Probabilistic Safety Assessment and Management

ESREL'96 — PSAM-III

Springer
*London*
*Berlin*
*Heidelberg*
*New York*
*Barcelona*
*Budapest*
*Hong Kong*
*Milan*
*Paris*
*Santa Clara*
*Singapore*
*Tokyo*

# Probabilistic Safety Assessment and Management '96

## ESREL'96 — PSAM-III

*June 24-28 1996, Crete, Greece*

## Volume 1

*Edited by*
Carlo Cacciabue and Ioannis A. Papazoglou

Springer

P. Carlo Cacciabue, PhD
I A PSAM, Joint Research Centre, The European Commission
Institute for Systems, Informatics and Safety, 1-21020 ISPRA (Va), Italy

Ioannis A. Papazoglou, PhD
I A PSAM, NCSR 'Demokritos'
PO Box 60228, GR-15310 Aghin Paraskevi, Greece

ISBN 3-540-76051-2 Springer-Verlag Berlin Heidelberg New York

British Library Cataloguing in Publication Data
Probabilistic safety assessment and management '96 : ESREL '96 - PSAM - III
1.Risk assessment 2.Risk management 3.Probabilities
I.Cacciabue, Petro Carlo II.Papazoglou, Ioannis A.
363.1'00151923
ISBN 3540760512

Library of Congress Cataloging-in-Publication Data
A catalog record for this book is available from the Library of Congress


Printed in Great Britain



Printed and bound at the Athenæum Press Ltd., Gateshead, Tyne and Wear
34/3830-543210 Printed on acid-free paper

# CONFERENCE ORGANISED
# By

**ESRA**
European Safety and
Reliability Association

**IAPSAM**
International Association for Probabilistic Safety Analysis and Management

## Sponsored By

European Commission Joint Research Centre EC - JRC
National Center for Scientific Research "DEMOKRITOS"
European Commission EC - DG-XII/D
European Safety, Reliability and Data Association (ESReDA)
Greek Atomic Energy Commission (GAEC)
Scandinavian Reliability Engineers
Society for Risk Analysis (SRA)

# CONFERENCE ORGANISATION

**Ioannis A. Papazoglou,** General Programme Chair

## Senior Advisory Board

**Volta G.,** JRC, EU, Chairman
**Ale B. J. M.,** RIVM, NL
**Apostolakis G.,** MIT, USA
**Bertrand Y.,** ISDF, F
**Garrick B. J.,** PLG Inc., USA
**Hsia D. Y.,** INER Rep. of China
**McQuaid J.,** Health and Safety Executive, UK
**Sato K.,** Nuclear Safety Commission, J

## Organising Committee

**Papazoglou I. A.,** N.C.S.R. " Demokritos", GR
**Cacciabue P. C.,** European Commission, JRC, EU
**Karvounakis G.,** N.C.S.R." Demokritos", GR
**Cojazzi G.,** European Commission, JRC, EU
**Parisi P.,** European Commission, JRC, EU
**Nivolianitou Z.,** N.C.S.R. " Demokritos", GR
**Anaziris O.,** N.C.S.R." Demokritos", GR
**Bonanos G.,** N.C.S.R." Demokritos", GR

## Conference Secretariat

**Cicognani C.,** European Commission, JRC, EU
**Finokalioti G.,** N.C.S.R." Demokritos", GR
**Magkia S.,** N.C.S.R." Demokritos", GR

## TECHNICAL PROGRAMME COMMITTEE

**P. Carlo Cacciabue,** Technical Programme Chair

**G. Cojazzi,** Associated Chair, Technical Programme

# PREFACE

## BACKGROUND HISTORY

This Conference represents the confluence of two successful traditions: the annual ESREL Conferences which themselves stem from a European initiative merging several national conferences into a Pan-European safety and reliability event, and the PSAM International Conferences which have provided a forum for the presentation of the latest developments in methodology and application of probabilistic methods in various industries.

**ESREL'96** follows ESREL'90 in Brest, France, ESREL'91 in London UK, ESREL'92 in Copenhagen, Denmark, ESREL'93 in Munich, Germany, ESREL'94 in La Baule, France, and ESREL'95 in Bournemouth, UK
**PSAM III** follows PSAM I in 1990 in Los Angeles and PSAM II in San Diego both in the USA. By moving outside the USA PSAM III establishes its international character not only in terms of participants' origin but also in terms of Conference Venue.

The objective of the ESREL series of conferences is to promote communication and integration of reliability and safety experts across Europe. PSAM's objectives include promotion of relevant experience among various industries. This joint ESREL - PSAM conference aims at enhancing communication, sharing of experience and integration of approaches not only among various industries but at a truly world wide basis by bringing together leading experts from all over the world.
**ESREL'96 - PSAM III** has been co-organised by the European Safety and Reliability Association (ESRA) and the International Association for Probabilistic Safety Assessment and Management (IPSAM). Sponsoring by the Commission of the European Union (Joint Research Centre and DG-XII/D Environment), and the Center for Scientific Research "DEMOKRITOS" is particularly appreciated and gratefully acknowledged along with that of the other sponsoring organisations.

# THE CONFERENCE THEME

*" We Athenians in our persons, take our decisions on policy and submit them to proper discussion. The worst thing is to rush into action before the consequences have been properly debated. And this is another point where we differ from other people. We are capable at the same time of taking risks and estimating them before hand. Others are brave out of ignorance; and when they stop to think, they begin to fear. But the man who can most truly be accounted brave is he who best knows the meaning of what is sweet in life, and what is terrible, and he then goes out undeterred to meet what is to come".*

from Pericle's Funeral Oration in Thucydides'
" History of the Peloponnesian War"
(started in 431 B. C.)

More than 2.400 years ago Pericles offered, as one of the reasons for the superiority of the Athenian political system, their capacity of estimating risks and evaluating them before taking decisions. In the last 30 years, contemporary scientists have been trying to provide modern societies with a systematic, self consistent and coherent framework for making decisions, concerning at least one class of risks, those stemming from modern technological applications.

Parts of this framework, are supported by methods and techniques developed in the scientific areas, known as Reliability Availability Maintainability (RAM) analysis of systems (recently referred to also as Dependability), Probabilistic Safety Analysis (PSA), and/or Quantified Risk Assessment.

Most of the effort in the last three decades has been spent in trying to develop techniques for assessing the dependability of technological systems, and assessing or estimating the levels of safety and associated risks. A wide spectrum of Engineering, Physical and Economic Sciences are involved in this assessment effort.

We believe that the time is ripe for putting these assessment methods at work, and use the results in relevant decisions. This is what we collectively call Risk Management.

Decision making and more specifically risk management and risk based regulation are the main themes of this Conference. A large class of decisions concerning technology, is being made through various forms of regulations. Creating an appropriate regulatory framework making proper use of the capabilities of RAMs and QRA techniques is, in our belief, the challenge facing nowadays the reliability and probabilistic safety analysis fields. Use and testing of existing probabilistic methods, will also indicate the areas in need of further research and development.

This Conference has widely covered and discussed both well established practices and open issues in the domain of Probabilistic Safety Assessment and Management, identifying areas where maturity has been reached and where more and new research is needed.

Can we now argue that the lesson of Pericles has been learned and the wisdom of the Athenians has been spread? We like to think that it has started to do so, although modern technology always presents us with new challenges and open questions demanding the continuous and passionate work of safety analysts and researchers.

**PROCEEDINGS ORGANISATION**

Three hundred and sixty three (363) papers, are included in these Proceedings covering a wide spectrum of topics in the areas of System Dependability and Probabilistic Safety Assessment and Management, and authored by scientists originating from 30 different countries.

One of the particular features in this Conference Organisation pioneered by the PSAM series, is that members of the Technical Programme Committee (TPC), are responsible for soliciting papers for the Conference and organising them into Sessions. All papers are then reviewed by the TPC and arranged in new sessions according to subject and final availability. In order to facilitate Conference attendance and the following of presentations, the papers in the Proceedings are arranged according to the presentation sequence, which of course is seriously affected by space and time constraints.

We have organised the papers in five major thematic categories. To facilitate the reader to locate papers in specific thematic areas, the following table is provided:

I. RAMs - DEPENDABILITY (71 papers)
Sessions : 3, 8, 13, 18, 28, 33, 38, 44, 47, 50, 55, 58, 60, 63, 64, 66, 72, 77.

II. PSA - QRA METHODS & APPLICATIONS (98 papers)
Sessions : 2, 7, 12, 17, 22, 27, 32, 37, 43, 49, 54, 59, 65, 68, 71, 74, 76, 79, 80, 81, 85, 86, 91.

III. RISK MANAGEMENT, RISK BASED REGULATION, DECISION MAKING (89 papers)
Sessions : 4, 9, 14, 19, 23, 24, 29, 34, 39, 45, 51, 56, 61, 67, 73, 75, 78, 82, 83, 87, 89, 90.

IV. HUMAN FACTORS (45 papers)
Sessions : 1, 6, 11, 16, 21, 26, 31, 36, 41, 42, 48, 53.

V. SPECIAL TOPICS : Data Analysis, Uncertainty Analysis, Expert Systems, Expert Judgment (60 papers)
Sessions : 5, 10, 15, 20, 25, 30, 35, 40, 46, 52, 57, 62, 70, 84, 88.

A reader interested in the general subject of Human Factors, for example, could record the numbers of the sessions in this category, and then through the Table of Contents (which provides the sessions in numerical order) identify the paper(s), of greater interest to him or her.

The editors would like to take this opportunity to thank all those who have contributed to the organisation of this Conference, in particular the Senior Advisory Board, the Technical Programme Committee and the Conference Organising Committee.

P. C. Cacciabue
Joint Research Centre
European Commission
Institute for Systems, Engineering and Informatics
I-21020 Ispra (VA), ITALY

I. A. Papazoglou
NCSR "Demokritos"
P.O. Box 60228
GR-153 10 Aghia Paraskevi
GREECE

# CONTENTS

## Emergency Planning
Chairs:Harvey, D.; Banda, Z.



## Learning from Operating Experience-I
Chairs:Amendola, A.; Sagar, B.



## Innovative Applications of HR Techniques-I
Chairs:Parry, G. W.; Kondo, S.

## PSA Studies/Applications-II

Chairs:Hauptmanns, U.; Soth, L.



## Systems Evaluations

Chairs:Hirschberg, S.; Ellia-Hervy, A.

## Risk Based Regulation: Multilateral Agreements

Chairs:Kyriakopoulos, N.; Avenhaus, R.



## Learning from Operating Experience-II

Chairs:Kortner, H.; Procaccia, H.



## Life Cycle Cost/Profit

Chairs:Ravindra, M. K.; Bot, Y.

## PSA Studies/Applications-III

Chairs:Mulvihill, R.; Macris, A. C.



## Reliability and Risk Based Maintenance

Chairs:Dykes, A. A.; Carlsson, L.

## Risk Based Regulation: Applications-I

Chairs:Spitzer, C.; Bley, D.



## Data and Reliability Analysis

Chairs: Kafka, P., C.; Senni, S.



## Human Error Tolerant Systems

Chairs:Hollnagel, E.; Mazet, C.

## Modelling Approaches to Safety Analysis

Chairs:Siu, N. O.; Dutuit, Y.



## Safety Concepts and Tools for Design and Operations

Chairs:Henneke, D. W.; Vestrucci, I. P.



## Environmental Risk Management Strategies

Chairs:Ravindra, M. K.; Bonano, E. J.

## Expert Judgement Probability Assessment

Chairs:Cojazzi, G.; Goossens, L. H. J.



## Innovative Applications of HR Techniques-II

Chairs:Bley, D.; Gertman, D. I.

## PSA in Industrial Applications

Chairs:Prassinos, P.; Petersen, K. E.



## Risk Management-I

Chairs:Lederman, L.; Siu, N. O.



## Agency Guidance

Chairs:Murphy, J. A.; Virolainen, R. K.

## Uncertainty Analysis-I

Chairs:Klaassen, D. C. M.; Devooght, J. E.



## HRA in Practice-I

Chairs:Hirschberg, S.; Cacciabue, P. C.



## Health Effects Modelling

Chairs:Lee, K. J.; Kelly, D. L.

## Testing Optimization

Chairs:Haim, M.; Krymsky, V. G.



## Risk Based Regulation: Methods

Chairs:Speis, T.; Cassidy, K.

## Uncertainty Analysis-II

Chairs:Hessel, P.; Wicks, P.



## Advances in Human Reliability Modelling-I

Chairs:Wreathall, J. W.; Cacciabue, P. C.



## Performance Assessment for Radioactive Waste Repositories

Chairs:Thompson, B.; Anderson, D. R.

## Total Quality Methods

Chairs:Surbone, G.; Pyy, P.



## The Decision To Launch: Perspectives on Aerospace Uses of PRA

Chairs:Frank, M. V.; Wu, J. S.



## Applications of Expert Judgement for Risk Analysis

Chairs:Cooke, R. M.; Cojazzi, G.

## HRA in Practice-II

Chairs:Furuta, T.; Pyy, P.



## Aspects of Safety Culture

Chairs:Cummings, G. E.; Maharik, M.



## Maintenance and Scheduling

Chairs:Guarro, S.; Lanore, J. M.

## Risk Based Regulation: Applications-II

Chairs:Ilderg, D.; Kaiser, G. D.



## Uncertainty Issues and Expert Judgement

Chairs:Goossens, L. H. J.; Frangopol, D. M.



## Learning from Operating Experience-III

Chairs:Minarick, J. W.; Cooper, S. E.

## Application of Simulator Data for HRA

*Chairs:Svenson, O.; Hollnagel, E.*



## Functional Analysis for PSA

*Chairs:Gheorghe, A.V.; Modarres, M.*



## Design for Reliability and Mainenainability

*Chairs:Papazoglou, I.A.; Kafka, P.*

## Environmental Decision Making

*Chairs:Kastenberg, W.E.; Apostolakis, G.E.*



## Analysis of Critical Software-I

Chairs:Guarro, S.; Bourgade, E.

## Computational Methods-I

Chairs:Izquierdo, J. M.; Cho, N. Z.



## Advances in Human Reliability Modelling-II

Chairs:Orvis, D.; Spurgin, A. J.



## External Events

Chairs:Ho, V.; Bari, R. A.

## Reliability and Risk Assessment Applications for Hardware

Chairs:Arueti, S.; Procaccia, H.



## Risk Perception Versus Risk Analysis-I

Chairs:Okrent, D.; Garrick, B. J.



## Bayesian Methods

Chairs:Papazoglou, I. A.; Atwood, C. L.

## HRA at the Turning Point?

Chairs:Hollnagel, E.; Wreathall, J. W.



## Risk Issues for Soviet-Designed Reactors

Chairs:Stamatelatos, M.; Holló, E.



## Common Cause and Dependencies

Chairs:Mosleh, A. M.; Vasseur, D.

## Risk Based Regulation: Statistical Methods

Chairs:Karydas, D. M.; Stavrianidis, P.



## Analysis of Critical Software-II

Chairs:Saglietti, F.; Bourgade, E.



## Reliability/Availability Analysis Methods-II

Chairs:Dutuit, Y.; Lydersen, S.

## PSA Methods in Transport Systems
Chairs:Evans, M. G. K.; Hessel, P.



## Reliability Prediction
Chairs:Bacivarov, I.; Bedford, T.



## PRA Based Decision Making
Chairs:Murphy, J. A.; Kaufer, B.

## Data Collection and Management

Chairs:Gheorghe, A. V.; Kazarians, M.



## Reliability/Availability Analysis Methods-III

Chairs:Ortiz, N.; Aldemir, T.



## Dynamic Reliability

Chairs:Matsuoka, T.; Aldemir, T.

## QRA Methods and Computer Codes-I
Chairs:Mosleh, A. M.; Talbott, D. V.



## Risk Assessment and RAMS Decision Support Methods for Space Applications
Chairs:Ferrante, M.; Lloyd, J.



## Risk Based Regulation: Verification of Non-Proliferation
Chairs:Kyriakopoulos, N.; Wakefield, D. J.

## Modelling of Physical Phenomena for Risk-I

Chairs:Park, C. K.; Cazzoli, E.



## Ariane 5: Ariane Reliability Studies

Chairs:Bianchi, S.; Preyssl, C.

## Structural Reliability Estimation/Analysis

*Chairs:Schuëller, G. I.; Kurisaka, K.,*



## QRA Methods and Computer Codes-II

Chairs:Camarinopoulos, L.; Lambright, J. A.



## Reliability Based Maintenance

Chairs:Holló, E.; Senni, S.

## Risk Perception Versus Risk Analysis-II
Chairs:Okrent, D.; Garrick, B. J.



## Atmospheric Modelling
Chairs:Kaiser, G. D.; Petersen, K. E.



## Decision Analysis in Risk Management
Chairs:Preyssl, C.; Fragola, J. R.

## PSA Studies/Applications-IV

Chairs:Kondo, S.; Schmocker, U.



## Reliability/Availability Analysis Methods-IV

Chairs:Kafka, P.; Wu, J. S.



## Risk Based Land Use Planning

Chairs:Ale, B. J. M.; Volta, G.

## Modelling of Physical Phenomena for Risk-II

Chairs:Spray, S. D.; Islamov, R. T.



## Computational Methods-II

Chairs:Sato, Y.; Cho, N. Z.



## Integrating Regional EHS Risk Assessment

Chairs:Gheorghe, A. V.; Camarinopoulos, L.

## Risk Based Regulation: Applications-III

Chairs: Apostolakis, G. E.; Brown, M. L.



## Risk Management for Space Systems

Chairs:Fragola, J. R.; Bianchi, S.



## Data Collection and Quantification

Chairs:Procaccia, H.; Patev, R. C.

## PSA Studies/Applications-V

Chairs:Bonano, E. J.; Wilson, D.



## Computational Methods-III

Chairs:Devooght, J. E.; Smidts, C.

## Risk Based Prioritisation

Chairs:Ortiz, N.; Bonano, E. J.



## Expert System in Reliability Analysis

Chairs:Marseguerra, M.; Dutuit, Y.



## Risk Based Regulation: Issues Facing the Regulator

Chairs:Thompson, B.; Apostolakis, G. E.

## Risk Perception vs Risk Analysis-III

Chairs:Hoegberg, L.; Bari, R. A.



## Environmental and Health Effects

Chairs:Budnitz, R. J; Fthenakis, V.

# The Use of the Dynamic Flowgraph Methodology in Modeling Human Performance and Team Effects

A. Milici, J. -S. Wu and G. Apostolakis
Advanced Systems Concepts Associates
2250 East Imperial Hwy., Suite 200
El Segundo, CA 90245

**Abstract**

In current probabilistic safety assessments (PSAs), the modeling of teamwork in the analysis of complex system operation is limited in many respects. One major limitation is that the operating team has been modeled as one operator performing an isolated set of tasks, rather than as a group of individuals working in an interactive environment. Studies conducted by behavioral scientists addressing team effects, on the other hand, are mostly qualitative and do not include system hardware failures. In this paper, we present the Dynamic Flowgraph Methodology (DFM) as an approach to model the operating team. DFM provides a tool with which the performance of individuals in the team and the system hardware can be modeled in an integrated fashion using a clear graphic representation of the causal relationships between physical variables and human characteristics. The advantages of using DFM models include its dynamic representation of the system operation and its modular nature, through which human and team characteristics can be modeled with much flexibility.

# 1. Introduction

PSA methodology has evolved over the years to a mature level that many industries have conducted PSAs to gain safety measures and insights of their complex engineering facilities. One major factor that contributes to the safe operation of a complex system is the performance of its operating team. While PSA methodology includes extensive human reliability analysis (HRA) models, in which the impacts of human performance on system safety are estimated quantitatively, the approach in these models is to treat the operating team as one entity conducting an isolated set of tasks [1]. Such a strong assumption precludes the modeling of communication, coordination, and group decision making, all of which are important factors that may affect the safe operation of a system.

Team performance has long been a subject of interest to behavioral scientists. In the past, many studies have been conducted to address team decision-making and team performance issues. These studies, however, provide mostly qualitative measures and treat human performance issues separately from the hardware. Since

PSA is used mainly to evaluate quantitatively the overall safety performance of the system, models used in PSAs would need to provide quantitative measures and to treat hardware and humanware as integrated parts.

This paper presents some preliminary results of an attempt in this direction. The DFM methodology [2, 3] was chosen on the basis that it provides a tool with which the performance of individuals in the operating team and the system hardware can be modeled in an integrated fashion using a clear graphic representation of the causal relationships between physical variables and human characteristics. The fact that DFM generates prime implicants also provides an avenue for quantitative analysis models to be developed in the future.

## 2. Description of DFM Basic Features

The basic DFM methodology can be viewed as a two-step process. The first step consists of building a model that expresses the logical and dynamic behavior of the system and its operational environment in terms of its hardware, software, and humanware elements. The second step uses the model developed in the first step to build timed fault (or success) trees that identify "prime implicants", which are logical combinations of hardware, software, and humanware conditions that cause certain specific system states of interest. Timed fault trees also show the time sequences in which these conditions come about. These system states can be desirable or undesirable, depending on the objective of the analysis. This is accomplished by backtracking through the DFM model of the system and its operational conditions in a systematic manner.

DFM takes the form of directed graphs with relations of causality represented by arcs that connect network nodes and special operants, called transfer boxes and transition boxes. The nodes represent important process variables and parameters, while the operants represent the different types of possible causal or sequential interactions among them. DFM's usefulness as a tool for the analysis of the entire operational environment derives from its direct and intuitive applicability to the modeling of causality driven processes. DFM can provide a comprehensive representation of the way an operational environment of interconnected and interacting hardware, software, and humanware elements is supposed to work and how their working order can be compromised to failure and/or abnormal conditions and interactions.

In DFM, the causal relationships between parameters are represented as logic transfer functions, or decision tables. Figure 1a shows a simple DFM Model, where A represents a variable, possibly a physical parameter, and B represents a variable whose value depends on variable A. Variable A consists of a set of discretized states $\{a_i\}$ and variable B consists of the states $\{b_j\}$. The box TF represents a transfer function that maps the states $\{a_i\}$ to the states $\{b_j\}$, i.e., $b_j = TF(a_i)$. In some cases the transfer function TF will depend on the value of another parameter C, which is represented by a set of states $\{c_k\}$, as shown in Figure 1b. Then TF is a set of functions $\{TF_k\}$, and each function in the set is associated with each state $c_k$ of C.

When the causal relationship has a time delay associated with it, then the transfer function is applied to the variables ("fired") at a future time step.

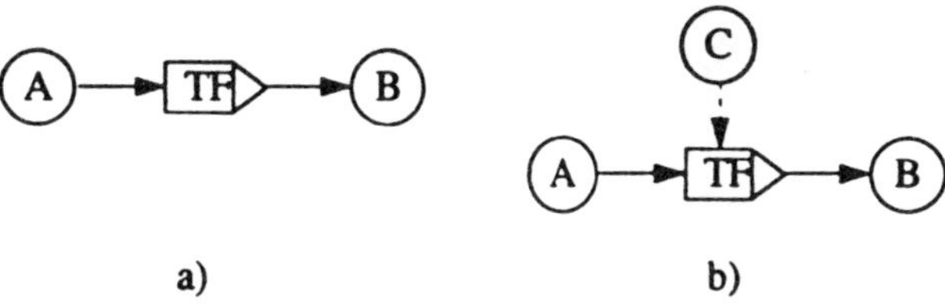


Figure 1: Simple DFM models

# 3. The DFM Approach

To illustrate the basic concept of how team performance in a plant operating environment could be modeled by DFM, consider the following simple example, in which the team related concepts of mental models and communication between team members are explicitly considered. In this example, we present a model of a simple system, with a team of operators consisting of two individuals. The system is a pressurized gas tank, and a diagram of it is shown in Figure 2. The operators are supposed to keep the pressure from becoming too high or too low, controlling the pressure by opening or closing a relief valve or by starting and stopping a pressurizing pump. One operator reads the pressure sensor to determine the value of the pressure in the tank, and decides what to do, i. e., whether to increase the pressure, decrease the pressure, or do nothing, and then communicates his decision to the second operator who performs the appropriate action.

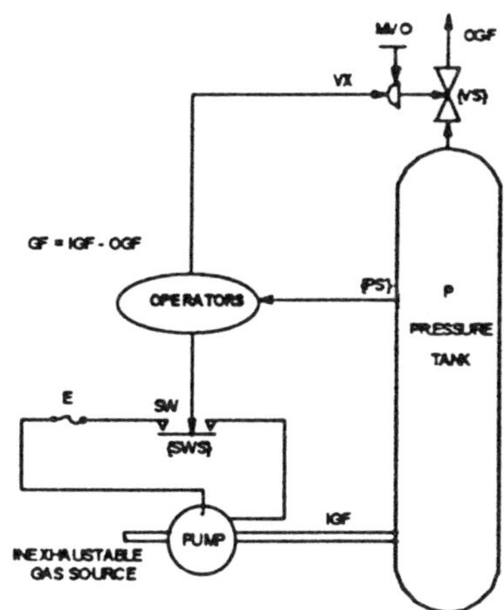


Figure 2: Pressurized gas tank example system

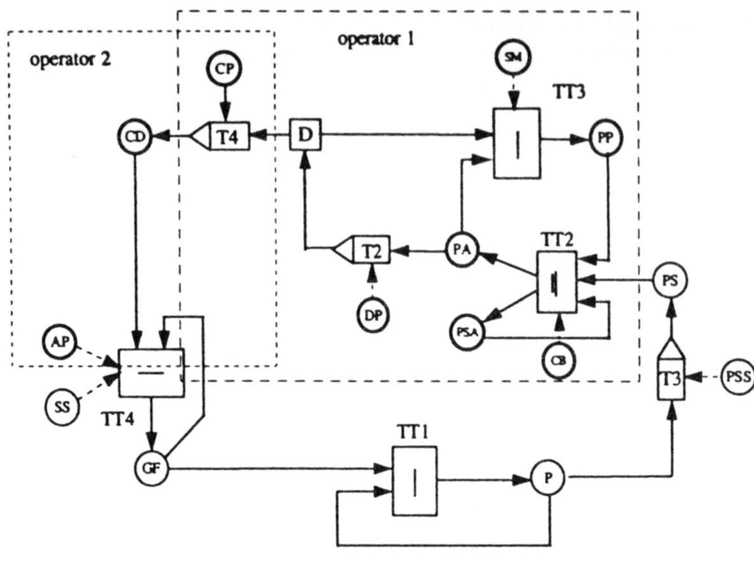


System Nodes:
P: Tank Pressure {very low (ll), low (l), normal (n), high (h), very high (hh)}
PS: Pressure Sensor Reading {very low (ll), low (l), normal (n), high (h), very high (hh)}
PSS: Pressure Sensor Status {reliable (r), unreliable (u)}
SS: Pressure Control System Status {available (a), unavailable (u)}
GF: Gas Flow {inflow (i), outflow (o), no flow (n)}

Operator Nodes:
PA: Pressure Assessment {very low (ll), low (l), normal (n), high (h), very high (hh), vague low (lln), vague normal (lh), vague high (nhh), completely vague (u)}
PP: Predicted Pressure {very low (ll), low (l), normal (n), high (h), very high (hh), vague low (lln), vague normal (lh), vague high (nhh), completely vague (u)}
PSA: Pressure Sensor Status Assessment {reliable (r), unreliable (u), vague (v)}
CB: Cognitive Bias {no bias (n), anchoring bias (a), overweighting bias (o), confirmation bias (c)}
SM: System Behavior mental model {correct (c), fast (f), slow (s)}
D: Decision {increase pressure (i), decrease pressure (d), do nothing (n)}
DP: Decison Process {correct (c), incorrect uncertain (iu), incorrect certain (ic), no decision uncertain (nu), no decision certain (nc)}
CD: Communicated Decision {increase pressure (i), decrease pressure (d), do nothing (n)}
DP: Decision Process {success (s), sender communication failure (sf), sender miscommunication (sm), receiver communication failure (rf), receiver miscommunication (rm)}
AP: Action Execution {correct (c), error of commission (c), error of omission (o)}

Figure 3: DFM model of example system

## 3.1 DFM Model

Figure 3 shows the DFM model for the example system and a team of two operators. Node GF represents the gas flow into or out of the tank, which is a direct result of the operator actions. The tank pressure (node P) at a future time step is determined by the gas flow and the current tank pressure. The transition box TT1 indicates that the pressure at the next time step depends on the pressure at the current time step and the gas flow at the current time step. If the gas flow is into the tank, then the pressure will be one state higher in the next time step. If the gas flow is out of the tank, then the pressure will be one state lower in the next time step. The transition box TT4 indicates that the gas flow in the next time step depends upon the operator action and the status of the system used to control the pressure, as well as the gas flow in the current time step. If the system is unavailable, then there will be no gas flow, regardless of the operator's decision.

Transition box TT2, together with nodes PS, PP, PSA and PA represent the first operator's mental model of the state of the system, and describes how his assessment of the pressure in the tank (node PA) and his assessment of the state of the pressure sensor (node PSA), is updated based on the sensor reading (node PS), what he thinks the pressure should be (node PP) and his own cognitive bias (node CB).

Transfer box TT3, together with nodes PA, PP, D and SM represents the first operator's mental model of the system behavior, i. e., it models how he makes a prediction about what the pressure should be in the next time step. His prediction (node PP) depends on his current assessment of the pressure (node PA), his decision about whether to increase or decrease the pressure (node D) and his mental model of how fast the pressure should increase or decrease (node SM).

Transfer box T2, together with nodes D, PA and DP represent the operator's decision making process. Node D represents the operator's decision, i. e., whether to decrease the pressure, increase the pressure or do nothing. Node DP represents possible faults in the operator's decision making process, possible resulting from confusion due to conflicting information, or time pressure.

Node CD represents the decision communicated from the fist operator to the second operator and has the same states as Node D, and node CP represents the communication process, which determines whether the first operator's decision is correctly communicated to the second operator. Node AP represents the second operator's action execution process, which determines whether the action is carried out correctly, and includes errors of commission and errors of omission.

## 3.2 DFM Analysis

Here we describe the results of the backtracking DFM analysis for the example model. Very high tank pressure, combined with gas flow into the tank, is the undesirable top event for which we generated prime implicants, where we have assumed that this condition may lead to failure of the tank. As a first step, we back traced one time step, and the analysis resulted in 262 prime implicants.

Of the 262 prime implicants, 68 are "single failure" prime implicants. By a "single failure" prime implicant, we mean that the prime implicant contains only one basic event that is considered to be a "failure" state. Failure states for system nodes are the unavailable or unreliable states of nodes SS (pressure controlling system status) and PSS (pressure sensor status), respectively. The failure states for the human process nodes consist of operator errors or mis-assessments, and are grouped into decision errors (failure states of Node DP), action execution errors (failure states of Node AP), communication errors (failure states of node CP) and wrong parameter assessments (where the state of Node PA is not consistent with the state of Node P). Of the 68 single failure prime implicants, 23 are of the decision error type, 11 are of the action execution error type, 23 are of the communication error type and 11 are of the wrong parameter assessment type. One of the single failure prime implicants is given below, where the failure is of the wrong pressure assessment type (the first operator believes the pressure is low when it is actually high, which causes him to increase the pressure when he should decrease it).

At time -1, PA: low AND
At time -1, DP: correct AND
At time -1, CP: success AND
At time -1, AP: correct AD
At time -1, SS: available AND
At time -1, GF: in AND
At time -1, P: high

It is interesting to see how the first operator could arrive at the incorrect pressure assessment given above. To do this, we backtracked one more time step with a top event of high tank pressure (Node P), low pressure assessment (Node PA) and gas inflow (Node GF). This analysis resulted in 66 valid prime implicants. Of these 66 prime implicants, there is one single failure prime implicant, 18 double failure prime implicants, 43 triple failure prime implicants and 4 quadruple failure prime implicants. In the single failure prime implicant, the lone failure (the sensor is unreliable) causes the tank pressure to reach a very high state from a normal state in two time steps, in the absence of any other failures.

**Acknowledgments**

The work presented in this paper has been funded by a Small Business Innovation Research (SBIR) grant, awarded by the SBIR Program Office of the U.S. Nuclear Regulatory Commission (NRC). The opinion and viewpoint presented herein are those of the authors and do not necessary represent the criteria, requirements and guidelines of the U.S. NRC.

## References

1. G.E. Apostolakis, V.M. Bier and A. Mosleh, "A Critique of Recent Models for Human Error Rate Assessment," Reliab. Engng. & System Safety, 22, 201-217, 1988.
2. C.J. Garrett, S.B. Guarro, G.E. Apostolakis, "The Dynamic Flowgraph Methodology: A Methodology for Assessing the Dependability of Embedded Software Systems," IEEE Trans. on Systems, Man, and Cybernetics, 25 (5), 1995.
3. S.B. Guarro, M. Yau, C. Garret, G. Apostolakis, J.-S. Wu, "Development of Tools for Software Safety Analysis in Aerospace Applications," ASCA Report AR-94-03, ASCA, Inc., El Segundo, CA, 1994.

# Man-machine interaction analytical approach

Prof. Ing. G. Surace, Ing. O.P. Casetta
Department of Aeronautic and Space Engineering
Politecnico di Torino, C.so Duca degli Abruzzi 10129 TORINO Italy

**Abstract**

Nowadays, the very high technological level otteined by aeronautic operative systems, together with a more complex structure of control procedures, increases the risk of human error in accident occurrence and evolution.

# 1 Introduction

In the aviation, safety systems incorporate many factors such as design and, aircraft maintenance, human professionality and training. In the last twenty years, safety has not done significant progress (a accident probability is stable at 10E-6), because, in spite of the high structural safety standards, human accounts for 70-80% of aviation accidents. The significans of these findings translate to an increase in absolute terms in the numbers of accidents which will occur as Air Transportation expands leading as a consequence to the deterioration of the image of aviation. Human Factors, in all respects, is currently a very important topic in aviation and represent "The new fronteer of Safety in Aeronautic". [1]

## 1.1 Safety Philosophy

The functioning of complex sociotechnical systems, such as in aviation, requires the appropriate coordination of many factors and the definition of guidelines for a philosophy which puts the man at the centre of man-machine binomial. Many accidents which in the past were simply referred to a "human error" hide a drastic reduction in man-machine system reliability.[2]

The introduction of computer systems in the new, so called, *glass cockpits*, has created new problems for pilots when they are called to modify suddenly their flight plans; working for long periods to modify too complex plan parameters, pilots could lose their exact space and temporal reference, when, instead, a simple action is required. For example this problem, for example, is true for a runway alteration as such during the final approach.

Human Factors (HFs) represent the fundamental technology and has to play a central role in the future developement of aeronautic systems.[3]

## 1.2 Purpose

The Probabilistic Safety Assessment (PSA) represents a systemic approach for the identification of the possible contributions to risk in an operative scenario, which requires the application of a number of systemic steps to reach the fixed objectives of evaluation. On the other hand, risk represents the probability grade measure that an hazardous event can occur.

A fundamental chapter of the PSA refers to HFs and the evaluation of the possible man-machine interactions in an operative scenario managment during emergency situations.

The purpose of this research has been to develope a methodology which permits the study and the reduction of human error probability in aeronautic systems by using simulation techniques of the man-machine system, based on models of physical phenomena and behavioural cognitive processes connected by stocastic nature relations. This methodology is compatible with other PSA analysis methods, from which it derives.

# 2 Description of Methodology

Some analytical steps are necessary for the developement of the methodology:

- Safety/Operational Hazard Analysis or Failure Mode Effect Analysis;
- existing requirements analysis;
- developement of a human reference behaviour model, to link to the physical model examined;
- developement and application of a taxonomy for human error analysis and isolation;
- analysis and developement of existing empirical correlations and formulations between real working environment and the behaviour of the operator, by the definition of the external factors which influence human behaviour;[4]
- procedures formalization for a prospective or retrospective analysis.

The starting point is given by the analysis of the possible modes in which an event, involving the possibility of human failure, can occur and then, research for error minimisation by the use of an adequate taxonomy. One of the ways to proceed could be as follows:

- *Accident detailed analysis* in its dynamic evolution and outcome.
- *Selection of a specific scenario* regarding human behaviour, in which input data come from previous safety analyses still stressed and which cannot be reduced by a simple data system implementing. The second step is the detailed description of the working environment and the interface system architecture, to have a clear picture of the operative scenario where HFs operate, and identification of the probable consequences which an error could lead to.
- *Detailed description of the human behaviour* for an action by action survey of the possible human behaviour.
- *Available time identification* as a fundamental element in human error probability and identification of the ways to reduce it. If necessary, determination of the specific time available and its importance, action by action, so to evaluate the influence in an operative scenario.
- *Methodology application* for the identification of primary causes (Root Cause Analysis) from the taxonomy used.
- *Error probability reduction* by the definition of some new instructions and the improvement these which already exist, referring to the actual scenario such as the replanning of existing software or old emergency procedures.

# 3 Operator Cognitive Model (OCM)

The introduction of the human component in man-machine system, needs the developement of a human behaviour model where the rules permit the simulation of the cognitive processes, i.e. the processes of logic links between operator knowledge and the control actions over a physical system.

Mental processes in the model proposed are not rigorous procedures executed by passing different mental states, but cyclic-active, characterized by a continuous perceptive cycle [5], and consisting of four fundamental functions: *identification/observation; interpretation/evaluation; planning/choice; action/execution* and Memory which plays the role of the human knowledge residence area.

The main characteristic of this cognitive model is the double value of the cognitive functions which may be, at the same time, either an action result or its cause.[6]

Automatic systems use thousands of rules through the comparison of several variables at the same time, while humans can only sequentially order, and with defined procedures, certain sequences, without considering the machine inner logic. Therefore in an emergency situation, the result can be different from that which the pilot desired because of the distance between computer logic and the brain. In these situations pilots' reflexes are very slow and space-time profile could be modified as a result. The control procedures have a dynamic nature related to the contingent evolution of the transient sequence so that, one of the most important factors in operator sensorial response is the available time. Functions which correlate the operator response time with the error probability, describing human response to a sensorial stimulus (SRF, Stimulus Response Function) and built by simulations and extrapolations, stress a high time-depending human behaviour. In particular, the SRF decreases with the time available while brain activity can increase.

These considerations have promoted the idea of dividing available time into four sections: independent time period IT, medium time period MT, reduced time period RT and zero time period ZT. Taking into account these standard reaction human times, for every phase : - sight time 0.1, warning recognition time 1.0, danger consciousness time 5.0, manouvering decision time 4.0, muscular reaction time 0.4, aircraft reaction time 2.0 -, for a total of 12.5 sec., we can assume a first reduced period of ten tofifteen seconds.( Aeronautic Medicine Data)

Examining one by one the four periods in order to isolate their characteristics it can be observed, like in the Rasmussen model, three behaviour typologies, each one related to a characteristic available time.[7]

The basis of the model is the "Skill Based Behaviour" SBB or Stimulus Response (SR) for which we can identify that actions istinctively occurring and based on own operator's well defined mental schemes. This case represents the first attempt to define the perceved reality by two fundamental modes: analogy research with other past experiences by memory (similary matching) and frequency analysis of past similar situations (frequency gambling) [8]; between the stimulus and the response, there is a direct connection, depending on the knowledges of the operator. If the time available is too short, about a second, this can be considered a-priori inhibited because the time available is shorter than operator reaction time (obtained by experts' judgements or simulations); in these cases a complete automation is necessary.

If the time available is more than a second, but not enough to organize adequate responses, the operator reacts to simple sensorial stimuli with a Planning function by-pass, that is the difference between deduction/induction and the observation of something evident.

The second level of the model is represented by the Rule Based Behaviour (RBB). The action composition is based on well defined procedures and between stimulus and response there is no direct bond. The cognitive process is based on a logic information acquisition derived by the system which allows the identification of appropriate procedure for the situation analyzed. The available time has to be sufficient to organize a valid strategy for emergency states and the reference behaviour model is that procedural with all the functions of the model. In these situations, the main problem is the identification of the first causes and the application of the most appropriate procedure, with a particular reference to the time "Time Tracking".

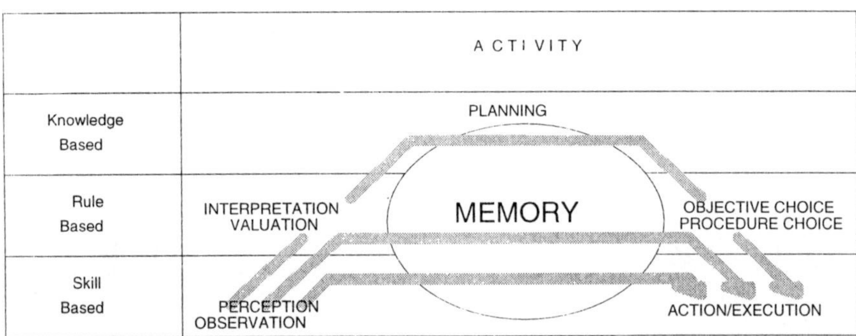


Fig.1

The most complex model is the Knowledge-Based-Behaviour (KBB), characterized by uncommon situations in which operator is called to organize a valid strategy to a problem without any procedural help. At this level there is no direct bond between stimulus and response and it indicates only the state of the system. In these cases, time available is sufficient for a valid strategy but there is a problem represented by the probable "attention failure" in that cases in which the operator is called to remain for long periods in a stand-by situation. During this phase the cognitive process is performed by many repeated steps.[9]

## 4 "Timing Error Taxonomy"

Taxonomy is a classification of a group of categories of possible erroneous human behaviour. It represents the reference model required to correlate human behaviour with the possibility of runnimg into dangerous situations by which it is possible to classify information from an accidental analysis through defined schemes based on a reference model (the considered cyclic cognitive model).

Taxonomy has been developed starting from an existing model, to which it has been referred for other clarifications and improvements [10], reorganized by the considerations above operator's available time and the possibility to by-pass the Planning function if the time is not enough to organize a valid procedure.

In the taxonomy, as in the cognitive model, the accidental cause, called genotype, can be introduced in one of the phases of which the model is composed, Perception/Observation, Interpretation/Evaluation, Planning, Choice and Action/Execution, in one of the columns "specific cause" or "generic cause"; this can refer to that specific directly connected, or one precedent, depending on when the primary erroneous action was committed.

The erroneous behaviour manifestations (fenotypes), appear only in the Action/Execution phase, while specific effects represent only the manifestation of the specific causes which cannot give action errors because they cannot rise up in the passage from the decisional process to the operative one. They all refer to psychophisic conditions of the operator and belong to the person related causes which are the main cause of error manifestations in the cognitive process and they occur spontaneously despite system related causes. These, could be activated by random events too, which activate person related causes and then erroneous behaviour.

As first, we must identify and choose the right time period in a operative sequence or in a single action and then, from the Action/Execution phase, we go back in the cognitive model , from the generic effect to the specific cause, to identify the true error cause and the possible effects of these.

External influence factors are grouped into six categories: comunication, interface, interference, procedures, working environment, operator available time.

# 5 Time factor in a human reliability analysis

The final step of a HRA is the numerical evalutation of the error probability. One of the more commonly used techniques, the THERP method (Technique for Human Error Rate Prediction), could be implemented if in the NHEP (Nominal Human Error Probability) the exact time period available is taken into account.[11]

In the THERP method, errors are modeled by the use of probability trees and dependence models between actions executed and considerations about the factors influencing performance. The NHEP are built from an enormous data bank obtained by experimental or expert-judgement information. The method is compatible with other PSA analysis but does not allow the evaluation of complex cognitive processes, such as diagnosis or knowledge-based planning, and to stress dynamic angles which characterize causes, manifestations and consequences of the possible pilot errors.

The first step for the general application scheme is the determination of the factors which influence operator activities and critical sequences where the error probability is very high. For this purpose, the developed taxonomy is a valid instrument. When critical actions has been determined, the ETs of the HRA can be developed by a top-down process characterized by a binary architecture of which the whose branches end with the success or failure of the procedure.

Action-by-action every NHEP nominal value has to be determined through expert judgement or a reference handbook (chapter.20). These tables have, for all data, an error factor and the NHEP value is the mean of a log-normal distribution of the 95th and 5th percentile, to which the upper and lower uncertainty limits are respectively associated . The evaluation is performed by:

NHEP*EF/EF

These nominal values have to be modified for the specific analysed case through the introduction of the possible factors influencing operator performances, one of which is the available time. The nominal value can be increased to a level NHEP/EF or decreased to NHEP*EF if conditions are better or worse than nominal. An Available Time Factor ATF, a variable in the range 0-1, can be introduced to modify further on the NHEP, by a scheme:

| Time available: | Reduced | NHEP*ATF |
|---|---|---|
| | Medium | NHEP |
| | Independent | NHEP/ATF |

To more correctly determine the conditioned probability of success or failure, in performing an action, it is fundamental to take into account the dependence levels (positive or negative) among operator actions, and modify again the NHEP first obtained.

To obtain cautionary values, in the NHEP handbook, only the positive dependence is considered. This is divided into five levels: short (<5%), medium (<15%), high (<50%) and complete (100%).

# 6 Error minimization directions

The main problem in HFs related accidents, has been identified in "pilots'deviation from operative procedures" (Lautman&Gallimore 1988). The lack of a general project philosophy of coherent procedures, could cause a progressive deviation from the operational standard procedures. It so important to develop a guide flight operation philosophy which establishes the right direction for group strategies.

The model by which we can minimize errors, distinguishes the possibility to process informations regarding the available time. When S-R function dominates, indications have to be too procedural and well defined; on the contrary, when more time is available, a greater freedom of choice for the operator is possible.

## 6.1 Reduced time period

*simple direct action*

Every required action in this phase has to be clear, short and well defined, expressed in a way to facilitate operator's answer and memory.

*null implementation time*

Every required action has to have an implementation time too reduced for minimizing operator time answer, related to the short available time

## 6.2 Medium time period

*time evolving*

Every command has to maintain a well defined time domain (chronological tracking) to prevent the risk of a loss of the time evolution.

*ambiguous directions prevention*

The phraseology adopted in operative procedures represents another element to reduce human error probability, averages and/or accidents, which has to be coherent with cockpit procedures to prevent misunderstandings between crew members and the flight control operators

*right sequentiality*

Sequentiality is the inner mechanism which guides many time dependent cockpit procedures, especially in emergency or abnormal situations. In these situations any error arising when following a checklist could have dangerous or irreversible consequences because of the limitated available time and of the working overload.

*procedural diagnosis*

For the diagnosis of abnormal system situations, error probability can be reduced by replacing an operator knowledge based diagnosis, with a more procedural one based on written rules and the S-R function.

TIME AVAILABLE - BEHAVIOURAL MODEL

| // | SKILL | RULE | KNOWLEDGE |
|---|---|---|---|
| ZERO | REDUCED | MEDIUM | INDEPENDENT |
| human action impossible | single direct action<br><br>no implement. time | cronological tracking<br><br>ambigous tests prevention<br><br>procedural diagnosys<br><br>right sequence | ambigous signals prevention<br><br>sequnce errors prevention<br><br>operative maintening |
| seconds | minutes | hours | |

available time

Fig. 2 Directions for human error probability reduction

## 6.3 Independent time period

*ambiguous signals errors prevention*
Although the time available is not defined, it is necessary that all the procedures are characterized by very different signals, so to prevent misunderstanding errors.
*sequential and time errors prevention*
As in the medium time period, the interface design plays an important role; it has to be deigned so as to prevent time-space sequence errors
*operative maintenance*
It is necessary that the operator has not to maintain very low focussing or attention levels for long periods. This could be made by periodical simulations and regular controls which prevent dangerous attention losses.

The planning of the cockpit procedures as an interface component, permits tte introduction of the HF as a central and not peripherical element. The engineer, using a bottom-up methodology, obtains directly which machine outputs are required and assesses which of these operator could probably make.

# References


-1- "Human Factor in civil aviation" international conference, September 21 1995 Rome

-2- Degani A., Wiener E.L. (1991), "Philosophy, Policies, Procedures and Pratices: The four P's of Flight Deck Operations", in Proceedings of the Sixth International Symposium on Aviation Psychology, Columbus, Ohio.

-3- Wiener E.L., Kanki B., Helmreich R. (1993). Cockpit Resource Managment, New York Accademic Press.

-4- Johnson, W.B. (1990), "Advanced Technology for aviation maintenance: research and development in the USA" , in Proceedings of the ICAO Flight Safety and Human Factors Seminar, Duala, Camerun.

-5- P.C. Cacciabue, Affidabilità dinamica e fattori umani in sistemi nucleari. Dip. Ing. Nucleare Politecnico di Milano 1994.

-6- Rasmussen J., Pejtersen A.M., Schmidt K., (1990). Taxonomy for cognitive work analysis. Proceedings of 1st Workshop of MOHAWC Esprit II Project, B.R. Action 3105, CEC. May15-16, Liege, Belgium

-7- Rasmussen J, (1986). Information processes and Human-machine interaction. An approach to cognitive engineering. North Holland. New York

-8- Rouse W.B., (1993). Model of Human Problem Solving: Detection, Diagnosis and Compensation for System Failures. In Automatica, 19 (6), pp 613-625

-9- Humphreys P.(Ed.) (1988). Human reliability assessors guide. United Kingdom Atomic Energy Authority, RTS88/95Q.

-10- Vestrucci P., (1990) Modelli per la valutaone dell'affidabilità umana. Franco Angeli Libri Milano.

-11- Bell B.J., Swain A.D., (1981). A procedure for conducting a Human Reliability Analysis for nuclear Power Plants. NUREG/CR-2254, SAND81-1655, Draft Report, USNCR, Washington, US.

# Simulating Interactions Between Operator and Environment

K. Furuta, T. Shimada and S. Kondo
QUEST, The University of Tokyo
7-3-1 Hongo, Bunkyo-ku, Tokyo 113, Japan

**Abstract**

A comprehensive simulation system including interactions between an operator and the environment has been developed. This system was then tested using a case problem of operating a nuclear power plant in an emergency using three control room designs.

## 1 Introduction

Comprehensive simulation of human-machine systems (HMS) is expected to be a useful method to assess performance of HMS and to point out problems in HMS design. Such simulation has been tried by combining a process model and an operator model, to which artificial intelligence (AI) technologies are often applied for implementation [1]. In classical AI approaches, it is assumed that a physical symbol system can construct a proper comprehension of the world by processing the symbolic representations within the system. Once the state and situation of the world have been encoded in physical symbols, information processing is performed almost independently of the external world, the environment.

The pure physical symbolism as described above, however, is facing several difficulties in understanding capabilities of human cognition. New concepts therefore such as direct perception by Gibson [2] or external knowledge by Norman [3] have been proposed. In these views, human cognition is an open and interactive process rather than a closed process within a physical symbol system. It is thought humans do not keep all contextual information in mind but get a large part of the information from the environment as required. Humans can thereby not only avoid cognitive overload but also respond sensitively to contextual changes. Interactions between humans and the environment occur when humans seek contextual information in the environment and these interactions determine a large part of human behavior. Such interactions therefore should be considered in assessing human performance for engineering purposes.

In this paper we discuss how such interactions can be considered in HMS simulation and propose an architecture of HMS simulation. Implementation of a simulation system based on the proposed architecture will be described, and then usefulness of the new system will be shown with a case study.

# 2 Architecture of HMS Simulation

In the previous architecture, a HMS simulator was constructed by coupling the plant model and the operator model as shown in Fig. 1 (a), and plant parameters and operational commands are exchanged between these two models. Here it is assumed that all contextual information is kept and dealt with in the operator model. According to late views in cognitive psychology it is not the reality, but contextual information should reside mainly outside of the operator model and it is sensed by the operator model as in Fig. 1 (b). Sensing of contextual information includes so called direct perception. Direct perception requires little cognitive resources so that humans feel unconsciousness and directness, though the amount of information actually processed seems enormous from a viewpoint of information theory. Since it is difficult to model this process as it is based on the current knowledge on the phenomena and current computer technologies, a trade-off between reality and practicality must be considered to select an architecture to be adopted.

Based on the above consideration, we introduced an environment model between the process model and the operator model as shown in Fig. 1 (c) to simulate operator-environment interactions in a phenomenological manner. The environment model contains definitive information of human-machine interface, i.e. design specs, as well as various kinds of simulation algorithms of plant-interface, operator-interface, and operator-operator interactions. This architecture enables us to consider operator-environment interactions without excessively complicating information transfer between the process and operator models.

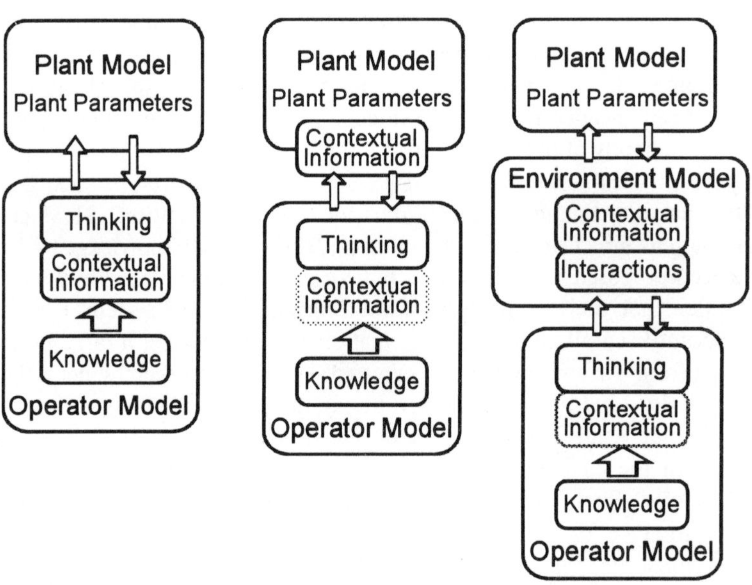


Figure 1 Configuration of comprehensive HMS simulation.

# 3 Implementation

A prototype simulation system was constructed based on the proposed architecture. The plant simulation module of this prototype is a simple simulator of a BWR-type nuclear power plant, and the operator simulation module is a multiagent knowledge-based system based on the blackboard model of control. The operator model in this module is a modified decision ladder, where diagnostic reasoning is modeled as hypothesis retrieval by similarity matching and hypothesis validation by observation. A more detail of the operator simulation module is explained in the previous paper [4].

The environment model was implemented as an interface simulation module (ISM) between the plant simulation module and the operator simulation module. All information exchanged between the plant and the operator modules passes through ISM. The configuration of ISM is illustrated in Fig. 2.

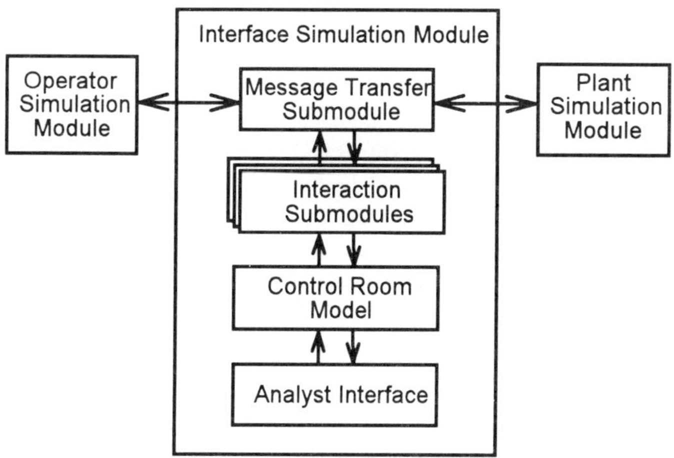


Figure 2 Configuration of interface simulation module.

ISM has a control room model where attributes of interface items such as control boards, control panels, annunciators, alarms, indicators, operational equipment are defined by object-oriented class definition. Some of these attributes are related to perceptual aspects or human factors such as the maximum distance within which an indicator can be read out or a manual switch can be operated. ISM also knows attributes of an operator such as position, height of eyes, direction of sight line, etc.

Several submodules are implemented in ISM to simulate interactions in the work environment. The vision analysis submodule calculates the vision of each operator and picks up visible interface items. It is assumed presently that operator's eye span extends within a cone along the sight line. The sight movement submodule controls the search of a target interface item. The sight line is moved in two phases: the group of items containing the target is caught in sight first and then the target is caught at the center of sight. The search is performed from the left to the right, from the top to the bottom direction in a control panel. The path planning submodule calculates operator's approach path to the point where a target comes into a readable or operable distance. A simple obstacle avoidance algorithm is included in this module. The judgment submodule judges whether or not a target is readable from the current position of an operator. The maximum distance of readability depends on the type and the size of indication to be read as well as operator's eye sight. The message transfer submodule controls the flow of messages exchanged between the submodules in ISM, the plant simulation module and the operator simulation module. The analyst interface presents a result of simulation in a graphical form and provides a platform for modifying the control room model to the analyst.

# 4 Case Study

Test simulations were performed using a case of plant operation in an emergency. Three types of control room designs, L-type, U-type, and integrated, were modeled in ISM and compared in the test. In the L-type and the U-type control rooms, 520 annunciators, 198 indicators and 451 manual switches are arranged on the main and two subsidiary control boards. In the integrated control room, 207 annunciators, 220 switches, and 7 CRTs are arranged on the huge display board as well as the main control board. The location of each interface item was determined after published materials, but neither it is exactly the same as the actual plants nor all of the items have connections with the plant simulation module.

The size of an operating crew was four for all cases, and each operator was located in front of the board in his charge at the beginning of simulation. A walk speed of 1.5 m/sec was used from observation of experiments with a full scope simulator [5]. The event scenario used for the test is a total loss of feedwater without initiation signals of reactor core isolation cooling (RCIC) and high-pressure core spray (HPCS). In this scenario the operators are expected to start-up RCIC and HPCS in manual.

The result of simulation for the L-type control room was compared with the previous simulation lacking ISM, and it was also compared with an experimental result that was obtained using a similar event scenario.

Some operator performance was simulated more realistically in the present simulation than the previous one. For example, all of the follower operators go to read an annunciator when they heard alarm sound, but those who are not close to the flickering item will stop after having seen another operator is already approaching it. This behavior could not be simulated with the previous system, because the simulated operators will not look at the situation in the control room. In stead the scope of business for each operator was defined and given to the simulator a priori. In addition to such behavior, what the previous simulation could not but the present one could simulate was that an operator detects an unexpected change of indication within his sight while looking at another target.

The improvements as above resulted in a more accurate prediction of a time line of operator performance, if time intervals between perception of some initiation signal and the related action were compared. For example, it took no less than 22 seconds from readout of the first annunciator and to certification of reactor shutdown in the previous simulation, while it took just four seconds in the present one, which is more comparable to three seconds observed in the experiment. Similarly the time required from detection of a unexpectedly low reactor level to manual start-up of RCIC was 8 seconds in the previous simulation, 19 seconds in the present one, and 16 seconds in the experiment.

Simulation of operator's motion enables assessment of walk distance of the operator in task achievement. Figure 3 compares the trajectories of operators' head positions between the three control rooms. The total distance walked by the three follower operators was 97 m in the L-type, 85 m in the U-type, and 11 m in the

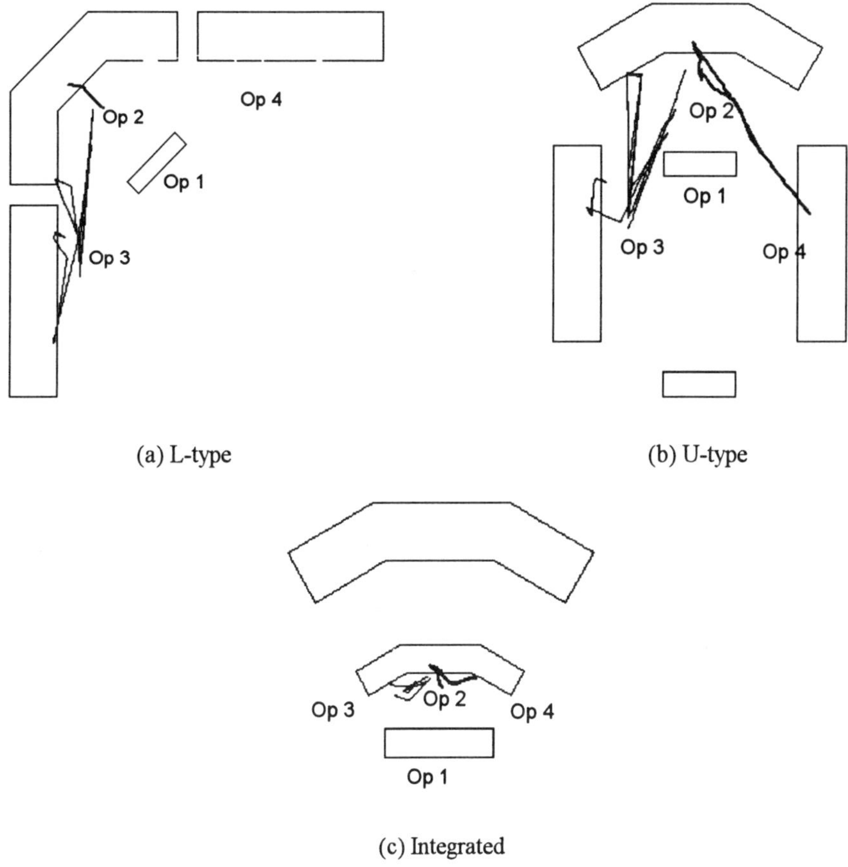


Figure 3 Head trajectories of operators in three control rooms.

integrated control room. It is apparent that the integrated design contributes greatly to reduce the walk distance of operators. The total distance walked by the subjects in the experiment were evaluated 107±28 m from video tape recording, and the interval includes the value predicted by the simulation.

# 5 Conclusion

Interactions between an operator and the environment should be considered in comprehensive HMS simulation. Taking a trade-off between reality and practicality into account, an architecture that consists of separate plant model, environment model, and operator model has been adopted. Based on this architecture, a prototype system was developed and tested using a scenario of nuclear power plant operation.

Merits of the new simulation system shown by the case study can be summarized as follows:

- A change of operator's behavior by interactions between an operator and the environment as well as by those among operators themselves can be simulated.
- A time line of operator performance predicted by the new system is more plausible than the previous one.
- Some measures that were unobtainable with the previous system such as walk path trajectory and eye motion trajectory are now available.

# Acknowledgment

The authors wish to express their thanks to Dr. Ujita of Hitachi Ltd. for his suggestions on the validity of the simulation in comparison with experiment.

# References

1. K. Furuta and S. Kondo. An approach to assessment of plant man-machine systems by computer simulation of an operator's cognitive behavior. Int. J. Man-Machine Studies 1993; 39: 473-493
2. J. J. Gibson. The ecological approach to visual perception. Houghton Mifflin, Boston, 1979
3. D. A. Norman. The psychology of everyday things. Basic Books, 1988
4. Furuta and S. Kondo. Computer simulation model of cognitive process in group works. Proc. 5th Int. Conf. on Human-Computer Interaction, 979-984, 1993
5. H. Ujita. Private communication

# Knowledge Dependency Analysis of Plant Operators using Consistent Protocol Formulation

Shinji Yoshikawa[1)], Kenji Ozawa[1)], Naoki Koyagoshi[2)], and Toshihiro Oodo[2)]

1)Frontier Technology Section, Advanced technology Division, O-arai Engineering Center,

2)Operations Engineering Section, Monju Construction Office,

Power Reactor and Nuclear Fuel Development Corporation.

Narita-cho 4002, O-arai-machi, Ibaraki, 311-13 JAPAN

Shiraki 2-1, Tsuruga-shi, Fukui, 914 JAPAN

## 1. Introduction

Safety and reliability of nuclear power plant(NPP) operation are now achieved by elaborate and systematic operator education. However, it is still difficult to completely deny the possibility of operators' cognitive overload caused by very complicated plant abnormal consequences. This cognitive overload may confuse the operators in important decision making processes and thus reduce safety margins in settling or recovering the target plants.

Nowadays, many efforts are devoted to human factor analysis and simulation, mainly to optimize man-machine interface design and cooperative task allocation within an operation crew in future generation nuclear power plants[1]. However, it is not easy to realize optimum rationalization of crucial information in every abnormal plant situation[2]. At the same time, methodology for enhanced reliability and safety must be sought also for the existing power plants. Therefore, it is important and essential to establish a methodology acting on minds of power plant operators to form a well organized and effective knowledge of the target plant.

The authors have launched a research program to establish a methodology to support human operators in forming this kind of integrated plant knowledge( hereafter referred to as "plant mental model"). The final goal of this research is to build an off-line intelligent computer-aided instruction(ICAI) system to monitor the subject operator's mental model, and to support plant operators in acquiring, correcting, and integrating plant knowledge.

This paper explains first an experiment with a training simulator and on-duty operators. A formulation scheme of operator protocols is described next. This formulation scheme has enabled some computerized analysis which is shown next. Future direction of both protocol formulation and analytical tool development are discussed in the last part in this paper.

# 2. Experiment

## 2.1 Criteria of simulated anomaly given to the subject operators

Observation of the current on-duty operators' behavior is undoubtedly the fundamental information resource for this kind of research. An abnormal situation was given to two on-duty plant operation crews in a computer-simulated control room, to obtain the behavioral data of the operators. The following criteria have been introduced for adequateness of the simulated anomalies:

(1) The anomalies should be more complicated than those included in the training curriculum

(2) The root cause should be uniquely derivable from the observable parameters, from the knowledge of plant functions and structure to be provided in the training curriculum, and from the available documents in the control room.

## 2.2 Characteristics of the selected anomaly

The selected anomaly is "Terminated steam supply to a feed water heater", where steam flow into a feed water heater as heat source is terminated due to erroneously closed check valve at the steam supply line from a high pressure turbine[3]. This anomaly is hard to diagnose because the operators can not observe the anomaly propagation sequentially. (Figure1)

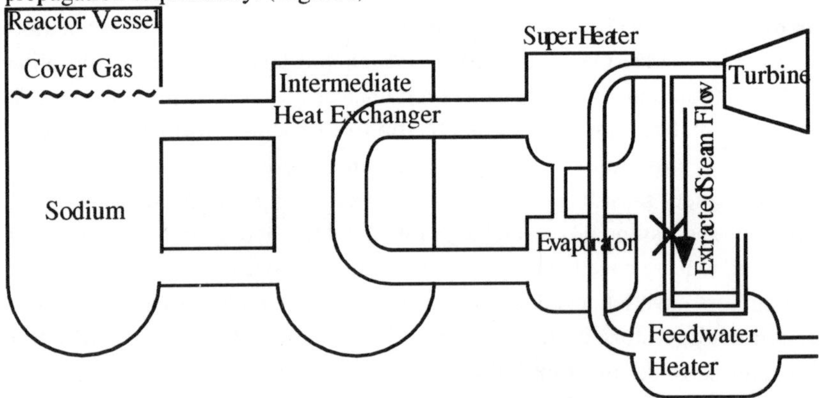


Figure1 Terminated Steam Supply to a feedwater heater

The authors would like to emphasize that the subject operators were asked to concentrate on cause identification. i.e., they were forced to maintain power generation until they become sure about the anomaly cause. This request is very important because, in the real plant operations, the operators can shutdown the reactor before cause identification, to minimize possibility of hazardous consequences.

# 3. FORMULATION OF OPERATOR PROTOCOLS

The authors developed a formulation scheme to transform the acquired protocols into analyzable database, based on an observation that the most of the recorded protocols can be viewed as a combination of a <verb part> and a <object part>.

## 3.1 Verb Parts

The following <verb> types have been identified:
*(1)statement*:Simple statements and confirmations of information as <object part>s
*(2)guess*:Presumption and supposition of information as <object part>s
*(3)question_w*:Question about some parameter value in the <object part>s
*(4)question_yn*:Asking for affirmation or negation of a statement.
*(5)proposal*:Proposal or question about the propriety of an assumed operation
*(6)prediction & Guess*:Predictions/guesses of plant behaviors/states
*(7)request*:Requests about plant operations (generated only by shift supervisors)

## 3.2 Object Parts

The information as <object part>s have been classified into:
*(a)plant state*:A set of parameter states including quantitative and qualitative values, and qualitative time derivatives, with clock information
*(b)plant state evolution (up to present time)&*
*(c)plant state evolution (from present time)*:A sequence of <plant state>s
*(d)causality*:Physical phenomena and engineering intention of control A set of <cause plant state> and <effect plant state>
*(e)operation*:A set of <forced plant state> and <intended effect as a plant state>
*(f)observability*:Description of parameter observability can be represented by a set of <parameter>, <observation point>, <expected accuracy> and <required time to get the value>
*(f)estimation*:Normality judgement of plant states and subsystem states, and propriety judgements of assumed operation.
This type of protocol is represented as:<<plant state|evolution|operation> <estimation value>>.

This formulation scheme is still a tentative one since new protocol categories can be introduced, as long as new experimental data will be available. Regarding the currently available behavioral data, this formulation method is applicable to most of the operator protocols, except some remarks crucial in important decision making and vocal readings of documents. The operator's ID(or the real name) is included in the trial formulation of the recorded protocol data.

Table 1:The proposed classification matrix for operators' protocols

| Verb Object | Statements | Questions _5w1h | Questions _yn | Proposal | Prediction & Guess | Request |
|---|---|---|---|---|---|---|
| Plant State | "This valve is fully open" | "Check the outlet coolant temperature" | "Is the valve closed?" | | "This valve is likely to be closed" | |
| Evolu-tion (1) | "This temperature decreases after hitting the upper limit" | "Has this flow been kept as it is?" | "has this level got lowered to this height?" | | "Looks like the temperature used to be lower" | |
| Evolu-tion (2) | | "How is this flow rate likely to behave?" | "Is this tempera-ture going up ?" | | "The outlet temperature is supposed to rise" | |
| Causali-ty | "The power increase is caused by the cooled core" | "Why this level rises?" | "Is this tempera-ture rise due to the decreased flow?" | | "This flow increase seems due to a closed bypass line" | |
| Opera-tion | "I am closing the valve" | "What should we do to decrease this tempera-ture?" | "Did you start up the backup pump?" | "How about decreas-ing the power demand?" | | "Please close the valve" |
| Observ-ability | "The valve position is not displayed at the panel" | "Where can I check the level on the supervisory console?" | "Is the recorder accurate enough to see 1% change?" | | "The valve position is not checkable in this room" | |
| Estima-tion | "This pump speed is abnormally low" | "What is the normal position of this valve?" | "Is this valve normally open ?" | | "It does not make sense to increase the flow for temperature control" | |

# 4. Analytical Approach

There is a common consens that a causal network among plant parameters plays a key role in operators' plant mental models. The proposed protocol formulation scheme is intended to describe operators' cognitive behaviors along with the causal network. The analytical tool generates recognized causal network(s) by combining all the expressed causalities and compares it(them) with the theoretical one.

Referrings to state evolutions are then drawn on the expressed and theoretical causal networks. One of the results obtained by this procedure is shown in Figure2. Thick lines and thin line show expressed causalities and evolution relation, respectively, on the theoretical causalities drawn by dotted lines. Numbers above each parameter show times of operators' refferences.

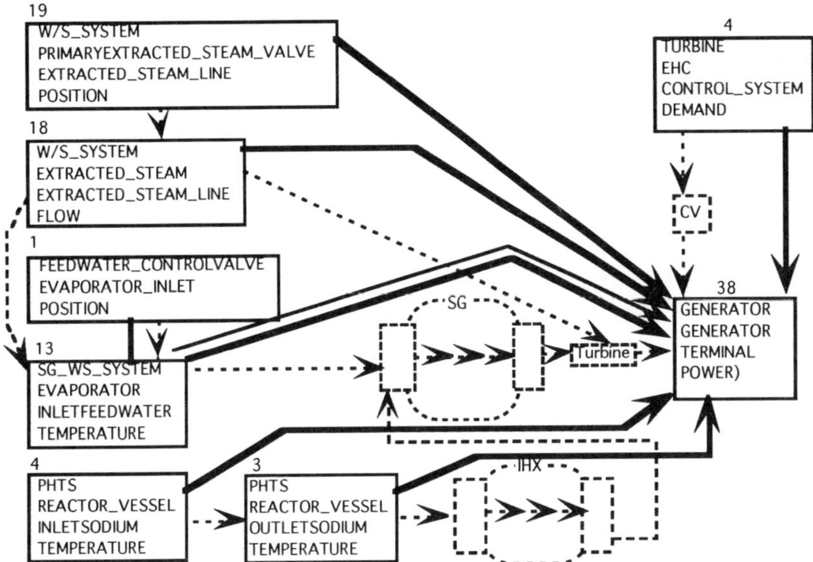


Fig 2: Derived Cognitive Chunks of Causal Network

# 5. Operators' Knowledge Dependency in Diagnosis

## 5.1 Plant As Assembled Subsystems

The operators understand the plant configuration by component groups-subsystems. The most important knowledge in this understanding for knowledge-level diagnosis is the relationship between subsystems and representative parameters. The operators judge normality of each subsystem by normal value of each representative parameter. The representative parameters are eminent within each subsystem for many times for being referred. In addition to the causal network, it is suggested that the operators have a vague causal network among subsystems, although it is hard to conclude before a wide variety of simulation experiments.

## 5.2 Causalities and Evolutions

Deviation propagation from feedwater temperature to electic power generation is also addressed as a plant state evolution as shown in Figure2. This tendency is common in case that deviation propagation is not too spontaneous nor too slow, i.e., the time delay is recognizable and number of nodes in the propagation root can be held simultaneously in operator's mind.

# 6. Future Directions

Observations of the obtained results described in this paper derived some necessary improvements of the protocol formulation and of the analytical tool.

It is suggested that operators have some approximated range of time constant for each deviation propagation. This attribute may contribute the enhanced precision of operators' mental models and this precision will be crucial in cases that given anomaly has stronger time stress than the simulated anomaly in this experiment.

More than one causal networks were derived from the expressed causal relations. The analytical tool needs some functions to suggest operators' cognitive process to connect together or switch between these seemingly separated causal networks.

# 7. Concluding Remarks

A protocol formulation scheme have been applied to a protocol set obtained from operators' diagnosis against a simulated plant anomaly. Some basic analytical functions have been developed to suggest the possibility for more sophisticated analyses in the future. Mutual feedback between protocol analysis and experiment planning is indispensable to establish an effective methodology to model operators' mental models and their forming processes.

# 8. Acknowledgements

The authors would like to express thanks to Mr. Yutaka Furuhama of Department of Quantum Engineering and System Science, University of Tokyo, for helpful suggestions, and to the people in the MONJU Construction Office of Power Reactor and Nuclear Fuel Development Corporation for friendly support.

# References

1. Ujita, H., Human Characteristics of Plant Operation and Man-machine Interface, Reliability Engineering and System Safety, 119 (1992).
2. N. Moray, Advanced Displays Can be Hazardous: the Problem of Evaluation, Post-HCI Conference, Hieizan, Japan (1995)
3. S. Yoshikawa, et.al., Observations and Trial Formulation of Knowledge Dependency of Plant Operator Behavior,Post-HCI Conference, Hieizan, Japan (1995)

# The Use of PRA in Designing the Westinghouse AP600 Plant

F. Wolvaardt
Westinghouse Energy Systems Europe, Brussels Belgium
C. Haag, S. Sancaktar, T. Schulz, J. Scobal
Westinghouse Electric Corporation, Pittsburgh USA

## 1 Introduction

A full-scope Level 3 Probabilistic Risk Assessment (PRA) has been performed for the Westinghouse-designed advanced passive light-water reactor (AP600) [1]. The PRA study estimates the core damage frequency and release frequency of the AP600 design for both power and shutdown conditions. The analysis also includes an internal fire PRA and an internal flood PRA. A seismic margins analysis has been performed to evaluate the plant's capability to withstand seismic events. Westinghouse designers and PRA analysts worked closely together for over five years to assess the safety of a variety of passive plant design configurations.

## 2 Use of PRA in the Design Process

The AP600 design has evolved over a period of years. PRA techniques have been used since the beginning in an iterative process to optimize the AP600 with respect to public safety. Each of these iterations has included:

- Development of a PRA model
- Use of the model to identify weaknesses
- Quantification of PRA benefits of alternate designs and operational strategies
- Adoption of selected design and operational improvements.

The scope and detail of the PRA model has increased from the early studies as the plant design has matured. This iterative design process has resulted in a number of design and operational improvements. The use of PRA in the AP600 consisted of five distinct stages including: conceptual design analysis (stages 1 and 2), PRA analysis as part of the design certification application (stage 3), and revisions of the PRA in support of further refinement of the design and modeling assumptions (stages 4 and 5).

### 2.1 Stage 1 - Use of PRA During the Early Design Stage

The initial AP600 design incorporated features that were intended to address leading causes of core damage and severe release. These features included passive safety-related core damage prevention and mitigation systems, active nonsafety-related systems, and other plant features. Passive safety-related core damage prevention systems require no support systems other than instrumentation and control, and need fail safe equipment for the most common events. Passive system mitigative features included a reduced number of containment penetrations compared to currently operating plants, the penetrations that are open at power are fail safe, improvement of the interfacing systems loss-of-coolant accident event, and hydrogen igniters are provided in the containment. Other plant features factored into the initial design include a physical separation of electrical and instrumentation and control trains, reduction in the number of flooding sources, and a diverse actuation system for anticipated transients without scram events.

Prior to 1989, several probabilistic scoping studies were performed on the AP600 conceptual design, which concentrated on quantifying the core damage frequency and large release frequency for internal initiating events during power operation. The early studies included detailed models of the passive safety-related fluid systems. They did not include detailed models of other systems such as instrumentation and control. The use of scoping studies was an iterative process at this stage of the design's evolution. The outcome of the scoping study provided insights into the AP600 conceptual design, which led to many design and operational enhancements. Examples of design enhancements include:

- Originally the depressurization system consisted of three stages, each stage contained two lines with two normally closed motor-operated valves. An alternate design was then analyzed which included a fourth depressurization stage off the hot leg with valve types diverse from the first three stages.

- Diverse automatic actuation for certain safety-related functions was introduced. In addition, separate and diverse manual actuation for certain safety-related functions was provided along with alarms and information to the main control room.

- The normal residual heat removal system, with piping routed outside of the containment, was designed with three containment isolation valves to reduce the probability of interfacing systems loss-of-coolant accident events that result in containment bypass.

- Protection system logic modifications are adopted to preclude steam generator overfilling during a steam generator tube rupture event reducing the need for full reactor depressurization.

- The number of onsite power supplies was increased to two nonsafety-related diesel generators.

In addition to plant design changes, some changes to the success criteria were made. In the early stages of the PRA, the success criteria were primarily based on engineering judgement or preliminary design basis analyses. However, during the iterative process of this stage of the PRA, some success criteria refinement was examined. An example is the success criteria originally did not credit the accumulators for mitigation of a small loss-of-coolant accident event. Further examination of the response of the accumulators to small and medium loss-of-coolant accidents indicated that, if the core makeup tanks fail, the accumulators will inject and the core will be cooled provided the operators manually initiate the automatic depressurization system. Thus, the small and medium loss-of-coolant accident success criteria were enhanced.

Operational changes were also evaluated at this stage of the PRA. Initial automatic depressurization system valve positioning is an example of an operational change. Originally, the first three stages each contained two lines with two normally closed motor-operated valves. The valve configuration was changed to one valve open and one valve closed in each line to allow for testing during refueling.

## 2.2 Stage 2 - Preliminary PRA

Beginning in 1989, a preliminary PRA was conducted in support of the Westinghouse AP600 application for design certification. The preliminary PRA was performed on the AP600 design that existed at the time of completion of the scoping studies along with design changes made as a result of the final scoping study. The scope of the PRA was also expanded to evaluate both at-power and shutdown conditions as well as external events. Because the AP600 design was evolving throughout this period, the success criteria were primarily based on engineering judgement derived from preliminary design basis safety analyses. The system and component dependency analysis and the data used in the preliminary PRA were deliberately conservative. The results of the preliminary PRA identified important areas of the AP600 design where the design effort would focus. Examples of specific system design changes made during this stage of the PRA include:

- The in-containment refueling water storage tank system initially consisted of one line containing a normally closed motor-operated valve and two series check valves. To improve the reliability of the injection phase of the system, a second parallel path of two check valves in series was added to the existing two series check valves. Additionally, the motor-operated valve is now normally open.

- To improve the reliability of the sump recirculation function, redundant and diverse recirculation valves were incorporated into the design. The AP600 conceptual design consisted of two parallel check valves from the sump. Diversity was modeled into the design by changing one of the check valves to a motor-operated valve; redundancy was incorporated by making each line contain two valves in series. Thus, the resulting recirculation path consists of one line of two motor-operated valves and one line of two series check valves.

- Alarms are provided in the main control room to inform the operator of mispositioned isolation valves of the passive core cooling system that have remote manual control capability. This reduces the probability of valve mispositioning.

In the first stage of the PRA, the success criteria were primarily based on engineering judgement. For this stage of the PRA, the success criteria were refined. Examples of more refined success criteria include:

- The more significant success criteria changes were for the depressurization system. Taking credit for a design change that increased the size of the fourth-stage valves and performing best-estimate loss-of-coolant accident calculations allowed the use of a success criteria that tolerated multiple failures.

- Analysis shows that the containment cooling system only requires air cooling to prevent containment failure.

Operational changes were also made as part of this stage of the PRA. The normal residual heat removal system and automatic depressurization system provide some examples of operational changes.

- To start the normal residual heat removal system, it was necessary for the operators to locally open three valves. To reduce the operator's burden as to when it was appropriate to actuate normal residual heat removal, an operation change was made so that the operator initiates the system whenever the automatic depressurization system is actuated, with the exception of cases when radiation could leak out of containment. Additionally, the system can now be manually actuated from the main control room instead of using local manual actuation.

- Evaluation of the automatic depressurization system configuration showed that the potential for spurious actuation had increased. The valve configuration was changed to two closed valves with quarterly testing.

## 2.3 Stage 3 - AP600 PRA Submittal to NRC (1992)

The third stage culminated with the submittal of the AP600 PRA report, along with the AP600 Standard Safety Analysis Report (SSAR), to the NRC on June 26, 1992. This stage included a complete Level 3 PRA. The PRA factored in design changes made as a result of the preliminary PRA findings. The success criteria assumptions were verified. Some of the conservative data and dependency factors were adjusted to be more realistic during this stage. Because of the extensive interactions during previous design/PRA studies, few plant changes resulted from this study. Two design changes that did result include:

- The core makeup tank can now be actuated on a low steam generator level plus high hot leg temperature indication. This was done to indirectly reduce the importance of operator actions to initiate passive feed and bleed.

- The scope of the diverse actuation system was expanded to include control rod insertion. The system was also expanded to include an actuation signal for opening of the in-containment refueling water storage tank motor-operated valves during mid-loop operations. This was done to provide automatic operation to reduce the dependence on operators to open the valves in the event of an accident during mid-loop operation.

## 2.4 Stage 4 - PRA Revision 1 (1994)

Stage 4 was the first revision to the AP600 PRA. The revision, submitted in July 1994, included the following major changes: introduction of phenomonology onto the Level 2 containment event tree and performance of the risk-based seismic margins analysis. This stage also included the focused PRA sensitivity study and initiating event evaluation as part of the regulatory treatment of nonsafety-related systems topic.

In September 1993, the focused PRA sensitivity study and initiating event evaluation were submitted to the NRC. The focused PRA sensitivity study evaluated the core damage and large release frequencies for AP600 without taking mitigation credit for nonsafety-related systems. The results of the study show that even with no credit taken for nonsafety-related systems, AP600 meets the regulatory goals.

The Level 2 PRA was revised to introduce the use of decomposition event trees and to incorporate phenomena onto the containment event tree. Six decomposition event trees were created to analyze the following phenomena:

- In-vessel retention of molten core debris
- Thermally induced failures of the reactor coolant system pressure boundary
- In-vessel steam explosion
- Ex-vessel steam explosion
- Ex-vessel debris coolability
- Hydrogen combustion analysis.

A risk-based seismic margins analysis was also performed as part of Revision 1 of the AP600 PRA.

There were no appreciable changes in the plant design as a result of this stage of the PRA.

## 2.5 Stage 5 - PRA Revisions 2 - 6 (1995)

This stage includes the updates leading to various revisions submitted to the NRC during 1995. The changes made to the PRA resulted from plant changes and NRC questions. Most plant changes incorporated into the PRA were made for other reasons than the PRA. The design changes resulted in small improvements to the core damage and large release frequencies. The primary emphasis of this stage of the PRA was to refine the success criteria calculations and the system and event tree modeling. Some of the PRA-related feedback to the design is summarized below.

- Further refinement of the PRA success criteria calculations resulted in making the automatic depressurization system success criteria more conservative.

- Based in part on a PRA sensitivity study, the automatic depressurization system stage 4 valves were changed from air-operated to explosive-operated (squib) valves .

- Service water blowdown procedures and sources of makeup water were evaluated as a function of service water heat loads during various plant conditions to ensure that the assumed success criteria will be met. The heat loads were also evaluated to assess the required number of cooling tower fans that must operate to ensure adequate cooling. In addition, the initial fault tree evaluated indicated the potential vulnerability to bypass flow occurring upon loss of a dc power supply causing an air-operated valve to open. Consequently, the power supplies to the equipment were reevaluated.

- Initial PRA modeling of the need to open the main generator breaker in the fault trees for the 4160 vac buses following a plant trip highlighted the importance of certain functions initially assumed to be performed by the plant control system. It was determined that the plant control system would not be fast enough to perform this action and that a reverse power relay would control opening this breaker.

# 3 Concluding Remarks

The total AP600 core damage frequency due to internal initiating events from all modes of plant operation is assessed to be less than $5x10^{-7}$ events per year. The AP600 design is more than one order of magnitude better than the safety goal of $1x10^{-5}$ events per year as required by the Utility Requirements Document and more than two orders of magnitude better than the current safety goal of $1x10^{-4}$ events per year as set forth by the U.S. NRC. This low AP600 core damage frequency can be attributed to the added redundancy and diversity which has been incorporated into the design. Not only does the AP600 meet the ALWR and NRC safety goals, but the analysis shows the plant design is not overly dependent on nonsafety-related systems or any one passive safety system.

By using PRA techniques such as initially performing scoping studies and a preliminary PRA, the initial AP600 conceptual design has been enhanced to the point of not only meeting but exceeding the nuclear industry safety goals. The results of the AP600 PRA demonstrate the benefits of integrating PRA into the nuclear plant design process.

# 4 References

1. AP600 Probabilistic Risk Assessment, Prepared for the U. S. Department of Energy, DE-AC03-90SF18495, Westinghouse Electric Corporation, Revision 6, November 1995.

# PSA-Based Optimization of Technical Specifications for the Borssele Nuclear Power Plant *

Ad Seebregts
ECN Nuclear Energy, Petten, The Netherlands
Herman Schoonakker,
N.V. EPZ, Borssele, The Netherlands

## 1 Introduction

The Borssele Nuclear Power Plant (NPP) is a Siemens/KWU 472 MWe Pressurized Water Reactor which has been in operation since 1973. In 1989, a Probabilistic Safety Assessment (PSA) program was initiated to complement deterministic safety studies and operational experience in forming a plant safety concept. In 1993, the PSA-MER model was completed and used to determine the effects a package of proposed modifications would have on plant safety and risks to the environment. This model was used to start retrospective risk profile and allowed outage times (AOTs) analyses, which both concerned the calculation of the change in total core damage frequency (TCDF) given a change in configuration. The main problems identified and reported in this paper are: (i) How to calculate the change in TCDF (ΔTCDF)? (section 3); and (ii) How to set practical decision criteria and how to use the PSA as extension to Technical Specifications (TS) AOTs? (section 4). Finally, a pilot study was conducted in order to optimize surveillance test intervals (STIs) which are also part of the TS (section 5).

## 2 Borssele NPP Technical Specifications

The Borssele NPP TS define limits and conditions to assure that the plant is operated safely and in a manner consistent with the assumptions in the safety analyses. The TS requirements are based on deterministic analysis and engineering judgement. A PSA can be used to improve or extend these requirements [1], in particular the AOT and STI requirements.

An AOT specifies the maximum allowed time for a component to be unavailable in a given plant state. If the availability cannot be restored within the AOT, the plant must be placed in a safer state, usually in a (hot or cold) shutdown state.

The Borssele TS AOTs are only specified for one component or combinations of components within one system at a time. So, simultaneous unavailability of components belonging to different systems is not addressed. The PSA can be used:

- to quantify the risk impact (i.e., impact on TCDF) of a TS AOT; and

---

* This work was partly sponsored by the Dutch Ministery of Economic Affairs

- to set AOTs for combinations of components not yet included in the TS AOTs.

Quantification of the risk impact means the calculation of ΔTCDF given a change in configuration. This is what is done in the retrospective risk profile analyses, see section 3. The next problem is how to define a criterion to bound ΔTCDF, which is addressed as part of the PSA-based AOT analyses, see section 4.

An STI requirement specifies the frequency with which the surveillance tests must be performed. By testing, the risk associated with undetected failures (of standby components) can be limited. The PSA can be used to optimize STIs and test arrangements (e.g., staggering of tests), see section 5.

# 3 Retrospective Risk Profile

In April 1994, EPZ started to use the PSA-MER to calculate off-line the monthly risk (i.e., TCDF) profile of the plant. To date, only the power operational state has been included. It is the intention to use the shutdown part of the integrated PSA-3 model (available since 1995) to generate the risk profile during the 1996 refuelling outage.

The objectives of the risk profile analyses were: (i) to gain experience with the PSA model (which was largely made by the external contractor NUS); (ii) to obtain an impression of the variation in TCDF during operation; and (iii) to assess the applicability of the PSA for plant operational activities.

In principle, for each configuration (i.e., combinations of unavailable components), the whole PSA model needs to be requantified because:

- an alternative configuration may require a change in the fault tree model in order to incorporate possible re-alignments because of unavailable components;
- unavailable components could be present in possibly previously truncated cut sets which should now enter the final TCDF cut set equation.

On the average, 10 to 20 changes in configuration occur per month. The computer solution time needed for one combination based on a total requantification of the PSA model is about two hours; modifying the model input and analyzing the results takes a few days. As a faster alternative, simple NUPRA (the PSA software used) sensitivity analyses based on a priori calculated TCDF cut set equations were performed. Two TCDF cut set equations were used: (i) based on normal truncation limits and (ii) a truncation limit lowered by a factor of 10. Generation of (i) takes approx. 2 hours; (ii) takes approx. 6 hours, while the number of cut sets increases by a factor of 2 compared with (i); the resulting TCDF is about 1% higher than (i). Figure 1 shows the results of these two cases compared with the total requantification of the model. It was concluded that such fast sensitivity analyses provide useful insights assuming that sufficiently low truncation limits are used (i.e., sufficiently comprehensive cut sets) and the analyst is aware of the limitations of such sensitivity analyses. However, total requantification is still preferred.

The risk profiles assessed sofar (April 1994 to December 1995) show TCDF increases well below internationally applied decision criteria (see also section 4).

The analyses have proven to be very successful in communicating possible uses and benefits of PSA to the plant operators. Originally, the control room personnel were reluctant to use additional tools to the TS. Because of the success of the risk profile calculations, this position is being reconsidered.

events are not included for cases in which the system is automatically realigned upon demand. These three types of unavailabilities are combined into one basic event (at a system or train level). Unavailabilities modelled in the fault trees can result in multiple system unavailabilities which may be a violation of the TS. In the accident sequence (event tree) quantification, minimal cut sets containing such violations are eliminated from the results (so-called Disallowed Maintenance Combinations). This modelling is very efficient and suited for average risk assessment and identification of relatively weak points in the design, but it has some drawbacks for optimization of the STIs or periodic maintenance, because:

- the different contributions (test, maintenance, and repair) are not clear from the PSA data base;
- it is not obvious from the PSA documentation which tests can be overrided in case of a demand;
- actual test and maintenance times cannot be taken into account.

Moreover, the NUPRA software does not contain time-dependent unavailability models, which is necessary for optimization of the STIs and actual test timings (e.g., see [2]). Therefore, a tool additional to NUPRA had to be developed which could be used the PSA database and the TCDF cut set equation, supplemented with information on actual test placings. The result is a time-dependent plot of the TCDF. Not only variations in the configuration can be taken into account (see also section 3), but also the time-dependent unavailability of periodically tested and maintained components between two instants of tests or maintenance. Due to a lack of data, several scoping analyses with simplifying assumptions were performed to:

- compare the average TCDF with a time-dependent TCDF;
- show the effect of using a time-dependent unavailability model for the periodically tested and maintained components in order to optimize test intervals and test arrangements (e.g., staggering tests).

Figure 2 shows examples of this last effect. The following situations are shown: (1) average unavailabilities; (2) time-dependent unavailabilities; and (3) time-dependent unavailabilities in combination with a perfect staggering scheme. The differences between the averages are 3-4%.

These preliminary findings indicate that the tool can be used for indicative optimization purposes and for comparison with the off-line risk profiles. However, data is needed on the actual timings of tests and periodic maintenance. In addition, the relative contributions of test, scheduled maintenance, and repair in the basic events for test and maintenance should be known.

# 6 Conclusions

The main conclusions of the work can be summarized as follows:

- The Borssele PSA is suited for off-line risk profile and AOT applications. For the optimization of surveillance tests and periodic maintenance, actual test and maintenance data should be collected and fed into the PSA model, which probably needs some modification.
- The calculated PSA-based single AOTs results show that the TS AOTs are generally more restrictive than the PSA-based AOTs using the TCDF increases from continued operation as a criterion.

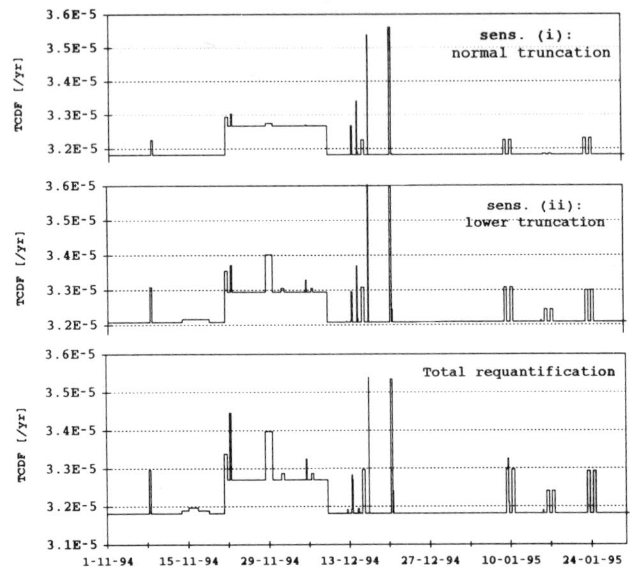


Figure 1 Retrospective risk profile results with TCDF calculated in three different ways

# 4 Allowed Outage Times and Decision Criteria

The PSA-MER was also used for a pilot study on PSA-based AOTs. The objectives were: (i) to perform an indicative evaluation of AOTs of the most important Borssele NPP systems and components; (ii) to compare these PSA-based AOTs with the TS AOTs; and (iii) to establish practical decision criteria.

The analyses were restricted to 'single' AOTs (i.e., one component/train unavailable) and primarily focused on a comparison with the deterministic TS.

The criterion used was based on the TCDF increase from continued operation [1]:

$$\Delta TCDF \text{ x } AOT \leq B$$

where ΔTCDF equals the increase in TCDF as a result of the unavailability of a component or system, and B equals a dimensionless fixed limit. The use of the limit value B is based on the reasoning that ΔTCDF above an accepted safe level would be tolerable for only a limited period. The higher ΔTCDF is above the acceptable level when equipment is unavailable, the less time the plant would be allowed to operate. The constant B represents the highest acceptable TCDF increase integrated over the duration T, which is specified by the AOT. Choosing a specific value for B is a matter of opinion (more or less subjective). Ref. [1] uses a value of 5E-7 for B, which was used as starting point for the Borssele PSA-based AOT analyses.

Because requantification of the PSA-MER model was too time-consuming for the large number of AOT analyses required (about hundred), sensitivity analyses based on the TCDF cut set equation were used to compute the ΔTCDFs. Requantification of the total PSA model confirmed the results of the most important AOTs.

The study concluded that the PSA-based single AOTs are generally much longer than the TS AOTs, except for the components given in Table 1. The short PSA-based AOT of the CM bus is explained by a conservative asymmetry in the PSA model. Normally, 2 of the 3 VF (Emergency Cooling Water) pumps are fed by CM, and the remaining VF pump by bus CL. This configuration is modelled in the PSA. However, one of the VF pumps, assumed to be connected to CM, can be switched to CL. This switching-over takes 0.5 hour, which is below the calculated PSA-based AOT. Therefore, the CM bus AOT requires no additional investigation.

Table 1 Components with PSA-based AOTs shorter than or close to TS AOTs

| Component | TS AOT (hours) | PSA-based AOT (hours) |
|---|---|---|
| RZ storage tank RZ001/2/3/4-B001 | 8 | 8.3 |
| AC 380V main emergency power bus CM | 8 | 0.6 |
| Valves TJ011-S007/6 and TJ012-S007/6 | 24 | 23.5 |

The results of these AOT analyses lead to a discussion on criteria and limits to be used for setting PSA-based AOTs and for application in the risk profile calculations (see section 3). The current opinion is to maintain existing TS AOTs. For situations not yet included in the TS (mainly simultaneous unavailability of different systems), the following PSA-based extensions are under consideration:

- use $\Delta$TCDF x AOT $\leq$ B, with B set to a percentage of the average TCDF (i.e., a relative limit);
- use practical multilevel instead of continuous AOTs (e.g., see Table 2);
- limit the cumulative yearly AOT contributions by a percentage of the average TCDF ($\Sigma_i$ $\Delta TCDF_i$ x $AOT_i \leq B$); and
- choose a smaller B for planned unavailabilities of safety system components, e.g., 10 % of B used for unplanned unavailabilities.

Table 2 Allowed $\Delta$TCDF per AOT class: unplanned unavailabilities

| $\Delta$TCDF (%) | > 2000 | > 600 | > 100 | > 20 | > 2 | $\leq$ 2 |
|---|---|---|---|---|---|---|
| AOT | 0 (Shutdown) | 8 hour | 1 day | 1 week | 1 month | 1 year |

Another alternative would be to base the criterion not only on the $\Delta$TCDF during operation, but also to compare this increase with the possible TCDF decrease when shutting down the reactor. The development of such a criterion is now feasible given the availability of the Borssele PSA-3 model for the shutdown operational states.

# 5 Surveillance Tests and Periodical Maintenance

Based on the PSA-3 power model, a pilot study of PSA-based optimization of STIs was initiated. The objective of this study was to assess whether the PSA-3 model is suitable for optimization of STIs and periodic maintenance.

Component test and maintenance unavailabilities are modelled in the system fault trees. Normally, test and maintenance unavailabilities are modelled separately from the hardware faults. This also applies for repair unavailabilities. Test unavailability

• Multilevel, relative, and cumulative criteria are currently being considered as most suited for setting PSA-based AOTs in situations not yet included in the TS.
• Conservatisms and asymmetry in the fault tree models need to be investigated in more detail for future PSA applications.

## References


1 B. Atefi, D.W. Gallagher. Feasibility Assessment of a Risk-Based Approach to Technical Specifications. NUREG/CR-5742, Washington D.C., 1991

2 G. Johanson, J. Holmberg (editors). Safety Evaluation by Living Probabilistic Safety Assessment - Procedures and Applications for Planning of Operational Activities and Analysis of Operating Experience. SKI Report 94:2, NKS/SIK-1(93)16, 1994


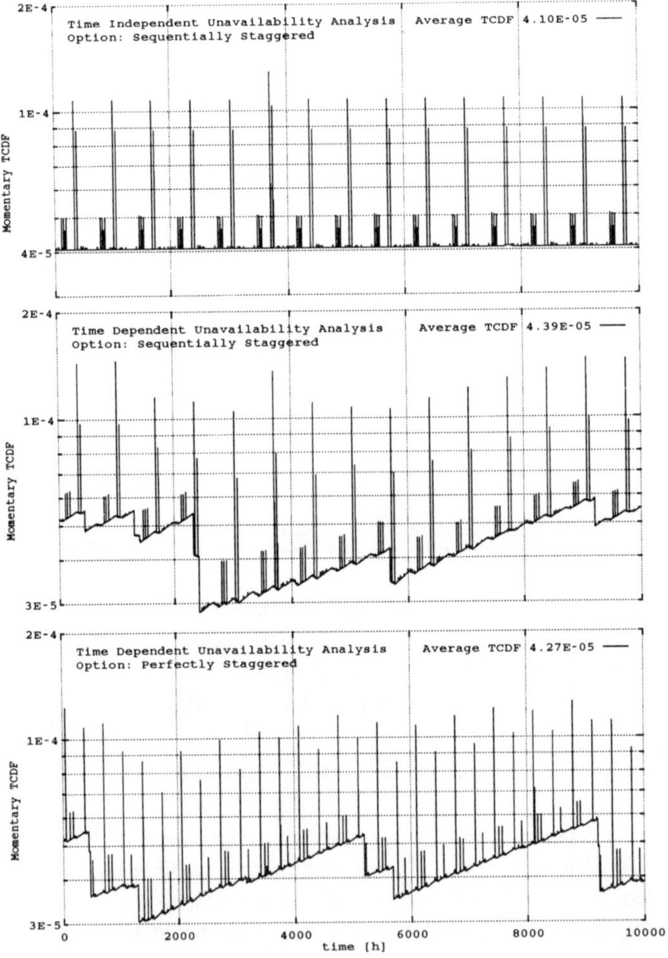


Figure 2 Time-dependent unavailability calculations for optimization of STIs

# Comparative Evaluation of Severe Accident Risks Associated with Electricity Generation Systems

S. Hirschberg and G. Spiekerman
Paul Scherrer Institute
Villigen, Switzerland

**Abstract**

This paper addresses experiences from a systematic investigation of severe accidents associated with electricity generation. The scope of the analysis covers full energy chains. The work has been focused on the actual experience data as reflected in a number of existing databases and in numerous other sources. As a result, a comprehensive database was established. It appears to have a significantly higher degree of coverage of energy-related accidents than other known sources. The evaluations concentrate on health effects associated with severe accidents; whenever feasible other types of consequences are also being addressed.

## 1 Introduction

Comparative assessment of the risks associated with electricity generation plays an increasingly important role in the strategic evaluations of national policy options. At the same time the topic is controversial and the statistical basis for such evaluations has been relatively poor. In particular, severe accidents have not been given the attention they deserve in the available, otherwise comprehensive, studies on external costs of power generation.

Early results of the present work were summarized in 1994 [1]. Since that time the accident database ENSAD (Energy-related Severe Accidents Database), established by Paul Scherrer Institute (PSI), has been greatly expanded and consolidated; at the same time a number of detailed data analyses were carried out. Apart from a much increased (by 44.7%) number of energy-related accidents included in the database, the most important advancements concern:

- extension of the spectrum of consequences considered for the different energy chains, particularly for oil and gas chains
- detailed examination of the risks associated with hydro power and of the consequences of the Chernobyl accident

In a parallel effort the Probabilistic Safety Assessment (PSA) technique was used to evaluate the economic consequences of nuclear accidents[2]. This was followed by a detailed review of other studies addressing this topic, including comparisons with own results [3].

In this paper some results examples are provided. We refer to[4] for the complete analysis and the full spectrum of results.

## 2 Some Facts about ENSAD

ENSAD is not limited to energy-related accidents. However, they are in focus of the data analyses carried out and having the highest priority are far more complete in comparison with the information on accidents in other sectors.

As of January 10, 1996 the total number of accidents included is 9844, of which 3312 are energy-related. Among the energy-related accidents 1239 are severe according to the definition of severe accidents used by us. The definition specifies a number of consequence types (fatalities, injuries, evacuation, ban on consumption of food, releases of hydrocarbons and other chemicals, enforced clean up of land or water, economic loss), and a minimum numerical level of damage for each of these categories, necessary for classifying a specific accident as severe.

The nature of the work on a severe accident database is such that the scope of this task can always be expanded. First, there exists an enormous amount of sources of information on past accidents. These sources are of varying quality and depth; they may be partially overlapping or may in some cases supplement and in other contradict each other. Second, there is a flow of new information concerning recent accidents. As a result, the evaluations of severe accidents are constantly changing. The result examples provided here cover accidents that occurred until the end of year 1992. There is always a substantial time lag with respect to the inclusion, accounting for and possible analysis of the accidents in the multitude of information sources used as the input. At this time, according to our judgment acceptable completeness and quality can only be achieved for the period until the end of 1992.

## 3 Examples of Energy Chain-specific Results

Different parts of the fuel cycles provide the dominant contributions to the consequences of accidents, which depending on the specific configuration of a national electricity supply, may affect prioritizations in the context of comparative evaluations. For example, most of the accidents in the coal cycle occur in mining while within the oil cycle long range transportation and local distribution are clearly dominant; in the case of nuclear and hydro (large dams), the risks are clearly associated with the power plants. Nuclear and hydro-based systems together supply 98% of the total electricity generated yearly in Switzerland and are, consequently, focused on in the context of the Swiss energy policy. These two energy sources exemplify the difficulties with respect to the applicability and transferability of accident data based on past world experience.

When evaluating hydro accidents a number of essential dam characteristics were considered. Large dams having purposes other than power production (such as irrigation, flood control, water supply, recreational) were included. This allowed a significant extension of the statistical material while many of the insights gained are independent of the purpose of the dam.

Figure 1 shows the estimated critical dam failures rates for the different types of dams and for two time periods, For the period 1850 - 1992 all known dam accidents were included while for the period 1930 - 1992 the accidents involving dams taken into operation before 1930 were excluded; the background for choosing these time boundaries is the fact that dams built before 1930 with few exceptions use masonry as the construction material while the structurally stronger concrete has become the dominant material after 1930. Furthermore, a separate evaluation was made excluding accidents that occurred within five years after the first filling of the reservoir. The results show significant differences between the various types of dams, lower failure rates for dams built from 1930 and on, and lower susceptibility to accidents after five years of operation. Other evaluations have shown differences between frequency-consequence (f-N) curves for dams in Asia and Africa on the one hand and dams in America and Europe on the other. As illustrated by Figure 2 the latter show significantly lower risks.

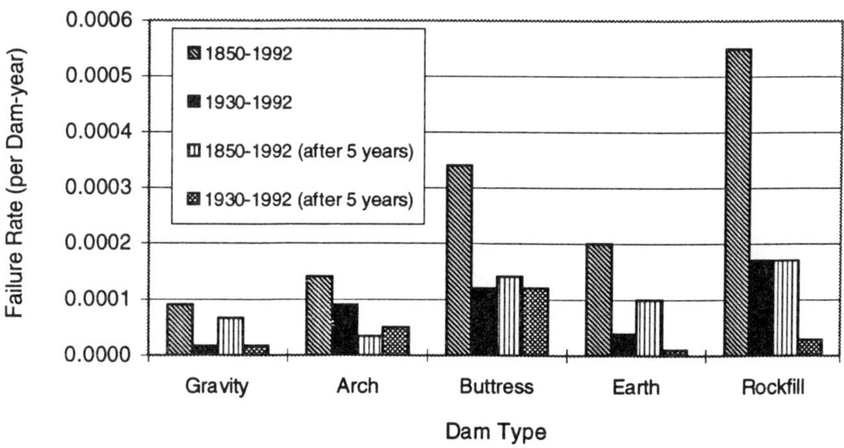


**Figure 1 Failure rates for different dam types and for different time periods.**

The Swiss dams were examined with respect to a number of characteristics that are important in the context of evaluating the potential severe accidents. Examples include: type of dam (dominance of gravity dams), height, capacity and quality of supervision. In most cases the prevailing characteristics of the Swiss dams are favorable from the risk point of view.

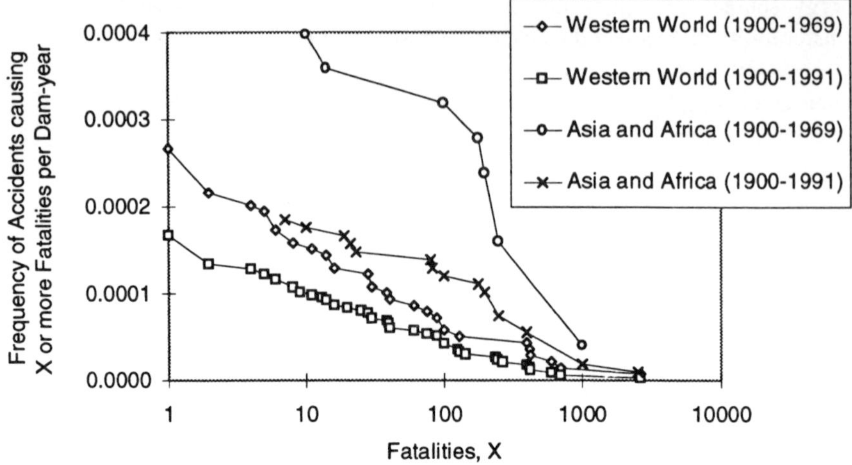


**Figure 2 Frequency of exceedance of number of fatalities due to dam accidents in Asia and Africa and in the western world, plotted for two time periods.**

The limitations in applicability of past accident data to cases that are radically different in terms of technology and operational environment are evident in the case of the nuclear fuel cycle. For the nuclear chain only two major accidents occurred in the past (Three Mile Island and Chernobyl). The latter led to catastrophic health and environmental consequences. Given the scarceness and very limited applicability of such data, representative results for the western plants with state-of-the-art safety standards can only be obtained through PSA applications. This aspect is highlighted by examining recent studies of external costs associated with severe reactor accidents, carried out in Switzerland and elsewhere. Figure 3 shows the results of the different studies. Although the values in the figure cover a range of some five orders of magnitude all recent studies (which as opposed to several older ones do not use Chernobyl as the reference for the consequence assessment), show results below 0.1 cents per kWh, unless risk aversion is included. Different approaches to the estimation of risk aversion and the problems encountered are briefly discussed in [3]. The same reference categorizes 11 published studies into one of the three types of analysis used in the context of the overall external cost assessment ("top-down", limited "bottom-up" and full scope "bottom-up" ). The full scope "bottom-up" approach, utilizing modern and comprehensive PSA represents the current state-of-the-art. However, among the published studies only two (including [2]) fully implemented this approach.

Regarding fossil fuel cycles substantial completeness improvements have been achieved for oil and gas chains. As an example, the database contains 60 on- and offshore oil spills exceeding 25000 tons in the period 1967 - 1992. Most of the severe oil accidents occur either during the transport to refinery or when the oil is regionally distributed. In the case of natural gas the accidents occur predominantly in the long distance transport and in the regional or local distribution.

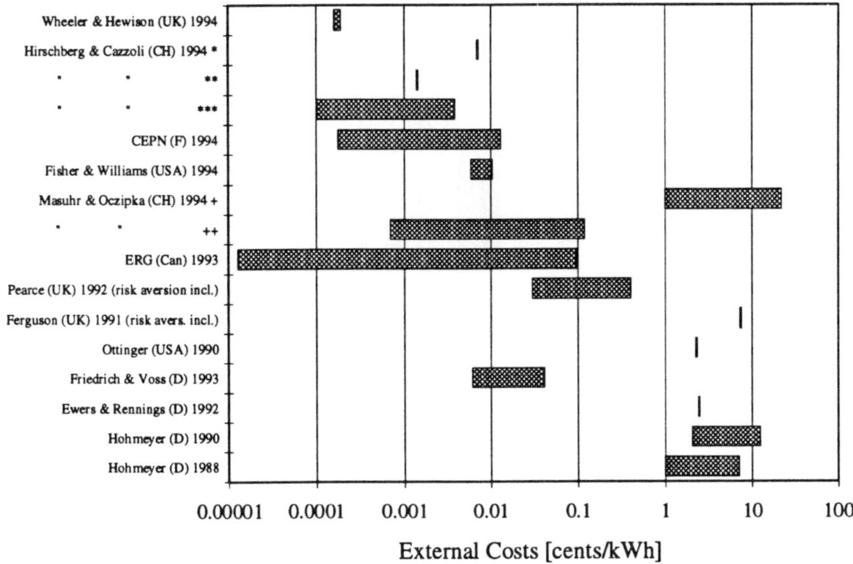


*) Zion (mean). **) Peach Bottom (mean). ***) Mühleberg.
+) Under "risk awareness". ++) Under "risk neutrality".

**Figure 3 Span of estimated external costs of severe accidents.**

# 4 Comparative Evaluations

Figure 4 shows the estimated number of immediate fatalities, injuries and evacuated persons per unit of energy for six energy chains; only accidents with at least 5 fatalities, 10 injuries and 200 evacuated, respectively, have been included. With the exception of Liquid Petroleum Gas (LPG), which is included for comparison, all other energy chains represent different means for electricity production.

It should be noted that the completeness of the data is much higher for fatalities than for injuries or evacuations; particularly poor is the information on evacuations associated with hydro power. Furthermore, since the figure only shows immediate fatalities, the delayed ones, particularly relevant for the Chernobyl accident, need to be treated separately. The current best estimate of the Chernobyl-specific delayed fatalities, primarily based on the assessed doses received by the emergency workers and by the public, is about 14 fatalities per GWe*a [4]. On the other hand, probabilistic plant-specific estimates of normalized number of latent fatalities

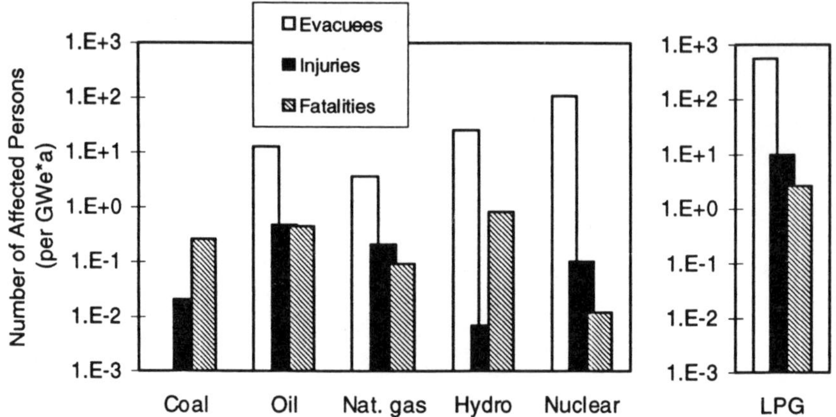


**Figure 4 Comparison of energy-related severe accident records for the period 1969 - 1992: immediate fatalities, injuries and evacuated persons per unit of energy.**

are in the case of nuclear power plants with good safety standards several orders of magnitude smaller than this estimate (typically, in the range 0.01 - 0.1 fatalities per GWe*a). This is due to differences in terms of technology and operational environment.

Figure 5 shows frequency-consequence curves obtained for the coal, oil, gas and hydro chains, based on world-wide accidents. As opposed to Figures 1 and 2 only dams used for power generation are included here.

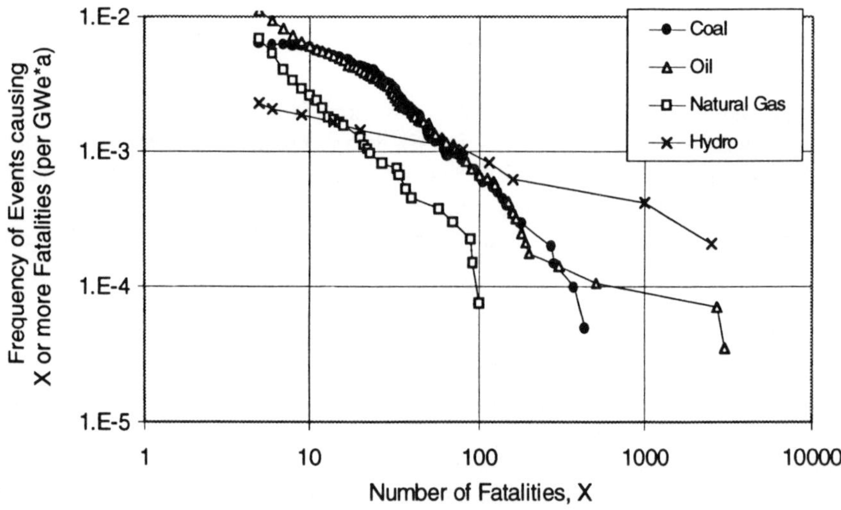


**Figure 5 Frequency-consequence curves for different energy chains in the period 1969 - 1992, normalized by electricity production.**

The coal and oil chains have the highest frequencies of accidents up to the level of about 70 fatalities while hydro has the lowest. For higher levels of consequences the situation becomes reversed. As opposed to the other energy carriers the fatalities in the coal chain are predominantly occupational.

# 5 Concluding Remarks

The completeness of the state-of-knowledge on past energy-related accidents has been significantly improved. Nevertheless, when using this information the applicability of the data to the country- , site- and plant-specific conditions must be carefully examined.

Use of PSA techniques may in some cases be the only relevant approach. In the context of the estimation of external costs associated with severe nuclear accidents the probabilistic approach has been employed to a different extent and with varying stringency.

Apart from the issue of the applicability and transferability of data, there are other matters which either remain open or present inherent practical difficulties in comparative assessment of severe accidents. These issues are elaborated in [4] and include:

- Non-uniform level of knowledge and limited scope of applications of risk analysis;
- Difficulties to cover a wide range of consequences in a consistent manner;
- Treatment of the distribution of impacts in time and space;
- Treatment of risk aversion and non-quantifiable social detriments associated with extreme accidents.

## References


1. Hirschberg S. and Parlavantzas C. Severe accidents in comparative assessment of energy sources: Current issues and actual experience data. Proceedings of PSAM-II, 20 - 25 March 1994, San Diego, California.
2. Hirschberg S. and Cazzoli E. Contribution of severe accidents to external costs of nuclear power. Proceedings of ENS Topical Meeting on PSA and Severe Accidents, 17 - 20 April 1994, Ljubljana, Slovenia.
3. Hirschberg S. External costs of nuclear reactor accidents. Contribution to the OECD/NEA Report on Methodologies for assessing the economic consequences of nuclear reactor accidents. To be published 1996.
4. Hirschberg S., Spiekerman R. and Dones R. Severe accidents in the energy sector. PSI Report, Paul Scherrer Institute, Würenlingen and Villigen, to be published 1996.

# Role of PSA in Improving Plant Safety During Shutdown Operation

Tae-Yong Sung, Jin Hee Park and Tae-Woon Kim
Advanced Research Group
Korea Atomic Energy Research Institute
Taejon, Korea

**Abstract**

Probabilistic safety assessment is a useful tool to evaluate and improve plant safety. During plant shutdown, operator action is the most important for mitigating the event after initiating event occurs. Because PSA can evaluate the reliability of human action, PSA technique is useful to evaluate and improve plant safety during plant shutdown. An overdraining event was evaluated and the results showed that an operator error was a dominant contributor. Two plant improvement options were selected to reduce the operator error probability. One was modification of abnormal operating procedure and the other was installation of additional level indicator independent of tygon tube. The quantification results of modification alternatives show that the modification options derived from the risk-based approach can indeed improve the plant safety. However many problems for LPS PSA methodology should be resolved to get reasonable results and to achieve efficacious plant safety improvement.

## 1. Introduction

The Probabilistic safety assessment (PSA) has been used widely in the process of meeting the requirement of the regulatory body, validating the design of nuclear power plant, and improving its safety. Most of the analyses had been performed for full power operation since the risk during shutdown had been regarded as low due to comparatively low decay heat. However, recent studies indicated that the risk during shutdown was not negligible and that the most vulnerable period for PWR was the mid-loop operation [1, 2].

Because most of the safety systems can not be actuated automatically during plant shutdown, operator action is the most important for mitigating the initiating event. Since the PSA technique can evaluate the reliability of human action, it can be used to evaluate and improve plant safety during shutdown operation.

To improve the plant safety and to obtain the insight for the low power and shutdown (LPS) PSA methodology, a PSA case study was performed for an overdraining event for the Yonggwang nuclear power plant units 3, 4 (YGN 3, 4) which are both Combustion Engineering 1000 MWe plants.

# 2. PSA Model

## 2.1 Plant Condition

Because the YGN 3, 4 have no experience of the midloop operation, the plant initial conditions were assumed based on the operating experience of the other PWRs in Korea and the technical specifications for the YGN 3, 4.

1) The overdraing event occurs at 86 hours after reactor shutdown. Decay heat is 0.38% of full power at this time.

2) The temperature of the reactor coolant system (RCS) is 60°C.

3) The level of the RCS is at midplane of hot legs.

4) The RCS has an opening of 3.81 cm (1.5") diameter through reactor vent, and the vent size is not enough for gravity feed operation.

5) The SGs are not isolated from the RCS and feed water is not available.

## 2.2 Event Tree Analysis

For the lack of abnormal operating procedures and the thermal-hydraulic analysis for the loss of SCS event during RCS reduced inventory operation for the YGN 3, 4, the event tree was developed based on the discussion of a group of thermal-hydraulic analysts and plant staffs. The success criteria were made by engineering judgments and assumptions.

Figure 1 shows the event tree for overdraining initiating event during drain-down for the first mid-loop operation. The heading of the event tree and the success criteria of top event are described as follows.

1) OVIE : overdraining initiating event. The event occurs when operator reduces the RCS coolant inventory excessively during drain-down process. If the RCS is being drained below the hot leg midplane, the loss of operating shutdown cooling pump due to air entrainment may be resulted. Review of overdraining event experiences revealed that incorrect level indication was the principal contributor to the events [3]. The frequency of initiating event was evaluated as 1.602e-2 /demand using US PWR experience data.

2) OP : operator diagnosis of the initiating event. After the operating SCS pump is tripped, the operator diagnose that the cause of pump trip is overdraining correctly. If the operator fails to diagnose the cause, the operator may consider that the cause is hardware related failure and will start the 2nd SCS pump. The started 2nd SCS pump will be also tripped due to air entrainment. In this analysis, it is conservatively assumed that the tripped pumps could not restart within 24 hours.

3) MU : coolant makeup. For the overdraining event, it is necessary to raise the coolant level above the hot leg midplane before starting the 2nd SCS pump. The makeup can be accomplished by high pressure injection system (HPSI), low pressure injection system (LPSI), charging and volume control system (CVCS),

and containment spray system (CSS). The success criteria of top events are operator's starting one pump and aligning the flow path from refueling water tank (RWT) to RCS.

4) REC : recovery of the 2nd SCS pump. This top event represents the restoration of one SCS using the 2nd SCS pump to remove decay heat.

5) F&B-I : feed and bleed I. This event represents primary feed and bleed operation using one HPSI train and one safety depressurization system (SDS) valve after RCS inventory makeup is succeeded.

6) F&B-II : feed and bleed II. This event represents primary feed and bleed operation using one HPSI train and one SDS valve after operator fails to diagnose the overdraining.

7) LTC : long term cooling. This event is succeeding operation of feed and bleed operation. As the feeding is continued, RWT will be empty. The suction of HPSI pump should be changed from RWT to containment sump.

## 2.3 Fault Tree Analysis

All fault trees for headings in the event tree were developed using the fault tree models of full power PSA for the plant analyzed [4]. The following modifications were made to these fault trees to fit the plant operating status.

1) All the automatic signals were not modeled.

2) Front-line and support systems train B were not available due to maintenance except SCS train B and related subsystem.

Some systems are used in many headings for different functions, as the HPSI is used for inventory makeup, feed operation, and long term cooling. Some systems share common components and piping, as the HPSI, LPSI, CVCS and CSS share the suction line and related components from RWT for inventory makeup. The alignments of systems used in later heading were changed by the conditions of previous headings. The fault trees for all conditions were developed in consideration of relation between systems and headings.

Human reliability analysis (HRA) has been performed mainly based on the ASEP (accident sequence evaluation program) HRA procedure [5] which has been used for HRA of full power PSA. Since the plant condition and work environment during shutdown are much different from the normal operation, some part of the procedure was modified to incorporate the situational differences.

## 2.4 Quantification and Results

The event tree was quantified by the fault tree linking method using KIRAP [6]. In the quantification, the relationships between the fault trees were treated by complement event.

There were 5 core damage accident sequences in the overdraining event tree. The accident sequences 7 and 8 are dominant sequences. The results showed that the major contributor of core damage was an operator error to diagnose the overdraining event (OP).

# 3. Sensitivity Study

## 3.1 Plant Modification Alternatives

Two plant improvement options were derived to reduce the operator error to diagnose the overdraining event (OP). Option 1 was modification of the operating procedures and operator training, by which operator could use the CSS as a backup of the shutdown cooling system (SCS). At the YGN 3, 4 the CS pump can be used as a backup of SC pump (Fig 2). However, abnormal operating procedure for the loss of shutdown cooling for the YGN 3, 4 does not describe the event during reduced RCS inventory, and normal operating procedure for drain-down of RCS does not describe the backup capability of CSS.

Option 2 was modification of the hardware, i.e. installation of additional level indication system independent of tygon tube. Additional independent level indicator can help operator to diagnose the initiating event and reduce the occurrence frequency of initiating event.

Three plant modification alternatives were made, case 1; the use of CSS for backup of SCS (option 1), case 2 ; additional level indication system (option 2), and case 3 ; simultaneous adoption of options 1 and 2.

## 3.2 PSA Models for Plant Modification Alternatives

PSA analysis for plant modification alternatives was performed to evaluate the effectiveness of the options. For cases 1 and 3, an event tree was developed to model the CSS as a backup of SCS. Figure 3 shows the event tree for cases 1 and 3. The headings which are different from Figure 1 and success criteria of them are described as follows.

1) OP-BAK : operator re-diagnose the initiating event. After the 2nd SCS pump is tripped as a result of mis-diagnosis of initiating event, the operator re-diagnose that the cause of pump trip is overdraining correctly.

2) RE-CSS : recover SCS using CSS. This top event represents the restoration of shutdown cooling function using CSS pump to remove decay heat.

The effect of option 2 was modeled in human reliability analysis for OP. In this analysis, the effect of reducing initiating event frequency is not modeled because of the difficulty in the evaluation of the effect . The human error probability of OP was reduced by a factor of 10 by option 2.

# 4. Results and Discussion

The core damage frequency (CDF) is used as a measure of risk for the base case (current design) and plant modification alternatives. The case-by-case quantification results are shown in Table 1 in the ratio of CDF to base case. The procedure change for using CSS could reduce the CDF by a factor of 20. The additional level instrumentation could reduce the CDF by a factor of 7 and the case 3 which adopted both options could reduce the CDF by a factor of 62. The results demonstrate that the modification options derived from the risk-based approach can indeed improve the plant safety.

However, the following problems for LPS PSA methodology should be resolved to get reasonable results and to achieve efficacious plant safety improvement. For evaluation of the human error probability, the inter-dependencies between human actions should be properly evaluated. To get more information on the phenomena after the accident, the best-estimate thermal-hydraulic calculations are needed. Since the available systems for mitigating initiating event are limited and used for many different functions during shutdown, system dependency between functions and systems is complicated to model the relationship and to quantify the models. Therefore, the dynamic approach for accident sequence analysis should be developed.

## References


1. CEA/IPSN, "Probabilistic Safety Study on French 900 MWe Plant", EPS 900, CEA/IPSN, 1990.
2. EdF, "Probabilistic Safety Study on French 1300 MWe Plant", EPS 1300, EdF, 1991.
3. T.L. Chu et al., "Evaluation of Potential Severe Accidents During Low Power and Shutdown Operations at Surry, Unit 1", NUREG/CR-6144, 1994.
4. Y.G. Jo et al., "Final Level 1 PRA Update for Yonggwang Nuclear Units 3 and 4", Korea Atomic Energy Research Institute, Jul. 1993.
5. A.D. Swain, "Accident Sequence Evaluation Program Human Reliability Analysis Procedure", NUREG/CR-4772, 1987.
6. S.H. Han et al., "KIRAP (KAERI Integrated Reliability Analysis Code Package) Release 2.0 User's Manual", KAERI/TR-361/93, 1993.


**Table 1 The comparisons of quantification results**

| Case | Option | CDF Ratio (Case / Base Case) |
|---|---|---|
| Base Case | Current design | 1 |
| Case 1 | Use of CSS for backup of SCS | 1/20 |
| Case 2 | Installation of additional level indication system | 1/7 |
| Case 3 | Use of CSS for backup of SCS and installation of additional level indication system | 1/62 |

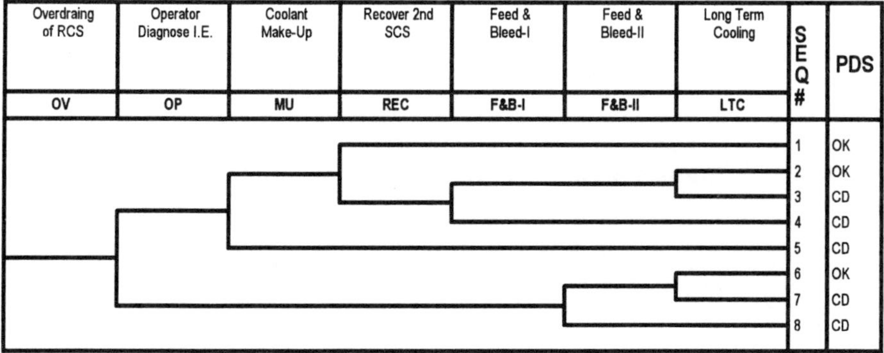


Figure 1 Event tree for overdraining event and case 2

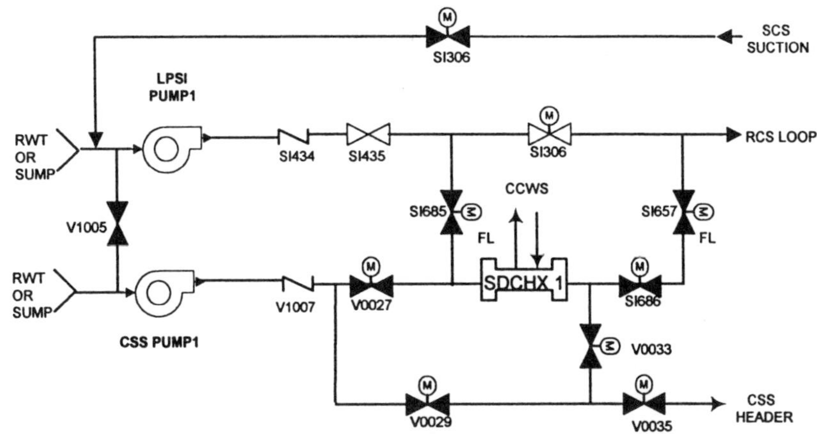


Figure 2 Simplified diagram for SCS train 1

| Overdraing of RCS | Operator Diagnose I.E. | Operator Re-Diagnose I.E. | Coolant Make-Up | Recover 2nd SCS | Recover SCS Using CSS | Feed & Bleed-I | Feed & Bleed-II | Long Term Cooling | SEQ # | PDS |
|---|---|---|---|---|---|---|---|---|---|---|
| OV | OP | OP-BAK | MU | REC | REC-CSS | F&B-I | F&B-II | LTC | | |
| | | | | | | | | | 1 | OK |
| | | | | | | | | | 2 | OK |
| | | | | | | | | | 3 | OK |
| | | | | | | | | | 4 | CD |
| | | | | | | | | | 5 | CD |
| | | | | | | | | | 6 | CD |
| | | | | | | | | | 7 | OK |
| | | | | | | | | | 8 | OK |
| | | | | | | | | | 9 | CD |
| | | | | | | | | | 10 | CD |
| | | | | | | | | | 11 | CD |
| | | | | | | | | | 12 | OK |
| | | | | | | | | | 13 | CD |
| | | | | | | | | | 14 | CD |

Figure 3 Event tree for case 1 and 3

# Modeling Components and Systems with Buffers Providing a Grace Period

J.K.Vaurio
Imatran Voima Oy, PL 23, 07901 Loviisa, Finland

## 1 Introduction

Most repairable systems have been modeled so far under the assumption that component failures have an immediate influence on the output reliability of a system. Modern technology provides many examples in which a failure of a component affects the system later than the failure occurs, and only if repair is not completed within a certain time T, a grace period. A clear distinction is made here between *component* failures and *output* failures which occur conditionally with delay T and often with a shorter downtime than the repair time. Some examples:

* A pumping system with an intermediate storage. The output fails only if and when component repair is not completed within the time it takes to empty the storage.
* A waste treatment system with buffer tanks. The output fails only if the buffer is filled before repairs are completed.
* Room cooling (or heating) systems. Failures are harmful only if repair is not completed before the temperature goes above (or below) a critical limit.
* Power generation systems with or without standby units. Severe consequences take place only when a power outage lasts longer than a critical time limit.
* Space missions: failures are repairable as long as repairs are completed within a grace period. The very first output failure is catastrophic, but several short component failures may precede such an event.

A grace period exists in these examples only if component failures are detected immediately and repairs are started before the output fails. Analytical relationships are developed here for the reliability characteristics of the *output*, given the corresponding characteristics of *components*.

### Notation

| | |
|---|---|
| $F(x)$ | failure time distribution of a good and new component |
| $f(x)$ | $F'(x)$, failure time density |
| $F_{e1}(t)$ | distribution of the time to the first output failure |
| $f_{e1}(t)$ | $F'_{e1}(t)$, density of the first output failure time |
| $G(y)$ | repair time distribution of a new component |
| $g(y)$ | $G'(y)$, repair time density |

| | |
|---|---|
| H(x) | hazard function, $\bar{F}(x)=\exp[-H(x)]$ |
| h(x) | hazard rate, H'(x) |
| K(y\|x) | probability of component repair duration > y, given failure at time x |
| M(y) | repair function, $\bar{G}(y)=\exp[-M(y)]$ |
| m(y) | repair rate, M'(y) |
| q | initial failure probability, u(o) = q; failure-to-start probability |
| t | time; the process starts at t = 0 |
| $\tau$ | repair time/outage time variable |
| $t_f$ | mean time to failure, $\int_0^\infty \bar{F}(x)dx$ |
| $\tau_r$ | mean repair time, $\int_0^\infty \bar{G}(y)dy$ |
| u(t) [$u_e(t)$] | probability of component [output] failed state at time t; unavailability |
| $U_{se}(t)$ | unavailability of system output at time t |
| W(t)[$W_e(t)$] | expected (mean) number of component [output] failures during [0, t) |
| w(t)[$w_e(t)$] | W'(t) [$W'_e(t)$], failure intensity at time t; w(t) = 0 for t < 0 |
| X* | Laplace transform or any function X(t) |
| $\bar{p}$ | 1-p, for any probability or fraction p |
| $\delta(t)$ | Dirac's delta function |

## 2 Single Component Analysis

Assuming that the output is restored immediately when the component is repaired, the following relationships can be established:

$$w_e(t) = K(T|t-T) \cdot w(t-T), \tag{1}$$

$$u_e(t) = \int_0^{t-T} w(t')K(t-t'|t')dt'. \tag{2}$$

The most common alternatives for K(y|x) are

Category $R_{new}$: $K(y|x)=\bar{G}(y)$

when repair times are independent and identically distributed, and

Category $R_{old}$: $K(y|x)=\bar{G}(x+y)/\bar{G}(x)=\exp\left[-\int_x^{x+y} m(t')dt'\right]$

when m(t')dt') is the probability of repair completion during (t', t'+dt'], given that the component is failed at t', and this probability is independent of events before time t'.

Eq. 2 yields also u(t) when T = 0. Thus, w(t) is the component function that needs to be solved separately. It depends on F(t), G(t), the initial condition, and the maintenance policy. Basic equations for w(t) and u(t) have been obtained [1] under two maintenance policies,

Category $F_{new}$: each repair makes the component "as good as new", i.e. the times to failure are independent and identically distributed,

Category $F_{old}$: a repair makes the component "as good as old but unfailed",

both in combination with any of the repair time categories $R_{new}$, $R_{old}$ [1]. (In case of a component with multiple failure modes: w(t), u(t) and G(y) are the intensity, the probability and the repair time distribution of a specific mode, respectively). Taking into account the initial condition u(o) = q, the following results can be obtained:

Category $[F_{new}, R_{new}]$: $w^* = [q+(1-q)f^*]/(1-f^*g^*)$.

Category $[F_{old}, R_{new}]$: w(t) is the solution of the integral equation

$$w(t) = q\delta(t) + h(t)[1 - \int_0^t \bar{G}(\tau)w(t-\tau)d\tau].$$

Category $[F_{old}, R_{old}]$:

$$u(t) = e^{-H(t)-M(t)}\left[q + \int_0^t h(t')e^{H(t')+M(t')}dt'\right],$$

$$w(t) = q\delta(t) + h(t)[1-u(t)].$$

The asymptotic results ($t\to\infty$) for the *output* are

$[F_{new}, R_{new}]$:

$$w_e(\infty) = \frac{1}{t_e + \tau_e},\ u_e(\infty) = \frac{\tau_e}{t_e + \tau_e},$$

$$t_e = (t_f + \tau_r)/\bar{G}(T) - \tau_e,\ \tau_e = \int_T^\infty \bar{G}(\tau)d\tau/\bar{G}(T)$$

$[F_{old}, R_{new}]$: $w_e(\infty) = \bar{G}(T)h(\infty)/[1+\tau_r h(\infty)]$, $u_e(\infty) = \tau_e w_e(\infty)$.

$[F_{old}, R_{old}]$: $u_e(\infty) = e^{-Tm(\infty)}h(\infty) / [h(\infty)+m(\infty)]$, $w_e(\infty) = m(\infty)u_e(\infty)$.

Exponential case: The case of exponential $\bar{F}(t) = e^{-\lambda t}$ and $\bar{G}(t) = e^{-\mu t}$ belongs to all categories $[F_{new}, F_{old}, R_{new}, R_{old}]$. It leads to (for $t \geq T$)

$$w_e(t) = e^{-\mu T}\left[q\delta(t) + \frac{\lambda\mu}{\lambda+\mu} + \lambda\left(\frac{\lambda}{\lambda+\mu} - q\right)e^{-(\lambda+\mu)(t-T)}\right],$$

$$u_e(t) = e^{-\mu T}\left[\frac{\lambda}{\lambda+\mu} + \left(q - \frac{\lambda}{\lambda+\mu}\right)e^{-(\lambda+\mu)(t-T)}\right].$$

## 2.1 Time to First Output Failure

The distribution of the time to the first *output* failure is of major interest if the output failure is catastrophic, or non-repairable. In most practical cases one can use the inequality

$$\underset{0 \le x \le t}{\text{Max}}\ u_e(x) \le F_{e1}(t) \le \min\{W_e(t), 1\}. \tag{3}$$

(A similar inequality has been pointed out for system unreliability [2]). Especially the upper bound is accurate when $W_e(t) << 1$. The exact derivation of $F_{e1}(t)$ is rather complex. For category $[F_{new}, R_{new}]$ the Laplace transform of $f_{e1}(t) = F'_{e1}(t)$ turns out to be

$$f^*_{e1} = \frac{q+(1-q)f^*}{1-f^*\tilde{g}^*}\bar{G}(T)e^{-sT},$$

$$\tilde{g}^* = \int_0^T e^{-s\tau}dG(\tau).$$

# 3 Systems with Independent Components

In many cases the system unavailability is completely determined by the unavailabilities of mutually independent components or basic events, often as a linear or polynomial function indirectly specified by a fault-tree or a block diagram. In case this mutual independence holds also for the output events, the *system output* unavailability is a similar function of the basic event output unavailabilities $U_{se}(t) = Q[u_{e1}(t), u_{e2}(t), ...]$, where different basic events can have different grace periods $T_i$. One can then show that the system output failure intensity is (for $t > \max\{T_i\}$)

$$w_{se}(t) = \sum_i \frac{\partial Q}{\partial u_{ei}} w_{ei}(t). \tag{4}$$

Here $\partial Q/\partial u_{ei}$ is commonly known as the Birnbaum importance (now for the *output* events), often easily determined by picking out those terms in Q which contain $u_{ei}$. The condition $t > \max\{T_i\}$ can be waived if all components are good at the beginning, $u_{ei}(o) = 0$. (One condition for Eq. 4 is that multiple failures do not occur simultaneously, e.g. initial failures).

For the unreliability of the system output, an inequality similar to Eq. 3 holds.

As a simple example consider a continuously operating waste treatment process consisting of an evaporator a and an incinerator b in series, with buffer tanks providing grace periods $T_a$ and $T_b$, respectively. Consider the stationary (asymptotic) state.

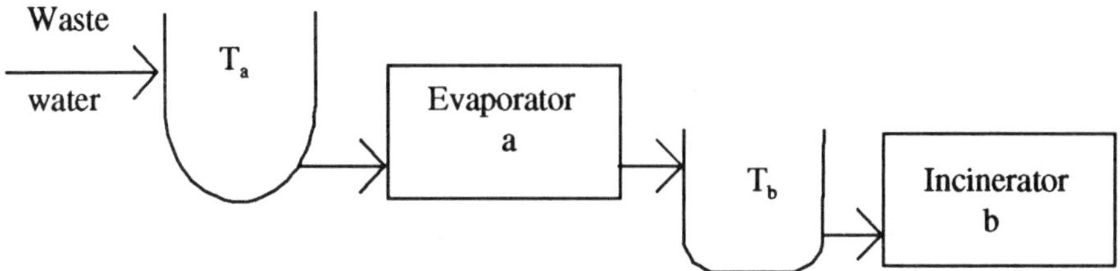


Without buffer tanks the system unavailability would be $u_s = u_a + u_b - u_a u_b$, and the failure intensity $w_s = (1-u_b)w_a + (1-u_a)w_b$. With buffers the grace period for the incinerator is actually $T_a + T_b$ because both tanks can be filled before the plant has to be shut down. The system function ("output") is up if no component output failure is present. Thus, $U_{se} \cong u_{ea}(T_a) + u_{eb}(T_a+T_b)$ and $w_{se} \cong w_{ea}(T_a) + w_{eb}(T_a + T_b)$, where the time parameters indicate the critical repair times for the components, a and b.

# 4 Systems with Standby Components

Consider a system consisting of one normally operating subsystem (the main system) with failure intensity w and repair time distribution $G(\tau)$, and n standby systems with failure-to-starat probabilities $q_j$, probability densities of the operating times $f_j(t)$, and repair time distributions $G_j(\tau)$, $j = 1, 2, ..., n$. When the main system ($j=0$) fails, the standby systems are started one by one, the next one if and when the preceding subsystem fails, until the main system or one of the standby systems is repaired. The quantity of interest is the probability that the total outage duration exceeds a grace period T. The mean repair times are assumed to be short compared to the mean operating times (multiple total outages are unlikely during one main system outage). The following conditional probabilities can be defined recursively:

$$P_0(\tau) = \bar{G}(\tau)$$

$$P_1(\tau) = \left[q_1 \bar{G}(\tau) + (1-q_1)\int_0^{\infty} f_1(x)\bar{G}(x+\tau)dx\right]\bar{G}_1(\tau)$$

$$P_j(\tau) = \left[q_j P_{j-1}(\tau) + (1-q_j)\int_0^{\infty} f_j(x) P_{j-1}(x+\tau)dx\right]\bar{G}_j(\tau),\ j=1,2,...,n$$

$P_j(\tau)$ is the probability that subsystems 0, 1, ..., j are simultaneously down for a period > $\tau$, given that the main system (j=0) fails. The intensity of total outages longer than T is

$$\mathbf{r}(T) = \mathbf{w}\mathbf{P}_n(T).$$

This formalism applies to many of the examples mentioned in the Introduction. It allows several repair policies (single or multiple repair crews). Explicit analytical solutions are possible with exponential repair time distributions [3]. Mutual dependencies of the subsystems can also be taken into account: because there is a definite order of failures (and starts), each $q_j$ (and $f_j$) can be defined as the *conditional* probability (failure time density) under the condition that the preceding subsystems 0, ..., j-1 have failed. It is also possible to use the formalism separately to different causes or modes of the main system failures, each with different parameters w, $q_j$, $f_j$ and $G_j$ [3].

## Acknowledgement

This work was financially supported by Lappeenranta University of Technology, Lappeenranta, Finland

## References

1. Vaurio J.K. Reliability and availability equations for multi-state components. Reliability Engineering 1984; 7:1-19

2. Murchland J.D. Fundamental probability relations for repairable items. NATO Advanced Study Institute, Liverpool, July 1973

3. Vaurio J.K., Tammi P. Modeling the loss and recovery of electric power. Nucl. Engng & Design 1995; 157:281-293

# Selecting And Implementing The Best Group Replacement Policy For A Non Markovian System[1]

Elmira Popova
Department of Mechanical Engineering, The University of Texas at Austin
Austin, TX 78712, U. S. A.

John G. Wilson
Babcock Graduate School of Management, Wake Forest University
Winston-Salem, NC 27109, U. S. A.

## 1. Introduction

One of the most important procedures in the process of designing a new piece of equipment which ages over time is to set up the optimal preventive maintenance schedule. The schedule is usually based on laboratory experiments during the design period. Once the equipment starts producing goods on the manufacturing floor or in the computer lab, managers and engineers schedule preventive or, if necessary, corrective maintenance over time to keep the system operational. In general they will follow the recommended procedure obtained with the new piece of equipment.

There are many different types of system. Each of them can be divided into series and parallel operating subunits. For instance, any microelectronics device is constructed of systems of series and parallel working components. Any machine in a manufacturing cell can be divided into parallel and series operating blocks - such as hydraulic, power source, etc. Knowledge of the behavior of the subunits is necessary to obtain a complete analysis of the whole system.

For the general case of n items operating in parallel, several group replacement policies are discussed in the literature. An m-failure policy requires replacement of a system of n machines at the time of the mth failure. The case of i.i.d. exponentially failing machines under the assumption that replacement was instantaneous is considered in [1]. It is proven that the class of m-failure policies is optimal. This model was extended in [2] to the case where replacement time is a random variable. An (m,T) - policy is one where the system is replaced at the time of the mth failure or time T, whichever occurs first. This policy for a single unit system was investigated in [3] while [4] considered a multiunit system. A T - age failure policy is one where the system is replaced every T units of time [5]. The importance of knowing the associated variance per unit time to make optimal managerial decisions is given in [6]. Each of the above analyses requires that the parameters of the failure time distribution be known with certainty.

The problem of integrating the group replacement decision with Bayesian estimation of the underlying parameters is presented in [7]. An algorithm for obtaining optimal Bayesian decisions within a certain policy class is obtained.

---

[1] This research was partially supported by the State of Texas Advanced Technology Program grant #0003658-472

Bayesian policies for the case of n=1 and a Weibull failure distribution has been recently given in [8]. They investigated both the cases when the system is repaired to "as good as new" and to "as good as old". In [9] and [10], the nature of optimal Bayesian replacement policies when the failure time distribution is exponential is investigated.

For many real systems the assumption of exponential failure times is an oversimplification. The use of phase distributions to model failure times is considered in [11]. Closed form formulae for the expected cost and variance per unit time for T-age, m-failure and (m,T) failure policies are obtained. However, for non exponential failure times, the class of optimal group replacement policies is not known.

In most manufacturing settings, the reliability engineer has failure time data and has formed an opinion about system behavior. This knowledge can be incorporated using Bayesian analysis. The object of investigation in this paper is a system with one operating unit which has a Weibull failure time with unknown scale and shape parameters. The optimal replacement strategy is the one which minimizes the expected cost per unit time. Sensitivity analysis regarding the behavior of the optimal maintenance strategy for different prior distributions is performed.

# 2. Weibull Lifetime With Unknown Parameters.

Assume that a system with n parallel items is operating. The failure time is Weibull with pdf f(x) given by $f(x) = \alpha\beta x^{\beta-1} e^{-\alpha x^{\beta}}$ where $\alpha$ is the scale parameter and $\beta$ is the shape parameter. Each time a replacement is performed, a fixed cost (which includes the cost of replacing the machines) $c_0$ is incurred. There is a salvage value of $c_s$ for each used but functioning machine that is replaced. The downtime cost per unit time equals $c_d$. Further assume that the scale and shape parameters are random variables with a given prior distribution. The goal is to find a replacement policy that minimizes the expected cost per unit time. From renewal theory, the long run expected cost per unit time equals $\{c_0 - c_s E[S] + c_d E[D]\} E[T]^{-1}$, where $E[S], E[D]$ and $E[T]$ are the expected salvage, downtime and length of a cycle. All the expectations are taken with respect to the prior distribution of the failure parameters.

The Weibull distribution is one of the most common distributions for modeling failure times. The two parameters allow a great deal of flexibility. Going from use of an exponential distribution to a Weibull distribution presents considerable mathematical difficulties. Results, even for the case for one machine, can provide insight into the structure of optimal policies for more complicated systems. Indeed, a measure of how complicated it is to mathematically model maintenance systems can be seen from the current state of the literature: most results are given for exponential machines (see, e.g., [1], [2]) or for systems with only one machine (see, e.g., [5] and [8]). Including prior distribution in this analysis makes it even more complicated. In both [12] and [8] a Gamma prior for the scale parameter and a discretization of the Beta density for the shape parameter are used. Bayesian estimators of the scale and shape parameters under uniform and exponential densities are obtained in [13].

# 3. Optimality Results For n=1.

Assume one machine is operating with a Weibull$(\alpha,\beta)$ lifetime. It is shown in [14] that the optimal replacement policy is of the form "replace at the time of the first failure or time T, whichever occurs first". As was assumed in [12], a gamma prior for $\alpha$ and a discrete prior for $\beta$ will be used. The robustness of the policy and its associated expected cost per unit time with respect to prior distribution is of interest. The mean and variance $\mu$ and $\sigma^2$ of a gamma prior distribution with parameters $v$ and $\lambda$ equal $v\lambda^{-1}$ and $v\lambda^{-2}$, respectively. Assume that the mean of this distribution is fixed at some value, then small values for the variance would indicate more certainty about the actual value for $\alpha$ while large values of the variance imply a greater degree of uncertainty about the actual value for $\alpha$. The case where the standard deviation equals zero, i.e. $\alpha$ is known with certainty was used as the benchmark against which to measure the expected cost per unit time as the prior variance was increased. Due to the nature of the optimal policy there is no downtime cost involved. Only two cost parameters need to be varied - fixed and salvage cost. Without any loss of generality, assume that $c_s$ equals 1.

Consider, for example, the curve labeled $\mu=10$ on Figure 1. The horizontal axis represents the standard deviation as a percentage of the mean. For instance, a value of 5 on the horizontal axis stands for $\sigma = 0.05\mu$. The height of the curve at this point represents the percentage difference in the objective function if the prior distribution satisfies $\mu=10$ and $\sigma=.05\mu$ compared to the case with a prior distribution satisfies $\mu=10$ and $\sigma = 0$.

In order to investigate the sensitivity of the expected cost per unit time, many problems with different values for $\mu$, $\sigma^2$ and $c_0$ were analyzed. Representative examples are presented in Figures 1-3. For a fixed value of $\mu$, increasing variance decreases the expected cost per unit time. For a fixed value on the horizontal axis, the expected cost per unit time decreased as the value for $\mu$ increased. Higher expected cost per unit time occurred for low values of $\mu$ and low values of variance.

## References


1. Assaf, D. and G. Shanthikumar, Optimal group maintenance policies with continuous and periodic inspections, Management Science 1987; 33, 11, 1440-1452

2. Wilson, J. G. and Benmerzouga, A. , Optimal m - failure policies with random repair time, Operations Research Letters 1990; 9 , 203 - 209

3. Nakagawa, T., Optimal number of failures before replacement time, IEEE Transactions on Reliability 1983; R - 32, 1, 114 - 116

4. Ritchken, P. and Wilson, J. G., (m, T) group maintenance policies, Management Science 1990; 36, 5 , 632 - 639

5. Okumoto, K. and E. Elsayed, An optimum group maintenance policy, Naval Research Logistics Quarterly 1983; 30 , 667 - 674

6. Wilson, J. G., A note on variance reducing group maintenance policies, to appear in Management Science 1996

7. Wilson, J. G. and A. Benmerzouga, Bayesian group replacement policies, Operations Research 1995; 43, 3, 471-476

8. Mazzuchi, T. A. and R. Soyer, Adaptive Bayesian replacement strategies, Working Paper, 1994, Dept. Operations Research, George Washington University.

9. Wilson, J. G. and E. T. Popova, Optimal Bayesian group replacement policies, working paper, 1996

10. ______________________, Adaptive replacement policies for a system of parallel machines, N. P. Jewell et. al. (eds.) Lifetime Data : Models in Reliability and Survival Analysis, Klewer Academic Publishers, 1996, pp 371-375

11. Popova, E. T. and J. G. Wilson, Group replacement policies for parallel systems whose components have phase distributed failure times, working paper, 1996

12. Soland, R. M., Bayesian analysis of the Weibull process with unknown scale and shape parameters, IEEE Transactions on reliability 1969; R-18, 4, 181-189

13. Canavos, G. C. and C. P. Tsokos, Bayesian estimation of life parameters in the Weibull distribution, Journal of Operations Research 1973; 21, 3, 755-763

14. Popova, E. T. and J. G. Wilson, Optimal replacement policies for one machine systems, working paper, 1996

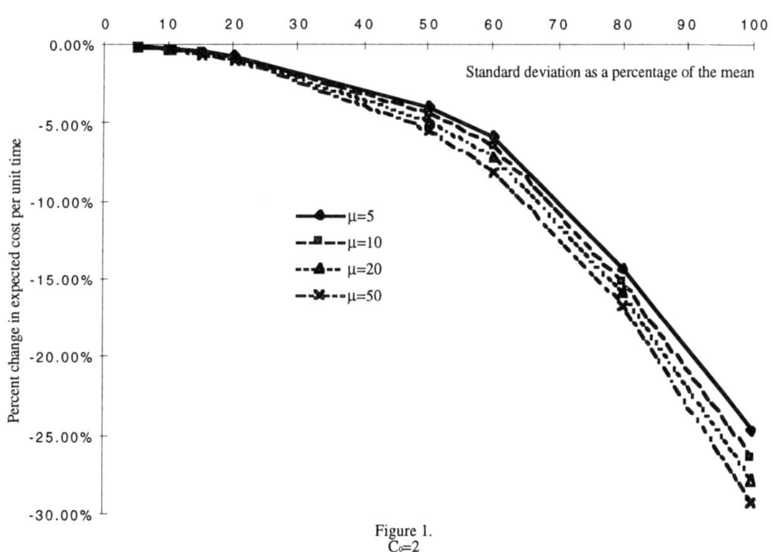


Figure 1.
$C_0=2$

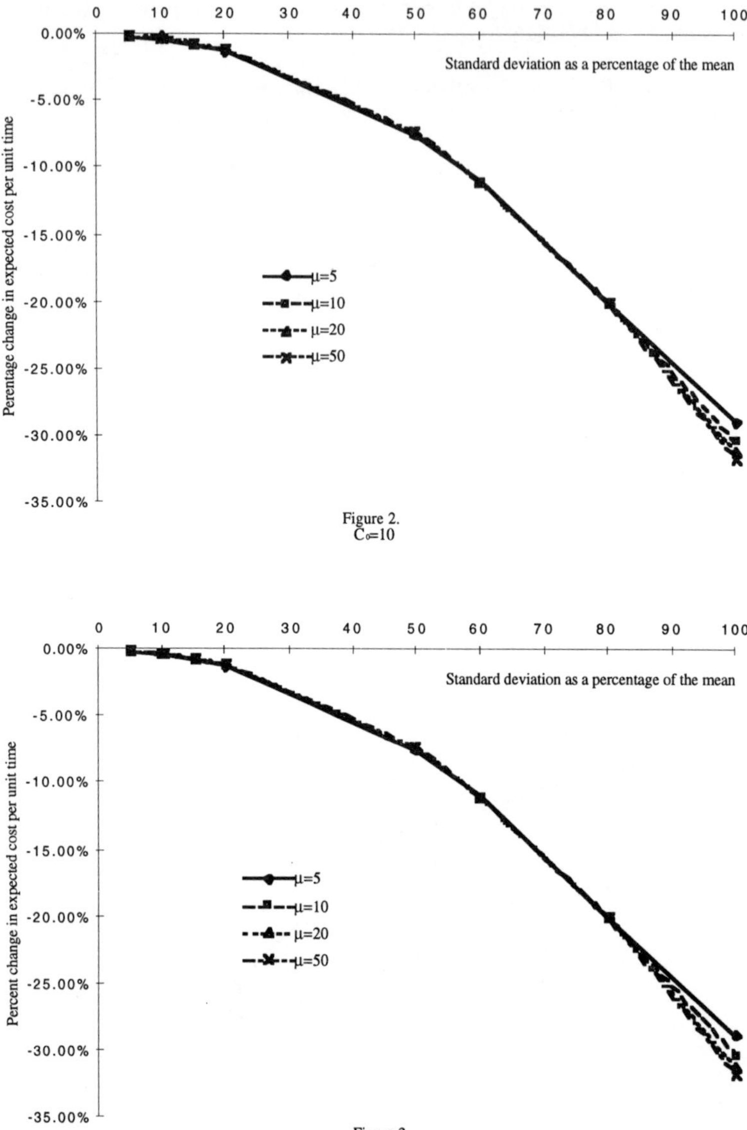


Figure 2.
$C_0$=10

Figure 3.
$C_0$=19

# A Tighter Upper Bound for System Unavailability

Winfrid G. Schneeweiss
Computer Engg., Fern University
D - 58084 Hagen, Germany

**Abstract**

It is shown that with a very small computational overhead the error interval between the first two Bonferroni inequalities can be reduced, thus allowing for very much simpler approximate analysis of mincut-based fault trees in many cases than by Hunter's tighter upper bound.

Key words: Fault tree, Mincut, Unavailability, Approximation, Upper bound, Bonferroni inequalities

## 1 Introduction

With most real-world engineering systems an approximate fault tree analysis is desirable, since an exact analysis is too expensive or/and the basic data given are not accurate enough for an exact analysis. A plausible method to find upper and lower bounds of system unavailability/unreliability is the method of exclusion/inclusion [5, 6], which is tightly coupled to the Bonferroni inequalities [5]. Two approaches based on given error bounds were discussed by this author in [1, 2].

The main aim of any approximate analysis is to get sufficient accuracy at low cost. The possibility of prescribing the error bound in [1, 2] has its price even though the fact that no sum-of-products (SOP) form of the fault tree's Boolean function is needed in [2] offers some recompensation. However in many cases from the very definition of a fault tree – typically with m-out-of-n systems – that SOP form is given and a tentative approximate analysis will be based on the first two Bonferroni inequalities. From Hunter's Theorem of 1976 [3] it is known that one can find tighter upper bounds also without higher-order Bonferroni inequalities. However the selection of the best of those bounds implies the search for a maximum spanning trees of the complete graph with as many nodes as there are system mincuts.

The approach to be discussed here avoids such a complex search procedure. Rather, in the Matrix of the probabilities for activating pairs of mincuts only the maxima of half-rows (lines or columns) have to be found which is of low computational complexity. Unfortunately, there is still some chance that the error interval found will be unacceptably big. So the new approach will not always work, but one should use it as a first step, with more sophisticated approximation procedures following only if need there be.

## 2 A basis for non-Bonferroni-type upper bounds for the probability of a union of random events

As is well known, the principle of inclusion/exclusion is based on the following identity for the probability of the union of the events $a_1, ..., a_m$:

$$P(\bigcup_{i=1}^{m} a_i) = \sum_{i=1}^{m} P(a_i) - \sum_{i=1}^{m-1} \sum_{j=i+1}^{m} P(a_i \cap a_j) + - \cdots \ . \tag{1}$$

The Bonferroni inequalities are [5]:

$$1) \quad P(\bigcup_{i=1}^{m} a_i) \leq \sum_{i=1}^{m} P(a_i) \ , \tag{2}$$

$$2) \quad P(\bigcup_{i=1}^{m} a_i) \geq \sum_{i=1}^{m} P(a_i) - \sum_{i=1}^{m-1} \sum_{j=i+1}^{m} P(a_i \cap a_j) \ , \tag{3}$$

...

with alternating inequality signs. Clearly, the difference between the bounds defined by (2) and (3):

$$\epsilon_{max} \equiv \sum_{i=1}^{m-1} \sum_{j=i+1}^{m} P(a_i \cap a_j) \tag{4}$$

is a maximum error interval. When it is small enough, one can stop any approximation procedure. Otherwise, the approach outlined in the sequel may prove to be helpful. A smaller maximum error can be found based on Lemma 1.

**Lemma 1: General second order upper bound for the probability of a union of events**

$$P(\bigcup_{i=1}^{m} a_i) \leq \sum_{i=1}^{m} P(a_i) - \sum_{i=2}^{m} P(a_i \cap a_{(i)}), \ \forall (i) \in \{1, \cdots, i-1\} \ . \tag{5}$$

(Proof in [3,4,9].)

Considering the matrix $(p_{i,j})_{mm}$, where $p_{i,j} \equiv P(a_i \cap a_j)$, the following is obvious. Lemma 1 yields the best result, i.e. the (relatively) tightest upper bound on $P(\bigcup_i a_i)$, if in each line $i$ of the $p$-matrix the biggest value of $P(a_i \cap a_j)$ is selected and all those $m-1$ biggest values are added, according to Lemma 1; see Tab. 1, which shows the $p$-matrix with a biggest value element in each line marked by a cross.

If the $p_{i,j}$ are determined line-wise, i.e. for any given $i$, first $p_{i,1}$, then $p_{i,2}, \ldots$, finally $p_{i,i-1}$, then it is extremely simple to determine the maximum $p_{imax}$ of line $i$ as follows. Initialize with $p_{i,1}$, and then, after determining $p_{i,j}$, just insert "if $p_{i,j} > p_{imax}$ then $p_{imax} \leftarrow p_{i,j}$".

Tab. 1: Sample matrix $(p_{i,j})_{m,m}$ for $m = 7$ with $p_{i,j} \equiv P(a_i \cap a_j)$. (The symmetrical other half of the matrix is left blank.)

| i \ j | 1 | 2 | 3 | 4 | 5 | 6 | 7 |
|---|---|---|---|---|---|---|---|
| 1 | | | | | | | |
| 2 | x | | | | | | |
| 3 | x | | | | | | |
| 4 | | | x | | | | |
| 5 | | | | x | | | |
| 6 | | | | x | | | |
| 7 | | | | | | x | |

Fig. 1 shows that the selected $p_{i,j}$, when used to select edges $e_{i,j}$ from $K_m$ (the complete graph with $m$ nodes), yields a spanning tree. In the following, Lemma 2 is used for a better understanding of Lemma 1.

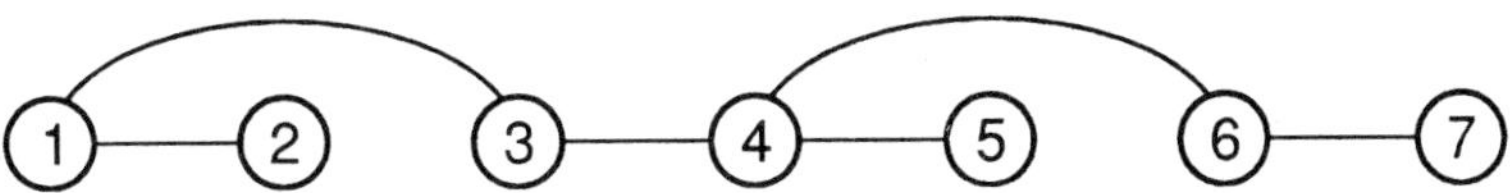


Fig. 1: Tree defining tightest upper bounds for $P(\bigcup_i a_i)$ according to Lemma 1; see Tab. 1.

**Lemma 2: Adjacency matrix of certain spanning trees**
An adjacency matrix, whose sub-main diagonal triangle has in each line exactly one entry 1 (otherwise 0), belongs to a tree. (Proof in [9].)

According to Lemma 2 and Tab. 1 the indices $j = (i)$ of the tightest bound via Lemma 1 are those of one of $2 \cdot 3 \cdot 4 \cdot \cdots \cdot (m-1) = (m-1)!$ spanning trees of a graph with nodes $1, \ldots, m$. Therefore, in order to determine the least upper bound of $P(\bigcup_i a_i)$ given via

$$P(\bigcup_{i=1}^{m} a_i) \leq \sum_{i=1}^{m} P(a_i) - P_{max} \; ; \; P_{max} \equiv \sum_{i=2}^{m} \max_{(i)} P(a_i \cap a_{(i)}), \tag{6}$$

it suffices to i) determine all the $\binom{n}{2}$ values $P(a_i \cap a_j)$, ii) select the largest of them for all given $i = 2, \ldots, m$, iii) add those largest values to get $P_{max}$.

This upper bound is possibly not quite as tight as Hunter's [3] is, but it has the advantage that it can be calculated with very little computational overhead, since instead of the $m^{m-2}$ spanning trees of the complete graph [7] only $(m-1)!$ of them have to be considered, whose construction is trivial and where only the maxima in $m-2$ lists of length 2 to $m-1$, respectively, have to be selected. Fortunately, due to [8] and similar well known results of graph theory and/or operations research, there is no need to identify all those $m^{m-2}$ spanning trees.

# 3 Application to mincut - based analysis

In mincut-based procedures to determine unavailability/unreliability the events $a_i$ of Section 2 are $a_i \equiv (T_i = 1)$, where the $T_i$'s are the terms of the SOP (sum-of-products [6]) form of the fault tree's Boolean function

$$\varphi = \bigvee_{i=1}^{m} T_i \ , \tag{7}$$

with $T_i = 1$, iff mincut $i$ is activated. Consequently, system unavailability/unreliability $U_S$ corresponds to $P(\bigcup_{i=1}^{m} a_i)$, and for evaluating the latter

$$P(a_i \cap a_j) = P(T_i T_j = 1) = E(T_i T_j) \ . \tag{8}$$

The main result of this short paper is then the following; see (6).

**Theorem 1: New upper bound for system unavailability**
A (tighter) upper bound for $U_S$ is given by

$$U_S = E(\bigvee_{i=1}^{m} T_i) \leq \sum_{i=1}^{m} E(T_i) - \sum_{i=2}^{m} \max_{j \in \{1,\dots i-1\}} E(T_i T_j) \ . \tag{9}$$

Note that this is true also for s-dependent system components.

# 4 Examples

An excellent, motivating example is the 2-out-of-3 system, where the tighter upper bound happens to be the exact result. The fault tree's function is

$$X_{2,3:F} = X_1 X_2 \vee X_1 X_3 \vee X_2 X_3 = T_1 \vee T_2 \vee T_3 \ . \tag{10}$$

Since $T_2 T_1 = T_3 T_1 = T_3 T_2 = X_1 X_2 X_3$ , (9) results in

$$U_{2,3:F} \leq E(X_1 X_2) + E(X_1 X_3) + E(X_2 X_3) - 2E(X_1 X_2 X_3) \ , \tag{11}$$

Notice that the r.h.s. of (11) is the well known [6] exact result of the general case of s-dependent components states.

The fault tree of the 2-out-of-4:F system, i.e. a 4-components system tolerating at most one faulty unit, has, by definition, the Boolean function

$$\begin{aligned} X_{2,4:F} &= T_1 \vee T_2 \vee T_3 \vee T_4 \vee T_5 \vee T_6 \\ &\equiv X_1X_2 \vee X_1X_3 \vee X_1X_4 \vee X_2X_3 \vee X_2X_4 \vee X_3X_4 \ . \end{aligned} \tag{12}$$

Tab. 2 shows all the $E(T_iT_j)$ for s-independent $X_1, ..., X_4$. Their sum is $\epsilon_{max}$ .

Tab. 2: Matrix $p$ with elements $p_{i,j} = E(T_iT_j)$ for $i > j$.

| i j | 1 | 2 | 3 | 4 | 5 | 6 |
|---|---|---|---|---|---|---|
| 1 | | | | | | |
| 2 | $\underline{U_1U_2U_3}$ | | | | | |
| 3 | $U_1U_2U_4$ | $\underline{U_1U_3U_4}$ | | | | |
| 4 | $\underline{U_1U_2U_3}$ | $U_1U_2U_3$ | $U_1U_2U_3U_4$ | | | |
| 5 | $U_1U_2U_4$ | $U_1U_2U_3U_4$ | $U_1U_2U_4$ | $\underline{U_2U_3U_4}$ | | |
| 6 | $U_1U_2U_3U_4$ | $U_1U_3U_4$ | $U_1U_3U_4$ | $\underline{U_2U_3U_4}$ | $U_2U_3U_4$ | |

Now let $i$ be chosen such that $U_i > U_j$ for $i > j$. Then by Theorem 1 and the underlined terms of Tab. 2:

$$\begin{aligned} U_{2,4:F} \leq \ & U_1U_2 + U_1U_3 + U_1U_4 + U_2U_3 + U_2U_4 + U_3U_4 \\ & -2U_1U_2U_3 - U_1U_3U_4 - 2U_2U_3U_4 \end{aligned} \tag{13}$$

and – see the non-underlined terms of Tab. 2, and (4) –

$$\epsilon_{max,new} = U_1U_2U_3 + 3U_1U_2U_4 + 2U_1U_3U_4 + U_2U_3U_4 + 3U_1U_2U_3U_4 \ . \tag{14}$$

Check: Comparing the easily derived exact result [9] (in the s-independent components case):

$$\begin{aligned} U_{2,4:F} = \ & U_1U_2 + U_1U_3 + U_1U_4 + U_2U_3 + U_2U_4 + U_3U_4 \\ & -2(U_1U_2U_3 + U_1U_2U_4 + U_1U_3U_4 + U_2U_3U_4) \\ & +3U_1U_2U_3U_4 \ . \end{aligned} \tag{15}$$

with (13) we find that the exact value of $U_{2,4:F}$ minus the "new" upper bound is negative, i.e. the new upper bound is really an "upper" bound, q.e.d.:

$$\begin{aligned} & 3U_1U_2U_3U_4 - 2U_1U_2U_4 - U_1U_3U_4 \\ & = -2U_1U_2U_4(1 - U_3) - U_1U_3U_4(1 - U_2) < 0 \ . \end{aligned}$$

By Tab. 2 the improvement of the maximum error interval amounts to

$$\epsilon_{max} - \epsilon_{max,new} = 2U_1U_2U_3 + U_1U_3U_4 + 2U_2U_3U_4 \ . \tag{16}$$

This is the best result from $5! = 2 \cdot 3 \cdot 4 \cdot 5 = 120$ spanning trees. For equal $U_i = U \ll 1$ this would mean a reduction from $\approx 12U^3$ to $7U^3$.

# 5 Conclusions

It was shown that with little computational overhead of the $m^{m-2}$ possible candidates for the least upper bound of $P(\bigcup_i a_i)$ according to Lemma 1, specifically of system unavailability, after all $(m-1)!$ of them can be checked. This can bring the error interval length down to an acceptable value, even though not always the biggest $m-1$ joint probabilities $P(a_i \cap a_j)$, i.e. $E(T_iT_j)$, will be subtracted from the upper bound $\sum_i P(a_i)$ of the first Bonferroni inequality (2). A simple example showed, how the procedure works and that a substantial reduction of the latter upper bound is nothing exceptional.

As to Hunt's work [3], which was used strongly here and its representation in [4], it is, at first sight, surprising that both authors appear to imply that the tightest upper bound to be got via (5) is found on examining all $m^{m-2}$ spanning trees of $m$ nodes; see for various proofs of the last formula [7]. In this paper we have searched only $(m-1)!$ of them. However any spanning tree of $K_m$ can be transformed to a tree as discussed in Lemma 2 on renaming its nodes; see [3] for details. And then Lemma 1 is applicable again. The number of spanning trees of our example system not covered by our approach is $6^4 - 5! = 1176$, i.e. only about 10% of the trees were searched for their maximum.

Finally it should be mentioned that there exist also tighter lower bounds inbetween those of the first two Bonferroni inequalities (2),(3) [4].

## References


1. Schneeweiss W. Approximate fault tree analysis with prescribed accuracy. IEEE Trans. R 1987, 36, 250-254.
2. Schneeweiss W. Approximate fault tree analysis without cut sets. Proc. Ann. Reliab. & Maintainab. Symp. 1992, 31, 370-375.
3. Hunter D. An upper bound for the probability of a union. J. Appl. Prob. 1976, 13, 597-603.
4. Kohlas J. Reliability and Availability (Zuverlässigkeit und Verfügbarkeit; German language textbook) Teubner, Stuttgart 1987.
5. Feller W. Probability Theory and its Applications. Vol. 1 (2nd. ed.). Wiley, New York 1957.
6. Schneeweiss W. Boolean Functions with Engineering Applications and Computer Programs. Springer, Heidelberg, New York 1989.
7. Moon J. Various proofs of Cayley's formula for counting trees. Ch. 11 in F. Harary (ed.) A Seminar on Graph Theory. Holt, Rinehart and Winston, New York 1967.
8. Kruskal J. On the shortest spanning subtree of a graph and the traveling salesman problem. Proc. Am. Math. Soc. 1956, 71, 48-50.
9. Schneeweiss W. A simple tighter upper bound for the probability of a union of events; with dependability applications. Informatik Tech. Rep. 2/1996, Fern University Hagen.

# Effects on system availability caused by non–exponential lifetimes

Siegfried Eisinger

Stord/Haugesund College, N–5500 Haugesund, Norway
e-mail: siegfried.eisinger@hsh.no

**Summary.** Reliability calculations normally assume exponential life– and repair times. The reasons are many, the most prominent of them might be mathematical simplicity and that the limiting availability is not dependent on the type of distributions for systems of independent components. This paper will discuss the effects caused by non–exponential distributions. The objective is a classification of systems and situations where the "exponential approximation" is not valid.

In order to avoid unreasonable complexity we consider repairable systems of identical and independent components with given life– and repairtime distributions. It is seen that the transient behaviour of the availability changes considerably with the distribution types. The monotonous form of the decrease to the limiting availability becomes oscillatory in some cases and in other cases it approaches the equilibrium from below. Generally, the time to reach equilibrium is considerably prolonged (a factor of 1000 is not unusual). This "time to relaxation" is taken as an important measure of difference and a simple approximate formula is derived giving an estimate of this "time to relaxation" for any system and life– repairtime distribution.

## 1. Introduction

Exponential life– and repairtimes are widely used in the calculation of system reliability. The reasons for this are many, the most prominent of them are:

Mathematical simplicity: The exponential distribution is easy to use in many analytical calculations. Its differential and integral is easy to calculate and it easily fits into Markov analysis.

Number of parameters: The exponential distribution needs only one parameter, namely the MTTF (or MTTR). This is desirable, especially if data are sparce. Many reliability data handbooks (e.g. [2]) only give a single parameter, limiting the use of non–exponential distributions.

Computer tools for the calculation of reliability measures often only support the use of exponential distributions.

Nevertheless it is well known that exponential distributions are often unrealistic approximations of reality. Light bulbs and batteries do e.g. not exhibit exponential lifetimes. For these components the Weibull distribution or the normal distribution would be more appropriate. For repairtime distributions the exponential distribution is often poor, because the time to repair often can be well estimated when the type of error is known. Therefore the lognormal distribution or even a constant distribution may describe the situation much better than the exponential distribution.

In spite of these facts little is known about the validity of the "exponential approximation". This article wants to address this subject and make a step towards a better understanding of where it is safe to use the approximation and where problems might arise.

The article is organized as follows: In section 2. the problem is defined and limitations are discussed. The one–component system is discussed in section 3.. In section 4. a suitable classificaton scheme for the structures of $n$–components systems is presented and a simple formula for an important difference measure is derived. Finally we give a conclusion and discuss future challenges in section 5..

## 2. Definition of the problem and limitations

In the following sections we will discuss differences in system availability measures when using non–exponential distributions instead of exponential distributions for both life– and repairtimes.

The systems we take into account are binary, koherent systems of $n$ components. We assume that the components are independent and identical. The first assumption is quite common in the analysis of binary systems (though it might also be unrealistic in many practical cases), but the second assumption is made to keep the number of parameters tractable. In addition we will often use the restriction $MTTF \gg MTTR$ which is seldom violated in practical cases.

## 3. One component systems

Any multi–component system of independent components can be mapped into a one-component system by choosing suitable life- and repairtime distributions. It is therefore worthwhile to consider these simple systems in order to get an imagination of which challenges that can arise for more complicated systems.

The availability $A(t)$ of one-component repairable systems can be calculated by numerical laplace transformation or by Monte Carlo simulation. In this paper the latter method is used.

Figure 3.1 shows the availability $A(t)$ of a component with exponentially distributed lifetime and three different repair distributions. First we note that the assymptotic availability is given by

$$A_\infty = A(t \to \infty) = \mu_T/(\mu_T + \mu_D) \tag{3.1}$$

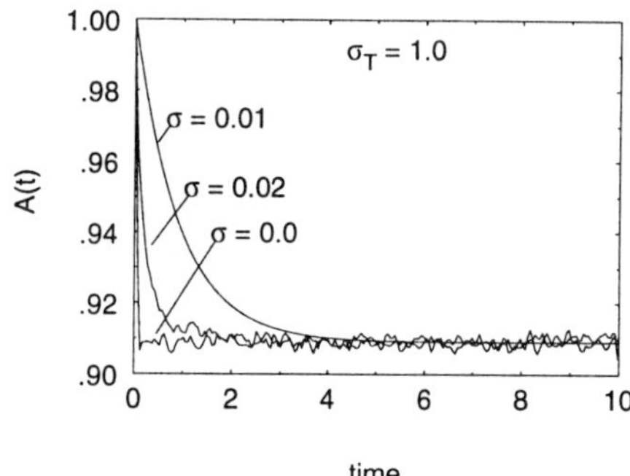


**Fig. 3.1.** Repairable component with exponential lifetime ($\mu_T = 1$, $\sigma_T = 1$) and three different repair-distributions

| $\sigma_D$ | distribution | $\tau^*$ |
|---|---|---|
| 0.0 | constant | 0.0 |
| 0.01 | exponential | 0.01 |
| 0.02 | weibull | 0.02 |

which is generally true for repairable components with non–lattice life–repair distribution [4].

Second, figure 3.1 shows that the transient behaviour of $A(t)$ depends on the "details" of the life- and repair distributions (compare 3.1 with figure 3.2 for more extreme examples of this). This fact is also rather well known [3].

To check the validity of the exponential approximation we should therefore concentrate on the transient behaviour of $A(t)$ which is conveniently given by the relative difference of $A(t)$ from $A_\infty$

$$\phi(t) \equiv \frac{A(t) - A_\infty}{A(0) - A_\infty} = \frac{A(t) - A_\infty}{1 - A_\infty} = 1 - \frac{\bar{A}(t)}{\bar{A}_\infty} \tag{3.2}$$

with $\bar{A} = 1 - A$. $\phi(t)$ is also called *relaxation function.*

We are interested in how long time relaxation takes for non–exponential cases relative to the exponential case. The relaxation time can conveniently be defined as the time $\phi(t)$ reaches some level $\phi_{\text{crit.}}$. Note that $\tau_{e^{-1}} \simeq \frac{1}{1/\mu_T + 1/\mu_D} \simeq \mu_D \simeq 0.0099$ if both the life- and repair distribution is exponential.

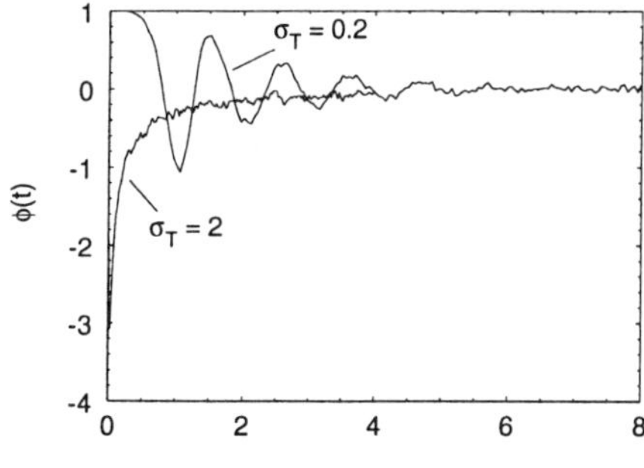


**Fig. 3.2.** Repairable component with IFR- and DFR-lifetime — $\mu_T = 1$, $\sigma_T = 0.2$ (2). The repair-distribution is rather unimportant for $\mu_D = 0.01$.

| $\sigma_T$ | distribution | $\tau^*$ |
|---|---|---|
| 0.2 | Weibull | 6.4 |
| 2.0 | Weibull | 1.0 |

The IFR–curve of figure 3.2 represents another example of transient behaviour. Here, an IFR–lifetime is chosen (weibull, with $\mu_T = 1$ as above and

$\sigma_T = 0.2$). Comparing figures 3.2 and 3.1 shows that the transient behaviour is crucially dependent on the lifetime distribution. Simulation of different repairtime distributions showed that the resulting curves are undistinguishable when $\mu_T \gg \mu_D$, therefore only one representative is plotted.

The relaxation function exhibits a damped oscillatory behaviour. The explanation for this behaviour also provides a clue towards the estimation of the relaxation time.

Let $TD_1, TD_2, \ldots$ denote the stochastic times between first, second, ...renewals. The time for the $j^{\text{th}}$ renewal with $j \gg 1$ is then approximately normally distributed with

$$\mathrm{N}(\mu, \sigma) = \mathrm{N}\left(j(\mu_T + \mu_D), \sqrt{j(\sigma_T^2 + \sigma_D^2)}\right) \tag{3.3}$$

For small $j$, $\mathrm{N}_j$ and $\mathrm{N}_{j+1}$ do not overlap appreciably and the availability exhibits marked top–points for each repair–time and bottom–points for each failure–time. The time, when these tops begin to become smeared out can therefore serve as an estimate for the relaxation time. In statistical terms, we define the time $\tau^*$ as the time when the distance between the expected values of $\mathrm{N}_j$ and $\mathrm{N}_{j+1}$ is the same as $\sigma_j$, or $(j+1)(\mu_T + \mu_D) - j(\mu_T + \mu_D) \simeq 2\sqrt{j(\sigma_T^2 + \sigma_D^2)}$. Solving for $j$ and multiplicating with $(\mu_T + \mu_D)$ yields

$$\tau^* = \frac{(\mu_T + \mu_D)^3}{4(\sigma_T^2 + \sigma_D^2)} \quad , (\sigma_T^2 + \sigma_D^2) \lesssim (\mu_T + \mu_D)^2/4 \tag{3.4}$$

where the restriction comes from the demand that $j \gtrsim 1$ for the sentral limit theorem to be (approximately) applicable. In figure 3.2 the calculated value for eq. 3.4 is included and is in good agreement with the plot of $\phi$.

The last example for non–exponential life– and repairtimes we shall discuss here, is the DFR–case, also shown in figure 3.2. We have chosen $\sigma_T = 2$ and the same repair–distributions as in figures 3.1 and the IFR–case. Also here, the repairtime distribution is rather unimportant, but the DFR–lifetime distribution causes the relaxation function to decrease below 0 and approach the x-axis asymptotically from *below*. Simulations with different repair–distributions (not shown) indicate that the bottom point is reached at about $\sigma_D$. It seems that $\phi$ first follows the same type of behaviour as the exponential case in figure 3.1 where we already mentioned the approximate decay time of $\mu_D (= \sigma_D$ for exponential repair–times). After that time, the components with very long life–times seem to affect the availability considerably, causing the assymptotic behaviour of $\phi$. This behaviour is not yet fully understood, but the following "intelligent guess" is in accordance with both the above discussion and the data:

$$\tau^* \simeq \sigma_D + |\sigma_T - \mu_T| \quad , \sigma_T \gtrsim \mu_T \gg \mu_D \tag{3.5}$$

Figures 3.1 and 3.2, where $\tau^*$ is included in the caption, show good agreement with formula 3.5. $\tau^*$ is somewhat underestimated for the case $\sigma_D = 0.02$ in fig. 3.1, showing that formula 3.5 can be improoved for exponential lifetimes.

Concluding the current section, we collect eq. 3.4 and 3.5 into one formula,

$$\tau^* \simeq \begin{cases} \sigma_D + |\sigma_T - \mu_T| & \sigma_T \gtrsim \mu_T \\ \frac{(\mu_T+\mu_D)^3}{4(\sigma_T^2+\sigma_D^2)} & \text{else} \end{cases} \tag{3.6}$$

which should only be used if $\mu_T \gg \mu_D$. It should be noted that the two cases of eq. 3.6 do not really exhibit a "clear cut" at $\sigma_T = \mu_T$.

## 4. $n$–component systems

To cope with $n$–component systems we should first classify the many different system types. A clue towards this goal is given by the well known cut–set approximation [1]. For identical components we can write

$$\bar{A}(t) = \sum_j n_j \bar{a}^k \simeq n_k \bar{a}^k \tag{4.1}$$

where $\bar{a}$ denotes the component unavailability, the sum runs over all orders of minimal cut–sets, $n_j$ denotes the number of cut–sets of order $j$ and $k$ is the order of the lowest cut–set of the system. The unavailability of a system of $n$ identical, independent and repairable components is therefore in a first order approximation given by the lowest order cut–sets. The order $k$ of this cut–set serves therefore as a very simple classification parameter for these n–component systems.

Inserting eq. 4.1 into 3.2 yields $\phi_k(t) \simeq 1 - \bar{a}^k(t)/\bar{a}^k_\infty$. Within this approximation it can be shown for the tail of $\phi_k(t)$ that

$$\phi_k(t) \simeq k \cdot \phi_1(t) \quad , \text{for } \phi_1(t) << 1 \tag{4.2}$$

Using this equation and fitting an exponential function to the tail of both $\phi_1(t)$ and $\phi_k(t)$ yields then an approximate relation between $\tau_k^*$ and $\tau_1^*$, namely

$$\tau_k^* \simeq \tau_1^*(1 + \ln k) \tag{4.3}$$

To test formula 4.3 we plot $\tau_k^*/\tau_1^* - 1$ versus $\ln k$ in figure 4.1. We use the simulated availabilities from figures 3.1 and 3.2. System relaxation functions are calculated and $\tau^*$ read of at the level of $\phi_{\text{crit}} \simeq 0.2$ This level is somewhat high to be truely conformant with approximation 4.2, but the level must be chosen above the noise–level of the simulations, which are already quite time–consuming. For eq. 4.3 to hold, the points of figure 4.1 should fit to the straight line $y = x$. Taking into account the above mentioned restriction, figure 4.1 supports eq. 4.3, but improoved checks should in future be carried out.

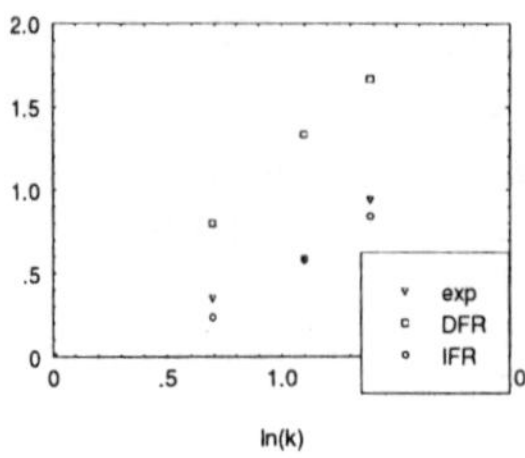


**Fig. 4.1.** Test of formula 4.3 against simulated systems.

# 5. Conclusion

In this paper we have investigated the effect of non–exponential life– and repairtime distributions on the system availability for systems of $n$ identical, repairable components. We found, that the time $\tau$ to reach the equilibrium value $A_\infty$ represents an important measure og difference. The following conclusions about $\tau$ are surprisingly simple and easy to use by engineers.

- For $MTTF \gg MTTR$ the details of the repairtime distribution are only important for the availability and the relaxation time $\tau$ when the lifetime is exponentially distributed.
- For any $n$–component system (with the above mentioned restrictions) the approximate relaxation time can be calculated with the formula 4.3 and 3.6 The exponential approximation is thus only applicable for times $t \gtrsim \tau_k^*$.
- For systems with non–identical components one can still use this formula if it is possible to find a "typical component" of the lowest order cut–sets, or (better) a worst case component. Generally one should pay special attention on the components with IFR lifetimes.

In future work, the lifecycles of DFR–components should be better understood, to give a better foundation of eq. 3.5. There are also plans to make progress into non–repairable systems and periodically tested systems.

# References

1. Terje Aven. *Reliability and risk analysis.* Elsevier, 1992.
2. DNV Technica. *Offshore Reliability Data Handbook*, 2 edition, 1992.
3. Arnljot Høyland and Marvin Rausand. *System Reliability Theory.* Wiley, 1994.
4. S.M. Ross. *Applied Probability Models with Optimization Applications.* Holden Day, San Francisco, 1970.

# RODOS Application on Complex Terrain Dispersion Problem using DETRACT

P.Deligiannis, J.G.Bartzis, N.Catsaros, S.Davakis, M.Varvayanni, J.Ehrhardt*
NCSR "DEMOKRITOS",
Institute of Nuclear Technology - Radiation Protection,
Attiki, Greece
*Research Center of Karlsruhe,
Institute of Neutron Physics and Reactor Engineering,
Karlsruhe, Germany

**Abstract**

RODOS system has been developed to provide integrated and comprehensive real-time on-line decision support for off-site emergency management of nuclear accidents. It has been established under the auspices of the European Commission's Radiation Protection Research Action. A large number of EC contractors participate in this project and advantage has been taken of existing developments at national level. Both West and East European Institutes are also involved together with Institutes in the CIS Republics which contribute to the present work.

RODOS provides protective actions and countermeasures in such a way to permit:
(i) the estimation of the extent of effects in both time and space,
(ii) the estimation of dose and health effects,
(iii) the quantification of the costs for the society and economy.

RODOS system allows for the use of available appropriate external codes for the evaluation of dispersion in the surrounding area of the nuclear site. Such a code is the *DE*mokritos *TRA*nsport for *C*omplex *T*errain (DETRACT ) system.

DETRACT deals with topography and land use simulation, meteorological data preprocessing wind flow and dispersion predictions.

In the present work the structure and main functions of RODOS are summarised. Furthermore for illustration purposes the TRANSALP field experiment [15] is simulated using DETRACT code.

## 1 Introduction

Following the Chernobyl accident, increasing resources were allocated in many countries to the improvement of a system capable of off-site emergency response in the event of a nuclear accident. Within the European Commission's Radiation Protection Research Action, a major project was initiated in 1990 to develop a comprehensive Real-time On-line DecisiOn Support system, RODOS, for nuclear emergency management [1]. Developments of a similar nature have been undertaken within the former Soviet Union and subsequently within individual Republics, taking into account the practical experience gained in responding to the Chernobyl accident.

In case of an accidental release from a nuclear plant, the domain in which human health and environment are affected may extend to several hundreds of kilometers (e.g. Chernobyl Accident). Thus it is obvious that decisions must have a more global character beyond national boundaries applicable and accepted in both West and East Europe. RODOS outcome reflects such a global character on its output which includes estimates, analyses and forecasts of accident consequences and protective measures and countermeasures which are consistent throughout all accident phases and distance ranges.

As stated before, the expansion of the accident is likely to cover a great continental area depending on topographic and meteorological characteristics. At such scales of interest, the terrain is usually complex and the characteristics of the planetary boundary layer may exhibit important temporal and spatial non-uniformity. Therefore it is essential for the atmospheric transport of radioactive pollutants, which strongly depends on the non-uniformity of these characteristics, to be predicted using a code capable of handling complex topography and simulating realistic meteorological conditions.

To this extend the Environmental Research Laboratory at NCSR "DEMOKRITOS" developed a transport code system that completely covers the needs of RODOS system up to the point of countermeasure evaluation. It can provide the subsequent RODOS modules with:(i) detailed topographical description, (ii) time dependent meteorological predictions under any topography and stability conditions and (iii) pollutant dispersion forecast (noble gas, aerosols, particles).

# 2 RODOS system

RODOS system provides all those tools required for the interconnection of all program modules, the input data transfer and data interexchange between the databases and the modules. All relevant environmental data, including radiological and meteorological information and readings, are processed, by means of models and mathematical procedures and understandable, interpretable pictures of protective actions and countermeasures are produced. This does not only permit their extension in terms of time and space to be estimated, but, together with dose, health effects and economic models, also to allow their benefits and disadvantages to be quantified.

The conceptual RODOS architecture is split into three distinct subsystems, which are denoted by ASY, CSY and ESY[1]. Each one of the subsystems consists of a variety of modules developed for processing data and calculating endpoints belonging to the corresponding level of information processing. The modules are fed with data stored in four different data bases comprising real-time data with information coming from regional or national radiological and meteorological data networks, geographical data, external program output and facts and rules reflecting aspects and subjective arguments.

If connected to on-line meteorological and radiological monitoring networks, the RODOS system provides decision support on various stages of information processing which conveniently can be categorised into four distinct levels. The functions performed at any given level include those specified together with those applying at all lower levels: Level0: Acquisition and checking of radiological data and their presentation, along with geographical and demographic information. Level1: Analysis and prediction of the current and future radiological situation based upon monitoring and meteorological data. Level2: Simulation of potential countermeasures. Determination of their feasibility and quantification on their benefits and disadvantages. Level3: Evaluation and ranking of alternative countermeasure strategies.

## 2.1 Operating System

RODOS *Operating Subsystem* (OSY) is responsible for the management of the whole system of data flow from external programs-modules, the database and the *Graphics Subsystem* (GSY) and the control of the system operation modes (interactive-automatic). OSY has been designed and developed on the *Client – Server* basis [2].

As mentioned above RODOS can be run in two modes of operation. In the automatic mode the system automatically presents all the information relevant to decision making and quantifiable in accordance with the current state of knowledge in the real cycle time. For this purpose, all the data entered into the system in the preceding cycle (either on-line or by the user) are taken into account.

Either in parallel to the automatic mode or alone, RODOS can be operated in the interactive mode. In the later phases of an accident, when longer-term protective actions and countermeasures must be considered and no quick decisions are necessary for emergency planning, this mode is more important.

## 2.2 Countermeasures

The countermeasure and consequence subsystem CSY incorporates the following module groups: *ECOAMOR* for calculating individual doses via all exposure pathways, in particular ingestion pathways, *EMERSIM* for simulating sheltering, evacuation and distribution of stable iodine tablets, *FRODO* for simulating relocation, decontamination and agricultural countermeasures, *HEALTH* for quantifying stochastic and deterministic health effects[2], *ECONOM* for estimating the economic costs of emergency actions, countermeasure and health effects[3].

In its present version, ECOAMOR[4] considers 31 basic food products, 22 foodstuffs and 55 processed foodstuffs. The models describe the dynamics of the different radioecological transfer processes, such as the seasonality in the growing cycle of plants, the feeding practices of domestic animals and human dietary habits. In the dose modules of ECOAMOR, the ingestion doses are calculated from the activity concentrations in foodstuffs, age and possibly season dependent intake rates and age dependent dose factors. Doses due to short term inhalation from the passing plume as well as long-term inhalation of resuspended material are also calculated. In addition, doses resulting from external exposure pathways are determined, such as irradiation from the passing plume and from deposited material on ground surface and the skin. Dose reductions from nuclide migration into deeper soil layers and by the shielding of houses are considered, as well as the influence of variable deposition patterns at different urban environments.

The early emergency actions considered in EMERSIM [4,6] can be defined indirectly by dose intervention criteria or directly by graphical input of areas. Important endpoints are areas and number of people affected and individual doses with and without emergency actions. EMERSIM will be completed by the optimisation module STOP [6], which is able to optimise routes for evacuation with respect to route length, dose saved, starting time and costs.

The relocation model in FRODO[4,7] uses criteria for the imposition and relaxation of permanent and temporary relocation in the form of dose levels. The endpoints evaluate relate to the areas of land interdicted, the time periods over which this occurs, the number of people relocated, the doses saved as a result of relocation, the doses received by those temporarily relocated following their return, and the doses received by individuals resettling in an area following the lifting of land interdiction after the permanent relocation of the original population.

The impact of decontamination on relocation can be evaluated for decontamination occurring either before of after relocation is implemented. The decontamination of agricultural land is included in so far as its impact on the need for or reduction in food restrictions is evaluated. The other agricultural countermeasures considered in FRODO are: banning and disposal, food storage, food processing, supplementing animal foodstuffs with uncontaminated, lesser contaminated of different foodstuff, use of sorbents in animal feeds of boli, changes in crop variety and species grown, amelioration of land and change in land use.

## 2.3 Action Evaluation

The evaluation subsystem ESY provides alternate countermeasure strategies under the aspects of feasibility in a given situation, public acceptance of the actions, socio-psychological and political implications and subjective arguments reflecting the judgments of a decision maker. These parameters can be taken into account in ESY using mathematical formulations as rules, weights, and preference functions.

The output of RODOS will be a short list of strategies, each of which satisfies the constraints implied by intervention levels, practicability, etc. together with a detailed commentary on each strategy explaining its strength and weaknesses.

# 3 DETRACT

*DETRACT*[14] is a code system designed and developed to predict airborne radioactive pollutant dispersion and deposition in highly complex terrain and under realistic atmospheric conditions at meso- or microscale level.

DETRACT is especially suitable for terrain of high complexity due to the fact that 2-D grid scale on the ground surface is completely independent from the 3-D grid scale of modeling domain in the air. This allows a quite detailed description of the ground and its land uses "ignoring" computation restrictions coming from the size of the 3-D air domain.

## 3.1 Topography and Land Use Simulator: DELTA/GAIA

The ground description is performed through the topography and land use simulator DELTA/GAIA [9]. Given the area under consideration and the cartesian discretization grid corresponding to the wind field description, DELTA/GAIA basically starts on one hand from a digitized topographic chart, containing the orographic data as a cloud of points given by their $3-D$ cartesian co-ordinates and on the other hand from the set of geometrical points describing the so called "regions". In the context of DELTA the "region" is defined as a continuous area of homogeneous land use.

The module divides the ground surface that corresponds to a particular horizontal grid DX*DY into $4^n$ (n=2,3,...) ground surface triangles. Each triangle has its own position and orientation in space as well as its own physical properties such as landuse, soil texture, deposition properties etc.

## 3.2 Weather Preprocessor: FILMAKER

The concept in this module is the use of any available useful weather observational or forecasting information coming from national or international weather services or individual meteorological stations and covering the greater area of interest. FILMAKER is based on the given usually incomplete set of meteorological parameters arbitrarily distributed in space and time and produces within the given grid the complete set of the necessary meteorological data directly recognised by the other modules of the system [10]. The completeness is obtained based on the basis of best scientific judgment and produces 3-D and 2-D meteorological information.

## 3.3 Prognostic/Diagnostic Module: ADREA-I

The diagnostic module ADREA-I generates mass consistent wind fields using least square methodology and utilizing Langrangian multiplier concept [11].

ADREA-I [12,13] performs small to medium scale atmospheric calculations under any atmospheric stability conditions. It is fully compressible and non-hydrostatic and uses equations for the conservation relations of mass, momentum and internal energy to describe the mass and energy field. The closure scheme is based on the eddy viscosity/diffusivity concept and it is obtained through the utilization of the one equation (default) or two equations turbulence model. Stability dependent length scales are allowed for a modified Richardson number copes with the multicomponent and multiphase air system.

## 3.4 3-D Puff Langrangian model: DIPCOT

For either continuous or instantaneous releases, the air and deposited concentration are computed by the DIPCOT module which simulates the release as a series of puffs. The complete trajectory of the individual puff is computed using the mean wind speed at its center, as it moves within the computational domain. The distribution coefficients estimation is based on the turbulent diffusion parameters provided by the ADREA I module. The concentrations are estimated assuming gaussian distribution of the spread around the center-line.

# 4 Application

## 4.1 The Experiment

The *TRANSALP* experiment refers to the Alpine region and it was designed to study the transport of atmospheric constituents over the Alpine barrier from the Western Po Valley to the Swiss Plateau and vice versa [15]. The release time resembles that of a real nuclear accident and this is why it was chosen for the present study.

The particular area in which the experiment (i.e. the release and the concentration measurements) took place, belongs to the complex topographical structure of the central Alpine area, including the upper part of the Italian Po valley with the first hills and lakes up to the main Alpine ridge and a large fraction of the Swiss Plateau.

The tracer was released near Giornico (G) at the beginning and near the base of the Leventina (L) valley (see Fig.2). The tracer used was perfluoromethylhexane PP2 (C7F14). The material was released at a rate of 8gr/s starting at 10:00 LST together with the development of the valley breeze and lasted for two hours.

## 4.2 Methodology

The calculation domain covers an area 47.5x47.5 $km^2$ as shown in Fig 1. This area includes the network of ground measurements and the two main valleys, located in the middle of the simulated area.

The horizontal grid selected for the atmospheric calculations is homogeneous (1158 m). In the vertical axis, a non-homogeneous grid spacing is set, with minimum interval of 50 m (over ground) and maximum 500m (near the top). The height of the domain is about 9.5 km.

In order for a real emergency case to be simulated it is assumed that there are meteorological data that it constantly flows from measurements in the surrounding area. The first step is assumed to be received at 10:00 LST the time at which the accident takes place. The time step of the meteorological update is every 30 minutes.

# 5 Results

Figures 1 and 2 give an overview of the RODOS environment and the complexity of the problem in hand. It is clear that the topography of the domain of interest is very complex (see figure 2) i.e. narrow valleys and steep slopes. From a meteorological point of view this affects the dispersion severely by introducing a number of different local meteorological systems (temperature and pressure gradients and in turn slope and valley winds).

As was expected the bulk of the pollution moves up the Leventina valley. At the top of Leventina valley, by the time the plume arrives, northern coming winds, from Gotthard pass, interact with Leventina channeled flow producing a plume split towards north and southwest directions. By the look of the dispersion evolution it is obvious that the wind flow is quite accurately diagnosed experiencing channeling characteristics. Thus due to ground heat exchange up going winds in the valleys are quite normal at this time of the day.

Figure 3 shows theoretical time integrated concentration. Qualitatively speaking this graph determines the trajectory of each of the puffs released from source. This again proves the quite accurate estimation of the wind flow and dispersion.

Finally figure 4 (scatter graph) illustrates the comparison between experimental and theoretical concentrations at the individual stations, with adequate degree of success. In more detail figure 5 shows time series of both predicted and measured concentrations seperately for each of the stations. From the above mentioned figure it becomes quite clear that there is a time lag between experiment and theory (about 1 hour) due to the underpredicted

# 6 Conclusions

*RODOS*: (i) Straightforward methodology and use of a decision support tool, (ii) Flexibility in the selection and use of the most appropriate set of mathematical models, (iii) Sophisticated utilization of European computer and station network and distributed databases.
*DETRACT*: (i) Complete set of mathematical models (***topography***, ***weather***, ***dispersion***), (ii) Intelligent treatment of complex topography simulation, (iii) Accurate modelling of wind flow and dispersion, (iv) Quite short CPU time: five hour modeling, two hour real time run on a 715 HP work station.

# References

[1] J. Ehrhard, J. Pasler-Sauer, O. Schule, G. Benz, M. Rafat, J. Richter, Development of RODOS, a comprehensive decision support system for nuclear emergencies in Europe - an overview, Radiation Protection Dosimetry, 50, 195-203, 1993.

[2] O. Schule, M. Rafat, V. Kossykh, The software environment ofODOS, 1st International Conference of the European Commission, Belarus, the Russia Federation and Ukrain on the Consequences of the Chernobyl Accident, Minsk, Belarus, 18-22 March, 1996.

[3] J. Pasler-Sauer, T. Schichtel, T. Mikkelsen, S. Thykier-Nielsen, Meteorology and atmospheric dispersion, simulation of emergency actions and consequence assessment in RODOS, Proceedings for Oslo Conference on International Aspects of Emergency Management and Environmental Technology, 18-21 June, Oslo, Norway, 197-204, 1995.

[4] J. Brown, K.R. Smith, P. Mansfield, J. Smith, H. Muller, The modelling of exposure pathways and relocation, decontamination and agricultural countermeasures in the European RODOS system, Proceedings for Oslo Conference on International Aspects of Emergency Technology, 18-21 June, Oslo, Norway, 207-213, 1995.

[5] V. Glushkova, T. Schichtel, Modelling of early countermeasures in RODOS, 1st International Conference of the European Commission, Belarus,

the Russia Federation and Ukrain on the Consequences of the Chernobyl Accident, Minsk, Belarus, 18-22 March, 1996.

[6] J. Brown, Y.A. Ivanov, L. Perepelietnikova, B.S. Priester, Modelling of agricultural countermeasures in RODOS, 1st International Conference of the European Commission, Belarus, the Russian Federation and Ukrain on the Consequences of the Chernobyl Accident, Minsk, Belarus, 18-22 March, 1996.

[7] V. Borzenko, S. French, Decision analysis methods in RODOS, 1st International Conference of the European Commission, Belarus, the Russian Federation and Ukrain on the Consequences of the Chernobyl Accident, Minsk, Belarus, 18-22 March,1996.

[8] N. Catsaros, D. Robeau, J.G Bartzis, M. Varvayanni, G.M. Horch, K. Konte, The DELTA code: acomputer code for simulating air/ground interaction zone - User's manual. DEMO 93/17, NCSR "DEMOKRITOS, 1993.

[9] P. Deligiannis, J.G. Bartzis, The weather pre-processor FILMAKER. DEMO report, NCSR 'DEMOKRITOS", 1995 (to be published).

[10] H. Yamazawa, H. Ishikawa, M. Chino, WIND04 and PHYSIC: Meteorological models comprised in the emergency dose information system SPEEDI. Proceedings of the Specialists Meeting on Advanced Modeling and Computer Codes for Calculating Local Scale and Mese-Scale Atmospheric Dispersion of radionuclides and their application, 1991.

[11] J.G. Bartzis, A. Venetsanos, M. Varvayanni, N. Catsaros, A. Megaritou, ADREA I: a transient three dimensional transport code for complex terrain and other applications, Nuclear Technology, 94, 135-148, 1991.

[12] J.G. Bartzis, Turbulent diffusion modeling for wind flow and dispersion analysis, Atmospheric Environment, 23, 1963-1969, 1989.

[13] J.G. Bartzis, M. Varvayanni, A. Venetsanos, N. Catsaros, C. Housiadas, G. Horch, J. Statharas, G.T. Amanatidis, A. Megaritou, K. Konte, ADREA I:a three-dimensional finite volume transport code for mesoscale atmospheric transport (the cartesian version) Part I: The Model Description. NCSR "DEMOKRITOS", DEMO 93/2 pt.1, 1993.

[14] N. Catsaros, D. Robeau, J.G. Bartzis, M. Varvayanni, A. Megaritou, K. Konte, A computer system for simulating the transfer of pollutants over complex tepa,ain-some recent applications, J. of Radiation Prot. dosimetry, 50, 257-263, 1993.

[15] Ambrosseti, P., Anfossi, D., Cieslik, S., Graziani, G., Grippa, G., R. Lamprecht, A. Marzorati, Stiengele A., Zimmermann H., The TRANSALP 90 Campaign. The Second Release Experiment in a Sub-Alpine Valley, EUR Report, 15952 EN, 1994.

# Risk Based Emergency Planning and Handling

Unni Nord Samdal and Olav Sæter
Norsk Hydro Research Centre
N-3901 Porsgrunn
Ola Brevig,
Norsk Hydro, Rafnes
N-3965 Herre
Norway

**Abstract**

Risk analysis and contingency planning have so far developed relatively independently, in spite of obvious common interests. This paper is based on the idea that co-operation can cause improvements in both areas. The intention of the paper is to show how the use of risk analysis methods can give useful input in an emergency situation.

# 1. Introduction

During the last 10-15 years, the field of risk analysis has been developed and refined. A systematic and scientifically based methodology has been developed, which is able to represent the physical evolution of an accident reasonably well. Parallel to this progress, methods for contingency planning has been developed and improved, resulting in more effective emergency response. Generally speaking, the two disciplines have developed independently, in spite of the common interest of making industry safer.

An advanced quantitative risk analysis results in numbers and figures describing the total risk from a combination of numerous events, each representing a certain probability and consequence, and a ranking of the risk contributions. These risk analysis results supply a background for decisions as to the need for remedial actions and information about which risk contributions to concentrate on. But they hardly give any direct input to emergency planning and handling. The work described in this paper is based on the idea that co-operation between the two disciplines will create increased understanding and improved methods for both parts. The paper describes how input from methods used in risk analysis can be utilised in emergency handling. It will concentrate on accidents in the process industry, i.e. accidents involving release of flammable or toxic material.

# 2. Information Needed in an Emergency Situation

## 2.1 General

The objective of the emergency staff in an accident situation is to minimise the extent of damage. This is done on one hand by treating injured persons and evacuating people in danger, and on the other hand by controlling the accident through limiting the release, collecting the released material, preventing escalation, etc. The treatment of injured people will always have a high priority. This paper will, however, concentrate on the control of the accident, since this is the part to which risk analysis methods can contribute.

An emergency situation is characterised by a high degree of stress, little time for decision-making and incomplete information available. Therefore, it is not feasible to perform calculations and modelling in such a situation. All information needed of this kind must be produced in advance and be easily available. Furthermore, it is not feasible to make use of a comprehensive manual with descriptions of numerous scenarios in various weather conditions, etc. Thus, the main challenge is to do enough (sensible) simplifications, rather than obtaining the most accurate and advanced analysis results.

When an accident occurs, the first areas in which there is a need to acquire information in order to limit further damage are:

- Release location and possibilities of isolating the release
- Operational data (material, pressure, volume, etc.)
- Resources (people, fire fighting equipment, cooling water, etc.)
- Observations (equipment leaking, size of visible gas cloud, pool size, gas concentration measurements if/when available, flame height, etc.)
- Weather conditions (wind direction, and possibly speed and stability class).

After having acquired as much as immediately possible of this information, the person/group in charge of the situation must decide on which actions to take. At this stage, easily available decision support can be utilised.

## 2.2 Accidents Involving Flammable Material

A release of flammable material has several potentials for further accident development. This is presented in figure 1 by an example of an "accident tree". The accident tree depicts how a situation can develop, and gives reference to forms (through the numbers indicated) with further information on what determines the development, potential consequences (both in time and space), possible actions and criteria for the actions to be successful. Note that such an accident tree should be developed for each installation separately, taking into account the special characteristics concerning the installation and material in question.

Information which would be useful for emergency response include:

- The area in which the gas concentration is flammable
- Degree of confinement
- Possible ignition sources
- Safe distance for personnel and equipment (such as vessels)
- Time to possible BLEVE

This is the sort of information contained in the forms referred to.

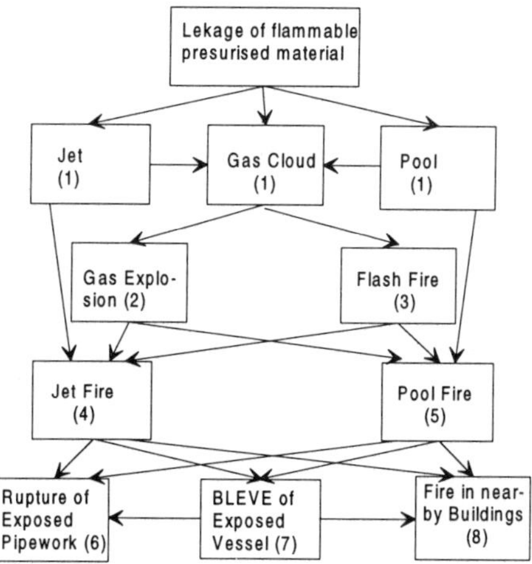


Figure 1: Example of "accident tree" for pressurised, flammable release

## 2.3 Accidents Involving Toxic Material

Releases of toxic material in general have a potential for severe consequences in much longer distances than flammable releases. On the other hand, there is limited potential for further development and escalation. The accident tree presented for flammable releases is therefore not relevant for toxic releases.

The main question in an accident involving toxic material is the distance to safe areas (low concentration), and how this distance can be reduced. Relevant actions are related to reducing the release or evaporation rate, or increasing the dilution.

# 3. How to Utilise Risk Analysis Information

## 3.1 General

The three main parts of a risk analysis are hazard identification, probability assessments and consequence calculations. In addition comes the task of evaluating the results.

The hazard identification part is of obvious use for the emergency planning. For existing plants, this part will often have been performed in one way or the other, independent of the risk analysis. However, the risk analysis approach will generally in more detail look into how these accidents can occur. This might be relevant information for the planning of accident training, but is of no value in an accident situation. The same applies to the probability assessments, which might be utilised when evaluating the personnel and resources needed for emergencies, but are of no value in accident situations. But when it comes to the consequence calculations, these are of major relevance for the emergency handling. The assessment of the potential consequences of an actual release is the key issue for the person/group in charge in an emergency situation.

## 3.2 Methods and Calculations

It is often possible to make discrete descriptions on the adverse effects of concern, for example whether or not there is a possibility of fatalities amongst the local population, whether or not nearby working places can be affected, etc. For the various types of accidents in question, three or (maximum) four categories should be defined, adjusted to the adverse effects in question. It is then possible to calculate the release rates or (in case of an instantaneous release) release quantities which will represent each category. This is used to identify what kind of events, related to equipment dimensions and release types, represent each category.

Note that the calculated release rate is only an intermediate result. The emergency and operation personnel have no possibility of estimating a release rate, but are often able to tell that the release originates from e.g. a pipeline of a certain dimension or a leaking valve.

Subsequently, detailed consequence calculations can be performed for each of the described categories, using the upper bound release rate or quantity of the category.

Some examples of information which is available from a quantitative risk analysis, or from using risk analysis tools, are given in table 1. Information on the actual hazard (fatality, house demolishment etc.) should be given rather than information on more abstract quantities such as heat radiation levels, concentrations, etc.

Table 1: Relevant information from risk analysis for emergency handling

| Fire | Explosion | Toxic release |
|---|---|---|
| - Distance to safe area for ignition of wooden houses<br>- Distance to given fatality rate (lower than x %) | - Distance to safe area for breaking of window glass<br>- Distance to safe area for demolishment of normal building | - Distance to a given concentration, for example IDLH<br>- Distance to given fatality rate (lower than x %) |

The calculations will depend on a set of assumptions such as weather conditions and release duration. It is not feasible to utilise consequence calculations for several weather conditions in an accident situation. At the most, the calculations could be split into a summer situation and a winter situation. A conservative but not unrealistic weather condition should thus be assumed. Likewise, when the consequences depend on the duration, a slightly conservative but not unrealistic duration should be assumed. The definition of the categories in terms of release rates or quantities will be different depending on whether the release is from gas phase or liquid phase. Also, if the process parameters and/or probable duration varies greatly between the process sections, the categories must take this into account. These few considerations, i.e. summer/winter, phase and process section, together with identification of the leaking equipment, are the upper limit of what can be tackled in an emergency situation.

# 4. Case Study, Flammable Release

## 4.1 Description of Case Installation

The studied installation is a 3700 $m^3$ spherical vessel for pressurised VCM (Vinyl chloride monomer) storage, including piping and loading equipment. There is a sloping bund designed to remove liquid VCM from directly underneath the vessel. The normal boiling point for VCM is -13 °C, and lower and upper flammability level is 3.8 and 29.3 vol-% respectively.

## 4.2 Developed Information for Case Installation

The accident tree presented in figure 1 is used as a basis. Examples of the reference forms developed follow in the appendix. Note that the examples only give indications on which information should be provided. The needed information is not fully developed. More examples should for instance be included in the categorisation logic in form 1.

# 5. Conclusion

When producing information to be used in emergency response, the challenge for the risk analyst is to present the results in a format suitable as a basis for quick decisions in a stressful situation. Development of the type of information indicated in the appendix will give the operational and emergency staff improved support for decisions with respect to alarm level, immediate actions and later actions when an accident occurs.

## Acknowledgement

The work presented in this paper is partly financed by the Research Council of Norway.

# Appendix: Examples of Reference Forms

**Form 1: VCM Release**

Categories of VCM release

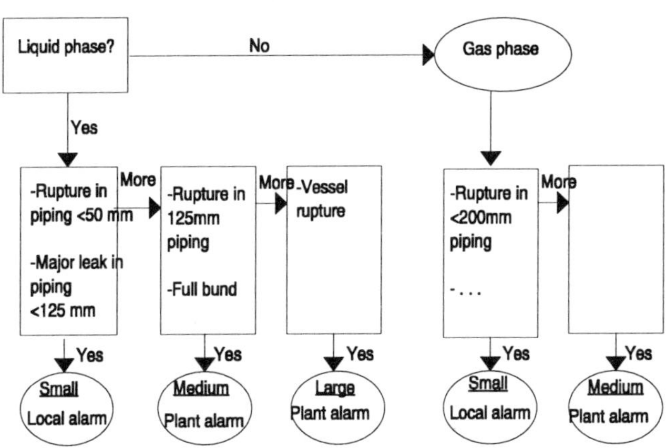

**Immediate actions:**
Evacuate obviously threatened area. Prepare cooling of obviously threatened vessels in case of ignition.

**Later actions:**
Flammable gas cloud. Evacuate people in areas covered by stable cloud:

| | Small | Medium | Large (Vessel rupture) | |
|---|---|---|---|---|
| Critical distances | Downwind length: 40m | Downwind length: 225 m | First couple of minutes: Downwind length: 1400m Upwind length: 500m Width: 1700m | Stabilised phase: Downwind length: 150m Width: 150m |
| Actions | Local alarm<br><br>Evacuate people | Plant alarm<br>Prepare foaming of bund<br>If vessels are threatened: Prepare cooling<br>Get local population threatened by BLEVE indoors (see form 7) | | |

## Form 4: Jet Fire

**Short term information:**
Jet fire closer than 12 meter to vessel: Critical! Few minutes to impairment of metal in vessel without cooling. Danger of BLEVE.

**Longer term information:**

| Category (see form 1) | Small | Medium |
|---|---|---|
| Size | Jet fire of length up to 60 m and diameter up to 9 m. | Jet fire of length up to 130 m and diameter up to 20 m. |

Rule of thumb: Safe distance for unprotected people from flame=flame length.

| Distance from flame to vessel / flame length | Time to beginning vessel impairment |
|---|---|
| 1/10 | 5 minutes |
| 1/3 | 15 minutes |
| 2/3 | 45 minutes |

Possible actions (avoid personnel in threatened area, see form 7):
Bend the jet fire through use of high pressure water.
Cool affected area of vessel.

## Form 7: BLEVE

| Observations | 10 % filled sphere | 50 % filled sphere |
|---|---|---|
| Critical distances<br>Ignition of wooden houses<br>Hazard distance for unprotected persons | <br>170 m<br><br>320 m | <br>300 m<br><br>540 m |
| Actions | Evacuate / warn people in threatened areas to move indoors | |

# Hazard Assessment for Emergency Planning

by

Zeferino Banda Jr.
Sandia National Laboratories, (SNL)
Albuquerque, NM 87185-1037 USA

## 1.0 Introduction

A key objective of the U.S. Department of Energy's (DOE) Emergency Management Program is to ensure that all DOE facilities and operations develop and maintain emergency planning, preparedness and response capabilities, as well as effective public and interagency communications to minimize consequences to workers and the general public from events involving the release of hazardous materials[1]. If planning and preparedness for emergencies is to be adequate and appropriate, then the hazards that are specific to each facility and operation must first be identified and understood. The Hazards Assessment process described herein forms the technical basis for such planning.

Hazards assessments are used to analyze the processes, chemicals, and radiological materials associated with facilities and/or activities, to determine the types and severities of accidents that could affect persons on-site and off-site of the facility. Hazards Assessment Documents (HADs) are used to formally establish and maintain emergency planning, preparedness, and response programs for all Sandia National Laboratories (SNL) facilities (and activities).

## 2.0 Hazard Screening

The first step in the SNL facility-specific hazard assessment process, is to identify the facility and site to be analyzed, and determine whether the hazards within the facility are significant enough to warrant a formal HAD. Sandia utilizes a "hazard screening process" whereby the chemical and/or radioactive material inventory for a given facility is compared against pre-determined screening criteria to ascertain whether the material(s) warrants further analysis.

The screening process utilizes the maximum chemical and/or radiological inventories available at any time in each facility. The process by which materials are either screened (identified as non-hazardous to the health and safety of personnel and the public) or kept (identified as potentially hazardous to the health and safety of personnel and the public) is detailed in Section 2.1. The SNL

Screening Criteria[2] is arranged in order of priority in an attempt to establish a programmatic and rational process with which to identify and/or screen hazards. The purpose of this process is to generate consistent screening of materials between all preparers of hazard assessments at SNL. This enhances the credibility and defensibility of the screening stage and reduces subjectivity or bias that would be introduced by individuals applying different criteria.

The screening criteria have been developed over the past several years as part of SNL's HAD program. Screening criteria for chemicals have been developed utilizing chemical toxicity, reactivity, flammability, dispersability, boiling point, vapor pressure, and other physical characteristics and the quantities involved. Radioisotopes are screened utilizing criteria in 10 CFR Part 30.72[3].

## 2.1 Chemical Screening Criteria

For chemical materials, the screening criteria are based on the following:

*(1) Standard Industrial Hazard* (SIH) In accordance with 40 CFR, Part 355.20, Hazardous Chemicals[4], materials found in a facility's chemical inventory that are, "...used for personal, family, or household purposes, or is present in the same form and concentration as a product packaged for distribution and use by the general public are considered SIHs and can be eliminated from further evaluation."

(2) *Toxicity of Material*-For those chemicals with maximum expected inventory quantities exceeding one pound, the Material Safety Data Sheet (MSDS) and/or the Hazardous Chemical Desk Reference, are reviewed to determine if the chemical is hazardous due to its toxicity. If a chemical of concern is found to exhibit acute or chronic toxicity it is kept for further characterization. These chemicals usually have an assigned occupational exposure limit. Only those chemicals that exhibit acute or chronic toxicity are kept for further screening.

(3) *Quantity of Material*-The screening criterion based on quantity is set at one pound. The Hazardous Substances and Reportable Quantities listed in 40 CFR Part 302, Table 302.4[5] and the Extremely Hazardous Substances and Threshold Planning Quantities listed in 40 CFR 355, Appendix A[6] are used as the bases for emergency planning. Neither of these lists contain chemical quantities of less than one pound. Chemical quantities of less than one pound will only be kept for further screening if identified as extremely dispersible and toxic.

(4) *Dispersability*-For chemicals in excess of one pound, that exhibit acute or chronic toxicity, and have a listed occupational exposure guideline, the MSDS and/or the Hazardous Chemical Desk Reference are reviewed to determine the dispersability into the atmosphere. In order for a chemical to be dispersible it must meet one of the following criteria: a) liquids that have a boiling point of less than 100° C; b) solids that are a finely divided powder (10 microns or less); or, c) materials that can be involved in a high energy event such as a fire or explosion.

(5) *Dispersion Modeling*-For those chemicals that exceed one pound, exhibit acute or chronic toxicity, and/or are found to be dispersible, a facility-specific plume dispersion is performed using an accepted dispersion model. A quantity of material producing less than Emergency Response Planning Guide (ERPG) 1[7] concentration at 30 meters can be screened from further consideration. (A 30 meter facility boundary is assumed to exist at all SNL facilities.) "Worst case" meteorological conditions are employed for the purpose of modeling (i.e., wind speed of 1 m/s, 10% cloud cover, F stability, 50% humidity, and 75 degrees F) to ensure conservatism in the analysis. Commonly used chemicals are listed in 40 CFR 355, Appendix B[8] relating dispersability and/or toxicity.

## 2.2 Radioactive Material Criteria

For radioactive materials, the screening criteria is based on 10 CFR, Part 30.72, Schedule C. Schedule C of 10 CFR 30.72 is a list of radioactive materials that require consideration for emergency planning. The list includes the maximum quantities, in curies, for specific radioactive materials, as well as mixed fission products, contaminated equipment (beta, gamma and alpha), irradiated material, any other beta gamma emitters, and any other alpha emitters. The quantities specified in Schedule C are based on the corresponding release fraction that the regulation has identified for a specific radioactive material. If the chemical or physical form of the radioactive material causes the release fraction to change either up or down, then the maximum quantity in curies will also change respectively. From an emergency planning perspective, the release fraction and quantity in curies is based upon 1 rem effective dose equivalent and 5 rem thyroid[9] exposures. Therefore, if the release fraction were to increase, then the quantity in curies that would give 1 rem effective dose equivalent under worst case conditions would decrease.

Any radioactive materials that exceed the quantities in Schedule C are kept for further evaluation and characterization. All other radioactive materials are screened out as being an insignificant hazard.

# 3.0 Hazard Characterization

After the screening process has been completed, the materials that are “kept” for further evaluation are characterized for use in the development of facility event scenarios. The characterization includes developing a detailed listing of those properties and characteristics that are specific to hazards identified at a given site or facility. These properties include, but are not limited to the following: physical state (e.g., gas, liquid), color, odor, boiling point, melting point, density, vapor density, vapor pressure, and level of concern as it applies to flammable, explosive, and toxic limits (e.g., Lower Explosive Limits, Time Weighted Average Exposures, Emergency Response Planning Guidelines). A safety profile is then developed (e.g., explosive, poisonous) with a list of reactive and or volatile conditions. Potential fire fighting responses are also provided for those situations that may warrant such responses.

# 4.0 Developing Event Scenarios

In developing the facility event scenarios, the preparer of the hazard assessment is required to identify and examine primary and secondary barriers and postulate failure modes, in addition to estimating the quantities of the material released for each cause of failure. Additionally, the hazard assessment preparer is required to systematically evaluate initiating events and scenarios (e.g., compressed gas cylinder valve damage, drum corrosion, etc.), for each set of barrier failures that could lead to the release of a hazardous material. Selection of initiating events and accident scenarios ranging from minor to severe for each combination of hazard and release type is required. Initiating events are assumed to occur (i.e., probability of 1).

A comprehensive list of credible "generic" release scenarios has been developed for identified hazards consistent with methods of shipment, storage, and use. The materials released in quantities consistent with a facility's minimum and maximum inventories provide a range of consequences for emergency planning and preparedness. The scenarios are assigned release designations and assembled in tabular form for quick reference.

# 5.0 Assessment of Event Consequences

Computer models (e.g., ALOHA, ARCHIE & HOTSPOT) are utilized to analyze the consequences based on ERPG thresholds. The hazard assessment preparer employs release scenarios postulated in the previous section as the basis for consequence assessment. Two meteorological conditions, "worst case" (defined by the Federal Emergency Management Agency (FEMA) of 1 meter per second wind speed with a Pasquill-Gifford stability class of F), and average (for the SNL/New Mexico site of 4 meters per second with a Pasquill-Gifford stability class of C), are used when performing these assessments.

The consequences that result from these events are based on calculated health effects on workers and/or the public. The American Industrial Hygiene Association's (AIHA) ERPGs are employed in the modeling effort to identify distances to reversible, irreversible, and life threatening health effects due to a chemical or radiological release. Emergency classes (i.e., Alert, Site Area Emergency, and General Emergency) are assigned to each event that can adversely effect the public or non-involved workers. An Alert is defined as an ERPG-1 concentration (reversible health effect) for chemical releases and/or exposures equal to the Protective Action Guide[10] (PAG) for radiological releases at 30 meters. A Site Area Emergency is defined as ERPG-2 concentrations (irreversible health effects) and/or exposures equal to the PAG at the site boundary. A General

Emergency is defined as $\geq$ERPG-3 concentrations (life threatening health effects) and/or exposures greater than the PAG at the site boundary.

# 6.0 Definition of Emergency Planning Zones (EPZs)

The results of the consequence analysis, along with the DOE Emergency Management Guide[11] (EMG), are used when recommending the facility EPZ. An EPZ is a geographic area surrounding a specific DOE facility for which special planning and preparedness efforts are carried out. The DOE requires that the choice of EPZ for each facility be based on objective analysis of the hazards associated with that facility and not on arbitrary factors such as historical precedent or distance to the site boundary.

The estimated consequences leading to the definition of areas of potential impact are depicted on a map large enough to accommodate the site and impact zones. This information is then used in initial EPZ development and later in establishing Event Classifications. The hazard assessment preparer then applies five tests of reasonableness to the initial EPZ in the following manner:

1) Identify accident scenarios and all event scenarios that constitute concentrations equal to Early Severe Health Effects, (ESHEs), (ERPG-3's for chemical hazards and the corresponding PAGs for radiological hazards) that fall within the defined EPZ.
2) Determine whether the defined EPZ allows for expanded response activities if the situation warrants.
3) Determine whether the defined EPZ is large enough to support an effective response.
4) Determine whether the defined EPZ conforms to natural and jurisdictional boundaries where reasonable, and whether the needs and expectations of offsite agencies are likely to be met
5) Determine whether site preparedness would be enhanced if the area extent of the EPZ was increased.

# 7.0 Development of Emergency Action Levels (EALs) & Protective Actions

The final step in the performance of hazard assessment is the development of EALs. Emergency Action Levels are specific, predetermined, observable criteria used to detect, recognize, and determine the emergency class of Operational Emergencies. The three classes of Operational Emergencies in ascending order of severity are Alert, Site Area Emergency, and General Emergency. The correlation of event scenarios and estimated consequences developed in accordance with the DOE EMG are used to determine the emergency classes and protective actions that are appropriate for the scenarios, as well as the observable indications (i.e. EALs) to trigger such emergency declarations and protective actions. Protective actions are physical measures, such as evacuation or sheltering in place, taken to prevent potential health hazards resulting from a release of hazardous materials to the environment from adversely affecting employees or offsite populations.

# 8.0 Conclusion

Hazard Assessment allows for adequate and appropriate emergency planning and preparedness. The hazard specific screening, results in a built-in conservatism that provides a measure of safety for those involved in emergency management and response activities. By assuming the worst, emergency managers can plan accordingly so as to limit the negative impacts on responders and the public.

# 9.0 References


1. DOE Order 5500.3A, *Planning and Preparedness for Operational Emergencies.* February 27, 1992

2. Sandia National Laboratories (SNL) *Hazard Assessment Screening Criteria,* August 1995.

3. 10 CFR, Part 30.72, Schedule C. *Quantities of Radioactive Materials Requiring Consideration of the need for and Emergency Plan for Responding to a Release.*

4. 40 CFR 355, Emergency Planning and Notification, Part 355.20 (b) *Hazardous Chemicals.*

5. 40 CFR 302, Designation, Reportable Quantities, and Notification. Part 302.4, List of Hazardous Substances and Reportable Quantities.

6. 40 CFR 355, Emergency Planning and Notification, Appendix A, *The List of Extremely Hazardous Substances, and their Threshold Planning Quantities (Alphabetical Order).*

7. *Emergency Response Planning Guidelines.* Emergency Response Planning Committee of the American Industrial Hygiene Association. February 1992.

8. 40 CFR 355, Emergency Planning and Notification, Appendix B, *The List of Extremely Hazardous Substances and their Threshold Planning Quantities (CAS Number Order).*

9. *U.S. Environmental Protection Agency, (EPA) EPA400-R-92-001.*

10. Ibid.

11. *DOE Emergency Management Guide, Hazards Assessment,* Section D.3.b. June 26, 1992.

# Crowd Movement Models: Comparing Agora Models to the Results of Evacuation Trials of the Shuttles in the Channel Tunnel

Dr Emmanuel Lardeux
SOFRETEN, Parc Saint Christophe
Cergy Pontoise, France

**Abstract**

Sofreten organised the evacuation trials of shuttles in the Channel Tunnel on behalf of TML and Eurotunnel. Beyond the interest of this contribution to this great project, it is an original opportunity to compare actual evacuation test results with the results provided by evacuation models, specially the ones of our software Agora.

## 1 Safety and Evacuation

### 1.1 Generalities

The safety level can be validate only if we consider the crowd movement inside a public building. Indeed, whenever an incident occurs, we may have to evacuate the whole building, requiring then a lot of staff.

The safety level of the building depends then directly on the organisation of this operation, in order to put these people out of danger either outside or under protected areas (for example asking people, in a burning high building, to go down inferior floors). To succeed this evacuation, it should be asked to be done early enough considering the risk level it represents (fire, propagation of smoke, bomb alert, technical problems...).

In this case, the means of detection intervene such as alert and alarm signals, and also, sometimes, staff participation or/and the safety staff present at the time on the site.

A second condition to succeed this evacuation is its rapidity. This is where the structure of the building is essential, considering the number of exits available, possible ways, there length and width, these last depending on the possible congestion of the paths used during the evacuation.

Finally, the last condition to succeed is that the behaviour of the concerned people is adaptated, means that their behaviour could add on the existing danger. An "adaptated" behaviour would be the one that make people use the evacuation means (free exits) or avoiding hustle other person.

On could call all the inadapted behaviour: "panic" if this term was not so reducing in the common language. Be that as it may, this "unadapted" behaviour has to be avoided during an evacuation. This is the aim of the French Fire Regulation named "Regulation against fire and risks of panic". The legislation is trying to impose rules that help a good organisation of the evacuation phase.

## 1.2 The Channel Tunnel

In the case of the Channel Tunnel, a test programme has been carried out contractually between TML and Eurotunnel, but MdO was also closely involved, as were the Intergovernmental Commission (IGC) and the Safety Authority. Indeed, Eurotunnel needed to receive operating certificates from the IGC for the different services to be provided (freight shuttles, tourist shuttles and Eurostar).
Among the tests, a series of tunnel evacuations exercises has been carried out. The aim of these tests was to check the effectivness of the procedures that have been put in place by Eurotunnel to handle different crises.
Sofreten company has been asked by TML and Eurotunnel to take the organisation of these tests in charge, as well as the writting of the reports. Concerning the analysis of human behaviour's part, Sofreten collaborated with two english companies: Building Use Studies and Four Elements.
The evacuation tests took place between February 1994 and January 1995. So fourteen in total have been carried out, involving more than two thousand different people.

# 2 Evacuation of Tourist Shuttle under Channel Tunnel

## 2.1 Evacuation procedures

The way through the Eurotunnel system of the cars and coaches is carried out inside specific shuttles with one or two decks according to the vehicles they will have to carry. A shuttle has generally two trains, one is a simple deck and the other is a double deck. They each have twelve carriages. Whenever an incident occurs inside one of the carriages, the evacuation towards the next carriage has been considered, in order to isolate and to confine the incident (possible Halon gas relief) while the train is guided on the emergency platform of the terminal if necessary.
Other types of incidents could lead a train to obstruct the tunnel (breaking of catenary, breakdown of a locomotive, break failure...). In this case, the procedure consists in evacuating the occupiers of the carriage through the service tunnel, where they will wait for a train, specially affected for the evacuation, to reach them using the second tunnel track. They reach then the terminal destination.
These are the first type tests. They have been carried out in February 1994. For these tests, Eurotunnel was using a three-carriage train (three coaches and one loader one each side) which was already set on french terminal platform.
Four trials were made, consisting in loading the three carriages with coaches. The central carriage coach, involved in the evacuation, is a simple or double deck depending on the test, and filled with a changeable number of people. These volunteers, recruited around Calais and Folkestone area, didn't know neither the system or the procedure tested. They just know they were participating to a safety exercise during their visit of the french terminal.

## 2.2 Organisation of the Exercise

Few minutes after the end of the loading of the coaches in the three carriage train, the signal is trigged in the central carriage. For two out of the four trials, we send

artificial smoke at the front of the first vehicle for one minute before releasing the alarm, in order to observe how the smoke influences the occupiers.
A recorded voice informs passengers an incident happened and they have to evacuate. The seven video cameras located in the coach and in the three carriages allow us to record and to follow the evacuation process.
After each test, a questionnaire is given to each participant in order to complete the informations and to have a better understanding of the human factor, specially concerning the perception of available informations.

| | Test 1 | Test 2 | Test 3 | Test 4 |
|---|---|---|---|---|
| Coach Type | Double deck | Double deck | Single deck | Single deck |
| Number of participants | 83 | 90 | 60 * | 62 * |
| Smoke | No | Yes | No | Yes |
| (*) on which 4 persons in two different cars | | | | |

*Table 1 : Description of tests parameters*

# 3 Evacuation Model

## 3.1 Shuttle Models

Sofreten, supported by the Direction de la Sécurité Civile of the French Ministery of the Interior, and in collaboration with the laboratory of Rennes University, initiated a research program on crowd movement models. These studies led to the elaboration of a software, named **Agora**, intended for the safety analysis of every kind of buildings [1].
The evacuation models we are going to confront with the observations made when the tests took place are:

- The Togawa model : it is a simple model, static and empiric, elaborated by Togawa [2];
- A queuing network model : this model, developped by the author, is mathematically resolved both in steady state and transient mode;
- A determisistic fluid simulation model : this model has been originally developped by Predtechenskii and Milinskii [3].
- A simulation model, called "Sensitive simulation model", developped by the author.

All these models have been gathered in Agora to ease their use. The modelisation of a building is made by "cutting up" the plan into discrete entities refered to as "blocks". Quantitative attributes (length, width, initial occupancy...) and qualitative attributes (kind of path, links with adjoining blocks) are assigned to these blocks. When the transition rates from blocks to the others are fixed, it defines a "routing". In the special case we used as a support to our comparison, there are two disctinct areas to modelise :

- Model 1 : corresponding to the modelisation of tests 1 and 2,
- Model 1 : corresponding to the modelisation of tests 3 and 4.

This is to show how these two sites have been modelised.

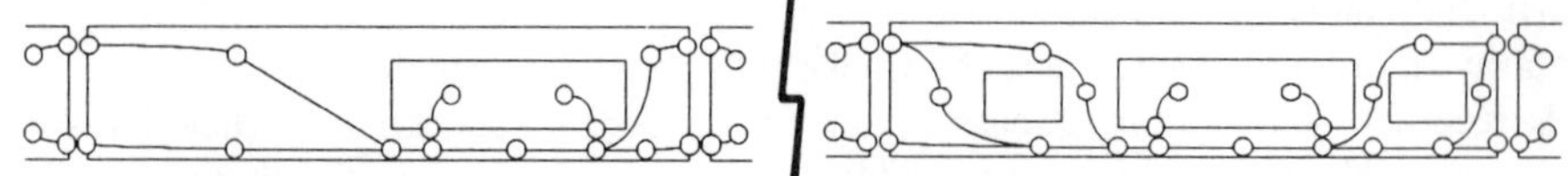

*Figure 1 : Modelisation of carriage evacuation trials - Model 1 and Model 2*

## 3.2 Results of Evacuation Trials

Evacuation paths of occupants and evacuation time for each trial are reported in the following figures.

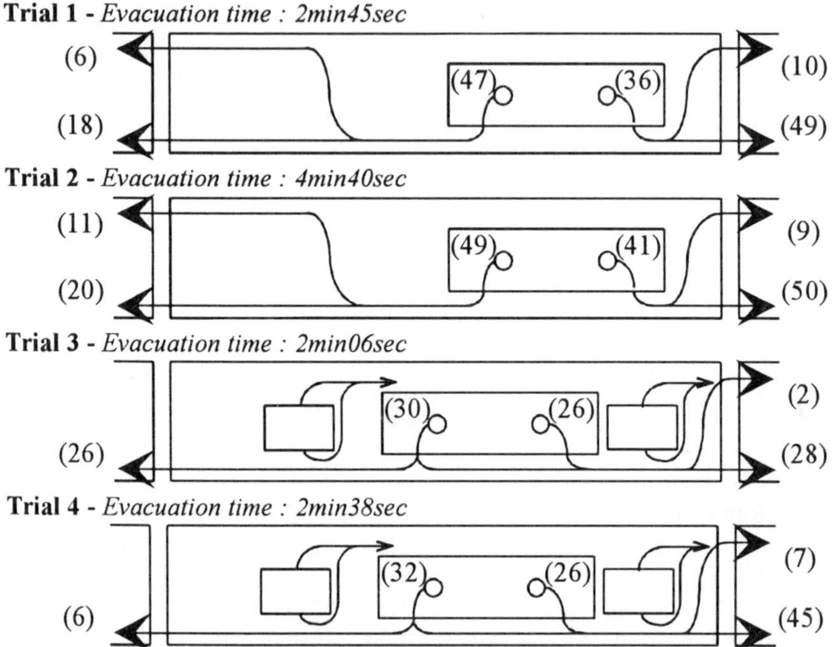


Concerning trial number 2, the video shows that 10 persons delayed to leave the coach, judging the situation not urgent. The 80 other persons had left the carriage after 3 minutes, which provides a second reference for our comparison (2b).

| Trial | Occupancy | Total duration | Actual duration | Latency time |
|---|---|---|---|---|
| **1** | 83 | 174 | 145 | 29 |
| **2** | 90 | 280 | 250 | 30 |
| **2b** | 80 | 180 | 150 | 30 |
| **3** | 56 | 126 | 99 | 27 |
| **4** | 58 | 158 | 128 | 30 |

*Table 2 : Actual evacuation times to compare with simulation times, in seconds*

Time origin correspond to triggering of the alarm signal. Evacuation message begins 5 seconds later.

Taking a latency time of about 30 seconds observed on the video into account, the basis of comparison between actual evacuation times and simulation is the table 2.

## 3.3 Comparison

In order to show what the modelisation would have led to without the informations we obtain from the observations, we will modelize two relevant routings.

*Rout. MIN*: This routing minimizes the distances to cover.

*Rout. UNIF*: It is the "optimum" scenario, where the occupiers share out uniformally through the four exits.

*Rout. REAL*: In addition to these scenarios, we modelize the one we really observed.

Evacuation durations are indicated in seconds in the following tables.

*3.3.1 Togawa model*

| **Trial** | ***Observed*** | *MIN* | *UNIF* | *REAL* |
|---|---|---|---|---|
| **1** | 145 | 72 | 51 | 85 |
| **2b** | 150 | 76 | 52 | 86 |
| **3** | 99 | 57 | 48 | 62 |
| **4** | 128 | 58 | 48 | 80 |

The Togawa Model is simple (few parameters) and gives rapidly evacuation durations for buildings. In this special case, the evaluation is lower than observed, even with the *REAL* routing.

*3.3.2 Random Generation of Routings Plus Togawa model*

| **Trial** | ***Observed*** | ***Minimum*** | ***Maximum*** | ***Mean Time*** |
|---|---|---|---|---|
| **1** | 145 | 72 | 124 | 90 |
| **2b** | 150 | 76 | 132 | 92 |
| **3** | 99 | 57 | 104 | 68 |
| **4** | 128 | 58 | 100 | 68 |

Agora generates randomly a sample of 500 routings, and evaluates evacuation times with Togawa model. This method provides interesting intervals of evacuation times (extreme routing may also be analysed), despite the imperfection of Togawa model.

*3.3.3 Queuing Network model - Steady state Analysis*

| **Trial** | ***Observed*** | *MIN* | *UNIF* | *REAL* |
|---|---|---|---|---|
| **1** | 145 | 61 | 64 | 67 |
| **2b** | 150 | 61 | 65 | 63 |
| **3** | 99 | 50 | 53 | 50 |
| **4** | 128 | 51 | 54 | 70 |

This steady state resolution is derived from an existing model [4] and has been improved before inclusion in Agora. However, this method provides optimistic results, due to steady state hypothesis, as soon as the conditions of evacuation are not very fluid.

*3.3.4 Queuing Network model - Transient Analysis*

The transient analysis is an original model developped by the author [1] and provides better results than the previous resolution method.

| Trial | *Observed* | *MIN* | *UNIF* | *REAL* |
|---|---|---|---|---|
| **1** | 145 | 114 | 159 | 255 |
| **2b** | 150 | 117 | 161 | 240 |
| **3** | 99 | 104 | 116 | 128 |
| **4** | 128 | 104 | 118 | 175 |

*3.3.5 Fluidic Deterministic Model*

| Trial | *Observed* | *MIN* | *UNIF* | *REAL* |
|---|---|---|---|---|
| **1** | 145 | 90 | 81 | 85 |
| **2b** | 150 | 93 | 93 | 75 |
| **3** | 99 | 72 | 72 | 72 |
| **4** | 128 | 74 | 73 | 53 |

This model generally provides relevant and detailled results (specially concerning congestion phenomena). In this case, the predicted evacuation times are lower than expected.

*3.3.6 Sensitive Simulation Model*

| Trial | *Observed* | *Minimum* | *Maximum* | *Mean Time* |
|---|---|---|---|---|
| **1** | 145 | 127 | 130 | 125 |
| **2b** | 150 | 146 | 173 | 160 |
| **3** | 99 | 124 | 142 | 128 |
| **4** | 128 | 118 | 131 | 127 |

This original model is a behavioural-based stochastic model and takes crowding phenomena and more human factor hypothesis into account. The results it provides for this study are the closest to the reality.

# 4 Conclusion

This paper shows that evacuation models may be relevant regarding to the prediction of evacuation times and progress. Among the several existing models, already implemented in the software Agora, the two most relevant models -in the special case studied above- are the most recent models, developped by the author. This result conforts a previous analysis [5]. Human factor hypothesis have not been very detailed in this paper, but they have an important contribution to the relative interest of each model.

**Reference list**

1. Lardeux E. La Modélisation des Mouvements de Foule, PhD thesis, University of Rennes, 1992
2. Togawa K, Study on Fire Escapes Based on Observations of Multitude Currents, Tokyo, Ministery of Construction, BRI, 1955
3. Predtechenskii V.M., Milinskii A.I. Planning for Foot Traffic Flow in Buildings, NBS, Amerind Publishing Co. Pvt Ldt, New Dehli, 1978
4. Smith J. An analytical queuing network computer program for the optimal egress problem, Fire Technology, vol 18, 1982, pp 18-37
5. Lardeux E. Agora: Un Outil d'Analyse des Mouvements de Foule, 8ème Colloque λ/μ, Grenoble, Octobre 1992, pp370-375

# Learning From Experience : The Major Accident Reporting System (MARS) in the European Union

G. A. Papadakis and A. Amendola
Major Accident Hazards Bureau,
Institute for Systems Engineering and Informatics,
Joint Research Centre, European Commission,
I-21020 Ispra (VA), Italy

## 1 Introduction

Industrial accident databases are designed and operate to meet various objectives among which the primary ones are to:

- contribute to evaluation of safety management and safety policy;
- develop statistical estimates and trends; and to
- validate assumptions, models and results of PSA or consequence assessment.

The first interest area addresses aspects such as the management of risk which is of high strategic importance in public authority and corporate policies. The management of the process of controlling the accident hazards is expected to improve with the lessons learned from accurately reported accidents. This includes the safe design & operation (immediate and root causes of the accidents), siting & authorisation, emergency preparedness & response for which both stake holders face diverse duties and responsibilities.

Statistical estimates and trends may be deduced from information provided by databases specially from those with a large number of events, concerning economic, environmental and social costs of accidents in relation to parameters such as activities and installations, dangerous substance and conditions. At this purpose aggregate indices and indicators may be developed. Furthermore statistical data are used to assign event frequency data in PSAs.

Quality of prediction of accident scenarios improves when considering field data on consequences for comparison with outcomes of predictive models, identification of fault dependency structures, human response and other significant issues derived from the data bases and the accident reports.

Databases become valuable when data provided are adequate and reliable to match their intended goals. The Major Accident Reporting System (MARS) and its database [1] set up by the European Commission (EC) within the framework of Seveso Directive (EEC/501/82) in 1984 has a priority aim: to support the major accident prevention policy developed from the EC and the national regulatory authorities in the European Union (EU). In addition MARS can provide significant input into accident scenarios and their possible consequences.

## 2. MARS and major accident hazards regulation

High quality information on industrial accidents has proved to be an important factor to improve prevention. Such an importance has become more vital recently since legislation is making a move to less prescriptive forms, where companies are required to demonstrate the safety of their plant, but are free to determine the means by which they will do this [2]. In addition national competent authorities may not be able to complete their control on safety cases unless they adopt criteria drawn on the base of reliable past accident information.

MARS has been established across EU member states as an obligatory notification scheme for major accidents by a reporting procedure which can guarantee a complete and accurate reporting protocol [1]. The efficiency of such notification procedure has been substantially improved by the adoption of a new two-step notification form which allows all available facts to be transmitted immediately after the accident. In a second step after the accident has been fully investigated a comprehensive report should be submitted. In this way data on major accidents can achieve high quality while the continuously updated database can promptly provide information to the EC and to the national competent authorities to support the evaluation of their regulatory activity.

A first feedback on the regulation has been derived by the study of the underlying causes of the accidents notified in MARS: approximately 90% of the accidents up to the end of 1991 were attributed to management faults. Similar findings came into light by a more recent review [3]. It appears that managerial and organisational omissions still comprise the predominant underlying causes identified and thus become principal factors in the prevention policy. In recognition of that the new Seveso Directive [4] includes improved consideration of management systems and human factors. This is one of the principal changes related to the introduced internal Major Accident Prevention Policy (MAPP) which now forms the basis of the new requirements for any major accident hazards establishment and is enhanced by a further requirement to implement Safety Management Systems (SMS) as appropriate to control the identified hazards [5].

The notification forms allows an easy examination of whether pre-accident evaluation has been performed and actual contingency and consequences have been correctly addressed. The objective to validate risk assessment and applied models by reliable data can thus be met.

One of the limitations of the reporting system for statistical and trend analysis, is the low number of cases in MARS (217 up to end 1995) because only major accidents are reported.

Information included in MARS was found useful for defining the set of accident notification criteria adopted now in the forthcoming SEVESO II Directive. As a consequence all member states will be obliged to report to the EC all accidents which will meet this set of criteria. It is interesting to note that individual criteria (i.e. substance released, casualties, environmental damages, material loss etc.) were given considerable attention while it was recognised that messages originated in and lessons learnt from incidents with minor consequences may be of great importance for a prevention policy. Therefore in addition to the adoption of well defined quantitative criteria Member States are encouraged to notify further incidents to MARS if regarded as "being of particular technical interest for preventing major accidents and limiting their consequences ...".

# 3. Analysis of MARS data

An interesting task for MARS with the EU accident data is to perform the identification of the accident causes (e.g. immediate and underlying) in addition to the common statistical analyses for particular purposes ; an example is given in fig.1 for notified accidents classified by activity. This has been the result of thorough and official investigation performed at several levels and different times after the accidents occurred and can be used to prevent similar incidents from occurring in relevant industrial sites.

The reporting procedures allow for causes identification in the vast majority of the cases ( >90 % ). Two samples of MARS accidents taken at the end of 1991 and 1993, were analysed as for their immediate causes (i.e. operator error, component failure, chemical reaction etc.) and also their root causes (i.e. managerial - organisational failings, design inadequacy etc.). A detailed picture of those factors to which the cases were attributed in 1991 and 1993 is shown in fig. 2 and fig. 3.

It is interesting to note that for the vast majority of cases it is not only one but numerous factors leading to an accident. Thus design errors for instance were frequently accompanied by managerial omissions (which were implicated in more than 70% of the accidents). The root causes show in general a remarkable preference primarily to inadequacies of the management system and secondarily to inadequacies in the design . Component failures are the most frequent immediate causes ( >40 % ), while operator errors and reaction runaways have a relatively high frequency. The level of these factors can well compare with the values obtained in other accident databases.

It is remarkable that despite the natural expansion of the MARS database (more accidents accumulate) the type of distribution of the causative factors remains the same over the years and also shows similarities with samples taken from other databases. The operator errors alone contribute (ca 25%) to the causative factors with comparable values when analysed either in 1991 or in 1993. Very similar values were obtained for these factors when analysed in samples from other databases i.e. with German accidents reported by ZEMA in 1993 and 1994 (fig 2).

Analysing the human errors involved in these cases it appeared that the majority of errors consisted of "representation errors" : that are mainly due to either incorrect risk evaluation or insufficient/biased use of information. Such findings can compare with those from other sources and dedicated projects i.e. human representation errors similar to those discussed in the Archimede project on recreating recorded aviation incidents, based on Rachel database [7].

Some of the experience gained from studying the accidents is related to specific fields such as specific processes, substances and to specific operations, emergency responses etc. Lessons learnt from emergencies after accidents in the individual member states involving dangerous substances has been the subject of extensive national studies. These reports successfully gathered the strong point in the analyses with the purpose to share knowledge and thus improve emergency preparedness and response. MARS accident data however allow for an analysis that lead to a wider spectrum of lessons related to design and construction, to operations and elements of Safety Management Systems, and finally to specific topics (i.e. substances). Each of these topics consist of a number of sub-categories as classified in previous reviews [1]. A distribution list of causative factors from MARS data which give insight and substantially contribute to lessons learnt is given below :

**Design and Construction**

- general and unclear factors 25 %
- process design 18 %
- substance related 5 %
- spacing requirements 3 %
- instrumentation - control 13 %
- electrical 8 %
- mechanical and piping 11 %
- emergency-safety systems 20 %
- fire fighting systems 8 %

**Operations and SMS elements**

- general-organisation factors 30%
- operations 9 %
- maintenance 7 %
- special operations 5 %
- inspection 11%
- training and staffing 11%
- emergency plann.-management 20%
- safeguarding 4 %
- communications and permits 2 %
- resources 3 %

**Specific topics (substances )** 20 %

It is clear from the list above that factors as design, safety systems, the organisation and finally the planning and management of emergencies receive more weight.
Integrated indicators for accident was an early concern the result of which is an accident gravity scale formally adopted in 1993 by the EU forum of competent authorities for a trial period of two years [8]. This gravity scale classifies the accidents in six levels (G = 1 to 6) according to the quantity of substance released or exploded (the hazards created by the event), and the accident actual consequences to : man, flora and fauna, material losses, property-production losses, water polluted and the cost of environmental damage recovery. To illustrate the gravity scale a distribution of the gravity classes corresponding to those MARS accidents with short-term consequences is given in fig. 4.
The retained notification criteria for SEVESO II will result in the obligation of reporting all accidents of gravity higher than an intermediate level lying between levels G2 and G3. The adoption of such criteria will eventually harmonise the obligation among EU members to notify accidents with serious consequences while allows the freedom for accidents without serious consequences to be also reported. Accident analysis in general has shown that equally useful lessons may be extracted from accidents with consequences of different gravity.

# 5. Conclusions

The implementation of the new Directive in the EU member states, considering the special attention given to the Major Accident Prevention Policy and to the Safety Management Systems, is aimed to improve industrial safety and minimise faults in the management system. Such improvement expected as feedback of the new regulation should be reflected in the distribution of causative factors of MARS accidents in the years to follow.

The quality of information and the functionality of the database is in general insured by the accident reporting system in use. The notification in turn should result from the need of a sound debriefing and comply with the principles similar to those of auditing [5]. The reporting forms could thus successfully serve as the formula for accident auditing schemes under several conditions and with various objectives in excess to those intended for the competent authorities (e.g. communication among companies, international HSE concern etc.). The recent reporting forms of MARS focus on the information necessary to evaluate preventive and mitigating measures but also on the compliance of management to safety standards and on the adequacy of existing legislation.

An industrial accident database including a number of parameters (as considered in the gravity scale), capable to describe the environmental effects may substantially contribute to identify the needs for the development of environmental risk assessment. The accident analysis may thus offer new inputs to R&D in the field of environmental risk.

## References


1. Drogaris G., Learning from Major Accidents Involving Dangerous Substances, Safety Science, 16 (1993) 89-113
2. Cole S.T. and Wicks P.J., Review of EU research in industrial safety 1987-1995, CEC 1995, DGXII/D1, SDME 7/47 Brussels.
3. Rasmussen K., The Experience with the Major Accident Reporting System from 1984 to 1993, (forthcoming EUR report 1996)
4. "Proposal for a Council Directive on the control of major-accident hazards involving dangerous substances", (COMAH). Europ. Commission COM(94) 4 final. Jan.26, 1994.
5. Mitchison N. and Papadakis G., Safety Management Systems - International Initiatives, 3ASI conference on Safety Management Systems, Milano, June 21-23 (1995)
6. Haastrup P. and Romer H., An analysis of the database coverage of industrial accidents involving hazardous materials in Europe, J.Loss Prev. Process Ind, 95, Vol 8, No 2, p79
7. Mancini Aw. S., and Ordonneau F.: How to improve safety by using a data base Rachel and Archimede Examples, ESReDA Seminar on Accident Analysis, JRC ISPRA Oct 94
8. Amendola A, Francocci F. and Chaugny M., Gravity Scales for Classifying Chemical Accidents, ESReDA Seminar on Accident Analysis, JRC ISPRA Oct 1994

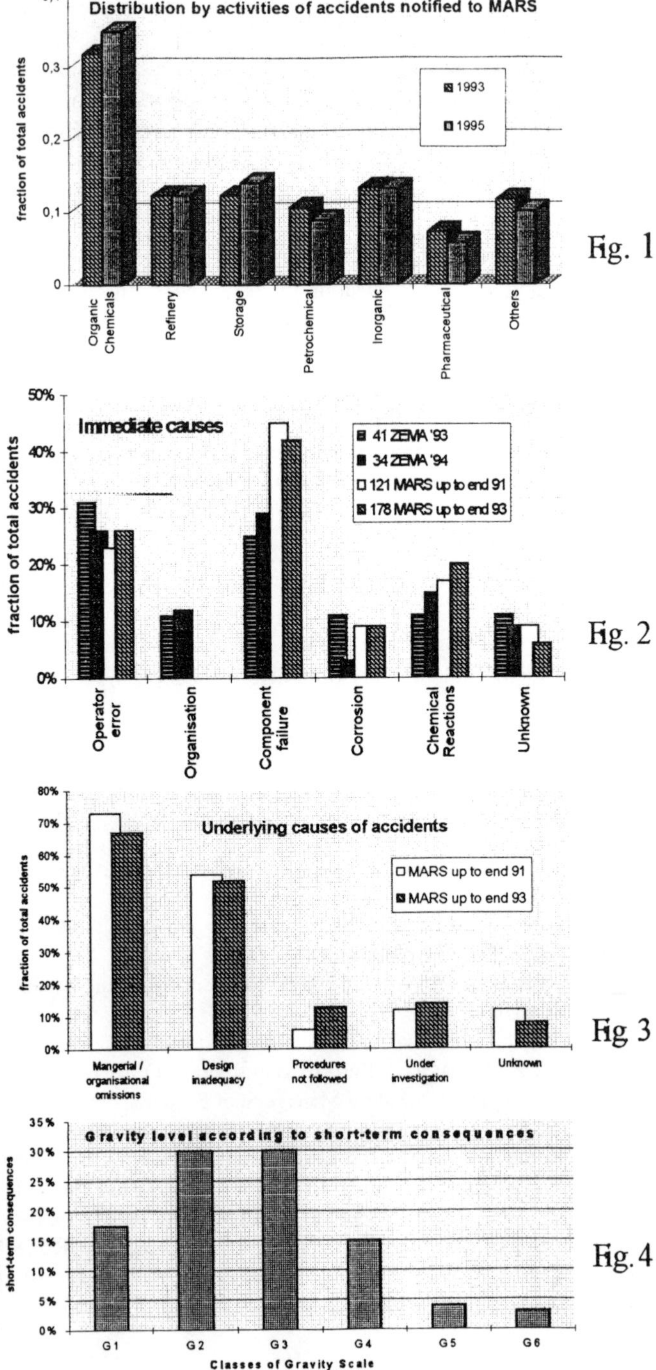


Fig. 1

Fig. 2

Fig 3

Fig. 4

# Operation feedback free text analysis

Authors

Silberberg S., Souchois T.
Electricité de France, Direction des Etudes et Recherches
Chatou, France

**Abstract**

EDF has to manage 54 Pressurised Water Reactors and databases are needed to follow operation and maintenance information.

The data collected are analysed for different purposes : operation, maintenance, safety, and one of the common difficulty in using this information is to analyse the free text in the reporting system. Until now, most of the realised analyses need to read the free text "manually". Furthermore, with the extension of databases, the free text becomes more and more extensive.

EDF started studies about automatic operation feedback free text analysis some years ago and we are to present with this paper a specific study : how to classify a set of information sheets by using statistical and clustering methods.

First part of the method is to measure proximity from one document to another one, in a given corpus. The corpus is represented by the different free texts coming from each document. The advantage of such a method belongs to the fact that the method is basically statistical and that means that few knowledge is previously necessary : few investment is asked before applying such a method.

Second part of the method is to work on the proximity measure results and to obtain a proposal of clusters based on the words belonging to the free texts.

A practical example is provided to fix the limits of such a method.

# 1 Introduction

This communication relates a study on how to classify texts with a proximity tool of literal data [1], associated with hierarchical cluster procedure.

Reports describing events in power plants constitute a corpus in which each report free text is an element. One goal of operation feedback is to analyse free text automatically for obtaining classes among the lot of reports by data similarity between different texts.

A proximity measurement tool has been developed by EDF and applied to event report free texts. A distance indicator represents the proximity for each report to all other reports. The proximity measure results are wooked on to obtain a proposal of clusters and the consistency of such a proposal is estimated.

# 2 The context

A lot of event reports provide operation feedback : about 8000 forms are released per year from the nuclear power plants for different reasons. How to take advantage of such an information, automatically as far as possible, for different purposes.

Each event form is partly coded but, for the time being, we are only interested by the event description. The description belongs to the writer and enables to add information the coded fields do not allow to have. The description has to be concise (less than 500 characters). That means that, most of the time, the words are specific to the operators and the writing style is telegraphic.

# 3 The method

Each form free text contains raw literal data which have to be cleaned. A cut-out extracts words from each description. It is possible to stem the different expressions of a same word. The proximities between all the different forms are computed. At last, the cluster procedure is applied to obtain classes.

## 3.1 The corpus preparation

The raw data have to be cleaned from typing and spelling errors. Some codes which are in the free text have to be written homogeneously (dates, system and compnent acronyms,...). Afterwards, the corpus is considered as cleaned.

## 3.2 The text cut-out

Not all the words are representative of a description content. The punctuation and the words which are considered without interest are not taken into account. The meaning of a sentence is not considered but only its content. As an example :

Plant shutdown for refuelling and maintenance works and works. Charging pump inspection as part of tech specs the following list of words represents the description:

- plant
- shutdown
- refuelling
- maintenance
- works
- charging
- pump
- inspection
- part
- tech
- specs

## 3.3 The stemming

The stemming is realised to improve the proximity measures. According to the language, the stemming can be more or less important. With the previous example, works and work, specs and spec have the same content.

## 3.4 The proximity computation

EDF (Direction des Etudes et Recherches) has developed a program (ADOC) to compute the proximity. ADOC is a tool to measure proximity between texts and it is based on G. Salton theory [2].

ADOC is used for different purposes and it has been used in this case for document clustering. ADOC measures proximity beetwen on text and other texts in the same corpus and these measures provide the texts which are the closest to the given text with there contents.

According to G. Salton, the formula to measure the proximity of two texts belonging to the same corpus is based on the angle of two vectors representing the texts.

Each word in a given text has a weight :

$$P_{ij} = n_{ij}\left(1+\log_2 \frac{N}{N_j}\right)$$

where:

$n_{ij}$ : number of times the word j belongs to the form i

N : number of forms (descriptions) in the corpus

$N_j$ : number of forms (descriptions) where the word j is found in.

The approach is statistical and provides weight to the more used words in a form and to the less used words in the corpus.

Supposed two texts represented by vector $\vec{R}$ and $\vec{L}$ :

$\vec{R}$ (leakage, shutdown, pump), $\vec{L}$ (primary, pump, shutdown)

The proximity of the two texts is measured by the following formula :

$$\cos\left(\vec{R},\vec{L}\right) = \frac{\sum_j P_{R,j}\, P_{L,j}}{\left\|\vec{R}\right\| \left\|\vec{L}\right\|}$$

Greater is the cosinus, closer are the texts. As a result of pratice, a cosinus value greater than 0.2 indicates a significant similarity betwen two texts.

When there are n texts, each one has a cosinus with the other texts. That means that a matrix which is symetrical is attached to a given corpus.

| | Form 1 | Form 2 | ........... | Form n |
|---|---|---|---|---|
| Form 1 | 1 | | | cos (1,n) |
| Form 2 | cos (2,1) | 1 | | cos (2,n) |
| .<br>.<br>. | | | | |
| Form n | cos (n,1) | cos (n,2) | | 1 |

This distance matrix is the starting point of the cluster procedure.

### 3.5 The cluster procedure

The clustering tree is built upward, using the CLUSTER procedure of SAS [3] which corresponds to the farest neighbours (COMPLETE LINKAGE).

In COMPLETE LINKAGE the distance between two clusters is the maximum distance between an observation in one cluster and an observation in the other cluster. COMPLETE LINKAGE is trongly biased toward producing clusters with roughly equal diameters and can be severely distorded by moderate outhiers (Milligan 1980). COMPLETE LINKAGE was originated by Sorensen (1948).

## 4 Result analysis

The cut level on the tree diagram fixes the number of clusters.

Most of the clusters are homogeneous but the procedure does not tell the user which words are representative of each cluster. On cluster represents forms which are not easy to group.

## 5 Conclusion

The advantage of such a method is to obtain results without important investment. With few more investment, by representing each text by expressions instead of simple word for instance, the proximity of texts should be more selective and more adequate.

We want to thank another group at EDF (Direction des Etudes et Recherches) for his help and more specifically Mr STA J.D. whose help was needed to implement ADOC.

# References


1. SOUCHOIS T. Proximity analysis of literal data on event forms for assistance with experience feedback analysis. Collection des notes internes de la DER - 1994-1995, No 0017
2. SALTON G., McGILL M. Introduction to modern information retrieval. Computer Science Series. McGraw, Hill book company, 1983
3. SAS/STAT. User's Guide, vol. 1. Procedure cluster, version 6, Fourth Edition, SAS Institute Inc

# Analysis of UK Offshore Accidents & Incidents

## 1.0 Introduction

Stephen Connolly, HSE (OSD) UK.

Accident and incident data for the UK offshore industry is collected via two reporting forms: Offshore Incident Report (OIR) forms OIR/9A and OIR12. OIR/9A deals with all offshore accidents and incidents, and OIR12 records details of "loss of containment" incidents, introduced specifically to cover recommendation 39 of the Cullen Report of the Piper Alpha disaster of June 1988. A loss of containment incident recorded on an OIR/9A triggers a complementary OIR12, which requests further technical details of the event.
The number of dangerous occurrences reported has increased significantly in recent years, with a corresponding increase in OIR/9A's, from 2-300 in the late 1980's to over 1100 in 1991/92. This may be equated to a higher awareness safety of due to the safety case requirements (Ref 1) in general and of hydrocarbon leaks in particular, following the Piper Alpha incident and the redefinition of dangerous occurrences.

## 2.0 Details of Incident and Accident Data Capture

Accident details are reported in four categories: fatalities, serious injuries, "over 3 day injuries" and dangerous occurrences and input into OSD's accident and incident database. Categories are recorded as one of 17 Broad Incident Types (BIT's), 15 Work Activities and 10 Operations which were being undertaken at the time of the incident. Examples of BIT's are, Falling Objects, Slips Trips and Falls, Use of Machinery, and Fire/Explosion. Examples of Operations are Production, Drilling, Maintenance. A full description of the incident is also included. Analysis of this description can yield important information relating to the underlying cause of the incident and may also highlight factors which are important but were not recorded in the "tick box" sections. This is seen as a good source of data for causation analysis which could be used to focus research effort aimed at providing a better understanding of the principle factors in accident causation.

### 2.1 Exposed Offshore Population

UK offshore accident statistics are derived from a small working population, typically 30,000 persons, exposed to a number of high risk scenarios, such as fires, explosions, ship collision and extreme environmental conditions. The size of the offshore worker population can vary considerably. Between 1989 and 1995 offshore population varied from a high in 1990/91 of 36,500 to a low in 1994/95 of 27,200, a variation of about 34% based on the 1994/95 population. Incident and accident rates are generally expressed per 100,000 employees for use as broad indicators of average accident rates as they do not take account of exposure times as a function of occupation (Ref 2, 3).

# 3.0 Offshore Incidents and Accidents Statistics

The combined fatal and serious injury rate for 1993/94 and 1994/95 show a significant downward trend. Using the data from 1989 onwards a comparison using a Chi squared test has been performed and it indicates that there is a significant difference between the observed rates for 1993/94 and 1994/95 and those expected, based on the 6 year data set.

## 3.1 Direct Observations of Serious Injury Rate & Work Activity

1991/92 was a period of particularly high activity with about 186 wells drilled, a worker population estimated at 33,200 and 22 serious injuries associated with Drilling/Workover. In the following year 1992/93, the number of wells drilled dropped to about 131 with a worker population of 29,500 sustaining 24 serious injuries. A fall of about 11% in overall exposed population and a rise of about 9% in serious injuries. In 1993/94 drilling activity fell below that of the two previous years, not significantly, but the serious injuries associated with drilling fell to 10; a significant reduction over the previous two years. Although the serious injuries associated with drilling activity show a downward trend over a 6 year period, the sudden increase between these two annual rates is difficult to explain. There appear to be other factors that have an effect on the serious injury rate that are not clear from initial observations. Construction and Deck Operations show accident rates that do not follow the overall downward trend. It is possibly linked to a high level of Construction or Commissioning of new installations following a year of relatively high well activity. The rate for 1992/93 is significantly different from previous years and "outlier" data sets of this type could be explained more accurately if workers exposed hours data as a function of occupation were available. Discrepancies of this type adds to the growing need for such information.

A factor which may have some influence over the disparity between overall accident trends and serious injuries is the accuracy of reported incident descriptions. Serious accidents recorded under "other" have fallen in recent years to a significantly low level. It is possible that the higher awareness of safety related activities has resulted in more accurate reporting of serious injury categories. Less accidents are recorded under "Other" and hence there is an apparent increase in specific activities due to more accurate reporting. Additionally, small worker populations are more "sensitive" to small increases in the number of accidents. Drilling, Maintenance, Domestic and Production all show a general downward trend. Production sourced injuries show an accident rate peak in 1992/93 with zero accidents reported in 1993/94, and again this is largely unexplained from direct observation. A similar picture emerges if injuries are considered as a function of their Broad Incident Type (BIT). A large number are categorised as Slips/Trips/Falls, Handling Goods and "Other". Most of the remaining BIT's show a downward trend from 1990 to 1995.

### 3.2 Seasonal Effects on Injury Rates

Activities requiring minimum weather conditions such as refurbishment, maintenance, shutdowns etc. are generally planned for the summer months, and a variation in incident and accident rates should be observable between summer and winter periods. A subset of data has been taken for the two years 1993-95 and split into 6 monthly sets. It is noted that the link is tenuous if the serious injuries for November and December 1993 are compared with the incidents recorded for November and December 1994, but detectable if over 3 day injuries are considered. The data indicates that there are a higher number of incidents in the summer period if over 3 day injuries and dangerous occurrences are considered. A suggested explanation is that the short timescales allowed for shutdowns are a contributing factor, as well as the higher level of equipment testing during start-up.

## 4.0 Interaction Analysis of Incidents and Accidents

Single, major accidents such as Sea Crest 1989, Piper Alpha 1988, Chinook helicopter 1985, Alexander Kielland 1980, are characterised by multiple fatalities. Statistically, large numbers occurring infrequently are extremely difficult to use for deriving meaningful messages. Statistical analysis of these rare events with associated multiple casualties, must be viewed with caution, particularly estimation of accident trends and causation. It is generally accepted that accidents and incidents rarely have a single cause. Almost all accidents and incidents, if analysed thoroughly, will produce several interrelated causes. In order to reduce accidents and incidents it is important to understand how and why they occurred, and what were the significant contributory factors. To provide meaningful and useful information from data analysis we need to discover "logical interrelationships" between contributing factors. Statistical relationships can be generated using linear regression to predict the value of one data field on the basis of values of one or more predictor data fields. When more than one variable is used as a predictor, this is called multiple linear regression. However, regression no matter how sophisticated has shortcomings. There are many non-linearities and discontinuities in cause and effect relationships, especially in socio-technical situations. Accident causation is a complicated inter-relationship that is almost certainly non-linear, and has many inter-dependant factors. This is known as an interaction effect, and an attempt has been made to identify these effects by using an "intelligent" software programme.

### 4.1 Data Fields used in the Analysis

The data fields used in the analysis are:-

**Dependant variables are:-** ***Accident category*** or ***Dangerous Occurrences***

**Independent variables are, respectively**:- i. Broad Incident type (BIT); ii.

Activity; iii. Occupational Experience; iv. Experience Offshore; v. Job at time of Incident. vi. Occupation of injured person; vii. Operation at time of incident; xiii. Time into Shift; xiv. Time into Tour; x. Time of Incident;
Incident and accident data from 1989 to 1995 has been analysed in two ways:- the 3 accident categories and Ill-health, and dangerous occurrences.
The **dependant variable** is the field against which all other data fields are analysed; injury categories used are fatalities, serious injuries, minor injuries. Dangerous occurrences are analysed as a function of Operations, Activity, and BIT only.

## 4.2 Interrelationships in Offshore Accidents

An initial analysis of 2477 data items covering the 3 injury categories has revealed 4 significance levels of principle contributing factors. Operations; Activity and Time into Tour; Experience in Occupation and Offshore; and Broad Incident Types. Operations is split into 6 statistically significant sub groups of which Diving has no further interrelationships. Construction, Production and Structural Modifications account for 638 (~26%) of injuries, Drilling and Deck Operations 759(~30.5%), Maintenance and "Other" 798(~32%), Domestic/Catering 171(~7%) and Transport 59(~2.5 %). Construction and Domestic groups show an approximate ratio of serious to over 3 day injuries of 1:10 whereas Maintenance and Drilling indicate a ratio of about 1:5. Transport shows a very different profile than the other groups with a high level (33.9%) of fatalities out of the 59 injuries reported. This data set is heavily influenced by the Cormorant Alpha helicopter accident involving 11 fatalities. An interesting relationship is highlighted as a function of occupational experience in 234 data items. There appears to be a 2:1 ratio of serious injuries between workers with 0-2 years experience and those with 2-6 years. Diving is shown to have a high number of serious injuries in comparison with over 3 day injuries, and a measurable level of occupational illness. The latter is a direct result of a close monitoring by OSD's Diving Section due to the nature of Diving activities and its recognised long term occupational hazards. Drilling and Deck Operations, and Maintenance reveals "time into tour" as a significant contributory factor.
Reporting inaccuracies account for 93 (~12%) of reported accidents in Drilling with about 120 (~15%) for Maintenance. Further investigation has revealed that "time into tour" is sometimes misinterpreted by Drilling Rig management as "time on location" rather than the time a worker has been onboard; a possible indicator for clearer guidance to this section of the offshore industry. However there is sufficient confidence in the quality of the 759 data items to indicate that "time into tour" plays a significant part in an injury being serious rather than "minor". A 2:1 ratio is revealed between "time into tour" of 13 days and over 13 days; for Drilling the same statistical significance is highlighted at 14 days. This an area of interest to OSD's Human Factors Section which will be investigated further. This interrelationship also highlights another issue, that of reporting accuracy. Within this data subset missing data is indicated as well as "90 years" occupational experience. Missing data and possible misinterpretation of reporting details occurs elsewhere and could indicate a requirement to review incident reporting guidance.

## 4.3 Interrelationships in Offshore Incidents

2862 reported Dangerous Occurrences have been analysed by Operations as a function of Activity and Broad Incident Type. Broad Incident Types appear to be the principle contributing factor followed by Activities. Initial analysis has revealed a complicated interrelationship with data clusters containing Activities repeated under different BIT's. Missing data indicators appear in all Activity sub-groups. Loss of containment accounts for 887 (~30.7%) of incidents, with Fire/Explosion 302(~10.7%), Crane Operations 344 (~12%), Using Machinery 376 (~13%) and "Other" 441 (~15.55). The latter figure indicates that a significant proportion of dangerous occurrences are not reported accurately, or alternatively cannot be placed into 1 of 16 pre defined categories. This point has been discussed with representatives from the offshore industry with a consensus option that serious consideration should be given to removal of "other" categories in incident reporting. 52 of the 887 Loss of Containment incidents are reported under "inspection" activities. This may be interpreted as 52 hydrocarbon leaks resulting from inspection activities. Inspection and Cleaning grouped under Fire/Explosion incident types indicate that 34% of dangerous occurrences appear under Maintenance operations. 54% of Inspection related incidents are also highlighted as "Maintenance" under Crane Operations and Handling Materials. Falling Objects and Slips/Trips/Falls recorded 195 (~6.8%) incidents with 50% reported under Drilling activities. A large number (65) of incidents reported as Falling Objects are sub-grouped under "Other" the same number as recorded under another sub-group "Operating Machinery". Falling Objects are an important category because in some circumstances they can become double jeopardy events. i.e. a falling object can cleave a pipe of vessel containing hydrocarbons and simultaneously supply an ignition source. Further analysis of Crane Operations and Falling Objects is planned to investigate the nature of these incidents in order to provide useful intelligence for technical assessment of safety cases particularly for new designs.

# 5.0 Discussion

The initial investigation into relationships within the data sets indicate several areas which require further investigation and research. Data or lack of it, on ship collisions has been recognised as being in need of improvements in several areas, and this view is supported by the outcome of the analysis. There are several research programmes underway to generate relevant intelligence including; installation damage assessment, vessel traffic collection, human factors contribution. A higher incidence of serious to over 3 day injuries for specific occupational categories is recognised and may be input into the offshore safety case assessment guidelines, to focus resources in these areas during technical assessments of installations' safety cases. Missing or inappropriate data i.e. "90 years" occupational experience, needs further investigation and could lead to a review of reporting guidance. Within OSD's three branches, data is used for differing purposes. Operations branch use data to target inspection programmes and investigations in support of reports of individual accidents. Analysed data is used to identify trends and patterns in accident causation and to identify particular problem areas. Policy branch require information to monitor the effectiveness of new and established policies, particularly before and after the introduction of new regulations. Data is also used to support proposals for new regulations. Technology branch use data for assessment of safety cases in the evaluation of risks to persons and consequence analysis of major accident events. Offshore regulations require safety cases to include information relating to identification of major hazards and initiating events. Technical assessment of safety cases needs valid data against which the safety arguments can be evaluated. Further benefit from accurate data is gained during contact with duty holder during assessment. Likely scenario's are considered and advice given based on information of incidents and accidents from a variety of offshore installations. Information is a crucial parameter for identifying areas where research or technology development may be required, and in formulating strategies for the development of technical guidance.

# 6.0 Conclusions

This initial analysis should be viewed as a broad indicator of areas that would benefit from further investigation, and not as an absolute measure of interrelationships between accident causation and offshore activities. The following conclusions are made:-

- Reporting inaccuracies, missing data points, and use of "Other" as a category needs to be reviewed.
- Injuries in activities such as Drilling appear to be sensitive to the length of time workers spend offshore.
- Worker exposed hours has been identified as a significant factor in highlighting interrelationships, and the need for more detailed exposed hours data is currently being addressed.
- There appears to be higher ratio of serious to over 3 day injuries in certain activities.
- Incidents recorded as related to Inspection as an activity should be further investigated.

# Acknowledgements

The contents of this paper are those of the author and may not represent those of the HSE. I would like to thank my colleagues for the advice and comments in preparing this paper.

Ref 1 The Hon. Lord Cullen, Nov 1990, Public Inquiry into the Piper Alpha Disaster, (The Cullen Report) in 2 volumes; HMSO.

Ref 2 OTO 94 010 available from HSE Information Services, Information Centre, Broad Lane , Sheffield, UK.

Ref 3 OTO 95 953 available from HSE Information Services, Information Centre, Broad Lane , Sheffield, UK.

# Risk-Based Approach to Analyzing Operating Events*

A. R. Marchese
U. S. Department of Energy
Washington, DC , U.S.A.
and
P. Neogy
Brookhaven National Laboratory
Upton, New York, U.S.A.

## Introduction

Existing programs for the analysis of operating events at the Department of Energy (DOE) facilities do not determine risk measures for the events. An approach for the risk based analysis of operating events has been developed, and applied to two events [1]. The approach utilizes the data now being collected in existing data programs and determines risk measures for the events which are not currently determined. Such risk measures allow risk appropriate responses to be made to events, and provide a means for comparing the safety significance of dissimilar events at different facilities.

## Risk-Based Approach

An overview of the approach is presented in Figure 1. Given an operating event, potential undesirable consequences (such as injury or fatality to workers or members of the public, property damage, impact on the environment, etc.) are identified. Qualitative estimates are made of the conditional probability of occurrence of the undesirable consequence(s), and of the magnitude of the consequence itself. The conditional probability is estimated on the basis of the residual protection, or the number of remaining barriers that provide protection from the undesirable consequence. If one (or fewer) barrier remains, the conditional probability is assessed as "high". It is assessed as "medium" with two barriers remaining, and as "low" with three or more barriers. The consequence is assessed as "high" if the potential exists for fatality or property damage in excess of $1,000,000, "medium" if the potential for severe injury or property damage in excess of $100,000 exists as a result of the event, and as "low" otherwise. The qualitative assessment of the conditional probability and the consequence allows an assessment of the conditional risk from the event. If this risk is less than "medium-medium", no further analysis is necessary, and the results of the qualitative assessment are documented.

---

*This work was supported by the U.S. Department of Energy under Contract DE-AC02-76CH00016.

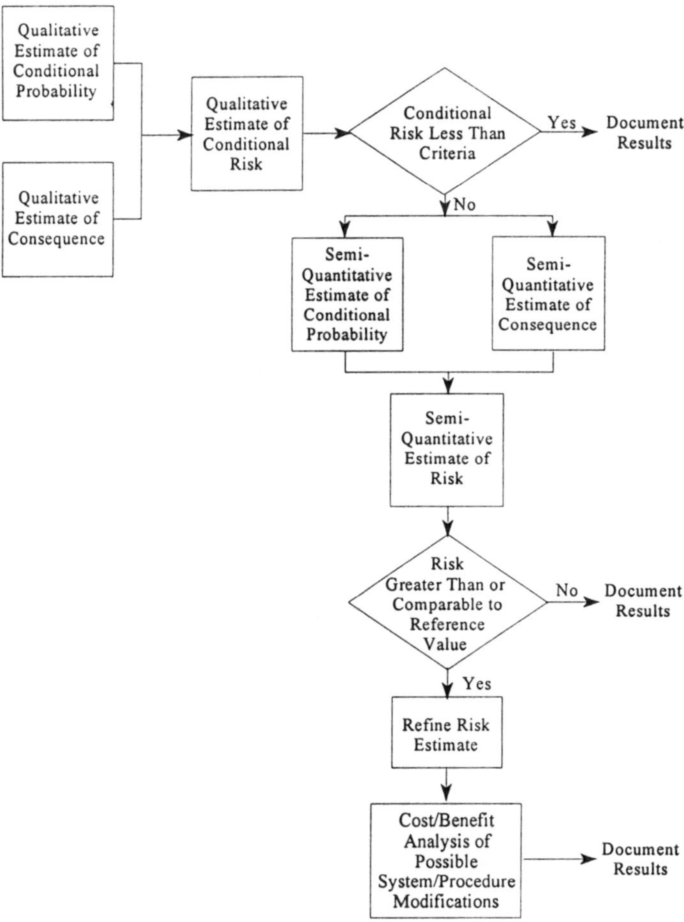


Figure 1 Overview of Risk Based Approach

A semi-quantitative estimate of the conditional probability is determined using a simplified event tree or a simplified fault tree that incorporates the barriers. The failure frequencies utilized are based on facility specific data if available, generic data or informed estimates. Similarly, a semi-quantitative estimate of the magnitude of the consequence is determined. For industrial hazards, the consequence is usually self-evident. For radiological and chemical hazards, an accident analysis approach similar to that developed within DOE's Defense Programs [2] is utilized to estimate the consequence. Since most events with a conditional risk of "medium-medium" or higher would have some likelihood of fatality to a worker, it is convenient to use the risk of fatality from the event as a risk measure to categorize the event. This risk measure gauges the safety significance of the event and helps in deciding what the appropriate level of response to the event should be. Before engaging in additional analyses, it is necessary to ensure that they are justified by the level of risk posed by the event. This is achieved by comparing a risk measure of the event, the fatality risk, to some reference value. If the fatality risk is not greater than or comparable to this reference value, then a risk appropriate response to the event may not include any corrective actions. In this case, further analysis of the event is unnecessary, and the results of the analysis performed thus far are documented. If the fatality risk of the event exceeds the reference risk value or is comparable to it, then the risk estimates already obtained are refined. A more detailed consideration of the function and efficacy of the systems, components, structures and procedures that play a preventive or mitigative role during the event, particularly those that are likely candidates for upgrading as a corrective measure, is also undertaken at this time. Finally, a cost-benefit analysis of possible systems and procedure modifications is undertaken.

## Applications

The analytical approach has been applied to (1) a glove box fire in a plutonium processing and fabrication facility, and (2) an electrical hazard event at a composite materials technology facility presently under construction. The results of the analysis are presented in the form of simplified event trees and fatality risks associated with the events. The choice of the events was made in part to demonstrate the applicability of the methodology to incidents involving radiological as well as non-radiological, industrial hazards.

### Glove Box Fire

In November 1994, contaminated rags drying on the floor of a glove box in a plutonium processing and fabrication facility were found to be undergoing spontaneous combustion. The glove box was successfully isolated from any source of oxygen, and the smoldering rags were subsequently allowed to burn to completion by controlling the flow of oxygen to the glove box. The rags are believed to have self heated due to contamination with $Pu^{238}$. No radioactivity was released as a result of this event. Given the observed event, the spontaneous combustion of rags in the glove box, several barriers existed to prevent the release of radioactivity. These are: (1) detection of the fire and intervention to contain it, (2) maintenance of glove box contamination despite the fire, and (3) the ventilation system which maintains the glove box at a negative pressure with respect to its surroundings and minimizes a release when the glove box containment is lost. Based on the three barriers that remained, the conditional probability of release was characterized as "low". The consequence of the release was judged to be in the "medium" category based on the large specific activity and the large inhalation dose conversion factor of $Pu^{238}$, although the amount of $Pu^{238}$ in the rags was presumed to be small. The conditional risk of the event is therefore "medium-low". At this point, the analyst may decide that no further analysis is necessary (in accordance with Figure 1). To illustrate the methodology, and determine that the conclusions based on qualitative analysis are valid, the remaining steps are described below.

Figure 2A presents a simplified event tree based on the barriers discussed above. The likelihood of a fire being detected depends on how frequently the room is checked by a worker during normal operational shifts or by a security personnel at other times. The likelihood of the fire being contained after detection will depend in part on the skill and training of the worker to perform this non-routine task. The fact that the fire was detected and contained in this instance indicates that the likelihood of detecting and successfully containing the fire is not considerably smaller than unity. In the absence of more detailed information or analysis, this likelihood was estimated to be about 0.5. If the fire escapes detection, there is still some likelihood that the fire would extinguish itself without breaching the glove box containment. This likelihood will depend on the size of the fire and its location (proximity to the flammable gloves). Since the amount of combustible material consisted of about a quarter pound of rags probably placed near the center of the glove box floor, the fire had the potential to be small and localized. The likelihood of the glove box containment to be maintained despite the fire burning undetected was again estimated to be about 0.5. At the facility in question, there have been incidents of release of radioactivity from a glove box after it is breached due to improper ventilation or improper worker response. The likelihood of the ventilation system to be defeated after a breach of glove box containment is estimated to be about $10^{-2}$. The conditional probability of a significant release was, therefore, estimated at $2.5\times10^{-3}$ ($0.5\times0.5\times.01$). Given a release, the maximum dose to a worker in the room was estimated at 19.2 rem (based on an estimated 18 g of $Pu^{238}$ in the rags, and assuming instantaneous, uniform dispersal of airborne $Pu^{238}$ particles within the room). The corresponding risk is presented in Figure 2B and compared to the threshold for significant risk adopted by the Occupational Safety and Health Agency (OSHA) in its final benzene rule ($10^{-3}$ fatality) and the average lifetime accidental fatality risk in U.S. industries ($4\times10^{-4}$ fatality per work life of 40 years).

## Electrical Hazard

In June 1994, an electrician working on a 480-volt main distribution panel in a composite materials technology facility received serious flash burns from an electrical fault and the subsequent electrical arc blast. The electrical fault occurred when a ground wire to be installed made contact with the exposed parts of energized incoming connections on the main breaker, which had been turned off. After an electrician removes the protective cabinet enclosure covering a distribution panel, several barriers exist in principle to protect him. The first of these is a work plan that acquaints him of the hazards involved and provides him with instructions to safely execute his task. A second barrier exists in the form of a procedure for electrical energy isolation and control (lockout/tagout). Lastly, protective equipment such as gloves, blanket and safety glasses provide a third barrier. For the event analyzed, as we shall see, the first barrier failed and, consequently, the second and third barrier failed as well. Because of the crucial role played by the failed first barrier, and the dependent nature of the subsequent barriers, the conditional probability of severe injury was judged to be "high". Since the potential for severe injury or fatality existed, the consequence was also judged as "high", leading to a "high-high" categorization of the conditional risk.

Figure 3A presents a simplified event tree for the electrical hazard incident. A work plan was generated for the activity but was deficient in several respects. The task was categorized as low risk based on considerations of public health and safety, not risk to the worker. The work plan was also deficient in that it did not require a high voltage lockout/tagout to completely de-energize the panel. The work plan also did not identify

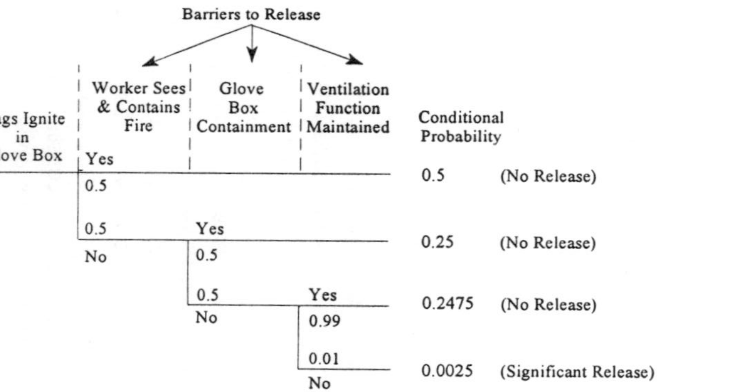


Figure 2A Simplified Event Tree for the Glove Box Fire

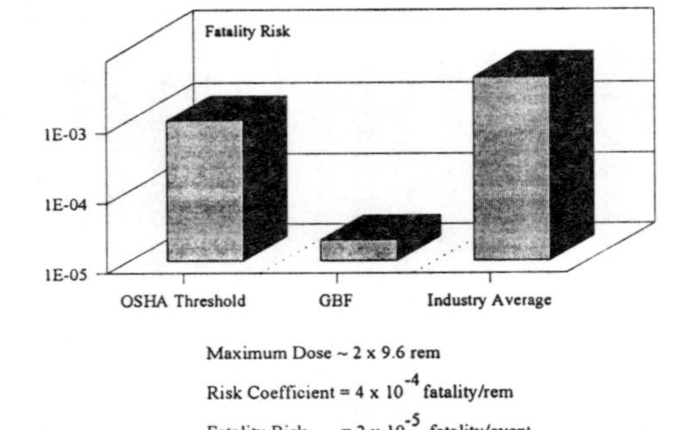


Figure 2B Comparison of Risk to Reference Risk for Glove Box Fire

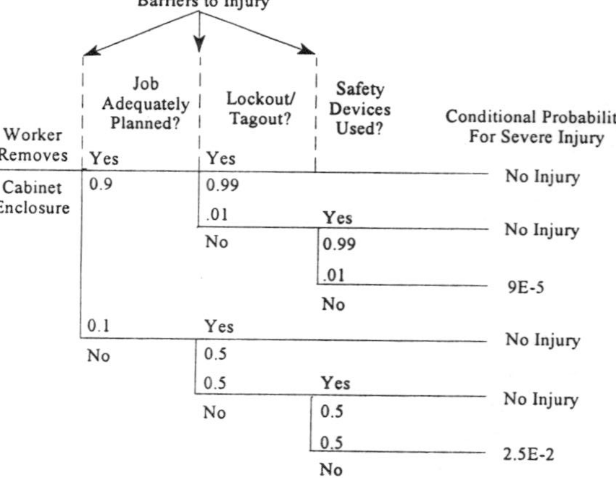


Figure 3A Simplified Event Tree for the Electrical Hazard Incident

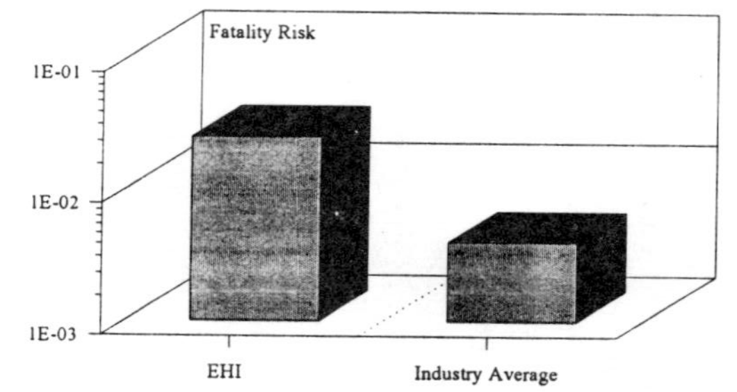


Figure 3B Fatality Risk from the Electrical Hazard Incident

which protective equipments, if any, were needed for the work, and did not make any provisions for making the equipment available to the worker. The deficiencies in the work plan were due to human errors. These human errors belong to the category of initiator actions, including slips and mistakes, that cause initiating events. This category of human error has nominal probabilities in the range of $10^{-2}$ to $10^{-4}$. The probability may be an order of magnitude higher if a need exists for systems knowledge or for the interpretation of indirect information, as existed in this case. The probability of an inadequate work plan was, therefore, taken to be $10^{-1}$. Because the work plan failed to specify lockout/tagout and protective equipment, these barriers were as likely not to be implemented as to be implemented. The conditional probability for a severe injury was, therefore, estimated at $0.1\times0.5\times0.5$ or $2.5\times10^{-2}$. Considering that fatal injuries are about an order of magnitude less likely than severe injuries, the conditional probability of fatality may be estimated at $2.5\times10^{-3}$, which is also the risk of fatality from this event. Figure 3B presents the fatality risk from the event and compares it to the average lifetime accidental fatality risk in U.S. industries. Clearly, the risk from the event is greater than the average accidental fatality risk, and further efforts are needed to reduce this risk. The following general observations are made here regarding the risks associated with this event and the benefits of reducing these risks. This event occurred due to human errors at two levels: (1) errors that led to an inadequate work plan, and (2) the failure on the part of the individual to take greater responsibility for his own safety and use appropriate safety equipment and safe work practices. Implementation of necessary actions to ensure that work plans are developed to take into account worker risks as well as public health and safety is a crucial step in reducing the frequency of similar incidents. Training the workers to take more responsibility for their own safety by using appropriate safety equipment and safe work practices will reduce both the frequency and consequences of such incidents.

## Conclusions

In this paper we have presented a step-wise approach to reviewing operating events for their safety and risk significance. The risk-based approach allows a quick determination of the appropriate level of response to an event, and the cost-benefit aspects of any contemplated corrective action. Reference risk values have also been suggested for comparison to the risks from individual events. The calculation of a quantitative risk measure such as the fatality risk associated with events also allows a meaningful comparison to be made of the safety and risk significance of dissimilar events. Although we have restricted ourselves to individual events, the method could be extended to examine the risk significance of a class or family of events. By aggregating and analyzing operating events of a similar nature, it would be possible to examine the risk implications of the underlying safety issues.

## References

1. Preliminary results of this analysis were presented at the 1995 Winter Meeting of the American Nuclear Society (P. Neogy and A.R. Marchese, Trans. Am. Nucl. Soc., **73**, 280, 1995).

2. U.S. Department of Energy Defense Programs Safety Survey Report, D. Pinkston, editor, DOE/DP/70056-H1, November, 1993.

# HRA for WWER 1000 NPP Temelín PRA Project - New Specific Issues to Be Addressed

**Jaroslav Holý**
**NRI Rez**
**Czech Republic**

## 1 Introduction

The Temelín WWER 1000 NPP PRA Project is a long term PRA project which includes complete Level 1 and Level 2 PRAs. The project started in September 1993 and is about to be completed. The project is controlled, organized and partially carried out by NUS Company from the U.S.A. Several organizations from Czech Republic have participated in this project, the principal ones being the Temelín NPP (Safety Analysis Group) and the Nuclear Research Institute Rez (Reliability and Risk Assessment Department).

As a part of the Level-1 PRA, a detailed HRA was carried out. The analysis directed by Garreth W. Parry from NUS followed the general HRA Guidelines approach presented in IAEA-TECDOC-499, which is based on the Systematic Human Action Reliability Procedure (SHARP). Besides this general approach, the EPRI decision tree methodology was used to quantify the most important procedure-driven operator actions.

There were two specific challenges that had to be addressed by the Temelín NPP HRA analysis. The first one is that NPP Temelín is under construction and much of the information needed for HRA was not available so that more assumptions than usual had to be made. The second is that Temelín NPP appears to be a special combination of "eastern" hardware controlled and driven by advanced "western" control and information processing systems. The information and control system COMPRO is an advanced computer system, which will be used for the first time to help operators to control the operation of NPP of "eastern" design. To address the specific features of operators′ communication with COMPRO, the EPRI decision tree approach had to be extended.

The paper describes the innovative approach adopted for the analysis of procedure (and COMPRO) driven operator interventions during the mitigation of the consequences of non-standard situations (accident conditions) and for the quantification of the probability of human failures connected with those interventions. The approach has been developed by the combination of several well-known HRA methods taking into account the unique operational conditions of NPP Temelín.

## 2 Representation of Human Error Event

Current PSA practice is to construct the plant logic model in such a way that the consequences of not performing a response action are clearly delineated. However, the consequences of the actions taken other than those required by the procedure, i.e. errors of commission, are generally not modelled. This is the practice in this PSA also. However, a recent study of a PWR with Westinghouse style EOPs /1/ showed that, assuming the control room crew was well-trained and well-disciplined, and that the instrumentation had a sufficiently high reliability with adequate redundancy, the likelihood of a significant error leading essentially to a misdiagnosis of the accident was extremely small. Thus, in this study following standard PSA practice, human error events are modelled as **errors of omission.**

The human error is regarded as having two contributions. The first is from the failure to **d**etect, **d**iagnose, or make a **d**ecision as to a plan of action (the **DDD** contribution). This is essentially the **cognitive** contribution. With symptom-based procedures, making a diagnosis is in fact a much less demanding task as the procedures themselves are structured to lead the operators to the appropriate procedure. The second contribution is from the failure to execute the planned action correctly. Different approaches are used to estimate the probabilities of the two phases.

## 3 Estimation of the Failure of DDD Phase Probability

In the DDD phase, a distinction was made between time-critical and non-time-critical responses. A time-critical response is one in which the decision on a plant action has to be made in a short time, on the order of minutes rather than hours. A non-time-critical response, on the other hand, is one in which there is an adequate time to reach the decision and the operators are not subjected to an undue time stress. It was decided that different HRA models would be used for the two regimes. In the time-critical regime, the time reliability curves (TRCs) in NUREG/CR-4772 would have been used. However, no time critical actions were identified for the Temelín PSA model. In the non-time-critical regime, a modification of the decision tree approach developed for EPRI /2,3/ was used.

The approach is based on the **decomposition** of each failure event into the contributions that represent different potential failure mechanisms. For each failure mechanism, a **decision tree** is constructed which has as its branches factors that are thought to influence the likelihood of an error through that mechanism. A probability is associated with each path through the decision tree to represent the analyst's assessment of the combined effect of the factors on that path. The probability of an error in the DDD phase for a particular scenario is then estimated by summing the contributions from the appropriate paths from the decision trees that are applicable to that scenario. The approach also suggests how to take credit for the possibility of recovery from mistakes when there are clear opportunities to do so, either as a result of a step in the procedure, or as a result of input by another crew member.

**Five basic failure mechanisms** have been identified for NPP Temelín operators' actions connected with information processing:

- information not available or hardly available
- failure of attention
- information misread/miscommunicated
- procedure step skipped
- procedure misinterpreted.

The final form of Temelín-specific decision trees results from applying Temelín-specific information to the general form of decision trees, which have been published in /2,3/. The original decision trees developed in the EPRI program are not directly applicable because of the use of the EOP tracking system COMPRO in the control room.

The most frequently used and well-known methods of quantification of basic attributes influencing operator's behaviour presented in /4,5/, for example, were used to quantify the individual branches of the decision trees for the case of Temelín. Two following examples can illustrate this process.

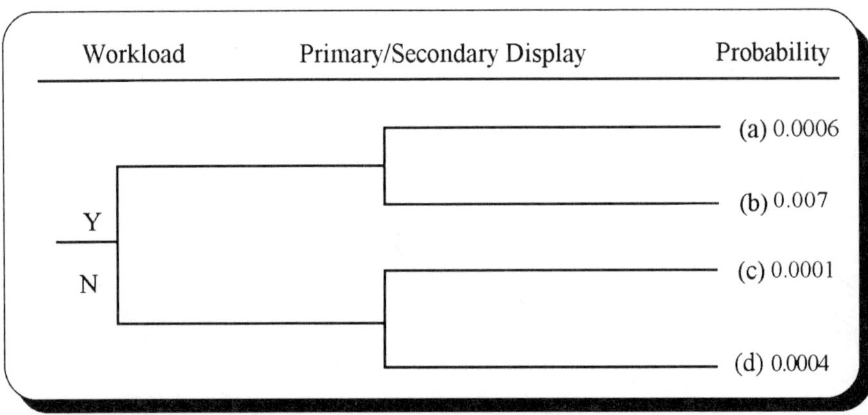


*Factor No.2: Failure of Attention (see the figure above)*

The numerical values of probabilities of human errors given the combination of conditions represented by sequences on the tree have been determined as follows:

- (c) - an approximate value from THERP
- (a) - the value for "normal load" multiplied by the factor of 6 - "*high overload*" in HEART (see /5/)
- (d) - this attribute is not directly addressed in THERP, the value has been obtained using two factors modelled by HEART:
  - "*A low signal/noise ratio*" - the maximum correcting factor of 10 is supposed in HEART, a factor of 2 has been used in the Temelín HRA

- "*A channel capacity overload particularly one caused by simultaneous presentation of non-redundant information*" - a maximum correcting factor of 6 is supposed in HEART; the factor of 2 has been used in Temelín HRA,

the resulting value was obtained by multiplication of the basic value (c) by both reducing HEART factors

- (b) - the estimation of the value was similar to (d), but higher HEART corrective factors have been used for the case of operator's overload (factors 4 and 3 respectively).

The following assumptions have been made when developing this decision tree:

- COMPRO provides the operator with assistance in monitoring the system status. Checking is the only type of activity of the operator.
- Using COMPRO, the operator is not influenced by the position of the panel as a source of information (Front/Back). On the other hand, two displays, primary and secondary, are at his disposal as the source of information about the plant, and the quality of the operator's action depends on the source of information it is taken from (it is event-specific).
- The operator communicates very intensively with COMPRO. It can be assumed that the critical values of the cues are signalled by an anunciator in COMPRO.

*Factor No.5: Procedure Step Skipped*

The questions asked in the original decision tree are: *Is the step obvious or hidden? Is the operator using a single procedure, or more than one? Is the step graphically distinct? Are there placekeeping aids?*

Only one branch from the decision tree No.5 in /3/ was used in the analysis. Firstly, the procedure step can not be skipped by being hidden in some way, because COMPRO delivers the description of the individual procedure steps to the primary screen and it even controls and checks carrying out the procedure step by step by the operator. Secondly, only one step of the procedure is performed during an arbitrary (small) time period of operator's action and it is not possible to violate this rule, because more than one step of the procedure is not readily available (at the disposal) in COMPRO. Thirdly, relatively effective means are used by COMPRO to make the important information graphically distinct (different colors). Fourthly, the placekeeping aids are inherent in COMPRO. As the result, the answers to the above mentioned questions indicated by the headings of the original decision tree are "*Obvious*", "*Single*", "*Yes*" and "*Yes*".

The probability value for the single branch of the reduced decision tree has been taken from THERP (/4/,Table 15.3). Because the "*not very short*" procedures will be typical for COMPRO, the subcategory "*more than 10 items*" or "*long list*" has been used conservatively from THER giving the resulting numerical value of 0.003.

# 4 Estimation of the Failure of Execution Probability

The **ASEP procedure** /6/ has been used as the basic methodical tool. The HRA analysis in frame of PSA Level-1 for NPP Temelín has been built on somewhat specific conditions, therefore the ASEP procedure could not be used in a complete and standard manner. In the following paragraphs, the Temelín specific points of using the procedure are discussed.

1. In the current PSA model of NPP Temelín, there is no example of skill-based or rule-based post-diagnosis action, which will not be described in the written procedure. The quality of the written procedures (computer system COMPRO) as well as the possibility of failure of the instrumentation supporting operator's activity is addressed in another part of the analysis.
2. The estimation of the maximum allowable time was performed for every important post-accident human interaction. However, because the Temelín plant has not been operated yet, no specific thermal-hydraulic analyses have been carried out for the purposes of HRA and even the operating and emergency procedures have not been completely developed, the individual estimates of the allowable times have been very approximate only ("*seconds*", "*minutes*", "*tens of minutes*", "*several hours*", "*enough time*"). Even these approximate estimates have been connected with a relatively high level of uncertainty in some cases.
3. The diagnosis is made by COMPRO. The probability of misdiagnosis is generally set to be 0.
4. For COMPRO performs the automatic (and immediate) diagnosis, almost all the time, which is at disposal, can be devoted to carrying out the individual executive steps.

In ASEP, the appropriate HEP(s) are derived for the individual human manipulations using the following rules:

1. The probability of failure of the original performer of the action is derived. In case that the recovery of the failure is possible (made by a second person), the probability of failure of recovery is derived and the final probability of the shift failing the manipulation equals to the product of the probability of failure of the original action and the probability of failure of the recovery action.
2. The probability of failure of either the original action or the recovery action depends on the *type of task* and on *level of stress.*
3. The probability of failure of the original action is a) 0.02 provided that the critical action is part of a *step-by-step* task done under *moderately high stress* b) 0.05 provided that the critical action is part of a *dynamic task* done under *moderately high stress* or a *step-by-step task* done under *extremely high stress* c) 0.25 provided that the critical action is part of a *dynamic task* done under *extremely high stress.*
4. The probability of failure of recovery action, i.e. verification of the correctness of the critical action, is a) 0.2 provided that it is connected

with *step-by-step task* under *moderately high stress* b) 0.5 provided that it is connected with *step-by-step task* done under *extremely high stress* or with the *dynamic task*.

The first part of ASEP methodology for this case, i.e. for the original action could be used for the purpose of analysis. However, the quantitative values for failing of potential recovery (the verification mainly) were not used, because the verification of correctness of the action is made almost exclusively by COMPRO system (not by another member of plant staff). The only possibility of failure connected with the verification is that the operator watching and following COMPRO screen fails to process, interpret and transfer the information about the failure to the manipulation with the elements of the panel board performed by another member of the shift. The decision trees, which have been used for assessing of the probability of failure of processing the information can also be used for assessing the probability of failure of the verification process.

For example, in the case of "failure of attention" it can be supposed that the information necessary for the verification is placed on the primary display. The second attribute, workload is not uniquely defined in this case, it is event-specific. For the situation typical with high workload, the probability of failure of attention is 0.0006. For low level of workload, the probability of failure of attention is 0.0001. For the attribute "procedure step skipped", the original decision tree for that case has one branch only characterized by the probability of 0.003. However, this probability is connected with the procedure consisted of a "long list" of steps. In case of verification of the action, the "list" consists of one step only, namely the verification itself, therefore the probability value has been modified using THERP value for the "short-list procedure" (0.001). The main contributors to the probability of failure of verification come from the attributes "failure of attention" (in this case the conservative value of 0.0006 was chosen), "information misread/miscommunicated" (0.002) and "procedure step skipped" (0.001). The total value of probability of failure of verification has been derived to be the sum of those three values, which equals to $3{,}6 \times 10^{-3}$.

Two basic scenarios resulting in the failures of the execution phase of some procedural step have been taken into account:

- the execution at panel board fails and the verification step at COMPRO screen fails
- the execution at panel board fails, the verification at COMPRO screen reveal the failure (does not fail), the execution step is repeated and fails (including verification).

The probability of failure of the execution phase equals to the sum of the values for both scenarios. The probability of scenario a) equals to the product of probability of execution failure and probability of verification failure. The numerical values of this probability are presented above. The probability of verification failure equals to 0.0036. Therefore the resulting probabilities of the individual variants of the scenario a) are: 1) $7.2 \times 10^{-5}$ provided that the critical action is part of a step-by-step task done under moderately high stress 2) $1.8 \times 10^{-4}$ provided that the critical action is part of a dynamic task done under moderately high stress or a step-by-step task done under extremely high stress 3) $9.0 \times 10^{-4}$ provided that the critical action is part of a dynamic task done under extremely high stress.

The probability of scenario b) equals approximately to the product of three factors: 1) probability of failure to execute the given step for the first time - the same as the probability of failure of execution in scenario a) 2) probability of failure to execute the given step after the successful verification of the failure in the first attempt - the value corresponding to the value in a), but high level of stress is suppoesed 3) probability of failure of the repeated verification, which is supposed to be one order of magnitude higher than the probability of the first verification. The related probabilities of individual variants of the scenario b) are: 1) $3.6 \times 10^{-5}$ provided that the critical action is part of a step-by-step task done under moderately high stress 2) $4.5 \times 10^{-4}$ provided that the critical action is part of a dynamic task done under moderately high stress or a step-by-step task done under extremely high stress 3) $2.3 \times 10^{-3}$ provided that the critical action is part of a dynamic task done under extremely high stress.

An observation can be made that in case of step-by-step task with moderate level of stress the probability of the scenario b) is smaller than the probability of the scenario a), i.e. in case of repeated execution, there is higher probability of success. On the other hand, in the remaining two cases, the influence of high stress is so significant that the probability of success of the repeated action is lower than the probability of success of the first execution.

The total values of probabilities of failure of execution can be obtained as the sums of probabilities of the scenarios a), b):

- **$1.1 \times 10^{-4}$** provided that the critical action is part of a step-by-step task done under moderately high stress
- **$6.3 \times 10^{-4}$** provided that the critical action is part of a dynamic task done under moderately high stress or a step-by-step task done under extremely high stress
- **$3.2 \times 10^{-3}$** provided that the critical action is part of a dynamic task done under extremely high stress.

## References

1. Parry, G.W., Julius, J.A., Jorgenson, E., and Mosleh, A.M., "A Procedure for the Analysis of Errors of Commission in a PSA", presented at PSAM II, San Diego, CA, 1994
2. Beare, A.N., Gaddy, C., Singh, A., and Parry, G.W., "An Approach for Assessment of the Reliability of Cognitive Response for Nuclear Power Plant Operating Crews", in Proceedings of Probabilistic Safety Assessment and Management, Beverley Hills, CA, February 1991, Elsevier.

3. Parry, G.W., et.al., "An Approach to the Analysis of Operating Crew Responses Using Simulator Exercises for Use in PSAs", presented at the OECD/BMU Workshop on special issues of level 1 PSA, Cologne, FRG, May 28, 1991
4. Swain, A.D., and Guttman, H., "Handbook of Human Reliability Analysis with Emphasis on Nuclear Power Plant Applications", NUREG/CR-1278, August 1983
5. Humpreys, P.(editor), Human Reliability Assessors Guide, October 1988, RTS 88/95Q, UKAEA
6. Swain, A.D., "Accident Sequence Evaluation Program Human Reliability Analysis Procedures", NUREG/CR-4772, February 1987.

# An Improved HRA Process for Use In PRAs

M.T.Barriere
Brookhaven National Laboratory, Upton, New York, USA
A.Ramey-Smith
United States Nuclear Regulatory Commission, Washington, DC
G.W.Parry
NUS, Gaithersburg, Md, USA

## Introduction

This paper summarizes an analytical process for performing an HRA in the context of a Probabilistic Risk Assessment (PRA), that addresses the major deficiencies of current HRA methods. This analytical process is being developed using the concepts captured in a multidisciplinary HRA framework [1], and is supplemented with the experience obtained from the analysis of historical nuclear power plant (NPP) events [2]. Both the framework and its application to the analysis of NPP events are earlier products of the HRA development project initiated by NRC in response to the recognized need for an improved, more realistic, approach to the modeling of human-system interactions. The analytical process is the application phase of a new approach to human reliability analysis called ATHEANA (A Technique for Human Error Analysis), which is based on an understanding of why human-system interaction failures occur, rather than on a behavioral, phenomenological description of operator responses, and represents a fundamental change in the approach to human reliability analysis.

There are several categories of human-system interaction failures that can impact the safety of a plant. ATHEANA has been developed to address important subsets of those failures; the human-system interaction failures that can result during operating crew responses following an initiating event (i.e., a disturbance to normal operation that would be expected to lead to a reactor trip), and subsequent system malfunctions.

## Objectives of the ATHEANA Project

The ATHEANA method is currently being developed to provide PRA analysts with an improved way of modeling human/system interactions, by consideration of the context under which their failure is likely. The impact of human/system interactions is incorporated in the PRA model in the usual way, by the inclusion of basic events, called human failure events (HFEs), in the structure of the model. Specifically the objectives of the new method, as identified in NUREG/CR-6350, Volume 1 [3] are to:

- identify and characterize important human-system interactions and their likely consequences under accident conditions,
- represent the most important severe accident sequences that could occur,
- estimate the frequencies of these sequences and the associated probabilities of human errors, and
- provide recommendations for improving human performance based upon characterizations of the causes of human errors.

# The ATHEANA Process

The ATHEANA process is developed so that it can be applied to an existing PRA model, but it is potentially most useful when used as an integral part of the development of the model. Application of ATHEANA presupposes that certain important elements of the PRA model have been determined, as this is what defines the modeling context. For example, it will be assumed that the identification of initiating events, a determination of the associated success criteria, and the initial construction of event trees, to at least the functional level, have been completed. In addition, the systems available to perform the functions necessary to satisfy the success criteria will have been identified, and their operational characteristics determined. While this stage of the process is not usually regarded as a human reliability analysis function, it is important that the initial PRA model and the accident scenarios it identifies are well understood by the HRA analyst.

## Description of the Major Elements of the Process

From a functional point of view, the process can be described as having five tasks. The technical objective of each of these tasks is described below:

### Task 1 Familiarization with the PRA model and Accident Scenarios

As discussed above, it is assumed that the starting point for ATHEANA is an initial PRA model. To analyze the human-system interactions within the context of that model it is necessary to become familiar with the definitions of all the elements of that model, and with the accident scenarios identified. It is also essential to understand the assumptions underlying the PRA model. Clearly, a key element in performing an HRA is to identify the role of the operating crew in mitigating, or controlling, the progress of the accidents represented by the PRA scenarios, by becoming familiar with the set of applicable procedures, and understanding which procedures are required, and under what plant conditions. It is also necessary for the human reliability analyst to develop a clear picture of how the plant responds to the functional failures represented in the scenarios.

## Task 2 Identification of Potential Human Failure Events and Associated Unsafe Actions

This task can be distinguished from the task that follows in that it does not require a knowledge of causes of human failures, but is based on an understanding of the different ways that the operators could interact with the plant systems, and relies on a knowledge of system design and operational practices. There are two levels of description of the impact of incorrect actions, differentiated by the level of detail they represent, namely, the human failure events, or HFEs, and the unsafe actions respectively. The steps are:

- Identification and definition of Candidate HFEs in the context of the PRA model. An HFE, as used in the multidisciplinary HRA framework, is defined in terms of its impact described as a mode of unavailability of a component, system, or function, e.g., system X rendered unavailable. The HFE is to be defined in such a way that it implies that the impact on the system is sustained long enough to lead to an unrecoverable change of state of the plant.

- Identification of potential unsafe actions for each HFE. An unsafe action is defined by specifying a specific action on the part of the crew that can lead to the mode of unavailability associated with the HFE. For example, a system could be terminated in many ways, such as placing the pumps in pull-to-lock, defeating automatic initiation signals, etc. It is important to identify the different unsafe actions that could contribute to an HFE, because the reasons for their occurrence may be different. The relevance of this will be apparent later.

The term HFE will be used for the basic event included in the PRA model. The final set of HFEs may include some whose impact is defined at the level of an unsafe action. Others may include contributions from several unsafe actions.

## Task 3 Identification of the Most Probable/Significant Causes of the Unsafe Actions.

This is essentially achieved by building simple plant-specific and scenario-specific models of the causes of the unsafe actions. The building of the models of causes may require several intermediate steps. As discussed in [2], from the practical perspective and for the purposes of quantification, the cause of an unsafe action in the ATHEANA method will be represented in terms of one or more error forcing contexts. An error forcing context (EFC) is a combination of challenging plant conditions (e.g., equipment and instrumentation failures), and deficient performance shaping factors (e.g., poor training, deficient procedures). In the following, two representative steps of this stage of the analysis are represented, namely:

- Identification of the most significant reasons for each unsafe action. A reason is a fairly high level description of why the unsafe action occurred. It can be characterized as being that level of description of cause that might be elicited from the operator following the incident, e.g., "Operators terminated SI because they believed that the termination criteria had been met". The reason should also address why the error was not recovered if that is appropriate for the HFE of interest. For example, the reason above may be expanded to "Operators terminated SI because they believed that the termination criteria had been met, and failed to notice indications that should have warned them that SI was in fact necessary."

- Identification of the most significant Error Forcing Contexts. The error forcing context is a scenario specific description of the factors that provide a plausible explanation for the operator behavior that is characterized by the reason given above. An example of an EFC for inappropriate SI termination, based on a mistaken belief that the conditions are met, is presented in Chapter 4 of NUREG/CR-6350, Volume 1 [3]. It comprises an incorrect, false high, RCS pressure measurement, an historically unreliable pressurizer level indication, and the occurrence of the incident at a time when operators are least ready to deal with an accident.

This stage of the process can be regarded as a search for the "root causes" of the unsafe actions. It has to be recognized however, that the "root cause" is not generally a single item, but a combination of items. Again, it is important to identify the factors that not only cause the error in the first place, but cause that error to persist, and thus prevent correcting the error in time.

As the title of this task of the process implies, there is a screening or prioritization involved in the identification of EFCs, so that only the most significant events and causes are included in the PRA model.

### Task 4 Refinement of HFE Definitions and Integration into PRA Logic Model

In constructing a PRA logic model it may be appropriate to define some HFEs as resulting from specific unsafe actions. If the impacts of several unsafe actions on the sequence development are identical, those unsafe actions can be grouped into one HFE for the purposes of economy in constructing the logic model However, if the consequences of a specific unsafe action on sequence development is sufficiently different from the consequences of other unsafe actions with the same initial functional impact, then it is necessary to model that unsafe action explicitly. For example, depletion of the source of injection will have the same impact on an injection system as terminating a system, but if the source is used for multiple systems, depletion could also impact a second system in a way that termination would not. This is a system oriented logic model structure issue.

It may also be necessary to modify the definitions of HFEs if the decision is made to explicitly include, in the PRA logic model structure, equipment failures that contribute to an error forcing context.

However, there is another concern which can lead to a redefinition of events. The concern is that of dependency. The reasons for some unsafe actions may also be the likely reasons for other unsafe actions appearing in the same scenario cut set. If a common reason can be identified for the occurrence of two HFEs in the same accident scenario description, it may be worthwhile modeling that reason explicitly, in a similar way that certain hardware dependent failure mechanisms are modeled. An alternative approach is to model the dependencies using similar techniques as those used in CCF analyses (NUREG/CR-4780) [4]. In any case, the final form of the PRA model and the definitions of the HFE events it contains will result from an analysis of the impact of the various unsafe actions on the accident development, both in terms of their impact on the plant and in terms of the contributing causes. There are two important steps:

- Searching for dependencies between HFEs/unsafe actions on a scenario basis, by, for example, identifying common EFC elements,

- Final definition of HFEs by, for example, identifying unsafe actions with unique consequences, grouping of other unsafe actions by consequence, choosing an appropriate decomposition of HFEs to address dependency.

**Task 5 Estimation of Probabilities of HFEs**

The approach to quantification is to first estimate the likelihood of the EFCs associated with each unsafe action, and for each EFC, to estimate the conditional probability of error.

- Estimation of likelihoods of EFCs. This will be based on constructing probabilistic models for the joint occurrence of the elements of the EFC. This entails obtaining estimates of the relative frequency with which the plant conditions and unfavorable PSFs in the EFC occur in the PRA scenario definition That this is feasible has been demonstrated by the trial application in NUREG/CR-6350.

- Estimation of conditional probabilities of error given an EFC. This will be a function of how specific the EFC definitions will be. In the case that the EFC is defined such that failure is almost guaranteed, then there is no need to estimate this conditional probability. However, in many cases, the EFC creates an environment in which the likelihood of failure is enhanced, and in this case, it will be necessary to propose methods for estimating these probabilities.

Once the HFEs to be included in the PRA model have been defined and incorporated into the logic structure in the appropriate way, the requantification of the PRA model is essentially trivial.

# Application of the ATHEANA Method

The application of ATHEANA is facilitated by using structured searches for identifying potential HFEs and their associated EFCs. It is planned that there will be two basic source documents comprising ATHEANA; the Frame of Reference (FOR) manual, and the Implementation Guidelines. The FOR manual is intended to be the knowledge base for the method and represents the ATHEANA model of human error causality. The Implementation Guidelines document will provide detailed guidance on how to apply ATHEANA.

# Acknowledgements

This support of the US Nuclear Regulatory Commission and Brookhaven National Laboratory is gratefully acknowledged. The opinions in this paper are those of the authors and do not necessarily represent those of the NRC, or BNL. The assistance of the other project team members, S. Cooper, D. Bley, J. Wreathall, W. Luckas, J. Taylor, and C. Thompson in the development of the work presented here is acknowledged.

# References

1 Barriere, M.T., W.J. Luckas, Jr., J. Wreathall, S.E. Cooper, D.C. Bley, and A. Ramey-Smith, "Multidisciplinary Framework for Analyzing Errors of Commission and Dependencies in Human Reliability Analysis," NUREG/CR-6265, August 1995.

2 Barriere, M.T., W.J. Luckas, D.W. Whitehead, and A. Ramey-Smith, "An Analysis of Operational Experience During LP&S and A Plan for Addressing Human Reliability Assessment Issues", NUREG/CR-6093, Brookhaven National Laboratory: Upton, NY and Sandia National Laboratories: Albuquerque, NM, 1994b.

3. Barriere, M.T., S.E. Cooper, G.W. Parry, D.C. Bley, J. Wreathall, W.J. Luckas, Jr., and A. Ramey-Smith, "Development of An Improved Approach for Human Reliability Analysis, Volume 1" Draft NUREG/CR-6350, December 1995

4. Mosleh, A. et al. "Procedures for Treating Common Cause Failures in Safety and Reliability Studies" NUREG/CR-4780, Vol. 1, 1988, Vol. 2, 1989.

# A Dynamic HRA Method Based on a Taxonomy and a Cognitive Simulation Model

P.C. Cacciabue, G. Cojazzi, P. Parisi

European Commission, Joint Research Centre,
Institute for Systems Engineering and Informatics,
21020 Ispra (Va), Italy

**Abstract**

This paper discusses the issue of Human Reliability Assessment (HRA) methods and the use of dynamic approaches for error probability assessment, based on cognitive simulation and related taxonomies of erroneous behaviour. A comparison between the proposed technique and the "classical" THERP method is shown.

## 1 Introduction

The consideration of Human Factors (HF) in safety analyses of complex systems demands that the human-machine interactions are carefully thought about and that some paradigms are formulated for describing the human activities.

These models are not necessarily transformed into computer simulation but, at least, they represent the correlations existing between the operators, the working contexts and the organisation, that ultimately affect the behaviour.

On the basis of these paradigms, it is possible to define the data for the predictive analysis of safety studies. Such data must then be compiled, usually, by a combined application of field studies, analysis of interviews and queries, expert judgement, and review of databases connected to real accidents.

There seems to be a general agreement amongst the practitioners and researchers that these paradigms and data are necessary for improving the methodologies on human factors [1, 2, 3]. Indeed, they are crucial for the quality and the reliability of the predictions during anticipated transients.

In practice, these requirements imply the use of *models of cognitive* behaviour and *data/parameters*, related to *classifications* of human erroneous behaviours, accounting for contextual conditions, personal factors and organisational influence. Moreover, for describing appropriately the humans interactions, some more elements are needed: a model of *the plant control* and emergency systems, a *structured representation* of the event sequences and, in some cases, a *model for the reliability* study. These characteristics make the advanced HF methods particularly flexible, so that they can be equally applied to PSA, to design basis studies, and to the evaluation of emergency procedures.

In this paper we will describe briefly the approach that we are following and then we will focus on the cognitive simulation model. We will show an application to a relatively complex case, with the aim of demonstrating the flexibility of the approach for PSA studies and for analysing procedures and design basis accidents.

## 2. The methodology for Human Factors Studies

The HF methodology under development, has been called HERMES, standing for Human Error Reliability Methods for Event Sequences and it is based on:

1. A cognitive simulation model, built on the theories of human error and contextual control of Hollnagel [2] and of Reason[4]. The model assumes that the operator carries out, dynamically and interactively with the plant, the loop "detection-diagnosis-planning-action", and, at each step, the cognitive processes are controlled by the effectiveness of the gathered information.
2. A classification scheme of erroneous behaviour, strictly correlated to these theories [5];
3. A model of the functional response of the plant, based on analytical and numerical treatment of differential equations and on the criteria of failure mode and effect analysis.
4. A method for structuring the interaction of the models of cognition and of plants and for controlling the dynamic evolution of events. This method generates the normal or inappropriate (failed) behaviour of the operator or of the components, according to logical and probabilistic considerations: the DYLAM approach [6].

The HERMES methodology, schematised as in figure 1, can be applied for prospective and retrospective type of studies.

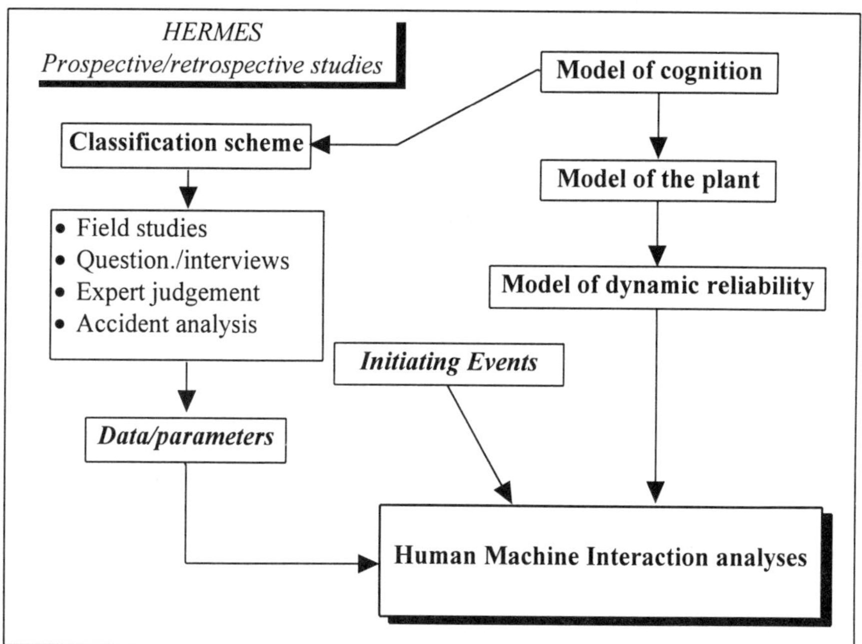


Figure 1 A Framework for Human Machine Interaction Methodology

The classification scheme, based on the model of cognition, guides the field studies, the development of questionnaires and interviews, the extraction of expert judgement, and the examination of accidents/incidents, with the aim of estimating data and parameters to be included in the analyses. A series of initiating events, selected by the analyst in accordance with the objective of the safety study, triggers the whole set of sequences. For brevity, no further details will be given of the classification scheme, of the dynamic reliability analysis methodology and of the plant models, which can be found elsewhere [7, 8].

*Retrospective* studies of real accidents lead to the identification of the root causes of human erroneous behaviour, while *prospective* analyses estimate the human-machine interaction during hypothetical accidents. This paper will concern the prospective use of the methodology, with particular reference to PSA applications.

# 3. Application to a real case study

The methodology has been applied to a study case which has been studied for quite sometime, the Auxiliary Feedwater System (AFWS) of a French type PWR nuclear power plant. The schematised representation of the AFWS (fig. 2) shows half of the system (train-A), feeding the steam generators 1 and 2 (SG1 and SG2), with train-B completely similar, but feeding the steam generator 3 and 4 (SG3 and SG4).

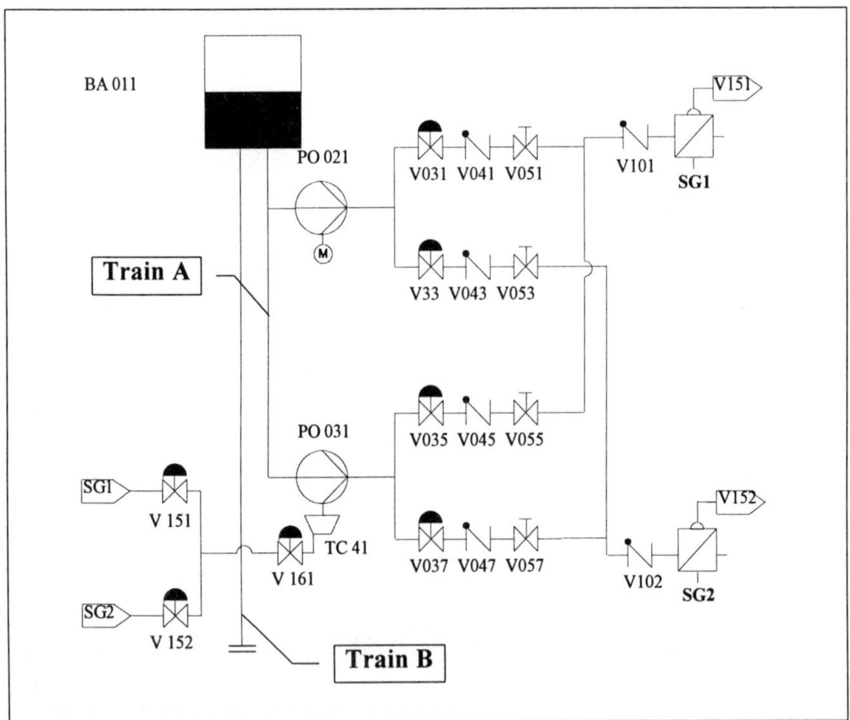


Figure 2. Schema of the Auxiliary Feedwater System

## 3.1 The cognitive simulation and other models.

The model of the operator has been developed as to reflect the expected reasoning processes, decisions and actions that the designer may demand to a trained nuclear power plant operator. In this sense the model is a very simple representation of actions, based on fuzzy decision rules [7]. The original Cognitive Simulation Model (COSIMO) has been taken as reference [9].

A fundamental feature of the model is the concept of the "salience", which represents the content of information attributed to certain indications either as the result of their physical appearance or as the cognitive relevance attributed to them by the reasoning process. In this case, the salience, *sal(i)*, has been calculated according to the content of *rule*, *trend* and *importance* associated with each indication: $sal(i) = \sum rule(i) + trend(i) + imp(i)$.

The possible control actions simulated by the model are: a) the isolation of a steam generator, by closure of the valves; b) the regulation of the mass flow to a steam generator, by setting of the automatic regulator to the demanded values of mass flow; and c) the progressive regulation of the mass flows during the accident progression. Recovery actions are also possible.

The weighted sum of the saliences associated to each quantity gives the "diagnosticity" of a steam generator, which is a measure of the diagnosis, i.e. the time dependent identification of the steam generator on which action has to be taken: $D(SG_j) = \sum sal^2(i_{SGj}) / \sum_{1,4} \sum sal^2(i_{SGj})$

The plant response to operator action has been calculated by a simple simulation of the thermohydraulic and fluid dynamic behaviour of the AFWS and Steam Generators. The logical models of the instrumentation available, as "actuators" and "indicators", have been developed, including automatic regulators, alarms, etc.

As mentioned above, the DYLAM technique is driving the human machine interaction, and, according to the data utilised, a reliability study or a more deterministic analysis can be performed.

## 3.2 The human reliability study: comparison with THERP

The objective of the human reliability analysis is the evaluation of the contribution of inappropriate behaviours to the overall probabilistic safety study of the system. A number of possible human errors and their probabilities have been postulated, differentiating between errors of perception, diagnosis/planning and actions.

The actual adherence to real data was here less relevant, as this was a demonstrative study case and, in order to reduce further the complexity of study, no system failures have been assigned and no interference of the environment on the operator behaviour has been considered.

While the robustness of HERMES to study human behaviours in case of accident with different initiating events and false alarms, has been studied elsewhere [7, 8], in this paper we discuss the comparison of HERMES with THERP. The case studied has been the Loss Of Feedwater (LOF) on the secondary side of the SG1.

### 3.3.1 The THERP analysis

The application procedure for the THERP [10] analysis implies:

1. Development of elementary Human Error trees;
2. Assignment of nominal human error probabilities
3. Evaluation of Performance Shaping Factors
4. Definition of dependencies;
5. Evaluation of failure and success probabilities;
6. Analysis of the error recoveries.

The analysis of the operator's task in case of LOF accident has been firstly developed and then the THERP procedure has been applied. Tables 1 and 2 show the task analysis and the evaluation of the Human Error Probabilities (HEP). In particular, the basic HEP have been identified in the database of THERP (Ch. 20) and it has been assumed that the errors A-D were affected by stress conditions, while errors D-E-G and errors H-I-J were linked by High Dependency. This has lead to the Modified HEPs finally used for the construction of the human error tree.

Table I. Task Analysis.

| Task Analysis LOF |
|---|
| Verify p, l, q in SG 1-4 |
| Diagnose initiating event |
| Identify failed SG |
| Close Regulation Valves |
| Close Safety Valves |
| Close Valves on Turbine |
| Verify Mass Flows to SGs |
| Identify available SGs |
| Initial reg.. of mass flows |
| Time dependent regulation |

Table 2. Evaluation of Generic and Modified Human Error Probabilities (HEP - HEP Mod)

| Error event | HEP | HEP-Mod |
|---|---|---|
| A: Instrument selection | .003 | .006 |
| B: Indicator Reading | .001 | .002 |
| C: Identification failed SG | .001 | .002 |
| D: Closing regulation valve | .003 | .006 |
| E: Closing safety valve | .003 | .006/.503 |
| G: Closing turbo-p. valve | .003 | .006/.503/.752 |
| H: Verifying of mass flows | .001 | .001 |
| I: Identifying available SGs | .010 | .010/.505 |
| J: Regulating of mass flows | .001 | .001/.144 |

As far as recovery is concerned, it has been considered that the closure of any valve (safety and turbine) leads to the recovery of previously omitted isolations and that the verification step of the procedure, also leads to the recovery of previously omitted procedural steps. The human error tree can then be developed (Fig. 3).

### 3.3.2 The HERMES analysis

The procedure for the application of HERMES [8] has been carried out by defining the error modes and types to be studied, the error probabilities and then by applying the DYLAM techniques to the models of the AFWS and of the operator.

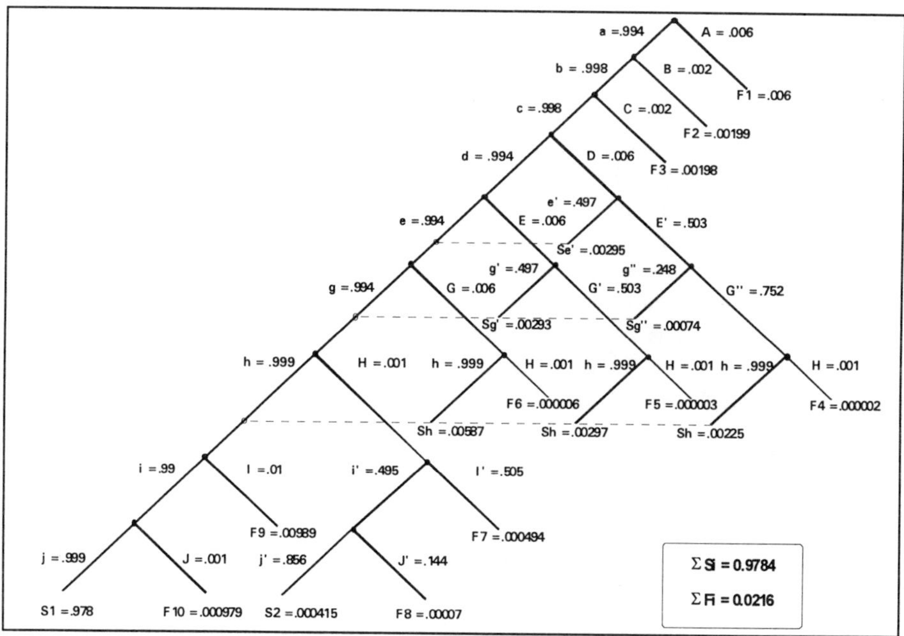


Figure 3. THERP human error tree for the Loss Of Flow to SG1.

In particular, the classification tables that define error manifestations and the related causes have been adopted [5]. The probabilities of erroneous behaviour have been derived from the THERP database.

Table III shows the behaviours, the effects, the manifestations and the associated probabilities. As an example, the erroneous regulation has been assumed to be of either 30 % above or below or 50 % above the required values.

Table III. Behaviours, effects manifestations and probabilities of operator model

| **Behaviours** | **Effects** | **Manifestations** | **Prob.** |
|---|---|---|---|
| Correct observat. | Wrong reading | 0. correct reading | .998 |
| Incorrect observat. | Missed reading | 1. missed reading of q, | .001 |
| | | 2. missed reading of q, p, | .0005 |
| | | 3. missed reading of q, p, l. | .0005 |
| Diagnosis/Planning | Correct diagnosis | 0. actions on SG1 | .99 |
| | Erroneous diag. | 1. SG1 <=> SG2 | .01 |
| Correct action | Correct regulation | 0. Correct regulation of *q* | .975 |
| Erroneous action | Erroneous regul. | 1. $\Delta q = + 30$ %, | .01 |
| | | 2. $\Delta q = - 30$ %, | .005 |
| | | 3. $\Delta q = + 50$ %. | .01 |

All the expected behaviour, including recovery and actions effected by alarms and other environmental conditions are implicitly included in the simulation. Moreover, all possible consequences of erroneous manifestations are calculated by only one single run of the combined program DYLAM-AFWS-COSIMO.

The case study has considered 32 possible accidental sequences, grouped in the following way: 1 "nominal case", with all correct behaviours, 7 cases of "order 1", i.e. with only one erroneous manifestation, 15 cases of "order" 2, i.e. with two combined erroneous manifestations, and 9 cases of " order" 3. With the error probabilities shown in table 3, the results for mission failure/success have been the following: ***S=.9776, F=.0224***.

For all 32 transients, the behaviours have been calculated for the mass flows, temperatures, pressures in all SGs. As an example Figure 4 shows the response of SG1 to the actions of the operator following the accident in presence of nominal behaviour, i.e. no human errors.

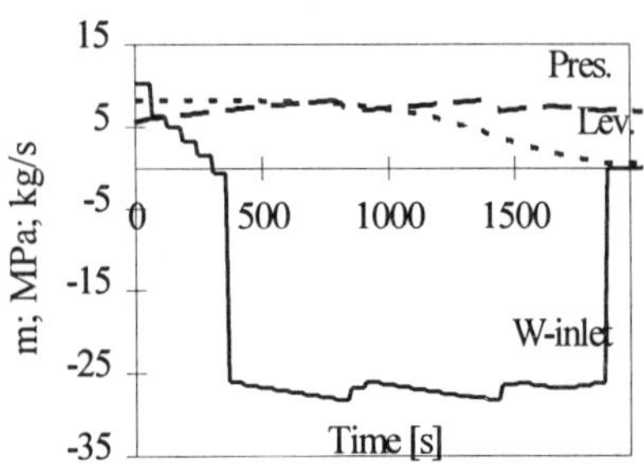


Fig. 4. Mass flows, pressures and levels in SG1. LOF - SG1-nominal case.

The study of the 32 sequences has enhanced situation of particular interest, such as the case of erroneous diagnosis with the systematic exchange of SG1 with SG2.

### *3.3.3 The Comparison between HERMES and THERP*

The numerical comparison of the results obtained by THERP and HERMES is very good. Indeed, the difference is negligible ($S_{HERMES} = .9776 \cong S_{THERP} = .9784$). This result in not surprising, as the basic probabilities of human errors have been assigned in a rather consistent manner.

However, important differences exists between HERMES and THERP from the methodological viewpoint. Firstly, the use of *simulation* represents an important difference, as HERMES needs to develop a model of the interaction between operators and systems, while THERP postulates only limited and fixed interactions.

This difference is crucial, as it leads to the intrinsic character of *dynamic* behaviour of HERMES versus the static response of THERP. It, also, implies that the recoveries are treated very differently in the two methods: in HERMES the recovery action is the results of the evolution of the interaction between operator and system and follows only the occurrence of certain events, like an alarm or a modification of the salience of specific indicators. In THERP, on the contrary, the recoveries are defined a-priori.

In HERMES the *manifestations* of operator behaviour are assigned by the analyst in input, by fixing the spectrum of possible actions associated with certain erroneous behaviour, and the *consequences* are calculated by the simulations. In THERP these aspects are completely ignored.

THERP, on the other hand, offers a very well structured and formalised data evaluation scheme, as well as a consolidated database to which reference can be made very easily.

# 4. Conclusions

This paper has presented the methodology HERMES for the inclusion of human factors in safety assessment studies, which is based on a combined effort of cognitive psychology and engineering skills and on the use of simulation for the dynamic representation of the accidental sequences.

The methodology has been compared with the more established THERP and, while good comparison has been found in the results, many methodological differences have been highlighted.

As in the case of all these type of approaches, the data are the bottlenecks of the method. The data evaluation demands an accurate and extended work of collection of information and analysis and can result in a sound of database or can unfortunately invalidate the entire result of the perspective studies. For these reason the HERMES methodology requires that, prior to the prospective studies, a preliminary retrospective analysis is performed for sound data retrieval.

The sample case shown in this paper has demonstrated that the method has the ability to analyse simple interactions as well as complex situations, requiring dynamic characteristics and human errors related to cognitive behaviour.

## References

1. Parry, G.W., Suggestions for an Improved HRA method for Use in Probabilistic Safety Assessment. *Reliability Engineering and System Safety*, RE&SS, **49**, pp. 1-12, 1995.
2. Hollnagel, E., *Human Reliability Analysis: Context and Control.* Academic Press, London, 1993.
3. Cacciabue, P.C., Cognitive Modelling: a Fundamental Issue for Human Reliability Assessment Methodology? *Reliability Engineering and System Safety, RE&SS,* **38**, pp. 91-97, 1992.
4. Reason, J., *Human error.* Cambridge University Press, Cambridge, UK, 1990.
5. Hollnagel, E. and Cacciabue, P.C., Reliability of Cognition, Context, and Data for a Second Generation HRA. Proceedings of International Conference on Probabilistic Safety Assessment and Management. San Diego, California, 20-25 March 1994.
6. Cojazzi, G., Cacciabue, P.C., Parisi, P., DYLAM-3: A Dynamic Methodology for Reliability Analysis and Consequences Evaluation in Industrial Plants. **EUR 15265 EN**, 1993.
7. Cacciabue, P.C., Cojazzi, G., A human factor methodology for safety assessment based on the DYLAM approach. *Reliability Engineering and System Safety, RE&SS,* **45**, pp. 127-138, 1994
8. Cacciabue, P.C., A Methodology for Human Factors Analysis for System Engineering: Theory and applications. To appear in *IEEE-System Man and Cybernetics,* 1996.
9. Cacciabue, P.C., Decortis, F., Drozdowicz, B., Masson, M., Nordvik, J.P., COSIMO: A Cognitive Simulation Model of Human Decision Making and Behaviour in Accident Management of Complex Plants. *IEEE Transaction on Systems, Man and Cybernetics, IEEE-SMC,* **22** (5), pp. 1058-1074, 1992
10. Swain, A.D., Guttmann, H.E., Handbook on Human Reliability Analysis with Emphasis on Nuclear Power Plant Application. **NUREG/CR-1278** Final Report, 1983.

# Procedure for the Analysis of Errors of Commission during Non-power Operations

by

J.A.Julius, E.J.Jorgenson
NUS, 1303 S.Central Ave., Kent, WA 98032
G.W.Parry
NUS, 910 Clopper Rd., Gaithersburg, MD, 20878
A.M.Mosleh
University of Maryland, 2135 Building 090, College Park, MD, 20742

## 1 Introduction

This paper presents a condensed version of a Probabilistic Risk Assessment (PRA) procedure to analyze the potential for errors of commission to have a significant impact on risk during the non-power operations at a nuclear power plant. The approach adopted is, in principle, the same as that for the analysis of errors of commission during full power operations[1]. However, in detail, it is different for several reasons. Firstly, in addition to addressing errors in the response to initiating events, it is more important in the non-power PRA to identify the errors that lead to, or contribute to the occurrence of, initiating events. In a full power PRA, errors resulting in initiating events are rarely modeled explicitly, since they are typically grouped with another initiating event such as reactor trip, and the safety systems required to respond to the trip are typically independent of the failures causing the trip. Secondly, the operator responses are often not as clearly guided by procedure as in the full-power case. For example, in a Westinghouse plant, there is generally no equivalent to the E-0 procedure to aid in the diagnosis of the event. Thus, misdiagnosis is potentially a much more significant concern because of the stronger dependence between what causes the initiator and how the operator responds to it. Thirdly, the plant configuration is constantly changing during an outage, and there are many different activities proceeding in parallel, both inside and outside of the control room, with many more opportunities for plant personnel to interact with the plant. In the full power case[1], the activities of interest were primarily associated with the control room crew responses to initiating events.

The analysis of errors of commission during non-power operations identifies specific error of commission events that are important at a given plant, evaluates the factors affecting the likelihood of these errors, and confirms the effectiveness of a plant's design, procedures, and practices in defending against errors of commission. The errors of commission analysis is thus of value in identifying and evaluating potential changes in the plant's design, procedures, or practices in order to further strengthen the defenses against errors of commission.

# 2 Overview

This analysis procedure for the identification of errors of commission during non-power modes of operation is conducted in three phases. These phases are: identification of activities that create the opportunities for error; identification of failure mechanisms of functions, systems, or components that could result from these activities (error expressions); and identification of the most significant of these based on consideration of consequences, recovery potential, and likelihood to decide if the events are significant. The PRA model provides a framework that identifies the context for the analysis in terms of the different plant and system configurations.

# 3 Procedure for Analysis of Errors of Commission During Non-Power Operations

As discussed above, the analysis consists of three phases, each of which is discussed below.

## 3.1 Identification of the Opportunities for Error

The purpose of the analysis is to identify, in as complete a manner as possible, human caused failures of equipment or functions that result in, or contribute to the occurrence of, initiating events, or have an impact on the operator response to initiators. In order to restrict the scope of the analysis to be reasonably tractable, only rational, purposeful human/system interactions are investigated. Therefore, there must be a reason why the activity associated with the human/system interaction that is postulated to result in an error is carried out.

Any activity that results from a requirement for, or permits plant personnel to, interact with the plant affords an opportunity for error. The activity might be responding to an unexpected change in plant conditions or performing an essential part of a specific evolution. In addition, there are maintenance, test, and calibration activities which can impact operating systems, but in addition, provide opportunities for errors which result in equipment being in an unrevealed unavailable state. From an understanding of each activity and the context within which it is taking place, specific errors expressions, corresponding to failure modes of equipment or functions, can be identified. The activities that take place in different plant operational states (POSs) affect different systems. Each POS defines the boundary conditions (temperature, pressure, decay heat level, and available mitigating systems) for the shutdown PRA model.

The identification of potential error expressions begins, therefore, with a search for opportunities for the plant personnel to interact with a system of importance, and, from an understanding of the activity, an identification of possible failure modes of the associated equipment. This is achieved through an understanding of the system functions, potential plant configurations, and operating practices and administrative procedures such as those to control maintenance, and shutdown and start-up.

## 3.2 Identification of Potential Error Expressions

From the point of view of the PRA model, it is useful to group the error expressions of interest into (1) those associated with the initiating events, and (2) those associated with the response phase. A further subdivision can be made by noting that errors may be latent i.e., they may not be revealed, or have an impact, at the time they are made, or they may be active i.e., they are the trigger events that result in the transition to the unwanted condition.

### *3.2.1 Error Expressions Associated with Initiating Events*

These error expressions either result directly in, or create a degraded state of a critical safety function and thereby contribute to, an initiating event. The initiating events identified in the shutdown PRA may be quantified in different ways. For some, such as the support system failure initiating events, fault tree models might be developed. For others, frequencies may be estimated based on extrapolations from full power conditions to those conditions specific to the different POSs. For others still, a list of credible scenarios can be constructed (e.g., reactivity addition). The latter is in a sense similar to the fault tree modeling approach but is a direct identification of the "cutsets".

When the initiating event frequencies are quantified with historical data, it is assumed, as in the full power case, that the frequency implicitly incorporates the contribution from any errors of commission. However, if an operator error can be identified as a credible mechanism for the initiating event, then it is analyzed in detail, especially if it may also impact the response to that initiating event. This analysis requires knowledge of the operational and maintenance practices and procedures throughout an outage.

For the fault tree and cutset approaches, it is assumed that the modeling has identified the failure scenarios in terms of equipment failure modes and human errors of omission. The identification process for errors of commission consists of determining whether any of the activities carried out in the various POSs could contribute to any of the hardware failure or functional unavailability states identified in the models for these events. Therefore, the first question asked is whether there is any activity conducted during a particular POS which calls for some interaction with a function, system, or component whose failure results in, or contributes to the occurrence of, an initiating event.

Error expressions identified during the analytical portion, are then supplemented by those identified during a review of plant-specific, historical data (where available) or generic, industry data.

### *3.2.2 Error Expressions Associated With Response*

There are two types of responses explicitly represented in the PRA models, namely recovery of the failed function, and the actuation and/or control of mitigating systems or alternate systems. Typically, failure to recover is modeled essentially as an error of omission. However, it is of interest to identify possible dependent failure mechanisms between initial failure of the function and failure to recover the function, and incorrect operation of alternate systems, in particular those resulting from misdiagnosis. Therefore, the error expressions of interest may include failing to initiate a system, but misdiagnosis as a cause of that error is of interest since it could be related to the cause of the initiator.

Principally, however, the non-power errors of commission procedure looks for failure mechanisms of systems that could be the result of errors of manipulation during alignment of alternate systems which lead to failure of remaining redundancies or failure of other alternatives. A search is made to identify what effect has to be created by the operators to do this. Examples of such effects are, flow diversion, run out of pumps, dead heading pumps, inadvertent isolation, etc..

## 3.3 Identification of Significant Error Expressions

As in the analysis of errors of commission in the full power phase, the next phase of the analysis is to identify the most significant of these error expressions. There are three different considerations: consequence, likelihood, and potential for recovery. Because of the nature of the shutdown analysis itself, an initial screening on consequence is performed during the search for error expressions. The only error expressions identified are those that either result in or contribute to an initiating event, or have a significant impact on the accident development or mitigation.

The errors are discussed in terms of the two major sub-groups, those resulting from control room related operational activities, and those resulting from maintenance activities. In the first of these subgroups, errors are identified as arising during a response to a plant/system perturbation, or as arising during a required manipulation, and are expected to exhibit cognitive modes of failure not seen in the more routine maintenance activities.

However, for both these groups, it is essential to have an understanding of how errors occur, i.e., it is necessary to have a model of error causes. The form of the model adopted for this project is discussed below.

### *3.3.1 An Influence Model of Error Causes*

The model of error causes used in the PRA for which this study was performed is an influence model, similar to that used for the full-power study. As in the previous analysis for errors of commission during power operations[1], the causes are expressed in terms of Error Producing Conditions, or EPCs, which are outside influences that have the capability to strongly impact the likelihood of failure. The model of error causes is constructed by identifying key EPCs for classes of high level failure modes, such as misdiagnosis, slips, skips, and different impacts of inadequately controlled maintenance. For the control room operator related errors in particular, it is important to choose some of the EPCs on the basis of a model of cognition, since the most significant errors of commission are likely to be errors of intent[2]. Potential defenses against these EPCs are also identified. Only those defenses which do not represent the simple negation of the EPCs are included at this stage. The resulting models of error causes are effectively represented as tables of causes/influences and corresponding defenses. The assessment of likelihood in this study was based judgementally on the quality of defenses against the occurrence of these mechanisms, as opposed to estimating the likelihood of EPCs directly.

### *3.3.2 Procedure for Identification of Significant Error Expressions*

#### 3.3.2.1 Errors During Responses to Perturbations

In general, for any mode of operation, there are three classes of scenarios that require operator response: (a) loss of a critical safety function, or reactor trip, (b) failure of equipment during particular evolutions that requires operator action to maintain the safety function, and (c) deliberate changes of plant status due to transitions such as those between POSs. It is clear that the likelihood and importance of these categories depend on the specific operational mode of the plant.

The procedure is essentially the same for all classes of perturbations, and is similar to the analysis of the errors during full-power operations[1]. The first step in this phase of the analysis is to identify what information is necessary to enable the operators to make an accurate assessment of the situation, and for each scenario of interest, to assess whether that information is distorted in such a way as to produce the inappropriate actions captured in the error expression. The cause-defence tables are used to identify the potential EPCs, and to identify any specific defenses. If, for a specific error expression, EPCs can be identified for which there is no clear defense, then the error expressions are ranked in importance on the basis of the consequences of the error, the likelihood of the scenario (including the EPCs), and the potential for non-recovery.

#### 3.3.2.2 Maintenance Errors that Result in System Failure

Listed below are classes of EPCs for maintenance errors.

- Poor Physical Environment
- Inadequate Job Specification
- Tools and Parts Unavailability
- Inadequate Communication Protocol and Means
- Insufficient Training and Experience

Psychological Environment

These EPCs were used to develop cause-defence tables for three different classes of maintenance related errors, those related to inadequate maintenance on the system itself, those which involve taking a wrong component/train out of service, and failure caused by maintenance on a nearby but otherwise unconnected component. For each maintenance error expression identify the relevant EPCs and possible defenses, using the cause-defense tables. The EPC may be evaluated either on a generic basis at the plant (e.g., the labeling is adequate and clear and should not be a significant cause of error), or on a specific basis. Those error expressions for which all EPCs cannot be eliminated as having appropriate defenses,are ranked on the basis of consequence, likelihood, or non-recovery as for the response errors.

# 4 Conclusion

This paper presents a condensed version of a procedure developed to analyze the potential for errors of commission as part of a low power and shutdown PRA. Details of this procedure will be presented in a subsequent paper that is in preparation[3]. This procedure, in full, has been successfully applied in the analysis of a pressurized water reactor and the portion pertaining to initiating events is being applied in several shutdown PRAs.

# Acknowledgement

The authors are indebted to Mario Van der Boorst of N.V. Elektriciteits-Produktiemaatschappij Zuid-Nederland (EPZ) for his support in the performance of the project.

# References

1. J.Julius, E.Jorgenson, G.Parry and A.Mosleh, "A Procedure for the Analysis of Errors of Commission in a Probabilistic Safety Assessment of a Nuclear Power Plant at Full Power", Reliability Engineering and System Safety, Vol. 50, (1995), pages 189-201.

2. Barriere, M.T., W.J. Luckas, Jr., J. Wreathall, S.E. Cooper, D.C. Bley, and A. Ramey-Smith, (1995) “Multidisciplinary Framework for Analyzing Errors of Commission and Dependencies in Human Reliability Analysis,” NUREG/CR-6265, BNL-NUREG-52431, August 1995.

3. J.Julius, E.Jorgenson, G.Parry and A.Mosleh, "Procedure for the Analysis of Errors of Commission During Non-power Modes of Operation" , in preparation.

# Insights and Benefits in Probabilistic Terms from Plant Modifications Implemented at Nuclear Power Plants in Germany, Sweden and the USA

Authors
Lennart Carlson and Hans Eriksson, Swedish Nuclear Power Inspectorate (SKI)
Wolfgang Werner, Safety Assessment Consulting (SAC)

## 1 Introduction

Selected modifications and backfits to PWR and BWR plants are presented and their effectiveness in probabilistic terms is discussed. The material is mostly from German, Swedish and US plants, and to a lesser extent from Swiss, French and Japanese plants. For a detailed description, see /1/.

## 2 Selected Modifications and Backfits at PWR Plants

### 2.1 Emergency Core Cooling and Containment Spray Systems

*2.1.1 Principal causes for failure of the ECCS before modifications:*

- Inadequacy or incompleteness of procedures for switching from HPI to LPI and for transferring from the injection to the recirculation mode of core cooling
- Early depletion of water sources
- Absence of automation in situations requiring quick alignment of safety systems
- Control logic problems preventing the actuation of safety systems

*2.1.2 Modifications and Backfits*

- Improved availability of the emergency core cooling (ECC) and containment spray (CS) systems by provision of alternate water sources and means of injection, e.g.:
  - ◊ Cross-ties between low pressure injection/recirculation (LPI/LPR) pumps and HPI pumps to enable high pressure recirculation (HPR) (older German plants).
  - ◊ Refilling of the refuelling water storage tank (RWST) with coolant from the sump by cross-ties to LPR pumps for continued HPI (some older German plants).
  - ◊ Cross-connecting to the HPI System, including RWST, of another unit (Surry).
  - ◊ Emergency refill of the RWST upon failure of LPR (Zion).
  - ◊ Use of external containment water injection for water supply to LPI/LPR and containment spray recirculation (Ringhals 2/3/4, Beznau 1/2)
  - ◊ Backup for LPR pumps by CS pumps (REP 900, 1300, Japan 1100 PWR)
  - ◊ Backup for CS pumps by mobile pumps (REP 900, 1300)
- Improved procedures and hardware for alignment and actuation of systems

- Automation of, or enhanced procedures for transfer from injection to recirculation.

*2.1.3 Evaluation of the Effectiveness of Modifications and Backfits*

The overall improvement of the ECC and CS functions is measured by: the unavailability on demand, given SLOCA situations. This was ~2 E-2 in the first PSAs, conducted before the implementation of the discussed modifications and backfits. The examination of recent PSAs shows three groups which are related to the design characteristics "redundancy/diversity" of the ECCS and CS trains:

1. Unavailability of 4 E-3 - 6 E-3 for 3-loop plants with intermeshed trains.
2. Unavailability of 2 E-4 to 6 E-4 for
   - ◊ 3-loop plants with intermeshed trains, and additional redundancy/ diversity,
   - ◊ 3 and 4-loop plants with separation of trains.
3. Unavailability of 5 E-5 for a 2-loop plant with two intermeshed trains, and one completely independent, diverse and self-sufficient train.

These figures demonstrate substantial improvement over the situation before the implementation of modifications and backfits. However, the unavailability of the ECC function remains to be of crucial importance because the application of bleed and feed procedures has reduced the relative importance of almost all other accident sequences, and, on the other side, increased the demand for the ECC function.

## 2.2 Systems for Mitigating Steam Generator Tube Rupture Events

*2.2.1 Background*

SGTR events were not analysed in the early PSAs. The inclusion of SGTR events in the DRS-B and NUREG-1150 studies revealed severe vulnerabilities of the systems for mitigation of SGTR events, primarily due to

- inadequate procedures for the isolation of the defective steam generator,
- problems in the control logic that prevented the HPI pumps from being shut off, with the consequence of overfeeding of the secondary side.

*2.2.2 Modifications and Backfits*

- Improvements of leakage detection, control logic for shutting off of HPI, and hardware and procedures for isolation of the faulted steam generator.

*2.2.3 Evaluation of the Effectiveness of Modifications and Backfits*

- Prior to the modifications, the contribution to both the core damage frequency and risk were high. With the implemented system improvements, the unavailabilities of the systems for coping with SGTR events are now generally ~ E-4 or lower, reducing the CDF contribution from SGTR to less than 10%. However, in all examined level-2 PSAs, SGTR remains to be a dominant contribution to offsite consequences.
- Significant mitigation is provided by primary side bleed/feed in situations with failed isolation of SG: the split-up of mass flow between pressuriser valves and leaking SG tube(s) reduces the amount of coolant lost to the outside. Besides the

mitigating effect of this AM action, the CDF contribution from SGTR is reduced by the factor 10 or more (German PWRs, Beznau 1/2, some US PWRs).

- Strategies for filling-up the faulted SG with water are being introduced (Beznau), as fission product scrubbing in the water column is seen as a possibility to significantly reduce the release of fission products. Studies indicate that reduction of releases by more than the factor 10 is achievable.

## 2.3 Auxiliary / Emergency Feedwater to the Steam Generators

### *2.3.1 Background*

- In the early PSAs, lack of diversity of the auxiliary/emergency steam generator feeding (AFW/EFW) was the main cause for the unavailability of this function.

### *2.3.2 Modifications and Backfits*

- At all plants, alternate, diverse water supply and/or injection to the steam generators has been provided.
- Check valve design and surveillance has been improved at plants susceptible to steam binding of AFW/EFW pumps (Westinghouse plants).
- Secondary side bleed/feed (preventive AM) has been implemented as back-up to AFW/EFW systems (German PWRs, Beznau 1/2, Maine Yankee).

### *2.3.3 Evaluation of the Effectiveness of Modifications and Backfits.*

- The modifications and backfits implemented at the plants included in this paper have reduced the unavailabilities from originally ~E-4 - ~E-3 per demand to ~E-7 - ~E-5 per demand.
- Together with the back-up by primary side bleed/feed, the CDF contribution resulting from loss of SG feeding is now insignificant at all examined plants.

## 2.4 Containment Systems

### *2.4.1 Background*

Loss of the retention capability of large dry PWR containments is mainly due to:

- Containment bypass sequences (SGTR, V-sequence, and isolation failure).
- Hydrogen combustion (early or late).
- Loads attending high pressure melt ejection (DCH).
- Slow overpressurisastion.
- Basemat penetration by molten corium.

### *2.4.2 Modifications and Backfits*

- The importance of the V-sequence scenario has been greatly reduced by improvements to the high pressure/low pressure interface of the ECCS.

- Numerous modifications and backfits to the systems for containment isolation, have practically eliminated the contribution to significant offsite consequences from failure of containment isolation.
- Hydrogen control: has been (or will be) improved by the
  - ◊ provision of high capacity catalytic recombiners, also in combination with igniters (German PWRs)
  - ◊ putting in place of procedures for early containment venting for removal from the containment atmosphere of hydrogen and oxygen (under study, Beznau 1/2)
  - ◊ provision of post- accident inertisation (under study, Borssele).
- The availability of containment fan coolers was improved by backup cooling using river water (Beznau 1/2).
- External water injection to the containment from fire trucks is provided for
  1. prevention of core damage, using the injected water as sources for LPI/LPR and/or CS (Beznau 1/2, Ringhals 2/3/4) , and for
  2. mitigation of core damage consequences by flooding of the containment when RPV failure is imminent, in order to reduce
     - ◊ basemat attack by molten corium,
     - ◊ production of combustible gases.

  To avoid late overpressurisation of the containment, the strategy is supported by high capacity filtered containment venting. (Beznau 1/2, Ringhals 2/3/4).
- Depressurisation of the RCS for prevention of high pressure melt ejection, for sequences involving loss of steam generator feeding (all examined plants).
- Filtered containment venting for avoiding catastrophic containment failure. (German PWRs, Beznau 1/2, Ringhals 2/3/4). It is also beneficial in combination with strategies for external water injection as means for ultimate containment heat removal, if all other systems have failed.(Ringhals 2/3/4).

# 3. Selected Modifications and Backfits at BWR Plants

## 3.1 Injection to the Reactor Pressure Vessel

### *3.1.1 Principal causes for failure of RPV injection before modifications*

Overheating of injection pumps due to high temperature in the suppression pool (caused by RHR failure).

### *3.1.2 Modifications and Backfits*

Making available alternate water sources and alternate means of injection. e.g.

- Control rod drive system as backup for HPCI (Gundremmingen) and RCIC (LaSalle), also with suction from the condensate storage tank (CST) (Peach Bottom, Grand Gulf), also from another unit (Browns Ferry).
- Standby lLiquid control (SLC) system as backup for HPCI and RCIC with suction from the SLC control tank (Browns Ferry).

- Pump seal water system as backup for HPCI (Gundremmingen).
- Addition of a diverse, independent motor driven high pressure system as backup for the steam driven HPCI system with suction from the demineralised water system, the liquid waste processing system, or the containment fuel pool via the fuel pool cooling and cleanup system (Forsmark 1/2).
- High pressure service water system as backup for LPCI via a cross-tie to the RHR system, with suction from the river (Peach Bottom).
- Condensate system (CDS) pumps as backup for LPCI via a bypass of the feedwater pumps, with suction from the condenser hotwell, also with the possibility for make-up from the CST (Peach Bottom, LaSalle, Grand Gulf Gundremmingen).
- Low pressure core spray (LPCS) system as backup for LPCI, with the CST as suction source in case of high suppression pool temperature (Peach Bottom).
- Fire water system as backup for low pressure injection (Grand Gulf, Perry, Browns Ferry).
- Cross-tie to the emergency service water system/RHR system for low pressure injection from the other unit (Perry).
- Standby service water basin as backup water source for low pressure injection via the SSW cross-tie to the LPCI (Grand Gulf).
- Direct injection of river water by the SW system as backup for low pressure injection (German BWRs).
- Additional diverse RPV-injection and heat removal system as backup for low pressure injection (Gundremmingen).
- Use of external water injection as water source for LPCI (Swedish BWRs).

*3.1.3 Evaluation of the Effectiveness of the Modifications and Backfits*

The unavailability per demand of the ECCS function, given transient situations with reactor scram, excluding LOSP events, was ~6 E-4 in WASH-1400.
Except for the newest designs, there is significant intermeshing among different systems which is now exploited for making available alternate water sources and means of injection. With credit for such alternatives, the unavailabilities range from 4 E-7 - 5 E-5. The variation results from differences in the availabilities of alternate coolant injection and from differences in the credit given to it.

## 3.2 Depressurisation of the Reactor Pressure Vessel

*3.2.1 Principal causes for failure of RPV depressurisation before modifications*

Common-cause-failures of multiply redundant valves.

*3.2.2 Modifications and Backfits*

- Addition of diverse bypass valves to 3 out of 7 S+R valves. The bypass valves can be manually actuated. (Gundremmingen, some other German BWRs). A similar modification is envisaged for Swedish BWRs.

*3.2.3 Evaluation of the Effectiveness of the Modifications and Backfits*

With available common-cause-failure data, unavailabilities below 5 E-5 - 1 E- 4 are hard to support for systems of highly redundant valves. Therefore, at most plants some diversity of the depressurisation function is or will be provided.

## 3.3 Residual Heat Removal and Containment Systems

*3.3.1 Principal causes for failure of RHR before modifications*

Common-cause-failures of pumps in the RHR system.

*3.3.2 Modifications and Backfits*

Diversity has been added to RHR systems, also by assigning heat removal functions to containment systems, e.g.:

- Primary containment venting, cContainment pressure relief for removal of residual heat and prevention of primary containment overpressure failure when the ECCS systems are operable, but the normal RHR systems have failed. As the core is not yet damaged, filtering is not required (Swedish and US BWRs).
- Filtered containment venting for ultimate heat removal and for avoiding catastrophic failure of the containment in the event of failure of core cooling and all other residual heat removal (Swedish and German BWRs).
- Accident mitigation by flooding of the containment to the upper core level, thus ensuring ultimate cooling of the core material (all Swedish BWRs).
- Additional RPV-injection and heat removal system (Gundremmingen)
- External water injection as additional suction source for the containment spray system. (all Swedish BWRs)

*3.3.3 Evaluation of the Effectiveness of the Modifications and Backfits*

- In the first PSAs, the contribution to core damage of RHR failure was dominant; the unavailability of the RHR function was 5 E-5 and higher.
- Improvements to the systems performing the RHR function have reduced the unavailability of this function by at least factor 10.

# 4. Conclusions

The presented examples show that a systematic and comprehensive probabilistic assessment of modifications and backfits can provide valuable assistance to the
(1) discussion of safety issues, (2) reconstitution of safety analyses reports, (3) focusing of further PSA applications, and (4) enhanced feedback from generic world-wide experience to safety evaluations for individual plants.

# 5. Reference

1. Compilation of Selected Modifications and Backfits at German, Swedish and US PWR and BWR Plants, SKI-Report 95-25.

# Performance Improvements for Electrical Power Plants : Designing-in the Concept of Availability

BOURGADE E., DEGRAVE E., LANNOY A., Electricité de France

*1, avenue du Général de Gaulle, 92141 CLAMART CEDEX FRANCE*

## Abstract

Availability plays an increasing role in operation and design of nuclear plants. This paper proposes means to build availability of a new plant right from the start of the design process. The emphasis is put on two tools, one for allocation and one for evaluation of availability. Finally, the main aspects of future developments are presented.

## 1. Context

Safety at nuclear power stations - which represent the bulk of the units built over the last twenty years - has been the overriding concern when considering the dependability of electricity generating plants in France. With medium-term forecasts pointing to stabilizing electricity consumption, and increasing competition between different sources of energy in the long term, this emphasis is gradually shifting toward considerations of power station profitability, which, among others, integrates the concepts of availability and maintainability.

So far, generating plant availability has been ensured by applying deterministic design rules (covering aspects such as redundancy and inherent component quality) and operating rules (covering issues such as maintenance control and spare parts control).

Experience shows, that we cannot expect to achieved improved power station availability through these methods alone. Among other things, we run up against obstacles concerning equipment accessibility and the impossibility of making accurate allowance during the design phase for the impact of future modifications and improvements. EDF is currently developing a specific methodology, CIDEM[1],[2], for integrating availability and maintainability considerations in the design of generating plant. CIDEM (Conception Intégrant la Disponibilité le retour d'Expérience et la Maintenance) means Design Integrating Availability Operating Experience and Maintenance. This methodology would allow the designer to design-in power station availability and maintainability from the start, and thus to ensure a better profitability.

In this paper, we give an overview of the CIDEM proposed methodology and a brief rundown on its main component factors concerning availability. We also discuss expected results and implementation problems.

## 2. CIDEM : Availability and Design

To design-in availability, we must first determine suitable measuring criteria, and the means of evaluating their values. For generating plants, we adopt two criteria : the numbers of days of **forced unavailability** (trips and forced outages) and the number of days of **scheduled outage**. This paper deals mainly with forced outages.

For each of the criteria, we set a target figure at the outset of the design process. Since our objective is to achieve plant profitability, these targets will naturally represent better performances than currently indicated by experience feedback. Regarding scheduled outages, if experience feedback indicates that present average annual duration is 40 days, the designer might set for example a target figure of 25 days.

Then, how can the designer take into account such targets during his work ? They are far too global to be directly used. So, to ensure that targets are met, we assign local objectives to each of the major functions and systems generated by the initial stage in the design process, which begins, or should begin by a functional description and is followed by a preliminary choice of systems and critical components. The designer of each plant system will thus have a clear objective to meet.

As the definition breakdown gets finer, the design process generates more complete architectures, down to equipment level. At this point, two situations may arise. Generally, experience feedback provides valid dependability data on identical or similar devices (pumps, valves, motors, for example). In some cases, rare, the technology of equiments is new, and the specifications of these equipments can be based on the assigned objectives, as on specific calculations based on experience feedback and innovation.

The next step consists of assessing the criteria associated with the systems suggested by the designer ; to do this we build a recursive process. At each iteration, recommendations are issued to help the de signer bring the system within the assigned availability target, if the target has not been reached already. To assess system availability, we evidently need data on the maintenance plan (stock control, repair times, preventive maintenance guides). It therefore appears logical for the design process to include development of a maintenance plan, which becomes more detailed as the design process advances. The final maintenance plan will provide a baseline document for operators during plant startup.

Similarly, final allocated availability values of components will be used in writing specifications for suppliers. Specifications will be easier to write, and the process of building the dependability specifications will be consistent.

A flowchart for the design process is shown in figure 1.

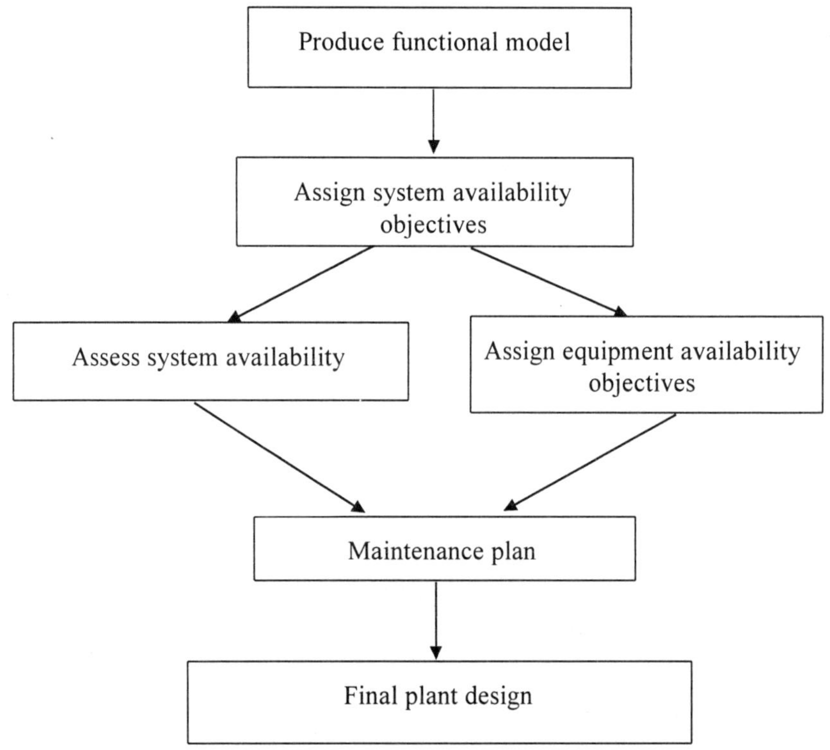


**Figure 1 - simplified CIDEM process**

## 2.1. Availability Allocation

Developments in dependability allocation methods are quite recent. From previous work and special developments for the CIDEM project, two methods for allocating forced outage targets have been defined [3]. Basically, they are both based on a dependability model, which links the unavailabilities of components to the unavailability of the plant caused by the system analyzed. One of these methods is based on importance factors deduced from previous experience and the dependability model and the other on optimization of an effort function. One of their most important characteristics is that they can be used with complex dependability models, i.e. models which are not series or parallel. Both methods are operational. They have been applied both to systems in nuclear power plant and to a whole 1300 MW power station.

## 2.2. Availability Assessment

**Characteristics**

Reliability assessment studies on high-risk industrial systems such as nuclear power stations have produced a number of sophisticated methods and tools. Availability assessment undoubtedly benefits from these developments, though recent work on the CIDEM project reveals a number of specificities in availability assessment that can have an important influence on dependability models. Furthermore, dependability models are conditioned by the nature of the undesirable event, and this can seriously limit their robustness. For example, a simple failure in a protection system (safety injection for example) produces unplanned power station outage, though redundancy means, that it will not cause overall plant failure during an accident. Availability assessments during design benefit from a great advantage over probabilistic safety (i.e. reliability) assessments after the design : in most cases, structural models (i.e. reliability block diagrams or fault trees) are sufficient. More than that, the models are much simpler than those developed for probabilistic safety studies.

Another specificity of forced outage assessment concerns the nature of relevant failure modes, which differ from those found in probabilistic safety assessment and during reliability centered maintenance work. Let us consider an auxiliary feedwater pump, which is supposed to be used to inject water in a steam generator when the main feedwater system is unavailable. If the system includes 3 redundant pumps, **during normal operation**, the cavitation of the pump will lead directly to the unavailability of the plant, due at least to technical specifications. On the other hand, **during an accident**, the same failure will be of no consequence if the 2 other pumps can provide the required flow.

**Tools : toward automated analysis**

The usual way to assess the dependability of a system is to **manually** build a model, in our case, a fault tree. The reliability analyst gathers information about the system, mainly provided by the designer, he builds the model, he gathers reliability data, and gets results through a dedicated software. This method is widely used, and will remain the only practical one for some time. Nevertheless, it has inconveniences, especially the two following : **lack of coherency** and **excessive duration of analysis**.

The lack of coherency is not a direct consequence of manual analysis, but a possible if not likely one : many systems are designed and have to be reviewed, i.e. assessed, by different analysts, implying a difficulty of clearly keeping track of modeling hypothesis, and consequently a difficulty of ensuring coherency between these hypothesis.

The excessive duration of analysis is a well-known of dependability, and it is especially unwanted during design : designers will not accept to wait several months before they know whether their system meets the allocated targets.

Thus, EDF has launched a program of automation of availability assessments. It has been started in 1994, and first concerns thermohydraulic systems. It is mainly based on experience gained in the development and use of EXPRESS [4], expert system built for reliability evaluation of systems of nuclear plants. A thorough analysis of specificities of availability analysis has been conducted, and we are now in the process of writing the technical specifications of a tool. A prototype will then be written, and tested on real systems.

The overall goal is not only to provide a tool used for assessments, but also for allocation : it would be used to build the structural model (fault tree) needed as an input of the allocation tool. The figure 2 illustrates the proposed process.

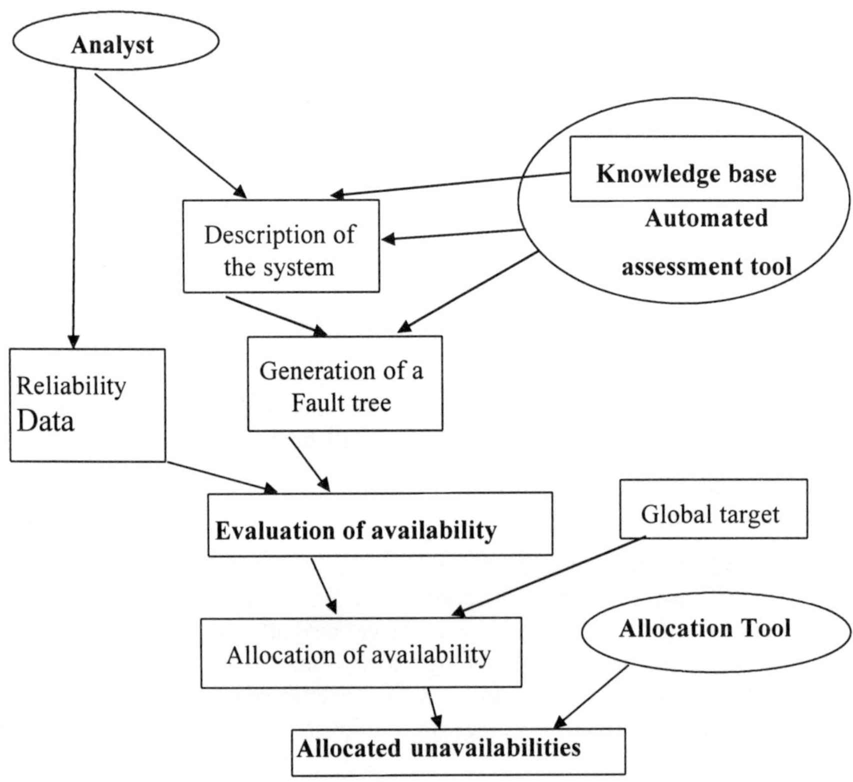


**Figure 2 - Process for availability assessment or allocation of objectives**

# 3. Keys for Future Developments

## 3.1. Maintenance costs and Dosimetry

So far, in this paper, we have mainly discussed availability, but two other fields, at last, are to be investigated, dosimetry and maintenance costs. The first task is to build models linking basic parameters, evaluated with experience feedback, to the corresponding parameter at the system level. Some encouraging progress has been made, especially for dosimetry, and it can be reasonably foreseen, that in the near future, we shall be able to build at least a statistical dosimetry evaluation model, when operating experience is available. The next step will be their use as an input for the allocation process.

## 3.2. Traceability and speed

To keep track of hypothesis and to be able to quickly allocate or assess availability, dosimetry and maintenance during design are important factors for acceptance of these types of methods and tools by designers.
It must be emphasized, that the first point, i.e. the traceability, is closely linked with the design process itself : functional and dysfunctional hypothesis assumed during allocations and evaluations are directly derived from design choices, either explicit or implicit. Then, it seems necessary to ensure transparency of the design process. Introduction of functional analysis as soon as possible during design, and even as a basis of this process could be a useful, if not necessary requirement to achieve such a goal.

# 4. References


[1] C. Degrave, M. Martin-Onraët, "Integrating Availability and Maintenance Objectives in plant design - EDF Approach", Third Joint International Conference of Nuclear Engineering, Vol. 3, pp. 1483-1488, Kyoto, Japan, 1995

[2] M. Martin-Onraët, C. Degrave, C. Meuwisse, "Integrated Logistic Support Concept in the Design of Nuclear Power Plants", ICONE 4, New Orleans, March 10-14 1996

[3] M. Bouissou, C. Brizec, "Application of Two Generic Availability Allocation Methods to a Real Life Example", ESREL'96

[4] M. Bouissou, H. Bouhadana, M. Bannelier, N. Villatte, "Knowledge Modeling and Reliability Processing : Presentation of the FIGARO Language and of Associated Tools", Proceedings of Safecomp'91, Trondheim, Norway, 1991.

# Modelling Industrial Systems using Fault Tree and Monte Carlo Analysis Techniques

P. Ashton, R. Leicht
NUKEM GmbH
D-63755 Alzenau, Germany

**Abstract**

The liberalisation of European markets and world-wide competition forces many manufacturers and operators caught in the cycle of lower costs, higher availability and maximum safety to continue to seek new ways to remain competitive. Against this background, optimisation of life-cycle costs can become a deciding factor in economic success. This paper looks at the fault tree and Monte Carlo methods for modelling industrial systems within the scope of ARM management and assurance programmes, using appropriate fault tree and simulation software tools, and contrasts the two approaches.

## 1 Introduction

The quantitative analysis of the availability, reliability and maintenance (ARM) as well as the safety of plant and systems is an established activity in various industrial sectors, particularly practised in the nuclear industry, but increasingly in various other sectors such as the chemical industry, transportation, waste treatment etc. Though the application of quantitative techniques to the analysis of the safety of plant and equipment is widespread, until recently, the wider use of such techniques for questions relating to plant and equipment reliability have been much more limited.

With the need to satisfy competing restraints, including stricter health and safety requirements, pressure to reduce costs etc. this situation has now changed and quantified analyses at various stages of the life cycle are much more widespread e.g. during the tendering phase of a project to demonstrate the meeting of ARM specifications, or during the operating life as a result of problems arising.

This paper concentrates on practical aspects of the two well known modelling techniques of fault tree analysis and Monte Carlo simulation for analysing plant availability.

Before the two modelling techniques are discussed a general remark concerning modelling is made. All quantitative ARM and safety analyses involve producing a model of the system under consideration with the model representing to a greater or lesser degree the real behaviour of the system. In some instances, for example in "living PSA's", the original model is refined in an iterative process of model improvements based on observations of operating behaviour during the plant's lifetime. Such techniques serve to reduce the distance between reality and the predictions of the model used. There exists a school of thought that the results of such analyses in absolute terms are less important than the values in relative terms (in particular when various alternative approaches are to be compared) as well as insights gained from actually performing the analysis. The logic of this argument contains a flaw, however, since if the model deviates fundamentally from reality what does this say about the meaningfulness of any conclusions drawn from the relative differences indicated by such a model. Hence, the aim must be to use realistic models.

# 2 Availability Modelling using Fault Tree Analysis

The main advantages of fault tree analysis for availability modelling are that the method has a high degree of acceptance (including acceptance by regulatory authorities) and allows both qualitative and quantitative analysis including detailed analysis of failure causes.

Fault tree analysis is, however, as the name suggests, fault orientated. In availability analysis, many more factors come into play, for example, factors affecting fault restoration may be as equally important in determining a system's availability as the failures themselves.

The general procedure followed when undertaking a fault tree analysis is depicted in Figure 2.1.

The tasks indicated by the boxes labelled (1) essentially fix the scope of the analysis, whilst the tasks indicated by the boxes labelled (2) involve determination of the reliability data requirements and their subsequent evaluation. These latter tasks run parallel to producing the initial fault tree and are in effect interdependent. Note, the preferred source of data should always be the operating experience of the plant or system in question (or similar plants in the case of design studies). Generic data is used only as a secondary source, as necessary, or for improving plant-specific data using the Bayes' method.

Limitations of fault tree analysis, in particular when modelling system availability, include:

- commonly used quantification techniques do not allow consideration of time dependence of failure or repair rates
- the repair model for the system can often not be adequately represented e.g. logistic constraints
- the repair of each basic event may not be statistically independent, particularly with limited repair resources
- difficulties occur in considering dependent failures during quantification

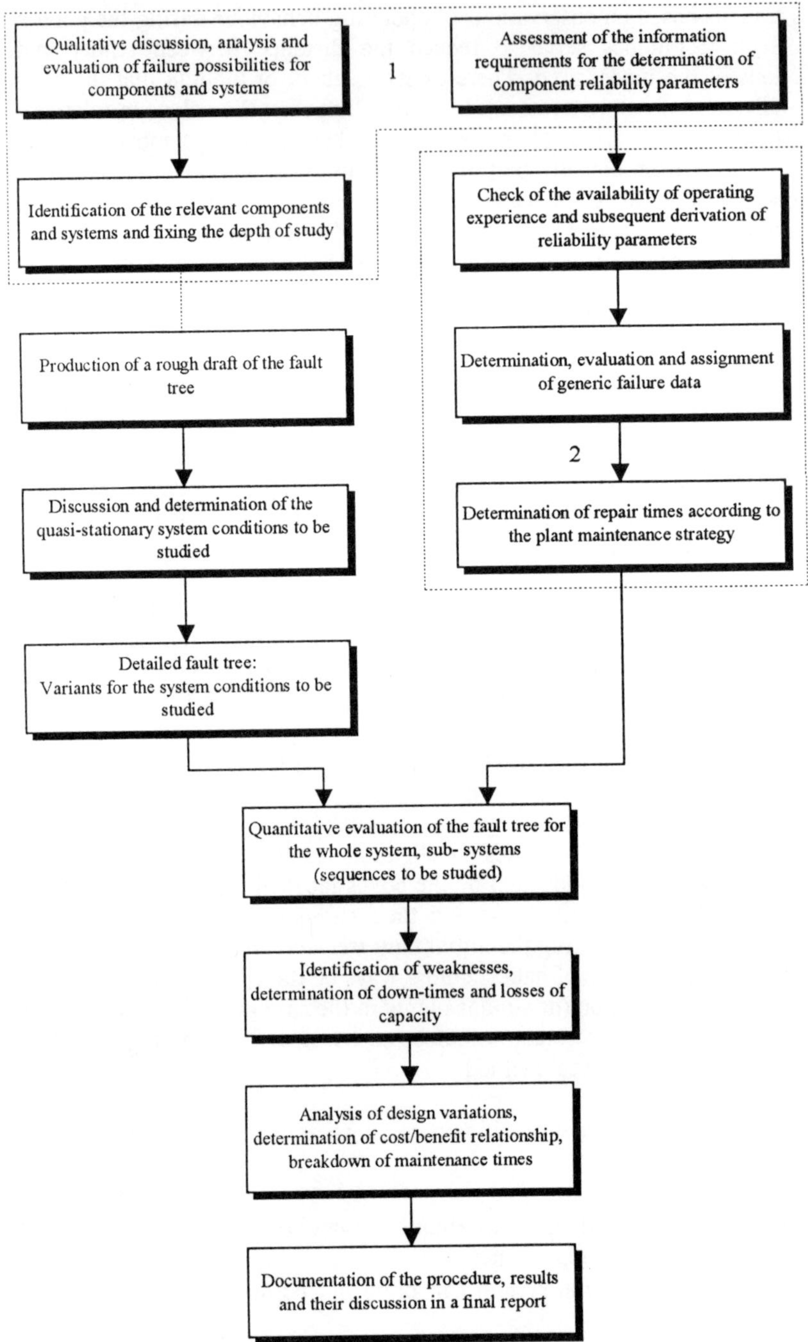


**Figure 2.1:**
**Procedure for the study of system availability - fault tree method**

- "feedback" effects can only be considered with difficulty, since the output of each gate is determined by the respective inputs
- the fault tree does not account well for operator corrective actions and other compensating factors which may prevent an undesirable outcome

Producing a fault tree requires detailed knowledge of the function and behaviour of the system under study. The value of a fault tree may be significantly reduced if all possible failure causes are not considered and the causality in event sequences not correctly portrayed. The high degree of subjectivity usually involved in producing a fault tree may be reduced by providing for several people to check a fault tree.

The following practical examples illustrate some of these limitations.

A recent study, in which the NUKEM fault tree code FTL [1] was used, required an estimate of the frequency of overfilling buffer storage tanks located between a production facility and an associated effluent treatment plant. The effluent treatment plant comprised essentially a multi-stage evaporator and concentrate incinerator. The analysis was required by the relevant authorities following an incident involving a controlled discharge of untreated effluent to a water course due to failures in the effluent treatment plant and a resulting overfilling of the buffer tanks. An essentially complete operating history of the effluent treatment plant over the previous three years including a daily measurement of the buffer tank levels was available as data input to the analysis.

The approach adopted was as follows:

- the average level in the buffer tanks over the previous three years was evaluated
- a max. tolerable downtime of the effluent treatment plant assuming a mean effluent arising rate and mean effluent treatment plant throughput was determined
- the frequency of events leading to unavailability of the effluent treatment plant for periods greater than the tolerable downtime was determined using fault tree analysis. As a number of failures only resulted in reduced plant throughput, three categories of unavailability (100%, 50% and 25% reduction in plant throughput) were defined and evaluated.

This approach involves a number of simplifications some of which limited the scope of the analysis. For example:

- repair rates were modelled using an exponential distribution whereas in reality a normal distribution may be expected to be more appropriate. When evaluating the frequency of exceeding maximum tolerable downtimes, this approach is only reasonable so long as the maximum tolerable downtime is similar to the mean repair times (as was the case)
- the operating data showed certain effects which could not easily be modelled, for example:

- a likely dependence between faults in the incinerator and blockages in the associated slag removal system
- certain failures, e.g. freezing of instrument air lines and inadequate cooling capability of a heat exchanger, were clearly dependent on weather conditions and time of year (the plant not being contained in a building)

- the reduced throughput categories only approximately reflected the true effect of the failures considered.

In short, to model the system using fault tree analysis, it was necessary to introduce simplifications to reduce a dynamic system to a system independent of time.

A second example concerns the analysis of the availability of a natural gas storage system during the plant design stage to optimise the provisions of redundant equipment. The system is required to pump natural gas from a distribution network into storage during summer months and then return this gas into the distribution network during winter months at periods of high demand. This latter operation involves a corresponding number of start-up sequences where failure behaviour may be expected to differ from periods of continuous operation. In addition, equipment requirements during gas removal is dependent on the gas pressure in storage. Clearly, for such a plant availability may be expected to be a function of the mission profile. Such an analysis does not easily lend itself to treatment by fault tree analysis.

# 3 Availability Modelling using Monte Carlo Analysis

The possibilities offered by Monte Carlo simulation for modelling complex systems have long been known, with the approach addressed in reliability text books and papers dating back to the seventies [2] and [5]. The main advantage of the approach over fault tree analysis is that a wider range of systems and associated boundary conditions may be taken into account. The ability to reproduce analytical results by simulation has previously been demonstrated so that no attempt to repeat such an exercise is made here. The reader is referred, for example, to [6] where the results of Monte Carlo simulation and an analytical solution to determine the optimal test interval for a system are compared.

A survey of literature reporting ARM and safety analyses using Monte Carlo methods showed - apart from a few exceptions [3] - [6] - no widespread use of the method. In [3] the application of Monte-Carlo simulation to fault tree quantification in assessing the availability of a fluid catalytic cracking unit is described. This paper also contains results of a review of literature pertaining to the reliability of refineries, ammonia plants and pressure vessels and also notes the relative sparcity of published information in this area. A reliability analysis of a compressor sub-system of a wider process system is described in [4]. In [5], flow sheet modelling of a 545 Mg/d ammonia plant to determine process performance is described, whilst [6] addresses the use of Monte Carlo simulation to determine optimal test intervals for industrial systems with a NPP standby liquid control system used as an example. The majority of applications identified in the literature survey concerned electrical power distribution systems e.g. [7].

With the increased demand for the application of quantitative techniques to questions relating to plant and equipment reliability noted earlier, the picture presented by the literature survey may be expected to change. A reason for the increased demand is the more general recognition of maintenance as a key strategic issue within the framework of system life cycle costs (LCC), with repair-based or breakdown maintenance no longer accepted as a viable option. The search to reduce costs is accompanied by a move to preventive regimes by the introduction of planned maintenance and reliability based philosophies as identified, for example, in the rail industry [8]. Thus the increased requirement for quantitative ARM analysis may also be expected to include a requirement for wider ranging analysis and more detailed treatment of issues affecting availability such as maintenance regimes, spares logistics, repair limitations etc. Monte Carlo simulation is particularly suited to such purposes.

The general procedure followed when undertaking a ARM or safety analysis using dynamic simulation is essentially the same as when performing an analysis using the fault tree method - this is not surprising since both methods involve representing a system as a number of events or line replaceable units in a Boolean network. Differences do, however, exist. In applying the Monte Carlo method the system model is "transported" through a hypothetical future covering possible system conditions that may be encountered over time (the mission profile). The process is repeated numerous times with the projected performance and malfunctions recorded each run. The result is a comprehensive statistical database of the entire life-cycle from which the required system parameters may be evaluated.

The mission profile is a way of combining the various sustained and peak requirements and involves defining a scenario covering an extended period of operation. From the mission profile the ARM analyst derives distinct phases of operation and then determines the upstate requirements in each phase.

Whilst the potential application of Monte Carlo simulation to ARM and safety analyses is not new, the accessibility of such techniques to analysts due to the ever increasing power of personal computers and the development of corresponding software tools, e.g. the SPAR stochastic simulation tool [9], is a recent development.

In comparison to fault tree analysis, dynamic simulation offers many more possibilities particularly relevant for analysing system reliability, availability, life-cycle costs etc. For example, features of the SPAR tool include:

- consideration of various maintenance philosophies (at system or LRU level)
- logistics; including spares strategy, failure of spares on the shelf, price considerations

- assignment of various attributes to the line replaceable units (LRU's - the "building blocks" of the system model) including failure distributions for active and passive operating states, repair distributions, age, effects of preventive maintenance, repair efficiency etc.
- LRU interactions - operational if-then rules that describe how LRU failure and repair affect other LRU's, sub-system etc.

Quantification is not limited to assumed constant failure and repair rates - all distributions commonly used in reliability modelling (Exponential, Normal, Lognormal, Weibull and user-defined distributions) may be applied.

Disadvantages of Monte Carlo simulation are principally associated with the required computing time. This may be prohibitively long when attempting to simulate highly reliable systems with an adequate degree of confidence due to the correspondingly large number of trials required or, for example, when in support of a sensitivity study.

# 4 Conclusions

1. The need to satisfy competing restraints, including stricter health and safety requirements, pressure to reduce costs etc. has given rise to an increased requirement for the quantification of system ARM characteristics. Availability modelling of plant and systems using appropriate modelling techniques is part of this process.
2. Fault tree analysis is a commonly used technique to model system availability and despite limitations of the method enjoys a high degree of acceptance.
3. Monte Carlo simulation has long been recognised as a useful method for modelling complex systems but apart from power supply applications, there is little evidence of it's widespread use for availability modelling.
4. The increasing power of PC's and the appearance of suitable software tools for Monte Carlo simulation has improved the accessibility of Monte Carlo simulation to ARM and safety analysts.
5. The increased requirement for quantification of system ARM characteristics may be expected to give rise to more detailed treatment of issues such as maintenance regimes, spares logistics, repair limitations etc. than is currently the norm. Monte Carlo simulation is more suited to such purposes than fault tree analysis.

# References


[1] FTL Ver. 3.0
Fault Tree Programme, NUKEM GmbH, 63775 Alzenau, Germany, 1994

[2] Gruhn G., Kafarov V.V.
"Zuverlässigkeit von Chemieanlagen", VEB Deutscher Verlag für Grundstoffindustrie, Leipzig 1979

[3] Thangamani, G.; Narendran, T.T.
"Assessment of availability of a fluid catalytic cracking unit through simulation", Reliability Engineering and System Safety (1995), vol. 47, No. 3, pp. 207 -20

[4] Neumann, W.; Gruhn, G.; Schön, H.G.
"A Monte Carlo Model for Analyzing the Reliability of Technological Installations", Theoret. Found. Chem. Eng., (1982), vol. 15, No. 3, pp 276 - 282

[5] Gaddy, J.L.; Culberson O.L.
"Prediction of Variable Process Performance by Stochastic Flow Sheet Simulation", AIChE Journal, vol. 19, No. 6 Nov. 1973, pp 1239 - 1243

[6] Wu, Y-F.; Lewins, J.D.
"Mechanical System Surveillance Test Interval Optimization", Journal of Nuclear Science and Technology, 30 [12], Dec. 1993, pp 1225 - 1233

[7] Massee, P.; Bollen, M.H.J.
Reliability Analysis of Industrial Electrical Supply, IEE Conference Publication no. 338, 3rd International Conference of Probabilistic Methods Applied to Electrical Power Systems, July 1991, pp 220 - 223

[8] "Life-Cycle-Costs sind heute schon wichtiger als der Fahrzeugpreis"
VDI Nachrichten Nr. 46, 17. November 1995, pp 23.

[9] SPAR
Malchi Science Ltd., 39 Hagalim Boulevard, PO Box 2062, Herzlia, 46120 Israel

# Periodically Overhauled Systems Availability

Stian Lydersen
Dept. of Mathematical Sciences
Norwegian University of Science and Technology
N-7034 Trondheim, Norway

**Abstract**

This paper concerns calculation of dependability (for example average availability) for systems with components subject to periodic overhaul in addition to failures and corrective maintenance. Traditional fault tree analysis does not directly incorporate this situation. Alternative ways to adapt fault tree analysis to overhauled systems are described.

## 1 Introduction

Some systems contain components that are overhauled regularly. For example, consider 2 feedwater pumps where sufficient power is obtained with 1 active pump. At regular intervals, say, 48 hours each year, each pump is taken out of service for overhaul. Such overhauls may include tightening bolts, seal change, lubrication etc. Hence, during 96 hours per year, this will be a *1oo1* system, the rest of the year a *1oo2* system. (A *koon* system consists of n components, and is functioning if and only if at least k of its components are functioning.) Other types of units may have longer overhaul durations, for example 26 days annually for a certain type of gas turbine.

If no failures occur during operation, the states of the two components will typically be as illustrated in Figure 1. The system functions as a *1oo2* system for time $T_O - 2T_V$, and a *1oo1* system for time $2T_V$, where $T_V$ is the duration of overhaul of one component, and $T_O$ the period between planned overhauls.

In addition, an operating component may fail at random points in time. The down time (corrective maintenance) is stochastic. Further, overhaul of a component will be started only if the other component is functioning at that time. For example, if component no 2 is undergoing corrective maintenance at time $T_O - 2T_V$, then overhaul of component no 1 will be postponed until the failed component is active again. The duration of overhaul, and the scheduled points in time for starting overhaul, are assumed to be fixed.

The aim of this paper is to present possible methods for calculating the average dependability of such a system, over a overhaul cycle $T_O$. This is not possible using straightforward, traditional fault tree analysis. Methods for adapting fault tree analysis are presented. More details are found in Myklebust (1994).

# 2 Proposed Methods

Basically, three different methods for fault tree analysis of overhauled systems are envisaged:

i) Including "Overhaul" as basic events in the fault tree

ii) Using separate fault trees for each phase

iii) Phased Mission Analysis (PMA)

## 2.1 Overhaul events

Let $C_i$ denote the event that component number i is in a failed state (under corrective maintenance). Further, let $V_i$ denote the event that component number i is being overhauled. Figure 2 shows the resulting fault tree for the example *1oo2* system, including "overhaul" as basic events. Traditional models, and hence, algorithms and computer codes, for fault tree analysis, have the following main limitations:

a) All combinations of basic events are assumed possible.
b) The probability of a basic event is assumed to be
   - independent of time, or,
   - modelled as a (repairable or non-repairable) component with stochastic time to failure (i.e. exponentially distributed), or
   - modelled as a hidden failure discovered at periodic testing, or
   - modelled as a renewal process (i.e. frequency event in CARA, 1991)
c) All sequences of basic event occurrences are permitted (not dynamic fault tree analysis)
d) Basic events are assumed to occur independent.

These limitations are violated if straightforward analysis of the fault tree in Figure 2 is performed:

a) The minimal cut sets are $\{C_1, C_2\}$, $\{C_1, V_2\}$, $\{V_1, C_2\}$, $\{V_1, V_2\}$. However, two components in a redundant system will not be overhauled at the same time, and the cut set $\{V_1, V_2\}$ should be omitted.

b) The time to (planned start of) overhaul, as well as its duration, are deterministic.

c) A component will not be taken out of service for overhaul if its parallel is failed. For example, the cut set $\{C_1, V_2\}$ may occur only if $V_2$ occurs before $C_1$.

d) The points mentioned above (a and c) create a significant, negative dependence between some of the basic events. Two examples are: $P(V_1|V_2) = 0$, and $P(C_1) = 0$ at the instant $V_2$ "starts".

Including impossible cut sets, such as $\{V_1, V_2\}$, gives incorrect results, up to several orders of magnitude. If the impossible cut sets are deleted, traditional fault tree analysis with some adaptions give approximately correct results. Deleting impossible cut sets may be done by

- Constructing an alternative fault tree without impossible cut sets. For example, construct a fault tree with (only) the minimal cut sets $\{C_1, C_2\}$, $\{C_1, V_2\}$, $\{V_1, C_2\}$ as shown in Figure 3.

or

- Compute all minimal cut sets in the original fault tree (Figure 2). Before quantitative analysis takes place, delete cut sets with more than one overhaul event. This may be done by the computer by allowing "marking" of overhaul events, or "by hand".

Note that occurrence of some possible minimal cut sets are disjoint events. For example, $C_1$ and $V_1$ may not occur simultaneously. Hence, the probability of the top event (system failure) is given by

$$P(TOP) = P(\{C_1, C_2\}) + P(\{C_1, V_2\}) + P(\{V_1, C_2\}) \tag{1}$$

This is equal to the first (upper bound) approximation in the inclusion - exclusion principle for fault tree with independent basic events.

*Example:*

*Consider two parallel components with $T_O$ = 1 year, $T_V$ = 26 days, constant failure rate $\lambda = 200 \cdot 10^{6}$ hours$^{-1}$, and MTTR = 24 hours. Then,*

$$q_C = P(C_i) \approx \lambda\ MTTR = 0.0048$$
$$q_V = P(V_i) = T_V/T_O = 0.071$$

*and equation (1) gives the average unavailability P(Top event):*

$$Q_0 \approx q_C^2 + 2q_C q_V = 0.000023 + 0.000684 = 0.000707.$$

*Including the impossible cut set $\{V_1, V_2\}$ would have given $Q_0 \approx 0.00578$.*

## 2.2 Several Fault Trees

The concept of this method is easily exemplified by considering the system in Figure 1. Perform separate fault tree analyses for the "full system" (*1oo2* system), and for the "reduced" system (*1oo1* system). For example, the average availability of the system will be the weighted average of the availability within each phase:

$$A_{av} = [(T_O - 2T_V)/T_O]\ A_{1oo2} + [2T_V)/T_O]\ A_{1oo1}. \tag{2}$$

*Example (continued):*

$$A_{av} = Q_0 \approx (1 - 2q_V)q_C^2 + 2q_C q_V = 0.0000198 + 0.000684 = 0.00704,$$

*which equals the answer in Section 2.1 up to two digits.*

This method does not take into account that overhaul will be postponed if a redundant component is in a failed state. For the rest, the limitations of fault tree analysis as listed in Section 2.1 do not create problems. The disadvantage of this method is the extra work due to construction, analysis, and weighted averaging of several fault trees.

## 2.3 Phased Mission Analysis

The system can be modelled in terms of Phased Mission Analysis (PMA), as already indicated in Section 2.2. Different models for PMA have been suggested in literature, and some of them are applicable to periodic overhaul. However, existing PMA computation models and methods suggested do not seem to be less cumbersome than the methods suggested in Sections 2.1 and 2.2 (Myklebust, 1994)

# 3 Concluding remarks

The examples in this paper concern average availability. For some systems, the average Rate of Occurrence of Failures (ROCOF) may be important, such as the expected number of outages per year for a power generating system. In RCM (Reliability Centred Maintenance) analysis, the computation of component criticalities (e.g. Improvement Potential) plays a central role.

The two methods with overhaul events and several fault trees, respectively, have been studied by this author and by Myklebust (1994) for several types of systems. Generally, using several fault trees give the most correct results, and is straightforward to carry out. However, it may be more laboursome than using overhaul events, especially if more detailed analysis like a study of component criticalities is to be carried out.

Markov models have been suggested for PMA by several authors. However, the number of system states may easily be quite large even for moderately large systems.

The examples in the paper deal with simple *1oo2* systems, to illustrate the methods. More realistic, practical systems would typically include a large number of components, with simple *koon* systems as subsystems. Further, the components in a *koon* subsystem need not be identical. For example, a turbine driven and an electrically driven pump may be installed in parallel. The methods described for the *1oo2* system in this paper are easily applied with straightforward generalizations.

## References

CARA User's Manual, 1991. "Computer Aided Reliability Analysis". PC program developed by SINTEF Safety and Reliability, N-7033 Trondheim.

Henley, E. J. and Kumamoto, H. (1981): "Reliability Engineering and Risk Assessment." Prentice-Hall, Englewood Cliffs, NJ. Reprinted and distributed by IEEE Press, 1991, with title: "Probabilistic Risk Assessment - Reliability Engineering, Design and Analysis").

Høyland, A. and Rausand, M. (1994): "System Reliability Theory. Models and Statistical Methods." Wiley.

Myklebust, K. I. (1994): "Systempålitelighet ved periodisk vedlikehold" (System Reliability under Periodic Maintenance). M.Sc. thesis, The Norwegian University of Science and Technology.

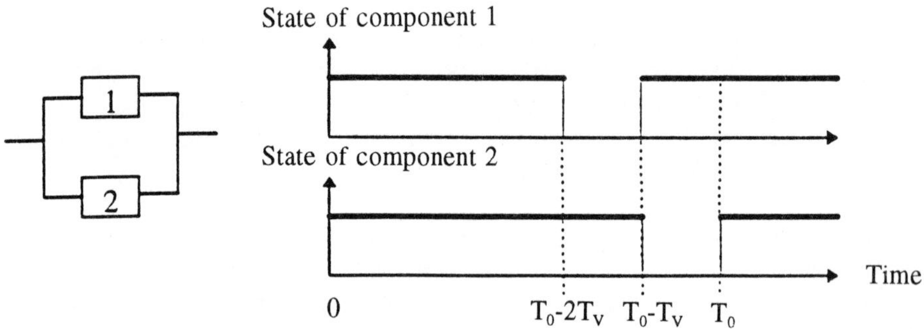


Figure 1: A parallel system of 2 components subject to periodic overhaul.

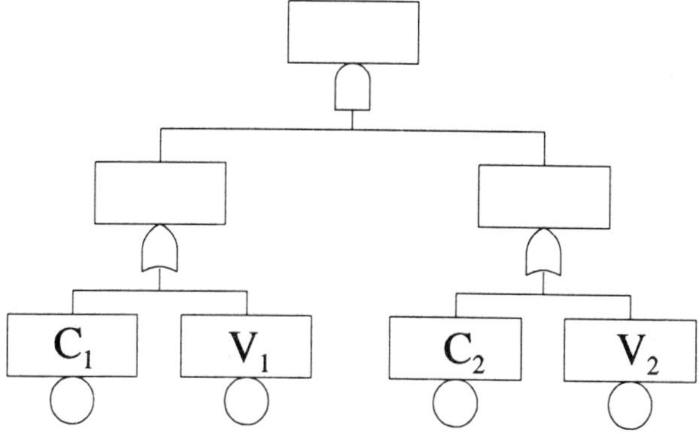


Figure 2: Fault tree for a 1oo2 system with overhaul events

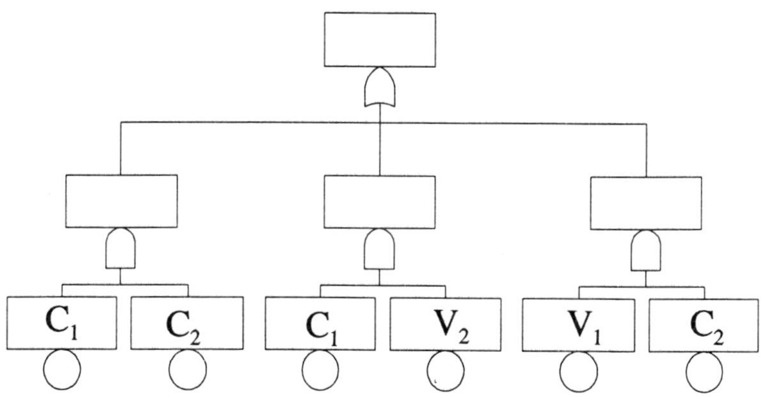


Figure 3: Fault tree with correct minimal cut sets for a 1oo2 system with overhaul events.

# Mitigation of Risks due to Pipeline Releases
## Integration of Risk Analysis and Leak detection design

G. Uguccioni[1], S. Belsito[2], J. Ham[3] , M. Silk[4]

(1) Snamprogetti SpA - SIAF - I 20097 San Donato Mil. (Italy)
(2) Consorzio Pisa Ricerche -TEA - I 56127 Pisa (Italy)
(3) TNO - Department of Industrial Safety - 7300 AH Apeldoorn (NL)
(4) National NDT Centre, AEA Technology, Harwell (UK)

### Abstract

The paper describes the objectives and some achievement of the Research Project DEPIRE sponsored under the CEC ENVIRONMENT Program, that is now in progress and will be completed by early 1997. The project is performed by Snamprogetti (Coordinator), Consorzio Pisa Ricerche, TNO and AEA Technology and is aimed at developing a procedure for the definition of the optimum leak detection system requirements based on the desired risk reduction and on the development of prototypes of leak detection systems based on innovative process based technology and on techniques not requiring process measurement.

### Acknowledgements

The sponsorship of the EC, DGXII (Contract EV5V-CT94-0419), is gratefully acknowledged.
The contribution of Prof. Sanjoy Banerjee (Univ. of California, S. Barbara) in the development of the Neural Network, of Prof. Paolo Andreussi (Project Manager for CPR) and of Ing. P. Lombardi (CPR) in the development of the process simulation, and of Fiona Ravenscroft (AEA) is acknowledged.

# 1 Introduction

While pipelines have proven an efficient and economic means of transporting hazardous fluids over long distances, the risks associated with accidental releases are still significant. Leak detection, at an acceptable cost, is of utmost importance in mitigating potential risks associated with pipeline transportation of hazardous fluids. Present leak detection systems are developed without explicit consideration of the desired risk reduction; to define the optimum characteristics of the system, integration with risk analysis concepts to define the desirable performance for the leak detection system(s) is necessary.

These considerations have led to a Project aimed at:

1. developing, systematising and assembling risk analysis methodology for determining the characteristics of leak detection systems as a function of the desired risk reduction goals for a given pipeline.
2. developing leak detection system concepts that can achieve the performance requirements arising from the risk studies, at an acceptable cost.
3. systematising techniques and software for reliability and unavailability analyses to confirm achievement of the performance targets.

The Project focuses its attention on two reference pipelines, carrying liquefied toxic (Ammonia) and flammable (LPG) gases and consists of four main tasks:

1. Risk Analysis of the reference pipelines
2. Development of a process-based leak detection technique
3. Development of a non process-based leak detection technique
4. Reliability Analysis of the proposed system

As a tentative threshold, a leak of 1% of the nominal flow in a liquefied gas pipeline is set as pushing the limits of sensing technology. Present systems can achieve performances lower than this order; e.g. performances of detection of a 5% leak in 45 minutes have been published [1].

The project is developing a system based on the measurement of process variables integrated with Neural Network technology and is studying alternative methods developed on the basis of experience in Non-Destructive Testing techniques. The framework within which these methods are being developed is given by a risk assessment that considers different population conditions, to assess specific requirements to the leak detection system in terms of minimum leak to be detected, detection time and probability of detection. The need for a leak localisation is also approached. The system reliability is also assessed, considering that the costs of spurious interruptions of pipeline operation are high, suggesting that the reliability of leak detection must be correspondingly high.

# 2 Risk Analysis

The two reference pipelines, representative of hazardous pipelines carrying toxic or flammable substances, are defined as follows:

| Pipeline | Substance | Diameter [in] | Length [km] | Pressure (MPa) | Temperature (°C) |
|---|---|---|---|---|---|
| A | LPG | 10 | 49 | 10 - 1.2 | 0 - 25 |
| B | Ammonia | 8 | 75 | 2.8 - 2.2 | 10 - 25 |

Both pipelines will be considered in the same demographical environment, to allow a comparison of different requirements to the leak detection systems posed by the population density in the vicinity of the line.

The risk analysis that is performed is aimed at quantifying Individual Risk and Societal Risk and is based on postulating four initial pipe leak dimensions, related to the release of 1%, 5%, 20% and 100% of the nominal pipeline flow rate. In addition, three flaw sizes (< 20 mm, < pipe radius and > pipe radius) commonly adopted in risk analysis studies [2] will also be analysed.

Following outflow of product, the following scenarios are analysed:

- flare of burning gas (LPG)
- fire ball (LPG)
- flash fire / deflagration (LPG)
- dispersion of toxic cloud (Ammonia)

For the calculation of the physical effects, the models implemented in EFFECTS-2 [3] are used. These include the Multi-Energy method for the determination of overpressures [4], Chamberlain model for the determination of heat radiation from jet fires [5]. The dispersion of Ammonia takes into account the effect of atmospheric humidity on the cloud density.

The consequences of each effect are assessed with the models given in the so-called 'Green Book' collected in the software package DAMAGE [6]. The risks are then calculated with the TNO software code RISKCURVES.

From the risk analysis, the performance criteria required to the leak detection system to maintain the risk to public below a certain level, and to what extent the risk is reduced by further improvement of the system are defined. The main parameters to judge the performance of the detection system are:

- bottom level of leak rate to be detected
- time between occurrence of the leak and detection
- time to activate intercepting actions, e.g. isolation

From this point of view, the criteria shall be based on acceptability limits for individual and societal risks; in the DEPIRE Project the limits applied in Dutch regulations will be adopted.

# 3 Artificial Neural Networks

One of the main tasks of the Project is to develop a detection system that shall be able to identify in reasonable time a flow leak as small as 1% of the nominal pipeline flowrate.

At present several promising applications of Artificial Neural Network (ANN) in different industrial contexts have started to be available. In theory, ANNs offer a valid alternative to classic deterministic and rule-based modelling approaches. The idea is to develop a methodology that, on the basis of as much field data as possible, allows to set up a knowledge-distributed system easily reproducible for various single pipelines and pipeline networks.

The development of an Artificial Neural Network (ANN) requires the availability of a large number of examples. During the learning phase the ANN is trained to recognise these examples. In the present case the examples should consist of field measurements of process variables (pressure, flowrate, temperature) performed in presence of a leak. Steady-state field data can be used for part of the training operation in addition to data measured in presence of leaks that are normally not available. To derive the cases necessary for ANN training a computer model simulating the flow in the pipeline in the presence of a leak has been set up [7]. The cases used for ANN training must be representative of the field conditions. For example, the instrumentation noise and drift must be included in the data.

The activities performed to set up a leak detection system based on ANN consisted of three tasks:

1) review of process variable technology;
2) development of a deterministic pipeline flow model;
3) development and training of a neural network for detection of the leaks.

In task 1) the existing technologies related to pipeline monitoring have been studied, with particular reference to four existing lines located in the north-eastern Italy. The instrumentation installed has been characterised. Data regarding typical noise and drift of the instrumentation have been collected.

In task 2) a pipeline flow model has been implemented in a code, based on the classical balance equations for mass, momentum and energy, solved using a finite difference scheme. The code simulates both single and two-phase flow and it is written in C++ in order to allow high modularity and reusability.

A qualification activity for the deterministic algorithm has been done against available field data. Several analyses have proven the numerical robustness of the code and showed sensible agreement with the data.

The code has been used to generate a database for training the ANN. Each case in the data set consists of the time evolution of pressure and flowrate calculated in the positions corresponding to measurement stations along the pipeline. The database includes also cases calculated without leaks. Different runs have been performed, in which the pressure and the inlet flowrate is varied, corresponding to the possible states of the pipeline. Considering all the possible combinations twelve cases without leak have been generated.

In addition, cases with leak with different dimensions (1%, 2%, 5%, 10%) have been simulated in different positions along the pipeline. Since the aim of the project is the detection of small leaks, only leak dimensions up to 10% of the flowrate have been considered in setting up of the database.

Considering the combination of the last two parameters with the pressure and flowrate values, a data set of 300 runs has been set up.

Nominal instrumentation noise and drift have been superimposed to the numerical prediction provided by the code.

The software SNNS (Stuttgart Neural Network Simulator), Version 4.0, recently released by the University of Stuttgart (August 1995), has been used for setting up of the neural network [8].

The final configuration of ANN for leak detection has been reached after various steps. The first attempt has been done by considering the data generated by the computer code without noise and drift being superimposed. The result of this activity was a fully connected feedforward neural network with 15 input nodes (corresponding to two flowrate signals -inlet and outlet-, plus thirteen pressure signals coming from the monitoring stations along the pipeline), 7 hidden nodes and 1 output node (configuration 15-7-1). The output of the network is the leak dimension. The ANN was able to correctly identify the dimension of the leak and to correctly discern situations where there was no leak from situations relative to the presence of the smallest leak we aim to detect (1%), thus indicating that the ANN should not give rise to spurious signals.

The noise and drift have then been superimposed to the signal provided by the code. The parameters characterising the noisy signals have been derived from commercial data. The use of noisy signals induced a strong deterioration on network capabilities: the ANN was not anymore able to correctly calculate the leak dimension and to distinguish a 1% leak from cases without leak.

At the end of a fairly extensive study, some actions effectively reduced the noise impact on the signals:

a) filtering of the data by using moving average techniques;

b) adding one further input to the ANN, namely the difference between the inlet and outlet flow measured 1800 s before the actual time, to give information on the flowmeter drift;

c) deriving of three different examples for training from a single data set. Superimposed noise and drift are randomly selected for the various cases (with the same values of maximum drift and noise).

Performances of the network (feedforward, fully connected, 16-6-1) in response to 269 test patterns (i. e. patterns different from those used for ANN training) are summarised in the following figure. It is important to note that errors in leak size evaluation are small enough to keep the no-leak histogram (total 5 cases are simulated) well distinct from the 1%-leak histogram (total 45 cases simulated).

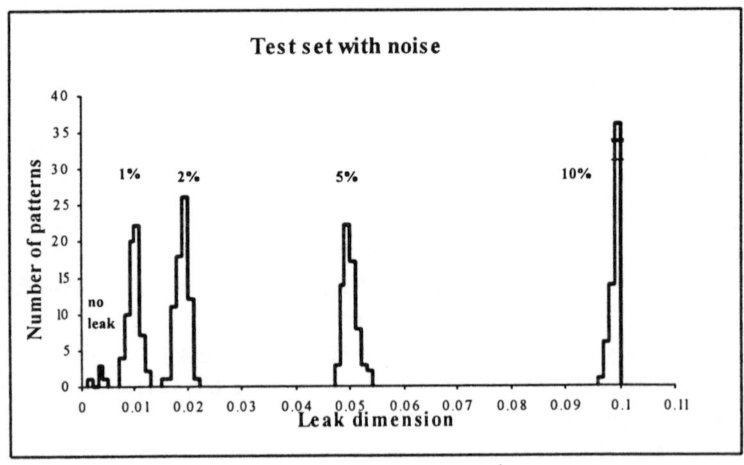

Next phase of the work will include leak-position determination and transient data analysis, to determine the maximum leak detection delay by the system. Extension of the method to pipeline networks could also be carried out.

# 4 Alternative Techniques

The Project is also considering techniques different from those based on process variable measurements that, in alternative or in conjunction with process-based techniques, can allow to reach the performance goals as identified by the Risk Analysis.

Three main categories of leak detection methods are envisaged:

- location of the leak site by patrols
- monitoring the pipeline performance by direct observation of pressure drop and volume loss
- collection/analysis of physical symptoms created by a leak

The final two categories are more appropriate to remote detection of leaks, and also offer the chance of rapid leak detection which obviously the first category cannot. However, even if the line patrol approach is not viable as a candidate technique in the context of the Project, it represents an existing baseline against which the cost and response time of the other techniques must be compared. In the review and appraisal of the techniques, the greatest interest has been directed towards the latter category.

Eight various possible methods have been found worthing of more detailed consideration for the leak detection task. In addition, the visual Inspection by patrolling has been considered as a baseline reference case.

The techniques have been assessed on the basis of:

1. Response Time
2. Capability to identify leak location
3. Probability of Detection
4. Invasiveness
5. Range of Application
6. Installation and Operating Costs

To each of these factors, a judgement based on available technical information has been given. These assessments are clearly not perfectly accurate, and it is also impossible to make accurate estimates of installation and service costs this far in advance. Nevertheless we believe that the costs and benefits identified are a fair reflection of the relative strengths and costs of the candidate techniques.

After assessing the values of the six parameters listed above, scores have been given to each parameter [9]. The overall benefit has been expressed as product of he scores, normalised to have a maximum value of 100. The total benefits have been found to vary from 100 (Sonic/Ultrasonic and Pressure wave transmission) to a minimum of 0.04 (Line Patrol, penalised by the high response time).

The cost to benefit ratios have been then assessed; these ratios, normalised to 100, has confirmed that the two techniques with higher benefit score can be considered the best candidates for further analysis, being a factor about 20 higher in score than the nearest rival.

It is also significant that none of the alternative techniques have been estimated significantly better than using a line patrol except in cost and response time.

Work is currently in progress on the two candidate techniques, these being acoustic noise and pressure wave detection, to answer the following questions:

- to what extent do these alternative techniques enhance leak detection?
- to what extent is new development required?

# 5 Conclusions

The activities carried out so far have shown the possibility of developing new concepts or major improvements to existing technologies that significantly enhance the performances of leak detection systems. Furthermore, the definition of performance requirements on the basis of risk reduction will allow a more balanced approach to pipeline safety. Further work is still necessary, partly within the DEPIRE Project, to test the prototypes on a real pipeline system and to extend the application to networks and to different products or systems (e.g. water transmission lines).

## References

1. A Hamande, V Condacse, J Modisette 'New Systems Pinpoints Leaks in Ethylene Pipeline', Pipeline & Gas Journ., April 1995
2. European Gas Pipeline Incident Data Group (EGIG) 'Gas Pipeline Incidents' - April 1988
3. 'Methods for the calculation of Physical effects of the Escape of Dangerous Materials (liquids and gases) - Yellow Book' - 3rd ed. (to be issued), CPR-14E
4. A C van den Berg 'The Multi-Energy method. A Framework for Vapour Cloud Explosion Blast Prediction' Journ. of Haz. Mat., 12 (1985)
5. G A Chamberlain, 'Developments in design methods for predicting thermal radiation from flares' - Chem. Eng. Res. Des., vol. 65, July 1987
6. 'Methods for the determination of possible damage - Green Book' - 1st ed. CPR-16E, 1992
7. S. Belsito, 'Development of a model for the simulation of leaks in pipelines' - DEPIRE Report TEC-T022-01, 1995
8. 'SNNS - Stuttgart Neural Network Simulator - User Manual' Report 6/95, University of Stutgart, Institute for Parallel and Distributed High Performance Systems - 1995
9. M G Silk, P Carter, 'A Review of Means of Pipeline Leak Detection Independent of Flow Measurement' DEPIRE Report TEC-T031-01, 1995

# Insights from the Wolf Creek Nuclear Generating Station Fire PRA

Dennis W. Henneke
NUS Corporation
107 Cumulus Court, Suite 100
Cary, NC USA 27513
Fax: (919) 469-5386

Vern Luckert
Wolf Creek Nuclear Operating Co.
P.O. Box 411
Burlington, KS USA 66839
Fax: (316) 364-4095

**Abstract**

The Wolf Creek Generating Station Fire PRA was completed in 1994 and 1995 by Wolf Creek Nuclear Operating Corporation with support from NUS Corporation. The Fire PRA was performed using an initial fire area screening employing the methodology, followed by the application of Fire PRA techniques on unscreened fire areas. The results of the Fire PRA showed that switchgear fires were the dominant fire scenario for Wolf Creek. The results, however, were considered conservative, due to the lack of a vigorous and accurate method for treating electrical cabinet fires, including propagation within a cabinet, propagation through cabinet penetrations, treatment of spurious valve operation and the ability of fire suppression to extinguish a cabinet fire.

## 1.0 INTRODUCTION

Wolf Creek Generating Station (WCGS) is a Standardized Nuclear Unit Power Plant System (SNUPPS) design located in Burlington, Kansas (USA). The primary system is a Westinghouse four-loop PWR design, and the plant received its operating license in 1985. Since the plant is of newer design and construction, the fire separation and suppression throughout the plant is excellent. In addition, the major pumps and components are compartmentalized, with very few large combustible loads contained within connecting corridors. Because of this, the Fire PRA results were expected to be very good. However, since the plant was designed with a two train Class 1E (or safety related) electrical system, it became evident early in the analysis that fires affecting an entire Class 1E AC power train, such as switchgear fires, would dominate the Fire PRA results.

The Wolf Creek Fire Risk Evaluation was performed in response to the NRC's request for Individual Plant Evaluation for External Events [1]. The Fire Risk Evaluation was performed using an initial fire area screening from the EPRI Fire-Induced Vulnerability Evaluation (FIVE) methodology [2]. FIVE uses a progressive screening approach based upon quantifying: (1) the frequency of the fire ignition in specific plant areas, (2) based upon quantifying: (1) the frequency of the fire ignition in specific plant areas, (2) the availability of automatic suppression systems, (3) the probability of having sufficient combustibles and heat release to cause damage to shutdown systems, and (4) the probability of manual suppression effectiveness. The FIVE methodology was used for the initial screening, ignition source frequency calculations, fire compartment interactions analysis, and walkdown portions of the fire analysis.

Following the initial FIVE screening, Fire PRA techniques were used on unscreened fire areas to determine a more accurate estimate of the fire-induced plant core damage frequency. The detailed fire PRA techniques included the use of COMPBRN IIIe for fire modeling, as well as event tree analysis for more complex fire scenarios. Additionally, probabilistic fire spreading from area to area was considered.

One of the main objectives of the fire PRA was to be able to use the analysis results for applications in a similar way to the Level I PRA results. One application was the planned inclusion of the fire PRA results in the development of a Safety Monitor™ for support of on-line maintenance scheduling. Therefore, the Fire PRA models had to be developed in sufficient detail to ensure accurate fire scenarios and reasonable core damage frequency results.

## 2.0 FIRE PRA METHODS AND RESULTS

A total of 133 fire areas were analyzed during the initial FIVE analysis. Twenty one (21) fire areas screened during the FIVE qualitative screening steps, with 105 fire areas screening during the FIVE quantitative screening steps. The remaining 7 areas included three switchgear rooms, two Auxiliary Building Corridors, and two electrical penetration rooms. These fire areas were then analyzed using detailed fire PRA techniques. A total of 62 ignition sources in the 7 fire areas were analyzed in detail using a combination of event tree and COMPBRN analysis.

An example of an event tree used for the detailed analysis is shown in Figure 1. This figure shows the analysis of a fire starting in an electrical cabinet, MCC NG01B. In this case, the cabinet contains cables associated with safe shutdown equipment (SSE). Additionally, there are SSE cables traveling directly overhead of the cabinet and still more SSE cables overhead and within approximately 20 feet of the cabinet. The cabinet is sealed with a fire retardant material, however, testing has not been performed to establish a fire rating for the installed configuration. Data from NSAC/178L [3] was used to derive a probability of fire propagation from cabinets such as these. A probability of non-propagation of 0.85 was derived from the data. Extinguishing a fire prior to propagation can occur from; 1) manual suppression, 2) self extinguishing of the fire, and 3) automatic suppression functions prior to propagation.

Once propagation from our example cabinet was determined to occur, the unavailability of automatic fire suppression was considered. In this case, a Halon fire suppression system was present in the room, and would provide protection for all cables not directly overhead of the cabinet. Manual suppression was not credited since it would be highly dependent on both the propagation probability and automatic suppression unavailability. Thus, the example cabinet fire results in three scenarios:

- The cabinet fire does not propagate. In this case all SSE in the cabinet is assumed failed.

- Propagation occurs but fire suppression actuates. In this case, all SSE in and directly above the cabinet is assumed failed.

- Propagation occurs and fire suppression fails. In this case, all SSE in the room is assumed failed.

A majority of the ignition sources (56 of 62) in the unscreened areas were electrical cabinets, including switchgear, MCCs, and distribution panels. Each of these cabinets was analyzed in a similar manner to MCC NG01B above. However, since no justification could be given for crediting non-propagation from an un-sealed cabinet (i.e., a cabinet with louvers or vents), these cabinet fires were assumed to propagate with a probability of 1.0.

The fire PRA resulted in an estimated fire core damage frequency value which was approximately 15% of the internal event core damage frequency value. The fire core damage frequency result was higher than initially expected when giving consideration to the excellent fire separation design for the plant. The dominant factor in these results was the conservative treatment of electrical cabinet fires, including the probability of propagation as well as treatment of spurious actuation of components (see below). This points to the need for additional fire damage (or probabilistic analysis) methodology development for cabinet fires. More developed analysis techniques, such as analysis of pump or oil fires, were not important to the Wolf Creek Fire PRA results, due to the high degree of compartmentalization and train separation employed in the design of the plant.

## 3.0 CONCLUSIONS

Treatment of electrical cabinets in previous fire PRAs and the EPRI documents [2, 4] was not consistent or clear. The EPRI Fire PRA Guidelines (Draft) provide some discussion on treatment of electrical cabinet fires, but only with regards to propagation and timing, not severity of damage. The major areas lacking consistent and accurate analysis techniques that affected the Wolf Creek Fire PRA results include:

1) Propagation from an electrical cabinet: cabinets come in all shapes and sizes. The EPRI FIVE methodology suggests that unless the cabinet is sealed with a rated seal, propagation is always assumed. The fire data suggests that

propagation may or may not occur. What is needed is a general methodology for treating all types of cabinets, especially those cabinets with louvers and vents that are not located at the top of the cabinet, and cabinets with seals fabricated from fire resistant, but not fire rated, materials.

2) <u>13.8 KV switchgear damage</u>: 13.8 kv switchgear is made up of cubicals that interconnect with a bus bar or common set of cables. The switchgear commonly include; 1) a supply breaker cubical, 2) a crosstie breaker cubical, and 3) equipment breaker cubicals. No single fire would necessarily fail the entire bus unless both the supply and crosstie breakers were failed, or the bus bar was affected. No data or analytical method was found that would treat this in a probabilistic fashion, and all fires were assumed to fail the entire switchgear.

3) <u>Spurious Operation of Equipment due to Hot Shorts</u>: NUREG/CR-2258 [5] recommends a conditional fire-induced hot short probability, including values for time-based recovery of hot shorts. These values should be applied carefully, as they may be non-conservative for un-recoverable hot shorts (e.g., a containment MOV where power is lost when the hot short is removed). Additionally, no guidance is given on how to analyze multiple hot shorts. For Wolf Creek, a method was applied, on a limited basis, where hot shorts were not considered for components in different parts of a cabinet. For example, an MCC containing 7 columns of cubicals with 4 to 5 cubicals per column, was analyzed assuming components in each column could hot short simultaneously, but components in the remaining columns would not spuriously operate due to hot shorts. This method was still conservative in that multiple hot shorts in a given MCC column were considered with a conditional probability of 1.0.

4) <u>Protection of Overhead Cables by Automatic Suppression</u>: For Wolf Creek, it was assumed that if a fire propagated from a cabinet, that all cables directly overhead (and within 5 feet) were assumed damaged prior to automatic fire suppression actuation. The problem with crediting suppression is the question of timing between propagation and suppression actuation. NSAC/178L data suggests that both propagation and smoke detection will occur at approximately 15 minutes after the initial fire ignition. In most cases, especially in vented cabinets, smoke detection would occur prior to propagation. However, at this time, there is insufficient data to support any probabilistic approach of suppressing cabinet fires prior to propagation.

For each of the above areas, insufficient guidance was provided on how to accurately analyze the Wolf Creek electrical cabinet fires. In all cases, a conservative approach was assumed. It is estimated that the approach utilized results in a final core damage frequency value due to fire initiators which is over-conservative by approximately a factor of three.

## References

1. U. S. Nuclear Regulatory Commission, Individual Plant Examination of External Events (IPEEE) for Severe Accident Vulnerabilities, June 28, 1991, Generic Letter 88-20, Supplement 4.

2. EPRI, Fire-Induced Vulnerability Evaluation (FIVE), TR-10030, Research Project 3000-41, Final Report, April 1992.

3. NSAC/178L, Fire Events Database for U.S. Nuclear Plants, Prepared by Science Applications International Corporation.

4. EPRI, Fire Risk Analysis Implementation Guide, Draft, January 31, 1994.

5. Kazarians, M. and Apostolakis, G., NUREG/CR-2258, Fire Risk Analysis for Nuclear Power Plants, Prepared for U.S. Nuclear Regulatory Commission, Office of Nuclear Regulatory Research, September, 1981.

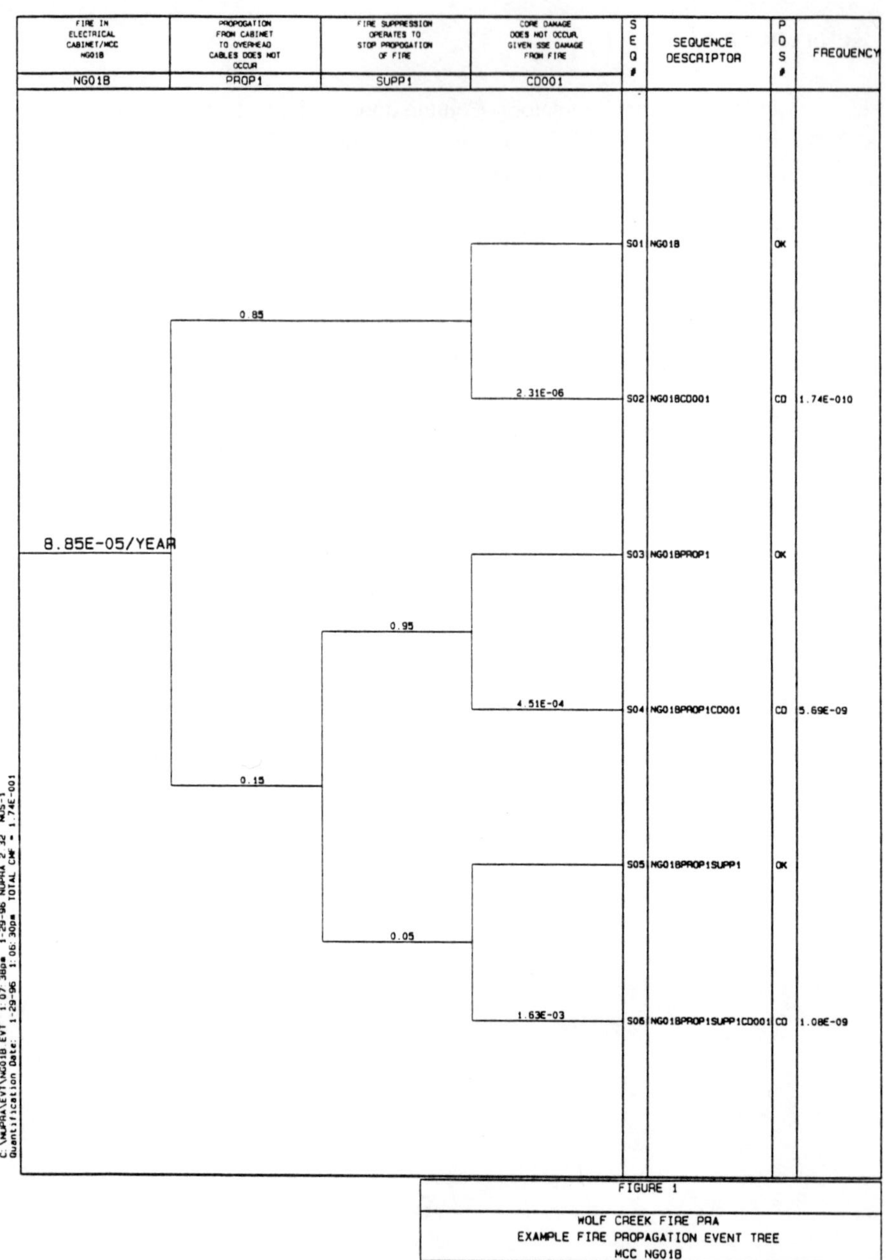


FIGURE 1

WOLF CREEK FIRE PRA
EXAMPLE FIRE PROPAGATION EVENT TREE
MCC NG01B

# A METHOD FOR OPTIMIZING THE SAFETY, AVAILABILITY, AND COSTS OF NEW PRESSURIZED WATER REACTORS

**Jean-Baptiste GOURDEL & Aline ELLIA-HERVY**
**SAFETY and RELIABILITY DEPARTMENT**
**FRAMATOME, FRANCE**

The design approach for a new pressurized water reactor (PWR) must take into account both the needs expressed by the utilities and the requirements imposed by the safety authorities. A design effort focusing on nuclear safety alone would risk leading to a reactor model that would not be very competitive (with too high a per-kWh cost), because it would not be sufficiently available, and would be too costly to operate and maintain.

It is therefore necessary to implement a method - an approach - that enables designing a new nuclear power unit while optimizing both availability and costs, both economic and dosimetric, while simultaneously maintaining the required safety level.

The purpose of this presentation is to show how designers can take these objectives into account.
The safety and availability requirements are expressed in the recommendations issued by the safety authorities and utilities (IAEA INSAG reports, joint recommendations for a common safety approach for future PWRs prepared by the French and German safety authorities, European or US utility requirements...).
As an example, the main safety and availability requirements expressed by the European utilities are the following ones :

- Core damage frequency: lower than $10^{-5}$ / year,
- Production availability: greater than 87% over the entire life of the plant unit,
- Frequency of unplanned reactor trips: less than 1 per 7000 hours critical,
- Unplanned capability loss factor: less than five days per year,
- Planned outage duration: less than 25 days per year.

To achieve these goals, a structured and logical methodology centered on reliability and risk analysis is used from the start of the design until the completion of the project.

The approach is based on:

- A thorough analysis of experience feedback from the existing PWR unit population,

- Probabilistic safety assessment models (PSA),
- Probabilistic availability assessment models (PAA); and
- Cost considerations, both economic and dosimetric.

Since the purpose is to design a PWR of the "evolutionary type," this optimization method, which involves multiple parameters and multiple solutions, is applied in accordance with an iterative process, using objective allocation at the systems level.

The process is implemented in four steps :

# 1. Step 1 : Functional Analysis and RAMS* Allocations

The first step of the analysis is to perform an allocation of the overall safety, availability and cost objectives. This allocation is system-oriented.

A functional analysis, carried out at system level, is performed in order to express (both the needs stated by the utilities, and the requirements imposed by the safety authorities) in terms of functions to be fulfilled by the system.
The newer the system design is, the more elaborated the functional analysis must be.

The functional analysis is supplemented by a detailed analysis of operating experience.
With these data, a first allocation can be made for each system for life cycle cost and availability (for the design of a new system, the allocation can be made using experience and engineering judgement).
Preliminary technical recommendations, to mitigate the impact of the system on safety, availability and dosimetry, are made.
This process of allocation is iterated during all the design phases.

# 2. Step 2 : Probabilistic Safety Assessment Models (PSA)

The PSA is used to verify that the level of safety for each proposed new system design is consistent with the overall safety objectives. The integration of PSA into the design is shown in Figure 1.

A preliminary PSA is performed to identify recommendations (operational and design), as well as items critical for safety. For these critical items, RAMS requirements for intrinsic MTBF (Mean Time Between Failure) and MTTR (Mean Time To Recover) are established.

The application of PSA methods as a decision making tool during the plant design phase, is an usual approach and is not detailed in this paper.

---

* RAMS : Reliability, Availability, Maintainability and Safety

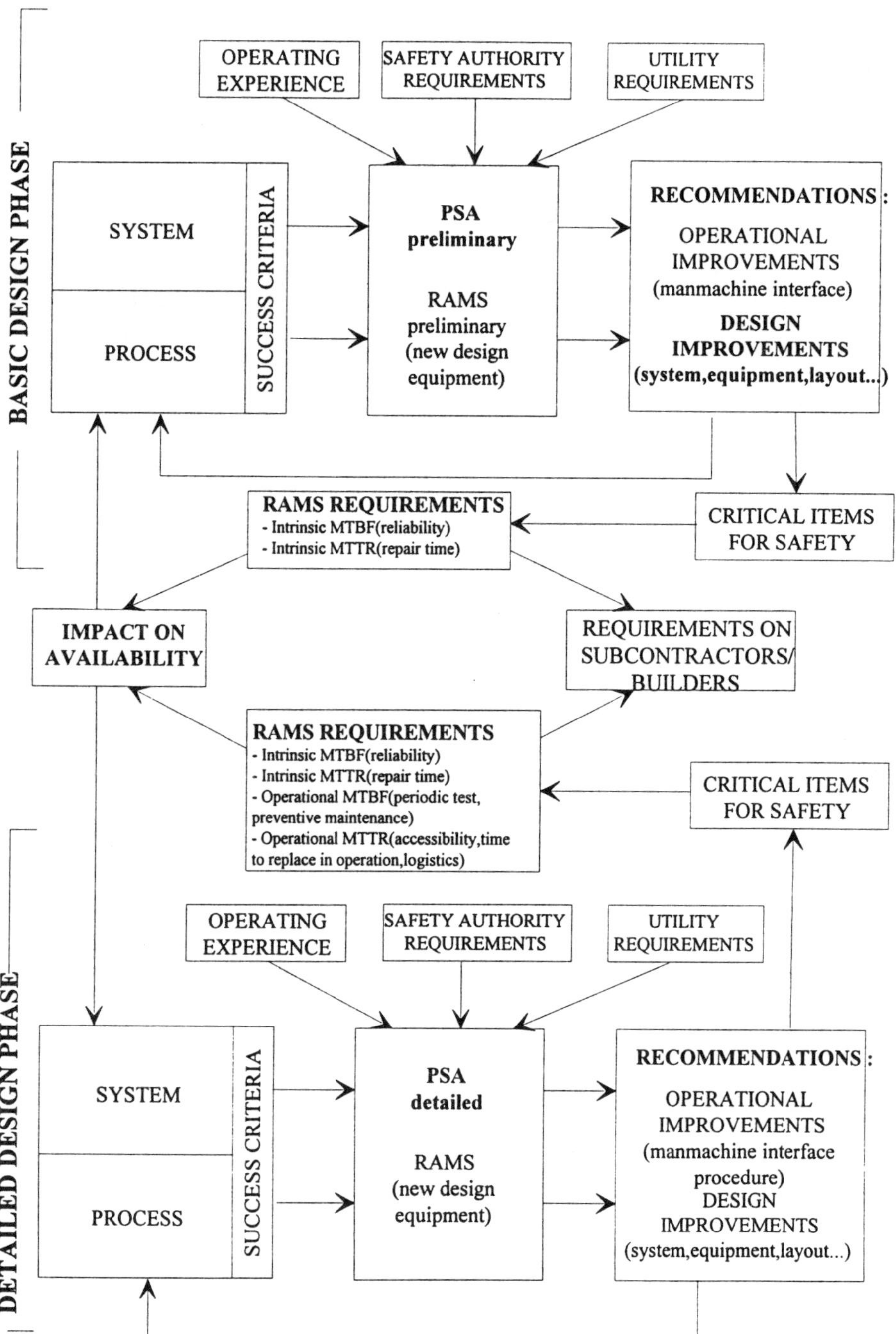


**Figure 1 : Integration of PSA into design (Safety goals)**

The systems or subsystems which are acceptable in term of safety are then selected for further analysis.

During the detailed design phase, a more detailed PSA taking into account outputs from Step 3 (for example, new maintenance or test programs) is performed.

# 3. Step 3 : Probabilistic Availability Assessment Models

PAA (Probabilistic Availability Assessment) is performed concurrently with the PSA program.

As the PSA model is developped in order to identify the combinations of failures leading to unacceptable safety consequences, the PAA model is developped in order to identify the failures or combinations of failures leading to plant unavailability.

The aim of PAA is to focus on availability and life cycle cost. The approach is different from PSA, because almost all systems have a direct impact on availability and cost. This approach is summarized in Figure 2.

## 3.1. Identification of critical equipment for plant unit availability

The aim is to identify what equipment is critical for unit availability. This equipment could induce forced outages, or be on a critical path during planned outages.
A simplified FMECA is performed to assign a criticality ranking for each equipment failure and select critical items for availability. Operating experience is also reviewed to identify past failure mode experience, and to determine equipment criticality.

## 3.2. Fault tree analysis

This analytical method is used to display the combinations of item failures that can result in an undesirable event. It takes into account the combinations of one or more item failures. (Undesirable events are forced outages or prolonged unit shutdowns.)
As a first step, only order 1 and 2 minimal cut sets, which are the most critical, are selected and critical items are identified for further analysis.

## 3.3. Decision logic process to select maintenance tasks (RCM-Reliability Centered Maintenance)

The critical equipment for safety and availability is then submitted to a decision logic tree process (RCM approach), to establish a maintenance plan.

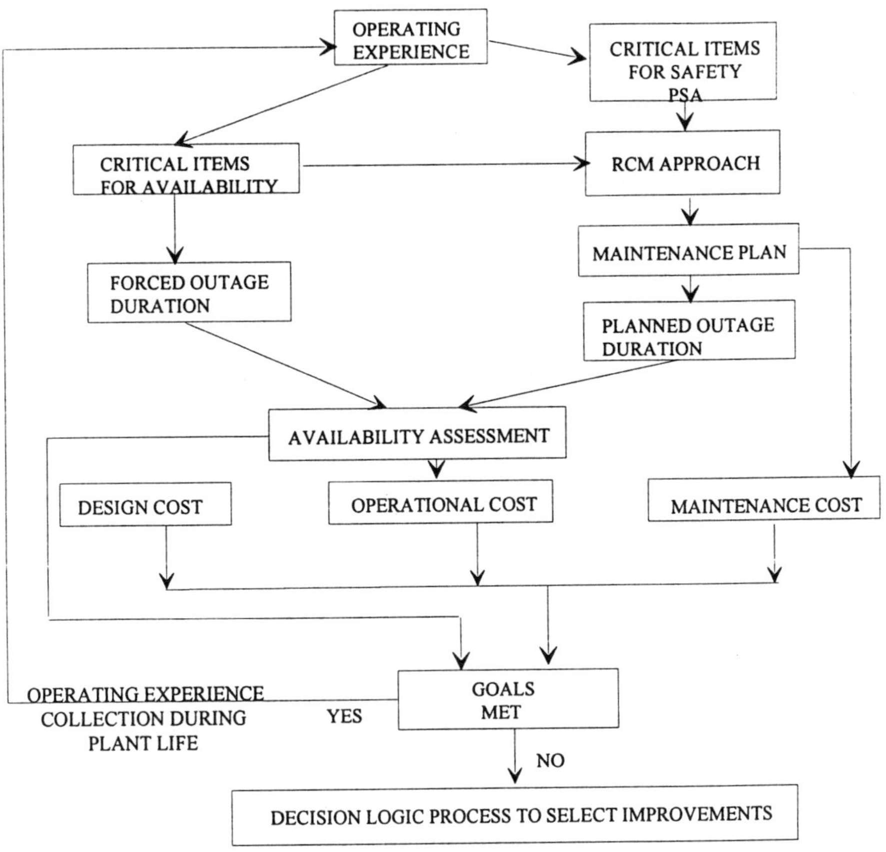


**Figure 2 : PAA process**

The MTBF and MTTR of equipment, taking into account preventive maintenance tasks, are assessed. These values are given to subcontractors or builders as RAMS specifications.

The cost of all the maintenance tasks, including logistic resources selected using the RCM approach, is assessed.

## 3.4. Impact of maintenance plan on availability

A planned outage schedule is established in a deterministic way. Operating experience is completed by an analysis of the maintenance to be performed on new items for which no experience record is available.

A detailed analysis of new critical maintenance tasks or logistic items is performed.

Critical paths of planned outages are identified. A path is defined by its tasks (of operation and maintenance) including the associated logistic items, and their duration. Critical maintenance tasks are those tasks on the critical paths.

The process used to identify critical items for availability is now used to define critical logistic items during an outage (e.g.: refuelling machine). These are the logistic items associated with critical tasks.

Critical items are added to the fault tree associated with the undesirable event: "prolonged facility shutdown", as potential causes of this event.

### 3.5. Assessment of availability

In this phase, operating experience, completed by the analysis of the newly designed items, is used to assess :

- In a probabilistic way (using fault trees, PERT diagram...) : the duration of the forced outages and the duration of the prolonged unit shutdowns (hazards during planned outages).
- In a deterministic way : the duration of an "ideal" planned outage.

## 4. Step 4 : Cost Evaluation (Comparison Between Different Designs)

The safety, cost and availability of the different system designs are now compared. To optimize the architecture of the selected system, a logic decision process to determine and select improvements can be used.

This process is only applied to the most critical items or tasks, until safety, availability or cost allocations are reached.

## Conclusion

The use of risk analysis methods to improve reactor design by directing effort at the most requiring attention points, has already permitted significant improvements in terms of safety, availability and operating costs on existing nuclear power plants.

Used early on, at the conceptual design stage, such methods are extremely helpful in optimizing the design and improving the performances of future plants.

But, this approach requires involvement very early on in the design process, and close collaboration, not only with the designers and manufacturers, but also with all the other participants (the utilities, for example). If the designers and manufacturers can optimize the intrinsic availability, maintainability, and safety of the facility that they are designing or manufacturing, by controlling the technical aspects and the costs of development and construction, it is the operators - the utilities - who are the best placed to optimize operational availability and safety, by controlling the organizational, logistic, and budgetary aspects of training, operation and maintenance.

Such an approach is ambitious and complex to completely implement, therefore it must be applied pragmatically. Nevertheless, this approach is the only means of optimizing the design of a new reactor to respect given objectives. In addition, the analyses performed to improve performance without sacrificing safety could be useful to demonstrate the achievement of these objectives to the utilities and the authorities, and to help gain public acceptance of these new reactors.

# Risk Based Approach to Define Allowed Outage Times of Electric Power Supply Systems at the TVO I/II Plant

Jari Pesonen, Mikko Kosonen
Teollisuuden Voima Oy
Olkiluoto, Finland

Tuomas Mankamo
Avaplan Oy
Espoo, Finland

**Abstract**

The Limiting Conditions for Operation (LCO) and Allowed Outage Times (AOTs) given in Technical Specifications (TechSpec) were originally set for Electric Power Supply Systems (EPSS) using deterministic criteria and qualitative analysis, following similar principles as for the frontline safety systems. Because of diverse functional consequences, the safety impact of EPSS failures has been difficult to assess in a balanced way prior to being able to use a plant specific PSA. This background led to the incentive to improve the consistency of the current requirements. An integrated approach to evaluate risk impact of single and multiple failures was developed. Our paper will describe the main features of the developed approach, including the applied working criteria for the risk-based setup of AOTs and action statements, which considers several risk measures in parallel.

## 1 Background and goals of the study

TVO I/II PSA program was initiated in 1984 by the utility. The level 1 analysis was in the first stage performed for internal initiating events. Later it was extended to include also flood, fire and shutdown mode analysis. Weather phenomena and level 2 analysis is ongoing. TVO's aim is to keep PSA living and thus make it possible to use it for instance to monitor safety level, to support safety improving modifications, to evaluate the safety benefit of modifications etc. [1].

The project for defining AOTs for Electric Power Supply Systems (EPSS) at TVO started 1992 as a pilot study. The background for the current definitions was explored, international practices and development in this area were studied. The results of pilot study showed that TVO I/II PSA model needed extensions, e.g. separation of external initiating events to train level and extensions of the controlled shutdown model. The pilot study was carried out as a joint effort by Avaplan Oy, ABB Atom AB and TVO, and later continuation between TVO and Avaplan Oy.

The aim of the project was to optimize and balance the LCOs for EPSS, but not to limit only to relaxations. Also the configuration contributions, i.e., situations when components in different systems are out of operation at the same time, was investigated. On the other hand the plant operators have found some of the requirements too restrictive and part of the requirements may not be beneficial for safety or the requirements seem not to be consistent from risk point of view.

Maintaining the connection to well-established deterministic framework was defined as a desirable goal when starting this study.

# 2 Electric Power Supply Systems of TVO I/II

Each unit is connected to the 400 kV net via one block transformer and to the 110 kV net via two startup transformers. The auxiliary power system is divided up into four separate sections (train A, B, C and D). Four diesel-generating units provide on-site standby power supply. In order to enhance the on-site standby power supply a cross-connection between diesel-backed busbars of unit I and II was installed in late 1980´s.

TechSpec requirements for EPSS are based on an US practices, namely RegGuide 1.93, version 1974, for a plant with 2•100% safety system design. These regulations have been modified based on engineering judgement and some probabilistic analysis to fit for the TVO I/II with 4•50% safety system design. RegGuide 1.93 is applied to EPSS which is designed according to general Design Criterion 17. It means that the following has to be fulfilled in the design:

- two power supplies from external grid and
- redundant internal power supply (both AC and DC).

Generally all the primary and secondary safety systems are included in the TechSpec with the minimum system capacity as defined in Final Safety Analysis Report (FSAR). This means normally that 2 out of 4 trains shall operate when necessary. A single failure together with preventive maintenance in another train still meets the FSAR criterion. AOTs for EPSS are generally shorter than for process systems, for example in case of DC bus failure the 8 hour AOT applies before the shutdown requirement expires; cold shutdown has to be reached within the following 8 hours.

The analyzed failure situations concern only power operation in this approach. All the basic points in TechSpec for EPSS were analyzed with exception of a few subcases. Selected cross-combination cases were analyzed including also combinations in current TechSpecs, which do not allow any AOT.

# 3 Using the Plant Specific PSA

The generic purpose fast PSA code called SPSA, developed by STUK (Finnish Regulatory Body) is used for modeling and solving TVO I /II PSA model [1].

Conditional risk increase was calculated in order to obtain risk increase factors from baseline for each LCO situation. Baseline risk was calculated by excluding events concerning surveillance testing, scheduled preventive maintenance and planned unscheduled maintenance during power operation. These kinds of events can supposed to be postponed if LCO state exists.

Risk increase factors were calculated for both continued operation and LCO shutdown. The risk of LCO shutdown was obtained by use of event trees for planned shutdown, loss of main feedwater and condenser. Shutdown risk includes the shutdown phase, hot shutdown (20 hours), and startup phase. Each LCO situation is analyzed by increasing temporarily basic event data (and CCF-data if exists) and making MCS search to get representative list of MCS in a given failure situation. This means for example that for four redundant train systems, the single failure probabilities are increased to 0.25. After MCS search original data were restored and train-specific risk increase factors were calculated for derivation AOT risk measures explained later.

The TVO experience were reviewed for the ten-year period from 1983 through 1992 in order to obtain plant specific data about the frequency and downtime length of LCO events. In total 116 events have been recorded, which corresponds to 5.8 events per reactor-year in the electric power supply systems. The bulk contributors are diesel generators, rotating DC/AC converters and rectifiers. Also partial loss of external grid connections have occurred few times, but these have mostly been shortly recoverable. The mean downtime has been about 10 hours.

For the systems with two or more experienced LCO events, the point estimate data were used. For the other systems, PSA data were used, being mostly generic data from the combined Swedish-Finnish data base [3]. Generally, the plant experience showed reasonable compatibility with the generic data.

The definition of risk measures are presented in more detail in Refs.[2,4,5]. Briefly summarized the following variables are central:

- Time to cumulate an incurred risk of 1% relative to annual Baseline Core Damage Probability (BCDP)
- Time to cumulate an incurred risk equal to LCO shutdown risk
- Delta-risk, which describes the addition from repair downtime occurrences to long term average risk (called also as yearly AOT risk contribution); the definition was extended to include also shutdown risk contribution of the repairs not completed within AOT

The first two measures describe the conditional, situation-specific risk, while delta-risk incorporates also the occurrence rate of a given failure class. Delta-risk is hence desirable to be separately controlled in order to bound the long term risk impact of repairs.

# 4 Results

## 4.1 Risk impact of the current LCOs/AOTs

The LCO events can be divided up into three categories, compare to Table 1. Double CCF events mean failures of identical redundant components. Double cross-combination events are coincident downtime of two different systems: e.g. during the repair of a system, another system is detected failured.

A bulk of the analyzed failure situations have small or moderate risk increase factor less than 2. Hence, the single-event risks are relatively small. This together with the relatively low occurrence rate and short mean downtime lead to minor delta-risk as well, Table 1. The failures of vital buses give higher risk factors up to about 10. In fact, the sum delta-risk is dominated by failures of vital buses, for which situations the current AOT is short or no AOT is given at all, and the risk is connected to the likely plant shutdown if such a situation would occur. The LCOs for these failure situations are not proposed to be changed from the current stringent requirements.

Table 1. Summary of the analyzed failure situations.

| Failure category | Number of analyzed LCO cases[1] | Occurrence rate [1/r-yr] | Number of proposed changes | Delta-risk for the | |
|---|---|---|---|---|---|
| | | | | current TS | proposed TS |
| Single | 20(72) | 6.51 | 10 | 2.76% | 2.04% |
| Double CCF | 25(129) | 0.12 | 6 | 0.043% | 0.037% |
| Double cross-combination | 57(650) | 0.037 | 49 | 0.061% | 0.013% |
| In total | 102(851) | 6.67 | 65 | 2.86% | 2.09% |

Notes: 1. In parentheses is given the number of all underlying combinations of trains; the combinations of a similar risk impact are grouped into LCO cases

In all analyzed failure situations, the risk of LCO shutdown is relatively large: time to cumulate a risk increment equal to LCO shutdown risk is ten days or larger. This is connected to small risk increase factors and to the fact that already baseline shutdown (no downtime event existent) means a risk which corresponds to 7.5% of BCDP or about the risk cumulating in 30 days of baseline power operation. Simplifications of the PSA models for shutdown modes may impose some overestimation in this respect, but anyway the LCO shutdown is in most cases very unfavorable.

The calculations revealed substantial variation between the individual trains and combinations of trains due to asymmetry of hard-wired dependencies on the electric power supply. This concerns especially influences via support systems and turbine plant systems. Furthermore, among the trains pairs AC and BD a physically close, and hence vulnerable to internal floods and fires, while those pairs are mutually separate in high degree. In case of large differences in risk impact, the different train combinations were treated as disjoint LCO cases. Compare to the number of analyzed LCO cases in Table 1, which shows also the total number of underlying combinations.

## 4.2 Working criteria and proposed changes

The suitable AOT was determined by using the following criteria:

A. Primary criteria; Incurred risk over AOT is clearly below 1% relative to BCDP
B. Secondary criteria; Incurred risk over AOT is at about 1% relative to BCDP and clearly less than the conditional risk of LCO shutdown

Because delta-risk proved minor, it did not play determining role in defining AOTs. The total delta-risk of a few percent (due to corrective maintenance in EPSS at power) can be considered fully acceptable, compare to Table 1.

The subset of LCO cases for which changes are proposed represent in the mean about three events per reactor-year. Their delta-risk impact would decrease from 0.8% to 0.05% by the changes, mainly because of the likelihood of LCO shutdowns would substantially decrease (currently about once in ten reactor-years for the subset). Bulk of the net impact comes from the cases presented in Table 2.

Table 2: Calculated AOT's for selected EPSS sections

| **TechSpec, Section 3.10** | | | | **EXPLANATION** | | |
|---|---|---|---|---|---|---|
| | Point / Failure criteria | AOT [days] | | Calculated AOT [days] | | Used criteria |
| | | cur-rent | pro-posed | **A.** incurred risk of 1% relative to annual BCDP | **B.** incurred risk equal to LCO shutdown risk | |
| **E** | a): 1/4 684 switch-over automation | 7 | 30 | 53 | 24 | A |
| **F** | a)&b): 1/4 666 rotating converter | 7 | 14 | 34 | 26 | A&B |
| **G** | new basic point f): 1/4 672 110 V DC rectifier* | 8 hours | 30 | year | 25 | A |

* condition; a transportable rectifier has to be connected as a substitute

The proposals include also several cases of more stringent requirements and more specific statements about starting up of diesel generators in reserve or connecting transportable substitute equipment for the time of the repair.

# 5 Summary of insights

Our experience show first of all that for this kind of PSA application, modeling of electric power supply should be extended in high degree of detail to support systems and turbine plant systems in order to correctly model Common Cause Initiator nature of failure situations, and to obtain realistic results about the risk impact. Lot of effort needs to be allocated to quality assurance of the results. This means primarily that the relevance of dominant sequences shall be checked for each calculation case. The training simulator was also used to verify that the functional impact of failures in EPSS is properly taken into account.

It is also important that the PSA models are based on realistic assumptions, because the relative results matter in these kinds of applications. The risk calculations were performed by using Level 1 PSA. Additional bounding analyses were performed in order to verify that the safety impact of failures in EPSS is not particularly pronounced due to causal unavailability in containment systems or other accident mitigation systems. The practical conclusions of our application were made with good margin against the uncertainties and limitations of the study. In other respect, the current LCOs based on deterministic framework was left as is - being, however, generally in line with the insights from the risk calculations.

The closer investigation of how to maintain the connection to earlier deterministic framework showed that a mixed approach in the way of replacing PSA modeling assumptions (mostly of best estimate character) by more conservative design basis assumptions according to plant FSAR does not make sense. The multiple conservative items tend to produce results which are hard to interpret: e.g. the relative importance of safety systems may be badly distorted [2]. On the other hand, taking into account the uncertainties in the risk-based approach led to following pragmatic conclusions:

- conventional well established AOT scheme defined by deterministic rules, with defense in depth goal, should be generally preserved
- cases where the risk-based analysis indicates a significant deviation should be carefully analyzed in order to understand the causes to the discrepancy

E.g. the plant specific PSA may provide new insights about the safety importance, or alternatively about non-importance of a system and its downtime situation, which has been difficult to determine earlier by qualitative reasoning or engineering judgment. Especially, fires, floods and other types of system interactions have been difficult to evaluate in a balanced way without PSA. Furthermore, the FSAR analyses lack proper way to consider the risk of LCO shutdown.

## References


1. Himanen, R., Use of PSA as a tool to monitor and enhance safety of TVO NPP, TOPSAFE '95 Conference, Budapest, September 24-27, 1995.
2. Kosonen, M. & Mankamo, T., Risk-based approach to define Limiting Conditions of Operation and Allowed Outage Times for electrical power supply systems. IAEA Technical Committee Meeting on Procedures for Use of PSA for Optimizing NPP Operational Limits and Conditions, Barcelona, 20-23 September 1993. Proceedings.
3. T-Book. Reliability data book of components in Nordic NPPs, 3rd edition. Prepared by ATV Office, Vattenfall AB and Studsvik AB, 1992.
4. Risk-based optimization of technical specifications for operation of nuclear power plants. IAEA-TECDOC-729, December 1993.
5. Handbook of Methods for Risk-Based Analyses of Technical Specifications. Prepared by Samanta, P.K. & Kim, I.S. (BNL), Mankamo, T. (Avaplan Oy) and Vesely, W.E. (SAIC) for USNRC, Report NUREG/CR-5995, November 1994.

# Modeling for Risk in the Absence of Sufficient Data: A Strategy for Treaty Verification

Nicholas Kyriakopoulos
The George Washington University
Washington, DC, USA

Rudolf Avenhaus
Universität der Bundeswehr München
Neubiberg, Germany

**Abstract**

This paper presents a methodology for selecting facilities to be inspected combining game theory with qualitative measures of risk. The Chemical Weapons Convention is used as an illustrative example.

## 1 Introduction

Multilateral arms control agreements establish a complex regime of objectives, monitoring, verification and enforcement. For example, the goal of the Treaty on the Non-Proliferation of Nuclear Weapons (NPT) is to encourage the use of nuclear energy and detect diversion of nuclear materials for other than peaceful purposes. Since diversion could occur at any stage of the nuclear fuel cycle, monitoring of the flow of the controlled materials is required for the entire cycle. To this end, the NPT requires the establishment of National Systems of Accountancy and Control of nuclear materials and a system of international safeguards administered by the International Atomic Energy Agency (IAEA) and designed to verify the proper operation of the national accounting systems[1,2]. International safeguards are, in effect, a set of auditing procedures.

In contrast, the Convention on the Prohibition of the Development, Production, Stockpiling and Use of Chemical Weapons and Their Destruction[3] (CWC), while setting multiple goals and monitoring activities consisting of declarations and inspections, it does not specify a comprehensive verification concept. For example, certain chemicals may be produced and used for legitimate purposes, but they may not be diverted for use in chemical weapons, yet a mechanism for detecting diversion is not specified. States will be subject to verification by the Organization for the Prohibition of Chemical Weapons (OPCW) to assure compliance with the terms of the Convention[4]. While it would not be difficult to establish a monitoring regime for detecting the presence of specified materials at specified places, detecting diversion would require definition and monitoring of the entire life cycle of each controlled material to construct a materials balance[5].

In the absence of a treaty-defined verification concept, the OPCW may adopt a monitoring approach combining risk analysis with game theory. This paper presents an example of how such a methodology can be formulated. The Chemical Weapons Convention is used as a model because it contains a very detailed formulation of objectives and some qualitative measures of risk without an equally detailed definition of the verification concept. The first part of the paper presents a classification of chemicals and facilities followed by the development of a set of risk indicators based on the treaty requirements. Absent any *a priori* quantitative information for developing probabilistic risk indicators, a qualitative approach based on fuzzy logic concepts is used. The second part of the paper applies a game-theoretic scheme to the problem of selecting facilities to be inspected.

# 2 An Example of Qualitative Measures of Risk

The primary goal of the CWC is to eliminate all chemical weapons by requiring the destruction of existing stocks and production facilities and prohibiting the acquisition of new ones. The major verification issues arise from the nature of the toxic chemicals which could be used as chemical weapons. These same chemicals also have legitimate uses in many areas of human activities and the Convention recognizes the right of the States Parties to acquire and use the chemicals for purposes not prohibited under the Convention. In effect, the verification regime is required to discriminate between two categories of uses for each of the chemicals subject to the control of the Convention: prohibited and non-prohibited.

Verification is based on monitoring materials (chemicals) and facilities by the OPCW, referred to as *inspectorate* for the remaining of the paper . Chemicals are classified into three categories *(Schedules)* using a combination of quantitative and qualitative measures.

*Schedule 1*: Chemicals which a) have been developed or used as chemical weapons, b) pose high risk through their high potential use as chemical weapons and c) have little or no use for purposes not prohibited by the Convention.

Schedule 2: Chemicals which a) pose significant risk through their potential use as chemical weapons, b) may be used as precursors for the production of chemicals in Schedule 1, c) pose significant risk on the basis of their importance in the production of chemicals in Schedule 1 and d) are not produced in large quantities for purposes not prohibited by the Convention.

*Schedule 3*: Chemicals which a) have been produced or used as chemical weapons, b) pose risk because they might be used as chemical weapons, c) pose risk on the basis of their importance in the production of chemicals in Schedules 1 and 2 and d) may be produced in large quantities for purposes not prohibited by the Convention.

Facilities are classified in terms of a) function as: *production*, *processing*, *storage* and *consumption* and b) content as *Schedule 1, Schedule 2, Schedule 3*. In addition, the Convention defines a fourth category of facilities associated with the production of discrete organic chemicals not included in the preceding categories. The fourth category of chemicals is referred to as *Other Discrete Organic Chemicals*. Facilities having any of the four functions, being associated with the four categories of chemicals and exceeding specified design characteristics and quantitative thresholds of the listed chemicals are subject to the verification provisions of the Convention.

# 3 Some Verification Issues

The major issues related to the duality of use of the chemicals and the facilities are identified in this section. The verification issues may be grouped into two major cases:

*Case I*: A controlled chemical is produced, *etc,* in a plant where it is not supposed to be produced. Different issues arise if the plant
a) has been declared as producing other controlled chemicals,
b) has not been undeclared under the terms of the Convention.

For this case, the verification objective is a) to select a facility and b) to detect the presence of an undeclared chemical at the given facility. Given that the set of chemicals is finite, detection of any one of the elements of the set at a given facility becomes technically feasible. For declared facilities the problem is how to select a given facility at a particular time. Monitoring of undeclared facilities presents a different set of problems, which are beyond the scope of this paper.

*Case II*: A controlled chemical is produced in a plant where it may be produced, but it is produced
a) when it is not supposed to be produced
b) in quantities exceeding those declared. Excessive quantities may be generated
1) by maintaining production for a period greater than that declared as production period, or
2) within the specified production period, but in quantities greater than those declared as being anticipated.

For *Case II* the verification problem has the following components:
a) If the chemical is not supposed to be produced during a given time period, the verification objectives would be to detect production during the given period.
b) If the chemical is produced during the given period, the verification objectives would be
1) to detect extension of the specified period
2) to detect production rate likely to result in the production of quantities greater than those having been declared as anticipated.

# 4 Fuzzifying and Defuzzifying Risk Measures

For the inspectorate to maintain its impartiality and credibility, the selection of facilities within a State must be based on technical, not political, criteria. The CWC provides the basis for developing such criteria by defining risk categories for chemicals and facilities. The risks, however, are expressed in qualitative rather than quantitative measures limiting the applicability of probabilistic risk assessment methodologies. At the beginning of the verification process there is no *a priori* information for assigning risk for any of the facilities under the domain of the Convention other than that specified by the Convention. As the verification process evolves, statistical models could be developed for each facility on the basis of corresponding historical data. However, treaty imposed limitations on the number of inspections and the amount of information that could be obtained would make the development of statistical models difficult if not impossible.

The approach proposed in this paper is to develop an inspection strategy based on the qualitative measures of risk specified in the Convention and the premise that a State intenting to perform an illicit operation at a given facility will act strategically. The first problem to be considered is the assignment of risk for each class of facilities on the basis of function and material.

Following the terminology of the previous section, the risk levels for the chemicals contained in the four previously listed categories are designated as:
{ *Very High* $(M_1)$, *High* $(M_2)$, *Moderate* $(M_3)$ and *Low* $(M_4)$ }.
A similar classification can be adopted for the facilities which are subject to inspections. For the four types of facilities, *production, processing, storage* and *consumption* the corresponding risk levels are:
{ *Very High* $(F_1)$, *High* $(F_2)$, *Moderate* $(F_3)$ and *Low* $(F_4)$ }.
In general, for any classification attribute $P_i$, the levels of risk, in descending order, may be represented as $\{ P_{ij}, i = 1, ..., p; j = 1, ..., q \}$, where $p$ is the number of attributes and $q$ the number of risk levels. The assignment of numbers to the qualitative measures of risk is akin to the concept of *defuzzification* encountered in fuzzy systems.

Given that, for the example of this paper, the set of risk levels is $\{1,2,3,4\}$, a transformation is required that will map the combination of the risk categories $\{M_1, M_2, M_3, M_4\}$ and $\{F_1, F_2, F_3, F_4\}$ into the set $\{MF_1, MF_2, MF_3, MF_4\}$, where $MF$ represents the combination of the two attributes. Absent any other information about the facilities and chemicals, the required transformation is given by (1).

$$R_k = Integer\left[\frac{M_i + F_j}{2}\right], \qquad i,j,k = 1,...,4. \tag{1}$$

For example, the risk of a storage (moderate risk) facility containing Schedule 2 ( High risk) chemicals would be *Integer* $[ ( F_3 + M_2 )/2 ] = 2 = MF_2$, which would designate a facility of *High* risk. The translation of the numerical values to the qualitative risk categories corresponds to the process of *fuzzification.*

## 5 A Game-Theoretic Strategy for Selecting Facilities

Having classified the facilities under consideration in a State, the question arises how to translate this classification into an inspection plan for these facilities. If a State intends to act illegally, it will do so according to a strategic plan. Accordingly, a *game theoretical analysis* involving the State and the inspectorate is appropriate. Following is such an analysis for a simple example; more elaborate models can be found elsewhere[6].

Consider a State with only two facilities. The inspectorate, can inspect only one with probability $p$. The probability of selecting the second becomes $(1-p)$. Furthermore, assume that the State is determined to act illegally in just one of these facilities, with probabilities $q$ and $(1\text{-}q)$, respectively. The payoffs to the inspectorate and the State are $(-a_i, -b_i)$, if the illegal action is performed in the $i$th facility which is inspected at the same time and $(-c_i, +d_i)$, if the illegal action is performed in the $i$th facility and inspection takes place in the other one, $i=1,2$.

Assume $a_i$, $b_i$, $c_i$ and $d_i > 0$ and, furthermore, $a_i < c_i$ for $i=1,2$. For later purposes, assume the payoffs to both the players to be zero, if the State acts legally. The second condition reflects the fact that the inspectorate's highest priority is legal behavior of the State; detected illegal behavior is obviously better than undetected one. Also, the payoff to the State is positive, otherwise there would be no incentive for the State to act illegally.

The problem is formulated as a non-cooperative 2 x 2 -bimatrix game which is represented by the following table:

| | | q | 1 - q |
|---|---|---|---|
| | 1 \ 2 | 1 | 2 |
| p | 1 | $-a_1$ / $-b_1$ | $-c_2$ / $+d_2$ |
| 1 - p | 2 | $-c_1$ / $+d_1$ | $-a_2$ / $-b_2$ |

It can be shown that the solution in the sense of Nash[7], *i.e.*, the equilibrium strategies $p^*$ and $q^*$ as well as the equilibrium payoffs $I_1^*$, $I_2^*$ to both players are given by (2).

$$p^* = \frac{d_1 + b_2}{d_1 + b_1 + d_2 + b_2}, \quad I_1^* = \frac{a_1 a_2 - c_1 c_2}{c_1 - a_1 + c_2 - a_2}$$
$$q^* = \frac{c_2 - a_2}{c_1 - a_1 + c_2 - a_2}, \quad I_2^* = \frac{d_1 d_2 - b_1 b_2}{d_1 + b_1 + d_2 + b_2} \tag{2}$$

Assuming, further, that the "sanctions" $b_1$ and $b_2$ are the same for both facilities, they can be normalized to one, and *the gains $d_1$ and $d_2$ become the risk factors of the preceding section.* Using (2), the ratio of the inspection probabilities $p^*$ and $(1\text{-}p^*)$ for the two facilities is given by (3). In the first case, the inspection probability in the $i$th facility is the larger, the larger its risk factor compared to the other is, while in the second case inspection of either facility is equiprobable.

$$\frac{p^*}{1-p^*} = \begin{cases} \dfrac{d_1}{d_2}, & for\ (d_1,d_2) \gg (1,1) \\ 1, & for\ (d_1,d_2) \ll (1,1) \end{cases} \tag{3}$$

Furthermore, since the State's payoff in case of legal behavior has been normalized to zero, the conditions for the State to act legally, in equilibrium, are given by (4).

$$I_2^* < 0 \ , \quad i.e., \quad \frac{d_1}{b_1} \cdot \frac{d_2}{b_2} < 1 \ . \tag{4}$$

In other words, the State will act legally, if the sanctions in case of detected illegal behavior are large compared to the gains in case of undetected illegal behavior. In such a case, both facilities should be inspected with approximate equal probability.

# 6 Conclusions

Some multilateral treaties have complete verification concepts imbedded in the treaty text while others do not. For the latter class, the development of technical monitoring criteria presents challenges because the foundation for constructing comprehensive verification procedures is incomplete. Nevertheless, as the example presented in this paper has demonstrated, it is possible to devise procedures combining qualitative and quantitative measures for monitoring such treaties without introducing political considerations.

**References**


1. *Treaty on the Non-Proliferation of Nuclear Weapons*, INFCIRC/140, International Atomic Energy Agency, Vienna, 1970.

2. *The Structure and Content of Agreements and States Required in Connection with the Treaty on the Non-Proliferation of Nuclear Weapons*, INFCIRC/153, International Atomic Energy Agency, Vienna, 1971.

3. *Convention on the Prohibition of the Development, Production, Stockpiling and Use of Chemical Weapons and on Their Destruction,* United States Arms Control And Disarmament Agency, Washington, DC, October 1993.

4. *Ibid.*, Articles IV, V and VI.

5. N. Kyriakopoulos, " A Model for Monitoring the Production and Distribution of Thiodiglycol", in S. J. Lundin (ed), *Verification of Dual-Use Chemicals under the Chemical Weapons Convention: The Case of Thiodiglycol*, SIPRI Chemical and Biological Warfare Studies, No. 13, Oxford University Press, 1991.

6. R. Avenhaus, M. Canty, *Compliance Quantified: An Introduction to Verification Theory*, Cambridge University Press, in print.

7. J. Nash, " Non-Cooperative Games", *Annals of Mathematics*, 54.2, pp. 286- 295, 1951.

# Arms Control Verification Synergies: Theory and Application in the United Nations Context

**D.Marc KILGOUR**
**Wilfrid Laurier University**
**Waterloo, ON N2L 3C5 CANADA**

**F.Ronald CLEMINSON**
**DFAIT Verification Research Program**
**Ottawa, ON K1A 0G2 CANADA**

**Abstract**

The application of synergies in monitoring and verifying United Nations arms-control regimes is reviewed. Theoretical characterizations of verification synergies, especially for on-site inspections, are interpreted and assessed.

## 1 Introduction

Without effective verification, arms control is almost pointless. First, verification is the key to accountability in international Non-proliferation, Arms Control, and Disarmament (NACD) processes, accountability is essential for ensuring compliance. Second, verification reduces international tensions by communicating the reduction of a state's military capabilities to its rivals. Yet, although it is crucial to NACD, the implementation of effective and efficient verification has been a great challenge.

An important new idea in verification, one that both increases effectiveness and reduces costs, is to exploit *synergies* across different verification techniques, rather than applying only one technique, or applying several but in isolation from each other. Synergies are positive effects arising from the combination of inputs--when the whole is greater than the sum of its parts. Perhaps the prototype example of synergy in verification is the use of aerial photographs to cue On-Site Inspections (OSIs). OSI constitutes a very accurate and therefore effective technique for Ongoing Monitoring and Verification (OMV), but is costly and often impractical because of personnel and equipment limitations. An alternative OMV strategy exploiting synergy is to carry out OSIs after overhead imagery (from satellites or aircraft) has pinpointed the locations of possible violations. Synergy can improve verification effectiveness and efficiency, or simply make it practical, in a wide range of NACD contexts [1].

Arms-control operations carried out under United Nations mandates have begun to utilize verification synergies. While national inspection agencies have relied on synergies, especially of OSI with human intelligence gathering, for some time [1], the use of innovative OMV policies by independent or multilateral inspectorates, has been less common, perhaps for institutional reasons. But recently, agencies associated with the United Nations have begun to exploit synergies, propelled by the twin needs of improving effectiveness and controlling costs. For example, the application of safeguards by the International Atomic Energy Agency (IAEA) under the Treaty on the Non-Proliferation of Nuclear Weapons (NPT) is in the process of being strengthened by the "93 + 2" initiatives, which incorporate many innovative synergies such as the use of environmental sensing to cue OSIs. [2] But most of the pioneering applications of verification synergies have occurred in the UN's operations in Iraq, and are especially evident in the design and operation of the Baghdad Monitoring and Verification Center (BMVC).

The purpose of this article is to review and comment on UN activities in Iraq, especially the BMVC and its strategies exploiting synergy in OMV, and to provide an assessment of these measures using several models of synergy. One of these models was developed to assess the use of OSI in these operations, especially those carried out by the BMVC.

# 2 United Nations Operations in Iraq

May 1996 marked five years since the United Nations Special Commission (UNSCOM) was created by the Security Council and authorized to develop an effective OMV capability relating to Iraq's obligations under UNSCR 687, 707 and 715(1991). Under the Security Council's mandate, much has been accomplished toward the identification, dismantling, and destruction of Iraq's programs to construct weapons of mass destruction and the means for their delivery. The scope and objectives of UNSCOM's operations are unprecedented. While UNSCR 687(1991) was imposed rather than negotiated, so this is clearly not arms control in the traditional sense, Iraq has formally acknowledged this regime and is cooperating with it.

The International Atomic Energy Agency (IAEA) has played an essential role in United Nations operations in Iraq, both independently and in cooperation with UNSCOM. An autonomous organization within the UN system, IAEA might well be termed "bureaucracy at its best." In the mid-1960s it initiated the world's first OSI program in support of the NPT, and has operated this "safeguards system" ever since. Currently, safeguards cost about US$67 million per year; resources include a staff of about 600, including 200 inspectors. It was IAEA's hands-on experience and immediately available expertise that permitted the United Nations to begin OSI in Iraq almost immediately after the passage of UNSCR 687.

In contrast, UNSCOM, created by the Security Council in 1991, can be seen as "ad-hockery at its best." UNSCOM organized and carried out its first OSI only six weeks after it came into existence. Subsequently, UNSCOM has demonstrated a remarkable ability to cobble together various techniques, to operate within severe time restrictions, to draw on the expertise of IAEA and many member states, and to develop innovative procedures in order to meet unforeseen challenges and to deal with in-country issues as they were encountered. In short order, its package of verification techniques expanded from on-site inspections and overhead imagery into the long list shown in Table 1. Together, UNSCOM and IAEA assembled the most effective tools in the OMV field to monitor activities and to support OSIs, at the same time fine-tuning OSI team compositions and inspection time frames. A unique and significant feature of the procedures developed by UNSCOM and IAEA in Iraq is their ability to capitalize on synergies among monitoring and verification tools.

| Methodology | | Source/Varieties |
|---|---|---|
| 1 | Satellite Imagery | NTM and commercial |
| 2 | High-Altitude Aerial Imagery | UNSCOM U-2 Reconnaissance (USA) |
| 3 | Medium-Altitude Aerial Imagery | AN-30 (Russia) (currently on hold) |
| 4 | Helicopter Aerial Imagery | UNSCOM C53 helicopters (Germany) |
| 5 | On-Site Inspections (OSIs) | 4 team models |
| 6 | Environmental monitoring | Air/soil/water sampling |
| 7 | IAEA Safeguards | NPT mandate |
| 8 | Ground Penetrating Radar (GPR) | Bell helicopters & GPR (France) |
| 9 | Radiation detectors | Multi-purpose |
| 10 | Remote sensors *in situ* | Cameras, sniffers, etc. |
| 11 | Collateral analysis | Literature search |
| 12 | Human information | Travellers, defectors, etc. |

**Table 1: OMV Methodologies used by UNSCOM/IAEA in Iraq, 1995**

## 2.1 The Baghdad Monitoring and Verification Center (BMVC)

The layered and packaged approach of UNSCOM and IAEA is most clearly evident in the organization and operation of the recently-established Baghdad Monitoring and Verification Center (BMVC). The BMVC, now functioning with a staff of about 80, is the centre of the in-country OMV process. For the first time, a multilateral organization, under a United Nations mandate, is directing the use of a wide variety of monitoring and verification methods, including on-site inspection, *in-situ* remote monitoring, and overhead imagery. Many aspects of BMVC's operations are quite sophisticated; for instance, images from remote cameras are being transmitted to BMVC headquarters where continuous video and computer scanning equipment "fuse" the images with data from other sources. The host country has a liaison office located within the BMVC, providing a variety of supporting mechanisms. Altogether, more than forty states have contributed personnel to the BMVC.

The importance of the BMVC from the standpoint of multilateral verification is difficult to over-emphasize. The operational experience accumulated by the BMVC, optimizing the use of cutting-edge monitoring technology, will be of immense value in future non-proliferation, arms-control, and disarmament regimes. The activities of the BMVC alone justify the recent conclusion of the Secretary General of the United Nations that the implementation of UNSCR 687(1991) constitutes a "'verification laboratory' for the testing, particularly in combination, of a wide variety of old and some new verification methods, procedures and techniques." [3, §196, p.60]

# 4 On-Site Inspections

The world's first On-Site Inspection system, the safeguards system of IAEA, was designed to verify the NPT in industrially advanced states. Its slow evolution during the 1970s and 80s accelerated remarkably when IAEA and UNSCOM adapted it to the verification of UNSCR 687(1991). Since then, the equivalent of a century's worth of OSI development has taken place.

One important aspect of the evolution of OSI in Iraq is the emergence of several team models. IAEA's past experience determined the original team complements and operating procedures. But that changed rapidly. To capitalize on the right to unimpeded access, the standard OSI team was expanded to 20-30 inspectors, typically staying in-country for 10-15 days at a cost of around $250,000. Later, smaller and longer inspections, typically involving 3-4 inspectors remaining *in situ* for 2-3 months, were found to be more effective at reduced cost. But there was at least one instance of a coordinated two-week mission by a "super team" of approximately 50 inspectors.

The BMVC's role in OSI is now expanding rapidly. Although "out-of-country" teams will continue to conduct specialized inspections from time to time, BMVC inspection groups are now established for each of four weapons categories. Each in-country group, which will consist of four inspectors resident in Baghdad for up to six months, will have primary responsibility for monitoring and verifying facilities and equipment in Iraq. The support of an aerial inspection team and an in-house imaging capability will permit a dramatic increase in small, short, minimum-notice OSIs. Furthermore, inspection teams will carry out preliminary analysis on the spot, and so will be able to react immediately to developments.

It is perhaps remarkable that the capabilities of the various OSI team models can be explained using synergy. Refer again to Figure 1, and recall that OSI typically falls around position B. It is a principle of classical statistics that changing the "strength of evidence" criterion in any test can produce a tradeoff of $\alpha$ against $\beta$, i.e. can move from one ($\alpha$, $\beta$) pair to another, along one of the continuous curves shown in Figure 1. Furthermore, more resources means more information in that the decision-maker can now choose a point on an ($\alpha$, $\beta$)-curve that lies below and to the left.

To distinguish among OSI team models, consider the following question: Beginning with a small (i.e. low-resource) OSI at point B and desiring to apply more resources so as to move toward point C, is it more effective

- to carry out the original procedure using a higher level of resources (thereby moving to a lower curve), or
- to combine the original procedure with another (also near point B) by paying an additional fixed (or set-up) cost?

Of course, procedures should be combined so as to achieve synergies, if possible.

This question was addressed by comparing inspection strategies using abstract models of OSI featuring variable resources, set-up costs, missed violation probabilities ($\beta$) that decrease exponentially as resources increase, and no false alarms ($\alpha = 0$). Surprisingly, the crucial variable is the extent to which the two processes are isolated from each other. Figure 2 illustrates the comparison. Suppose that there is exactly one violation, and that it has actually been identified by one of the two procedures.

If the probability that it is also identified by the other procedure is zero, then this is an Isolated (or Two-Haystack) problem; as illustrated in Figure 2(a), both procedures should be invoked provided resources are adequate. But if the focal probability is unchanged, this is an Independent problem, as illustrated in Figure 2(b), and it is never optimal to use both procedures at once.

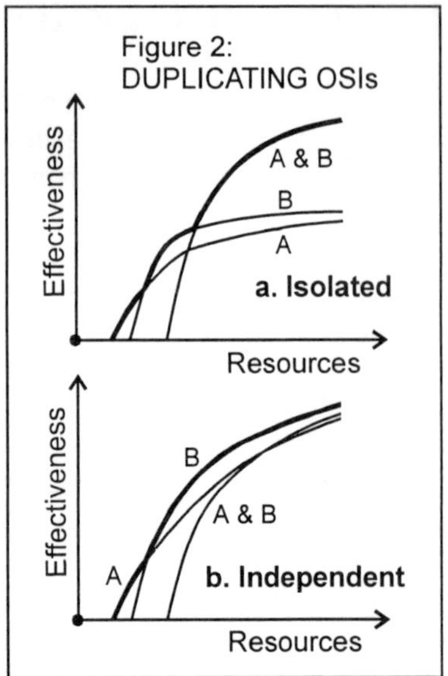


Generalizing from this example, the lesson for OSI seems to be that if a search area is known well enough that it can be subdivided into more-or-less isolated cells (sites, say, or facilities), then small teams should be used to inspect as many cells as possible. But if the search area cannot be subdivided in this way, then all available resources should be applied within a single optimally-designed search team.

# 5 Conclusions

In order to verify UNSCR 687, 707 and 715(1991), UNSCOM and IAEA have improved verification procedures dramatically. Despite financial and time constraints, unforeseen problems, and occasional crises, they used their relative freedom from operating restrictions to find ingenious solutions. The exploitation of synergy across verification techniques was, in many cases, crucial. This article has given an indication of how such synergies work, and why they are important.

Sometime in 1996, UNSCOM and IAEA will focus on the development and integration of the export/import control mechanism that must be in operation before the Security Council can consider lifting the sanctions and embargo on Iraq. OSIs directed by the BMVC, reinforced by aerial surveillance, air inspections, and remote monitoring, will be essential to meet this challenge.

The importance of export/import control in NACD extends well beyond Iraq. Only global mechanisms that are based on well-designed and well-integrated procedures will be able to regulate supplier/customer interfaces and thereby combat proliferation.

## References

1. McFate, P.B., S.N.Graybeal, G.Lindsey, and D.M.Kilgour (1993), "Constraining Proliferation: The Contribution of Verification Synergies," Arms Control Verification Studies No.5, Foreign Affairs and International Trade Canada, Ottawa, Canada.

2. Keeley, J.F.(1995), "Verification, On-Site Inspection, and '93+2'," University of Calgary, Calgary, Canada.

3. United Nations (1995), "Verification in All its Aspects, Including the Role of the United Nations in the Field of Verification,"Report of the Secretary General, UN Document A/50/377.

4. Kilgour, D.M.(1991), "New Research In Arms Control Verification using Decision Theory," Foreign Affairs and International Trade Canada, Ottawa, Canada.

# 3 Verification Synergies

Synergies across techniques are a readily available means of increasing the effectiveness of verification, while simultaneously decreasing its cost. To capitalize on synergies, sequencing and co-ordination are crucial. The usual strategy is to begin with a verification method that makes broad or general observations, identifying as many "potential violation sites" as possible. Then a follow-up at each of these sites can be carried out by a very accurate but expensive method. The process of following-up may have more than two steps; in fact, very complex conditional response sequences are entirely possible.

In this way, synergies has been exploited extensively in the OMV program of IAEA/UNSCOM in Iraq. The mainstay of the program is the use of OSIs cued by aerial surveillance. UNSCOM has a variety of aerial assets, and makes substantial use of them; to date, over 250 missions have been undertaken by the high-altitude surveillance aircraft (U-2) and over 600 by the Baghdad-based Aerial Inspection Team. The CH-53G helicopters provide fast, immediate transportation for inspection teams, without which no-notice inspections would be impossible. They are also used for low-altitude aerial photography, as well as medical evacuation and vehicle airlift. The BMVC has a photographic development laboratory to facilitate rapid processing and review of aerial photographs. This capability has proven crucial to the success of ground OSI teams.

A general notion of how synergies operate can be obtained from Figure 1. Any monitoring procedure can be located on the axes shown--the horizontal axis represents false alarm probability ($\alpha$), and the vertical axis missed violation probability ($\beta$). A good monitoring procedure minimizes both error probabilities, i.e. is located near C. Aerial inspection, which tends to identify actual violations as anomalies, but find many other anomalies besides, is typically located near A. OSI, which rarely identifies violations incorrectly, but may miss some violations because of limited team size, inspection time, etc., is located near B. It can be proven (see [1, Appendix A] and [4]) that the use of a procedure near A to cue a procedure near B produces a combined procedure near C.

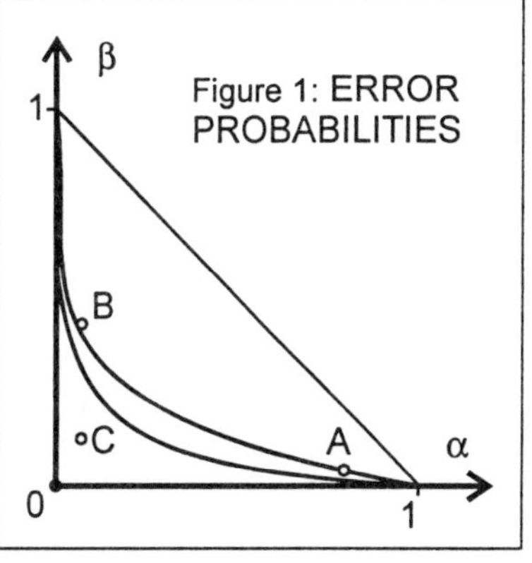

Figure 1: ERROR PROBABILITIES

The combination of aerial and OSIs has been particularly effective when underpinned with a full array of interlocking monitoring activities. The overall system in Iraq has now achieved a level of confidence far beyond what any single element could alone provide. It is without doubt the most comprehensive international monitoring system ever established in the history of arms control.

# Providing Assurance on the Absence of Unknown Activities

Jack L. King
Safeguards Department
International Atomic Energy Agency, Vienna, Austria

## 1.0 Introduction

After the startling discovery of a clandestine nuclear weapons program in Iraq, the effectiveness of the verification methodologies used for international agreements on nuclear non-proliferation was called into review. Traditional techniques such as surveillance, containment, and material accountancy [1] are designed to provide assurance that no nuclear material has been diverted for military use. Clearly such techniques can only be applied to known nuclear activities, where direct measurement and observation is possible.

The goal of IAEA safeguards is to provide assurance on the use of nuclear material to its Member States [2]. For nuclear material associated with undeclared activities such as were present in Iraq, new techniques are needed. One approach has been to use improved information analysis to develop a nuclear profile for each Member State which can provide assurance on the absence of undeclared activities [3]. This profile presents a complete and consistent argument for the belief (or disbelief) in all nuclear activities for each Member State. Where incomplete or inconsistent information exists, appropriate safeguards actions can be undertaken to gather additional information to improve the analysis.

## 2.0 Approach

The impetus of our effort was to develop a tool for providing assurance on the possibility of undeclared nuclear activities. We began with no preconceived notion of probability interpretation or statistical inference, but rather looked for the techniques which we felt most applicable to the problem. The tool should help to determine responsibility and accountability for a Member State's actions regarding its nuclear program and our associated responses in terms of safeguards efforts. As such, its development includes an ethical responsibility since it is to be part of an impartial process equally applied to all Member States. We expressed this requirement as "transparency", using the vernacular of the time. It was not our intention to try to determine the probability of the existence of nuclear proliferation, or to maximise the expected utility of the safeguards effort. The following requirements were established initially and guided the development [3].

1.) a model based on physical processes
2.) an analysis resulting in a fair and impartial application of safeguards
3.) a tool for the analyst

## 3.0 Physical Process Model

The method pursued is aimed at establishing a belief in the possibility of undeclared activities using relevant evidence on each process comprising the activity. Since nuclear activities consist of physical processes, they require certain items in order to be undertaken (equipment and materials) and generate characteristic outputs (products and by-products). If an item is necessarily associated with a process, then it is an indicator for it, and information about the indicator provides indirect evidence on the possibility of the existence of the process. We have therefore compiled a detailed domain model including all known nuclear processes in the nuclear fuel cycle and their indicators.

Since there was little existing theory for assurance on unknown activities, we developed the basis for a theory of evidence for providing assurance [14]. New terminology with associated semantics was developed for indicators, evidence, credibility of the sources of evidence, and belief in the hypothesis of an activity. Much of the nomenclature for indicators was adopted from existing documents regarding nuclear materials and equipment, such as INFCIRC 254 [4]. The development of the physical model, its processes, and relevant indicators culminated in a comprehensive document describing nuclear processes and indicators for them and an associated database.

## 4.0 Dealing with Uncertainty

### 4.1 Ethical Considerations

The method of dealing with uncertainty can have a profound effect on the ethical interpretation of the inference or decision when it implies accountability or liability. Consider the example given by [5] of a rock concert with 1000 attendees, 400 of whom paid and 600 gate crashers. Selecting any one attendee with no other information (i.e. no tickets), their probability of being a gate crasher is clearly 0.6, and if this classical probability is used to make a decision, the individual would be found guilty and forced to pay. However, this situation somehow offends our sense of equity, since we can imagine ourselves as being the suspect and knowing that we had paid the entrance fee. Furthermore, if this decision is allowed, the sponsors would be able to collect from all 1000 attendees, since each would be found guilty using the 0.6 prior probability. Most would find this situation unacceptable and seek some kind of alternate solution. Although the example may seem contrived, the idea of using classical probability and a betting scenario or maximizing utility using Bayesian analysis can also error on the side of efficiency and offend ones sense of fairness [5]. This is especially true if, as was pointed out earlier, the issue is an ethical one, such as the question of treaty compliance with which safeguards is concerned.

Consider also the case of information providing evidence for indicators which comes from a wide variety of sources, such as public media (newspapers and journals), special topical databases, import/export reporting, and inspection reports. They must be examined using the indicators from the physical process model, not through a "data mining" technique. Since the information obtained could be the basis of an inquiry into a Member State's activities, it must be supported ethically by a predetermined list of possible relevant data and not the result of a random discovery. Scientific investigators have long been aware of the ethical responsibility of generating hypotheses first, and then, only after having generated the hypotheses, testing them with data. These ideas are well developed in the ethics literature [6] as the stakeholder view wherein it is held that the process by which the decision is made is more important than finding the "correct" decision.

## 4.2 Structural Issues

Indirect evidence involves uncertainty, since information sources vary in their credibility and some indicators for processes may be present in other (non-nuclear) activities. A newspaper report of special-use high-strength steel has less credibility than an inspection report documenting its presence. But even if information about such an indicator were completely credible, it is still not sufficient to deduce unequivocally the presence of an activity. By rating the amount of evidence each indicator provides for a process we create an inference network for the physical model.

Given the above argument, measures are needed for the amount of assurance which could be provided through various safeguard activities. These measures would represent the belief, disbelief, completeness, and consistency in the nuclear profile. They would enable us to allocate safeguards resources to provide the best efforts for assurance under a given situation. They would also provide a means for evaluating new techniques and alternative methodologies given a specific context. The questions were:

1.) How do we measure uncertainty?
2.) How do we combine (and propagate) uncertainty?
3.) How do we represent the results?

## 4.3 Bayesian Analysis

The classical techniques of Bayesian analysis and utility theory incorporated into them as loss functions [7] was not appealing from an ethical standpoint. Techniques of dealing with uncertainty using causal models [8] were known to us, but the basis for probability we had was the importance of indicators for nuclear processes, not the probability of indicators given the process exists.

Our main problem with classical Bayesian analysis was the prior probabilities which are required to produce useful models. It is simply not possible to assign a prior probability to nuclear proliferation no matter how small it might be or no matter how "fair" in terms of noninformative prior it can be argued. Until the Iraq experience there was no evidence of any proliferation, and lacking any frequentist or classical interpretation for the probability, one is left with a subjective assumption which is hardly defensible in the international community as part of a fair and impartial process.

Another problem we had with the Bayesian approach is the concept of additivity. While it is true that a nation either proliferates or it doesn't, the idea that a piece of evidence can supply information for proliferation, and that the lack of the evidence therefore supplies evidence against proliferation was not workable. Much of the evidence is opportunistic in nature, and the lack of it provides no information at all. Based on our objectives, we needed a measure of the amount of evidence, not a probability of the process existing.

The final problem with the Bayesian approach was that we just don't have sufficient information about conditional probabilities, nor do we have a source from which to draw them. The probabilities available to us are about the import and export of nuclear-related equipment, environmental sampling laboratory analysis results for radio-nuclides, and other measurements which, although related to nuclear proliferation, do not provide any direct measure of the probability of proliferation.

We believe the problem is more of a jurist decision process in which the evidence is accumulated to provide assurance of the declared innocence of a defendant, and with additional evidence gathered to resolve anomalies which may occur. This corresponds more with the ideas of evidence and hypothesis expressed in [5] and [9] with the added condition that the decision is revisited on a periodic basis.

## 4.4 Dempster-Shafer Theory

We found the Dempster-Shafer theory [10] deals with uncertainty in a way which satisfies the ethical responsibility for a fair and impartial process in the application of safeguards to provide assurance on the absence of undeclared activities. This theory provides a systematic method of calculating belief functions using basic probability assignments from a given probability space onto a space where probabilities do not exist using compatibility relations. It has been interpreted as providing upper and lower probabilities, and entropy-like measures have been developed for the basic probability assignments [11].

The use of belief functions allows us to provide useful information about anomalous situations and conflicts in the data we receive. Using this information, we can direct the safeguards efforts in order to increase the level of assurance on an undeclared activity. It overcomes the problems stated above because it 1.) does not require priors, 2.) relaxes the additivity axiom of classical probability, and 3.) allows us to map a probability space (e.g. equipment shipments, laboratory results) onto a space where we have no probabilities (proliferation) using the physical relationships which exist in nuclear processes.

## 4.5 Fuzzy Set Theory

Although fuzzy set theory deals with a basically different kind of uncertainty, we found the fuzzy measures to be applicable in the assessment of evidence. Linguistic variables and their calibration [12] are used in the determination of uncertainty in order to help remove the misrepresentation of accuracy caused by using numbers and propagating them to an arbitrary precision. Of course the assessments are converted to real numbers for combination and propagation, but this conversion is subject to critical review and calibration.

# 5.0 Analytical Tool

The quality of the nuclear profile and the level of assurance provided by it will depend heavily on the work done by the information analyst in response to feedback received. We have used visualization techniques to provide a tool which can help the analyst readily identify conflict in the information (improve consistency) and guide the information search (improve completeness). A graphical user interface provides an overall view, details on demand, and access to an extensive array of help and reference documents [12]. The system is controlled using direct manipulation techniques with essentially no keyboard operations. Iconic measures of belief, disbelief, consistency, and completeness can be displayed for each activity with a variable color intensity representing the strength of the measure for that activity.

In the same manner as fuzzy measures were used in building the inference network for the physical model in order to avoid the misrepresentations of accuracy, visualization allows a limited number of intervals for the representation of the resulting measures. Dynamic queries [12] provide the ability to rapidly change the parameters of the analysis, and are used to present the variability of the information. Figure 1 shows a screen from the current prototype system -- VENAS (Visualization of Evidence on Nuclear Activities). The nuclear activities overview, dynamic query, and HTML client are shown in the upper left, lower left, and right sides respectively. See [3] for a more detailed description of the implementation of VENAS and early experience with the prototype.

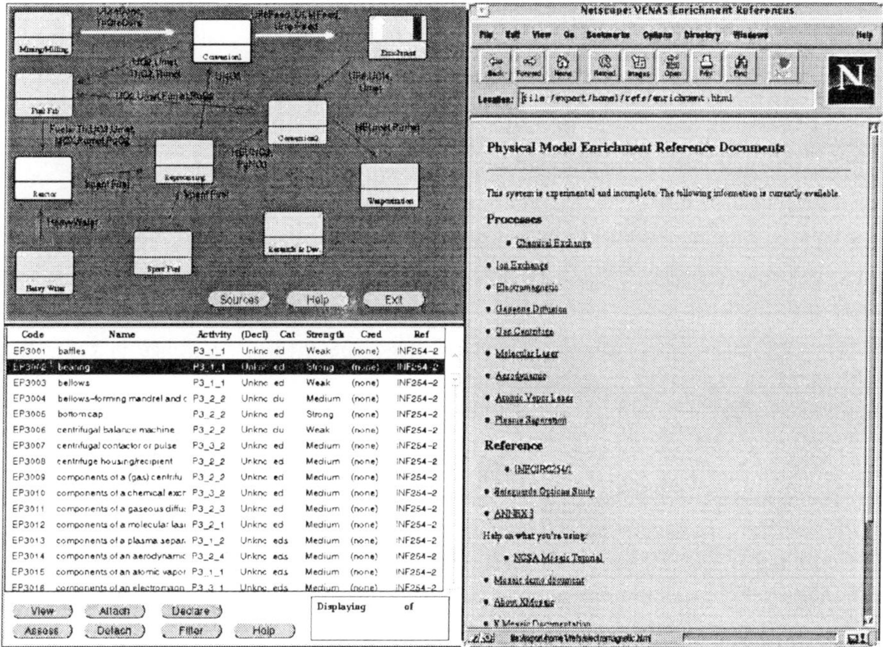


Figure 1. VENAS top level computer screen.

## 6.0 Conclusions

We have developed a general framework and desideratum for a system which can provide assurance on multiple related hypotheses using indirect evidence. This framework includes a domain model, evidence-hypothesis network, uncertainty measures, and visualization techniques. Its design has taken into account the ethical considerations of the IAEA safeguards mission, and we believe it can serve as an effective tool for providing assurance on the absence of undeclared nuclear activities. Our implementation of this approach (VENAS) is currently under review and we expect scenario and usability testing to take place this year.

## References

1. IAEA *Safeguards Manual of Operations*, Vienna, Austria, (internal document), 1990

2. Avenhaus, R. *Material Accountability: Theory, Verification, and Applications.* Wiley, Chichester, 1977

3. King, J. L. Improved Analysis of Information on States' Nuclear Activities. Proceedings of the 35th Annual Meeting of the INMM, Naples, Fla., July 199414

4. IAEA *Guidelines for Transfers of Nuclear-Related Dual-Use Equipment, Material and Related Technology.* INFCIRC/254/Rev. 1/Part 2., Vienna, Austria, rev. 1992.

5. Cohen, L. J. *The Probable and the Provable.* Ashgate Publishing Co., Brookfield, Vermont, 1991

6. Mason, R.O. Mason, F.M., and Culnam, M.J. *Ethics of Information Management.* Sage, Thousand Oaks, Calif., 1995.

7. Berger, J.O. *Statistical Decision Theory and Bayesian Analysis*, 2 ed. Springer-Verlag, New York, 1993

8. Pearl, J. *Probabilistic Reasoning in Intelligent Systems: Networks of Plausible Inference.* 2nd ed. Morgan Kaufman, San Mateo, Cal., 1988

9. Schum, D. A. *Evidential Foundations of Probabilistic Reasoning.* Wiley, New York, 1994

10. Shafer, G. *A Mathematical Theory of Evidence.* Princeton University Press, Princeton, 1976

11. Klir, G. Measures of uncertainty in the Dempster-Shafer theory of evidence, in Yager, R. Fedrizzi, M. and Kacprzyk, J. Advances in the Dempster-Shafer Theory of Evidence. Wiley, New York, 1994

12. Zadeh, L.A. A computational approach to fuzzy quantifiers in natural languages. *Computers and Mathematics* 1983; 9:149-184

13. Shneiderman, B. *Designing the User Interface: Strategies for Effective Human-Computer Interaction.* Addison-Wesley, Menlo Park, 1987

14. King, J.L. A Theory of Evidence for Undeclared Nuclear Activities. (to be published in *Proceedings of the 36th Annual Meeting of the INMM.* Palm Desert, CA, Jul 1995.

# Strategy Development, Risk Assessment and Resource Management: Some Issues Related to the Implementation of the Verification Regime of the Chemical Weapons Convention

John Gee, Ralf Trapp and Maurizio Barbeschi

Provisional Technical Secretariat of the Preparatory Commission for the OPCW
Den Haag, The Netherlands

**Abstract**

The Chemical Weapons Convention (CWC), expected to enter into force in 1996, poses a number of new conceptual problems. It establishes a number of verification sub-regimes, only some of which follow conventional, well-tested concepts. Other verification sub-regimes defy simple analogy with methodologies developed for bilateral arms control accords or nuclear safeguarding. This paper analyses the two following aspects of verification under the CWC: routine inspection of the chemical industry and challenge inspection. Routine inspection of the chemical industry poses the problem of managing limited inspection resources against a very large number of inspectable facilities. We argue that, instead of aiming at the impossible (high detection probabilities and short prewarning times), building and maintaining confidence in the *verification strategy* will be the key to the credibility of the verification regime. At the same time, routine inspection, while using a different procedural framework, will have to be conducted in such a way that the ability to implement a credible *challenge inspection* is retained. The definition of the *strategic intent* of the Technical Secretariat will be critical if the conflict between these potentially conflicting roles is to be avoided.

## 1 Introduction

The Chemical Weapons Convention (CWC) is a unique disarmament agreement containing the most extensive and intrusive verification provisions of any multilateral disarmament or arms control agreement concluded thus far. It will pose a number of novel problems for the concept of verification, a concept developed in the past around either bilateral/regional security frameworks or multilateral agreements of more limited scope. Most of the work done so far in relation to the development of the CWC verification regime, during the negotiations in the Conference on Disarmament in Geneva as well as in the work of the Preparatory Commission, relates to the *procedural* aspect of verification. To the extent that problems of verification methodology and efficiency have been addressed, the focus has often been on the individual inspection. With the entry into force (EIF) of the Convention approaching, more information, if still incomplete, has become available that now allows an assessment of verification efficiency. At the same time, extrapolations of the numbers of inspectable facilities are becoming more reliable, and assessments can be undertaken against the boundary conditions largely set by the Preparatory Commission.[1]

The building of confidence in a regime's stability is the crux of all disarmament treaties. A major element in this confidence-building process is the establishment of confidence in the *verification regime* itself. To pursue this objective and to maintain that confidence can be considered as the OPCW's *strategic intent*. For this, the intrinsically dualistic nature - Technical/Quantifiable (**T**-dimension) and Political/Managerial (**P**-dimension) - of the verification process has to be borne in mind.

While the **T**-dimension dominates the CW-related verification regimes of the CWC, awareness of the P-dimension becomes essential for designing strategies for the Convention's other sub-regimes. In fact, for routine inspections of the chemical industry and for challenge inspections, it can be argued that a substantially new concept is needed: these verification sub-regimes cannot/will not rely on attempts to *systematically* verify all *potential* capabilities. Being predominately qualitative in nature, they establish *non-systematic* inspection regimes sufficiently intrusive to demonstrate any significant violation if targeted against the correct locations. Their effectiveness rests on the deterrence value provided by the intrusiveness of the individual inspection as much as on the confidence of the States Parties in the ability of the Technical Secretariat adequately to distribute its inspection resources in time and space. Consequently, verification *strategy* becomes crucial for the realisation of the CWC's objectives.

## 2 Routine Verification Strategy and Risk Assessment

The CW-related verification sub-regimes represent a well-defined, easily quantifiable problem. The inspection strategy for these sub-regimes can be approached by any of the resource optimisation methodologies developed for nuclear safeguarding or bilateral arms control accords, which build on accountancy methodologies. CW verification under the CWC is conceptually not a "new" problem. Some traditional

---

1 While the Preparatory Commission took account of the relationship between resources and inspection effort in its internal planning, other decisions on related issues such as the size of the OPCW building, the number of inspectors to be recruited and trained, the amount of inspection equipment to be procured, and so on, have effectively imposed constraints on the size of the future verification structure, quite irrespective of any changes in the initial assumptions in relation to actual requirements.

methodologies for resource optimisation are: queuing theory, management science algorithms and linear programming, and game theory.

Routine inspection of the chemical industry under the CWC will pose different problems. During the initial three years after entry into force, the inspection planning processes will be largely driven by explicit Convention requirements on the timing of initial inspections. Subsequently, inspections will be conducted at so-called Schedule 2/3 plant sites and, at a later stage, at DOC plant sites in accordance with some sort of qualified random selection process constrained by explicit Convention restrictions on the maximum number of inspections, as well as by resources.

Beyond the initial phase, the Convention requires the planning of inspections to be *risk-based*. For Schedule 2/3 facilities, the selection process is based on what have been called "weighing factors". Risk assessment methodology will thus be of critical importance for supporting resource optimisation management and for demonstrating the effectiveness of the verification regime at large.

These risk assessments will be conducted at the Technical Secretariat, on the basis of the information available on the declared sites, including declarations, as well as reports from initial and any subsequent inspections. Two points need to be stressed:

(a) the information, on the basis of which risk assessments are to be conducted, will differ for the different verification sub-regimes of the Convention. Even within the same sub-regime the information basis may not be the same for different facilities. Risk assessments for verification purposes may, consequently, have to be conducted against what is potentially an unequal information base for different facilities;

(b) the quality of the information available for the risk assessments will occasionally be uncertain. For example, it may depend on the interaction between inspection teams and inspected States Parties during the initial inspections. Also, different States Parties may take different attitudes in relation to how generous they are in providing information during inspections. This problem is of particular significance, as differences in approaching plant site/plant relationships may generate a non-homogeneous data set.[2] The Technical Secretariat will have to provide for consistency in its risk assessment process notwithstanding such fluctuations.

Ironically, for the largest category of inspectable facilities, i.e. those covered under Part IX of the Verification Annex ("DOC/PSF plant sites"), the quantum of information available per facility will be least. While this category of facility is often mistakenly assumed to present the lowest risk to the object and purpose of the Convention, the key difference between these and other declarable industrial facilities

---

2 The basis for the risk assessment of Schedule 2 (3) facilities is the plant site as a whole. Unimpeded access is, however, limited to the declared Schedule 2(3) plants within the plant site, while any further access into the plant site would be subject to the clarification procedure under paragraph 51 of Part II of the Verification Annex, and can be managed by the inspected State Party. The degree of access into the rest of the plant site is thus discretionary and depends on the attitude taken by the inspected State Party as much as on the negotiation skills of the inspection team.

is that, while the chemical(s) manufactured at these sites are of no particular concern to the Convention, the technological base of these facilities could provide a stand-by CW capability. A comparatively low risk posed by the average individual facility of this category thus turns into a substantially different risk perception for the "odd one out". The challenge for any verification regime, let alone for one confined by resource limitations and procedural constraints, results from the sheer size of this facility category (countable in tens of thousands).

Under such non-homogeneous conditions, the crux of the matter lies in the Technical Secretariat's ability in rapidly to evaluate potentially incoherent information and to plan its inspections in an uncertain environment. The credibility and effectiveness of the routine verification regime will depend critically on the approaches taken by the Technical Secretariat in relation to resource utilisation. Verification strategy design, development and execution, together with risk assessment methodologies, will be paramount for a credible CWC verification regime.

## 3 Knowledge management versus process management

In the case of the CW-related, Schedule 1 and Schedule 2 sub-regimes, the requirement for a "*knowledge* management" tool within the Technical Secretariat is apparent. The sub-regime in place for Schedule 3 and, later, for DOC/PSF facilities, by comparison, requires a sophisticated "*process* management" based on the continuous monitoring of the sources of information and the development of evaluation tools for cost-benefit analyses related to the value of each individual item of information.[3]

Economists and strategists have developed a clear understanding of the principal difference between these two managerial concepts: the impact of the time variable on each phenomenon. "Process management" implies a linear time dependence. The complexity is mainly due to the large number of "events" which are considered unrelated. When *knowledge* is the managerial objective, the impact of time on the phenomenon is non-linear (in fact, "accumulation steps" may well become critical). The time-relationship among events may result in a derivative or even non-stochastic impact. Therefore, in most cases, it is not possible to predict *ex ante* (i) when a single event will occur, and (ii) how and when all events will mutually influence each other. Thus, the crucial factor to manage and to optimise is "learning". This learning process is crucial for the building of regime confidence, and the

---

3 This assumes that the Technical Secretariat will have to select inspectable facilities under this sub-regime predominantly on a random basis. It should be noted here that for DOC facilities, paragraph 11 (c) of Part IX of the Verification Annex establishes the option of selections based on State Party proposals. If implemented, that would (a) link the selection process, at least in part, to *actual* security perceptions of States Parties and would thus overcome the dilemma inherent in the large number of facilities, and would (b) re-establish the requirement of knowledge management, for risk assessments *during* individual inspections.

OPCW's "knowledge management" process can be depicted in terms of this learning process.

"Improving decision making" in risk assessment conduct can be approached through:

(a) an improvement of the decision itself (the *capability* to improve in the case of "process management"); and

(b) an improvement in the decision-making process (the *core competence* needed for establishing a "knowledge management" infrastructure)[4].

This consideration leads to the following formal expression, taking account of both the above aspects of decision improvement:[5]

$$|N(I,t)| = \Sigma_i \, n_i(I,t) = \xi_1(t)^*\Sigma_i \, U_i(I) + \xi_2(t)^*\Sigma_i \, \{V[T(I)] - C(I)\}_i$$

where:

I = single information
N or n (I,t) = net value of I (absolute value)
C (I) = cost of providing I
T (I) = time and effort saved if I is used
V [T(I)] = "opportunity" value
U (I) = value of I to the decision
$\xi_{1,2}(t)$ = relational factors (empirical constants, time function)

Accordingly, a resources optimisation (i.e. linear programming) capability becomes less and less reliable as the values of the relational factors $\xi_{1,2}(t)$ increase. Operationally, the risk assessment has to be performed at the Technical Secretariat's headquarters by the same personnel also in charge of long-term inspection planning in order to maximise the strategic understanding of the links between single inspections. And, as the regime's stringency increases from DOCs/Schedule 3 to Schedule 2 and Schedule 1/CW, the increasingly dominant question for regime-confidence-building is "how good will the Technical Secretariat be in strategically planning routine inspections, thereby enhancing its core competence?"

---

4 The term *capability* is used in the economic sciences to indicate "a set of skills mandatory to master a technicality/procedure" while a *core competence* requires long-term knowledge accumulation and implies some relational elements embedded in the decision-making process.

5 "Method (a)" is dominant for Schedule 3/DOCs-PSFs verification strategies (quasi-random selection of the facility, "symbolic" regime, poor quality and disaggregated information: the relational factor $\xi \Rightarrow 0$), while both of these relational factors substantially drive the Schedule 2 regime strategy. The CW sub-regimes are predominately driven by "method (b)".

# 4 Challenge Inspection and Chaos Management Theory

Challenge inspection is often characterised as the most revolutionary element in the CWC verification regime, and as the cornerstone of its deterrence effect. In terms of effectiveness assessment, it poses problems quite unlike those presented by the routine verification regime. One of the main reasons is that the initiator of a challenge inspection is different from its executor. The effectiveness of challenge inspection will mostly be determined by the following two factors:

(a) the likelihood that a State Party, having gathered intelligence about possible non-compliance, will actually resort to challenge inspection; and

(b) the capability of an international inspection to gather evidence that would allow a definitive judgement to be made concerning compliance.

The challenge inspection device is most critical for the credibility of the verification regime of the Convention as a whole. However, assessing the effectiveness of verification strategy and resources management optimisation by the Technical Secretariat is complicated by the fact that the target selection and the aims of each challenge inspection are left to the requesting State Party. The Technical Secretariat's function is largely reduced to execution of the inspection itself. Competence does not relate to any long-term performance of the Technical Secretariat other than in relation to retaining the *capability* to mount a successful challenge inspection at short notice. Strategy and risk assessment become (tactical) elements of individual inspection planning/conduct rather than of developing a credible inspection pattern. Regime confidence is further determined by the management of individual inspections, for example through the history of team/team leader selection, appropriate time-management, support provided to the team from headquarters, the processing of the information gathered *in situ*, and so on.

In an actual challenge inspection, the main aspect under observation by States Parties will be the way in which the Technical Secretariat applies its core competence to ensuring the proper execution of the particular inspection in question. Optimum use of Technical Secretariat expertise, in its broad exegesis, will drive the "credibility" of the whole verification castle. In essence, the question is not one of developing a strategy to find a resource utilisation *optimum* over time and space, but one of swift, credible and effective resource re-allocation as a reaction to an unexpected event. From the point of view of the long-term *modus operandi*, a challenge inspection is a something of an anomaly, which reduces the effectiveness of the resource optimisation planning. At the same time, any substantial failure to execute it in a credible manner will compromise the system as a whole.

A crucial problem is where and to what extent the Technical Secretariat will retain its ability to react in a timely and credible manner to a challenge inspection request when its resources are routinely used in a fundamentally different inspection mode. The option of having "specialised" challenge inspection team core constituents is limited by resource constraints and the Convention's requirement that inspectors designated for challenge inspection need to be drawn from amongst inspectors involved in routine inspection. In other words, the OPCW strategy makers face the potential dilemma that:

(a) the most intricate inspection type needs to be staffed with personnel that on a daily basis conducts routine inspections, and

(b) that such human resources are "trained by conduct" to operate in an environment which will in the main be more predictable than that obtaining in a challenge inspection.

Thus, in an inspection scenario where flexibility, creativity and negotiating ability are needed most, the Secretariat will need to rely on staff that the system itself has a tendency to bias towards routine behaviour in foreseen circumstances.

That dilemma poses serious conceptual and operational problems for the Technical Secretariat. On the strategic side, the routine inspection effort will have to be geared, to the extent, feasible in such a way that the ability to implement a credible challenge inspection is retained. To attain this principles from "chaos management theory" could, for example, be used by the Technical Secretariat, coupling mathematical consistency with managerial practices. Chaos management theory was originally used by large, transnational petrochemical corporations as well as by venture capital enterprises. It is aimed at providing those organisations with (i) a day-by-day managerial infrastructure for, *inter alia*, the optimisation of resources *and* (ii) at retaining the capability of implementing contingency plans whenever a large and unpredictable (in time and space) event occurs, without disrupting the entire potential of the managerial infrastructure.

The crux of the methodology lies in the acute management of "buffer" resources and in their "marked" planning.[6] A mixture of skills, equipment and *ad hoc* training coupled with a system of "just-in-time" resource monitoring has to be developed[7]. This maintaining of a "virtual inspection team" will be utilised for routine inspection with an eye on the "critical time" that could be tolerated for assembling a challenge inspection team if and when necessary. The "critical time" is a key managerial decision with a high political overlay, and becomes a regime attribute of fundamental importance for the regime's credibility.

The definition of the *strategic intent* (very long-term set of objectives for non-compliance deterrence and detection) becomes critically important. This strategic intent has to be agreed by all Member States, as the basis for the Secretariat to be able to define and build its *core competence*: by managing routine verification resource optimisation while maintaining its inspection (including challenge inspection) *capability*.

---

6 "Buffer" must not be confused with "idle".

7 "Just-in-time" techniques, originally introduced by some Japanese enterprises, are nowadays widely used to deal with those cases presenting the time variable as the driving force of the competitive advantage.

# Accident Sequence Precursor Lessons for PRA

J.W. Minarick
Science Applications International Corporation
Oak Ridge, TN U.S.A.

**Abstract**

A reduction in the incidence of unusual initiating events and total system failures observed in the Accident Sequence Precursor program indicates an improvement in the risk-related performance of U.S. commercial nuclear power plants during 1969-94. However, a number of historic system failures are inconsistent with PRA models, an indication of the need for more careful consideration of operational events when developing these models.

## 1 Introduction

The Accident Sequence Precursor (ASP) Program, conducted for the U.S. Nuclear Regulatory Commission by Oak Ridge National Laboratory with support from Science Applications International Corporation, analyzes operational events that have occurred at commercial light-water reactors (LWRs) in the United States in order to identify and categorize precursors to potential severe core damage accident sequences.

Accident sequences of interest in the program are those that, if sufficient failures occur, can result in inadequate core cooling and potentially lead to core damage. Accident sequence precursors are important elements in such sequences, such as an infrequent initiating event or concurrent equipment failures. Approximately 590 precursors from 1969-81 and 1984-94 [1-6] have been documented.

This paper examines changes in the incidence of certain precursors and draws conclusions related to changes in risk-related plant performance over the 1969-94 time period. Analysis of 1982-83 events for precursors, initially deferred, was scheduled for completion before the end of 1995. Unfortunately, since this effort has been delayed, the results presented must be considered preliminary.

## 2 Approach

Two consistently identified groups of precursors are unexpected initiating events and failures-on-demand of safety-related systems. Changes in the incidence rates for these two groups of events are observed in the ASP data. Since the frequency of these initiating events and the failure probabilities for these systems contribute to the frequency of many core damage sequences, changes in their incidence may be used to infer changes in risk-related plant performance.

The nature of the two groups of precursors facilitates estimation of their occurrence rates. Changes in the occurrence rate for initiating events can be estimated by comparing the number of such events over fixed reactor-year periods. Changes in failure-on-demand probabilities of safety-related systems can be similarly compared, provided the number of challenges can be determined. For most systems, the number of system-level demands is dominated by the number of surveillance tests. The frequency of such tests is specified in the plant Technical Specifications, and it is assumed that the number of demands over fixed reactor-year periods has remained reasonably constant. Some systems, particularly auxiliary feedwater (AFW) in pressurized water reactors (PWRs) and high-pressure coolant injection (HPCI)/reactor core isolation cooling (RCIC)[1] in boiling water reactors (BWRs), are also demanded after many reactor trips, and the number of additional demands due to reactor trips must also be considered.

Differences in the models used in different time periods prevent comparison of the *significance* of events identified in the yearly precursor assessments without reconciliation of differences, but do not impact a comparison of changes in the incidence of initiating events and total system failures.

## 2.1 Identification of Constant Reactor-Year Periods

The dates of initial criticality and, when applicable, final shutdown for all U.S. commercial LWRs were used to calculate the cumulative number of reactor years. Six reasonably equal periods between 1969-81 and 1984-94 were defined in which the observed number of unexpected initiating events and failures-on-demand of safety-related systems were compared [Table 1]. Because the fraction of BWRs and PWRs has remained relatively constant since the mid-1970s, these periods are also reasonably constant for BWRs and PWRs, except for the first period, where the percentage of BWR reactor-years is greater.

**Table 1. Constant Reactor-Year Periods in 1969-81 and 1984-94**

| Period | Date Range | Reactor Years<br>LWRs | PWRs | BWRs |
|---|---|---|---|---|
| 1 | Jan 1, 1969 - Jun 30, 1977 | 271 (.16*) | 144 (.13) | 127 (.20) |
| 2 | Jul 1, 1977 - Dec 31, 1981 | 295 (.17) | 187 (.17) | 108 (.17) |
| 3 | Jan 1, 1984 - Mar 31, 1987 | 295 (.17) | 190 (.18) | 105 (.17) |
| 4 | Apr 1, 1987 - Dec 31, 1989 | 295 (.17) | 194 (.18) | 102 (.16) |
| 5 | Jan 1, 1990 - Jun 30, 1992 | 279 (.16) | 184 (.17) | 95 (.15) |
| 6 | Jul 1, 1992 - Dec 31, 1994 | 279 (.16) | 184 (.17) | 95 (.15) |

* Fraction of the total number of reactor-years

[1] These single-train BWR systems are considered one redundant injection system for the purposes of this paper.

## 2.2 Precursors Involving Unusual Initiating Events and Total System Failures-on-Demand

All precursors were reviewed and those that occurred at power (or could have impacted a reactor trip from power) and involved the following initiating events or total failures-on-demand[2] (failures associated with a trip or surveillance test) were identified:

- loss of offsite power (LOOP) or small-break loss-of-coolant accident (LOCA),
- emergency power (EP), AFW, high-pressure injection (HPI), or high pressure recirculation (HPR) in PWRs, and
- EP, HPCI/RCIC, or the automatic depressurization system (ADS) in BWRs.

These initiating events and system failures were chosen because (1) they have been observed multiple times within the 1969-81 and 1984-94 time periods and (2) they are potentially associated with almost all plants (a few early BWRs do not have HPCI and RCIC). Changes in occurrence rates for these events can be observed and compared without further subdivision of the six time periods.

Initiating events and failures-on-demand identified[3] in the six reactor-year periods are shown in Table 2. These events include closure of the refueling water tank valves at Maine Yankee, incorrect reduction of HPI flow at Three Mile Island 2, and failure of two relief valves at Harris, all of which resulted in failure of HPI.

There has been a general reduction over time in the number of failures-on-demand and small-break LOCAs, particularly when compared to the July 1, 1977 - December 31, 1981 period. With one exception, PWR EP, this period includes the greatest number of failures for each initiating event and system studied.

## 2.3 Tests for Reduced Incidence of Initiating Events and System Failures

To determine if the reduction in the number of events in later years was anything but random, two statistical tests were employed. First, a Run Test was used to determine if the number of initiators and system failures observed in each period exhibited significant nonrandomness. This test [7] compares fluctuations in the sequence of observed events above and below the median value with the expected number of such fluctuations if the sequence was the result of random variation. The incidence of both PWR and BWR small-break LOCAs and AFW and HPCI/RCIC failures can be shown to be non-constant (decreasing) at the 90% confidence level.

The second test compared those initiating events and systems that exhibited a reduced number of occurrences or failures in the later three periods (April 1987-1994) with the expected number if no change occurred in frequencies or failure probabilities.

---

[2] The failures-on-demand addressed in this analysis would have impacted plant response to a variety of initiating events. Failures that impact plant response to only one initiating event are occasionally observed, but are not addressed.

[3] While many of the events considered in this paper were potentially recoverable, for example, by manually starting initially failed equipment, recovery was not addressed in the analysis; only the incidence of initial failures is compared.

**Table 2. Initiating Events and Total System Failures by Period**

| Initiating Event/System | Period 1 | 2 | 3 | 4 | 5 | 6 |
|---|---|---|---|---|---|---|
| PWR LOOP | 7 | 17 | 7 | 11 | 5 | 6 |
| PWR Small-break LOCA | 2 | 6 | 2 | 1 | 0 | 1 |
| PWR EP | 3 | 0 | 1 | 3 | 1 | 0 |
| AFW | 4 | 4 | 3 | 1 | 0 | 1 |
| HPI | 1 | 3 | 0 | 0 | 1 | 0 |
| HPR (given successful HPI) | 1 | 0 | 0 | 0 | 1 | 0 |
| BWR LOOP | 3 | 4 | 4 | 2 | 4 | 2 |
| BWR Small-break LOCA | 2 | 5 | 2 | 1 | 1 | 0 |
| BWR EP | 2 | 6 | 0 | 2 | 1 | 0 |
| HPCI/RCIC | 3 | 7 | 3 | 0 | 2 | 0 |
| ADS | 3 | 0 | 0 | 0 | 0 | 0 |

For the eleven initiating events and systems considered, 5.5, on average, would be expected to show a higher number of occurrences and failures in the later three periods, compared with the first three, if the failure rates were constant. Nine initiating events and systems exhibited a reduced number of occurrences and failures, while two systems (PWR EP and HPR) had the same number of failures in the first three and the last three periods (there were no increases in the number of failures in the last three periods). Using a Chi-square test [7], an assumption that the observed reduction in the number of events is the result of random fluctuations is rejected at the 95% confidence level [comparison of the number of events in the middle (1984-89) and last two periods (1990-94), results in a similar conclusion, but only at the 80% level].

# 3 Results

These two tests demonstrate that the reduction in the number of initiating events and total system failures-on-demand listed in Table 2 is most likely not a result of random fluctuations in the data, but is instead an indication of improved performance during the 1969-94 period. (Changes in operating procedures plus the recognition that other systems can provide core protection are expected to further improve risk-related plant performance. However, the impact of these changes cannot be readily discerned in the precursor data.)

Retrospective industry-wide frequencies and system failure-on-demand probabilities for 1984-94 were estimated from the data in Tables 1 and 2 [Table 3]. The 1984-94 time interval was chosen as indicative of current industry performance while providing a long observation period with a consistent data set. This is the longest period in which no substantial change is discerned in the ASP data (see Section 2.3).

**Table 3. Industry-average Initiating Event Frequencies and System Failure Probabilities Developed from 1984-94 Precursors**

| Initiating Event/System | Frequency/Probability[4] |
|---|---|
| PWR LOOP | 3.9E-2/yr |
| PWR Small-break LOCA | 5.3E-3/yr |
| PWR EP | 4.4E-4 |
| AFW | 3.4E-4 |
| HPI | 1.1E-4 |
| HPR (given successful HPI) | 1.1E-4 |
| | |
| BWR LOOP | 3.0E-2/yr |
| BWR Small-break LOCA | 1.0E-2/yr |
| BWR EP | 6.3E-4 |
| HPCI/RCIC | 9.9E-4 |
| ADS | no failures observed |

Much of the data in Table 3 is consistent with estimates developed in contemporary PRAs; the frequencies of LOOP and PWR small-break LOCA are, in fact, lower. This is considerably different from the results of the early precursor assessments [1 and 2], in which frequencies and failure probabilities greater than those predicted in PRA were observed.

Combinations of the initiating events and system failures in Table 2 comprise important core damage accident sequences, e.g. losses of secondary-side cooling and feed and bleed in PWRs (loss of feedwater,[5] AFW, and HPI), station blackout in BWRs (LOOP, EP, and HPCI/RCIC),[6] and small-break LOCAs with failure of reactor coolant makeup in both PWRs (LOCA and HPI) and BWRs (LOCA and HPCI/RCIC). Given the reduction in initiating events and system failures in the later years of the 1969-94 period, one may conclude that a reduction in the frequency of these sequences also occured.

An earlier qualitative comparison of important 1969-81 precursors with those that occurred in 1984-86 [8] showed the serious 1984-86 precursors were more consistent with events typically modeled in PRAs than was the case in 1969-81. Complicated

[4] Twelve surveillance test demands/reactor-year were assumed for system failures, except for AFW (12 surveillance tests + 3.7 shutdowns/reactor-year) and HPCI/RCIC (12 surveillance tests + 0.7 losses of feedwater/reactor-year).

[5] Incidences of loss of feedwater, the initiating event for this sequence, are not consistently identified in the ASP program. However, other studies have indicated a significant reduction in the number of such events since the 1970s.

[6] Station blackout in PWRs involves failure of only part of the AFW system (turbine-driven pump train). These failures are not consistently identified in the ASP program.

events involving electric power and instrumentation interactions were not seen in 1984-86 nearly to the extent they were in 1969-81, and both AFW and HPCI/RCIC system performance appeared to have improved.

These observations are also applicable to the period following 1986. Both AFW and HPCI/RCIC performance appear to have improved, small-break LOCA frequencies decreased, and complex, system-interaction events were infrequently seen.

While system failures probabilities developed from precursors are consistent with PRA estimates, many of the system failure causes remain inconsistent with those modeled in PRAs; the HPI and HPR failures that have occurred since 1981 were both the result of a common-cause failure of two similar relief valves—a failure mechanism that is not typically modeled.

PRA models still suffer from a lack of comparison with historically observed failures. Confidence must exist that the models are representative of actual performance if they are to be used for risk-based decision making. This can only be achieved by comparison with actual operation as documented in the ASP program.

## References

1. Minarick J and Kukielka C. Precursors to potential severe core damage accident sequences: 1969-1979, a status report. NUREG/CR-2497, Vols. 1 and 2, Oak Ridge National Laboratory, 1982
2. Cottrell W, Minarick J, *et al.*. Precursors to potential severe core damage accidents: 1980-1981, a status report. NUREG/CR-3591, Vols. 1 and 2, Oak Ridge National Laboratory, 1984
3. Phung D, Minarick J, Harris J. A review of comments on the 1969-1979 accident sequence precursor program status report: NUREG/CR-2497. ORNL/NRC/LTR-85/14. Oak Ridge National Laboratory, 1985
4. Minarick J, Cletcher J, *et al.*. Precursors to potential severe core damage accidents: 1985, a status report. NUREG/CR-4674, Vols. 1 and 2, Oak Ridge National Laboratory, 1986 and subsequent reports for 1984 and 1986-91 (Vols. 3-16)
5. Copinger D, Cletcher J, *et al.* Precursors to potential severe core damage accidents: 1992, a status report. NUREG/CR-4674, Vols. 17 and 18, Oak Ridge National Laboratory, 1993
6. Vanden Heuvel L, Cletcher J, *et al.* Precursors to potential severe core damage accidents: 1993, a status report. NUREG/CR-4674, Vols. 19 and 20, Oak Ridge National Laboratory, 1994 and a subsequent report for 1994 (Vols. 21 and 22, 1995)
7. Guttman I and Wilks S. Introductory engineering statistics. Wiley, New York, 1965
8. Minarick J. The US NRC accident sequence precursor program: present methods and findings. Reliability engineering and system safety 1990; 27: 23-51

# Novel Designs, How Do We Obtain Reliability Data?

Odd Fagerjord, Ms. Sc.
Det Norske Veritas
Høvik, Norway

**Abstract**

Presentation of results from a data survey after reliability data for valve-trees on subsea completed wells for oil and gas production. The data have been modified to be usable for special equipment which are constructed somewhat different from the equipment behind the generic data.

## 1. Introduction

Components used subsea reveal that they are more reliable than when used above water. There may be several reasons for this. First of all it is time consuming and very expensive to perform maintenance and repair subsea. Therefore great effort and costs have been put into testing and refinements of the design to make it as reliable as possible. Due to the high repair costs one will also try to keep the system "running" even in degraded performance condition.

When first "diving subsea" with the equipment one tended to use the failure rates as experienced from similar equipment used topside at the platforms. A typical failure rate (critical failure) for only one of the 4 to 6 valves on a valve-tree topside is according to ref. 1 in the order of 30 failures per $10^6$ hours or about 2.3 years mean time to failure. Due to supposed improvements and some subsea experience, MTTF values in the range 10 to 20 years have been typical values used for the entire valve-tree when evaluating subsea production reliability etc. Experience so far from subsea oil and gas production in Norwegian waters indicates that a subsea valve tree performs better than this too.

### 1.1 Acknowledgement

Most of the data presented in this paper has been obtained in a study sponsored by Norsk Hydro. We like to thank Norsk Hydro for their willingness to let us use data from this report (ref. 2).

## 2. Subsea Reliability Data

In order to improve the basis for making decisions with regard to e.g. engagement of maintenance resources (rigs etc.), and having indications that the designs used in the North Sea have better reliability than average, Norsk Hydro[1] took the initiative to collect experience data from subsea production in the Norwegian waters, ref. 2.

Totally the survey did not cover more than 167.5 well-years (which represented almost all the Norwegian subsea completed wells to that date) and very few failures as shown below:

**Table 2-1 Number of failures on subsea valve-trees installed in Norwegian waters**

| Equipment Failure | Degraded | Critical | Critical failures per mill. calendar-hours |
|---|---|---|---|
| Downhole failures | 1 | 1 | 0.68 |
| Downhole Safety Valve | 0 | 3 | 2.05 |
| Wellhead | 2 | 0 | - |
| Control pipings and seals | 7 | 1 | 0.68 |
| Controls, electrical | 0 | 0 | - |
| Valve Tree | 0 | 1 | 0.68 |

This data material is too small to make predictions with regard to distribution of various failure modes and failure mechanisms for individual equipment. In particular it was of interest to obtain failure rates for in-depth failure modes for subsea valves. Therefore the Norwegian data were merged with a larger database , ref. 3. This database gives statistics on interventions and not necessarily failures, but after a review of each intervention, the number of failures as shown in Table 2-2, were found for a subsea valve tree. As can be seen there it is almost no difference in the valve tree failure rate obtained from Norwegian waters compared with the world-wide data. None of the world-wide (W/W) data were recorded on Norwegian wells, and the two databases could be "merged" directly without counting a failure twice. I. e. the failure rates were derived as follows: $f=(n1+n2)/(T1+T2)$, where n is the number of failures from data-group 1 or 2, and T is the exposure time for group 1 or 2.[2] . Thus the data-material contains 8 critical failures on a well valve-tree for a total of 1029.5 well-years calendar time. The merged failure rate, $0.9 \cdot 10^{-6}$/h, for the two datgroups is recommended to be used for the valve tree.

---

[1] Norsk Hydro is one of the larger oil and gas operators at the Norwgian shelf

[2] Actually this is identical to a Bayes approximation to modify the Norwegian data with the w/w data as priori distribution.

**Table 2-2 Number of failures on subsea valve-trees recorded world-wide**

| Failure | Number of failures | | Distribution |
|---|---|---|---|
| | Degraded | Critical | |
| Tree Cap Leak | 3 | 0 | 0 |
| Tree seals leak (external) | 0 | 1 | 14% |
| Valve actuator failure | 1 | 5 | 72% |
| Valve leakage (internal) | 0 | 1 | 14% |
| Valve corrosion | 1 | 0 | 0 |
| Total | 5 | 7 | |
| Failure rate (per mill h. calendar time) | 0.68 | 0.93 | |

# 3. Break down of total failure rate into component failure rates

Almost all well valve-trees consist of minimum 4 operable emergency shut down valves of the "gate" type. It is therefore reasonable to estimate that the failure rate on an individual valve is less than 1/4 of the failure rate of the tree. Indeed a significantly less value than the 30 failures per million hours as found from ref. 1.

Several projects are now modifying the designs of subsea valve-trees and also individual valves to fit specific purposes, and one needs to have means to modify the failure rates as well.

In order to obtain a qualified guess, the valve tree has been broken down into single components with individual failure modes which have been combined in a fault tree. The failure rates for the individual components combined in the fault tree should then give the generic answer as found for the well valve-tree. Various assumptions have to be taken in order to make this "break-down" of the failure rates. A proposal on how to do this is given in the following.

A typical valve tree consists of various components which may fail. Mainly the components are valves or seals/gaskets. The Figure 3-1 shows the main components schematically. The "names" used are explained below:

**Valves** (Fail safe: closed)

PMV Process Master Valve (regularly tested, 6, 12 months.)
PWV Process Wing Valve (regularly tested, 6, 12 months.)
PSV Process Swab Valve (Closed, Stab-in hydraulics at WO)
AMV Annulus Master Valve (regularly tested, each 6 or 12 months.)
ASV Annulus Swab Valve (Closed, Stab-in hydraulics at WO)
MIV Methanol Injection Valve (Normally closed, used before shut in)
SIV Scale Inhibitor Injection Valve (Mostly open)
XOV Cross Over Valve (Normally closed,)

**Connections**

WHC Well-Head Connection
PWC Process Wing Connection
AWC Annulus Wing Connection

External leak paths:
All valves (stem / spindle and flanges) and flow line connections.

**Figure 3-1 Schematic drawing of Valve-Tree with components and leak paths**

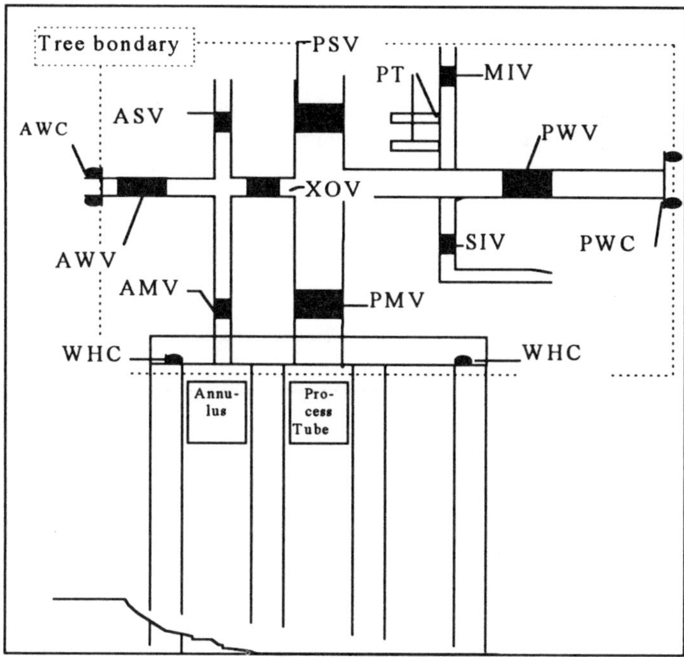


## 3.1 Valve-Tree, critical failure modes

In order to evaluate the relevance of a failure mode, it is important to know that all the valves that are operable during normal process (production or injection) are of "fail safe close" type, which means that they will close if the control system power is lost. The causes for the failure modes are described in the following.

**SPO** **Spurious operation.**
May be caused by failures with the control system, either electronic or hydraulic. Electric: Faulty signal, or power outage (power outage may not cause closure if the pilot system is "hydraulically latched). El. system is not inside valve boundary. Hydraulic: Loss of hydraulic pressure so that spring in actuator will move the valve in "fail safe" position. Can be caused by loss in hydraulic supply or leak actuator piston. The actuator is inside the valve boundary.

**FTC** **Fail To Close.** El. failure in control system, not inside valve boundary (remedial action: manual relief of hydraulic pressure). Hydraulic pressure retained (remedial action: manual relief of hydraulic pressure). Blocked hydraulic return, stuck actuator piston, stuck valve, spring failure.

**FTO** **Fail to open**. Stuck solenoid or pilot valve, stuck actuator, large leak across actuator piston. Stuck valve

**INL** **Internal leakage** (Large leak across valve ). Bad sealing, contamination, wear, hose deformation.

**EXL** **External leakage** (loss of valve containment). Seal failure at stem/spindle or flanges/connections.

Whether the failure modes are critical or not depends upon what function the individual valve has. By counting together the contribution from each valve and it's critical failure modes one can obtain the distribution of the failure modes for the valve-tree.

**Table 3-1 Number of possible occurrences of critical failure modes from individual component on a valve-tree**

| | **Actuator failures** ("Stuck valve" is assumed negligible) | | | **Sealing failures** | |
|---|---|---|---|---|---|
| Failure modes | SPO | FTC | FTO | INL | EXL |
| PMV | x | x | x | x | x |
| PWV | x | x | x | x | x |
| PSV | | | | | x |
| AMV | | x | x | x | x |
| AWV | | x | x | x | x |
| MIV | | x/2 | | x/2 | x/2 |
| SIV | | x/2 | | | x/2 |
| XOV | | | | x/2 | x/2 |
| Leaks: (WHC, PWC AWC+2PT) | | | | | 4x |
| Count: | 2 | 5 | 4 | 5 | 10.5 |
| Failure rate | | 0.64 | | 0,13 | 0,13 |

Assuming that all valves are identical with regard to reliability one has the following relations:

| | | | |
|---|---|---|---|
| Actuator failures per million hours: | 2·SPO+5·FTC+4·FTO= | | 0.64 |
| Internal leakages per million hours: | 5·INL=0.13 | INL= | 0.026 |
| External leakages per million hours: | 10.5·EXL=0.13 | EXL= | 0.0124 |

The various actuator failure modes are not solvable by this method. But based on experience from other valves one may assume a distribution. Some of the causes are the same for the three failure modes, SPO, FTC and FTO, and due to the relation: 2SPO+5FTC+4FTO=0.64 it can be shown that the relation between SPO,FTC and FTO can not be selected freely. Based on i.a. assumptions that stuck acuator or stuck solenoid is more likely than blocked hydraulic return or actuator spring breakage, we found that the relations had to be within the following limits:

FTC/SPO > FTO/SPO-1 and FTC/SPO < 2(FTO/SPO-1)

With basis in this and the judgement that FTO is larger than FTC, the relation: 1; 2.5; 3 was chosen. Thus one can obtain the failure rates for the three failure modes: SPO=0,024·$10^{-6}$/h, FTC=0,060·$10^{-6}$/h, FTO=0,072·$10^{-6}$/h.

Recommended failure rates for the individual valves used in a modern subsea valve tree can then be distributed according to it's individual contribution, (Table 3-1).

**Table 3-2 Recommended failure rates for subsea valve-tree components**

| Valves | SPO | FTC | FTO | INL | EXL | All |
|---|---|---|---|---|---|---|
| PMV | ,02415 | ,06038 | ,07245 | 0,026 | 0,0124 | 0,195 |
| PWV | ,02415 | ,06038 | ,07245 | 0,026 | 0,0124 | 0,195 |
| PSV | | ,0 | ,0 | 0,0 | 0,0124 | 0,012 |
| AMV | | ,06038 | ,07245 | 0,026 | 0,0124 | 0,171 |
| AWV | | ,06038 | ,07245 | 0,026 | 0,0124 | 0,171 |
| MIV | | ,03019 | 0, | 0,013 | 0,0062 | 0,049 |
| SIV | | ,03019 | 0, | 0,0 | 0,0062 | 0,036 |
| XOV | | | | 0,013 | 0,0062 | 0,019 |
| Leaks: | | | | | 0,0496 | 0,050 |
| Total | 0,05 | 0,30 | 0,29 | 0,13 | 0,13 | 0,90 |

The principle as shown here can be further utilised in order to quantify and distribute e.g. contribution from causes for failure modes and so on. This was done in order to establish failure rates for a gate valve system to protect a pipeline for overpressure (HIPPS). This system had some similarities with a valve-tree as it was built into an integrated block and the same type of technology as for the tree was used. The allocated space for this paper, however, prevent us from presenting details of the evaluation of this valve system.

## References

1 OREDA 92, Offshore Reliability Data Handbook, 2nd Edition, Det Norske Veritas Industry Norge A.S, Høvik 1992

2 Norsk Hydro, "Intervensjoner og feilhendelser på undervannsanlegg på norsk sokkel". DNV report no. 93-3550, February 1994

3 A survey of operation and maintenance requirements for subsea wells. Subsea Data Services, Houston Texas, February 1991

# RELIABILITY DETERMINATION OF TELECOMMUNICATION LINE EQUIPMENT BASED ON RETURN OF EXPERIENCE

Z.CHERFI
Université de technologie de Compiègne, dept de génie mécanique, BP649, 60206 Compiègne, France.

G.CHIESA
France-Télécom, O.N.S lignes et vidéocommunications, site d'Arcueil, 67, avenue Lénine, 94112 Arcueil, France.

## 1 Introduction

Today, FRANCE TELECOM has to remain competitive in the face of the international competition. In the area of lines and the telecommunication network, the economic rigor imposes the optimization of the quality of equipment on the basis of the reduction of expenses by report to services rendered. More, the park of equipment is in a constant evolution. So, It is necessary to know it well and to anticipate its behavior, in order to improve the maintenance, the control of costs and to favor the benchmarking.

## 2 Definition of the return of experience

As a rule, one can define the return of experience as follows: human and technical ways adapted to a best knowledge of the equipments behavior, by the intermediary of technical expertise of the events and their analysis. These expertises allow to improve the comprehension of modes of failure or degradation of products observed.
Thus, the return of experience provides ways to improve as much the reliability of equipments that that of the network.
The return of experience comprises four phases: observation, memorizing, analysis and restitution.

## 3 Return of experience at France Telecom

The relative study to management of the return of experience is called "PANEL DES MATERIELS" . It is a system of observation and of data analysis. One of its aim is to be a good tool of decision making. Its finality is to observe in permanence the behavior of the new equipments in exploitation and to estimate their reliability.

## 3.1 Advantages

In a first time, we determine the characteristic regions (of climatical viewpoint) which will take part in the study. These regions are called "ZONE PANEL". In each of them, we define the type of equipment to study, who is in charge of this work and particularly, what is the representative sample size. This choice presents several advantages:
- The number of followed equipments being weak, the number of persons concerned by the study in each region is therefore reduced. This allows to put on site an efficient, short and bilateral system of information.
- According to climatical characteristics of each zone, one can compare the evolution of equipment and to discern propitious environments to the failures.
- By the general knowledge of the zone and the concerned person know-how, it is possible to deepen locally analyses of failures.

Each new equipment is studied by the team of « reliabilists », beforehand to include it in the « Panel des Matériels », so as to determine a human and technical organization which is specific and negotiated.
- This flexibility improves the rigor and the exhaustiveness of data collected on the terrain.

## 3.2 Organizational Structure of the return of experience

The team of specialists of the « panel » centered in Paris, devotes to preliminary studies, management of informations and their analysis. In each" zone panel", a correspondent insures the permanent relationship between agents of the terrain and this team of specialists.

These agents inform forms of failures and send them to the service pilots. informations contained in these forms are seized in the database that we have developed for management of the return of experience.

This structure is the key of vault of the return of experience. Its pertinence is stemming from appropriate the management of aptitudes and capacity of each of participants.

## 3.3 Sampling Technique

One « zone panel » is characterized by a particular factor of environment which is different of that of the other zones.

Each of them comprises a sample of equipments whose size is determined thanks to a technique of sampling that we have computerized. The purpose is to determine a size of sample "n" allowing to validate a contractual failure rate from a given duration of tests and an admissible number of failed equipments.

Necessary data for the determination of "n" are:
- the estimation of the average failure rate of the equipment to follow: L.
- the percentage of precision linked to the L value.
- the duration of the study: T
- the number of returns of information desired: N
- the level of confidence associated to ulterior reliability calculations: G.

The obtained value is the minimal size of the sample to follow in each « zone panel ».

The procedure of estimation of the average failure rate is based on data of various origin:

- Either the equipment to study is of new generation: in this case, the failure rate estimation is obtained from data of anterior version coupled to a coefficient of improvement.

- Or the equipment is entirely new: in this case, it is necessary to make call to judgment of experts.

Furthermore, we propose a method allowing to determine the minimal duration for keeping track of different types of equipments (electronics or mechanics).

# 4 Ascent and processing of the informations

In a first time, the team of « reliabilists » studies the equipment thanks to data contained in the functional and technical specifications notebook.
This first study will allow, thanks to the technique of sampling above-mentioned, to determine the number of equipments to follow in different regions.
The totality of agents concerned by the study receive a training. It is obviously fundamental to motivate them and to make them responsible in order to have an ascent of exhaustive informations.
They are loaded to insure the expertise of the equipments and to fill the forms of failures.

As a rule, one distinguishes seven essential points for analysis in an expertise:
- moment of the discovery of the failure
- identification of the equipment
- description of the environment
- description of the failure
- degree of the consequence
- origin of the failure
- nature of the intervention.

Others criteria can be added in case of need.

All forms of failures are transmitted to the team of the « reliabilists ». the Informations are seized in the database. It allows to structure the brute data and insures to the user a research and a well-off statistical processing.

Realized under ACCESS, this relational database is very convivial. It allows also to keep track of many types of equipments. The user can question it to its manner, to undertake sortings and requests according to its needs. It constitutes a unique and reliable data pole to the service of FRANCE TELECOM, avoiding thus all drift oral information.

# 5 Optimization of the reliability

From collected data then seized in our database, several levels of analysis are possible.

## 5.1 Safety Indicators

- The statistical processing allows a periodic evaluation of the rate of failure by region and global and of others specific indicators (MTTF...).
- The graphs render account rapidly of the evolution of the behavior of the studied equipments and to compare it to the initial reliability objective.
- The failure modes are formed in a hierarchy according to different criteria. This allows us to be focused on the most critical.

## 5.2 Analysis of data

Thanks to capacities of extraction of the database, a more thorough study emanating from expertise of failures brings a more precise vision, that, coupled to statistical tools (analysis of data), lightens the comprehension of the events observed: origin of the failure, most propitious environments, consequences on the system...

A such step gives a net overview of the impact of the external environment on the equipments. An extreme external environment can not be taken in account in the course of the elaboration of specifications notebook and be put in obviousness in the course of this study. All analyses are undertaken in collaboration with all actors implied in this work. Stemming processing conclusions and statistical data analyses are confronted with the knowledge and to the experience of these persons. The results bring a best knowledge of the behavior of the equipment and its characteristics of reliability. It results some a teaching being able to improve the design of the equipment, its ergonomic and the methods of implementation...
A preventive maintenance plan is then established so as to reduce risks of degradation of the network fathered by failures of the equipments.

## 5.3 Human Reliability

The analysis of human reliability data is one original parts of our work. The interest aroused by this aspect is appeared when we have noticed that a certain number of breakdowns had for origin human failures.

A good appreciation of the interface man/ equipment allows to take necessary dispositions to limit human errors.
This allows to improve as much the global reliability of the network that the conditions of work.

# 6 Conclusion

This study has allowed us to learn a lot of things on the duration of equipment life of lines of communication. It has especially allowed us to realize the importance of the human factor, so much a human failure viewpoint, that a viewpoint of the management and the communication. We have invested in the training of agents and their awareness to the importance of the project.
We are about the obtained results and also pleased to have led to term, a work that associates a technical aspect to the human factor.

## References


1. Pages A., Gondran M. Fiabilité des systèmes, Ed Eyrolles 1980.
2. Leplat J. Erreur humaine, fiabilité humaine dans le travail, Ed.Collin 1988.
3. Villemeur A. Sûreté de fonctionnement des systèmes industriels, Ed Eyrolles, 1988.
4. Moura E.C. How to determine sample and estimate failure rate in life testing, ASQC 1990.
5. Procaccia H., Piepszownik L. Fiabilité des équipements et théorie de la décision statistique fréquentielle et bayésienne, Ed Eyrolles 1992.
6. Gardarin G. Maîtriser les bases de données, Ed Eyrolles 1993.
7. Marcenac P. SGBD relationnel, optimisation des performances, Ed Eyrolles 1993.
8. ISDF, Etat de l'art dans le domaine de la fiabilité humaine, Octarés 1994.
9. Lannoy A., Procaccia H. Méthodes avancées d'analyse de base de données du retour d'expérience industriel, Ed Eyrolles 1994.
10. Aupied J. Retour d'expérience appliqué à la sûreté de fonctionnement des matériels en exploitation, Ed Eyrolles 1994.

# The Application of the Accident Sequence Precursor Methodology for Marine Safety Systems

Louis H.J. Goossens[1] and Cees C. Glansdorp[2]
[1] Safety Science Group, Delft University of Technology, Delft, the Netherlands
[2] Marine Analytics, Maritime Consultancy Bureau, Rotterdam, the Netherlands

**Abstract**

Maritime safety has always been given high priority in the maritime countries of the European Union. From an inventory of marine safety highlights an in-depth study of incidents in relation to safety systems appeared to be highly relevant under the European Communities EURET 1.3 program in order to contribute to the cost-effectiveness of (combined) marine safety systems, given an accepted level of safety. A basic element in such a study is the analysis of ship incident data, for which the Accident Sequence Precursor Methodology (ASP Method) has in principle a suitable framework.

The objectives of the study was to test whether the ASP Method can discriminate in the effectiveness of safety nets and can describe them to find an optimal mix of safety nets for each navigable area used by certain flow of ships of known composition, and to investigate whether human failures may be modelled using the ASP methodology.

The paper will explore some of the results, but will mainly deal with the methodological aspects of using the Accident Sequence Precursor methodology for complex systems as navigation.

# 1 Introduction

Maritime safety has always been given high priority in the Netherlands and other countries of the European Union. This led to the use of various types of marine safety systems like markings, piloted waterways, patrol boats and so on. It also led to the development of traffic guidance systems like VTS (vessel traffic services). Policy evaluations indicate that advanced, newly developed safety systems are added on to the existing system configurations without sufficient in-depth knowledge of the added-on effectiveness of the new systems in relation to already existing systems.

From an inventory of marine safety highlights an in-depth study of incidents in relation to safety systems appeared to be highly relevant in order to contribute to the cost-effectiveness of (combined) marine safety systems, given an accepted level of safety. A basic element in such a study was the analysis of ship incident data, for which the Accident Sequence Precursor Methodology (ASP Method) has in principle a suitable framework. This paper presents the results from the research program EURET 1.3 (Commission of the European Communities) [1].

## 2 Accident Sequence Precursor Methodology - General Aspects

The ASP Methodology originated at the Oak Ridge National Laboratory shortly after the accident at Three Mile Island as a part of the Accident Sequence Precursor program for light water nuclear reactors in the United States [2]. As a general conclusion, the precursor program has made a very significant contribution to the field of risk assessment.

The Dutch Ministry of Housing, Physical Planning and Environment has sponsored, taking its responsibility for developing risk management methodologies, a research program into the feasibility of the ASP Method for the chemical process industries in the Netherlands [3,4].

Whereas the traditional probabilistic risk analysis methodology was developed to predict risks in systems for which no or scarce operational experience was available, the ASP methodology is specifically designed for situations for which operating experience is available, though insufficient to allow statistical predictions of serious accident probabilities. This is typically the case in the chemical process (and nuclear) sector.

In those cases, the ASP calculations of the overall accident probability is done by summing conditional probabilities for accidents, given data available for the accident sequence precursors. Normally, incident data are not avaialble, but data of failing safety systems are. For marine systems, the overall accident probability can be deduced from incident registrations.

The contributions of the various precursors, for which no data are readily available from the databanks, can be deduced from an analysis of the incidents data and descriptions. This procedure is explored by defining generic accident sequence paths within a generic event tree developed for marine accidents. Each path consists of a subset of relevant precursors, indicative for the safety systems' effectiveness.

# 3 Human Errors in Marine Systems

Marine incidents are generally a result of a large sequence of events, in which human actions are alternated with technical failures and (un)successful human recovery actions. It appears that these sequence paths in the generic event tree are heavily centered around the actions of the navigator (with or without pilotage). This intrinsic modelling dependency leads to complications in the modelling of the sequences of the individual sub-events in the event tree. In order to provide a sensible set of generic sub-events additional modelling assumptions were necessary.

The analysis procedure applies the ASP Method which entails that each incident will be analysed according to a scheme, leading to the building up of the generic event tree. This requires:
- identification of the initiating event for each incident
- identification of the incident path for each incident.

The initiating events are defined in several groups. For the safety effectiveness analysis three types of initiating events are distinguished:
- the macro-strategic initiating events
- the meso-strategic initiating events
- the tactic initiating events.

Initiating events are defined as conditions which do lead to an incident unless the deviation is recovered to a non-incident situation (usually before the point-of-no-return). The macro-strategic initiating events contain those conditions which provide inadequate information to the navigator at a much earlier stage prior to the time of the incident, such as voyage preparation. Once the ship enters the "incident area" it arrives at the stage where meso-strategic initiating events may appear. The navigator then has to consider inadequate information for determining concrete actions on board for adjusting the ship's course and speed. At this level, active and passive safety systems are intended to prevent collisions and strandings. In particular, systems like VTS, have a task in providing additional meso-strategic information. Tactic initiating events generally occur as a result of a human error close to the point-of-no-return. In most cases, only human recovery onboard the vessel may help out.

A special database was developed to accommodate all items which were thought to be important to this investigation. The ASP methodology has proven its worth in the nuclear field and appears to be promising in the field of marine safety. In marine environments, this method
- uses operational data from incidents which occurred in fairways, and
- processes these operational data into a generic event tree model that results in numerical information on the contribution of safety systems to incidents.

# 4 Approach to Marine Safety

The following constraints are present in marine situations:
- complex maritime routes (like estuaries of big rivers and harbour entrances, or dense sea routes like the English Channel and parts of the North Sea) are considered
- course and speed of the ships are important factors in determining the cost-effectiveness of ship manoeuvres and the risk-benefit-ratio
- the expected behaviour of the navigator with respect to course and speed and his response to deviations determine the failures and recovery actions made by the navigator
- the expected behaviour of the ship (mechanical failures) and its surrounding elements (other ships and fixed objects) determine the physical constraints
- passive safety measures have a function in increasing the attention level of the navigator and to provide information on the conditions of the maritime route
- active safety measures also have a function in increasing the attention level of the navigator and to provide information on the conditions of the maritime route, plus they have the ability of providing concrete time-constrained advices and communications.

As indicated above, the incidents are identified as specific paths in the generic event tree [1]. The basic elements of the generic event tree (Figure 1) are:
- the identification of the initiating event
- the complexity of the "incident area", whereby complexity is defined by the characteristics (lenght and width) of the vessel relative to the navigable space, whether other ships are in its vicinity and by meteorological conditions
- the corrective effect of the VTS, whther or not it failed in correcting the navigator's decisions on course and speed
- whether or not the point-of-no-return has been reached (all incidents are a result of having no ability of prevention after the point-of-no-return has been reached, otherwise there would have been no incident at all).

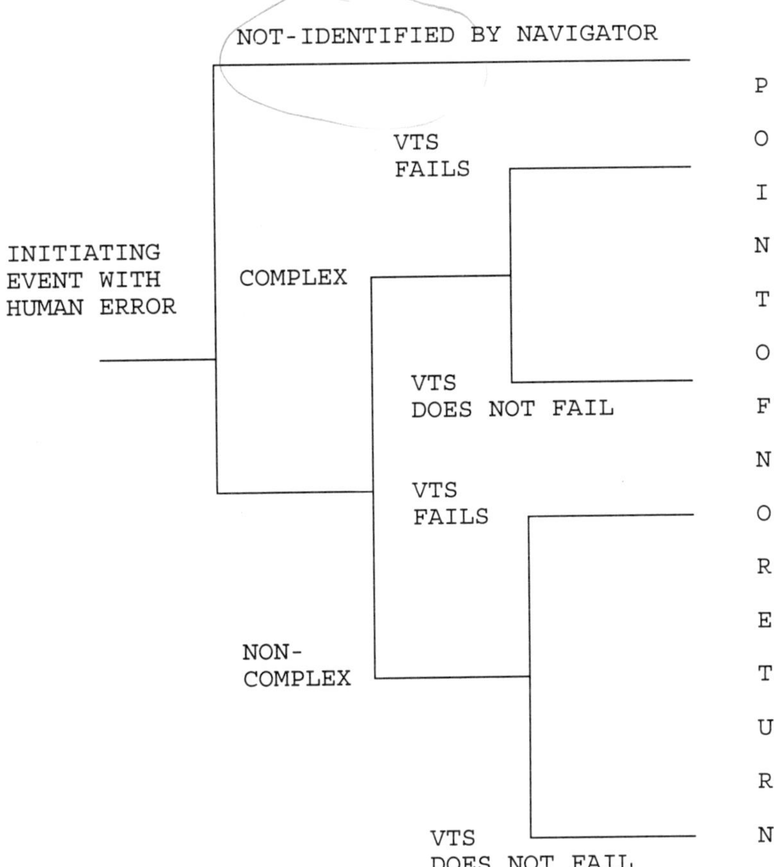


**Figure 1**

Generic event tree representing navigation through safeguarded waterways for identified deviations only

Figure 1 shows the generic event tree for use in marine safety analyses. It only explores initiating events with a human error at the level of the navigation control. Not included are initiating events defined by mechanical failures 9roughly less than 20 percent of all incidents).

Another important aspect is whether the safety systems have a task in all situations. For this reason, complex and non-complex situations were distinguished. VTS's mainly have a task in providing adequate meso-strategic information for complex situations. [Otherwise VTS's could be useful in every possible waterway.] The ASP-analysis structure aimed at identifying whether a VTS (or any other combination of safety systems) had an effective performance on the safety records.

# 5 Results

From the generic event tree in figure 1 it may be clear that the ASP analysis enables incidents to be put into the sequence paths of the generic event tree and therefor has the potential to compare the number of incidents occurred through the various paths. Two typical examples of results are discussed in this paper: the effectiveness of VTS's with respect to the complexity of the marine system and the potential for VTS support for the various types of initiating events.

The first issue (effectiveness of VTS and complexity) can be analysed with the use of the complexity ratio defined as the ratio of incidents taken place under complex conditions and incidents taken place under non-complex conditions. The ratio shows the impact of changes in VTS configurations. Table 1 shows such changes for the Western Scheldt VTS configuration. In the first period the VTS consisted of a simple radar system with no ship labeling facility built in. In the second period the radar system was advanced and was capable of labeling sea going vessels (only).

**Table 1**

Complexity ratio for accidents at the Western Scheldt

| | RADAR NO SHIP LABELING | RADAR WITH SHIP LABELING |
|---|---|---|
| # ACCIDENTS | 94 | 126 |
| # SHIPS | 130 | 179 |
| SEA GOING VESSELS PILOTED | 1.60 | 0.43 |
| SEA GOING VESSELS NOT PILOTED | 0.67 | 0.27 |
| INLAND GOING VESS. | 0.30 | 0.40 |

Table 1 shows a drastic reduction of the complexity ratio for sea going vessels (both piloted and non-piloted vessels). There occurred less incidents under complex conditions relative to the incidents under non-complex conditions. [In absolute terms, the complex incidents dropped from 20 per year in the first period to about 8 per year in the second period, while the number of non-complex incidents did not change significantly (from 24 to 20 per year)].

Another important result from the ASP analysis is the relative contributions of the three types of initiating (human error) events. Table 2 shows the overall results.

**Table 2**
Human error incidents (in percentages)

| NAVIGABLE SPACE | MACROSTR. | MESOSTR. | TACTIC |
|---|---|---|---|
| **SEA GOING VESSELS** | | | |
| NORTHSEA/EUROGEUL | 18 | 26 | 56 |
| NORTHSEA/CONT.SHELF | 15 | 13 | 72 |
| EMS ESTUARY | 7 | 10 | 83 |
| WESTERN SCHELDT | | | |
| * NO SHIP LABEL | 5 | 23 | 72 |
| * WITH SHIP LABEL | 13 | 15 | 72 |
| **INLAND GOING VESSELS** | | | |
| EMS ESTUARY | 25 | 12 | 63 |
| WESTERN SCHEDLT | | | |
| * NO SHIP LABEL | 25 | 21 | 54 |
| * WITH SHIP LABEL | 14 | 22 | 64 |

Table 2 shows that two-thirds of the incidents initiated with a tactic event, which means that they commenced close to the point-of-no-return. Vessel Traffic Services are not designed to address this type of incidents as the time to respond is very short, and it is very difficult, if not impossible, for VTS operators to identify tactic initating events onboard ships. Only changes in the ship's course could be detected from the radar screens, but the resolution for significant changes is too low in the short time still available prior to an incident.

VTS systems are typically designed to address mesostategic initiating events. This type of events relates to the planning of the vessel's speed and course, for which there is ample time if relevant information from the VTS operator is appropriately picked up onboard the vessels. Table 2 shows that less than one quarter of the incidents initiated with this type of human errors, and could therfore in principle be addressed by (improvement of) VTS systems.

The macrostrategic initiating events are related to what is generally called substandard conditions onboard ships. Again, for VTS operators these conditions are difficult to identify (how can they know that the vessel has outdated charts for the fairways?). Pilots can overcome these situations sometimes, as they are (very) familiar with the specific navigable spaces.

# 6 Conclusions

The Accident Sequence Precursor Methodology has shown to provide insight into the mechanisms under which ship incidents in densely navigated fairways may occur. The defined incident sequence paths lead to typical representations of the complexity of the situation and the response of active safety systems like Vessel Traffic Services. Examples of results show how the ASP Method can be usefully applied to identify for which types of incidents which type of VTS actions can be effective.

# 7 Acknowlegdements

This paper reports the results under the CEC research program EURET 1.3, co-sponsored by the Ministry of Traffic and Waterways in the Netherlands. The authors want to thank Reinout Leurink for his contribution in setting up the data-bank and for performing all necessary analyses.

## References

1. Goossens L.H.J., Glansdorp C.C., Leurink R. Operational benefits: Risk reduction analysis with Accident Sequence Precursor Sequence Methodology. Report for EURET 1.3 (CEC and Dutch Ministry of Traffic and Waterways), Marine Analytics b.v./Rotterdam and TUDelft, March 1992
2. Minarick J.W. "The US NRC Accident Sequence Precursor Program: Present method and findings", Reliability Engineering & System Safety, 1990, Vol.27, No.1, p.23-51
3. Cooke R.M., Goossens L.H.J., Hale A.R., van der Horst J. Accident Sequence Precursor methodology - A feasibility study for the chemical process industries. Report for the Dutch Ministry of Environment, TUDelft/TNO Apeldoorn, March 1987
4. Cooke R., Goossens L.H.J. The accident sequence precursor methodology for the European Post-Seveso era. Reliability Engineering & System Safety 1990; 27:117-130

# Cost Analysis of an Air Transport System Model with Critical Human Errors and Two Repair Policies

Sławomir Klimaszewski
Henryk Smoliński
Józef Żurek
Air Force Institute of Technology
Warsaw, Poland

**Abstract**

The paper deals with a certain mathematical model of an air transport system. The model refers to a single aircraft and takes the impact of other system components on the aircraft reliability and operating costs into account.

# 1 Introduction

Many publications have discussed the reliability of man-machine systems [1-4]. Most of these publications have concentrated their attention on on-surface systems. However, one may be interesting in analysis of reliability and cost of an air transport system.

Every air transport system includes the following components:

1. aircrew,
2. aircraft,
3. flying control and support subsystem,
4. flying ground support subsystem,
5. environment.

The model represents the case when the aircraft is liable to two levels of maintenance. i.e. preventive and corrective ones. It has been assumend that there is no period of waiting for the preventive maintenance, which is consistent with what operating practice shows. There is, however, a period of waiting for the corrective maintenance to be executed. Times of executing both the preventive and corrective maintenance processes show generalized distributions described with specific probability density functions.

# 2 Assumptions underlying the model

1. The system shows only one state of fitness for use (i.e. state 0), other states are the states of unfitness for use;

2. In t=0 the system is in the state of fitness for use (i.e. performs its tasks according to the predetermined operational use requirements);
3. There are two levels of maintenance: preventive maintenance (state 1), and corrective maintenance (state 3);
4. There is a state of waiting for the corrective maintenance to be executed (state 2);
5. Time until preventive maintenance (repair) of aircraft becomes necessary shows exponential distribution with parameter $\lambda_1$ ;
6. Time of waiting for the corrective maintenance to be executed shows exponential distribution with parameter $\lambda_2$.
7. Time until corrective maintenence becomes necessary shows exponential distribution with parameter $\lambda_3$;
8. The system may become unoperational due to an unrepairable (catastrophic) failure ;
9. The system's working life to catastrophic failure shows exponential distribution with parameter $\lambda_4$;
10. Time to catastrophic failure of the system subject to preventive maintenance shows exponential distribution with parameter $\lambda_5$;
11. Time to catastrophic failure of the system that waits for corrective maintenance shows exponential distribution with parameter $\lambda_7$;
12. Time to catastrophic failure of the system under corrective maintenance shows exponential distribution with parameter $\lambda_6$;
13. Time of carrying out preventive maintenance (repair) shows optional distribution described with the probability density function $F_1$ (x);
14. Time of carrying out corrective maintenance (repair) shows optional distribution described with the probability density function $f_3(y)$;
15. At any optional instance of time only one of the elementary events can occur, i.e.transition from one state to another;
16. The failures are statistically independent;
17. Having been serviced/repaired the system is as operational as a new one.
18. Random variables that denote: times to failures, times of waiting for maintenance/repairs, times of executing maintenance procedures (repairs) are independent.

# 3 Notations

i i-th state of the system

s Laplace transform variable

$\lambda_j$ constant hardware failure rate of the system; for $j = 1, 2, 3$

$\lambda_j$ constant critical human error rate; for $j = 4, 5, 6, 7$

$\mu_1(x)$ time dependent system repair rate when the system is in state 1 and has an elapsed repair time of x

$\mu_2(x)$ time dependent system repair rate when the system is in state 2 and has an elapsed repair time of y.

$P_i(t)$ probability that the system is in state i at time t; for $i = 0, 1, \ldots, 4$

$P_i(s)$ Laplace transform of $P_i(t)$

$p_1(x,t)$ probability density (with respect to repair time) that the failed system is in state 1 and has elapsed repair time of x

$p_3(y,t)$ probability density (with respect to repair time) that the failed system is in state 3 and has an elapsed repair time of y

$f_1(x)$ probability density function (pdf) of the system repair time when the system is in state 1 and has elapsed repair time of x

$f_3(y)$ probabilty density function (pdf) of the system repair time when the system is in state 3 and has elapsed repair time of y

$C_s(t)$ total expected cost of the system

$G_s(t)$ total expected gain of the system

$C_j$ cost per unit time or earning rate; for $j=1,\ldots,4$

# 4 Time dependent system cost analysis

Fig.1 shows the diagram of space of states. The differential and integral equations for the system are:

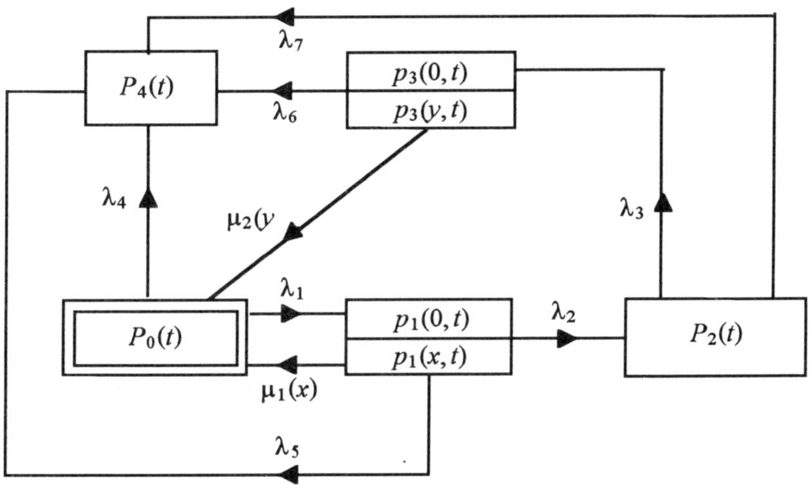


Figure 1 Diagram of space of states

$$[\frac{d}{dt} + \lambda_1 + \lambda_4] P_0(t) = \int_0^\infty p_1(x,t) * \mu_1(x)\, dx + $$

$$+ \int_0^\infty p_3(y,t) * \mu_3(y)\, dy \tag{1}$$

$$[\frac{\delta}{\delta t} + \frac{\delta}{\delta x} + \mu_1(x) + \lambda_2 + \lambda_5] p_1(x,t) = 0 \tag{2}$$

$$[\frac{d}{dt} + \lambda_3 + \lambda_7] P_2(t) = \int_0^\infty p_1(x,t) * \lambda_2 \, dx \tag{3}$$

$$[\frac{\delta}{\delta t} + \frac{\delta}{\delta y} + \mu_2(y) + \lambda_6] p_3(y,t) = 0 \tag{4}$$

$$[\frac{d}{dt}] P_4(t) = P_0(t) * \lambda_4 + \int_0^\infty p_1(x,t) * \lambda_5 \, dx + P_2(t) * \lambda_7 +$$

$$+ \int_0^\infty p_3(y,t) * \lambda_5 \, dy \tag{5}$$

The associated boundary conditions are as follows:

$$p_1(0,t) = P_0(t) * \lambda_1 \tag{6}$$

$$p_3(0,t) = P_2(t) * \lambda_3 \tag{7}$$

At time t = 0,

$$P_0(0) = 1; \qquad P_1(0) = 0 \qquad dla\ i = 1,2,3,4 \tag{8}$$

Using Laplace transforms in Equations (1) - (5) and than solving them, we get the following Laplace transforms of state probabilities $P_i(s)$:

$$P^*_0(s) = \frac{1}{B^*(s)} \tag{9}$$

where:

$$B^*(s) = s + \lambda_4 + \lambda_1 [1 - f^*_1 (s + \lambda_2 + \lambda_5) -$$

$$- \frac{\lambda_2 * \lambda_3 * N^*_1 (s + \lambda_2 + \lambda_5)}{s + \lambda_3 + \lambda_7} f^*_3 (s + \lambda_6)] \tag{10}$$

$$N^*_1 (s + \lambda_2 + \lambda_5) = \frac{1 - f^*_1 (s + \lambda_2 + \lambda_5)}{s + \lambda_2 + \lambda_5} \tag{11}$$

$$P^*_1(s) = \frac{\lambda_1 * N^*_1 (s + \lambda_2 + \lambda_5)}{B^*(s)} \tag{12}$$

$$P_2^*(s) = \frac{\lambda_1 * \lambda_2 * N_1^*(s + \lambda_2 + \lambda_5)}{B^*(s) * (s + \lambda_3 + \lambda_7)} \tag{13}$$

$$P_3^*(s) = \frac{\lambda_1 * \lambda_2 * \lambda_3 * N_1^*(s + \lambda_2 + \lambda_5) * N_3^*(s + \lambda_6)}{B^*(s) * (s + \lambda_3 + \lambda_7)} \tag{14}$$

$$P_4^*(s) = \frac{1}{s * B^*(s)} [\lambda_4 + \lambda_1 * \lambda_5 * N_1^*(s + \lambda_2 + \lambda_5) +$$

$$+ \frac{\lambda_1 * \lambda_2 * \lambda_7 * N_1^*(s + \lambda_2 + \lambda_5)}{s + \lambda_3 + \lambda_7} +$$

$$+ \frac{\lambda_1 * \lambda_2 * \lambda_3 * \lambda_5 * N_1^*(s + \lambda_2 + \lambda_5) * N_3^*(s + \lambda_6)}{s + \lambda_3 + \lambda_7}] \tag{15}$$

System reliability can be obtained by inverting Equation (9). Total expected gain of the system is given by

$$G_s(t) = C_1 * \int_0^t P_0(t)\, dt - C_3 * t \tag{16}$$

Total expected cost of the system is given by

$$C_s(t) = C_2 * \int_0^t P_0(t)\, dt - C_4 * t \tag{17}$$

Plots of the total expected gain of the system are shown in Figure 2.

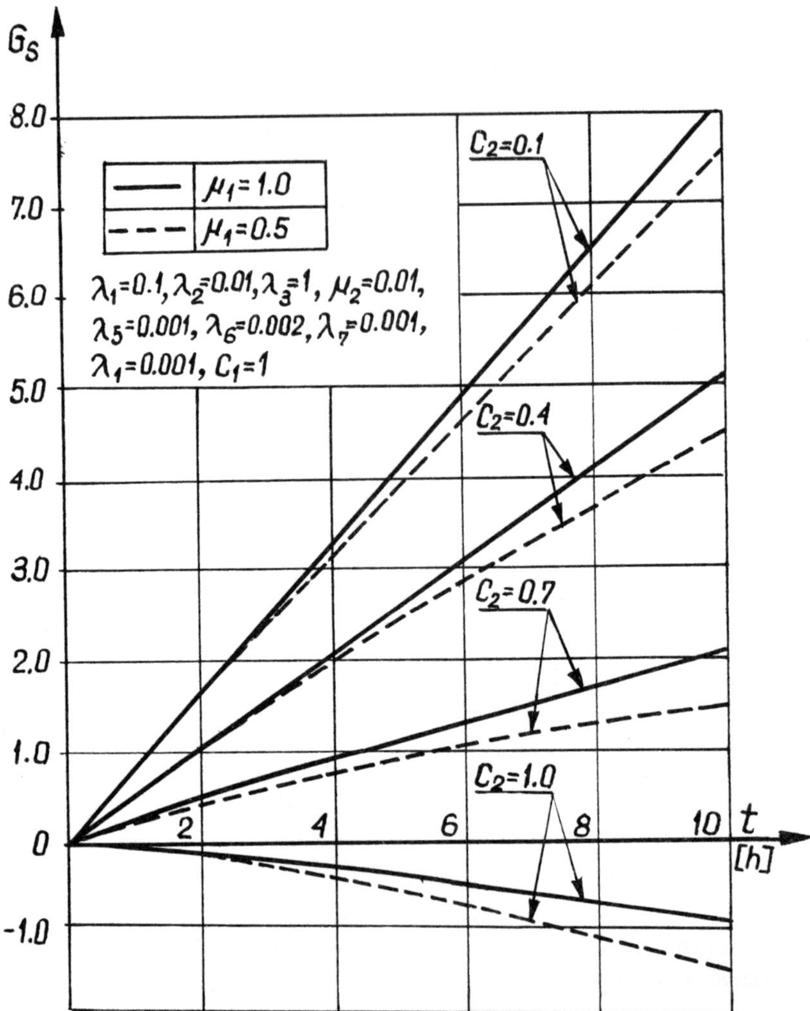


Figure 2. Total expected gain of the system.

## References

1. Dhillon BS, Rayapati SN. Reliability and availability analysis of transit system, Microelectronics and Reliability 1985; 25:1073-1085.
2. Dhillon BS, Rayapati SN. Reliability modeling of on-surface transit systems with human errors, Microelectronics and Reliability 1985; 25:1087-1098.
3. Gupta PP, Gupta RK. Cost analysis of an electronics redundant system with critical human errors, Microelectronics and Reliability 1986; 26: 417-421.
4. Dhillon BS, Yang N. Availability analysis of a repairable standby human - machine system, Microelectronics and Reliability 1995; 35: 1401 - 1413.

# Reliability and Cost Optimization

Yizhak BOT, BQR Reliability Engineering Ltd.
P.O.Box 208 Rishon LeZion 75101 Israel
Tel : 972 3 966 3569, Fax : 972 3 969 8459

## 1. Introduction

One of the problems that Reliability Engineers are faced after defining the redundancy techniques is to select the one which is the cheapest in Life Cycle Cost term. In this article we shall present an example of redundancy that increase the cost of the system but reduce the LCC and we shall introduce CARE (Computer Aided Reliability Engineering) software tool which was used.

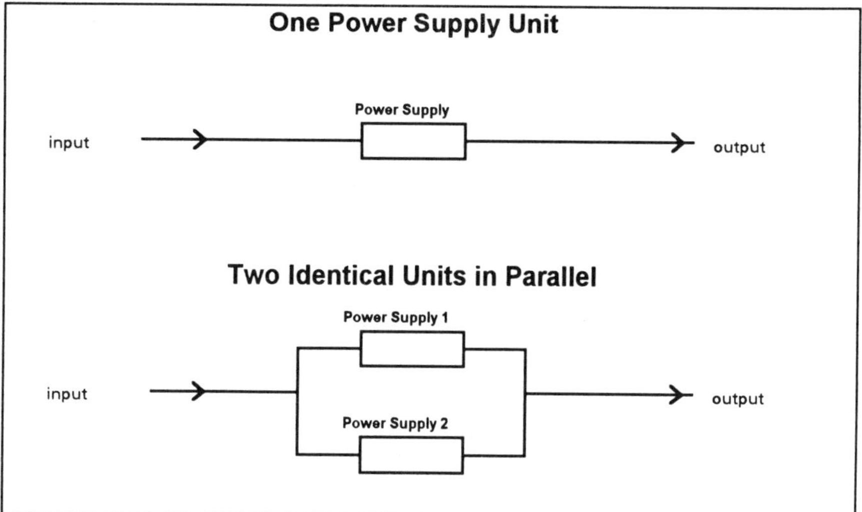


Figure 1. Reliability Block Diagram for Both System Configurations

The two configurations represent the analysis of a situation that might well arise in practice: the requirement to provide a very reliable power supply at optimal cost.

## 2. System description

A single unit is considered that has a MTTR of 72 hours and in the operating conditions has a MTBF of 10,000 hours and costs $5000.

When two are used in parallel in partial active redundancy, the MTTR remains the same but as the operating load is halved, the MTBF increases to 40,000h and the initial cost is $10,000 for the pair.

## 3. The Reliability Model

To make a meaningful comparison it is necessary to consider the availability and overall costs during the operating life of 22 years. To do so the effective MTBF for the system with redundancy and reparability needs to be computed. The Markov chain in figure 2. below represents the states and transitions of the system. Using it ensures that all possible states are considered in the analysis. With it, the state of the system and the operational mode of the units can be tabulated as in table 1 below.

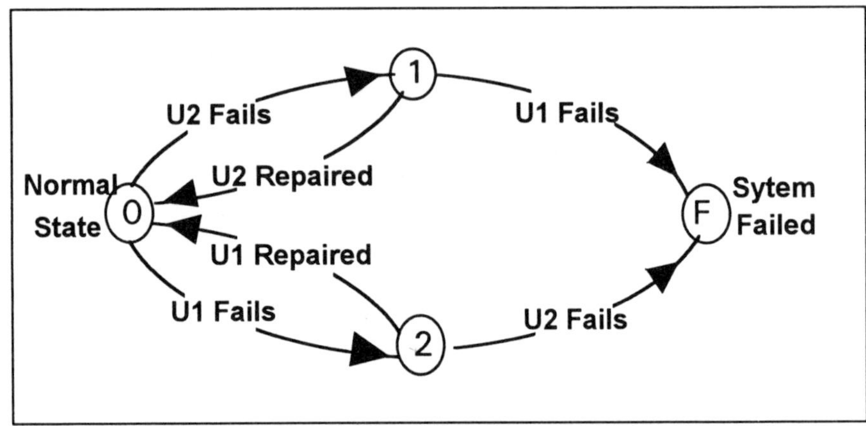


Figure 2. State Transition Diagram for the System

| States | Reliability Block - Units | | | | |
|---|---|---|---|---|---|
| | Currently Operating | Active Redundant | Partial Active Redundant | Stand-By Redundant | Under Repair |
| 0 | | | U1, U2 | | |
| 1 | U1 | | | | U2 |
| 2 | U2 | | | | U1 |
| F | | | | | U1, U2 |

Table 1. The States Tabulated

This is then calculated by considering the possible transitions from each state and the instantaneous probability of such change. A mathematical model representing each state and the probability of transition to another as a matrix is formulated, and from this, vectors to represent the probability of remaining in any of the non-failed states are derived. These then form a series of differential equations that will describe the probability of the system being in a non failed state that is then solved, and from this the effective MTBF is evaluated.

# 4. The LCC Model

The overall costing throughout the life cycle including the initial capital costs, maintenance costs with all staffing training, equipment and other overheads is evaluated for the Life Cycle Costing. Part of the input to this analysis and the results therefrom is set out in the tables 2 and 3 below, for a Life Cycle of 22 years with 100 units in the field.
The input was the same in every respect other than the MTBF, and the fact that where there are two units, there is a possibility for preventative maintenance, which cost is accounted for, whereas when there is only one unit there cannot be preventative only corrective maintenance. The overall cost of operation during the full life cycle shows a saving by a factor of nearly 2, despite the fact that the availability is significantly increased.

LCC SUMMARY in (x1000 $ )

| COST ELEMENT | % | Current |
|---|---|---|
| R & D | 0.3 | 260.000 |
| Production | 1.1 | 1000.000 |
| Investment | 0.5 | 517.580 |
| Operation | 10.5 | 9914.130 |
| Support-CM | 0.7 | 688.844 |
| Support-PM | 86.9 | 81938.700 |
| Support-SW | 0.0 | 0.000 |
| Disposal | 0.0 | 0.000 |
| Total | 100.0 | 94319.254 |

Table 2. LCC with redundancy

LCC SUMMARY in (x1000 $ )

| COST ELEMENT | % | Current |
|---|---|---|
| R & D | 0.2 | 260.000 |
| Production | 0.3 | 500.000 |
| Investment | 0.1 | 215.780 |
| Operation | 5.4 | 9054.000 |
| Support-CM | 94.1 | 159129.714 |
| Support-PM | 0.0 | 0.000 |
| Support-SW | 0.0 | 0.000 |
| Disposal | 0.0 | 0.000 |
| Total | 100.0 | 169159.494 |

Table 3. LCC for single Unit configuration

# 5. Results

The Cost the MTBF and the Availability for the two configurations are:

| | **Single Unit** | **Two Units** |
|---|---|---|
| **MTBF hrs.** | 10,000 | 2,807,175 |
| **Availability** | 0.9928514694 | 0.9999743521 |

Table 4. Results for the two configurations

# 6. The Optimization Process

## 6.1 Decisions Types

For the optimization we need to make a few decisions. The decisions are based on comparison of a few options which affect the cost and the reliability.
The decisions are :

- ♦ To select a more reliable (high Quality Level) but more expensive components', if there are alternatives

- To improve the system reliability by reducing stress versus the cost of such implementation (for example : adding fan's for cooling)
- To select the redundancy degree and type for critical blocks (for example : parallel, K out of N, standby etc.)
- To make changes in system design for simplifying repairing (for example : replace failed assembly during operation)
- BIT implementation and ATE design to detect and isolate failures to reduce repair time
- To make changes in maintenance policy (for example : Repair level, preventive maintenance etc.)

Implementation of all those options usually increase the initial cost of the system and the logistic support, but decrease the maintenance cost. The total Life Cycle Cost and the reliability is calculated by CARE software and allow the Reliability and Design Engineers to estimate their decisions for selecting the optimal one.

## 6.2 CARE Software Involved

In the process, 6 software modules from CARE tools are involved, for the corresponding data flow see Figure 4.

CFW (CARE Frame Work) : This module prepares the project tree including assemblies, sub-assemblies and components, describes also the components stresses, environment and other conditions for failure rate prediction;
PREDICTION : This module predict the failure rates bottom up starting from components, sub-assemblies, assemblies up the system level according to different prediction methods;
ORLA (Optimal Repair Level Analysis) : This module optimize and select the most cost effective repair solution according to the various maintenance cost elements;
VLCC (Visual Life Cycle Cost) : This module makes the final LCC calculation;
RBD : (Reliability Block Diagram) This module calculate the reliability, MTBF and the MTTR of each option and for each assembly in the project tree according to the selected redundancy type, repair policy and switching policy;
REL - COST (Reliability - Cost Optimizer) : This module collect the results from the previous modules and present all options in a PARETO table were the best option is in the first line.

## 7. Conclusion

This optimization process is very complicated but can be very easy if the user uses the CARE software. CARE provide the user with cost and reliability estimations for each option.

BQR has successful experience using this technique with many consulting projects which the customers saved a lot of money.

Figure 3. CARE main window

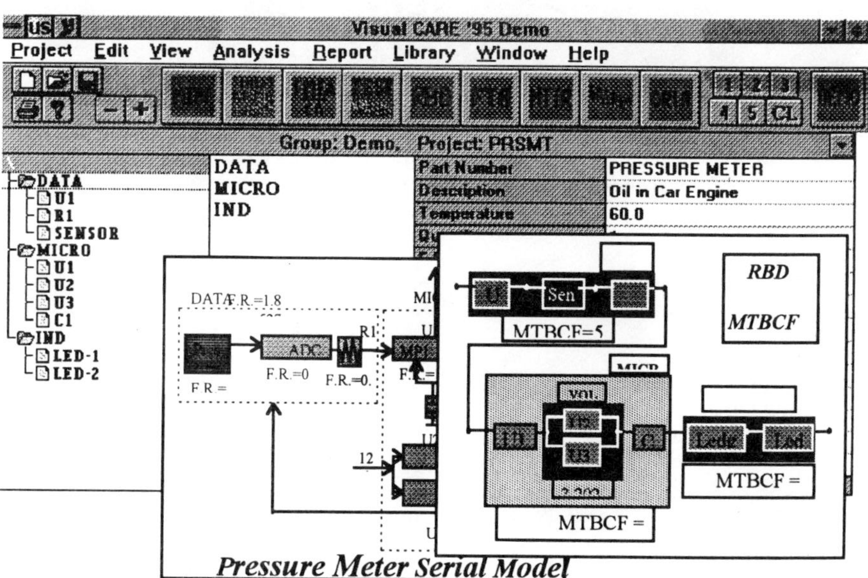

# Figure 4. RELIABILITY - COST OPTIMIZATION

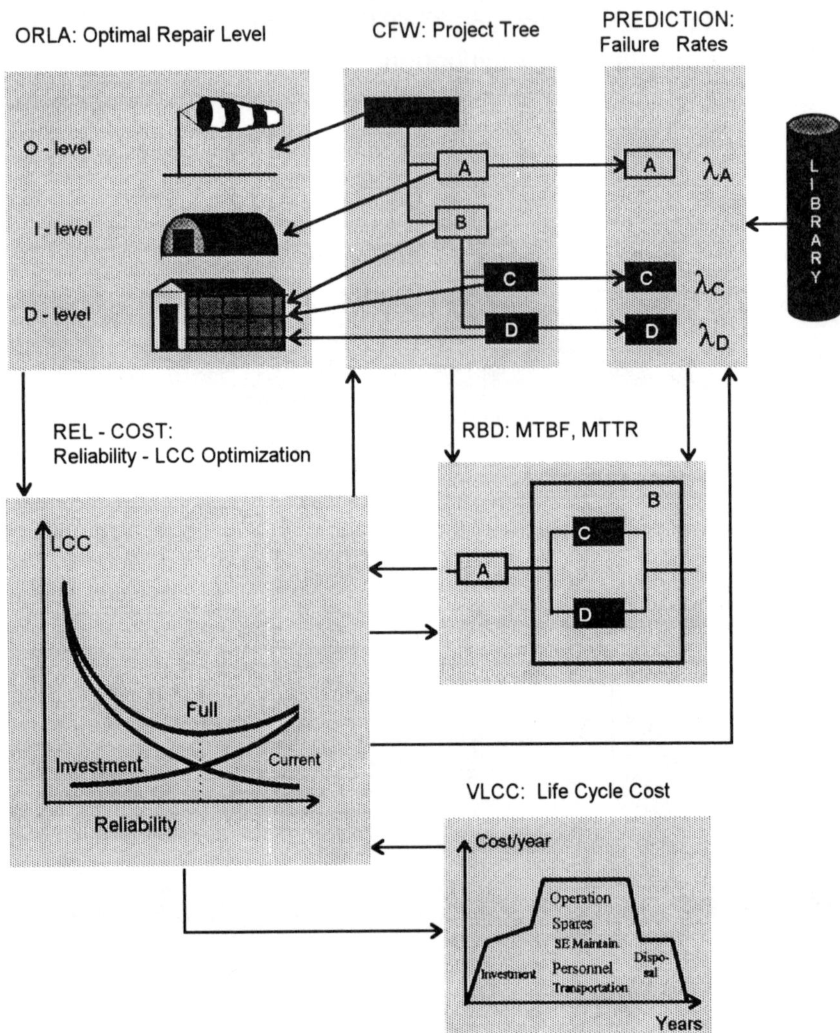

# Seismic IPEEE of US Nuclear Power Plants: Lessons for NPPs in Other Countries

**M.K. Ravindra**
**EQE International, Inc.**
**Irvine, California USA**

**Abstract**

This paper summarizes the lessons learned in the seismic IPEEE of a large sample of plants in the US of different reactor types/vintages and site conditions and reviews their applicability to plants outside of the US. These encompass the aspects of seismic hazard, seismic fragility/margin and systems analyses. Both seismic PSA and margin methods are reviewed.

# 1 Introduction

As part of the implementation of its policy statement on severe accidents, the USNRC issued Generic Letter 88-20 requesting each nuclear power plant owner to conduct an individual plant examination (IPE) for internally initiated events. An extension of this request is the IPEEE, the individual plant examination of external events. All the operating plants in the US (about 110 reactors) are conducting this examination since 1992.

Seismic IPEEE is considered an essential part of this search for severe accident vulnerabilities. Two methodologies - seismic PSA and seismic margin assessment - have been developed to address the seismic events in the IPEEE. These methodologies are summarized in the IPEEE guidance document NUREG-1407 [1]. The seismic IPEEEs of US nuclear power plants are nearing completion. Results of the walkdowns and fragility/margin evaluations along with the preliminary estimates of the seismic core damage frequencies are available. Identification of the dominant contributors to seismic risk is underway.

Seismic IPEEEs of nuclear power plants are being conducted in Canada, Czech Republic, Finland, Spain, Sweden, Slovakia, Hungary, Slovenia, Korea and Japan. Some of these countries have adopted the seismic PSA as the IPEEE method while others are conducting seismic margin assessments.

The objective of the paper is to summarize the lessons learned in the seismic IPEEE of a large sample of plants in the US of different reactor types/vintages and site conditions and review their applicability to plants outside of the US. These will encompass the aspects of seismic hazard, seismic fragility/margin and systems analysis. Both seismic PSA and margin methods will be examined for this purpose.

# 2 Seismic PSA Methodology

## 2.1 Overview of Methodology

The key elements of a seismic PRA are:

1. Seismic hazard analysis - estimation of the frequency of various levels of seismic ground motion (acceleration) occurring at the site

2. Seismic fragility evaluation - estimation of the conditional probabilities of structural or equipment failure for given levels of ground acceleration

3. Systems/accident sequence analysis - modeling of the various combinations of structural and equipment failures that could initiate and propagate a seismic core damage accident sequence

4. Evaluation of core damage frequency - assembly of the results of the seismic hazard, fragility and systems analyses to estimate the frequencies of core damage and plant damage states; and assessment of the impact of seismic events on the containment performance and integrity.

This process is described in detail in Reference [1,2]. About half of the nuclear power plant owners in the US chose to perform the IPEEE using seismic PSA. Seismic fragility evaluation of structures and equipment is a unique procedure in the seismic PSA and is described in the following.

## 2.2 Seismic Fragility Evaluation

The methodology for evaluating seismic fragilities of structures and equipment is documented in References [3, 4]. Seismic fragility of a structure or equipment item is defined as the conditional probability of its failure at a given value of the seismic input or response parameter (e.g., ground acceleration, stress, moment, or spectral acceleration). Seismic fragilities are needed in a PRA to estimate the conditional probabilities of occurrence of initiating events (i.e., loss of emergency AC power, loss of forced circulation cooling systems) and the conditional failure probabilities of different mitigating systems (e.g., auxiliary feedwater system).

The objective of fragility evaluation is to estimate the ground acceleration capacity of a given component and its uncertainty. This capacity is defined as the peak ground motion acceleration value at which the seismic response of a given component located at a specified point in the structure exceeds the component's resistance capacity, resulting in its failure. The ground acceleration capacity of the component is estimated using information on plant design bases, responses calculated at the design analysis stage, as-built dimensions, and material properties. Because there are many variables in the estimation of this ground acceleration capacity, component fragility is described by a family of fragility curves; a probability value is assigned to each curve to reflect the uncertainty in the fragility estimation, (Figure 1). This family of fragility curves may be described by three parameters; the median acceleration capacity Am, and logarithmic standard deviations, $\beta_R$ and $\beta_U$, for randomness and uncertainty.

In seismic margin studies, an indicator of the seismic margin of the component, HCLPF, is used. HCLPF is an acronym for high-confidence-of-low-probability-of-failure. It is defined mathematically as 95% confidence of less than 5% probability of failure. If the fragility curve is described by the median, $A_m$, the randomness, $\beta_R$, and uncertainty, $\beta_U$, the HCLPF may be computed from:

$$\text{HCLPF} = A_m \exp\left[-1.65\left(\beta_R + \beta_U\right)\right]$$

HCLPF could be calculated using the fragility analysis or could be estimated from deterministic procedures. The latter approach is called the Conservative Deterministic Design Margin (CDFM) method and is described in EPRI NP-6041 [5]

# 3 Seismic Margins Methodology

Two seismic margins methods with enhancements can be used for the seismic IPEEE. One acceptable methodology was developed for EPRI, and another was developed for NRC. The NRC provides guidance (NUREG-1407) for both the EPRI and NRC seismic margins methodologies. The objective of a seismic margins review of a plant is to assess if the plant can safely withstand an earthquake larger than the SSE. The margins methodology (both NRC and EPRI) utilizes two review or screening levels tied to peak ground accelerations of 0.3g and 0.5g. The review level earthquake (RLE) for different plants in the U.S. were assigned by the NRC based on the Lawrence Livermore National Laboratory (LLNL) and EPRI seismic hazard estimates, sensitivity studies, seismological and engineering judgment, and plant design considerations. The level of effort has been further divided into three scopes: a reduced-scope margins methodology which emphasizes plant walkdowns, a focused-scope methodology and a full-scope methodology. The level of effort in the analysis of relay chatter is the major difference between the focused-scope and full-scope methodologies. Discussion presented in the following is primarily applicable to the focused and full-scope seismic margin studies.

The EPRI methodology is based on a system "success path" approach. This approach defines and evaluates the seismic capacity of those components required to bring the plant to a stable condition (either hot or cold shutdown) and maintain that condition for at least 72 hours following an earthquake larger than the SSE. Several paths may exist. GL 88-20 Appendix 1 requires that an alternative success path be selected so that it involves to the maximum extent possible systems, piping runs and components that are different from the preferred success path.

The seismic margin review conducted for the seismic IPEEE follows the guidelines outlined in the EPRI seismic margin report [5]. Additional guidelines and enhancements for conducting the seismic margin assessment are contained in NUREG-1407 and the NRC Generic Letter No. 88-20 Supplement 4. The important elements of the seismic margins assessment are listed below:

1. *Selection of the Review Level Earthquake*
2. *Selection of Assessment Team*
3. *Preparatory Work Prior to Walkdowns*
4. *Systems and Elements Selection ("Success Paths") Walkdown*
5. *Seismic Capability Walkdowns*
6. *Seismic Margin Calculations*
7. *Documentation*

# 4 Results of Seismic IPEEEs

Most of the utilities have completed the IPEEEs of their nuclear power plants and the results are being reviewed by the NRC. In the following, some of the preliminary results that are generic in nature are discussed.

## 4.1 Seismic PRA

The mean core damage frequency for the plants studied so far is in the range of $10^{-5}$ per year; this represents about 25 - 30 % of the overall core damage frequency from internal and external events. Most plants report no unique/significant seismic vulnerabilities Plant walkdown was helpful in identifying low capacity components which could be upgraded with minor cost/effort.

Recent seismic PRAs have focused much attention on support systems including room cooling capabilities, relay chatter and human reliability for unique external events caused by seismic events. The seismic PRAs also confirmed the previous finding that seismic hazard uncertainty does not change the rank ordering of sequences. In terms of the risk contribution, it was concluded that structures rarely contribute to seismic risk; the electrical components and diesel systems dominate the seismic risk contribution. The contribution to seismic core damage frequency comes from earthquakes mainly in the range of 2 to 5 times the SSE

## 4.2 Seismic Margin Studies

Seismic margin studies conducted so far have shown that the plant seismic margin is generally above the SSE after certain minor upgrades have been implemented. It is to be noted that the plant margin could be below the RLE. The plant walkdowns were able to identify the seismic vulnerabilities in most cases. The dominant contributors to the calculated margin of the plants happened to be yard tanks and electrical components.

# 5 Applications for Plants Outside of US

Based on the IPEEEs conducted so far, the following lessons have been learnt for application to plants outside of US.

1. For plant sites with low seismic hazard, a seismic PSA is more beneficial since it can minimize the upgrades.

2. Seismic margin method draws upon the screening procedures developed for US designs and state of practice; applicability to other countries should be assessed. Walkdown by engineers with knowledge of the background and limitations of the seismic margin method and screening criteria is important.

3. Relay chatter evaluation could be significant because data on relays may not be available and the cabinet may not have been tested with the relays inside them; reliance on operator action in seismic events is not fully justified.

4. Uncertainty in seismic hazard is not typically considered in other countries since there is more consensus.

5. Although the US methods could be used in other countries; it should be done carefully. Seismic PSA method is more universally applicable compared to the seismic margin method.

6. Examination of the seismic IPEEEs of US plants would help in understanding the relative level of effort and the resources needed and the schedule of activities. Plant walkdown would take place around the plant outage.

# 6 References


1. Procedural and Submittal Guidance for the Individual Plant Examination of External Events (IPEEE) for Severe Accident Vulnerabilities" NUREG-1407, US Nuclear Regulatory Commission, June 1991.

2. "PRA Procedures Guide" NUREG/CR-2300, US Nuclear Regulatory Commission, January 1983

3. Kennedy, R.P., and M.K. Ravindra, "Seismic Fragilities for Nuclear Power Plant Risk Studies, Nuclear Engineering and Design, Vol. 79, No. 1, pp. 47-68, May 1984

4. "Methodology for Developing Seismic Fragilities," EPRI TR-103959, Electric Power Research Institute, Palo Alto, California, June 1994

5. "A Methodology for Assessment of Nuclear Power Plant Seismic Margin," Electric Power Research Institute, Palo Alto, California, EPRI NP-6041, June 1991.


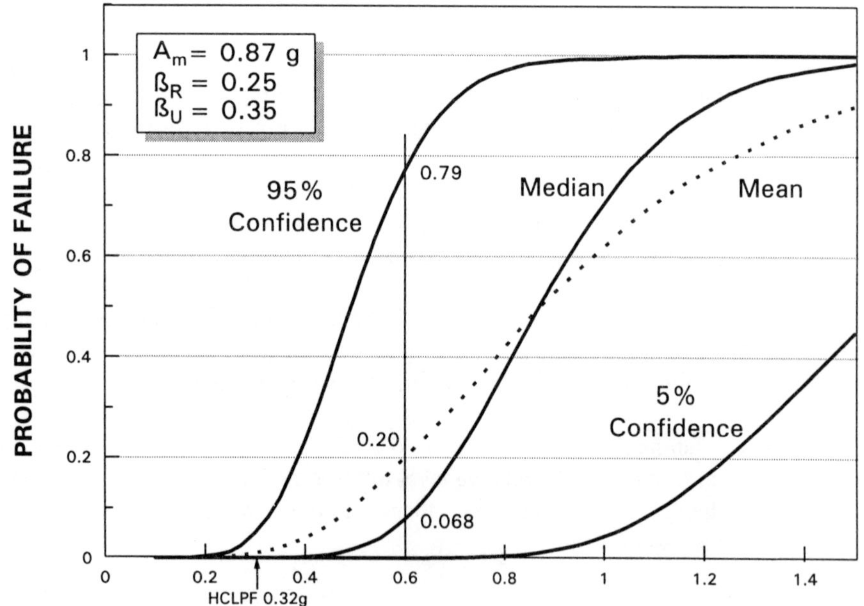


Figure 1: Typical Family of Fragility Curves for a Component

# Incorporating Uncertainty into Cost Estimation at the Time of Preliminary Design

Dan M. Frangopol
University of Colorado
Boulder, Colorado, U.S.A.

Ahmed Banafa
King Abdulaziz University
Jeddah, Saudi Arabia

**Abstract**

A fuzzy logic approach is proposed for incorporating uncertainty into cost estimation at the time of preliminary design. The study is exemplified on pile foundation projects at the feasibility stage. However, the results presented can be generalized to cover other types of projects and other project stages.

## 1 Introduction

Cost estimation is made prior to the detailed design stage. This can lead to large variations between the estimated and actual costs. Several attempts have been made to deal with the uncertainties associated with cost estimation. Most of these attempts deal with objective (i.e., random) uncertainties, in which historical data permits a quantitative assessment of the risks involved. Cost uncertainty, however, is not limited to the variability associated with randomness. Uncertainties associated with imperfect modeling and estimation play a major role in the estimation of project cost. In many cases, these uncertainties are more significant than those associated with randomness.

At the preliminary design stage, prediction of several variables affecting the cost of the project have to be made using linguistic terms, such as "high" and "good". Therefore, a cost estimation model able to include linguistic variables at the time of preliminary design is necessary. Such a model is proposed in this paper. In the proposed model, regression analysis is used to treat the historical data in order to construct the relationship between independent and dependent variables. The fuzzy set theory is then used as an appropriate tool to treat the uncertainties described in linguistic terms. Owing to space constraints, the paper is limited to the development of the cost membership function and the brief description of a numerical example.

## 2 Fuzzy Approach

The cost estimation methods may be classified into detailed and preliminary estimating methods. Detailed methods rely on quanitities obtained from a complete set of drawings and specifications of a project. Preliminary cost estimation methods are performed at the earliest stages of a project. At these stages less information is available. Moreover, this information is also less accurate. The only thing known about some variables affecting cost may be "verbal descriptions" such as: the price may be "high", the weather may be "good", the quality may be "low". Usually, variables like these are omitted from the model. These omissions result in errors.

The objective here is to use fuzzy logic to handle these types of variables. This logic was implemented in a computer program called DREC (Design Reliability for Estimating Cost). DREC is an interactive program developed in FORTRAN 77. It handles both continuous and discrete variable membership functions. The flowchart of the Main Program was described in [1, 2]. Some of its subroutines are described below.

## 2.1 Membership

The flow-chart of the subroutine membership is shown in Fig. 1. This subroutine generates the predicted cost values for each set of the historical data using the linear $L$, power $P$, and multiplicative $M$, cost function forms given in [2]. The calculation of the cost membership value for the data point $i$, $\mu_{Ci}$, is made using the extension principle, as follows:

$$\mu_{Ci} = \max\left[\min\left(\mu_{i1}, \cdots, \mu_{ij}, \cdots, \mu_{im}\right)\right] \tag{1}$$

where $\mu_{ij}$ = membership value for the variable $j$ ($j = 1, \cdots, m$) at the data point $i$ ($i = 1, \cdots, n$). Inputs from this subroutine are used in the following subroutines.

## 2.2 Sorting

This subroutine sorts the cost data, $C_i$, in ascending order and then rearranges the membership values, $\mu_{ij}$, accordingly.

## 2.3 Grouping

This subroutine (a) groups the cost values in a narrow range $R$ (i.e., $|\hat{C}_1 - \hat{C}_2| \leq R$) into a single value (i.e., max $(\hat{C}_1, \hat{C}_2)$), and (b) calculates the fuzzy probability for each cost value, $P(\hat{C}_i)$ as follows

$$P(\hat{C}_i) = \mu_{Ci} / \sum_{i=1}^{k} \mu_{Ci} \tag{2}$$

where $k$ = number of grouped cost values.

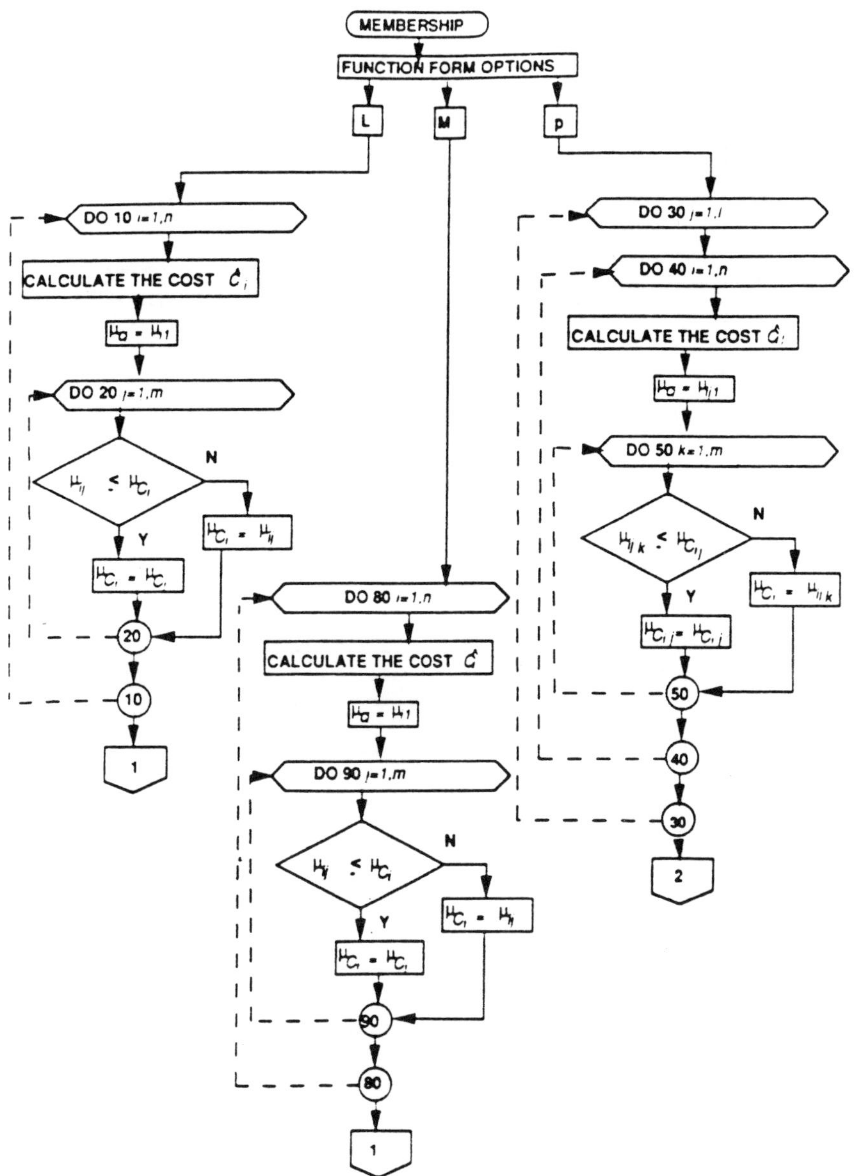


Figure 1 Flow-chart of Subroutine Membership

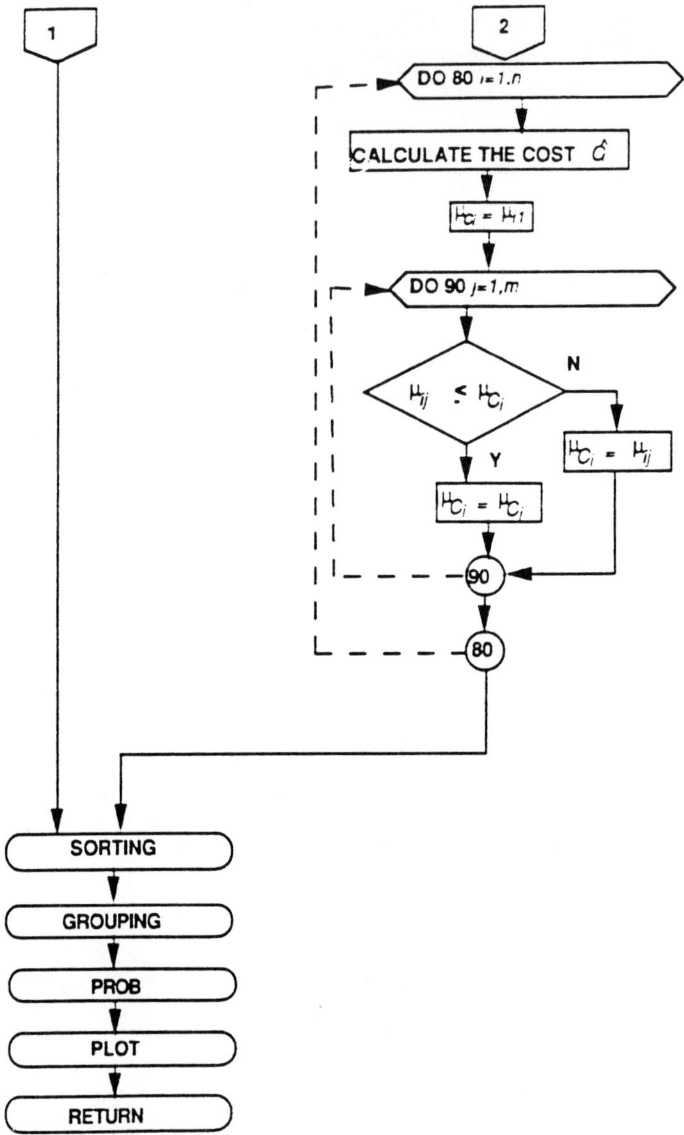


Figure 1 Flow-chart of Subroutine Membership (continued)

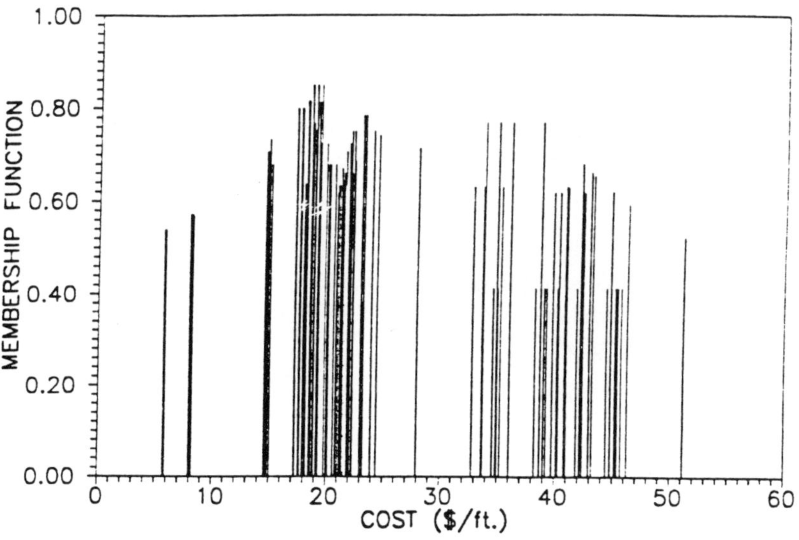


**Figure 2 Membership Function of Cost**

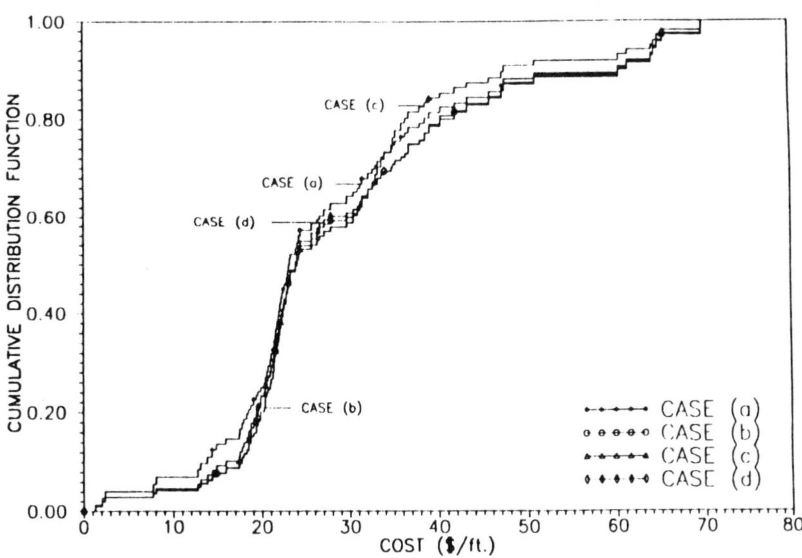


**Figure 3 Cummulative Distribution Function of Cost**

### 2.4 Probability (Prob)

This subroutine calculates (a) the project cost $C_N$ associated with linear, power, and multiplicative cost functions, (b) the project cost fuzzy probability, (c) the cost that corresponds to any given probability, and (d) the cost fuzzy probability that corresponds to any given cost range.

### 2.5 Plot

In this subroutine plots of the cost mass probability, cost membership, and cost fuzzy probability functions are generated. A detailed description of the above subroutines is presented in [2].

## 3 Results

To demonstrate the validity of the proposed approach, the cost estimation of pile foundation projects at the preliminary design stage was investigated. Examples of results are presented in Figs. 2 and 3, which show examples of the membership function and the cummulative distribution function of the cost. The cases (a), (b), (c), and (d) in Fig. 3 are associated, respectively, with low, average, high, and a combination of low, average, and high membership functions of all variables considered in computations.

## 4 Conclusions

An efficient fuzzy approach for incorporating uncertainty into cost estimation at the time of preliminary design was presented. Further effort has to be made to consolidate the proposed approach. Especially, reliable data for creating membership functions of the fuzzy variables affecting cost must be assembled.

## Acknowledgements

The support from the U.S. Army Corps of Engineers through Contracts DACW 39-89-M-4488 and DACW 39-90-C-0064 is gratefully acknowledged. It made it possible for the second author to pursue graduate research work towards a Ph.D. degree in the Department of Civil Engineering, University of Colorado, Boulder.

## References

1. Banafa A., Frangopol, D.M., Mosher, R.L., Reliability model for cost estimation of pile foundations. In: Uncertainty modeling and analysis. IEEE Computer Society Press, Los Alamitos, California, 1993; 129-135.

2. Banafa, A., Design reliability for estimating cost of pile foundations: From theory to application. Ph.D. thesis, University of Colorado, Boulder, 1991.

# An Operating PSA Application at EDF: the Probabilistic Incident Analysis

A. Dubreuil-Chambardel, P. François, H. Pesme, B. Maliverney
Electricité de France, DER and DEPT
Paris - France

## 1. Why Analyze the Potential Consequences of Incidents?

Operational nuclear safety is defined as all the steps taken in a nuclear plant to prevent the release of radioactive substances into the environment. This definition poses a practical problem: can the efficiency of these steps be measured? In other words, how is safety measured?

The answer to this question is not a simple matter, for the absence of serious accidents does not constitute proof that the level of safety is good.

Simply enumerating incidents, however, does give an indication of the level of safety of a plant (although the question of detection of incident situations and their declaration by the operator is an essential factor). But this approach is not sufficient in itself, for it oversimplifies the diversity of incident situations encountered.

The following questions are therefore raised:

- how could an incident have degenerated into an accident with more serious consequences?
- is it possible to measure what separates an incident from the potential accident with more serious consequences?

This is the purpose of probabilistic incident analysis.

This problem is not specific to the field of nuclear power. But nuclear power plant (NPP) operators are rather special in that they have been developing Probabilistic Safety Analysis (PSA) for quite some years. These analyses constitute models of the operation of plants in accident situations. They have a set of accident scenarios which can result in unacceptable consequences. These accident scenarios are quantified by probabilities. These probabilistic models constitute the data base from which it is possible to build possible scenarios for aggravation of a real incident.

Probabilistic analysis based on the consequences of an accident is therefore different to analysis of causes (which remains indispensable, however). Of course consequences have always been analyzed to some extent, but not with any real method. Probabilistic analysis provides a method. The procedure developed by EDF therefore consists in applying the models of scenarios from PSAs to the most significant incidents.

In the early 1980s, the US Nuclear Regulatory Commission (NRC) made the first attempt at systematically analyzing incidents in American nuclear power plants. However the analyses were highly simplified due to the great diversity of American plants. The standardization of French units means our operational models and reliability data are closely adapted to the incidents to be analyzed.

# 2. The EDF Programme

The programme for probabilistic analysis of incidents in EDF NPPs is referred to as the "precursors" programme. It was launched by EDF in 1993. It was initiated after exploratory studies carried out at the international level, especially with the NRC. It involves identifying and analyzing all the incidents that might degenerate and result in more serious consequences such as reactor damage (called "core meltdown") or radioactive release.

This programme is based on a method for systematic selection of incidents and on a method for analysis of the selected incidents. The selection method is based on a set of qualitative criteria which identify "safety-significant" incidents. The purpose of the analysis method is to assess the conditional probability of core meltdown subsequent to the incident and to identify its main modes of possible degradation.

At the end of 1995, we were able to weigh up the systematic analysis of the incidents that occurred in 1994 throughout EDF NPPs. The methods have been stabilized and we are now trying to incorporate them in all the methods for analysis of operating feedback.

# 3. Varied Objectives

The objectives of the analysis method can be classified in four main categories.

- Measuring the seriousness of the incidents : this is called quantitative incident analysis. The risk measure is used to detect the most important incidents, to prioritize treatment, and to assess the level of safety of an individual unit or of EDF NPPs as a whole.
- Finding the qualitative lessons to be learnt from the real incidents used as precursors of incipient, potentially more serious accidents: this is what we call qualitative incident analysis, or accident feedback.
- Promoting a risk analysis approach among operators. This risk analysis approach looks at the potential consequences of incidents and involves probabilistic methods. This objective is part of the setting up of a "probabilistic safety culture".
- The fourth category more particularly concerns PSA designers. Incidents on NPPs are a means of comparing PSAs with real events. Incident analysis is therefore a means of validating and enhancing PSAs.

Special stress is put here on the value of accident feedback: by prolonging incidents to their potentially serious accident consequences, a lot can be learnt about accidents, on the basis of common incidents, without suffering their real consequences. The method thus makes it possible to learn from non-events in the same way as we usually learn from real experience.

# 4. Principle of the Analysis Method

Probabilistic analysis is a method for assessment of the probability of potential unacceptable scenarios that might have occurred after a real incident. It is important to understand what is meant by "incident" and "potential consequence". An incident may have no real consequence on the operation of a plant: the unavailability of a backup system has no real effect on operation as long as no accident transient occurs. The risk then lies in the potential consequences of the incident: if the plant were to reach a position where the unavailable system would be necessary for safety, the consequences would be more serious. Probabilistic analysis sets out to envisage such eventualities: it takes account both of real transients and, when they aggravate the situation, potential transients.

A nuclear power plant is continually evolving in a risk space which is limited by a lower threshold (everything is available, safety is optimum) and an upper threshold (unacceptable accident). The level of risk fluctuates slightly above the lower threshold and deviates considerably from it in the course of an incident. Probabilistic analysis involves identifying the potential scenarios of aggravation of the situation and quantifying the variation in risk.

The method consists in applying event trees derived from PSAs to an actual incident (in this case the incident is considered to be an initiating event : this is the case of operating transients), and/or in modifying the existing trees to take account of the effect the incident has on the degradation of safety functions (this is the case when equipment is unavailable). The analysis is then performed up to the point of assessment of the conditional probability of core meltdown, thus identifying the seriousness and main sequences of the incident, i.e. the most probable degradation modes. Probabilistic incident analysis therefore involves reasoning in terms of conditional probabilities: "given that the incident has occurred, what is the probability of core melt?" This results in calculation of the increased risk due to the incident, i.e. the conditional probability of core meltdown once the incident has occurred

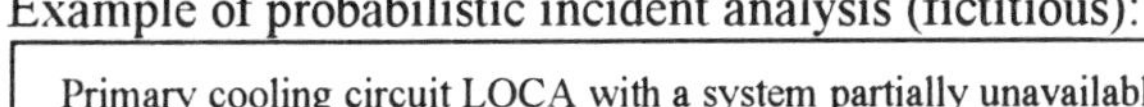
Example of probabilistic incident analysis (fictitious):

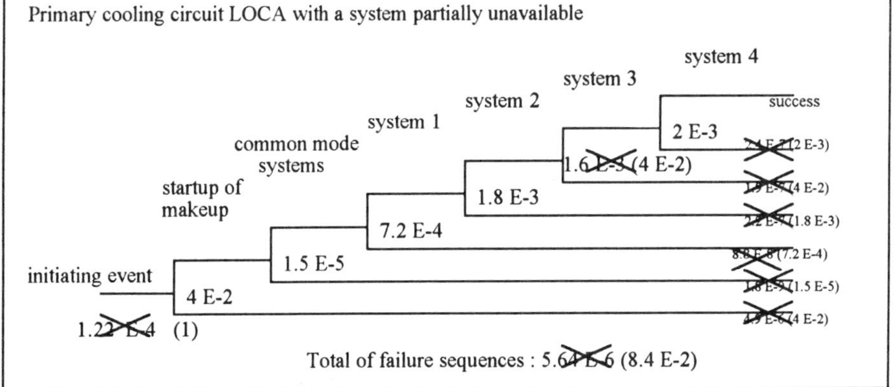

The incident (fictitious) described here is a small Loss of Coolant Accident occurring in a shutdown state with one safety injection system unavailable (system 3). The appropriate tree is taken from the EPS 900 PSA. Two probabilities are then modified to represent the incident:

- the probability of the initiating event (1.22 E-4/year) is replaced by 1
- the probability of the event occurring with two available trains (1.6 E-3) is replaced by a value corresponding to a single available train (4 E-2).

The total risk rises from 5.64 E-6 per reactor-year for this type of initiating event to 8.4 E-2, which is the conditional probability of core melt of the incident.

To assess a given risk, all the factors that describe the incident are examined, and all those that are likely to have an appreciable aggravating effect on consequences are identified. For example, a real equipment failure may have no effect on the state of the plant at the time it occurred, but could have had a detrimental effect if the plant had been in another state at the time. The initial state of the plant is therefore an important parameter that has to be varied in order to envisage the potential consequences of the failure.

Analysis has identified nine types of parameters in the PSA that can be varied in accordance with the actual situation:

i) hourly rate of occurrence of an event ($\lambda$)
ii) probability of occurrence of an event ($\gamma$)
iii) probability of occurrence of human error
iv) frequency of an initiating event
v) rate of common cause failure
vi) hourly rate for repair of an equipment item
vii) type of reactor (900 or 1300 MW)
viii) initial state of reactor
ix) chronology of events occurring in the incident.

With the exception of unfavourable events which actually occurred in the course of the incident, and whose probability is set to 1, each variation in the parameters is weighted by its usual probability of occurrence. For example, if a transient is more detrimental for a reactor in a state which represents 10% of the annual duration, the probability of being in that state is taken as 0.1, except if the reactor was indeed in that unfavourable state, in which case the probability is 1.

The sum of probabilities on all accident sequences is therefore equal to the conditional probability of unacceptable consequences once the incident has occurred.

An "accident precursor" is any incident whose conditional probability of core melt is greater than $10^{-6}$ when all the parameters are adjusted.

# 5. Some Significant Results

Among the 400 significant incidents declared by EDF power plants in 1994, 114 were selected, analyzed, and classified in accordance with their conditional probability of core meltdown. The graph hereafter shows the ten most significant incidents for the year, classified in accordance with the conditional probability of core meltdown due to the incident.

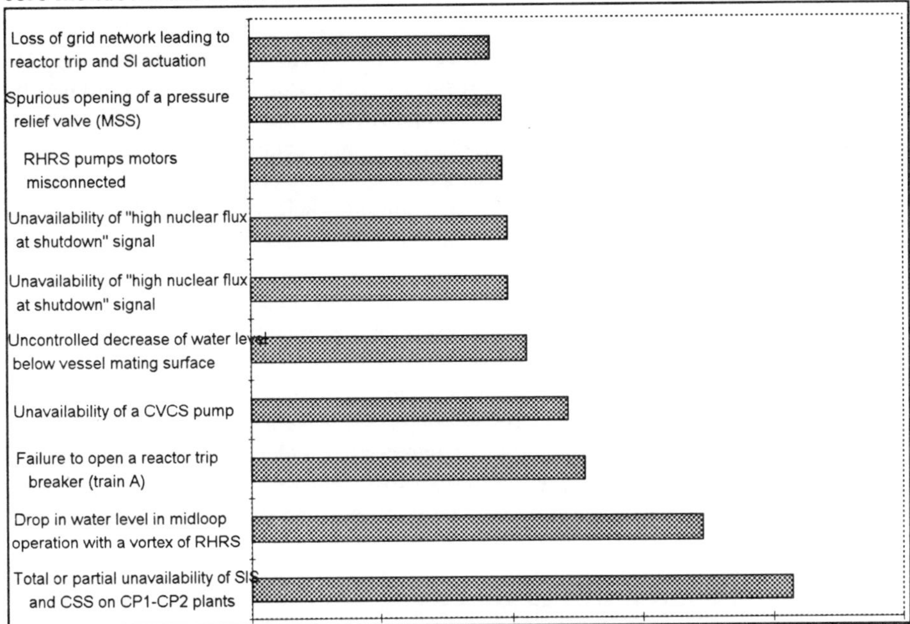


The sum of these probabilities can give an indication of the level of safety of the power plants, but this indicator is of a markedly random nature: the figure obtained is representative only for the 2 or 3 most significant incidents in the year, which means there is no guarantee of representativity of the level of safety of all power plants, since these incidents could occur a few months earlier or later and thus completely change a yearly indicator, without that variation being of any relevance with respect to the real safety of all the power plants.

In addition, some results are very sensitive to the calculation assumptions. This is the case when it is necessary to assess the probability of failure of a component whose characteristics are impaired (cracking of a weld for example), or when it is necessary to assess the rate of common cause failure for several components (probability of failure of a sound component, when an identical component has already failed). These problems are encountered when calculating the probabilities of failure of manual operations. In all these cases solutions have been proposed, but will have to be "fine tuned" with experience. In addition, we have models developed for PSAs (like Probabilistic Human Reliability Analysis). Despite this, the conversion of a real incident into a set of probabilities still involves some major difficulties.

However, assessment of seriousness does make it possible to rank incidents. Using this method, therefore, it is possible to attenuate the importance of incidents that have been highlighted by the media (loss of cold sources, wiring anomalies, etc.) and, on the contrary, to draw attention to incidents which were previously judged minor on the basis of conventional assessment criteria (e.g. failure to close of a valve on steam lines, unavailability of certain alarms such as the "high flux on shutdown" alarm, etc.). Moreover, the scenarios highlighted by probabilistic analysis sometimes contain sequences that surprise specialists in accident operation (certain cumulative failures).

The value of qualitative analysis of impairment sequences can be seen in practically all incidents: it is clear that announcing a numerical value associated with an incident is not enough: the numerical result in each case must be substantiated by describing at the very least the main impairment sequence. In most cases, a description of one to three sequences is enough to appreciate the essential elements of the risk. This qualitative analysis can be used to determine the most relevant "operating feedback" actions and to substantiate their urgency. This form of operating feedback can take on a variety of forms: design modifications substantiated by the results of analysis (automatic interruption of dilution by the RCV Chemical & Volume Control System, automatic makeup of primary coolant in midloop operation), modifications to operation documents specifying the control requirements in the event of incidents or equipment unavailability, etc.

The use of probabilistic analyses in power plants in order to enrich safety culture is considered to be a particularly interesting objective. Risk analysis can benefit from knowledge of the potential aggravation sequences of incidents (sequences described in PSAs and used in probabilistic analyses). Probabilistic analysis gives operators the possibility of better identifying risk situations and thus preventing certain incidents.

# 6. Conclusion and Perspectives

In relation to this analysis method one might ask "What good is it to give a weather forecast for the day before?". What is the point of producing a method that determines risk after the event? It could even be said that the *post hoc* probability of core melt is always zero since it has not occurred. In fact this method gives another point of view on safety: with respect to lessons learnt, it is decided to attribute as much importance to "near misses" as to accidents that actually happened. This makes use of the large knowledge base in the Probabilistic Safety Assessment models.

This method can be used to understand "how serious it is", but also, and above all, "why it is serious". Whereas one is limited to a risk indicator when one is not in a position to understand "why it is serious", the probabilistic method should make it possible to understand what sort of risk is involved.

Probabilistic incident analysis is therefore seen as an additional means for using operating feedback from NPPs: it provides a method for qualitative analysis of potential consequences. Through analysis of "impaired lines of defence" or of the frequency of the initiating events observed, it makes it possible to propose operating feedback actions specifically related to the lessons learnt from this type of analysis, or to confirm the urgency of the actions proposed by other analyses.

The experimentation carried out in 1994 is therefore conclusive and should be extended: the method will evolve and will gradually be incorporated into EDF's operating feedback structures.

# The Weibull distribution in Control Charts and in some Maintenance Strategies

M.F. Ramalhoto
Maths Dept.; Instituto Superior Técnico; Technical University of Lisbon
Av. Rovisco Pais, 1000 Lisbon, Portugal (Fax +351-1-8499242)

## 1 Introduction

This paper emphasizes the increasing need to construct adequate control charts linked to holistic maintenance schemes, in modern quality management. The importance of the Weibull distribution to construct adequate control charts is stressed.

The most important results in the current literature on the three parameter Weibull distribution are reviewed in section 2. In section 3 it is shown the importance of the Weibull distribution assumption in control charts: (a) as the distribution of the quality characteristic; (b) as the distribution of the time to the occurrence of a special cause. A short review of the current literature concerning (a) and (b) is given. Section 4 deals with control charts linked to maintenance schemes; it is suggested to look at control charts also as the sampling part of holistic maintenance schemes, in order to, for instance, help to set up more adequate inspection times. In section 5 a brief summary of recent development in control charts/maintenance schemes using the Weibull distribution, is given. The need to develop new models and methods in control charts/maintenance schemes and its quick transfer to industry is stressed in section 6.

### Control Charts

In any organization, there are always too many signals, concerns and wish lists. Prioritization and focus are extremely useful. The control charts provide reassurance to the managers and also the basis to prioritization of special improvement efforts. Control charts enable the operators (or different shifts) to aim at a common target value, thus reducing the final product-to-product variation. Furthermore, control charts provide a common language through the organization.

Use of control charts serves to divide and define the responsibility between the operators and management: (a) it is the operators' responsibility to keep the process on target (largely by using a X control chart type); (b) the R control chart type is used by the operator to alert the management, when the variation of the process (which is the responsibility of the management) is getting excessive.

### What is a control chart?

Control charts are simple graphical devices which give a way to obtain a process variability information by perspective observation of the process.

According to [1] and [2], the design of a control chart should be such that, when and only when, the control chart statistics falls outside its control limits, it will be possible to find a "special cause of variation". (According to the same authors, a "special cause of variation" is one that can be found by experiment, without costing more that it is worth to find (clearly, a special cause today might not be one tomorrow)). By construction control charts provide an operational definition of special causes.

### What Do We Hope to Achieve by Using Control Charts?

- To detect process changes more quickly so that timely action can be taken to restore the process to its best performance (short term process improvement).
- To achieve a long-term relation in variability (long term process improvement).

In fact, control charts are most powerful when used as tools for continuous improvement and when used close to the production process (they are well suited to shop-floor applications). They enhance process knowledge and quickly reveal opportunities for improvement, when adequately designed. So, the potential user of control charts may have two objectives: - the detection of process change and the long term reduction of variability. Surely, it is reasonable to assert that either or both of these objectives will be more easily achieved if the action lines on the control charts are appropriately placed.

## 2 The Weibull Distribution

The three-parameter Weibull distribution is very rich, it can be used to simulate various practical situations by varying its location (threshold), scale and shape parameters. The two-parameter exponential distribution is a special case of the three-parameter Weibull distribution with $\delta=1$.

If X is a three-parameter Weibull distribution ($X\sim W(\eta,\delta,\alpha)$) its distribution function is given by:

(1) $$F(x)=P(X\le x)=\begin{cases}0, & x\le\eta\\ 1-e^{-[(x-\eta)/\delta]^{\alpha}}, & x\succ\eta\end{cases}$$

$$-\infty\prec\eta\prec+\infty;\delta\succ 0;\alpha\succ 0$$

where $\eta$, $\delta$ and $\alpha$ are the location, scale and shape parameters, respectively [3, p.250] (though, there is no mathematical reason for $\eta$ not to be negative, in most of the applications $\eta\ge 0$). Its failure rate function, $\lambda(x)=f(x)/[1-F(x)]$, is constant if $\alpha=1$, decreasing failure rate (DFR) if $\alpha>1$ and increasing failure rate (IFR) if $\alpha<1$.

For $\alpha>1$ and for $0\prec\alpha\prec 1$, the mode ($m_0$) is, respectively

(2) $$m_0=\eta+\delta[(\alpha-1)/\alpha]^{1/\alpha};\ m_0=\eta.$$

The median ($m_e$), the mean and the variance are given, respectively, by

(3) $$m_e=F^{-1}(0.5)=\eta+\delta[\ln 2]^{1/\alpha}.$$

$$(4) \qquad E(X) = \eta + \delta\Gamma(1/\alpha + 1),$$

$$(5) \qquad Var(X) = \delta^2\left[\Gamma(2/\alpha + 1) - \Gamma^2(1/\alpha + 1)\right].$$

In [3; p.273] it is mentioned that - a Gumbel distribution of minima can be considered as a Weibull distribution with "a large" shape parameter. Considering that the location parameter is known, confidence intervals for the parameters, $\delta$, $\alpha$, based upon maximum likelihood estimates are given by [4] and by [5],[6] and [7]. These make use of the fact that the pivotal quantities $\hat{\delta}/\delta$ and $\hat{\alpha}\left(\ln\hat{\delta} - \ln\delta\right)$ have densities that do not depend upon the values of $\delta$ and $\alpha$.

[8] and [9] consider inference procedures based upon maximum likelihood estimator conditional upon the ancillary statistics. The resulting confidence intervals are just slightly shorter that the simpler ones based upon maximum likelihood estimators [10] shows that approximately

$$0.822n\alpha^2/\hat{\alpha}^2 \sim \chi^2(0.822(n-1)); \sqrt{n-1}\hat{\alpha}\left(\ln\hat{\delta} - \ln\delta\right)/1.053 \sim t(n-1)$$

these leads to convenient test of hypothesis and confidence intervals.

Sample sizes needed to choose the better of two Weibull distributions are given by [11]. [12] and [13] considered the probability of choosing the better of two Weibull distributions assuming a common shape parameter, $\alpha$.

A characteristic value of $\alpha$ may indeed be known in many situations. If this is the case, it was shown by [14] and by [15] that the maximum likelihood estimates for $\eta$ and $\delta$ always exist and are unique. If either $\eta$ or $\alpha$ are assumed to be known the maximum likelihood estimates of the remaining parameters always exist and are unique.

When all these parameters, $\eta$, $\delta$ and $\alpha$ are unknown, the likelihood function is known to be unbounded as $\eta \rightarrow min(X_1,\ldots,X_n)$. If it is assumed that $\alpha > 1$ then the likelihood function is bounded and often has a maximum. However, one should not rush into solving the likelihood equations because if they have one solution, they will for sure have at least two,[14]. [16], [17], [18] and [19] study the three-parameter Weibull distribution. The last paper gives confidence intervals for $\alpha$ and $\eta$ and $\delta$ unknown.

Models that generally compete with the Weibull are the exponential (one and two parameters), lognormal and gamma. [20] also indicates procedures for testing the Weibull distribution against its above competitors.

# 3 The Weibull Distribution in Control Charts

In the theory of control charts the Weibull distribution assumption is important in the flowing two situations:

1. The underlying distribution of the quality characteristic (under control consideration) is assumed to be a Weibull distribution.
2. The time to the occurrence of a special cause follows a Weibull distribution.

**Situation 1**

The consultation of the usual reference sources in the field of statistics in industrial and management applications, the only reference found which considers the Weibull distribution as the probabilistic model of the quality characteristic under study is [21]. This paper considers a two parameter Weibull distribution with $\eta=0$.

The normal distribution is generally acknowledged as a suitable probability law for many examples of continuous scale quality characteristics. However, the normal distribution has a symmetric density which is positive on the entire real line. These properties are inadequate for situations which require a threshold value $\eta$ (density 0 left of a $\eta$) or a skewed non-symmetric density. These requirements hold, for instance, for lifetime quality characteristics. An appropriate and very flexible distribution for such situations is the three-parameter Weibull distribution.

**Situation 2**

Models for choosing control chart design parameters according to economic considerations, specifically - expected total cost minimization- (i.e. the so called economic control charts) can be found in the quality control literature, see for instance, [22].

Economic control chart models typically assume that the time to the occurrence of a special cause follows an exponential distribution. Which is all right, if it is reasonable to assume that the associated shock model follows an homogeneous Poisson process (perhaps, it should be noticed that, when the shock model is a non-homogeneous Poisson process with intensity $\lambda(t) = \alpha t^{\alpha-1} / \delta^{\alpha}$, the time to first occurrence has a Weibull distribution of parameters $\delta$ and $\alpha$).

Some papers have been published concerning economic control charts where the underlying process failure mechanism is assumed to be a Weibull distribution (i.e. addressing situation 2).

[23] and [24] are some of the important papers found in the current literature that use the Weibull model instead of the exponential model. They concluded that if a constant sampling interval (FSI) is used the difference in expected loss is negligible under exponential and Weibull process failure mechanism. However, if a variable sampling interval (VSI) is used, the assumptions regarding the distributions of the process failure mechanism (exponential or Weibull) significantly affects the expected loss.

[25] consider an economic control chart, $\overline{X}$ - chart with a Weibull distribution for the failure mechanism and control chart design parameters varying over time (called the dynamic $\overline{X}$ - chart), where the Duncan's (1956) approach to the economic design of a $\overline{X}$ - chart was employed. They reported that the dynamic $\overline{X}$ - chart performs always better than Duncan's $\overline{X}$ - chart (in [26]), when the underlying distribution of the process failure mechanism is Weibull instead of exponential.

# 4 Control Charts and Maintenance Schemes

Control charts give indications about the reason for out-of-control situations and might be used to prioritize areas where, for instance, preventive maintenance should be used. Further, the effect of a preventive maintenance program can be seen in the control charts as a decrease in the machine failure(s) targeted by the program. Also, before a machine is returned to production from maintenance, a capability test should be performed to ensure that the machine is ready for production (i.e. an "acceptance test").

Therefore, because usually sampling is much cheaper than inspection, it is believed that, it makes sense to use control charts as the sampling part of a holistic maintenance scheme: (a) *to establish convenient times of inspection* (i.e., an alarm can lead to an inspection which might lead to a repair or even a renewal of some of the machinery and equipment in use); (b) *the maintenance and reliability known results about the equipment and machinery in use, might help to create adequate control charts with appropriated sampling interval policies.*

# 5 Recent Developments in Control Charts/Maintenance Schemes using the Weibull Distribution

*a)*The integration in the construction of control charts of probabilistic reliability results and other results coming from maintenance schemes are somehow already in the set up of the economic design of control charts. In [27] it is presented a complex process model, which generalizes the Collani's economic control chart model. The objective is to give a general framework to develop more satisfactory (useful in practice) economic control charts, able to properly incorporate more reliability results and more sophisticated maintenance schemes into its construction. However, many obstacles remain in the theory of economic control charts, for instance - they ignore statistical performance of the control charts and include costs that are very difficult or even impossible to obtain.

*b)*The classical control chart models for continuous scale quality characteristics has always concentrated on normal quality characteristics. Quality characteristics of lifetime type, which typically follows a Weibull distribution, have been widely ignored in spite of its practical importance (as already mentioned, in the literature only [21] seems to be dealing with this situation). On the other hand, in destructive testing situations, it is reasonable to take only one or two units in each sampling, (i.e., it is not acceptable to replace the Weibull distribution by the normal distribution). Therefore, control charts for quality characteristics following a Weibull distribution, are needed. [28] presents simple control charts for the location parameter of the three-parameter Weibull quality characteristic with fixed and variable sampling intervals (when the two other parameters are assumed to be known) .In this paper, among other results, it is shown that for lower sided control charts of the two parameter exponential quality characteristic ($\alpha$=1),

$$(6) \qquad ATS_{VSI}(\theta) = ATS_{FSI}(\theta) \qquad\qquad \infty \prec \theta \leq 0.$$

However, if, $\alpha \succ 1$ we have $ATS_{VSI}(\theta) \prec ATS_{FSI}(\theta)$; where $ATS_{FSI}(\theta)$ and $ATS_{VSI}(\theta)$ mean the average time to signal for a shift of magnitude $\theta$ of the location parameter , $\eta$, under fixed and variable sampling intervals, respectively. [29] presents simple control charts for the scale parameter of the three-parameter Weibull quality characteristic with fixed and variable sampling intervals (when the other two parameters are assumed to be known).

*c)*Process variability has to be taken very seriously because it implies usually process deterioration. To properly study process variability it is very often required the implementation of a broader data analysis strategy, i.e. a time series approach which moves away from the simplicity of the simple control charts. [30] recommended the exponential weighted moving average (EWMA) model as an all-purpose model that will work reasonably in most situations, reducing the need for broad time series training. [31] presents EWMA control charts for the scale parameter of the three-parameter Weibull quality characteristic with fixed and sampling intervals (when the other two parameter are known). It is perhaps worthwhile noticing that, when the control charts are integrated in a holistic maintenance scheme, for instance: (i) a switch, in a variable sampling interval control chart ($d_1$, $d_2$, $d_1 < d_2$), from the longer sampling interval ($d_2$) to the shorter sampling interval ($d_1$) might be followed by a quick simple inspection in the most sensitive equipment areas without causing production interruption; (ii) reliability and maintenance results about machinery and equipment might be used to choose the adequate lengths of $d_1$ and $d_2$.

# 6 Conclusions

The study of new models and methodologies in the field of control charts/maintenance and its quick transfer to industry, are important due to the following reasons:

*With adequate control charts/management schemes properly implemented:*

- it becomes possible, for instance, *to start work on improving material flow and to reduce the work-in-progress* (this in turns leads to a tidier environment);
- *the outcome of the process becomes more predictable* (this is a basic requirement for "just-in-time" production which involves extremely short lead times and basically an absolute minimum amount of work in process);
- it is possible *to identify and quantify problem areas objectively, enabling priorities for quality improvements and investments to be set.*

(This is why control charts are more and more regarded as an important element of total quality management (TQM)).

## References


1. Shewhart, W.A. Statistical method from the view point of quality control. New York: Dover Publications, 1939.
2. Deming, W.E. The new economics. Boston: Massachusetts Institute of Technology Press, 1993.
3. Johnson, N.L. and Kotz, S. Distributions in statistics: continuous univariate distributions, I. Wiley, New York, 1970.
4. Thoman, D.R., Bain, L.J., and Antle, C.E. Inferences on the parameters of the Weibull distribution. Technometrics, 1969, 11:445-460.

5. McCool, J.I. Inferential Techniques for Weibull Populations. Tech. Reps. 74-0180 and AD A-009645, ARL, Wright-Patterson, AFB, OH, 1974.
6. McCool, J.I. Inferential Techniques for Weibull Populations II. Tech. Reps. ARL 76-0233, ARL, Wright-Patterson, AFB, OH, 1975.
7. McCool, J.I. Analysis of single classification experiments based on censored samples from the two-parameter Weibull distribution. J. Statist.Plann Inf.., 1979,.3, 39-68.
8. Lawless J.F. Construction of tolerance bounds for the extreme-value and Weibull distributions. Tecnhometrics, 1975, 17:255-261.
9. Lawless J.F. Confidence interval estimation dor the Weibull and extreme value distributions. Tecnhometrics, 1978, 20:355-364.
10. Bain, L.J. and Engelhardt, M.E. Simple approximate distributional results for confidence and tolerance limits for the Weibull distribution based on maximum likelihood estimators. Technometrics, 1980, 23:15-20.
11. Rademaker, A.W., and Antle, C.E. Sample size for selecting the better of two populations. IEEE Trans. Rel., 1975, R-24:17-20.
12. Qureishi, A.S. The discrimination between two Weibull processes. Technometrics, 1964, 6:57-75.
13. Thoman, D.R., and Bain, L.J. Two sample tests in the Weibull distribution. Technometrics, 1969, 11:805-815.
14. Rockette, H., Antle, C.E., and Klimko, L.A. Maximum likelihood estimation with the Weibull model. J. Amer. Statist. Ass., 1974, 69:246-249
15. Peto, R., and Lee, P.N. Weibull distributions for continuous carcinogenesis experiments. Biometrics, 1973, 29:457-470.
16. Lemon, G.H. Maximum likelihood estimation for the three parameter Weibull distribution based on censored samples. Technometrics, 1975, 17:247-254.
17. Cohen, A.C. Multi-censored sampling in three-parameter Weibull distribution. Technometrics, 1974, 17:347-352.
18. Zanakis, S. A simulation study of some simple estimators for the three-parameter Weibull distribution. Journal Statistical Computation and Simulation, 1979, 9:101-116.
19. Wyckoff, J., Bain, L. and Engelhardt, M. Some complete and censored sampling results for the three-parameter Weibull distribution. Journal Statistical Computation and Simulation, 1980, 11:139-151.
20. Kotz, S., Johnson, N. and Read, C. (editors). Encyclopedia of Statistical Sciences, John Wiley & Sons, New York, vol.8, 1982.
21. Johnson, N.L. Cumulative sum control charts and the Weibull distribution. Technometrics, 1966, 18:481-491.
22. Ho, C. and Case, K.E. Economic design of control charts: a literature review for 1981-1991. Journal of Quality Technology, 1994, 26:39-53.
23. Banerjee, P.K. and Rahim, M.A. Economic design of $\overline{X}$ control charts under Weibull shock models. Technometrics, 1988, 30:407-414.
24. McWilliams, T.P. Economic control chart designs and the in-control time distribution - a sensitivity analysis. Journal of Quality Technology, vol.21, 1989, 2:103-110.
25. Parkhideh, B. and Case, K.E. The economic design of a dynamic $\overline{X}$ control charts. IIE Transactions, 1989, 21:313-323.
26. Duncan, A.J. The economic design of $\overline{X}$ charts used to maintain current control of a process. Journal of the American Statistical, 1956.
27. Göb, R., Beichelt, F., Dräger, K., Ramalhoto, M.F. and Schneidemann, H. Process maintenance from the point of view of reliability theory and of statistical process control. Res. Reps. of the Würzburg research group on quality control, 51, March 1995.
28. Ramalhoto, M.F. and Morais, M.J. Simple control charts for the location parameter of a Weibull distribution with fixed and variable sampling intervals. Tech. Reps. 18/95, Maths Dept., Instituto Superior Técnico, 1995.
29. Ramalhoto, M.F. and Morais, M.J. Simple control charts for the scale parameter of a Weibull distribution with fixed and variable sampling intervals. 1996 a) ( submitted to publication).
30. Montgomery, D.C. and Mastrangelo, C.M. Some statistical process control methods for autocorrelated data. Journal of Quality. Technologyl, 1991, 23:179-193.
31. Ramalhoto, M.F. and Morais, M.J. EWMA control charts for the scale parameter of a Weibull distribution with fixed and variable sampling intervals.1996 b) ( submitted to publication).

# Reliability evaluation of a periodically tested system by using different methods

Pierre CAILLIEZ *, Eric CHATELET **, Jean–Pierre SIGNORET *** – ISdF

* Régie Autonome des Transports Parisiens (RATP), ESE/QSF 7, square Félix Nadar 94684 – Vincennes CEDEX France

** Université de Technologie de Troyes, LM2S/SURFAS 13, boulevard Henri Barbusse 10010 – Troyes CEDEX France

*** ELF Aquitaine, DSE/MRT 64014 – Pau CEDEX France

## 1 Abstract – Introduction

The litterature related to the calculation of the reliability of periodically tested systems is rather poor. Then, the Methodological Research workshop of the ISdF (French Institute for dependability and safety) has worked in order to solve such problems. In this paper we intend to present an actual case concerning the reliability of a safety–related electronic equipment used in the Paris Metro. This case has been solved first in RATP by using an analytical method which is described in the paper. As this kind of method is rather complicated to handle without mistake, we began at ISdF, within the "test–cases activity", to identify and try several more flexible methods as the multi–phases Markovian approach which is presented hereafter.

As the Methodological Research workshop of the ISdF is going to become the Methodological Research workshop of ESRA, we intend to extend the "test–cases activity" at the european level. Then we are interested in gathering the various approaches usable to model the periodically tested systems and we welcome everybody who wants to participate.

## 2 System description

The system under analysis is a one–out–of–three active parallel system including three non identical components, whose characteristics are as follows:

- the failures of the first two components are hidden failures which are revealed only when the next test occurs;
- the third component is neither tested nor repairable.

# 3 Hypotheses and objectives

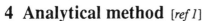

$\lambda_1, T_1$ | $\lambda_2, T_2$ | $\lambda_3$

In order to simplify we assume that:
- each component n° i has a constant failure rate $\lambda_i$ (i=1 to 3);
- the repair or replacement durations are negligible;
- the test intervals are respectively $T_1$ and $T_2=kT_1$, where k is an integer.

Our target is to calculate the system reliability R(t) at a given time t and the Mean Time To (first) Failure : $\mathrm{MTTF} = \int_0^{\infty} R(t)\,dt$

# 4 Analytical method *[ref 1]*

## 4.1 Principles of calculation

### *4.1.1 The reliability of a partly periodically tested set*

This method is aimed at giving accurate literal expressions of R(t) and MTTF, and simplified formulae of these for some sets of data.

We have better to calculate the probability $\Pi(t)$ of a system failure between times zero and t, than to calculate R(t) directly ($R(t)=1-\Pi(t)$). This enables us to separate failures of never tested components from failures of components which don't influence the system failure because a later periodical renewal will restart them.

As the calculation with three components is difficult to manage straight forward, we are going to proceed by steps and to calculate first two more simple scenarios, the results of which will be used for the complete system.

### *4.1.2 Scenario 1 (of reliability $R_1(t)$)*

The scenario deals with the reliability of a two components system, component 1 being periodically tested (test interval $T_1$) and component 2 being not tested nor repaired when failed.

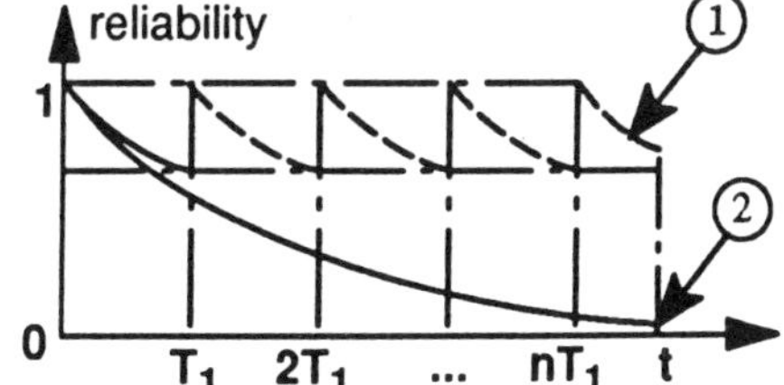


Let us consider $t \in [nT_1,(n+1)T_1]$. Then the period [0,t] is divided in n+1 periods: $[0,T_1]$, $[T_1,2T_1]$, ... $[nT_1,t]$. If the system is failed at t, that means that the non–repairable component 2 has failed before within one of the n+1 above periods, for example at time $\tau$ of the period n° i.

From the whole system failure, two cases have to be considered:
- component 1 has failed before component 2, i.e. within the interval $[(i-1)T_1,\tau]$
- component 1 has failed after component 2, i.e. within the interval $[\tau,t]$.

Thus, the probability of the whole system failure due to a failure of component 2 within the period n° i is: $\mathcal{P}\{$ "(2) failed during $[(i-1)T_1, iT_1]$" & "(1) failed during $[(i-1)T_1, t]$" $\}$

Components being independent, this reasoning applied to all periods yields:

$$\Pi(t) = \sum_{i=1}^{n} \underbrace{\mathcal{P}\{\text{(2) failed during } [(i-1)T_1, iT_1]\}}_{=e^{-\lambda_2 (i-1)T_1} - e^{-\lambda_2 iT_1}} \cdot \underbrace{\mathcal{P}\{\text{(1) failed during } [(i-1)T_1, t]\}}_{=1-e^{-\lambda_1 [t-(i-1)T_1]}}$$
$$+ \underbrace{\mathcal{P}\{\text{(2) failed during } [nT_1, t]\}}_{=e^{-\lambda_2 nT_1} - e^{-\lambda_2 t}} \cdot \underbrace{\mathcal{P}\{\text{(1) failed during } [nT_1, t]\}}_{=1-e^{-\lambda_1 (t-nT_1)}}$$

*4.1.3 Scenario 2 (of reliability $R_2(t)$)*

This scenario deals with the reliability of a three components system, component 1 being periodically tested (test interval $T_1$) and components 2 and 3 being not tested nor repaired when failed.

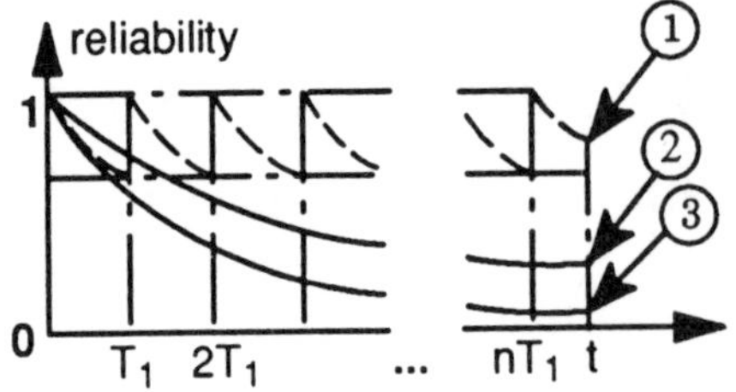


The time is also divided into the same n+1 periods as in scenario nº 1. The previous reasoning can be repeated, substituting component 3 for component 2 and set of components 1 and 2 for component 1. Thus:

$$\Pi(t) = \sum_{i=1}^{n} \underbrace{\mathcal{P}\{\text{(3) failed during } [(i-1)T_1, iT_1]\}}_{=e^{-\lambda_3 (i-1)T_1} - e^{-\lambda_3 iT_1}} \cdot \underbrace{\mathcal{P}\{\text{set } [(1) \,\&\, (2)] \text{ failed during } [(i-1)T_1, t]\}}_{=p_1}$$
$$+ \underbrace{\mathcal{P}\{\text{(3) failed during } [nT_1, t]\}}_{=e^{-\lambda_3 nT_1} - e^{-\lambda_3 t}} \cdot \underbrace{\mathcal{P}\{\text{set } [(1) \,\&\, (2)] \text{ failed during } [nT_1, t]\}}_{=p_2}$$

As the same reasoning applies to the set of components 1 and 2,

$$p_1 = \underbrace{\mathcal{P}\{\text{(2) failed during } [0, iT_1]\}}_{=1-e^{-\lambda_2 iT_1}} \cdot \underbrace{\mathcal{P}\{\text{(1) failed during } [(i-1)T_1, t]\}}_{=1-e^{-\lambda_1 [t-(i-1)T_1]}}$$
$$+ \sum_{j=i+1}^{n} \underbrace{\mathcal{P}\{\text{(2) failed during } [(j-1)T_1, jT_1]\}}_{=e^{-\lambda_2 (j-1)T_1} - e^{-\lambda_2 jT_1}} \cdot \underbrace{\mathcal{P}\{\text{(1) failed during } [(j-1)T_1, t]\}}_{=1-e^{-\lambda_1 [t-(j-1)T_1]}}$$
$$+ \underbrace{\mathcal{P}\{\text{(2) failed during } [nT_1, t]\}}_{=e^{-\lambda_2 nT_1} - e^{-\lambda_2 t}} \cdot \underbrace{\mathcal{P}\{\text{(1) failed during } [nT_1, t]\}}_{=1-e^{-\lambda_1 (t-nT_1)}}$$

and $p_2 = \mathcal{P}\{\text{(2) failed during } [0, t]\} \cdot \mathcal{P}\{\text{(1) failed during } [nT_1, t]\} = (1-e^{-\lambda_2 t})[1-e^{-\lambda_1 (t-nT_1)}]$

*4.1.4 Scenario 3 (of reliability $R_3(t)$)*

The basic scenario deals with the reliability of a three components system, components 1 and 2 being periodically tested (respective test intervals $T_1$ and $T_2$; $T_2$ multiple of $T_1$: $T_2 = vT_1$) and component 3 being not tested nor repaired when failed.

The above process is re–used and benefits by the reliabilities of both previous intermediate scenarios. Its steps are detailed below:

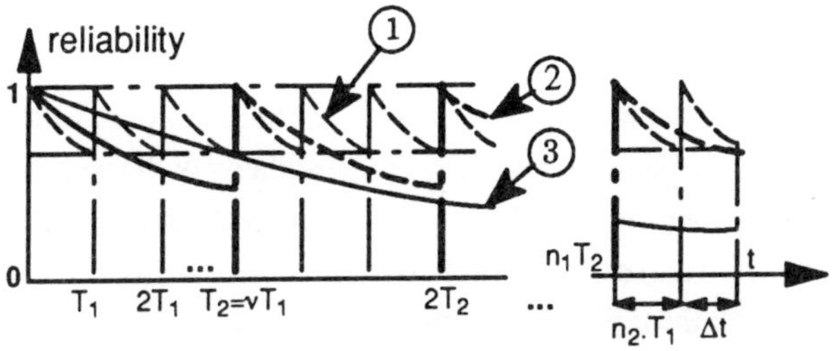

$t = n_1 T_2 + n_2 T_1 + \Delta t$

where : $n_1$ and $n_2$ = integers $\geq 0$
$0 \leq \Delta t < T_1$
$n_2 T_1 + \Delta t < T_2$
$n_2 < T_2 / T_1$

$$\Pi(t) = \sum_{i=1}^{n_1 v} \underbrace{\mathcal{P}\{\text{(3) failed during } [(i-1)T_1, iT_1]\}}_{= e^{-\lambda_3 (i-1)T_1} - e^{-\lambda_3 i T_1}} . \underbrace{\mathcal{P}\{\text{set } [(1) \,\&\, (2)] \text{ failed during } [(i-1)T_1, t]\}}_{= \pi_1}$$

$$+ \underbrace{\mathcal{P}\{\text{(3) working at time } n_1 T_2\} . \mathcal{P}\{\text{set } [(1) \,\&\, (2) \,\&\, (3)] \text{ failed during } [n_1 T_2, t]\}}_{= e^{-\lambda_3 n_1 T_2} [1 - \mathbf{R}_2(n_2 T_1 + \Delta t)]}$$

Whereas the interval $[(i-1)T_1, iT_1]$ is located inside the interval $[(j-1)T_2, jT_2]$,

$$\pi_1 = \pi_2 + (1 - \pi_2) \; [\; \mathcal{P}\{\text{set } [(1) \,\&\, (2)] \text{ failed during } [jT_2, t]\} = 1 - \mathbf{R}_1(t - jT_2) \;]$$

where $\pi_2 = \mathcal{P}\{\text{set } [(1) \,\&\, (2)] \text{ failed during } [(i-1)T_1, jT_2]\}$

$$= \mathcal{P}\{\text{(2) failed during } [(j-1)T_2, (i-1)T_1]\} . \mathcal{P}\{\text{(1) failed during } [(i-1)T_1, jT_2]\}$$
$$+ \mathcal{P}\{\text{(2) working at } (i-1)T_1\} . \mathcal{P}\{\text{set } [(1) \,\&\, (2)] \text{ failed during } [(i-1)T_1, jT_2]\}$$
$$= \{1 - e^{-\lambda_2 [(i-1)T_1 - (j-1)T_2]}\}\{1 - e^{-\lambda_1 [jT_2 - (i-1)T_1]}\}$$
$$+ e^{-\lambda_2 [(i-1)T_1 - (j-1)T_2]} \{1 - \mathbf{R}_1[jT_2 - (i-1)T_1]\}$$

## 4.2 Results

By calculating the above sums by means of geometrical series we obtain analytical formulae of reliability, and of MTTF by integration. In the general case (if $\lambda_1 \neq \lambda_2$, $\lambda_1 \neq \lambda_3$ and $\lambda_1 \neq \lambda_2 + \lambda_3$) they are rather complicated; the most simple one is, for example:

$$\mathbf{R}_1(t) = e^{-\lambda_2 t} + \left\{ \frac{1 - e^{-\lambda_2 T_1} - e^{n(\lambda_1 - \lambda_2)T_1} [\, e^{(\lambda_1 - \lambda_2)T_1} - e^{-\lambda_2 T_1} \,]}{1 - e^{(\lambda_1 - \lambda_2)T_1}} \right\} e^{-\lambda_1 t} - e^{n\lambda_1 T_1} e^{-(\lambda_1 + \lambda_2)t}$$

These formulae become simplified owing to truncated series expansions in particular cases (if $\lambda_1 = \lambda_2$, $\lambda_1 = \lambda_3$ and $\lambda_1 = \lambda_2 + \lambda_3$). Because of their complexity they have not been written here but they are available from the autors of the paper *[ref 1]*.
Most of them bave been verified thanks to the formal software MAPLE.

# 5 Reliability calculations of multi–phased systems based on homogeneous Markovian modelling *[ref 2]*

## 5.1 Principles

The first step consists in assessing the states of the system, i.e. 8 states for 3 components if each component has two states. The transition matrix is generated by using the Kro–

necker algebra [*ref 3*]. The reliability is the sum of the operating state probabilities at each time $t$, resolving the differential equations without the failure states – for the case of parallel components there is only one failure state. For each renewal of at least one component, we consider that a new phase starts. Then, the periodicity of the elementary phase is the minimum of all the renewal periods. For this phase, it is possible to calculate numerically or analytically (Laplace transformation method with MAPLE software) the state probabilities with the initial conditions of the phase. The principle is to iteratively calculate the probabilities needed to estimate the reliability at each elementary phase taking into account the renewal conditions – initial conditions $\vec{P}_0$ – and the final probabilities of the precedent phase. Then the reliability $R(t,\vec{P}_0)$ depends on $t$ and $\vec{P}_0$, and the Mean Time To Failure is:

$$\text{MTTF} = \sum_{i=0}^{+\infty} \int_{iT_1}^{(i+1)T_1} R(t,\vec{P}_0)\, dt$$

Therefore, the exact MTTF calculation is directly connected to the convergence of the series before, controlled by the overestimation of the error: $\dfrac{S_{i-1}\, S_i}{S_{i-1} - S_i}$ with:

$$S_i = \int_{iT_1}^{(i+1)T_1} R(t,\vec{P}_0)\, dt.$$ This error is obtained from an overestimation of the decreasing ratios $S_{i-1} / S_i$.

## 5.2 Analytical expressions associated to the scenarios 2 and 3

This is the case of three parallel components with one or two renewals. The system states are identified by the states of the components, the operating states being $\{(1,2,3), (\bar{1},2,3), (1,\bar{2},3), (1,2,\bar{3}), (\bar{1},\bar{2},3), (1,\bar{2},\bar{3}), (\bar{1},2,\bar{3})\}$. The analytical solutions for an elementary phase are easily obtained in the scenarios 2 and 3. The MAPLE software can be useful when the number of components increases or some components are repaired with constant repair rates. For more complicated cases, the numerical solution is the best one, and especially an O.D.E. integration method as the Runge–Kutta–Fehlberg one – more efficient: good accuracy and fast computing time [*ref 4*] – and Monte–Carlo simulation [*ref 5*].

The corresponding initial conditions are :

$$P_0^{1,2,3} = P^{1,2,3}(T_1) + P^{\bar{1},2,3}(T_1)\,[+ P^{1,\bar{2},3}(T_1) + P^{\bar{1},\bar{2},3}(T_1)]$$

$$P_0^{\bar{1},2,3} = P_0^{\bar{1},\bar{2},3} = P_0^{\bar{1},2,\bar{3}} = 0 \quad \text{and} \quad P_0^{1,\bar{2},3} = P^{1,\bar{2},3}(T_1) + P^{\bar{1},\bar{2},3}(T_1) \; [0]$$

$$P_0^{1,2,\bar{3}} = P^{1,2,\bar{3}}(T_1) + P^{\bar{1},2,\bar{3}}(T_1)\,[+ P^{1,\bar{2},\bar{3}}(T_1)] \quad \text{and} \quad P_0^{1,\bar{2},\bar{3}} = P^{1,\bar{2},\bar{3}}(T_1) \; [0]$$

The equalities are given for one renewal; in the case of two renewals the expressions inside brackets have to be added or they indicate an identity to zero if they are [0].

An algorithm has been developed to generate these expressions according to the renewed components.

## 6 Discussion and conclusions

When it is usable, the ***analytical method*** is very fast and efficient to provide accurate numerical values of both reliability and MTTF, however wide the range of parameters is. Moreover, its "white box" shape gives us a good visibility of the parameters (number and reliability of the components, duration of the test intervals) which are more influent on reliability, specially if the formulae are not too complex, for example when the $\lambda_i$ are equal, or with truncated series expansions of complicated expressions (when $\lambda_i T_i <<< 1$). With merely numerical methods, which are "black boxes", the same understanding would require the processing of a lot of various data sets.
However, the rapidly growing complexity of the calculations must be taken into account facing a new configuration with several components and test procedures. Thus, the analytical method could be quickly out of reach without the help of a formal software as MAPLE.
The ***actual case of the Paris metro*** consists of an electric supply service ($\lambda_3=10^{-5}$/hour, no periodical test), of a computer ($\lambda_1=10^{-7}$/hour, $T_1=1$ day) which controls a relay ($\lambda_2=10^{-6}$/hour, $T_2=3$ years). The analytical method provides a high level of safety for this equipment : R(300 years)=0.9967, MTTF=87646 years.
For this benchmark the ***Markovian method*** implemented by MAPLE yields accurate values of R(t) – with reasonable running times to compute even if $t$ is very great, but doesn't succeed in calculating the high level of the MTTF with enough precision because the series $S_i$ converges too slowly. It would be interesting to explore mathematical techniques for accelerating this convergence in such exceptional circumstances. In most cases this generic and flexible method is very powerful with an efficient software and furthermore, it can apply to non markovian models and to periods which are not multiple of each other.
The results that we have just presented above have been performed within the ***"test–cases activity" of the ISdF*** which aims at gathering all the methods usable to deal with periodically tested stand–by systems. Several members of the workshop are presently working on other methods able to handle a lot of others parameters of interest like MTTR, test duration, test coverage, probability of failure because of the tests, probability of bad configuration after a test has been performed, test policy (is the component available during the test or not ?), etc...
Among these, a ***stochastic Petri net*** model *[ref 6]* has processed the above benchmark and has succeeded in calculating reliability with a good accuracy, and MTTF with a reasonable approximation (14% lower than the exact value with 100 histories). These findings show that conventional Monte Carlo simulation is manageable in order to

compute such periodically tested systems. Nevertheless, some problems of computation time could occur for very rare events related to high security systems and convergence acceleration techniques as the importance sampling could be a solution in that case.
Otherwise an original method *[ref 7]* has been designed for producing ***analytical expressions of underestimated and overestimated values of R(t) and MTTF***. Its analytical and non iterative feature – making these calculations always possible – and the rather small difference observed between both values are quite attractive. This method doesn't allow to process periods which are not multiple of each other.
In that connexion, ***the "test–cases workshop" of the ISdF is open to anybody*** wanting to contribute to it, please just contact Mr E. CHATELET or Prof. Y. DUTUIT to participate.

## Acknowledgments

The authors wish to thank Mr Y. ROUCHAUD (RATP) for formulating the basic problem, Prof. Y. DUTUIT and Mr P. THOMAS (Technological Institute of Bordeaux University) for solving it by Petri net models, and Prof. M. ROUSSIGNOL (Marne–la–Vallée University) for both underestimation and overestimation methods.

## References

1. P. Cailliez: "Probabilité d'occurrence d'un scénario de panne mettant en jeu des maintenances périodiques", *Presentation at the Methodological Research workshop*, Institut de Sûreté de Fonctionnement, France, november 1994.
2. K. Kim and K.S. Park: "Phased–mission system reliability under Markov environment", *IEEE Transactions on reliability*, Vol. 43 nº 2, 1994, pp. 301–309.
3. V. Amoia, G. De Micheli and M. Santomauro: "Computer–oriented formulation of transition–rate matrices via Kronecker algebra", *IEEE Transactions on reliability*, Vol. 30 nº 2, 1981, pp. 123–132.
4. E. Châtelet, M. Djebabra, Y. Dutuit and J. Dos Santos: "A comparison between two indirect exponentiation methods and O.D.E. integration method for RAMS calculations based on homo–geneous Markovian models", *Reliability Engineering & System Safety*, to be published.
5. B. Tombuyses and J. Devooght: "Solving Markovian systems of O.D.E. for availability and reliability calculations", *Reliability Engineering & System Safety*, Vol. 48, 1995, pp. 47–55.
6. E. Châtelet, M. Djebabra, J. Dos Santos and Y. Dutuit: "Les réseaux de Petri stochastiques et interprétés", *Proceedings 9th International Conference on Reliability and Maintainability ($\lambda\mu$9)*, La Baule, France, june 1994, pp. 742–751.
7. V. Kalashnikov and M. Roussignol: "Fiabilité d'un système avec maintenances à intervalles réguliers", Report nº 29, Equipe d'Analyse et de Mathématiques Appliquées, Université de Marne–la–vallée, France, october 1995.

# Level 1 PSA Study for Unit 3 of J. BOHUNICE NPP

Z.Kovács, H.Nováková
RELKO Ltd, Engineering and Consulting Services
Bratislava, Slovak Republic

**Abstract**

The paper describes the results of the PSA study performed for Unit 3 of Bohunice V2 NPP.

## 1. Introduction

The Level 1 PSA study for unit 3 of Bohunice V2 NPP was undertaken in two phases. The first PSA model was developed for the status of the unit to September 30, 1993 by VÚJE and RELKO Ltd. This model identified that the emergency power supply system (category 2) has dominant contribution to the CDF. The start of the Load Sequencing System and DG on the technological reason (Large and Medium LOCA, Steam header break) decreases the reliability of the self consumption. Therefore, modification of this system was suggested.

The second PSA model was developed by RELKO Ltd, which included the modification of the emergency power supply (Category 2). In addition, the second model includes recovery actions and common cause failure of equipment in the Turbine Hall (elevation 14,7m) following secondary steam line break, steam header break and feedwater header break. The model was developed for the status of the unit to May 31, 1995.

The recovery actions modelled in the second PSA are limited to:

- drain of bubble condenser following interfacing LOCA and SG tube rupture,
- aggressive secondary cooling following loss of all HP pumps in case of small and very small LOCA
- realign of LP suction line to prevent overflow of LP tanks following small and very small LOCA

Modelling of the above mentioned pipe breaks is performed due to the possible high energy release of steam and water into the Turbine Hall. There is a potential for complete loss of secondary cooling systems (MFW, AFW and EFW) following these events. As there was no available analysis to determine the expected effects, a point estimate was used to represent the possibility of common cause failure of systems for each initiating event.

# 2. Scope of the Study

The study was prepared in accordance with the IAEA procedure for conducting level 1 PSA of NPP [1]. The following activities are included:

1. Document Collection and Plant Familiarization
2. Initiating Event, Accident Sequence Analysis, Success Criteria
3. Systems Analysis
4. Data Analysis
5. Human Reliability Analysis
6. Internal Fire Analysis
7. Core Damage Frequency Quantification and Interpretation of the Results

# 3. Description of the Results

The mean core melt frequency for unit 3 of J. Bohunice NPP is

$$6{,}41.10^{-4}\,.y^{-1}$$

The results of uncertainty analysis are as follows:

| | |
|---|---|
| * median | $5{,}36.10^{-4}.y^{-1}$ |
| * 5th Percentile | $2{,}14.10^{-4}.y^{-1}$ |
| * 95th Percentile | $1{,}88.10^{-3}.y^{-1}$ |

The next part of the section present the contribution of initiating events, category of primary events, systems, accidents sequence and minimal cat sets to the core damage frequency.

## 3.1 Contribution of Initiating Events

| | | |
|---|---|---|
| 1. Loss of offsite power | LOP | - 30,50% |
| 2. Steam line break outside confinement | SLBO | - 19,00% |
| 3. Steam header break | SHB | - 14,50% |
| 4. Fire in the turbine hall | FIRE-490 | - 10,80% |
| 5. Feedwater header break | LMF(FWHB) | - 7,72% |
| 6. Interfacing LOCA | IFSL | - 5,06% |
| 7. Very small LOCA | VSL | - 4,87% |
| 8. Medium LOCA | ML(20-40) | - 3,40% |
| 9. Pressurizer steam LOCA | PSL | - 1,34% |
| 10. Steam generator tube rupture | SGTR | - 0,70% |

## 3.2 Dominant Accident Sequences

| N. | Sequence<br>N.: Description | Event tree | Frequency | Contribution to CDF |
|---|---|---|---|---|
| 1 | 3.:LOP-B&F*MC(PV-B&F) | LOP-B&F | 1,40E-04 | 22,47% |
| 2 | 3.:SLBO-B&F*MC(PV-B&F) | SLBO-B&F | 1,09E-04 | 17,50% |
| 3 | 2.:SHB-B&F*MC(PV-B&F) | SHB-B&F | 6,68E-05 | 10,70% |
| 4 | 3.:FIRE-490-B&F*MC(PV-B&F) | FIRE-490-B&F | 6,28E-05 | 10,10% |
| 5 | 2.:LOP-B&F*HP(LL,CP,LOP) | LOP-B&F | 4,79E-05 | 7,70% |
| 6 | 3.:LMF(FWHB)-B&F*MC(PV-B&F) | LMF(FWHB)-B&F | 4,51E-05 | 7,20% |
| 7 | 7.:VSL.8*LP(OVERFL) | VSL.8 | 2,61E-05 | 4,20% |
| 8 | 9.:IFSL.7*SS(DRAIN) | IFSL.7 | 2,39E-05 | 3,80% |

Description of the events:

1. Following the loss of offsite power failure of emergency feedwater system occurs and operator fails to initiate feed and bleed. The sequence leads to core damage.

2. Following the loss of steamline break outside the confinement the MFW, AFW and EFW systems are lost, operator fails to initiate feed & bleed and core damage occurs.

3. Due to the steam header break MFW, AFW and EFW systems are lost and operator fails to initiate feed and bleed. The sequence leads to core damage.

4. Following the fire in the turbine hall and loss of primary to secondary side heat removal (MFW and AFW are lost due to the fire and failure of EFW) the operator fails to initiate feed and bleed and core damage occurs.

5. Loss of offsite power occurs, AFW and EFW fail to operate and due to the HP system failure the feed & bleed can not be performed. The sequence leads to core damage.

6. Following the steam header break the MFW, AFW and EFW systems are lost and operator fails to initiate feed & bleed. The sequence leads to core damage.

7. Given an unisolated very small LOCA and operator fails to prevent overfilling of LP tanks. In recirculation phase of HP pump operation the water from the confinement sump will be lost, HP pumps will cavitate and core damage occurs.

8. Given an unisolated interfacing LOCA and operator fails to drain the bubbler tower to the confinement floor and core damage occurs.

## 3.3 Contribution of Basic Event Categories

Post trip operator actions are represented in cut sets that account for 82,7% of the total CDF. The partial contributions are the following:

- 68,3% - operator fails to initiate feed & bleed
- 6,4% - operator fails to prevent overfilling of LP tanks
- 4,5% - operator fails to drain bubbler tower into the boxes
- 4,1% - operator fails to open valves of EFW system
- 2,0% - operator fails to depressurize primary side below secondary side
- 0,6% - operator fails to start demi-water system 1MPa,
- 0,6% - operator fails to initiate aggressive depressurization of primary side
- 0,5% - operator fails to isolate very small LOCA

Another important category of primary events is regarding turbine hall effect. The loss of primary to secondary side heat removal due to:

- steam line break outside confinement (19,1%)
- steam header break (11%)
- main feedwater header break (7,6%).

The contribution from hardware failure is 56,6%. The dominant contributors are:

- HP pump TJ61D01 fails to run - 9,5%
- Dieselgenerator QW fails to run - 9,1%
- HP pump TJ21D01 fails to run - 8,2%
- Failure of battery of category I - 6,9% (EE02 in standby state)
- Dieselgenerator QW fails to start - 6,7%
- HP pump TJ41D01 fails to run - 6,6%
- Dieselgenerator QV fails to run - 5,3%
- Control valve 04.7.309.1 in demi-water 1 MPa - 4,7% system fails to function
- Dieselgenerator QV fails to start - 3,9%
- AFW pump 04.2.03.1 fails to run - 3,0%

The contribution from I&C components is 6,1%.

The contribution from maintenance is 2,2%.

The contribution from common cause failures is 2,8%.

The contribution from test is neglectable.

## 3.4 Contribution of Systems

1. Primary circuit - 9,46%
2. EFW system - 4,29%
3. Demi water system 1MPa - 1,09%

4. HP system - 0,10%

## 3.5 Contribution from MCS

Minimal cut set for CD

| N. | Freq. | % | MCS | |
|---|---|---|---|---|
| 1 | 1,10E-04 | 17,22 | THEF-SLBO | Loss of primary to secondary side heat removal |
| | | | SLBO | Steam line break outside the confinement |
| | | | OA(B&F) | Operator fails to initiate feed & bleed |
| 2 | 6,72E-05 | 10,48 | THEF-SHB | Loss of primary to secondary side heat removal |
| | | | SHB | Steam header break |
| | | | OA(B&F) | Operator fails to initiate feed & bleed |
| 3 | 4,41E-05 | 6,88 | THEF-FWHB | Loss of primary to secondary side heat removal |
| | | | LMF(FWHB) | Main feedwater header break |
| | | | OA(B&F) | Operator fails to initiate feed & bleed |
| 4 | 2,75E-05 | 4,25 | PUS2BTA-EE02 | Failure of battery |
| | | | LOP | Loss of offsite power |
| | | | OA(B&F) | Operator fails to initiate feed & bleed |
| 5 | 2,10E-05 | 3,28 | FIRE-490 | Fire in the turbine hall |
| | | | OA(EF) | Operator fails to open the valves |
| | | | OA(B&F) | Operator fails to initiate feed & bleed |
| 6 | 1,60E-05 | 2,50 | OA(LP-OVERFL) | Operator fails to prevent overfilling of LP tanks |
| | | | ML(20-40) | Medium LOCA |
| 7 | 1,16E-5 | 1,81 | EFO0PMR-18.2.06.1 | Pump fails to run |
| | | | EFO0PMR-18.2.06.2 | Pump fails to run |
| | | | FIRE-490 | Fire in the turbine hall |
| | | | OA(B&F) | Operator fails to initiate feed & bleed |
| 8 | 8,74E-06 | 1,36 | OA(B&F) | Operator fails to initiate feed & bleed |
| | | | LOP | Loss of offsite power |
| | | | PEO2DGR-QW | DG fails to run |
| | | | DWS0VRO-04.7.309.1 | Control valve 04.7.309.1 fails to function |
| 9 | 6,60E-06 | 1,03 | HPO1PMR-TJ21D01 | Pump TJ21D01 fails to run |
| | | | HPO2PMR-TJ41D01 | Pump TJ41D01 fails to run |
| | | | HPO3PMR-TJ61D01 | Pump TJ61D01 fails to run |
| | | | SHB | Steam header break |
| 10 | 6,43E-06 | 1,00 | OA(B&F) | Operator fails to initiate feed & bleed |
| | | | LOP | Loss of offsite power |
| | | | PES2DGS-QW | DG fails to start |
| | | | DWS0VRO-04.7.309.1 | Control valve 04.7.309.1 fails to function |

# 4. Conclusions and Recommendations

Based on the results the following conclusions are given:

* Steam line, steam header and main feedwater header break on the elevation +14,7m has the most dominant contribution to the risk. This is a common problem of VVER 440 type reactors. Following these events the secondary cooling is completely lost. The solution for this problem is development of procedures for feed & bleed and operator training to perform this action.

* Dominant contribution is from post accident human-errors (82,7%). To minimize this contribution improvement of existing and development of new procedures for liquidation of failure states is needed. The pre-accident human error contribution is small (maintenance-2,2%, testing - neglectable).

* Contribution of ATWS to CDF is small. Provided that all ATWS sequences will lead to CD, the CDF would be increased only by the value of $9{,}84.10^{-6}$.

* Rapid overcooling (SHIVER) of primary circuit by secondary circuit is modelled for the no TG trip following the reactor trip, for cold feedwater supply by EFW system and for no closing of SG safety valves, valves of steam dump station to the atmosphere and relief valves of turbine bypass. In this field another thermal hydraulics analysis are needed, but it is expected, that the influence on CDF will be neglectable.

Recommendations to Improve Operational Safety:

To develop procedures for :
- feed & bleed, given loss of secondary cooling,
- overfilling prevention of LP tanks during small and medium LOCA,
- aggressive depressurization of the primary side by secondary side (by opening of a relief valve of turbine bypass) in case of small LOCA and simultaneous failure of all HP pumps and
- drainage of bubbler condenser into the confinement floor in case of interfacing LOCA.

To perform system modifications on the basis of PSA results .

Following the modifications the new CDF is 7,85E-5/y.

**References**

[1] Procedures for Conducting Probabilistic Safety Assessments of Nuclear Power Plants (Level 1), Safety Series No.50-P-4, IAEA

[2] Level 1 PSA Study for Unit 3 of Jaslovské Bohunice V2 NPP , Main report of the Study, RELKO Report, Sept. 1995

# Use of PRA to formulate a fresh set of bases for ALWR Passive Plants' Technical Specification

Dr A. Bassanelli
ENEL S.p.A., Nuclear Energy Division (ATN)
Rome, Italy

## 1.0 Background and Goals

ENEL has contributed to perform PRAs for the next generation of passive nuclear power plants jointly with General Electric for SBWR project and with Westinghouse for AP600 project.
All plant systems have been modeled in detail up to major components (i.e. pumps, valves, circuit breakers, logic cards etc.) and also outages for test and maintenance have been considered.

Next step is to establish a widely accepted methodology which, basing on the PRA model, allows to issue a set of Technical Specifications which, starting from those used in the actual operating plant, bridges for an innovative concept of Technical Specifications.
The nature of passive systems offers a unique opportunity to formulate a fresh set of bases, since Technical Specifications for existing plants cannot be adopted directly.
To supplement traditional approaches based on deterministic considerations, insights from probabilistic risk assessment (PRA) will intend to be used to guide the selection of limiting conditions of operation (LCOs), allowable outage times (AOTs), and surveillance intervals (SIs) for safety systems.
Consistent with the design goals of simplification, we intend to obtain a simplified specifications with logical and consistent bases that provide for operating and maintenance flexibility as well as high safety levels for all plant conditions including shutdown operation.

## 2.0 Bounded Conditions

The purpose of Technical Specifications is to establish a set of limits for safe plant operation (LCOs) designed to maintain plant operation within an envelope of analyzed conditions.
Analyzed conditions must include all operational modes of plant operation starting with a flooded refueling pool and progressing through the operation evolution necessary to bring the plant to 100% power, then returning the plant to refueling conditions with flooded refueling pool.

# 3.0 ALWR Passive System Design Basis and Technical Specification Requirements

## 3.1 ALWR Passive System Design Basis Requirements

ALWR passive safety systems shall meet licensing requirements for both prevention and mitigation of Licensing Design Basis (LDB) events without reliance on active systems to:

- maintain core inventory,
- remove decay heat,
- provide diverse reactivity control,
- maintain containment integrity,
- provide fission product control, and
- provide long-term post-accident core cooling.

Licensing requirements shall be met without taking credit for large continuously rotating machinery, multiple active valves, and AC powered divisions other than inverter supplied components. Therefore, Technical Specification for this type of equipment should not be required. The passive ALWR shall not require Technical Specifications for onsite AC power generators other than DC power used to generate AC power for instrumentation and control functions.

## 3.2 ALWR Passive System Technical Specification Requirements

Designed according the above requirements, the passive safety system :

- makes the operator procedures and Technical Specification less complex and less interdependent, particularly for emergency conditions;

- minimizes the number of realignments and related operations required to accomplish the functions;

- reduces the need for protective interlocks; and

- minimizes the required equipment consistent with achieving functional, safety, and reliability goals.

For purpose of Technical Specification analyses, no operator action should be assumed for 72 hours following DBA. Operator action before 72 hours, even if possible, should not be assumed. Operator action after 72 hours should be limited to simple, preplanned, unambiguous actions that can be performed without restoration of AC power.

# 4.0 Technical Specification Equipment Selection Criteria

The following criteria should be used to identify systems and equipment to be considered for Technical Specification control.

- Installed control room instrumentation that is used to detect, and indicate a significant abnormal degradation of the reactor coolant pressure boundary (RCPB). This criterion is intended to ensure that Technical Specifications control those instruments specifically installed to detect excessive reactor coolant system leakage, as basic concept to prevent accidents allowing operator actions to either correct the condition or to shutdown the plant safely, thus reducing the likelihood of a LOCA.

- Process variables that from analyses can be used to detect a challenge to the integrity of a fission product barrier. As basic assumption for providing adequate protection of the public health and safety is that the plant is operated within the bounds of the initial conditions assumed in the existing DBA analyses or transients analyses

- A system or component that is part of the primary success path preventing core damage and which, by design, functions or actuates to prevent or mitigate a DBA or transient. Primary success path can be defined from the event trees sequences, where the frontline systems, function and components and related supporting and actuating systems are sufficiently identified.

# 5.0 PRA Applications

Among the several Technical Specification requirements, only Limiting Conditions for Operation (LCOs) and Surveillance requirements are amenable to PRA. Even within these two categories, PRA techniques are suitable for addressing only portion of the current Technical Specification requirements. Many of the Technical Specification requirements in these categories such as shutdown margins, control rod response times, pressure, temperature, and power distribution limits are not the type of requirements that PRA is suited for.
Despite these limitations, the application of PRA techniques to Technical Specification has the potential to provide benefits such as application of a systematic approach for establishing AOTs, and surveillance intervals (SIs) and a more technically supportable basis for them.
Additionally, a consistent set of criteria developed using these technique would provide the basis for optimizing both utility and regulatory resources. It would also allow concentration on those items that are most risk-significant while simultaneously allowing more operational flexibility when risk is shown to be acceptable.

The PRA level 2 analysis, covering only sequences beyond the design bases, will be used as verification to impose additional Technical Specification requirements to systems or components which contribute to mitigate sequences having very low frequency, but very large consequences to the population (for example standby liquid control system in BWR plants).

# 6.0 Technical Specification Requirements for AOT Determination

The primary objective of the Technical Specification requirements for AOTs, is to define a position for the relationship between Technical Specification and defense-in-depth systems acceptable to both industry and regulatory body.

Defense-in-Depth systems are those active systems whose operation could terminate a DBA or transient without requiring operation of a passive safety system or whose operation could compensate for unavailability of a passive safety function, but should not be treated as redundant or back-up safety system functions.

As example, for purposes of Passive PWR Technical Specification, the following systems are defined as passive safety systems: Core Makeup Tanks (CMT), Automatic Depressurization System (ADS), Accumulators
Passive Residual Heat Removal (PRHR), In-Containment Refueling Water Storage Tank (IRWST), and Passive Containment Cooling System (PCCS).
while the following systems are defined as defense-in-depth systems: Startup Feedwater System, CVCS Makeup System,
Normal Residual Heat Removal System (NRHR), and Diesel Generators.

A regulatory body criterion requires:
*"the Technical Specifications include limitations and conditions for reactor operations that address significant contributors to the plant's overall core melt probability and risk as identified in risk evolution (i.e. PRA)".*

From a PRA perspective the unavailability of certain Defense-in-Depth systems (e.g. CVCS charging, back-up feedwater, NRHR and Diesel Generator for Passive PWR) cause about the same increase in Core Damage Frequency (CDF) as the loss of a train of a passive safety system.
In addition from a functional perspective, defense-in-depth systems have the capability to terminate certain transients and accidents before plant parameters exceed actuation setpoints for passive safety system actuating, thereby preventing a depressurization event or other challenge to the passive safety system.

An approach could be that:
defense-in-depth systems would not require Technical Specification consideration if a CDF < 1.0E-4 events/year is obtained from a PRA sensitivity study performed without credit for the defense-in-depth systems.

AOTs for passive safety systems should be evaluated based on availability of defense-in-depth systems. That means the current PRA should be used.

The availability for a defense-in-depth systems should be evaluated considering also the availability of the supporting systems. For example for NRHR availability evaluation also the availability of its support systems such Service Water and Component Cooling Water should be considered.

In PRAs recently performed, AOTs are based on functional similarity between passive plant safety systems and systems included in standard technical specifications for current plant. For example, the passive PWR CMT performs a function similar to a high pressure safety injection system train for existing PWR plants. Therefore the same AOT required in Standard Technical Specification for high pressure safety injection was used.
Risk evaluation should be performed to verify that the assigned AOTs do not collectively cause an undue increase in risk determined by PRA.

Defense-in-depth systems are not included in Technical Specification, however, unavailability of defense-in-depth systems concurrent with inoperable passive safety systems should be appropriately accounted for in technical specifications by reducing the Technical Specification AOT allowed for passive safety systems when defense-in-depth systems are not available.

## 6.1 AOT Determination

Technical Specification AOTs for passive safety systems action statements shall be limited by functional capabilities of non-safety defense-in-depth systems in accordance with the following guidelines:

*a)* Fixed maximum AOTs shall be determined for appropriate passive system LCO action statements:

- The maximum allowed AOT shall require availability of all defense-in-depth systems,
- The reduced AOT will be in effect whenever defense-in-depth systems are not available,
- The reduce AOT will be determined by using the combination of the single passive safety system train which is unavailable and assuming the two most limiting defense-in-depth systems are unavailable.

*b)* If the total number of unavailable defense-in-depth systems considered collectively is more than two, then the AOT for the applicable passive safety system LCO action statement shall be zero.

*c)* The AOT associated with the unavailability of a systems or components will be determined using the following guidelines:

- An allowable accumulated downtime is determined using the following relationship:

$$\mathrm{CDF} \cdot \mathrm{T} < \mathrm{L}$$

Where:

T is the cumulative AOT during a one year interval, i.e., the duration of time that the system or component is permitted to remain out of service expressed as fraction of a year.

CDF is the increase in the plant's core damage frequency as result of the unavailability of a system or component,

L is a dimensionless fixed limit representing as allowable accumulate risk due to downtime of a system or component.
The constant "L" represents the highest acceptable core damage frequency integrated over the duration "T" which is specified by the AOT. The limit "L" should be < 1.0E-6.

- To account for more than a single downtime for an system or component, the technical specification AOT will be determined by the following relationship:

$$AOT = T / r$$

where:

r is the number of annual downtime occurrences on which the AOT is based. This relationship provides a methodology of converting "T", which is a measure of cumulative time, to a traditional AOT.
Experience data will be utilized to estimate an appropriate value for "r".

*d)* AOTs for passive safety systems shall not exceed 30 days or be less than three days. Discreet time increments of 3 days, 7 days, 14 days and 30 days shall be used for AOT. For calculated AOTs between two of these discreet time increments, the lower value shall be used.

## 6.2 Conditional AOTs

Some AOTs determined by PRA methods will also depend upon which other systems or components are unavailable at the time a system or component is declared inoperable. To account for cases when other system or components are unavailable, a single, AOT specific, combination of systems or combinations should be used for determining each conditional AOT. The event trees from PRA will be used to determine the minimum sets of operable systems or components. This will minimize the number of combinations to be evaluated and seek to establish a single bounding set that provide reasonable operational flexibility.

## 6.3 AOT Extension: Process for Change or Temporary Relief

A process for change or temporary relief to the Technical Specification would allow utility management to respond to unique plant situations without excessive delays. The criteria need to ensure that such changes or relief would not adversely affect plant safety, e.g.., they would be infrequent, and would involve an increase in risk beyond a CDF defined goal (1.0E-4 event/year including internal and external events).
The process shall provide a sufficient basis for identifying those changes or temporary relief which can made with notification to the Regulatory Body but without prior concurrence of the Regulatory Body.

# 7.0 Surveillance Intervals

Initial baseline SIs have been established using reliability experience data. As gross rule SIs of one month, three months and refueling time have been applied in the recent PRAs.
Definition of longer SIs is a large time consuming task, where the PRA tool is extensively. To find a good balance between longer SI and longer AOT, a long trial-error process with many sensitivity computer runs are necessary.

# 8.0 Support systems

Because of the extensive use of passive features, reliance on support systems is significantly reduced. As a rule, where support system considerations are important, they will be contained in the specification associated with the supported system.
In most cases, supported system parameters will cover the required functions of the support system.
In other cases, such as where a more significant system provides a primary support function, such as DC electrical power, a separate specification will be retained. However, such specification will be self standing. For instance, the safety DC electrical power specifications will take into account their effect on safety systems, therefore, if a safety battery is declared inoperable, additional specifications need not be entered.
The definition of operability embodies the principle that a system can perform its function(s) only if all necessary support systems are capable of performing their relate support functions.
Support functions which are always required will be contained in the Technical Specification for the supported system.

# 9.0 Technical Specification for Shutdown and Refueling Conditions

Technical Specifications should include all modes of plant operation and therefore also shutdown and Refueling events.
A dedicated PRA analysis has been performed to identify shutdown risk and vulnerabilities, and a minimum set of requirements for equipment availability and cooling water inventories have been determined.
The same methodology beforehand described for AOT determination should be applied in presence of shutdown and Refueling conditions.
In PWR plants, for example, Technical Specification limitations should be placed on operations at reduced RCS coolant inventories, and requirements should be established for operability of passive safety systems.
No Technical Specification Surveillance Requirements will be established to determine the functional capability of active decay heat removal and its support systems.

# Plant Maintenance Advisory System Using 3D Configuration Model and PSA Technique (MAS3)

Kwang-Sub Jeong, Jae-Joo Ha, Seung-Hwan Kim, Sun-Yeong Choi
Nuclear Information Technology Development Team, Advanced Research Group
Korea Atomic Energy Research Institute
P.O.Box 105 Yusung, Taejon, Korea 350-600

**Abstract**

In general, maintenance activities consist of (1) the identification of maintenance tasks, (2) the acquisition of necessary information, and (3) the conduction of proper maintenance procedures. This paper describes the concept and the current effort to develop the computer aided maintenance advisory system using the state-of-the-art computer technology, "Plant Maintenance Advisory System using 3D Configuration Model and PSA Technique (MAS3)". The key concept is, first, to prioritize and determine maintenance activities using living PSA technique, to retrieve relevant maintenance information through integrated database, and to simulate maintenance procedures using 3D CAD model to reduce potential problems during maintenance. The necessary information is stored in database, i.e., plant configuration data in 3D configuration model, living PSA data, maintenance history and procedures in plant operation database, respectively, and those data are incorporated each other for the effective information retrieval. MAS3 gives an experience for maintenance, so the maintenance efficiency can be increased.

# 1. Introduction

Effective maintenance of equipment systems and structures at nuclear power plant is essential for its safe and reliable operation [1]. Maintenance at nuclear power plant is the programme which should assure that plant equipment (both safety related and production equipment) remains within original design criteria and that is not adversely affected during the plant lifetime.

Successful maintenance programme is the focal point of reliable operation of plant. Lack or substandard maintenance sooner or later results in unplanned outages and reduced plant availability. At the same time, maintenance activities are major single contributor to operational costs of a nuclear power plant, but also a programme where savings are possible. Increasing attention which is paid to safe operation of nuclear power plants worldwide also increases attention to maintenance as one of important contributors to safety. There are some measures for the effective maintenance: which system or component requires maintenance, which procedure should be performed, and how should the maintenance tasks be performed.

Korea Atomic Energy Research Institute is developing "Nuclear Integrated Database and Design Advancement System (NuIDEAS)" [2] for Nuclear Steam Supply System of Korean Standard Nuclear Power Plant. All design data including all design parameters, drawings and images are stored in the master database. In the project, the design data for components are stored in the main component database by re-organizing the data, and the plant 3-dimensional configuration model is established. These component data and configuration model can be utilized for the maintenance activity.

There are some major measures for maintenance: reliability, pathway, radiation level, available staffs, necessary tools, and so on. In this paper, the reliability measure is used to rank the maintenance priority among failed components of safety systems. In general, maintenance activities consist of (1) the identification of failed component, (2) the acquisition of necessary information, and (3) the conduction of proper maintenance procedures.

By combining the component master database, the living PSA technique [3], and 3-dimension configuration model, "Plant Maintenance Advisory System using 3D Configuration Model and PSA Technique (MAS3)" is being developed to assist the maintenance staffs.

In the next section, the basic concept of MAS3 will be described with the plant database for maintenance procedures and for plant 3-dimensional configuration model. The identification concept using living PSA technique will be followed. In section 4, the maintenance simulation will be described.

# 2. Development of MAS3

## 2.1 Basic Concept of MAS3

For the effective maintenance, the reliability of plant, available cost (man-hour and tools), radiation level, pathway, and so on. This paper incorporates all these to establish the maintenance policy, i.e. maintenance priority, for mailed components of safety system, and to assist the maintenance staffs by experiencing maintenance work by computer simulation. Figure 1 shows this basic concept.

For the identification of component, the technique of living PSA is used in terms of plant safety. The proper procedure for the component maintenance is retrieved from the database. Once the proper procedure of maintenance is chosen, it starts its conduction. Before the actual task, the 3-dimensional simulation can be performed to figure out the potential problem for actual maintenance, for example, radiation level around the working area and/or pathway to working area, for maintenance can be checked. These enable advisory for efficient maintenance in the aspects of safety and economic prior to the actual maintenance task.

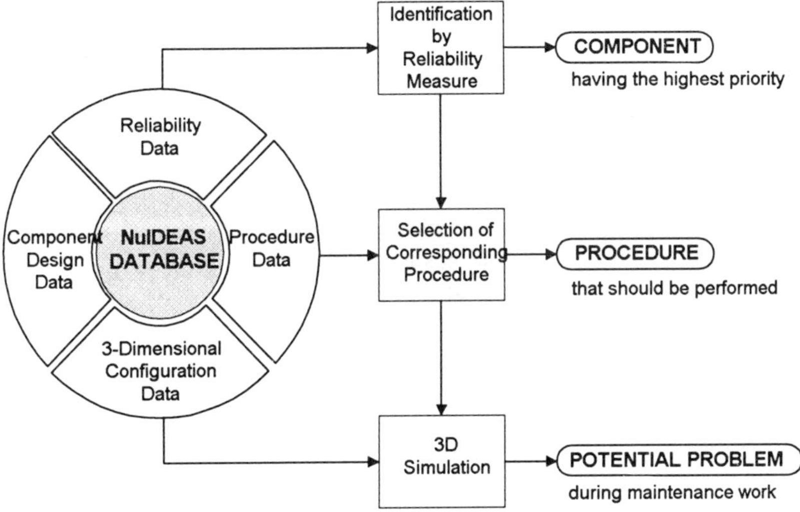


Figure 1. MAS3 Concept

## 2.2 Maintenance Database

Korea Atomic Energy Research Institute is performing the NuIDEAS project which integrates all design data into the master database. In NuIDEAS, all design-related data are stored in the master database and are retrieved from various client computers via computer network.

The component design data including design parameters, tables, figures, and images are stored in the master database according to the component hierarchy. The database structure for design document is established in the master database, too. The 3-dimensional configuration model for component is established in NuIDEAS project. This model is utilized for walk-through simulation prior to actual maintenance.

### 2.2.1 *Procedure Database*

There are a number of procedures in the nuclear power plant : normal/abnormal operating procedures, maintenance procedures, test procedures, and other procedures. Currently, all these procedures are written in a type of document and are utilized by the unit of document.

As computerizing these documents with the hyper-text technique and database concept, one can get the necessary information necessary for maintenance including graphical view from the personnel computer instead of written document via network. The computer can be located either in the maintenance office or working area, and the computer can provide additional information about the maintenance work to the maintenance staff. The procedure database is a part of NuIDEAS master database.

### 2.2.2 *Reliability Database*

KAERI developed a PSA level 1 code, "KAERI Integrated Reliability Analysis Code Package (KIRAP) ", [4] and it currently is utilized for many PSAs to evaluate the core damage frequency. KIRAP has its own reliability database [4,5], and this is used for prioritizing the components in terms of reliability/safety. The reliability database is a part of NuIDEAS master database.

## 2.3 Identification of Maintenance Task

### 2.3.1 *Basic Methodology*

The PSA technique is used to determine the maintenance policy such as maintenance priority for component of safety system The maintenance priority for failed components can be determined by applying the living PSA concept to the existing level 1 PSA results. In general, the level 1 PSA generated the core damage frequency (CDF) with averaged probability of components ($CDF_{base}$) :

$$CDF_{base} \quad = f\,[\,MCS_{plant}\,|\,p(\text{all components}) = \text{average}\,]$$

The living PSA can incorporate the change of plant configuration, and it calculates new CDF based on the changed status of components. Assume that there are n failed components in plant safety system, and the current CDF ($CDF_{current}$) can be calculated by assigning the failure probabilities of all these failed component to 1.0 :

$$CDF_{current} \quad = f\,[\,MCS_{plant}\,|\,p(\text{all failed components}) = 1.0\,]$$

### *2.3.2 Prioritization Algorithm*

For each failed component, assume the component is repaired and is operating, then the new CDF ($CDF_{new}$) can be calculated :

$$CDF_{new,comp\_1} = f\,[\,MCS_{plant} \mid p(comp\_1) = 0.0\,]$$
$$CDF_{new,comp\_2} = f\,[\,MCS_{plant} \mid p(comp\_2) = 0.0\,]$$
$$CDF_{new,comp\_n} = f\,[\,MCS_{plant} \mid p(comp\_n) = 0.0\,]$$

The component having the maximum decrease in CDF among the new CDFs has the highest priority in maintenance in the aspects of reliability measure.

To regenerate CDF by changing the probability in plant minimal cutsets, the fast running algorithm should be adopted for MAS3. As mentioned earlier, there are another measures for the effective maintenance, but in this paper, only reliability measure is used.

# 3. Simulation of Maintenance Task

## 3.1 Acquisition of Information

Once the component having the highest priority in maintenance, the corresponding procedure is retrieved from the procedure database. The procedure can be executed step-by-step by the operator prior to the actual maintenance, and the maintenance staff can use the networked computer to check the whole maintenance steps instead of written document. At the same time, all maintenance-related information can be retrieved from the master database by the user's selection. The component design data, drawings for manufacturing and sectional drawings, scanned image, 3-dimensional configuration model view, and the maintenance history data can be retrieved from the database.

## 3.2 Simulation of Maintenance Task

For the selected maintenance procedure, MAS3 can simulate the whole maintenance tasks step-by-step. While simulating, all available component design data can directly be retrieved from the database. These data make a role of the additional and necessary information for the effective maintenance.

The pathway can be checked during simulation by using the virtual reality (VR). The route from entrance to working area can be simulated, and the unexpected obstacles can be detected during the maintenance simulation.

The radiation level, if available, which is a important measure of maintenance prioritization, can be monitored from the local sensor, and it is very important to the working cost (time and man-hour). If the pathway including working area includes the high radiation area, the working cost is high. That is, it requires shorter maintenance time more man-hour in aspects of cost. Therefore the alternative pathway can be found by another simulation prior to the actual maintenance.

# 4. Discussion & Conclusion

The master database includes the component design data, procedure data, and 3-dimensional configuration data. The priority of maintenance task is determined by using the reliability measure in this paper by using the living PSA technique. The corresponding maintenance procedure is retrieved from the database, and the simulation for maintenance is performed by computer with 3-dimensional plant model.

The maintenance procedures are computerized with hyper-text technique, and they can be retrieved from the networked computer. The simulated maintenance gives an experience of maintenance to the staff. It increases the confident level of the operator for maintenance, and finally it increases the efficiency of maintenance task and it decreases the possibility of human error due to the maintenance.

# References

1. IAEA, Handbook on Safety Related Maintenance, IAEA Training Course Series, Draft, Nuclear Training Centre, KAERI, 1993.
2. J.J. Ha, K.S. Jeong, S.H. Kim, S.Y. Choi, and et al., Nuclear Integrated Database and Design Advancement System, KAERI/RR-1586/95, Dec. 1995.
3. S.C. Chang, T.W. Kim, S.H. Han, and C.K. Park, Development of Dynamic PSA Tool (PEPSI) for the Risk Management of NPP, IAEA TCM on Numerical Safety Indicator, Vienna, Austria, Nov. 1990.
4. S.H. Han, T.W. Kim, and C.K. Park, KIRAP Release 2.0 User's Manual, KAERI/TR-361/93, 1993.
5. T.J. Lim, C.K. Park, J.H. Park, T.W. Kim, Survey and Comparison of Generic Component Reliability Data and Establishment of Preliminary Generic Database for use in PSA, KAERI/TR-364/93, 1993.

# TURBINE DRIVEN PUMP SURVEILLANCES: "WHAT DO WE LEARN?"

Bob Christie
Performance Technology
P. O. Box 51663
Knoxville, TN, USA, 37950-1663
(423) 588-1444
(423) 584-3043 FAX

## 1 BACKGROUND

Most nuclear electric generating units in the United States use turbine driven pumps in a safety related function as a driving force to supply cooling water to selected equipment (reactor vessels or steam generators) in accident and transient conditions. These turbine driven pumps are normally "standby equipment" in that they are not running during power operation but are "function upon demand". These turbine driven pumps are generally found to be quite important pieces of equipment in the Unit's Probabilistic Risk Assessment because in some cases (loss of all ac power) the turbine driven pumps may be the only power source to provide water to cool the reactor core.

These turbine driven pumps are configured in essentially single train configurations with very little redundancy except for an alternate source of cooling water. All the systems are provided with recirculation piping that enables the system to be checked during surveillance testing without injection into the steam generators or reactor vessel. Only in the case of an emergency signal are the injection valves opened.

Mathematical modeling is done to determine the "Conditional Probability" of the turbine driven pump responding to a emergency demand. Most of the performance information available about these standby turbine driven pumps comes from test (surveillance) data. However, there are occasions that Non-Test data is collected. This paper describes a comparison of the performance information collected by the author from five turbine driven pumps at four different commercial nuclear electric generating units in the United States. The total performance information is listed and then is subdivided into surveillance testing and Non-Test actuations.

# 2 HISTORICAL RECORD

The total period of time covered by this paper is from January 1, 1988 to November 30, 1995. Not all of the units were tracked over the entire period. The time the turbine pumps were in either planned or forced maintenance was tracked and compared to the time the turbine pumps might have been demanded in an emergency function. All applicable starts of the turbine pumps were tracked and the ratio of successful starts to total starts was calculated for various time periods. All run hours and run failures of the turbine pumps were also tracked and a "failure rate to run" was calculated by dividing the run failures by the run hours and updating this information using Bayesian techniques. This failure rate to run was then used in a constant failure rate model to calculate the Probability of Run for a specified mission time.

# 3 RESULTS

The total values for starts and runs for the period are shown in Table 2:

# 4 COMPARISONS

As can be clearly seen in Table 2, most of the performance information comes from the surveillance testing.

| | Total | Surveillances | Non-Test |
|---|---|---|---|
| | | | |
| Starts | 666 | 647 | 19 |
| Successful Starts | 648 | 631 | 17 |
| Failures to Start | 18 | 16 | 2 |
| % Successful Starts | 97.3 | 97.5 | 89.5 |
| | | | |
| Run Hours | 349 | 341 | 8 |
| Run Failures | 3 | 3 | 0 |
| | | | |

While most of the information (approximately 97%) comes from the surveillance testing, the Non-Test data is consistent with the overall data if one looks at the overall "Conditional Probability." The overall Conditional Probability from the mathematical models for three of the turbine driven pumps (the ones with the longest period of tracking) are shown in Figures 1, 2 and 3. The Conditional Probability for the three turbine driven pumps ranges from 80% to 96%. The Non-Test data indicates that the overall Conditional Probability is in the same range.

There are a number of points to be noted in this comparison.

4.1. During the 19 Non-Test actuations, the turbine driven pumps were "Standby Available" in all cases; that is, they were not in maintenance at the time of the demand.

4.2. The turbine driven pumps are not used for long periods of time in the Non-Test actuations. The longest run was approximately 2.5 hours. Most Non-Test actuation runs are seconds or minutes.

4.3. It appears that the problems during the Non-Test actuations are the same as the problems during the surveillances. Most of the failures during the surveillance demands are overspeed trips of the turbine during the start. Both failures during the Non-Test actuations were overspeed trips during the start.

4.4. As the nuclear units have fewer reactor trips, the number of Non-Test actuations will decrease. For example, between May 1, 1995 and November 30, 1995, there were a total of 44 starts of the five turbine driven pumps and these starts were all surveillance test starts. There were no Non-Test actuations from May 1, 1995 to November 30, 1995. As the performance of the nuclear units gets better, a higher percentage of the performance information for standby equipment will come from surveillance testing.

# 5 SUMMARY

For five turbine driven pumps used in a "standby" role at four different nuclear units during the time period from January 1, 1988 to November 30, 1995, the percentage of performance information from Non-Test actuations is much smaller than the percentage of performance information from surveillance testing. In spite of the smaller amount of performance information from Non-Test actuations, the performance information gathered during Non-Test actuations appears to be consistent with the performance information gathered during surveillance testing.

# TABLE 1

NON - TEST ACTUATIONS

8/88 Reactor trip. Turbine driven pump was secured after 5 minutes of run time.

12/88 Reactor trip. Turbine driven pump was secured after 17 minutes of run time.

1/89 Reactor trip. Low level in steam generator started turbine driven pump. Secured within 10 seconds by control room personnel.

4/89 Reactor trip. Turbine driven pump failed to start because of overspeed trip.

11/89 Lost dc bus. RCIC automatically isolated. Then RCIC automatically unisolated and auto initiated. Operators quickly stopped RCIC.

8/90 Manual reactor trip. Turbine driven pump automatically started and quickly secured as not needed.

2/91 Reactor trip. Turbine driven pump was secured after 7 minutes of run time.

4/91 With the reactor at hot shutdown during power ascension, received low steam generator level signal. Turbine driven pump auto started. Reactor was manually tripped. Turbine driven pump was secured after 10 minutes of run time.

5/91 Reactor trip. Low level in steam generator started turbine driven pump. Turbine driven pump was secured after 35 minutes of run time.

5/92 Inadvertent actuation of RCIC during instrument and control surveillance. RCIC was stopped by operators after about 13 seconds of injection into the reactor vessel.

11/92 Lost all main feedwater pumps. Manual scram. Both turbine driven pumps were actuated by low level in the reactor vessel. Both pumps were secured within one minute.

8/93 False high level water signal scrammed reactor. Both turbine driven pumps were actuated on low level in the reactor vessel. Both pumps were automatically stopped on high level in the reactor vessel within two minutes.

8/93 During power ascension, reactor feedwater pump turbine tripped on high level in the reactor vessel. RCIC automatically started on low water level. RCIC was secured after 11 minutes of run time.

12/93 Following a reactor scram, RCIC was manually started and used for reactor vessel water level control. RCIC was secured after approximately 2.5 hours of run time.

9/94 Inadvertent reactor high level signal caused scram. Reactor low water level signal was generated. RCIC auto initiated but overspeed trip occurred on start.

4/95 Manual reactor scram was followed by very short duration reactor low water level signal. Both RCIC and HPCI auto actuated but only RCIC was necessary. RCIC was secured after 11 minutes of run time.

# TABLE 2

| | Turbine 1 | Turbine 2 | Turbine 3 | Turbine 4 | Turbine 5 |
|---|---|---|---|---|---|
| Total Starts | 119 | 410 | 62 | 42 | 33 |
| Successful Starts | 117 | 400 | 59 | 41 | 31 |
| Failures to Start | 2 | 10 | 3 | 1 | 2 |
| Run Hours | 111 | 138 | 45 | 30 | 25 |
| Run Failures | 1 | 1 | 0 | 1 | 0 |

The surveillance test values for starts and runs for the period are:

| | Turbine 1 | Turbine 2 | Turbine 3 | Turbine 4 | Turbine 5 |
|---|---|---|---|---|---|
| Surveillance Starts | 116 | 405 | 60 | 36 | 30 |
| Surveillance Successful Starts | 114 | 396 | 58 | 35 | 28 |
| Surveillance Failures to Start | 2 | 9 | 2 | 1 | 2 |
| Surveillance Run Hours | 110 | 136 | 44 | 27 | 24 |
| Surveillance Run Failures | 1 | 1 | 0 | 1 | 0 |

The values for Non-Test actuations are described in Table 1 and summarized below.

| | Turbine 1 | Turbine 2 | Turbine 3 | Turbine 4 | Turbine 5 |
|---|---|---|---|---|---|
| Non-Test Starts | 3 | 5 | 2 | 6 | 3 |
| Non-Test Successful Starts | 3 | 4 | 1 | 6 | 3 |
| Non-Test Failures to Start | 0 | 1 | 1 | 0 | 0 |
| Non-Test Run Hours | less than 1 hour | less than 2 hours | less than 1 hour | less than 3 hours | less than 1 hour |
| Non-Test Run Failures | 0 | 0 | 0 | 0 | 0 |

EFW TRAIN 'A' CONDITIONAL PROBABILITY
MODES 1, 2, 3

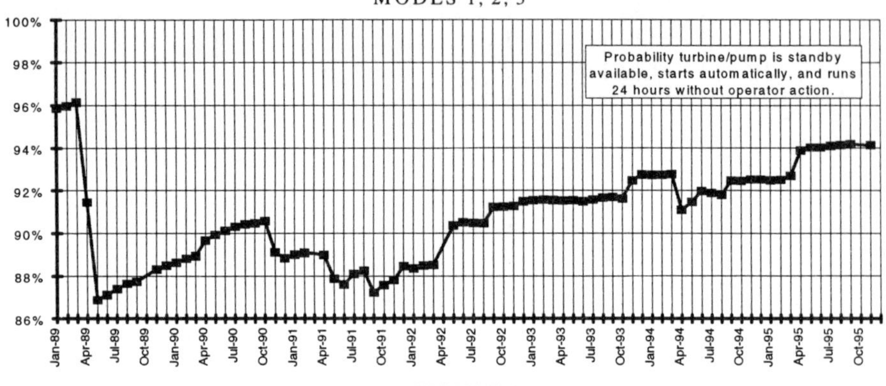


FIGURE 1

RCIC CONDITIONAL PROBABILITY
MODES 1, 2

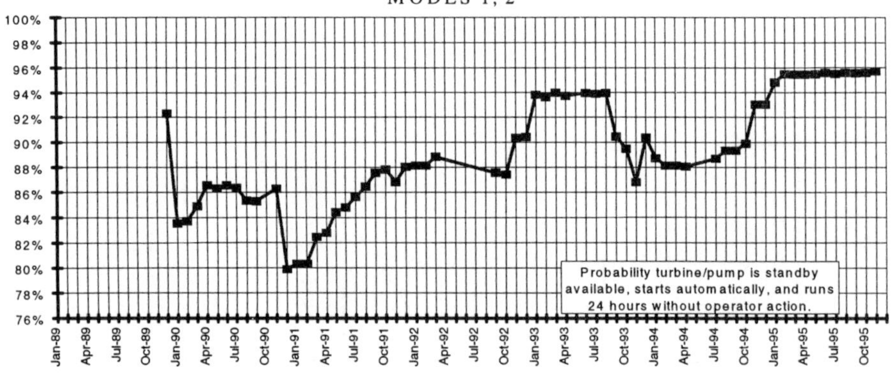


FIGURE 2

EFW TRAIN 'A' CONDITIONAL PROBABILITY
MODES 1, 2, 3

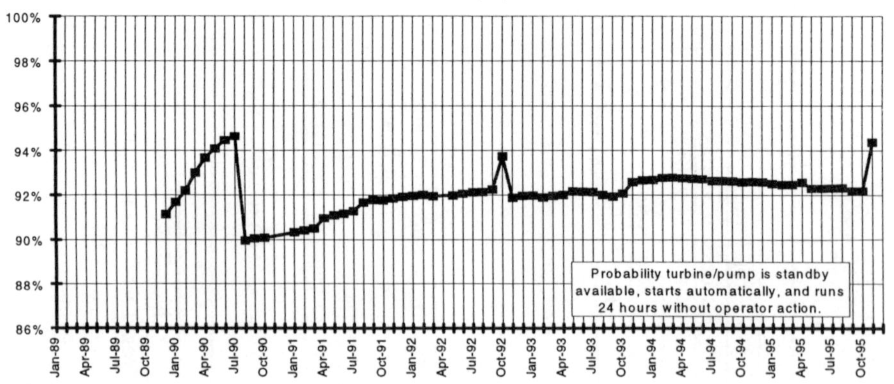


FIGURE 3

# FRAMS : A software prototype for incorporating Flexibility, Reliability, Availability, Maintenance and Safety in Process Design

T.V. THOMAIDIS, A.P. MELIN and E.N. PISTIKOPOULOS*
Centre for Process Systems Engineering
Imperial College, London SW7 2BY, U.K.

**Abstract**

A general framework for the systematic incorporation of operability and safety aspects at the design stage of process systems is presented in this work. Based on a stochastic model representation for the continuous and discrete state uncertainty, a set of analytical tools for the quantification of operability aspects and the identification and ranking of critical parts of equipment and/or critical events, a methodology supported by a software package is described for design and operability optimization of process systems. It is shown that the incorporation of operability measures at the early design level has economic benefits, while resulting in inherently flexible, reliable and safe design configurations.

# 1 Introduction

The integration of operability objectives (such as flexibility, controllability, reliability, maintainability and safety) in the design stage of process systems requires a common mathematical framework accounting for [1,2]:

- Continuous uncertain parameters either internal to the process (such as kinetic and heat transfer parameters, rates and temperatures of streams, etc) or external to the process (such as product demands and prices, feedstock compositions, etc) —usually described by continuous probability distribution functions.
- Discrete uncertainty such as equipment availability, unexpected events/ faults —here, reliability models and discrete probability functions are commonly used.
- A process model, i.e. a set of equality and inequality constraints representing mass/energy balances, reaction model equations, design equations/specifications, product specifications, etc; typically a differential-algebraic model is employed.

Based on this information the following set of analytical tools for the quantitative assessment of the various operability objectives has been developed [3,4]:

- Stochastic Flexibility index ($SF$), to measure the ability of the process system to absorb the continuous uncertainty.
- Combined Flexibility-Reliability index ($FR$), to measure the overall performance of a process system in the presence of both continuous process variations and equipment deterioration.

*Tel.:xx-44-171-594-6620 Fax.:xx-44-171-594-6606 Email:e.pistikopoulos@imperial.ac.uk

- Combined Flexibility-Availability index ($FA$), to evaluate the overall process performance when maintenance considerations are involved in the process model.
- Criticality index ($FRC$) for the identification of reliability and safety bottlenecks in a process system as well as a relative ranking of components/subsystems according to their operability significance.

These tools have been efficiently implemented in the software prototype package, FRAMS, which is described in some detail in the next section.

# 2 Overview of FRAMS

The basic structural blocks of this package, shown in figure 1, are as follows :

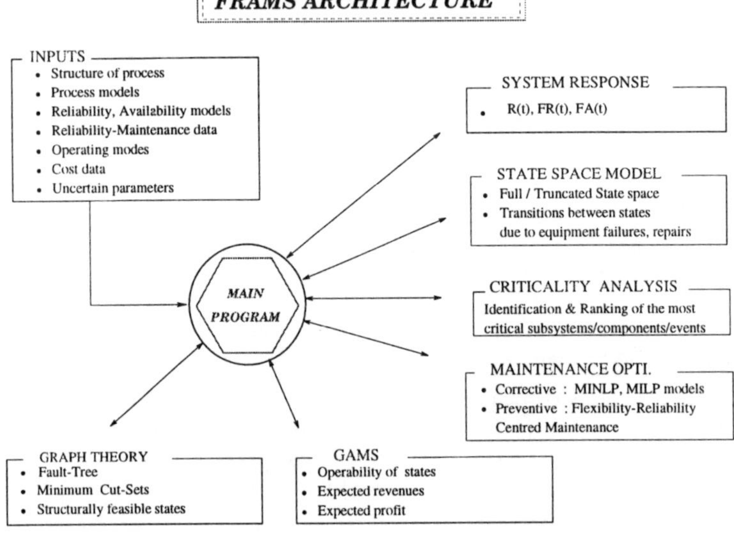


Figure 1: FRAMS - Block diagram of the software prototype

**System response.** For the evaluation and design optimization of the stochastic flexibility, Combined Flexibility-Reliability index and Combined Flexibility-Availability index (as functions of time), FRAMS is interfaced with GAMS (General Algebraic Modelling System, Brooke et al., (1988)), through which a number of commercial solvers (ZOOM,MINOS,CONOPT etc) can be used for the solution of the underlying (stochastic) optimization problems.
**State space representation.** This block automatically identifies all the discrete states involved in the system (according to operating/failure modes assumed for the equipment and subsystems participating in the process) as well as the transitions between them. Reduction options of the size of the state space utilising structural properties of the equipment (i.e. aggregation of symmetrical structures) are also available.
**Graph theory.** This block enables the automatic creation of the fault-trees and the associated minimal cuts. The required information for this block is embedded in the directed logic functional AND/OR digraphs (DFLD) and

the reliability/events data base. Structurally feasible states are then determined amongst the system states.

**Criticality block.** This part of the software deals with the identification of the most critical pieces of equipment/subsystems or events at any point in time. A criticality index is assigned then to each one and a criticality ranking is performed in order to obtain their relative importance. Its computation relies on minimal cut set and fault-tree concepts and reliability approximation schemes.

**Maintenance optimization.** Based on the information obtained from the flexibility analysis and the state space representation blocks, the associated maintenance superstructure model is formed [4], the solution of which leads to the optimal corrective maintenance policy. From the results of flexibility, reliability and criticality analyses, a second optimization procedure can be executed to yield a preventive maintenance plan over time.

Note that the necessary data regarding reliability and maintainability occurrence of desired events and process economics are provided via a database. Recent developments aim at extending this facility to commercial databases. FRAMS can be effectively used for the operability analysis of complex chemical process systems, especially when multiple operating modes are considered and a number of uncertain parameters are involved either in process operations and/or in equipment availability; hazard analysis is also included.

FRAMS can also be used as a design tool in order to obtain an inherently flexible, reliable and safe design configuration at maximum expected profit.

In this respect, FRAMS can be viewed as a systematic decision support system for the identification of the optimal design parameters and operating conditions in order to increase the capability of complex process systems to account for expected process variations, equipment unavailability and maintenance activities as well as safety regulations. Some basic features of FRAMS will be illustrated in the following example of the design of a multiproduct batch plant.

# 3 Illustrating Case Study

Consider the following multiproduct batch plant (typical in food, pharmaceutical and fine chemicals industries), as shown in figure 2(a) with the data of table 1. It involves three main processing stages (mixing, reaction and purification) for manufacturing two different products (A,B).

An interesting feature of batch plant models is that the feasible operating region in the space of the product demands $Q_i$ is linear (for fixed volumes and number of units) with the horizon time constraint limiting the plant capability. This is illustrated in figure 3(a).

The variations of the product demand are here assumed normally distributed following a joint probability density function, (see table 1), with which the evaluation of the stochastic flexibility index in FRAMS yields a value of $SF = 0.994608$ (integrating over the statistically significant area of $\mu \pm 4\ \sigma$), implying that the design can accommodate 99.4% of the expected product demand variations, Figure 3.(b).

If one of the two mixers fails (assume failure rates described by Weibull function, $weif(\frac{t}{\alpha};\beta)$, table 3) the feasible operating region, figure 3(c), and the operational characteristics of the plant significantly change, resulting in a reduction of the stochastic flexibility index value to 0.641703. For the different

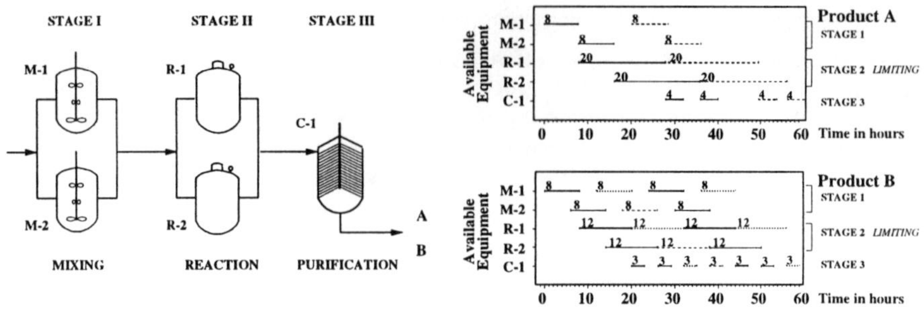


Figure 2: (a) Three stages multiproduct batch plant, (b) Gantt chart for Products A, B

operable system states, depicted in figure 5a, a corresponding stochastic flexibility index can be obtained. The estimation of the optimal expected profit, taking also into account the contribution of the degraded operating states, yields an expected profit value of 238,300 for a time horizon $\mathcal{H}$ of $6000hrs$, with an associated combined flexibility-reliability index of 0.826632. This implies that the system in average, without any maintenance action, can withstand 82.66% of both continuous demand variations and equipment failures. In this case the impetus for executing maintenance policies is clear, since there is a substantial deterioration of the system's performance (from 99.46% to 82.66%).

Next, the results of the design procedure in FRAMS are summarized in table 2 and shown graphically in figures 4(a) and 4(b).

Note as the design cost increases allowing for more investment cost, the volumes of the units increase and the corresponding time dependent flexibility, figure 4(a), and combined flexibility-reliability index values increase, figure 4(b). As a result, the more expensive design here has an average system performance (in absorbing the demand variations and equipment failures) which is almost identical to the maximum performance that the system can achieve when all the equipment is available (i.e. the stochastic flexibility). On the other hand, there is a trade-off, as revealed in figure 4(b). The design that maximizes expected profit corresponds to a less conservative design (smaller volumes, see table 2) with an associated combined flexibility-reliability index of 0.883744; this optimal design requires an investment cost of \$134,800.

| | Processing Times, hr | | | Size factor L/tn | | | Demand | | Price |
|---|---|---|---|---|---|---|---|---|---|
| | Stage | | | Stage | | | $tn/\mathcal{H}$ | | $\$/tn$ |
| Product | 1 | 2 | 3 | 1 | 2 | 3 | $\mu$ | $\sigma$ | |
| A | 12 | 20 | 4 | 2000 | 3000 | 4000 | 40.0 | 10.0 | 460 |
| B | 8 | 12 | 3 | 4000 | 6000 | 3000 | 60.0 | 15.0 | 520 |
| Times Horizon, $\mathcal{H} = 6,000\ hrs$ | | | | | | | | | |

Table 1: Processing times, size factors, demand-cost data and time horizon for the example problem I

| Case | $C_D$ | E{Profit} | $FR_{\mathcal{H}}$ | $V_{M-1,M-2}$ Lit | $V_{R-1,R-2}$ Lit | $V_{C-1}$ Lit |
|---|---|---|---|---|---|---|
| | 140,000 | 250,500 | 0.889873 | 1041.3 | 1561.9 | 2000.0 |
| Optimal | 134,800 | 252,500 | 0.883744 | 968.5 | 1452.7 | 1937.0 |
| | 120,000 | 238,300 | 0.826632 | 797.5 | 1196.3 | 1595.1 |
| | 90,000 | - | 0.624041 | 500.0 | 742.5 | 970.6 |

Table 2: Design Optimization - Summary of Results

When corrective maintenance is involved, with the maintenance cost data as shown in table 3, FRAMS results in an optimal design configuration with values for the capacities of the five units $d_1^* = d_2^* = 500\ Lit$, $d_3^* = d_4^* = 741\ Lit$ and $d_5^* = 988\ Lit$ respectively, (table 2), and an optimal maintenance plan as shown in figure 5b and table 4, involving one service crew ($\mathcal{NC} = 1$) with a maximum expected profit of \$ 260,080. While fixing to this optimal design, an optimal preventive maintenance scheduling is then determined.

## References

1. Van Rijn C.F.H. A System Engineering Approach to Reliability, Availability and Maintenance. Conf. on Foundation of Computer Aided Operations, FOCAPO, Salt Lake City, 1987

2. Grievink J.,Smit K.,Dekker R. and Van Rijn C.F.H., Managing Reliability and Maintenance in the Process Industry, Conf. on Foundation of Computer Aided Operations, FOCAPO, 1993

3. Pistikopoulos E.N. and Mazzuchi T.A., A novel flexibility analysis approach for processes with stochastic parameter, Comp. Chem. Engng., 1990; 14(9):991

4. Thomaidis T.V. and Pistikopoulos E.N., Optimal Design of Flexible and Reliable Process Systems, IEEE Trans. Rel. 1995; 44(2) 243-250

| $C_{\text{Design}} = \sum_j A_j V_j^{B_j}$ A/Lit | B | Weibull para. $weif\{\frac{u}{a};\beta\}$ | | Repair R. $\mu\ (hr^{-1})$ | Failure R. $l\ (hr^{-1})$ | Maint. Cost (\$/task) |
|---|---|---|---|---|---|---|
| 350 | 0.6 | 65,000 | 0.50 | 0.225 | 0.0070 | 300 |
| 650 | 0.6 | 80,000 | 0.60 | 0.090 | 0.0080 | 150 |
| 450 | 0.6 | 110,000 | 0.8 | 0.150 | 0.0110 | 450 |
| Time Horizon = 12,000 hr | | | | Fixed maintenance cost $\delta = \$50,000$ | | |

Table 3: Reliability, Maintenance and Cost Data of the Equipment

| Current State | Operable Eqpmnt Stage | State Probability | Stochastic Flexibility | Expected Revenue | Maintained To |
|---|---|---|---|---|---|
| 1 | 2 2 1 | 0.711326300 | 0.994608 | 493,325 | - |
| 2 | 1 2 1 | 0.084838227 | 0.957393 | 493,325 | 1 |
| 3 | 2 1 1 | 0.112958817 | 0.994606 | 467,663 | 1 |
| 6 | 1 1 1 | 0.033466687 | 0.957393 | 467,665 | 2 |

Table 4: Optimal Maintenance Plan and Expected Profits

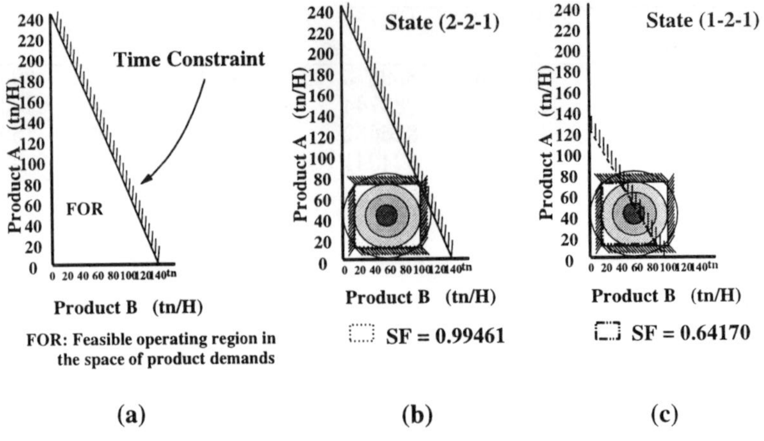


Figure 3: Effect of equipment failures on feasible operating region.

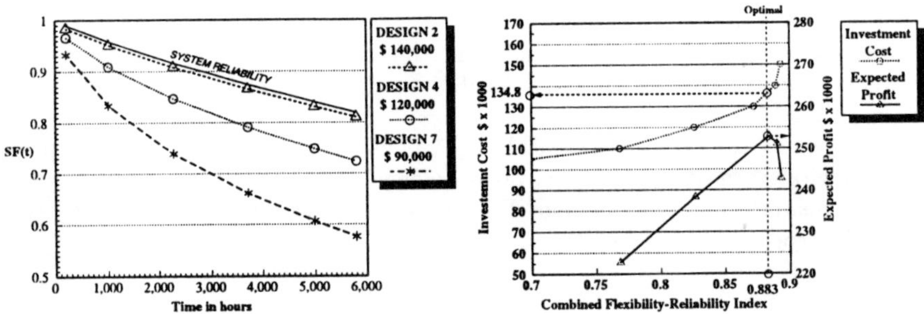


Figure 4: (a) Time dependent Stochastic Flexibility Index for different designs, (b) Investment Cost and Expected Profit vs. $FR_{\mathcal{H}}$

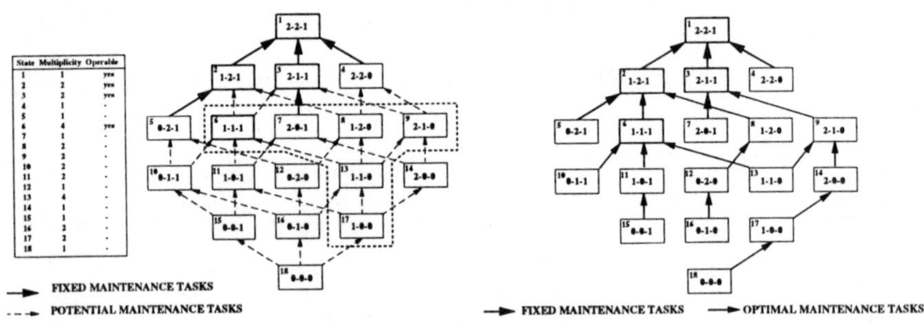


Figure 5: (a) State network and transitions due to maintenance, (b) Optimal Maintenance Plan

# The Use of RAMS Probabilistic Analysis for On-line Maintenance and Operational Decision Support

S. Contini, M. Wilikens, M. Masera
Joint Research Centre - ISEI
Ispra (VA), Italy

**Abstract**

This paper describes issues involved in the provision of RAMS support functions integrated in an existing on-line plant supervision tool called FORMENTOR[a] [1]. The FORMENTOR project aimed at offering real-time decision support to operators of complex industrial plants, acting as a supervision system helping in coping with abnormal or hazardous plant situations.

An on-line tool like FORMENTOR, after integrating the necessary extensions, can also be deployed as a useful tool for maintenance support by taking advantage of its monitoring and diagnostic capabilities, its knowledge bases and the type of support offered [2].

This paper presents an evolution of these concepts, taking advantage of a particularly efficient computer implementation of plant probabilistic RAMS models.

# 1. Introduction

In complex plants the correct management of Maintainability is a fundamental issue, not only from the production point of view, but also for the satisfaction of safety requirements. Maintainability relies on the relationship between operators and the maintenance crew: e.g. permits to perform maintenance work, which entail components unavailability with effects on production , product quality and safety.

The effective support of plant Maintainability means giving the operator a guide for deciding on the opportunity for maintenance interventions on the basis of the current status of the plant and on the pre-defined plant objectives to be respected. This can only be achieved by the use of on-line models, which should be flexible enough as to follow the modifications in the installation (e.g. component failure, maintenance activities, design modifications, production variation, etc.), and fast enough as to offer real-time support.

The solution presented here gives an adequate answer to these problems. It is based on the intelligent on-line decision support tool FORMENTOR, and the integration of probabilistic plant models developed during the plant design and the quantitative safety analysis.

---

[a] FORMENTOR is a project in the EUREKA program of collaborative international R&D projects. The partners are: APSYS, Cap Gemini Innovation, Det Norske Veritas and the Institute for Systems Engineering and Informatics of the Joint Research Centre of the European Commission.

The probabilistic information is associated with plant abnormal behaviors, i.e.: significant accidents, safety-related systems failures, production loss, component failures.

Users who may benefit from this information are:

- plant managers, who would be able to make decisions on how to improve or maintain the productivity according to the plant safety level and the availability of plant components;
- maintenance staff, who would be able to better plan interventions, according to the plant production needs, in the frame of reliability and availability considerations;
- plant operators, who need to know the consequences on safety, availability and product quality of any component taken off-line for maintenance/repair operations.

## 2. The FORMENTOR system

The FORMENTOR project developed methodologies, models and software tools, aimed at offering real-time decision support to operators of complex industrial plants. FORMENTOR as a supervision system helps operators in coping with detected abnormal plant situations that may degenerate into unwanted states: plant shut-down, too much product out of specification, or even a hazard-related incident.

The underlying objective is mainly oriented towards safely production by preventing risks, reducing operator stress in crisis, etc., but also other production objectives have been considered like process stability and product quality. The technological approach is generic to the application domains whose operators can benefit from this type of system but has so far mainly been applied in the petro-chemical industry.

The successful operation of technical systems depends on the accomplishment of a suitable maintenance strategy and this is particularly true for complex plant in which potential failure of a large amount of critical components can jeopardise safety and in which extension of the run times between outages is of utmost importance for its profitability.

The type of maintenance support system we are discussing differs from the typical software tool available, which normally only covers one aspect of the problem: specific diagnostics for one type of equipment, etc. FORMENTOR complements such tools by providing the capacity taking a global plant-wide view when treating failures, rather than dealing with particular problems on a piecemeal basis. In particular, it can combine information from several different sources for guiding the decision making process. For example it enables to identify several different plant anomalies as arising from one common underlying failure. A typical configuration integrated in overall plant control and in a FORMENTOR system is depicted in Figure 1.

The maintenance related actions proposed must of course correspond to the overall objectives of plant operation. In general, a certain number of goals for the operation of the plant are defined which can be classified into three broad categories: safety goals, process goals and economy goals. The fulfilment of these goals is reflected in the criticality of the equipment. It is therefore essential that any maintenance support system be capable of offering advice according to changing plant situations and of taking into account the relative urgency of the situation.

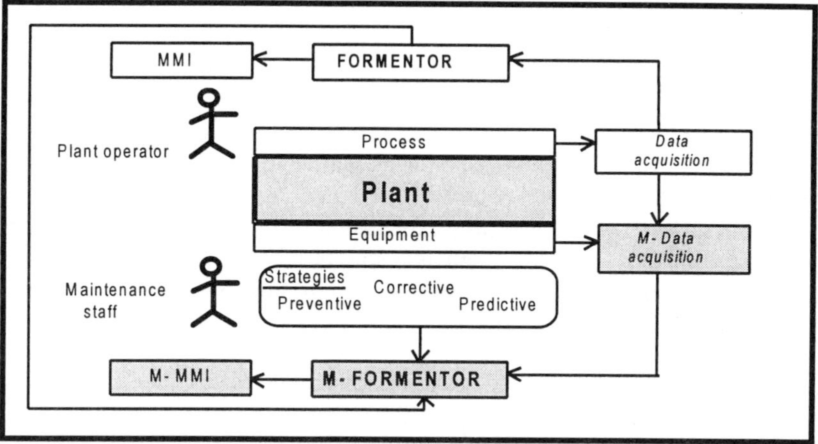


*Figure 1. Maintenance support configuration*

Our conceptual architecture for maintenance support is based on task specificity. The problem solving case is so divided into functions derived from the analysis of generic maintenance tasks. A number of generic and well defined functions have been identified for the maintenance support system. Figure 2 shows the functionalities in an architecture that integrates the FORMENTOR functions.

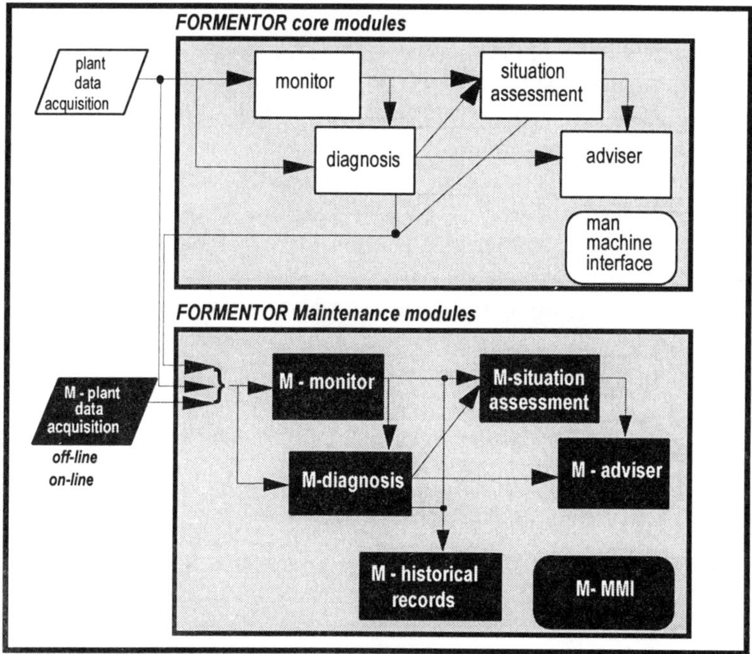


*Figure 2. FORMENTOR Maintenance functionality*

This set of functions, from which a particular application can draw functionalities according to its needs, includes:

- data acquisition - interfaces the retrieving plant data. In the case of maintenance these data can be on-line (available from diverse computerised systems) and off-line (the input of which will have to be done specifically by maintenance operators).
- monitoring - processes the raw plant data with the purpose of detecting any symptom indicating a potential equipment malfunction or precursor signs of degraded failures.
- diagnosing - tries to find a plausible explanation of the abnormal behaviour by identifying its causes (either based on heuristics or on behavioural models)
- assessing the current situation - determines and represents the instantaneous state of the plant with respect to the achievement of plant goals and functions.
- maintenance action advising - determines feasible interventions -for the given plant situation- and advises the most workable one; based on urgency, the type of intervention can be devised (immediate corrective, or delayed preventive).
- analysis of historical records - computes historical and statistical parameters useful for supporting the other tasks.
- man-machine interface - communicates with the maintenance staff and presents the information in an appropriate way, ensuring the fidelity to guidelines established for information presentation and interaction management.

The FORMENTOR approach is based on the modelling of the target system. The pursued advantage is to allow a methodical and exhaustive knowledge acquisition, which should allow a consistent representation and an easy maintenance and modification of the resulting knowledge bases.

The models used in a typical FORMENTOR application are [2]:

- The Goal Tree/Success Tree (GTST) Model, which gives a functional decomposition of the system into goals, and the functions needed for satisfying them; it is used for situation assessment, and for supporting action planning.
- The Plant Structural Model, which gives a structural decomposition of the system into components and subcomponents, incorporating structural and functional connections between them.
- The Plant Behavioural Models, which describe, either quantitatively or qualitatively, the behaviour of the process and components by appropriate models, supporting model-based diagnosis and prediction.
- The Heuristics Causal Models, which give a decomposition of the problem space into causal states, used for supporting the confirmation and diagnosis of abnormal states.

## 3. Usefulness of a Plant Probabilistic Model (PPM)

The models listed above are deterministic and the information they supply, in real time, comes from the analysis of the status of working components as well as of those under maintenance.

The information generated by deterministic models can be fruitfully integrated with probabilistic models, provided that these can give answers in real time. That is, that they can be recalculated any time a variation in the plant structure occurs.

Indeed, such models, for any plant status caused by component failures and subsequent repair, can give not only an updated evaluation of the plant safety, but also the variation of probability of each specific accident (e.g. fire, explosion, release of toxic substances), the on-demand unavailability of protection systems, etc., the criticality ranking of working components, and the degree of redundancy left for all accidents.

Furthermore, failed components can be ranked according to their relative priority for repair, with the aim of maximising the resulting availability and safety. With this probabilistic information available, the plant managers and operators can better decide the actions to take to maintain a pre-established level of safety, availability and product quality.

Another possibility offered by the probabilistic model is the simulation of a component failure before taking it out for preventive/corrective maintenance: if any plant goal cannot be satisfied the maintenance operation can be postponed until the plant will be working in more appropriate conditions.

The construction of the plant probabilistic model is made easier by the existence of the deterministic ones, the Goal-Tree-Success-Tree (GTST) already employed in FORMENTOR. The links and transfer of information between the two is under investigation, with the aim of generating a support module for fault tree construction.

# 4. The PPM approach in FORMENTOR

Real-time systems for operational safety monitoring, based on plant specific PSA, have been developed in the nuclear industry to improve the operational safety of NPPs [3].

In this paper, the approach used in the nuclear industry has been put under examination and an efficient computer implementation of the probabilistic model of a generic plant has been studied for FORMENTOR and tested on a potentially dangerous unit of a real petrochemical plant.

For the system specification, the following objectives have been considered :

- the quantification of the probabilistic plant model is to be possible in real time, i.e. taking not more than some tens of seconds to run it;
- the plant model is evaluated any time a component fails, or when the operator requests it;
- the quantification of the probabilistic model should give the occurrence probability of top events, the minimum degree of redundancy, the criticality of working components and the ranking of components for maintenance, for each defined plant goal (i.e. safety, quality, etc.);
- the operator should have the possibility to simulate the failure of one or more components and run the model accordingly;
- all results should be accessible and presented to the operator in an easily readable graphical form.

The most critical of these features is the real time constraint. Furthermore, the quantification should use plant specific data, thus entailing the use of a data base and a Bayesian module for parameters' estimation.

For the sake of simplicity only the plant safety goal is discussed, but all considerations are fully applicable to the others.

The safety goal is defined in terms of sub-goals, from which potential accidents are easily derived. For each of these hazardous situations the corresponding fault tree is generated. The set of fault trees FT1, FT2, ....FTn are therefore used to calculate the occurrence probability of the significant accidents (fire, explosion, dispersion of gases). These trees are not generally independent, since some component failure modes can appear in two or more of them.

When a system component fails, the fault trees containing it are automatically re-analysed and the new results displayed in a graphical form (histogram) showing to the operator the degree of the deviation of the accidents occurrence probability, from the previous values and from the "acceptable predefined levels". Consequently the operator, at a glance, has immediately a clear understanding of the risk variation and an indication of the level of protection for each potential accident through the knowledge of the components criticalities and of the minimal cut set (MCS) structure.

In order to obtain these results in real time there are two possibilities: either to store the set of pre-manipulated fault trees (in order to reduce the computation time), or to pre-analyse all fault trees and to store all MCSs (if manageable) or the most significant ones. Any of these solutions present advantages and disadvantages. The first one is adopted in FORMENTOR, as in off-line decision support tools [3].

The main difference with previous applications is that FORMENTOR stores the probabilistic models as compact graph-based structures, which allows the reduction of the computation time and working memory requirements.

# 5. The FORMENTOR prototyping solution

The probabilistic model of the plant is described in terms of fault trees derived from the GTST model. Fig. 3 shows the PPM in FORMENTOR.

The main on-line modules are: the data acquisition on component states, the graph-based probabilistic model, the reliability data base with associated reliability parameters estimation module. The results are integrated into the existing man machine interface.

The construction of the graph-based probabilistic model starts from the generation of fault trees for each dangerous plant state; the fault tree construction is facilitated by the available information on the plant working and failed conditions contained in the GTST model. Each tree is then transformed into a graph; failure data are taken from a reliability data base automatically feeded with new plant specific data; these data are updated by means of a Bayesian module each time a component fails and enters the restoration phase.

The set of graphs (one for each fault tree) are organised in a way suitable for rapid quantification. This solution allows the analysis of the set of fault trees in a time significantly shorter than that required by classical methods based on the manipulation of cut sets. The advantage is due to the use of binary decision diagrams [4] which allow the storage of complex Boolean expressions in a very compact way.

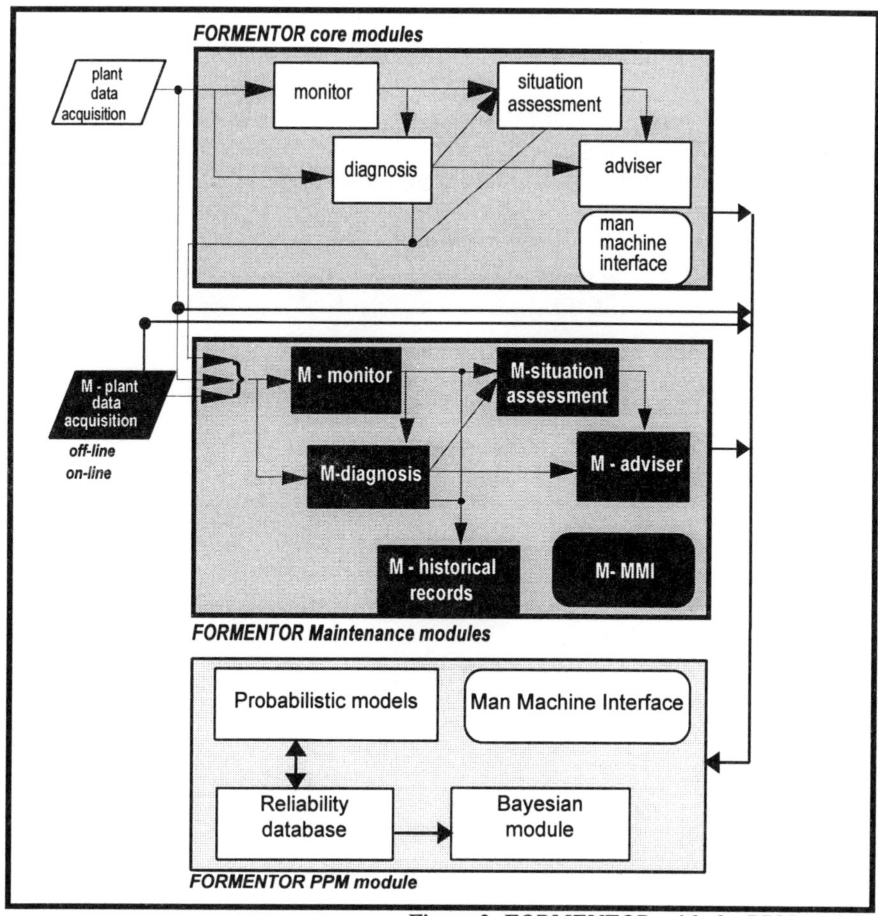


*Figure 3. FORMENTOR with the PPM module*

The analysis module developed has been derived from [5] where the heaviest part of the programme dealing with the determination of MCSs, without cut off techniques, has been changed using the data structure of binary decision diagrams.

# 6. Conclusions and further developments

The approach followed is very promising as it increases the real time understanding of the plant RAMS conditions.

The experimental results obtained so far are positive, fostering the completion of the FORMENTOR system and its application to more complex plants.

## References

1. FORMENTOR M.Wilikens, JP.Nordvik, A.Poucet. FORMENTOR: A Real-Time Expert System for Risk Prevention in Hazardous Environments - A Case Study. In Control Engineering Practice, Vol. 1, No. 2, pp 323-328, Pergamon Press, 1993.

2. M.Wilikens, C.J.Burton. FORMENTOR: Real-Time Operator Advisory System for Loss Control. Application to a Petro-Chemical plant. International Journal of Industrial Ergonomics., Elsevier Science Publishers.
3. W.J.Puglia, B Atefi. Examination of Issues Related to the Development and Implementation of Real-time Operational Safety Monitoring Tools in the Nuclear Power Industry. In Reliability Engineering and System Safety, Vol. 49, n.2, 1995
4. R.E.Bryant. Graph-Based Algorithms for Boolean Function Manipulation. In IEEE Transaction on Computers, Vol C-35, 1986
5. S.Contini, ISPRA-FTA, A Fault Tree Analysis Tool for Personal Computers, EUR 13997, 1992

# Toward Risk Based Regulation

Author
Philippe P. HESSEL
Reliability and PSA Section Head
Atomic Energy Control Board
OTTAWA - CANADA

**Abstract**

Moving from a purely deterministic regulation to the consideration of probabilistic arguments in the licensing decisions is not an easy task. Probabilistic techniques are characterized with uncertainty and are not able to encompass the whole safety picture of a Nuclear Power Plant. Deterministic criteria also have some weaknesses. Therefore, a rigorous process is necessary for deciding the conditions for a Risk Based Regulation policy without jeopardizing the effectiveness of the Regulatory Body.

## 1 - Introduction

Historically, Canada was the first country regulating Nuclear Power Plants with probabilistic unavailability targets for the so-called Special Safety Systems (2 Shutdown systems, Emergency Core Cooling and Containment). However, the present regulatory process does not formally allow crediting other probabilistic arguments in licensing decisions. The worldwide trend toward the use of PSA-based decisions leads the Atomic Energy Control Board (AECB)[1] to consider a move toward Risk Based Regulation. This paper describes the present regulating policy and its sources, the main features of deterministic and probabilistic criteria, then the process being set up for the determination of what could be the future policy of the AECB.

## 2 - The Present Canadian Regulatory System

The present Canadian regulatory system arose in the late sixties under the direction of Dr. Lawrence, then president of the AECB. His main rationale was that a nuclear power plant should not pose more risks to the public than a fossil fueled plant.

### 2.1 Development of the Present Regulations (before 1980)

The resulting target was, according to the then-present statistics, a probability of an accident lower than 10-5 per year. This value gave birth to the present probabilistic requirements, first published in the AECB-1059 document in 1972: "Each of the Special Safety Systems, Shutdown, Emergency Core Cooling and Containment, must have a probability of not meeting their design intent, when required to do so, lower than 10-3 per year" and "No serious process failure (failure of the process system requiring the action of a Special Safety System to avoid

---

[1]The AECB is a five-member board, each of them nominated by the Canadian government, and will be named "the Board" in the text. The Board is assisted by the AECB staff, generally named "the AECB" in the text.

damaging to the fuel) must have an occurrence rate greater than 3*10-1 per year." Meeting these targets is deemed to ensure that the global target is met.
Other requirements were issued, related to the so-called DEL (Derived Emission Limits), setting regulatory limits on the emission of radioactive releases during the normal operation of the plants.
A set of deterministic requirements supplemented these numerical targets, such as adherence to standards (for example ASME codes and CSA standards) and design rules (diversity, separation between process and safety systems, etc. . . . )
The set of requirements was updated with the occurrence of events in the plants and the work of national and international bodies such as the ACNS (Advisory Committee on Nuclear Safety) and the ICRP.

### 2.2 From 1980 to Present

The first important update of the Canadian regulations was the publication, in 1980, of Consultative document C6, used, as a trial, for the licensing of Darlington Power Plant. C6 requires a systematic design review and establishes five dose limits for different classes of initiating events, described in a list. The list is sorted according to the judgment of the AECB specialists, without direct reference to probabilities. In that way, C6 is, as said in its foreword, "of a deterministic nature."
Ontario Hydro answered to C6 by the publication of DPSE (Darlington Probabilistic Safety Evaluation) in 1987, considering that the analysis of initiating events met the requirement of a systematic design review from C6 and used the PSA's techniques to address the dose limits. DPSE was the first PSA produced in Canada and, while not a really Level 3 PSA, addressed the off-site effects.

### 2.3 The Present Status

While probabilistic arguments are not considered in the licensing decisions, the Board already used some risk-based rationales. For example the consideration, in radioprotection-related topics, of the ALARA (As Low As Reasonably Achievable) concept. Also, without using numerical figure, the AECB sometimes used risk-based arguments.
The ACNS proposed, in 1994, the draft of a new document, ACNS-20. In that document, the ACNS proposed a regulation based mainly on probabilistic targets at plant level, as much for normal operation as for accidents. ACNS-20 was not accepted by the AECB staff on the basis that probabilistic criteria were not sufficient on their own to ensure the overall safety of a plant. However, the staff acknowledged that probabilistic criteria give an insight into plant safety that deterministic ones could not.
Ontario-Hydro is presently carrying out PSAs for all his plants: Bruce Heavy Water Plant PSA, which was considered in the 1995 license renewal, Pickering-A Risk Assessment (PARA), was issued in December 1995 and BBRA (Bruce-B plant), BARA (Bruce-A plant) and DARA (Darlington plant) are to be issued within 2 years. Ontario-Hydro is eager to use these PSA for licensing purposes.
Supplementing these facts, the international trend toward an increasing use of PSAs in the regulatory process leads to consider Risk Based Regulation as a necessary evolution.

## 3 - Deterministic and Probabilistic Criteria Features

Probabilistic criteria are often associated with their weaknesses, such as, for example, uncertainty on the results, inability of coping properly with human behavior. Conversely, deterministic criteria are seldom analyzed in that way. The following sections try to describe the main strengths and weaknesses of both criteria.

## 3.1 Characterization of Deterministic Criteria

A deterministic criterion can be defined as defined fixed design rules whose fulfillment give a sufficient confidence that the design intent is met. These rules are based on the experience and expertise of the regulatory body's staff. They can be qualitative or quantitative.
Deterministic analyses were generally developed for ensuring that specific systems, important to safety, are designed with enough margins to tolerate the worst single failure that can be envisaged. As such, conservative assumptions have traditionally been assumed. The method was used as it made tractable the problem of dealing with a large number of failures. So, it reduced the number of failures to be analyzed to a small number.

### *3.1.1 - Strengths of Deterministic Criteria*

The first strength of deterministic criteria is that the associated decision making process is made simple. The answer is "Go" or "No-Go." Controlling their achievement is an easy task.
On many topics, they simplify the engineers' work by replacing complex and costly assessments or experiments with existing correlations or prescribed design practices (redundancy, adherence to ASME codes, etc.) They are therefore cost-effective.
They allow the designer and the regulator to deal with non or hardly quantifiable issues, for example security, protection against sabotage, etc.
The "Go" "No-Go" feature leads to simple instructions to the operators.

### *3.1.2 - Weaknesses of Deterministic Criteria*

Deterministic criteria's adaptation can be challenged as they are often based on expert judgment. This can be illustrated, in Canada, by the utilities' claims about the classification of events in the list of C6 (see 2.2).
They are determined according to the state-of-the-art that exists at the time. As technology evolves, they can become outdated and require then an over-conservative (or an under-conservative) design or impede innovation. No way exists to assess their adaptation.
The models are generally based on experiments, and use simplifications to allow an "easy" computation. The published model is rarely accompanied with the complete set of arguments, assumptions, experimental dispersion, method of fitting the data, etc. As a consequence, little is known on the validity of the analysis and the uncertainty on the results cannot be assessed. The behavior of an item beyond its design basis cannot be described.
All criteria are given the same importance. Ranking the issues according to their criticality is not possible.
The conservative assumptions result in analyses that do not usually represent the real plant behavior. Therefore, they can be misleading if they are used to assist the operator in understanding the actual reaction of the plant to upsets. They also give no indication on the relative importance of different systems as successive lines of defense.

## 3.2 Characterization of Probabilistic Criteria

According to their name, probabilistic criteria lead to results expressed in terms of probabilities (of failure, of success, etc.) As far the deterministic criteria, meeting the target is considered as giving a sufficient confidence in the result. The main difference is that probabilistic criteria recognize that an item can fail and defines the frequency of failure that can be accepted. Thus, they are generally quantitative.

### *3.2.1 - Strengths of Probabilistic Criteria*

The main strength of probabilistic criteria is that they describe the real world, where perfection does not exist. Also, they don't have to be conservative and they give realistic answers[1]. They then allow an optimization process.
They set out an acceptability level and allow to limit the design or operating procedures effort at a point where the result can be accepted. They allow to rank problems and to prioritize efforts.
As a PSA is based on "best estimate" data, it gives useful insights to the operators in a situation where decisions have to be made.

### *3.2.2 - Weaknesses of Probabilistic Criteria*

Probabilistic criteria's assessments are based on models and computations. They need numerical figures that can be very difficult to evaluate as, for example, failure rates come from experience. They are thus associated with uncertainties. The bottom line figures have to be considered with caution[2]. Also, their completeness cannot be assessed.
They are generally limited to quantifiable issues and can only be used when a model can be developed. Their adequacy is bounded by that of the model. One example of such limitations is the assessment of human performance.

## 3.3 Deterministic Vs Probabilistic

The weaknesses of probabilistic criteria prevent their use as a primary licensing method. However, they provide the only way to assess the real capability of a design and allow a decision making process based on risk comparisons. Thus, they can be considered as a powerful supplement to the deterministic criteria.
A balanced use of the two criteria would accept that a probabilistic argument can be challenged by a deterministic one and conversely. We will name such a policy "Risk Based Regulation."

# 4 - What is Risk Based Regulation?

Risk Based Regulation can be defined as a regulation process based not only on deterministic criteria, but also on an assessment of risks. The licensing decision is made to the solution that leads to the lower risk, or to a lower cost (the word "cost" including here all the consideration like time, doses, resources, impact on the public, etc. . . . ) while keeping the risk at an acceptable level, as the case may be. This is practically the application on the larger scale of the ALARA concept. One has to note that, depending on the issue, numerical values for the risk levels are not always mandatory, and that qualitative criteria can be sufficient.

## 4.1 The Benefits of Risk Based Regulation

The first characteristic of Risk Based Regulation is that the use of probabilistic techniques allows a ranking of the issues. It is then possible to answer to the question "What is important?". The regulatory body can focus the efforts on the most safety critical issues and ease or release the requirements on the less important.

---

[1]In that way, a PSA can be considered as the most deterministc analysis as it describes the plant "as it is", and not by reference to external models.

[2]This is a common feature between deterministic and probabilistic criteria. However, the uncertainty fields are generally more easily identified in a PSA than in a deterministic analysis where they are hidden in the correlations.

The benefit for the utilities would be a relief of the burden on the less important issues, allowing him to concentrate his resources on the more important. As a result, some savings can be foreseen, along with an improvement of the plant's safety.
For the regulatory body, the benefit would be a more effective regulating process, as for focusing on the important issues and better, less subjective, decisions. Therefore, the regulatory process would be more easily understandable and scrutable by the public and better accepted.

### 4.2 Risk Based Regulation Prerequisites

The benefits for the utilities are not without some obligations: Basing decisions on an assessment of risks needs a high level of confidence on the presented analysis, particularly on the relative ranking of importance. A strict commitment to use Risk Based Regulation would need formal guidelines for the utilities about the availability, the controllability and the quality of the data used and on the adequacy of the analysis and the methods with the issue.
Also, Risk Based Regulation does not mean that deterministic criteria are no longer valid. Most of these criteria are effective, as proved by the present results, and defining what are those that could be challenged by probabilistic criteria is not an easy process.
That means that such a policy is not immediately applicable and that a progressive process is necessary.
A plan "Toward a Balanced Regulation" was proposed and accepted by the AECB executive committee, then presented to the Board, on the beginning of 1996. Its main features are described in the following section.

## 5 - A Plan Leading to a Balanced Regulation

The process for moving from the present Deterministic Based Regulation toward a Risk Based Regulation must be such than the resulting regulatory system is free from weaknesses. Therefore, this process must be rigorous and the Public must be able to check the validity of each of the resulting options. A plan was then defined and presented, for approval, to the Executive Committee of the AECB, then to the Board.
The Plan aims at the establishment of a regulatory policy making benefit of probabilistic as much as deterministic arguments. It describes the activities to be performed to reach this target. It defines the following elements:

- Definition of the conditions and the criteria for the use of deterministic arguments, probabilistic arguments, or both,
- determination of a suitable set of Safety Goals and derived Safety Criteria,
- discussion of Value Impact / Cost Benefit analysis,
- determination of the suitability to require of a PSA as a part of the licensing file,
- definition of the way to cope with PSA applications,
- definition and preparation of the necessary regulatory documents framework,
- Organization to setup as to perform these activities,
- Schedule, periodic reports and milestones associated with each activity, and
- Parallel actions for training the staff and enhancing its reviewing and interpreting capabilities.

The Plan is organized in three sections.

### 5.1 Proposition for a Policy

The target of the first section is to provide the Board a proposition for an evolution of the regulatory process, supported by a comprehensive file. Three main topics have to be addressed in order to constitute a sound basis for the Board's decision:

- Deterministic Vs Probabilistic - Strength and weaknesses of each criteria, domain of application, where do they challenge and what to do then, setting up rules and criteria for the use of each of them and proposition of a policy.
- Safety Goals and Derived Criteria - ALARA and safety goals, determination of the optimum level for setting them, proposition of Safety Goals suitable to the CANDU design and compatible with the past policy, definition of the selective applicability to existing plants, definition of the areas of responsibility of the AECB and the Utilities in setting them and setting up the regulatory importance of meeting a Goal or a Criterion
- Requirement and regulatory use of a PSA - Benefits of requiring a PSA in the licensing file, what level and what purpose for the PSA, when in the design process, application to existing plants, criteria for the acceptability of PSA Applications in the licensing process and use of Value Impact and Cost Benefit analysis

As these topics are dependent, a Steering Committee will coordinate the progress of the working groups.

### 5.2 Preparation of the Regulatory Framework

The policy defined from the first section of the Plan can only be enforced via the issue of several regulatory documents and the update of several other ones. This framework will be organized into three tiers:

- A Policy Statement, giving general statements, not directly enforceable. It will define the policy of the Board concerning the use of both probabilistic and deterministic arguments, and will set out high-level Safety Goals (their level and values according to the results of the first section.)
- Top Tiers regulatory documents, setting out the general licensing requirements, including a set of Derived Safety Criteria to be met and, possibly, the status given to PSAs.
- Detailed or technical requirements: Requirements at system level (such as the present R7, R8, R9, R10), reporting requirements (such as the present R99) and methodological guidelines (such as C42, PSA guidelines, in preparation)

### 5.3 Training the AECB Staff

At the end of the first section of the Plan, each working group will propose the training curriculum to be followed by the AECB staff in order to enforce and make a correct use of the resulting policy. This section is a long term one, and has already begun, with seminars by worldwide authorities like Alan Swain and Bill Vessely, courses provided to the AECB staff by recognized consultants and courses at the IAEA and the US-NRC.

## 6 - Conclusion

The result of the described process would be an improvement in the effectiveness of the AECB, by focusing the resources of both the regulator and the utilities on the Safety Critical issues. Also, the rigor of the process would allow to a better understanding of the regulatory policy and decisions by the Public.

# References


Ref 1* ACNS 2

Ref 2* ACNS 24

Ref 3* ACNS 20 (draft)

Ref 4* AECB 1059

Ref 5* AECB regulatory requirements R7, R8, R9, R10 (requirements for the Special Safety Systems)

Ref 6* AECB regulatory documents R99 (Reporting requirements)

Ref 7* AECB consultative document C6

Ref 8* AECB Research project 2.315.1, "PSA application" contracted to Bill Vessely

Ref 9* AECB Research projects 2.316.1, "State of the art of PSA in UK" and 2.316.2 "State of the art of PSA in 4 European countries"

Ref 10* AECB INFO-0566 "The assesment of the Cost and Benefits of Regulatory Decision Making"

* These documents are available from the AECB Office of Public Information
280 Slater Street
PO Box 1046 - Station B
OTTAWA - CANADA - K1P 5S9

Ref 11 IAEA-J4-TC-697.6 Working material of technical committee meeting November 1994 "Use of safety indicators by regulatory bodies"

Ref 12 IAEA-J4-TC-803.3 Working material of Technical Committee meeting December 1994 "Use of PSA in the Regulatory Process"

Ref 13 IAEA-TECDOC-729 Risk based optimisation of technical specifications for operation of nuclear power plants.

Ref 14 NUREG-1489 A review of NRC staff uses of PRA

Ref 15 NUREG/CR-6141 BNL-NUREG-52398 Handbook of methods for Risk-Based analysis of technical specifications

Ref 16 OECD-NEA/CSNI/R(94)15 The use of Quantitative Safety Guidelines in Member Countries

Ref 17 Papers from Bill Vessely "Criteria for Risk-Based decision making" and "Definition of a framework for Risk-Based applications and Risk-Based regulation"

Ref 18 Forrest J. Remick, Implementation of PRA and Safety Goals in the USA for NPP safety. Paper presented at the IAEA Peer discussion on regulatory practices - Safety Goals - Vienna 1994

Ref19 IAEA CB-2 Reactor safety reevaluation - A common basis for judgement. Redraft January 1994

Ref20 IAEA Safety Series 106: The role of PSA and Probabilistic Safety Criteria in Nuclear Power Plants Safety

Ref21 UK-HSE T647 The tolerability of risk from nuclear power stations

Ref22 US-NRC SECY-94-176 Issuance of proposed rulemaking package on shutdown and low power opeartions for public comments

Ref23 US-NRC Safety Goals for the operation of NPP, Federal register, vol 51, pp 30028-300033, August 21, 1986

Ref24 US-NRC SECY-94-219 (Proposed Agency-wide implementation plan for PRA) 7590-01 (superseding SECY-94-218 :Use of PRA methods in Nuclear Regulatory Activities)

Ref25 US-NRC Comments from the Advisory Committee on Reactor Safeguards on the draft Policy Statement on the use of PRA, May 11, 1994

Ref26 US-NRC PRA Implementation Plan - June 14,1994

Ref 27 Villemeur - Reliability, Availability, Maintainability and Safety Assessment - 1991

Ref 28 Working material of the IAEA Specialist's meeting, April 26-29 1993 "Use of PSA in the Regulatory Process"

Ref29 IAEA-TECDOC-831 Policy for setting and assessing regulatory safety goals.

Ref30 Engineering Risks by Ulrich Hauptmanns and Wolfgang Werner. Edited by Springer-Verlag. Chapter 6.5 "Potential of the combined use of deterministic and probabilitic methods"

Ref31 P. Hessel - "The role of Reliability and Risk Assessment in the Canadian licensing procedures". ESRA Technical Committe Publication 2/95: "Proceedings of the workshop on Reliability of Mechanical Components and Structures" - La Baule 1994

# PSA applications for NPPs at HEW

H. Ohlmeyer, B. Schubert
Hamburgische Electricitäts-Werke AG
22286 Hamburg, Federal Republic of Germany

**Abstract**

PSA methods for NPPs were introduced to German utilities in the late 1980s. PSAs of the level 1 type have been implemented due to a request from the reactor safety commission. Using these PSAs as a basis, utilities have begun to use PSA methods in other areas to optimize safety and operation. This paper gives details about this type of applications as well as the lessons learned from them.

# 1 Introduction

In 1988 the German Reactor Safety Commission (RSK) requested periodic safety reviews for all German Nuclear Power Plants (NPP). For most German utilities and regulators, Probabilistic Safety Analysis (PSA) methods were fairly new at that time, especially with regard to the size of complete level 1 PSA. Therefore, most utilities faced the following problems:

1. Size and analytical depth of a PSA was not fully clear.

2. There was only little knowledge in the utilities and also in the regulators on how to prepare PSAs.

3. Software and its capability to support the PSA was almost unknown.

The first problem was solved by the rather fast publication of a PSA guide [1] which was based on international guidelines and was compiled by German experts under the auspices of the BMU (Federal Minister of Environment and Reactor Safety). To overcome the second problem, the utilities had to form groups of experts which had to be involved in the PSA construction work but also to contribute plant specific know how. The necessary software tools also had to be introduced to the utilities. HEW decided to develop the $RISA^+$ code [2].
It should also be mentioned that most utilities could not see the usefulness of the tremendous expense they had to invest for relatively little gain in new findings about the safety status of the NPPs. Therefore, there is still a strong request to PSA experts to show the usefulness of the method for plant operating purposes. This paper will discuss the areas of applications of PSA methods including the expected outcome at HEW (Fig.1).

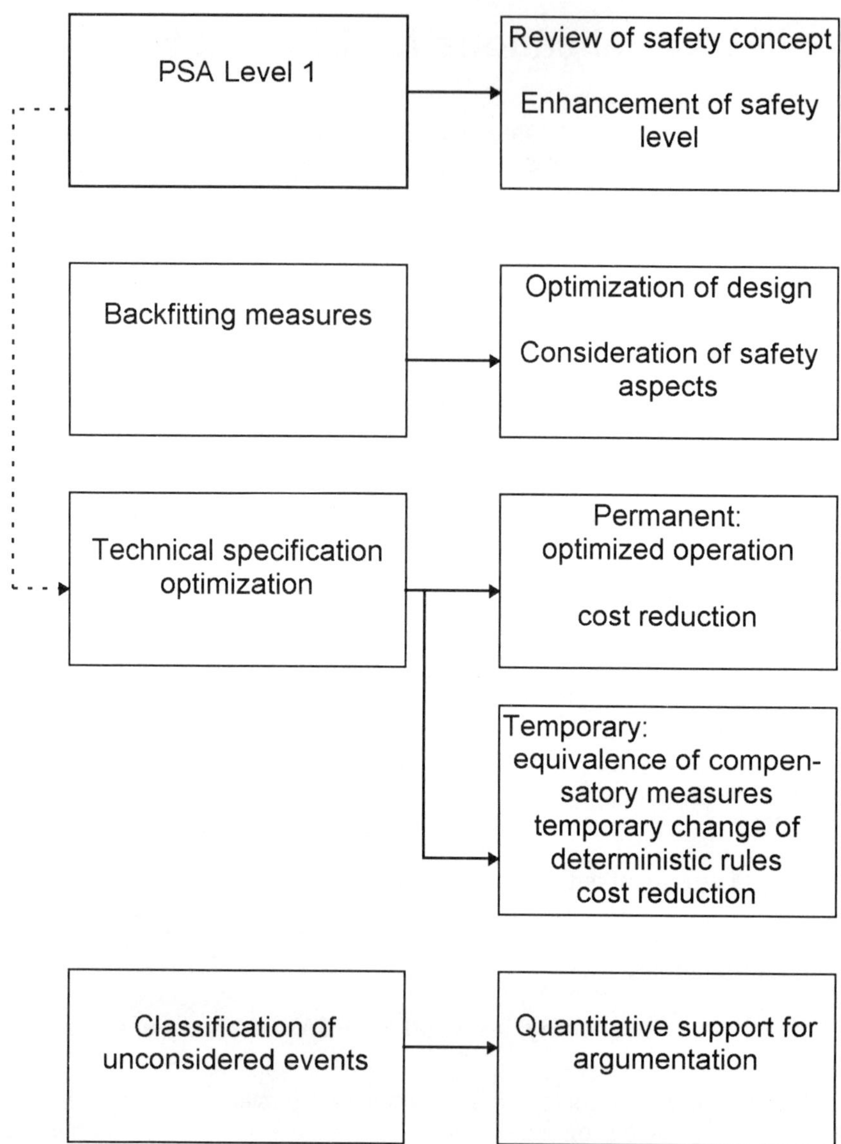


Figure 1: HEW applications of PSA methods with the expected result of the method

# 2 PSA applications at HEW

HEW operates two Boiling Water Reactors (BWRs). For the Brunsbüttel plant, the preparation of a PSA of the level 1 type was suggested to the local authority. While for the Krümmel plant, only parts of the plant have so far been analysed to answer specific questions concerning safety, design and operation of safety systems.

## 2.1 PSA Level 1

For the Brunsbüttel PSA, a process of reviewing the analysis parallel to the construction work was chosen in agreement with the controlling authority. The aim was to have an ongoing review process which could include all review findings directly in the PSA. The result would have been a completed and reviewed PSA at the end of the process. But there were some obstacles not foreseen when the PSA process started:

1. Capacity which should have been used to establish the PSA was used to consider the review outcomes and to reconstruct the PSA where this was necessary.

2. The HEW and the reviewer had differing concepts about the depth and the philosophie of the analysis. This led to unresolvable problems for the progress of the PSA.

3. From the utility's point of view, the reviewer costs became as high as the costs of the construction process, and were therefore unacceptably high with the risk of an open end.

These problems, and the consensus between all participants for the need of getting final PSA results led to the agreement to finish the process of ongoing review and to complete the PSA, taking into account the suggestions already made, in the responsibility of the utility.

The results for the Brunsbüttel PSA show no significant differences to the results of similar plants of SIEMENS-KWU BWR Type '69, which was expected on the basis of several other studies for this reactor type. Nevertheless, the advantages of the PSA process for the utility lay in the fact, that the plant's personal were now aware of the PSA tool and had a basis for further applications which can support the plant operation.

## 2.2 PSA applications in backfitting measures

For backfitting measures, it has become common to use PSA methods already in the design process if necessary and if safety systems are involved. Moreover, the licensing procedure for backfitting measures calls for such analysis in order to show and quantify the safety gain or to reach specific goals.

At HEW, backfitting measures for the main steam safety relief valve system (SRV) are an example of the intensive use of PSA methods from the beginning of the request for backfitting measures with the aim of reaching a final goal of an unavailibility of $<10^{-7}$ until the implementation of the solution in the plant:
In KKB, the SRV system consists of 7 steam controlled valves with 2 solenoid pilot valves for each main steam valve (in KKK there are 11 SRVs). The success criteria for this system is 1-out of-7 (1-out of-11 for KKK). Due to safety considerations the RSK postulated the complete loss of this system by common cause failure with an unavailability of $10^{-5}$. Due to this assumption the event sequence with loss of all SRVs became a dominant contributor for the core damage frequency.
The requested diversity was introduced by 4 motor driven valves in addition to the already existing 7 SRVs at KKB. The success criterion was defined to be 2-out of-4. It was shown that a combination of only 3 valves of the same type could not reach the goal of $10^{-7}$. The same applied to KKK where a design of 3-out of-5 motor driven valves was chosen. During the design process review experts postulated another common cause failure in the control system. They applied a common cause failure of $10^{-5}$ to the 13 pressure transducers in KKB, which would cause the loss of the trip signal to all controlled valves and the sequence under discussion to be of the same magnitude as it was at the beginning of the discussion. To overcome this problem, a diverse trip signal had to be introduced which was found in the selfactuating spring loaded pilot valves. To activate this feature, the so-called pneumatic load, which is installed for operational reasons, was reduced, causing the pilot valves to open independent of the control system at an acceptable reactor pressure level allowing sufficient injection of water.

The described process shows the strong influence of PSA methods on the design of safety systems and the dependence of the results on the application of probabilistic methods by different experts with different tasks. It also shows that rather uncertain values, like the postulated common cause values, often lead to rather costly backfitting measures.
On the other hand examples exists at HEW showing that PSA methods can support the daily work in the plants. Especially in the area of the redesign or extension of the existing control system, sometimes there are several design possibilities, which would produce the desired output but are different in expense and unavailability. In such cases PSA methods are useful for the plant to find the optimum between safety and cost considerations.

## 2.3 Technical Specification Optimization

As already mentioned, it is essential for utilities to draw benefits from the expense they had to invest in PSA including the necessary surroundings. Technical Specification Optimization (TSO) is one of the possible applications of probabilistic methods and an available PSA.
The technical specifications of the plant which are meant here determine the conditions and limitations under which the plant is allowed to operate. In

detail parameters like the surveillance testing intervals, the number of available redundancies of a system or of systems in general and the allowed outage times of systems are in the centre of the TSO process.
In the past the mentioned parameters were defined deterministically between utilities and regulators. Only sometimes probabilistic methods have been used for this task. Therefore, a fully probabilistical analysed definition of technical specifications does not yet exist and is now possible with the availibility of PSAs. It should be mentioned, that not all defined technical specifications of the discussed type can be handled with PSA methods or only with an sometimes unacceptable extension of the available PSA.

Applications of PSAs or probabilistic methods for TSO concerns to different areas distinguished by

- permanent modifications of technical specifications or

- temporary modification of deterministic rules for plant operation.

Permanent modifications of technical specifications are undertaken to optimize plant operation with regard to the above mentioned parameters. Optimization of plant operation in this context is defined as optimization in terms of cost. This does not mean that the safety level is reduced. From the utility's point of view it should be maintained and if even possible can be enhanced. Utilities are especially interested in the allowance for preventive maintenance. This has the advantage of shifting time consuming and rather sensitive work with respect to safety from the yearly shutdown to time periods during operation with less stringent time windows.

Temporary modifications of technical specifications are carried out if unforeseen or temporary plant conditions occur which will lead to violation of plant operation rules. In such cases it should be possible to take compensatory measures which keep the defined safety level and allow further operation with only minor restrictions.
A typical example of such applications was an event in KKK, which led to an unavailability of the current supply to one train out of four of the residual heat removal system (RHR system). The repair of the component concerned would have taken more than 100 h, which was the criterion to shutdown the plant if repair was not possible within this time period. Application of probabilistic methods could show that a reduction of the test intervals in conjunction with an reduction of power to 50% could have even reduced the unavailability of RHR system in this situation. Due to an unclear licensing situation, almost no benefit could be gained from the analysis.

At the moment, the application of these ideas and their acceptance is not fully clear between the participants of the process in Germany. For the cases mentioned, no risk oriented criteria or guidelines exist to review the probabilistic analysis and to determine the influence of its results on the determinstically oriented plant operation. There also exists no common consensus about the role of PSA methods in regulation and licensing. Therefore the outcome of such a review is unpredictable for the utility.

## 2.4 Support for event classification

During the daily discussions about the plant safety level events and accident sequences are often postulated which have not yet been considered in safety analysis of the plant. To decide wether these events should be analysed in more detail or not, probabilistic criteria are usually taken into consideration in order to estimate the contribution of the case under consideration to the overall core damage frequency of the plant. In most cases rough estimations of accident sequence frequencies have shown to be sufficient for decisions on consequences. Usually the decision is oriented at a cut criteria of $10^{-7}$/a which is usually used for the neglection of sequence frequencies in the PSA.

# 3. Conclusion

At HEW, PSA and probabilistic methods are used for PSA Level 1, design and cost optimization of backfitting measures, technical specification optimization and support for decision making. From these applications, the following conclusions can be drawn:

1. Plant operation is determined by deterministic rules, and PSA results can only support the operation at any single moment.

2. No regulation exists about the use of PSA in plant operation.

3. For some areas of PSA, no consensus in the PSA community exists in Germany. This leads to uncertainties concerning review results from the point of view of the utility.

4. For some areas of PSA application, uncertainties are large and therefore these applications should be avoided.

From the utility's point of view the practice of accepting only bad PSA results as a tool for new requirements by regulators and never giving credit for good PSA results for plant operation must be changed. To overcome the mentioned problems, methods of PSA must be enhanced and, which is essential, a consensus about the role of PSA for plant operation must be found in the PSA community.

## References

1. Methoden der Probabilistischen Sicherheitsanalyse, Guide of the Bundesminister für Umwelt und Reaktorsicherheit, November 1993, unpublished

2. RISA+ - Einführung und Referenzhandbuch, RISA Sicherheitsanalysen GmbH, Berlin, 1993

# Regulatory View on PSA for an Older-type PWR

Dietmar Keil
Ministry of Environment Baden-Württemberg
Department of Reactor Safety, Radioactivity of Environment
Kernerplatz 9
70182 Stuttgart
Germany

## 1 Introduction and Objectives

At present, Probabilistic Safety Assessments (PSA, level 1+) are carried out in the framework of the Periodic Safety Review (PSR) for all operating nuclear power plants (being in commission at least 10 years) in Germany. The PSR and PSA are basing on the recommendation of the German Reactor Safety Commission (November 1988), which has been a result of its safety examination of all German NPPs after the severe accident in Chernobyl. The concept for the performance of the PSR takes pattern by the safety concept of NPPs consisting of 4 levels: level 1 "normal operation", level 2 "disturbed operation", level 3 "accidents" and level 4 "very rare beyond-design events". The PSR comprises the following parts: plant description (summarized), safety status analysis, probabilistic safety analysis, security analysis and assessment of the results.

Essential goals of the submitted PSAs are:

- examination and qualification of event sequences leading to hazardeous states for the fuel element cooling. For the performance of the PSA the active functions of the containment have to be taken into consideration (level 1+),
- quantification of the expected conditional occurrence frequencies for the transition of an initiating event into the final (hazardeous) events,
- reviewing of the balance of the safety concepts and determination of deficiencies,
- specific optimization of system technique and operating mode,
- assessment of the effectiveness of realized and planned backfitting measures.

## 2 Procedure and Results

After carrying out of the German risk study for the reference plant Biblis B the probabilistic methodologies elaborated have been applied by performing a reference study "reliability analysis of safety-relevant systems for an older-type PWR" in 1989. About 30 improvements referring to the plant concerned resulted from this reference study. The first probabilistic analysis proved to be very valuable with respect to the enhancement of the safety level of the older-type PWR, especially from point of view of the regulatory body.

Moreover, a PSA level 1+ has been worked out by order of the regulatory body basing on the a. m. reference study. The performance of the PSA started in 1991 taking into consideration the German PSA guideline of October 1990. The spectrum of initiating events analyzed was selected on the basis of the guideline. In addition, the initiating events

- ATWS
- flooding of safety-relevant buildings (parts) due to internal causes
- leaks in a connection pipe (referring to primary circuit) outside containment

have been examined whether they can be neglected due to their low contributions (cut-off criteria: conditional occurrence frequency of a single event (transition into hazardeous states) < $10^{-7}$/a, total contribution of all neglectable events < $10^{-6}$/a).

A spectrum of 14 initiating events has been analyzed in the recent PSA level 1+. The expected total occurrence frequency (preliminary result) of plant hazard states for the older-type 2-loop-PWR examined has been calculated to 6 x $10^{-5}$/a. An additional contribution to the total occurrence frequency of plant hazard states results from the failure probability of the long term recirculation from the containment sump into the primary loop regarding a "loss of coolant accident" (LOCA) event. In case of a duration of three months (assumption) of a long term recirculation mode the expected total occurrence frequency of plant hazard states is about 21 % higher in comparison of assuming a duration of one month.

The main contribution to the total occurrence frequency of plant hazard states results from the initiating event "steam generator tube rupture" (SGTR). This is caused by the fact that the older-type PWR examined is equipped only with 2 steam generators. In case of a SGTR (resulting in the failure of one steam generator) only one other (intact) steam generator is available to control this initiating event. I. e. there is no further redundant intact steam generator available like in a 3-loop or 4-loop plant. The necessary countermeasures regarding the intact steam generator have therefore to be performed with a high level of reliability to control this event SGTR. But the results of the PSA revealed a relatively high mean unavailability of the system functions demanded to control the SGTR. This is primarily caused by the highly automated countermeasures being necessary due to the short waiting periods (< 30 minutes) for countermeasures regarding a 2-loop-PWR. The actuating and control of the automated countermeasures therefore demand a highly reliable performance of the corresponding instrumentation and control system; at present this is only insufficiently fulfilled in the estimation of the regulatory body. In particular the signal forming is carried out disadvantageously, leading to the circumstance that already a single failure of an assembly causes the failure of the system function. This insight is remarkable in so far as the instrumentation and control system has been backfitted a couple of years ago in order to improve the control of the initiating event SGTR. The backfitting measures were realized in the framework of a comprehensive regulatory procedure following purely deterministic and engineering principles. Only the probabilistic analysis revealed the deficiencies of the new deterministically developed concept for the instrumentation and control system. At present various measures are examined to achieve an improvement of the availability of the system functions demanded to control the event SGTR. The following considerations are at present in discussion:

- reduction of the unavailability of the system functions by shortening the test intervals, especially regarding instrumentation and control equipment (immediate action)
- improvement of the concept for the instrumentation and control system.

Unplanned human actions, emergency procedures and accident management (AM) measures - with a certain probability preventing core damage in case of the initiating events analyzed - have not yet been considered for the determination of plant hazard states. A significant improvement of the total result is to be expected if such measures are taken into account.

# 3 Insights of Regulatory Body

The PSA for an older-type PWR has proved to be very valuable with respect to the plant specific identification of vulnerabilities of the safety concept (e. g. SGTR) and the enhancement of the safety level of the plant. The regulatory body thereby gained essential insights relevant for the future performance of PSAs as well as for the implementation of a "Living PSA". The most important issues are addressed hereafter:

On principle, the PSA should be performed by the utility itself. The elaboration of PSAs by a reviewing or regulatory body includes not negligible "friction losses" caused by necessary information and data transfer.

It proved to be very useful that the utility reacted adequately to preliminary results of the analysis **during** the performance of the PSA (corresponding modifications) but did not wait for the final results.

The relative contribution of the single initiating events analyzed is of particular importance, because deficiencies not recognizable in the deterministic and engineering design of the plant can so be revealed.

In contrast to this the absolute quantity of the total occurrence frequency of plant hazard states should not be overemphasized, because large impacts on the results and on their uncertainties can be observed for data used as well as the assessment of common cause failures and human factors. The total result has to be analyzed and discussed carefully between regulatory and reviewing bodies as well as the utility concerned **before** exceeding backfitting measures are carried out.

It proved to be necessary that the elaboration of the PSA by the utility should be accompanied by the regulatory and reviewing bodies in a regular exchange of work packages performed and corresponding pre-assessments. In following such a procedure, time consuming discussions and extensive modifications and supplements of the PSA in the reviewing procedure can be avoided.

The examination of plant hazard states is important with respect to the assessment of the plant safety. Due to the political relevance of the PSA results - and in order to avoid a possible misuse of the results - plant damage states should be examined. In the course of this, additionally emergency procedures and AM measures are taken into account beside engineered safety systems coping with design basis accidents.

A prerequisite for a successful PSA represents also the necessity that utility staff performing the PSA is convinced of the benefit and usefulness of PSA; the staff has therefore to be provided with corresponding competence within the utility organization.

# 4 Conclusions and Prospects

From the authority's point of view the following conclusions can be drawn based on the experiences from the recent PSA and their results presented up to now:

- The data used for the reliability measures needed have a major influence on the results of a PSA. Referring to the PSA concerned conservative generic data for reliability measures have been used. Moreover, plant specific data have been applied for key components. These key components have been determined by importance and sensitivity analyses using generic data. The plant specific data for key components have been evaluated by analyzing the existing operational experience (like maintenance and test records).
- Furthermore, a very important topic concerning the results of the PSA refers to the correct representation of state and position of all relevant components and systems as well as the adequate application of success criteria depending on the event paths. Moreover, possible backfitting measures realized or planned have to be taken into consideration.
- Concerning the older-type PWR (2 steam generators) the main contribution to the total occurrence frequency of plant hazard states refers to the initiating event SGTR. In this context the instrumentation and control system has been revealed as essential deficiency.

Due to the further results and useful insights gained from the recent PSA level 1+ for the older-type 2-loop PWR the regulatory body intends to order further analyses with respect to:

- backfitting measures referring to the control of the initiating event SGTR
- determination of additional plant specific initiating events
- examination of the initiating event "internal fire"
- examination of non-full power states: start-up and shutdown phases, low power states
- examination of plant damage states, i. e. probabilistic assessment of emergency procedures and AM measures.

# On the Regulatory Review of the TVO I/II Low Power and Shutdown Risk Assessment

J.V. Sandberg, R.K. Virolainen, and I.M. Niemelä
Finnish Centre for Radiation and Nuclear Safety (STUK)
P.O.Box 14, FIN-00881 Helsinki, Finland

## 1 Introduction

TVO I and II at the Olkiluoto site are 710 MW(e) ASEA–ATOM boiling water reactors operated by Teollisuuden Voima Oy (Industrial Power Company) owned by major Finnish industrial corporations.

TVO made a shutdown and low power mode risk assessment in 1990 - 92 to supplement the full power operation PSA including internal events, floods, and fires [1,2]. The scope of the project comprised operational states with reactor power level less than 8 %. The loss of the external grid, or external initiating events like fires and floods were not included in the project. The small event tree/large fault tree method was used in the shutdown PSA as in the full power PSA.

TVO analyzed operating and maintenance instructions to gain information on human activities which might lead to initiating events or safety system unavailability. The possibilities to observe and recover errors were also analyzed. The following operations and events were considered: operations in the reactor hall, heavy lifts, pipe, pump, and valve operations, main circulation pump maintenance, power monitor work, fuel reloading, control rod replacement and actuator maintenance, plant shut down and start up from refuelling shutdown, isolation valve tests, tests of reactor and turbine plant protection systems, and criticality experiments.

A procedure was developed for the quantitative analysis of human actions and corrective actions which are especially important when the containment is open. Since the results are sensitive to the frequency of human errors, the success of recovery operations, and the accuracy of physical success criteria, error estimates and sensitivity analysis are essential for the interpretation of shutdown PSA results..

In the analysis of local criticality incidents, a task interaction matrix was used to describe the possibility of confusion among work phases.

Chronological phase diagrams were used for special analyses, such as cold over-pressurization of the reactor pressure vessel due to excess filling before opening the reactor lid.

To define initiating events and plant level models, the duration of low power level was divided into periods during which the configuration of the plant was essentially unchanged. The initiating events were then classified according to the plant response. This procedure gave a finite number of conservative classes of initiating events which were positioned in the specific periods of the shutdown when they were possible.

The preliminary findings of the shutdown PSA lead to some modifications of operating procedures. The changes were taken into consideration in the final version of the study. In some cases significant reduction of the core damage frequency was achieved. After the changes the mean frequency of severe fuel damage was calculated as $3.4 \times 10^{-6}$/outage which is not insignificant compared with the full power core melt frequency $3.6 \times 10^{-5}$ 1/a.

The most important initiating events were extra large bottom leak, medium small bottom leak, medium sized leak above the reactor core, loss of shutdown cooling system while the reactor pressure vessel lid is closed.

Probability distributions defined for the frequencies of the initiating events and basic events were used in the uncertainty analysis which gave the error factor 8.1 for the core damage frequency.

# 2 Main Review Results

The most important initiating event is an extra large bottom LOCA, diameter more than 189 mm. An extra large bottom LOCA may occur during main circulation pump maintenance if the plug of the shaft penetration is accidentally removed when the pump has been opened.

An extra large bottom LOCA cannot be compensated with the systems required to be operational by the technical specifications. The core will be uncovered in 20 minutes if the reactor pool gates are closed or in about an hour if the gates are open. If all applicable systems are operational, water level will stay slightly above the top of the core and the radiation level in the reactor hall will increase significantly.

If the air lock at the bottom level of the containment is open, the water inventory will be lost from the containment. The air lock should be closed within one minute of the initiation of the leak, because a later closing would be prevented by the high water flow. Core damage frequency due to extra large bottom LOCAs is $1 \times 10^{-6}$/outage.

Medium small bottom LOCAs correspond, for example, to control rod drive penetrations. The associated core damage frequency is also about $1 \times 10^{-6}$/outage.

Medium large LOCA above reactor core leads to core damage with frequency $2 \times 10^{-7}$/outage. This type of leaks can be caused, for example, by failures in the pool water treatment system. If the water inventory is lost through the lower air lock or if water recirculation is prevented, the core will be finally uncovered.

The loss of the shutdown cooling system when the reactor pressure vessel lid is closed, leads to core damage with $4 \times 10^{-7}$/outage frequency. The excess filling of the closed reactor pressure vessel associated with the loss of the shut–down cooling system also causes core damage with the frequency $4 \times 10^{-7}$/outage.

The five initiating events described above contribute about 80 % of the total core damage frequency $3.4 \times 10^{-6}$. The six most important accident sequences are discussed below.

**Extra large bottom LOCA and loss of water inventory through the open lower air lock**, frequency $9.0 \times 10^{-7}$/outage. The cooling systems are not sufficient to compensate the leak. The closing of the lower air lock fails and the water inventory will be lost from the containment. The core starts to degrade in about two hours from the initiating event.

**Medium small bottom LOCA, the auxiliary feedwater system and the core spray system fail to start**, frequency $7.8x10^{-7}$/outage. The core starts to degrade in about 11 h after the initiating event.

**Medium small bottom LOCA and loss of water inventory from the lower air lock**, frequency $5.0x10^{-7}$/outage. The core starts to degrade in about 18 hours after the initiating event.

**Loss of residual heat removal systems (feedwater, auxiliary feedwater, and core spray) while the reactor pressure vessel is still closed but the pressure is less than 12 bar**. Frequency of the sequence is $1.4x10^{-7}$/outage. The core is uncovered in about half an hour, but the containment is closed.

**Extra large bottom LOCA, the covers of the pipe ends connecting the upper drywell and the wetwell are not removed, and the upper air lock is not closed.** Frequency of the sequence is $1.1x10^{-7}$/outage. Water inventory will be lost through the upper air lock. Core starts to degrade in about 9 hours.

**Failure of the feedwater system, auxiliary feedwater system, and reactor pressure relief while reactor pressure is over 12 bar**. Frequency of the sequence is $1.1x10^{-7}$/outage. The core will be uncovered in about one hour after the initiating event. The unavailability of the safety and relief valves would be due to blocking of the valves for reasons of personnel safety. This is a high reactor pressure sequence with a closed containment.

The above six sequences account for 70 % of the total core damage frequency. The most important sequences involve human errors with highly uncertain probabilities. The bottom LOCAs are caused by human errors only. The closing of airlocks and the removal of wetwell connection covers also depend on human actions. The closing of the lower air lock has the largest influence on the core damage frequency. If it were always successful, the core damage frequency would be reduced by 80 %.

Modelling uncertainty plays an important role in almost all of the aforementioned sequences. A special review on the modelling uncertainties has been conducted recently [3].

There is a group of extra large bottom LOCA sequences with total frequency $2.5x10^{-7}$/outage where the water level is quite close to the top of core or the core is uncovered temporarily, but no core damage is supposed. Rough estimates indicate that uncovery time of less than one hour would cause, at worst, some fuel rod leakages, and uncovery time of more than three hours would cause extensive core melting.

Analysis of detailed behaviour of water level requires knowledge of the exact pumping capacity, size of the water storage, the cross sectional area of the containment at each level, the hydrodynamic characteristics of the leak etc. However, according to our 'quick and dirty' analyses, the period of core uncovery would be short, and no significant fuel leakage would be expected in the majority of these sequences. For some sequences the rough conservative estimates seem to be insufficient, but their frequency is very low, less than $1x10^{-10}$/outage.

Another issue connected with the temporary core uncovery, but not treated in the shutdown PSA, is the recriticality of a partially degraded core. According to the current understanding based on experiments, control rods may be liquefied due to the formation of eutectics and relocate away from the core regions before extensive fuel relocation. Recovery of cooling systems and filling of the core after the loss of control material might lead to recriticality, increased fuel damage, and high fission product release.

# 3 Level 2 PSA Aspects

The shutdown PSA had level 1 scope, and the release of radioactive materials and damage to the reactor building were not analyzed. During most of the shutdown the containment is open, and radioactive releases from melting fuel to the environment are prevented only by the reactor building.

The environmental consequences of a fuel melt may be more serious during shutdown than during power operation. Therefore the risk significance of the refuelling outages may be higher than implied by the shutdown core damage frequency compared to the full power operation core damage frequency.

## 3.1 Nuclide Inventories

There are three important groups of nuclides with different characteristics: noble gases, iodine, and cesium/strontium. Noble gases have relatively short half–lives and their release during fuel melt is practically complete. Their release to the environment can be prevented only by a leak–tight containment. The Kr–87 equivalent noble gas activity decreases by a factor of 10 in 24 hours after the shutdown, and again by a factor of 10 during the next two weeks.

The noble gas inventory during the early stages of the shutdown is only by a factor of 3 smaller than the noble gas inventory at the time of a containment failure in the majority of full power core melt sequences. Regarding the environmental consequences due to noble gases, a core melt accident at the early stages of a shutdown with an open containment would correspond to a full power core melt accident with late containment failure.

The 8 day half–life of the most important iodine isotope (I–131) is comparable to the length of the refuelling shutdown and cesium and strontium have longer half–lives. Iodine, Cs, and Sr are retained in filters and on the surfaces of the reactor building. The I–131 equivalent iodine activity decreases by 30 % in the first 24 hours after the shut–down and in two weeks it decreases by a factor of 5. Cs and Sr activities are practically the same during a shutdown and full power operation.

## 3.2 Factors Affecting Release

In many important shutdown core melt sequences reactor fuel might remain uncovered for a long period and might melt in air in the open reactor vessel. In this respect a shutdown core melt may have more serious environmental consequences than a full power core melt where the release takes place through the suppression pool and/or through the flooding pool in the lower containment and effective scrubbing is present.

When the containment is open, the reactor building is required to be closed. If radioactive substances are detected in the air in the reactor building, ventilation takes place with the filtered auxiliary ventilation system. The system is equipped with droplet separators, and two stages of particle and charcoal filters. However, the reactor building and the systems have not been designed for severe core damage conditions. Fission product aerosols may overheat or block the filters, and the reactor building may fail due to temperature increase or fire.

Physical blocking of the filters could be caused by the release of large quantities of aerosol particles. The mass of iodine in the core is about 15 kg, other volatile fission products include about 200 kg cesium (boiling point 678 °C) and 100 kg barium (boiling point 1640 °C). Aerosol particles could also be released from structural materials, for example, zircaloy contains about 700 kg tin with melting point 232 °C and boiling point 2270 °C.

The decay heat of radioactive aerosols is significant, for example, the decay power of iodines is hundreds of kilowatts. If even a moderate fraction of the decay power is deposited in the filters they may overheat and burn.

If the reactor cooling systems are not operable, the 10 MW residual power will heat reactor building structures and air. Ventilation could remove only about 600 kW if the temperature rise were 100 °C. The threats to the reactor building due to steam generation and heating have not been evaluated. In the case of serious core damage with high zirconium oxidation hydrogen burns could also destroy the reactor building.

In the case of an extra large bottom LOCA and an open lower air lock, the reactor vessel would become quite dry. The air circulating freely in the reactor vessel and heated by the residual and oxidation heat could create a hot air column which could damage the reactor building and enhance fission product release and transport.

Oxidation of some fission products in air may lead to of higly volatile oxides. This phenomenon contributed to the formation of hot ruthenium particles in the Chernobyl accident.

# 4 Measures Taken by TVO

The preliminary results caused the following changes in the equipment and procedures at the plant during the shutdown PSA project. The event frequencies were updated to correspond to the new situation.

- New plugs with overloading pins were procured for main circulation pump maintenance. The fracture of the pins prevents accidental removal of the plugs with a miscalibrated crane when the lower flange is not in place. The change caused a reduction of $2x10^{-6}$/outage in the final core damage frequency.

- During the maintenance of the main circulation pumps a permanent guard was positioned at the lower containment air lock for the fast closing in case of a bottom leak (assumed 50 % success probability). This change reduced the final core damage frequency by $1.4x10^{-6}$/outage. Later it was also required that the lower air lock be closed at the riskiest stages.

- To avoid cold overpressurization of the reactor vessel as a result of overfilling before removing the vessel lid, the safety and relief valves of the reactor will be kept operational during the risky stages. Operating procedures were also improved, e.g., the use of auxiliary feedwater pumps for reactor vessel filling was restricted. The $8.6x10^{-7}$/outage reduction of core damage frequency was taken into account in the final results. In the future the risk may also be reduced by two additional relief valves which will be designed for operation with both steam and water.

- Operating procedures during shutdown will be revised to minimize possibilities of equipment failures and errors. Emergency operating procedures will be prepared for low power and shutdown states and problem areas will be treated in the training of the emergency organization.

## References


1. Himanen R, Pesonen J, Sjövall H, Pyy P. Experience from shut down event PRA (SEPRA) for TVO I/II, in the proceedings of the IAEA – TCM on modelling of accident sequences during shutdown and low power conditions, Stockholm, 30 November – 3 December 1992.

2. Shutdown and low power safety assessment – A Status Report, NEA/CSNI/R(93)19, 1993

3. Pulkkinen U, Huovinen T. Model uncertainty in safety assessment, Finnish Centre for Radiation and Nuclear Safety report STUK–YTO–TR 95, 1995.

# Qualification Procedure for PSA Software Referring to STARS-FTA

Cornelia Spitzer
TÜV Südwest
Department of Nuclear Technology
and Radiation Protection
Dudenstraße 28
68167 Mannheim, Germany

## 1 Introduction

For the performance and in particular the evaluation of PSAs (level 1) a variety of software approaches is available originated from different countries. Strong and weak points of these tools have been described and discussed in several publications without carrying out a qualification procedure (at least not for the tools used in Germany). In the PSA-guideline [1] e. g., various software tools are mentioned with respect to the performance of a comprehensive NPP-PSA in the framework of the Periodic Safety Review; in the further developments of this guideline with appendices and corresponding documents (drafts) minimal requirements and features are listed which have to be fulfilled by software used.

In the domain of thermohydraulic or criticality analyses e. g., the qualification and validation of software can be performed by recalculating various experimental arrangements (benchmark exercises). Such possibilities do not exist with respect to software tools concerning the evaluation of PSAs, because the corresponding safety systems (and/or structure functions) generally do not provide observable "experimental" values due to high systemic reliability.

Results of comparisons only carried out with respect to the evaluation of selected and/or extensive fault trees are in general not satisfactory. Therefore, a detailed clarification of principal items like the correctness of the mathematical approaches and adequacy of approximations has to be performed in the framework of a qualification procedure regarding established as well as new software tools not yet qualified.

In this paper the aspects concerning a general qualification procedure for PSA software are described; the corresponding issues are addressed in different detail due to the limited length of the paper. The qualification procedure proposed is based on the actual state-of-the-art represented by guidelines, federal research/development projects as well as the discussion in the relevant working groups and boards. Further on, insights gained during the course of extensive PSA reviews - regarding plant related subjects and methodologies applied - performed within regulatory procedures are taken into consideration. The performance of quality assurance of software - with respect to the development process as well as the final product - following ISO 9000 e. g., is not explicitly discussed in the present paper regardless of its importance in a certification procedure.

Beside the general description of a qualification procedure, the initial validation and qualification of the software tool STARS-FTA [2], which has been developed at the Joint Research Centre in Ispra (JRC/ISEI), is presented.

# 2 General Qualification Procedure

Prime objective of the carrying out of a qualification procedure for PSA software is the provision of qualified tools for a - as far as possible - "realistic" modelling and representation of a real industrial installation, because a plant in operation has a well defined operating mode with particular test, repair and maintenance strategies. To a certain extent plant specific reliability data are available evaluated from the operational experience observed. These aspects have to be taken into account carefully to achieve the essential goals of PSAs, e. g. submitted for the Periodic Safety Reviews, namely the reviewing of the balance of the safety concepts, the specific optimization of system technique and operating mode as well as the assessment of the effectiveness of backfitting measures carried out and/or planned. I. e. the availability of appropriate modelling features has to be checked beside the possibilities of performing adequately the necessary analyses. The general qualification procedure proposed can be carried out for different applications (slightly modified regarding the requests in the phases, see below), in detail the present paper refers to the performance of an extensive NPP-PSA.

The performance of the work to be done to achieve the qualification of a software demands the carrying out and traceable documentation in successive work packages and phases matching with the below listed issues. Basically, most of the individual work packages (except the theoretical and mathematical parts) can be performed stepwise in phases. First the accomplishment of basic requirements in a software is examined and assessed. The respective fulfilment represents a necessary prerequisite for the qualification regarding the issue considered. In the next phase the realization of additional useful features is examined with respect to the performance and evaluation of an extensive PSA. The fulfilment of these additional features represents no necessity for the qualification, but it provides the possibility for a categorization of different qualified tools. Finally, possibly existing unusual or innovative features of a software are assessed with respect to their usefulness and correctness. Evidently, the qualification of a software presupposes the correctness of the mathematical approaches and approximations as well as their adequate realization in the software.

## 2.1 Work Packages

The following work packages - namely the examination and assessment of the following issues - have to be performed stepwise (see above) for the qualification of a PSA software:

*A Handling Characteristics*
**A-1** User Interface
**A-2** Features (like Graphics)
**A-3** Data Input / Output
**A-4** Data / Structure Management

**A-5** Interface between Tools
**A-6** Check Routines
**A-7** Documentation

*B Modelling of Safety and Operational Systems*
**B-1** Logical Operators / Connections
Basic requirements regarding Boolean logical operators represent the following gate types: OR, AND, K-of-N and transfer gates (no explicit logic). Additionally useful is: The provision of the NOT operator, NOT OR, NOT AND, XOR, INH and comment gates (no logic).
**B-2** Logical Switches / Boundary Conditions
Logical switches and boundary conditions to modify and/or adapt conveniently existing fault tree structures are in principle useful; an unequivocal and traceable application and documentation thereof has to be provided.
**B-3** Technical Scope / Dimensions
The maximum number of descendants from a gate has to be sufficiently high (about 80); the maximum number of gates and/or basic events allowed for an overall PSA has to be large too (for a real NPP in operation about 8000). Preferably, the corresponding variables should be parameterized.
**B-4** Basic Events
The respective reliability models are basically required for the description of the independent failure behaviour of: continuously monitored, repairable components / periodically tested components (including the possibility of taking into account staggered testing and not negligible repair times possibly needed after test) / unrepairable components / components with constant unavailability / components with fixed mission time (not necessarily separate model) / components with constant frequency (if event tree analyses can be performed).

An additional useful feature represents the possibility of taking into consideration a component unavailability unequal 0 at time equal 0. Preferably, exact formulas should be used. In general, a technique of differentiation between required reliability model parameters and optional ones (corresponding terms disappear if not applied) proved to be useful.

In general - and especially with reference to the work packages C, D and E below - the methods and approaches used should be as precise as possible, if approximations are used, the degree of approximation should be quantifiable and quantified. The adequacy and/or conservativeness of methods used is an important topic within regulatory procedures. These issues have to be examined and assessed theoretically for **all** applications performable with the PSA software under consideration. If need be, restrictions and constraints existing have clearly to be stated within the qualification procedure. In order to ensure an adequate and proper use of a software the aspect application range has to be "supported" by sufficiently precise documentation (see corresponding work package A-7) and preferably by the tool itself (e. g. error / warning messages in case of leaving the covered application range). This is of particular importance in case of possibly inconservative results. Focus of these considerations are points of view exceeding well known issues, like treatment of negations or higher order approximations regarding structure functions.

*C System and/or Structure Functions*
**C-1** Fault Tree Modularization
**C-2** Determination of Minimal Cut Sets (MCSs)
**C-2.1** Theoretical Basis
**C-2.2** Approaches / Approximations / Algorithms
**C-2.3** Realization in Software
**C-2.4** Quantification of Approximations / Neglects

***D** Probabilistic Parameters to be Evaluated*
The evaluation of the following probabilistic parameters (regarding basic events, MCSs and TOP events) is basically required: time dependent (maximum) unavailability / mean unavailability (correct and/or adequate numerical integration taking into consideration discontinuities) / unconditional failure intensity / expected number of failures in a time interval (adequate numerical integration, see above).

Additionally useful are: A "steady state mean" unavailability of TOP (and/or MCSs) based on mean unavailabilities for each basic event / conditional failure intensity / unreliability (probability of at least one failure in a time interval) / mean time between failures / mean time to failure / mean time to repair. Assumptions and boundary conditions regarding these parameters have clearly to be stated.

Esary-Proschan approximation regarding TOP evaluations is to prefer; second and third order approximations are useful from a general point of view.

**D-1, D-2, D-3, D-4:** see work packages C-2.1, C-2.2, C-2.3, C-2.4.

***E** Event Tree Analyses*
The provision of a systematic performance of event tree analyses represents a very useful feature; a precise and traceable description of the realization in a software is indispensable.

**E-1, E-2, E-3, E-4:** see work packages C-2.1, C-2.2, C-2.3, C-2.4.

*F Importance Analyses*
Basically, the Fussel-Veseley importance measure regarding basic events and MCSs is required. Additionally very useful are (including derived measures): risk achievement worth / risk reduction worth / Birnbaum importance / Barlow-Proschan importance / fractional, contributory and diagnostic importance. A precise definition of importance measures calculated in a software is indispensable.
**F-1** Theoretical Basis
**F-2** Approaches / Approximations
**F-3** Realization in Software
**F-4** Results to Present

*G Uncertainty Analyses*
The following uncertainty distributions regarding probabilistic parameters are basically required: Normal, Log-normal, Beta, Uniform, Log-uniform, Discrete. Additionally useful distributions are: Gamma, $\chi^2$, Empirical (histogran). Generally, the possibility of taking into account existing dependencies between probabilistic parameters has to be provided. The provision of possibilities to analyze further uncertainties (e. g. regarding modelling) is supplementary useful.

**G-1, G-2, G-3, G-4:** see work packages F-1, F-2, F-3, F-4.

***H*** *Sensitivity Analyses*
The provision of possibilitites to modify - individually or systematically - values of input probabilistic parameters represents a very useful feature; sensitivity measures defined and calculated in a software are to be described precisely.
**H-1, H-2, H-3, H-4:** see work packages F-1, F-2, F-3, F-4.

***I*** *Evaluation of Sample Cases*
Selected and representative fault/event trees with corresponding MCSs and results provided are additionally useful to be evaluated.

Further on, the consistent performance and updating of an extensive PSA, the quality assurance and the review are efficiently supported and simplified by an integral software package providing a modular basis representations for the different topics arising from PSA performance and assessment. Beside the already addressed issues, these topics refer to plant description, systematic data collection, evaluation of operational experience, provision/management of reliability data, common cause failures and human factors. The representations (i. e. modules) of such an integral tool should be executable independently of each other, but they are to be connected on a common data base with a well-defined input/output. A further categorization of tools can be carried out by performing a qualification of software matching with the above listed very useful issues of an integral software package.

# 3 Initial Validation of STARS-FTA

Prime objective of a contractual collaboration between the ISEI of the Joint Research Centre (Ispra) and TÜV Südwest is the qualification of the STARS software package [3, 4] as an integral tool for the performance and assessment of PSA and for the establishment of a Living PSA. At the beginning of this collaboration - under contract to JRC/ISEI - a preassessment of STARS-FTA (an implementation of ISPRA-FTA [2]) concerning practical requests was performed by TÜV Südwest in order to adapt the tool developed for chemical/petrochemical installations, whose probabilistic evaluations are therefore based on a conservative simplified modelling. With regard to the performance and assessment of a PSA for a real NPP in operation recommendations have been worked out classified corresponding to the priority for their realization.

Actually, the technical scope referring to an extensive PSA as well as the possibility of taking into account staggered testing and not negligible repair times possibly needed after test regarding periodically tested components have been recommended to be realized short-term. Preferably, exact formulas should be used. Furthermore, the evaluation of the time dependent unavailability as well as the resulting mean unavailability (correct and/or adequate numerical integration) have strongly been recommended. The same recommendation has been stated concerning the evaluation of the unconditional failure intensity and the expected number of failures in a time interval. Further modifications/supplements (referring to importance, uncertainty and sensitivity analyses e. g.) have been recommended to be realized medium-term.

Following the recommendations given by TÜV Südwest STARS-FTA will be adapted. The work to be done short-term has been specified by JRC/ISEI [5] and commented by TÜV Südwest. After the adaption of STARS-FTA the qualification of this module of the integral software package STARS will be carried out. It is to be expected - in consideration of the intended work program - that STARS-FTA after being extended will fulfil the basic requirements as well as most of the additional useful features presented in section 2 of this paper (without event tree analyses and uncertainty analyses).

# 4 Conclusions

In general, a qualification procedure for PSA software has to be performed. Issues to be addressed, examined and assessed while carrying out such a procedure are presented in this paper.

A preassessment of the JRC/ISEI software STARS-FTA has been performed by TÜV Südwest. The tool will be adapted following the recommendations given. It is to be expected that STARS-FTA after being qualified will be a powerful tool (especially concerning analytical fault tree analyses) also efficiently usable in German regulatory procedures. In particular, the hybrid method [6] implemented for the determination of the MCSs will provide more realistic (but still conservative) cut-off errors in comparison to those estimated by other analytical approaches. Moreover, in context with the entire software package STARS an integral tool will be available for a consistent and traceable performance and assessment of PSA over time (e. g. with regard to Living PSA).

# Acknowledgements

The author gratefully acknowledges the contract given by JRC/ISEI and would like to thank P. C. Cacciabue for many valuable comments and suggestions as well as G. Cojazzi, S. Contini and J.-P. Nordvik for numerous fruitful and beneficial discussions.

## References

1. PSA-Leitfaden, Facharbeitskreis "Probabilistische Sicherheitsanalyse für Kernkraftwerke". Federal Ministry of Environment, Nature Conservancy and Nuclear Safety, October 1990, Hab/Jab
2. Contini S. ISPRA-FTA. Fault Tree Analysis Tool for Personal Computers, Methodological Aspects and User Interface Description. Commission of the European Communities, Joint Research Centre, Ispra. EUR 13997 EN, 1992
3. Poucet A. STARS: Knowledge Based Tools for Safety and Reliability Analysis. Reliability Engineering and System Safety **30** (1990) 379-397
4. Nordvik J.-P., Carpignano A., Poucet A. Computer-Based System Modelling for Reliability and Safety Analysis and Management. In Proc. Topical Meeting on Computer-Based Human Support Systems: Technology, Methods and Future, 211-217. Philadelphia, Pennsylvania, June 25-29, 1995. Published by American Nuclear Society
5. Contini S., Cojazzi G., Sardella, R. Technical Specifications and Algorithms to Adapt Ispra-FTA for Nuclear Applications. Commission of the European Communities, Joint Research Centre, Ispra. Technical Note No. I.95.62, ISEI/IE/2936/95, May 1995
6. Contini S. A new hybrid method for fault tree analysis. Reliability Engineering and System Safety **49** (1995) 13-21

# Some Problems of Identification of the State for the Purpose of Steering the Reliability of the Object

Jan Borgoń
Jerzy Jaźwiński
Józef Żurek
Air Force Institute of Technology
Warsaw, Poland

## 1 Formulating the Problem

Systems are described with some specific features $\mathbf{C}\{C_i,\ i=1,2,...,n\}$ (system parameters). These features usually include: reliability, safety, availability, repairability, life, survivability, etc. Requirements $\mathbf{W}\{W_i,\ i=1,2,...,n\}$ are determined with reference to individual features. It is said that the system is in the state of ability, when a set of features is embodied in a set of requirements $\mathbf{C} \subset \mathbf{W}$. From the above statement it appears that the state of ability is the initial state for the system, i.e. $\mathbf{C} \subset \mathbf{W}$.

The nature of individual features can be different. Determinated values can be features, e.g. aircraft dimensions; let us denote them with "**c**". A random variable can also be a feature, e.g. a number of aircraft failures/damages per year (discrete random variable) or time of removing failures/damages (continuous random variable); let us denote it with "**C**". A given feature can be a stochastic process, e.g. a process of renovation; we denote it with **C(t)**. Requirements are determined with regard to individual features; they can be the determinated ones "**w**", the randomly variable ones "**W**", and the stochastic processes **W(t)**. Fig. 1 illustrates various interrelations between the features and the requirements.

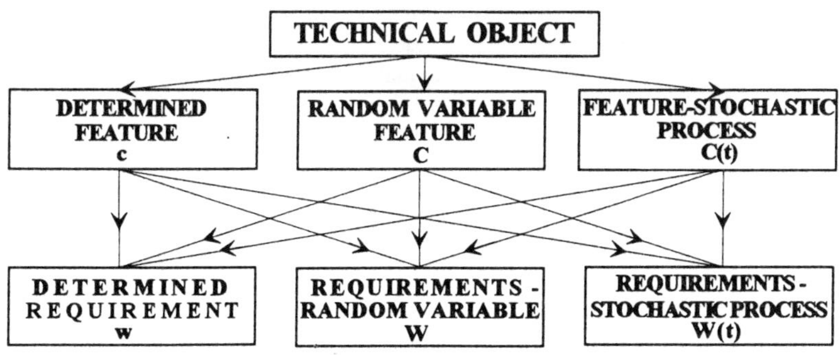


Fig. 1. Correlation of features and requirements of technical object

In the process of operational use of the system the individual features undergo changes that can be favourable and/or unfavourable. Experienced specialists need less time to maintain the system, which is a favourable tendency. In the course of operational use there are ageing processes proceeding within the system to result then in some increase in number of failures/damages - this is an unfavourable tendency of the feature. Finding the favourable and unfavourable tendencies of features and then counteracting them constitute significant steps of controlling the process of operating the system.

Finding the tendency of a feature to change consists in establishing the direction of changes with reference to requirements accepted as standards. Numerical values of a given feature assumed by a manufacturer can be the requirements (the determinated requirements **w**). A value of the feature accepted in the previous period of operational use can also be assumed to be the requirement (a random variable **W** or a stochastic process **W(t)** can be requirements in a given case).

The features under consideration can be included among the favourable and unfavourable ones:

- a feature is unfavourable if its increase is recognized to be undesirable (e.g. failure rate),

- a feature is favourable if its increase is admitted to be desirable (e.g. reliability).

Possibility of creating a measure of tendency to change(s) $Y(C,W,y_1,y_2) \subset [0,2]$ is given consideration. The measure should be such that for an unfavourable feature the tendency is: favourable when $Y \subset [0,y_1)$, unfavourable when $Y \subset (y_2,2]$, and there is no tendency to change when $Y \subset [y_1, y_2]$.

In the case under consideration the principle "the less the better" is compulsory. For the favourable index the relations proceed contrary, i.e. the principle "the more the better" becomes obligatory.

The measure **Y** is determined on the basis of statistical analysis. Indices of evaluation of tendency to change can be divided into two groups, i.e.: informational indices $\mathbf{Y_I}$ and determinating indices $\mathbf{Y_D}$. The unfavourable tendency of a feature that establishes the informational index does not influence evaluation of the tendency of the system to change (e.g. repairability). The unfavourable tendency of a feature that establishes the determinating index condemns the whole system (e.g. safety).

Besides the division into informational and determinating indices, specific weights **g** can be attributed to individual indices.

Evaluation of tendency to change quality of the whole system is a function of indices established on the basis of individual features.

For informational features:

$$Y_I = \frac{1}{n} \sum_{i=1}^{n} \cdot Y_{Ii} \tag{1}$$

For deterministic features:

$$Y_D = \sqrt[m]{\prod_{j=1}^{m} Y_{Dj}} \tag{2}$$

For the whole system:

$$Y_S = \sqrt{Y_I \cdot Y_D} \tag{3}$$

Having the weights **g** for deterministic indices introduced one can write down:

$$\lg Y_D = \frac{\sum_{j=1}^{m} g_j \cdot \lg Y_{Dj}}{\sum_{j=1}^{m} g_j} \tag{4}$$

For the informational index:

$$Y_I = \frac{\sum_{i=1}^{n} g_i Y_{Ii}}{\sum_{i=1}^{n} g_i} \tag{5}$$

where: m - number of features of deterministic meaning,
n - number of features of informational meaning.

Let us consider now some ways of establishing the index to evaluate tendencies of changes.

# 2 Establishing Indices of Evaluation of Tendencies of Changes

## 2.1 Formulating the Problem

Index **Y** describes the tendency of a specific feature to change and has been determined on the basis of relation between the feature and requirements. Fig. 1 shows nine types of such relations. Analyses of individual relations have been based on methods of probability mathematics and mathematical statistics.

Value of the index **Y** can be determined by means of verifying the hypothesis $\mathbf{H}_0$. Another hypothesis, alternative to the $\mathbf{H}_0$ one, has been denoted with $\bar{\mathbf{H}}_0$ . If the alternative hypothesis $\bar{\mathbf{H}}_0$ says the tendency is unfavourable, it is denoted with $\bar{\mathbf{H}}_0 = \mathbf{H}_1$. In case the alternative hypothesis says the tendency is favourable, it is denoted with $\bar{\mathbf{H}}_0 = \mathbf{H}_2$.

Having gained the results of analysing the specific feature one can define the index **Y** in the following way:

$$Y = \begin{cases} 1 - \Delta_1 & \text{when} \quad \bar{H}_0 \equiv H_1 \\ 1 & \text{when} \quad H_0 \\ 1 + \Delta_2 & \text{when} \quad \bar{H}_0 \equiv H_2 \end{cases} \tag{6}$$

Values $\Delta_1$ and $\Delta_2$ numerically determine the degree, to which the results of the conducted analysis are for the hypothesis $\mathbf{H_0}$.

Let us consider - by way of example - two from among nine various relations shown in Fig. 1:

1. a random variable feature **C**, requirements: determinated, **w**,
2. a random variable feature **C**, requirements: a random variable **W**.

## 2.2 Determination of Index of Tendency of Changes in Case of Unfavourable Random Feature and Determinated Requirements

Let us consider index $\lambda_i$ that represents rate of events (**u** - aircraft failures in the course of flight, **pwl** - a prerequisite for an aircraft accident, **wl** - aircraft accident, **kl** - air crash, **bp** - human (pilot's) error, **bo** - maintenance error, etc.).

Let $Z_i$ denote a number of events $\mathbf{i} \equiv \mathbf{u, pwl, wl, kl, bp, bo,}$ etc., respectively.

Let $\mathbf{T_i}$ denote total time of reference (total flying time, total calendar time of operating the aircraft, total number of take offs/landings, etc.).The rate of events is given with the following formula:

$$\lambda_i = \frac{Z_i}{T_i} \tag{7}$$

Let the required rate of events be denoted with $\mu_{0i}$ where:

$\mathbf{i} \equiv \mathbf{u, pwl, wl, kl, bp, bo,}$ etc.

It can be shown that:

$$\Delta_{1i} = \frac{P\left[1 < \chi^2_{2z} \le \frac{\lambda_i}{\mu_{0i}}\right]}{P\left(\chi^2_{2z} > 1\right)} \tag{8}$$

$$\Delta_{2i} = \frac{P\left[\frac{\lambda_i}{\mu_{0i}} \le \chi^2_{2z} < 1\right]}{P\left(\chi^2_{2z} < 1\right)} \tag{9}$$

where:

$\mathbf{i} \equiv \mathbf{u, pwl, wl, kl, bp, bo,}$ etc.

$\chi^2_{2z}$ - a random variable of chi-square distribution with 2z degrees of freedom

## 2.3 Determination of Index of Tendency of Change in Case of Unfavourable Random Feature and Random Requirements

In the case under consideration the rate of events $\lambda_i$; $\mathbf{i} \equiv \mathbf{u}, \mathbf{pwl}, \mathbf{wl}, \mathbf{kl}, \mathbf{bp}, \mathbf{bo}$, etc. is determined from the research work carried out at present.

The $\mu_i$; $\mathbf{i} \equiv \mathbf{u}, \mathbf{pwl}, \mathbf{wl}, \mathbf{kl}, \mathbf{bp}, \mathbf{bo}$, etc. has been determined from the research work conducted earlier (i.e. last year) under different operational conditions, etc.

$$\mu_{0i} = \frac{Z_{\alpha i}}{T_{\alpha i}} \tag{10}$$

It can be shown that:

$$\Delta_{1i} = \frac{P\left(1 < F_{2s_{\alpha i}, 2s_i} \le \frac{\lambda_i}{\mu_{\alpha i}}\right)}{P\left(F_{2s_{\alpha i} 2s_i} > 1\right)} \tag{11}$$

$$\Delta_{2i} = \frac{P\left[\frac{\lambda_i}{\mu_{\alpha i}} \le F_{2(s_{\alpha i}+1), 2(s_i+1)} < 1\right]}{P\left[F_{2(s_{\alpha i}+1), 2(s_i+1)} < 1\right]} \tag{12}$$

where:

$\mathbf{F}_{k_1,k_2}$ - F-Snedecor's statistics with $k_1$, $k_2$ degrees of freedom;
$\lambda_i$ - determined with formula (7);
$\mathbf{i} \equiv \mathbf{u}, \mathbf{pwl}, \mathbf{wl}, \mathbf{kl}, \mathbf{bp}, \mathbf{bo}$, etc.

## 2.4 Conclusions

1. To determine index **Y** for simpler cases one needs tables of Chi-square and F-Snedecor's distributions. The access to such tables is rather difficult, that is why we have to use computer techniques.

2. It is easy to determine index **Y** for special cases, e.g. when the feature shows normal distribution $N(\nu, \sigma)$ or when specific functions of the feature show normal or asymptotically normal distributions.

3. In general cases there is a need for theoretical studies to determine index **Y**.

# References

1. Firkowicz S. Statystyczna ocena jakości porównawczej lamp, Prace Przemysłowego Instytutu Elektroniki 1965; 3: 51-56 (in Polish)
2. Firkowicz S. O liczbowej ocenie statystycznej jakości wyrobów, Archiwum Elektrotechniczne 1967; 4: 32-39 (in Polish)
3. Jaźwiński J, Ważyńska-Fiok K. Bezpieczeństwo systemów, PWN Warszawa, 1993, (in Polish)
4. Jaźwiński J, Ważyńska-Fiok K, Zurek J, Intelligent systems in failure analysis, 5th International Conference SPT-5 Vienna, 1995, pp 121-122

# Reliability Study of the Nuclear Reactor BR2 SCRAM system*.

Yvan Pouleur, Bernard Verboomen
CEN/SCK Belgian Nuclear Research centre
Mol, Belgium

## 1 Description of the BR2.

The BR2 is a high flux material testing reactor (MTR) built in the fifties. It went critical for the first time in 1963 and has operated continuously on a regular basis since then. The only break in operations was when its Be matrix was replaced (1978-1980). The specific core array of the BR2 sets it apart from other comparable MTR. The core is composed of hexagonal beryllium blocks with central channels. These channels form a twisted hyperbolical bundle and hence are close together at the mid-plane but further apart at the lower and upper ends where the channels penetrate through the covers of the reactor pressure vessel. With this array, a high fuel density is achieved in the middle part of the vessel (reactor core) while leaving enough space at the extremities for easy access to the channel openings .

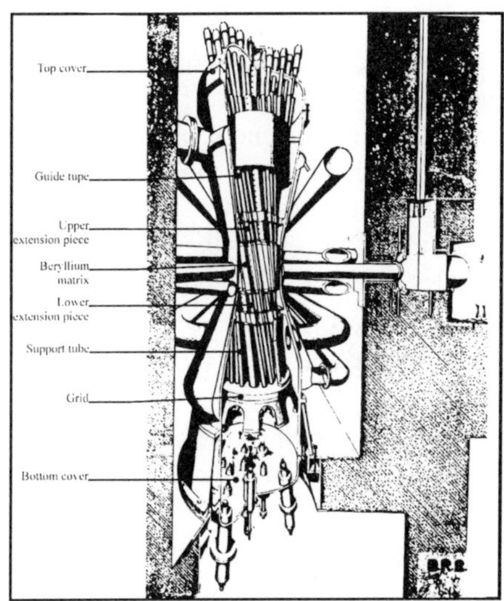


## 2 The Refurbishment Program.

To ensure the life extension of the reactor to ≈2010, it was decided to start a complete refurbishment program. Its tasks are mainly focused on the second beryllium matrix replacement, functional implementations and safety improvements. This process is closely dependant on the PSA level 1 process which was realised from '93 to '94 and the complementary studies that followed. One can easily understand that the SCRAM system occupies a major place in the safety defence of the reactor. It is present in almost all the accidental sequences envisaged in the PSA study. Therefore, it was decided to achieve a large analysis which would indicate the sensitive areas, suggest appropriate hardware modifications and judge of the adequacy of the actual test procedure.

# 3 Functional Description of the SCRAM System.

A functional diagram of the SCRAM system is to be found in annexe 1. The high neutron flux transient is detected by "L" chambers. Further, the signal goes through an amplifier, a comparator and enters in the voting logic module (VL) working in 1 out of 3, or, 2 out of 3. In this latter case, the VL module acts like a filter, if only one line presents a SCRAM demand (0), it will produce 1,1,1 at the output. For all other cases, output=input. From the VL module, the signals follow two paths, the first one to the logic interface which triggers the latch control mechanism, the second one to the relay line which acts as a backup (see below).
The B chambers also detect neutron transient but only trigger if the neutron period exceeds the alarm level. The path followed by the signals is very similar to the one of the "L" chambers.
Classical sensors detect all other phenomena which require SCRAM like overpressure, drop of flow... Each accidental event is detected by two or more sensors which open the contacts of the relay serial line. The chambers backup modules are included in that line.
By the opening of one contact of the relay line, a scram signal is sent to the logic interface which triggers the latch control mechanism. The hardware is composed of modern electronics and fast relays for the chambers paths in order to speed up the triggering. The rest of the system is mainly composed of common relays. Many modules are internally redundant.

## 3.1 The Reactor Cycles and the Tests.

The BR2, as a testing reactor, has its cycles adapted to the requirements imposed by the experiments . These cycle vary from a few days to more than a month, the average being three weeks. Therefore, two types of tests are realised:

1) On line tests ($T_{1i}$, $T_{2i}$): test of the chambers line, from the detector to the voting logic modules (not included).
2) Off line tests ($T_{3i}$): from the detector to the latch of the control rod (functional test).

# 4 Analysis.

First, the entire system was submitted to a Failure Mode and Effect Analysis (FMEA) and a dependant failure analysis of the control rod latch mechanism was realised. The FMEA specifies the types of component failures to be introduced in the fault tree and shows the weakness of the system due to its complexity. The dependant analysis is important because the scram effectiveness relies on the fall of several control rods. It clearly states the minimal value achievable for the scram system reliability. With the prerequisites it is possible to start the fault tree.

## 4.1 Fault Tree Modelisation.

The choice of the top event is based on the safety report which is related to neutron calculations. It is established that the antireactivity effect of the fall of 5 control rods out of the 6 compensates the ejection of an experimental device or a control rod. Therefore, the top event is: 2 control rods out of 6 do not fall on demand.
The following assumptions are made:
(1) the system is coherent [1], (2) basic events have a constant failure rate ($\lambda$), (3) components can be in 2 states: working or not working, (4) the system state depends unequivocally on the components state (no sequential dependency).
Basic events can either be a single component failure, or a whole group failure. In this latter case, the components are supposed to be serially connected, the failure of one component implies the failure of the group (conservative "black box" approach).
The final cutting out of the system is the result of a feed-back process based on the system topology, the sensitivity with regard to the top event success, and the knowledge of the sub-systems.
Concerning the common modes, a $\beta$ factor is frequently adopted, especially for electronic components situated on the same chip.

## 4.2 The Problem of Redundant Systems.

Many sub systems include redundant groups of components. The tests described above embodies some of these groups, it is therefore important to evaluate the effective test impact on reliability.
For a single component, after each test the component can be considered as good as new (for a limited period of time). This leads to the well known periodic curve. For a redundant system, only one assumption can be made: after a test, the system will work. At this moment its reliability will be 1, but its internal state will not be known, except if the test has revealed a failure. In this case, all the internal components have failed and will be repaired. Since that information is not available a priori, it is not possible to take it into account on a deterministic way. The adopted method consists in supposing that the system evolves as if it was never tested or repaired, but to reset its reliability to 1 after each test. Its availability can be easily calculated:

$$R(b+t) = R(b) * R(t)$$

R(b+t): probability that the system does not fail in [0,b+t].
R(b): probability that the system does not fail in [0,b].
R(t): conditional probability. Probability that the system does not fail in ]b,b+t], knowing it has not failed in [0,b].

If one considers b as the last test time point, R(t) is the availability of the system on]b,b+t], and:

$R(t) = R(b+t)/R(b)$, the formula is illustrated in the diagram below.

This calculation provides all the required information to assess the risk follow up and to suggest an adequate test period of the redundant sub-systems.

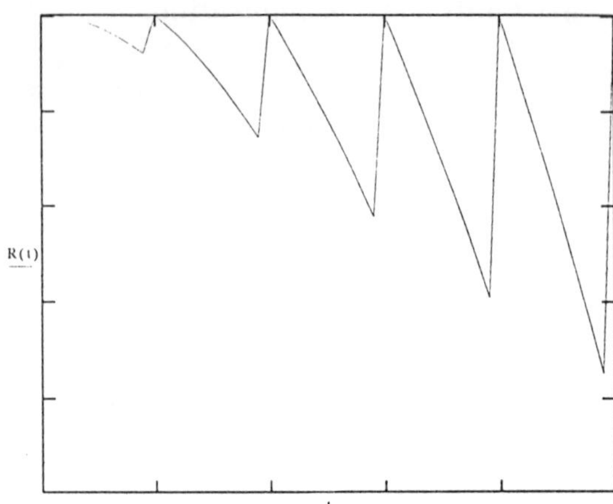


## 4.3 Scenarios.

Each accidental initiating event requests the working of a particular part of the system, leading to a specific reliability. Therefore, the fault tree, which has a size of approximately 500 gates, is designed to enable the runs of different scenarios. The main scenarios are: fast neutron transient - "L chambers" path response (with or without backup), start up transient - "L+B chambers" path response (with or without backup), classical sensors signal or manual demand - relay scram path response.

# 5 Results and Comments.

Although quantification has been extensively used, it is integrated in the reliability evaluation as an indicator of trend, reflecting the systems parameters influences. It plays a significant role in the improving process. This paragraph provides, beside the Minimal Cut Set (MCS) comments, the result of this process and the comparison of the reliability minimal value with international standard.

## 5.1 Minimal Cut Sets.

*5.1.1 L chambers path.*

The MCS analysis reveals three double failure, all included in the voting logic module. Since, each of them results from the failure of two AND gates located on a same chip, they should be conservatively considered as three single failure ($\beta$=1 assumption ). The other principal MCS are real double failure and concern the chambers line from the detectors to the VL (L1-C2, L3-CH1...). These last sequences result from the 2 out of 3 working regime.

*5.1.2 Relay line path.*

There is one MCS of order one. It concerns the failure of a protective R-C filter which by-pass the serial relay line. The next MCS are of order 2, they concern the failure combinations of sensors and relay contacts.

## 5.2 Sensitivity and Improvements.

By realising the independency of the voting logic components, which is a minor operation, the "L path" single failures will be suppressed. Quantitatively, this modification corresponds to a risk decrease factor of 10. Also, the replacement of the RC filter by electronics which have less short circuit failure mode will significantly improve the total reliability.
Once these two improvements are introduced into the model, no more dominant failure modes remain. The failures sources are homogeneously spread over the whole system which tend to show that the topology of the SCRAM system is quite healthy. Nevertheless, the sensitivity analysis also indicates that the backup modules of the neutron chambers do not increase significantly the reliability.
The test politics is adequate, there is no need to increase their frequency neither to adapt their position except for the redundant sub-system which should be extensively tested once a year.
Concerning quantification, the maximum mean unavailability is $3\ 10^{-5}$ per demand. The 90% confidence limit is $9\ 10^{-6} <$ Unavail. $< 9\ 10^{-5}$ . As the study methodology follows international recommendations, the calculations results can be considered consistent with the standards.

# 6 Conclusions.

The study of the BR2 SCRAM system by means of FMEA and Fault Tree techniques has demonstrated its efficiency. It has been possible to significantly improve the reliability of the system with minor modifications. Also, the study has produced precious information for the maintenance team.

* The SCRAM system is the complete device that activates the latch of the control rods in the reactor, to stop the neutron reaction in case of an emergency. It begins at the detectors and ends at the final latch mechanism.

## References.

1. A.Kaufman, D.Grouchko, R.Cruon, *Mathematical models for the study of the reliability of systems*, Academic Press, 1977.

2. Bernard Verboomen, *The BR2 Probabilistic Safety Assessment Project*, Proceedings of The 3rd JSME/ASME joint International Conference on Nuclear Engineering (ICONE-3), Volume 3, n° 5310-1, page 1361-1366, April 1995.

3. Yvan Pouleur, *The Probability Safety Assessment Impact on the BR2 Refurbishment*, Proceedings of The 3rd JSME/ASME joint International Conference on Nuclear Engineering (ICONE-3), Volume 3, n° 5310-2, page 1367-1372, April 1995.

4. IAEA-TECDOC-478, *Component reliability data for use in probabilistic safety assessment*, 1988.

# Annexe 1: BR2 SCRAM system

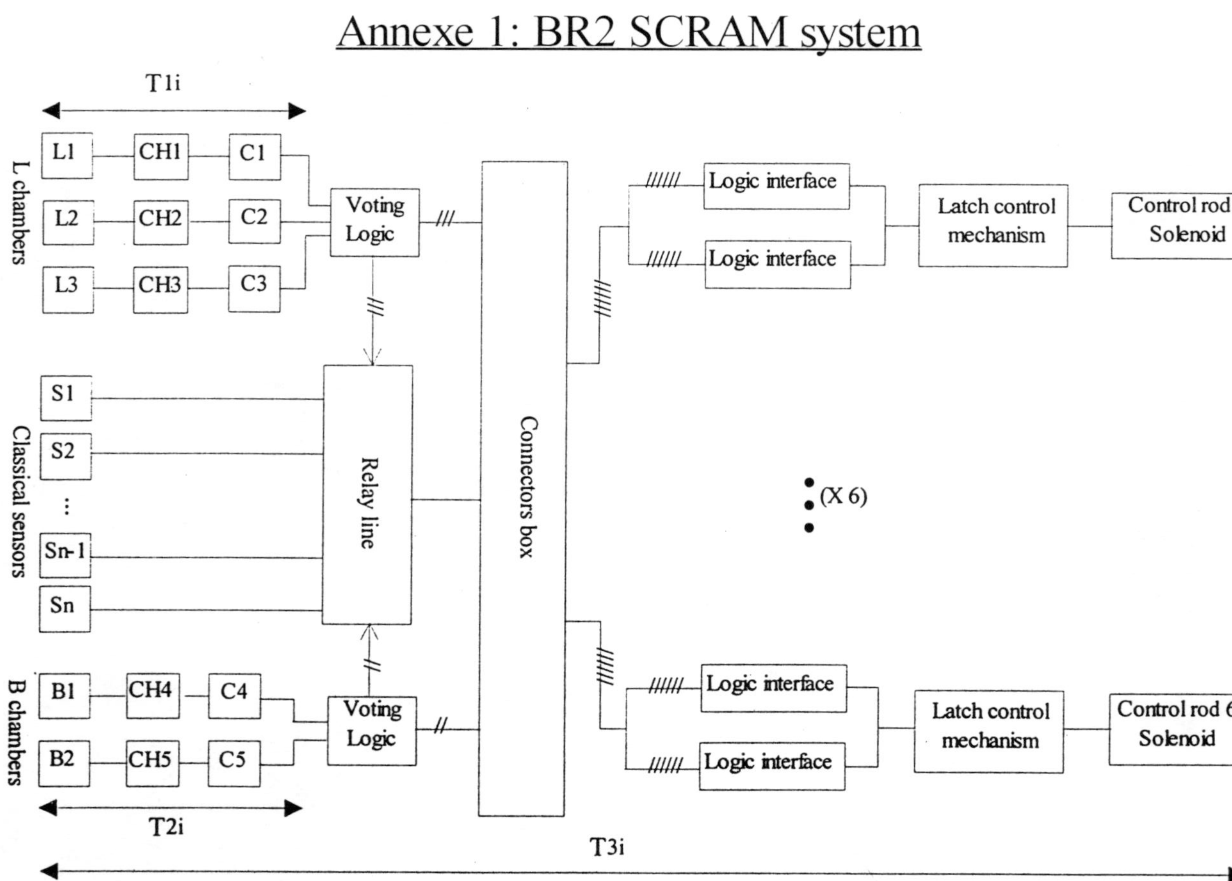

# Safety and Reliability Activities in the Design to reduce Operation and Environmental Risks

S. Senni Buratti, G. Uguccioni

Snamprogetti SpA - SIAF - I 20097 San Donato Mil. (Italy)

## 1 Introduction

Risk Analysis techniques are more and more frequently applied in the design of industrial plant. The growing complexity of the systems together with the need to develop increasingly reliable solutions make it essential that the plant design decisions are backed up by an in-depth, global knowledge of the most important specific problems of the plant itself.

In general, the application of reliability-based methods at design and operating levels in industrial plant has the main objective of identifying and characterising all the situations in which undesirable consequences may occur.

These consequences may take the form of simple unprogrammed shutdowns or more serious incidents capable of causing significant damage to the plant. They may even reach a level of gravity that has effects on the operators or the population in the surrounding zoneand have negative environmental repercussions.

There are three typical main objectives to be pursued at the design and operating stages: the minimisation of the risk of shutdown, the safeguarding of the people and environment and the identification of solutions which present a correct compromise in terms of safety, reliability and economy.

If it is not possible to privilege the economic dimension at the expense of operating safety, the adoption of highly reliable solutions, which are in general more costly, should be justified by concrete safety demands.

The decision-making process for the selection of the optimum plant system alternative should therefore be backed up by in-depth analyses which assess the specific problems of the plant under examination and take into account the implications on the area where it is located, thus enabling to obtain a global and unified vision of all the safety problems.

## 2 Project Safety and Reliability Activities

The possibility of managing the safety of a plant presupposes the setting up of a "safe" plant, whose reliability is maintained in the course of time by means of management and technical processes.

The first requirement that we have to satisfy if the plant is to be considered "safe" is that it is well designed. To ensure this, however, the first thing we have to do is decide to what extent we want it to be safe, that is, we have to decide on a reference point for the level of safety required for persons, the environment and property. It is only if we can demonstrate that the technical selections we have made guarantee this safety level that we can conclude that the plant has been well designed.

In the case of problems connected with the safety of persons, property and/or the environment, the definition and checking of the objectives are developed through activities of Safety Management, or "Project Safety and Reliability Activities", which involve Safety and Reliability Reviews, to be carried out independently of the specific actions of the designers.

The aim of these activities is to ensure that each designer has taken into account the specific problems concerning the system he is designing, that the safety objective has been well defined and is effectively reached, and that, in general, risk reduction is obtained for the particular installation we are dealing with.

To design well, with the consequence that the operator can run the plant safely and continuously, therefore means that the suitability of the design in safety and reliability terms is confirmed at design level. This suitability check, in an engineering company, is carried out by the Safety and Reliability Division, which has the task of assessing and giving guarantees on the safety, functional and operational levels of the plant.

The Project Safety and Reliability Activities can be implemented at different levels of depth, depending on the complexity and intrinsic risk factors of the plant, legislative or regulatory requirements, and any special restrictions - for example, those deriving from insurance or financing needs. These activities should always include the definition, at least in qualitative terms, of the objectives, the identification of the general safety criteria and regulations to be adhered to, and the sequence of the multidisciplinary audits and safety reviews, aimed at checking and certifying that the established safety criteria have been implemented in the design.

The performance of specific analyses and studies, including risk analysis, appears with increasing frequency in the Project Safety and Reliability Activities, not only because of the greater attention paid to risk control by the regulatory bodies, but also due to the awareness by design managers of the benefits to be derived for the overall development of the design from systematic studies of this kind.

In order to tune the safety activities in relation with the specific project as well as to develop these activities in an integrated manner within the design process, the "Project Safety and Reliability Plan" (figure 1) is defined each time a new project is to be undertaken. This is a programming document that sets out the level of safety and reliability the specific design is to adhere to, the criteria and regulations to be followed and the specific analysis and study activities required to obtain the safety and reliability standards we have set.

As the project progresses, specific studies, risk and reliability analyses and periodic reviews are then carried out to identify the problems and optimum safety measures and check that the safety criteria and requirements derived from the safety and reliability analyses are integrated into the design. Risk Analysis should therefore be seen as a design back-up activity, included within a broader plan, defined case by case on the basis of the specific design requirements and characteristics.

# 3 Risk Analysis in the Project Safety and Reliability Activities

The safety and reliability activities take into account not so much the single component, but the integrated plant within which the interaction of the different components leads to the possibility that the system beyond the expected operating conditions, may develop anomalies of varying degrees of seriousness. This makes it necessary to take into consideration methods for dealing with events in the realm of uncertainty, based on the evaluation of probability.

The tool that should be used to assess the level of safety or reliability is, therefore, Risk Analysis, which takes on the role of a decision-making tool, from whose results it is possible to identify any necessary adjustments to the plant in terms of hardware or management.

Risk analysis may be defined as a systematic and documented collection and processing of information, with a view to supplying decision-making support. To reach this stage, we have to consider the 'risk' as a function of the probability of occurrence of reference events and their consequences.

From this simple definition, we can set out the various stages that have to be developed to perform a Risk Analysis:

- The identification of the possible incidents
- The assessment of the probability of these events occurring
- The assessment of the consequences of an incident
- The identification of the risk function that represents the specific risk of the problem under analysis
- The comparison with the established acceptability criteria
- The decision-making process: the identification of the need for measures that will reduce the frequency and/or consequences of incidental events and the assessment of the most appropriate measures for the given project.

Let us now briefly examine the tools and procedures applied for the various steps set out above.

Identification of possible Incidents. The anomalous events that may take place in a plant can be identified by means of several techniques. The most important of these are:

- Historical data (incident data banks)
- Check list application
- Failure mode and effect analysis
- Operability/HAZOP analysis

The hazard identification techniques should, most importantly, guarantee the completeness and correctness of the identification of undesirable events. The selection of the technique to use should therefore be made in such a way as to satisfy these two criteria in each specific case.

The use of historical data and check lists, for example, is advisable when simple, well known and standardised systems are under analysis, as previous experience is already wide ranging and applicable in the case in hand.

Where the systems are more complex, or where there are significant variations from previous situations, the use of past experience alone is not advisable. In such cases we have to use more structured, formal techniques based (such as FMEA [1], Operability Analysis [2], Hazard and Operability Analysis (HAZOP) [3]) on the detailed analysis of the system in the interrelations of its component parts.

By applying the HAZOP technique, it is possible to carry out a kind of simulation on paper of the behaviour of the plant in anomalous situations and directly identify possible critical points (most typically, inadequately protected or detected anomalies) on which we can intervene by way of plant or procedure modifications.

This type of analysis is more complex and time consuming than those described previously. It has to be carried out by a team led by an expert in the application of the technique, known as a HAZOP Leader, and containing specialists in the main disciplines involved in the project. The technique may require several man-months of work. However, thanks to the completeness of the information and the benefits in terms of project improvement, this has become the most important technique, at least in the field of process engineering, and it is also widely used as a project revision technique independently of the subsequent application of quantity risk analyses.

Assessment of the Frequency of Occurrence. The assessment of the frequency of occurrence of incidental events enables us to measure the credibility of an event, by attributing a mean return time to it.

This can be done by means of statistics (data banks) or analytical methods based on the theory of probability, by which we can associate a probability of occurrence to each event that takes into account the specific operating and design specifications of the system.

As in the incident identification techniques, the use of statistics is advisable here only when we are certain that the historical information is based on a sample of systems which can be effectively compared with the one in hand, in terms of both the project and running characteristics. In general, the use of statistics without checking their effective applicability may lead to large-scale errors in the estimate of the incident probability. For example, if we take the probability analysis of leaks in a pipeline, the use of the leak statistics of piping in general may lead to over-estimates of several times the real probability if we do not take into account the fact that the thickness of the pipes plays a primary role in the possibility of leaks, which means that we are not using statistical information relating only to pipes of the same thickness of those we are dealing with. In the same way, the use of statistical information based on American pipes for those designed using European standards will lead to over-estimates of the probability of leaks caused by corrosion, as there is a considerable difference between European and US design conditions in terms of protection against corrosion.

In general, for relatively complex and not completely standardised systems (as is virtually always the case in process engineering), the use of analytical techniques based on the theory of probability is essential.

The main technique used is the Fault Tree [4]. This is a graph of the logical relationships of the events which, when they occur, may lead to the incident under analysis.

It is important to emphasise that the fault tree technique should be used not only to obtain the probability value of an event, but also to acquire information on the trends of the most critical breakdowns and any dependencies between systems caused by common components. It is also important to note that an elementary application of the fault tree technique, which makes direct use of the probability calculations, therefore without taking into account the possible presence of events common to more than one breakdown trend, may lead to large scale errors in probability assessment.

In special cases, different types of techniques are used, especially when we have to consider the interdependence of events, such as Markov diagrams, the Event Tree, the Montecarlo simulation or ad hoc analytical methods.

The Decision Tree, as implemented by Snamprogetti in the ADMIRA programme [5], requires a special mention. This is used not only for the calculation of the probability of events, but also for the complete induction analysis for the identification of all the possible operating modes and malfunctions of the system.

Analysis of consequences. Consequence analysis is the part of Risk Analysis that assesses the physical effects of the phenomena that follow an incident.

To carry out a consequence analysis, we have to follow up the evolution of the incident situation by means of:

- the definition of the source
- the identification of the phenomena subsequent to a leak
- the assessment of the damage caused by specific physical effects.

The definition of the source is strictly related to the identification of the incidental events. In particular, we have to define the expected dimensions of the leak, the process conditions when it occurs, the substance concerned and its phase (liquid, gas or mixed). On the basis of all this, and using appropriate models, we can calculate the flow rate and duration of the leak.

The phenomena typically taken into consideration in an analysis are:

- evaporation (on land or water)
- gas dispersion (continuous or instantaneous, at low or high speed, of gases lighter or heavier than air)
- fire (pool, flare or cloud)
- explosion (of vessels, clouds, confined or unconfined).

The application of simulation models enables to calculate the seriousness of the effects of an incident (for example, radiation or over pressure) on the surrounding territory, in accordance with the distance from the point of the incident and the atmospheric conditions at the moment of the leak.

This normally requires the performance of several simulations for the different weather conditions that may occur on the site, each of which is associated to a probability of occurrence.

The estimate of the effect of an incident is therefore correlated to the associated damage. This is achieved by using effect-damage correlation tables (for example, the assessment tables for radiation damage based on API standards [6] or Probit equations [7] which supply the damage probability associated with the doses for the main effects and toxic substances.

Risk Assessment and Acceptability Criteria. All the analyses mentioned above are geared towards the separate assessment of the two terms (probability and consequences) that form part of the risk definition.

The risk definition to be used also depends on the regulations in force in the area where the analysis is made. For example, to obtain authorisation in some European countries, documents that testify to the achievement of precise Individual and Social Risk values are necessary. In Italy these criteria are not laid down by the authorities, which means that it is the company that has to set the safety standards on its own initiative. There are various methods for the setting of these acceptability criteria; a practical way is the frequency/consequence matrix. If we place each event studied in the matrix, in the sector defined by the level of probability and consequently assessed, we can obtain a visual pattern of the risk associated with each event. Where an event is situated in an unacceptable risk zone, it may be taken into an acceptability zone by acting on the probability of occurrence, using measures aimed at increasing the level of protection or reducing the frequency of the initiating event or associated consequence. Typically, this involves making modifications to the location plan, the maximum quantities of the substance concerned and the emergency measures.

A representation of the risk level, requiring the calculation of the individual risk value in the area surrounding the plant, is the plot of iso-risk curves on the area map.

Further processing of results of this kind with, for example, the integration of information from several risk sources, enables us to evaluate the overall risk of the zone in question (figure 2).

This type of approach is undoubtedly essential for a "rational" management and planning of the territory or the drawing up of forecasting and protection programmes in the case of hypothetical risks connected with natural calamities or events linked to human actions.

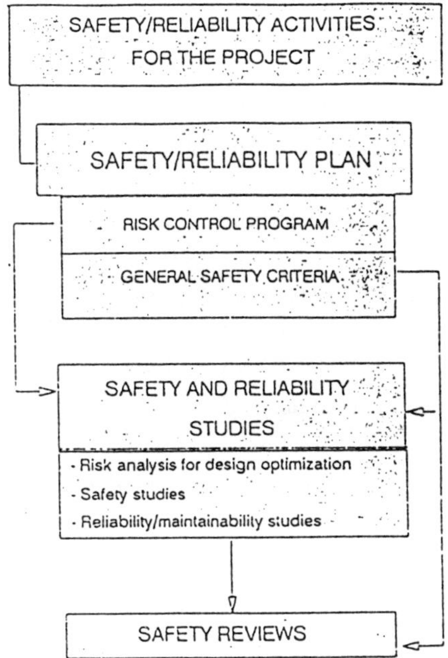


Figure 1 - Schematic of the 'Project Safety and Reliability Plan'

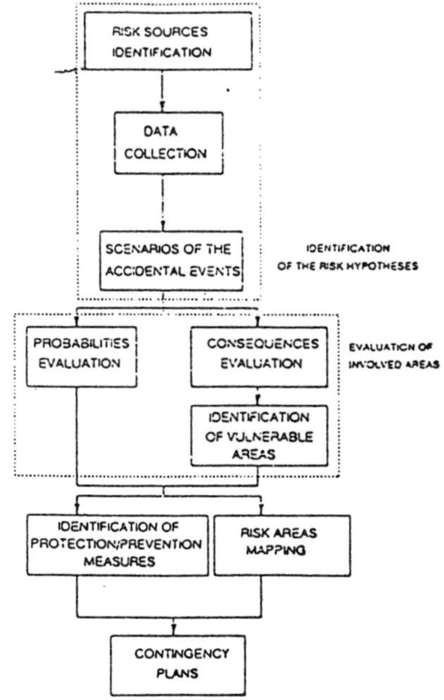


Figure 2 - Risk assessment for several risk sources

## References


1. 'Guidelines for Chemical Process Quantified Risk Analysis' - AIChemE, 1989
2. 'Basic Concepts for Operability Analysis ' Snamprogetti Repor INT-SI-GEN-25, 1985 (in Italian)
3. 'A Guide to Hazard and Operability Studies' - Chemical Industry Association, 1985
4. Vesely, 'Fault Tree Handbook' - NUREG 0492, 1981
5. Senni, Semenza, Galvagni, 'ADMIRA - An Analytical Dynamic Methodlogy for Integrated Risk Assessment' - PSAM 1, 1991, pagg. 407-412
6. American Petroleum Institute RP 521 - Guide for pressure relieving and depressurizing systems
7. 'Method for the determination of possible damage' - CPR 16E - TNO 1992

# Modeling of Physical Failure in Microwave and Optical Components For Assessment of System Risk

**N. Strifas, P. Yalamanchili, C. Pusarla, and A. Christou**
**Department of Materials and Nuclear Engineering**
**University of Maryland, College Park, MD 20742**
**Phone : (301) 405-5208 Fax : (301) 314-2029**

**Abstract**

This paper aims to compute electronic system risk in terms of the failure rate of individual microwave and optical components. The systems under consideration include a microwave multichip module and an optical multichip module. The system failure rate of a microwave or an optical multichip module depends on the architecture and failure rate of individual devices. The significant failure mechanisms have been identified and the appropriate physical models are analyzed to understand the physics of the failure. The physical model predicts the failure rate of the device in terms of the operating environment (temperature, humidity, and voltage) and the material system.

## 1 Introduction

The estimation of components reliability is a major importance to the equipment manufacturer. As the complexity of equipment and components increase so does the demand for higher reliability. Estimation of reliability is obtained via various modeling methods based on laboratory and field (operational) data. Knowledge of the failure mechanisms is an essential requirement of the analysis. Some failure mechanisms may result in either a catastrophic as wear out degradation such as the case of die-debonding due to excessive mechanical/electrical stress. The primary failure mechanisms include, debonding between the device and substrate (die attach) interface and low cycle thermo-mechanical fatigue of solder joints. This debonding reduces the dissipation of heat from the device and hence increases the device operating temperature. This could cause either catastrophic burn out of the device or the degradation of the device which could reduce the reliability of the device and the entire module.

Optical interconnects are finding rapid applications in electronic packaging. The use of optical interconnections between processors, board, chips, and even gates has been proposed to overcome the interconnection problem. The reliability issue has become a major concern for the development and insertion of the interconnect technologies into reliable commercial products. Of the many failure mechanisms associated with the new technologies, the effect of corrosion and solder fatigue have been studied on the performance degradation of electrical and optical interconnects respectively.

In the following section two different failure mechanisms will be presented and discussed in terms of their effects on limiting MMIC and Optical Interconnect (OIC) system reliability. The failure mechanisms to be presented are die bond failure and solder joint fatigue. The MMIC and OIC risk analysis will then be discussed using a system reliability approach. Typically a MMIC or OIC system however large or small contain a number of circuit or optical components and a substrate as backplane. It is natural then to subdivide such a system into a collection of subsystems and then organized into a reliability block diagram.

## 1.1 Die-Attach Evaluation

Voids can occur for various reasons: air entrapment, uneven spreading of the adhesive or solder over the backside of the die, preform outgassing, contamination of the bonding layer, etc. The presence of voids reduces the reliability of the devices. The voids may affect operating temperature of device and may induce thermal and mechanical failures. The gross voiding in the die bonds can significantly increase the junction temperature of a powered device, thereby impacting long term reliability.

Acoustic microscopy has been applied to GaAs MMIC packages subjected to temperature cycling in order to determine the die attach integrity, void density and die cracks. In order to investigate voiding, temperature cycling tests were performed by cycling the packages from room temperature to 225 °C for 80 cycles, maintaining the temperature constant for one hour. From the acoustic microscopy results, the void density as a function of number of temperature cycles is plotted in Figure 1a. Thermal measurements of the devices subjected to temperature cycling were carried out using Scanning Infrared Microscope (SIM). A fitting of thermal resistance vs. volume of die attach void shows a linear increase in the thermal resistance of the GaAs material with die attach void volume. This increase in thermal resistance may result either in an electrical wear-out failure or a catastrophic failure if local heating becomes excessive.

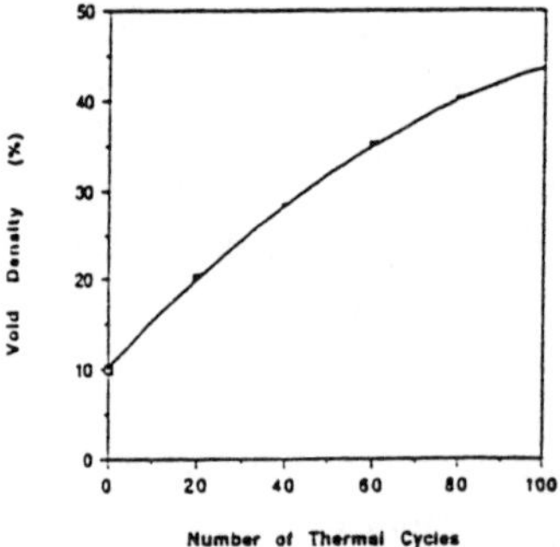


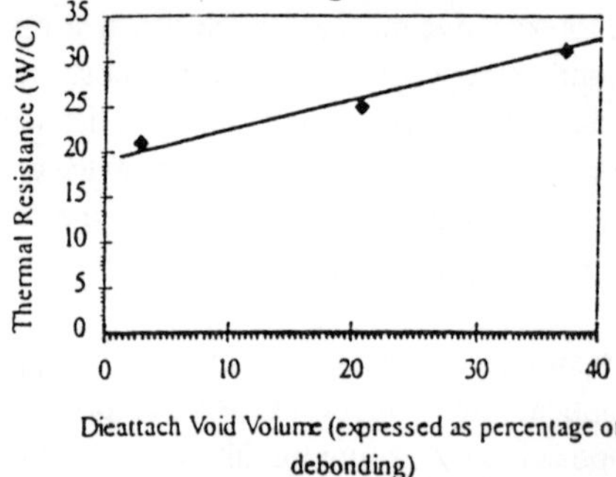


**Figure 1. (a)Void density as a function of temperature cycles. (b) Thermal resistance vs. Die attach void volume.**

## 1.2 Solder Joint Low Cycle Thermal Fatigue

Flip chip bonding using low temperature solder is commonly used to attach receiving and transmitting components to an optical backplane in an optical interconnect system. Low cycle thermal fatigue occurs at the solder joints due to the thermal coefficient of expansion mismatch between the opto-electronic component array and the silicon integrated circuit during temperature cycling. Temperature cycling occurs due to the chip powering cycle and the environmental temperature variations. Fatigue due to the shear stresses induced by the temperature cycling usually results from a low number of cycles to failure (1 - $10^4$ cycles) and is referred to as low cycle thermal fatigue.

The most commonly used solder materials for flip-chip applications are 95Pb/5Sn and 50Pb/50In solder. The process of fatigue failure at a solder joint is typically characterized by three distinct stages: (I) crack initiation, wherein a small crack forms some point of high stress concentration on the surface of the solder (ii) crack propagation, during which the crack advances incrementally with each stress cycle and (iii) final failure, which occurs very rapidly once the crack has reached a critical size. The low cycle fatigue life $N_f$ (number of cycles) of the solder follows the coffin-manson low cycle relationship which is given by [1]:

$$N_f = \frac{1}{2}\left[\frac{\Delta\gamma}{2\epsilon_f}\right]^{\frac{1}{c}} ; \qquad N_f = f(\Delta T, T, geometry, materials) \tag{1}$$

where $\Delta\gamma$ is the maximum cyclic strain range, $\epsilon_f$ is the fatigue ductility coefficient of the solder material, and c is the fatigue ductility exponent.

The predicted life of the solder joint at other temperatures and frequencies can be determined by the following relation:

$$N_{2f} = N_{1f}\left(\frac{f_2}{f_1}\right)^{\frac{1}{3}}\left(\frac{T_{1\max}}{T_{2\max}}\right)^2 \tag{2}$$

The performance failure due to the low cycle thermal fatigue is a catastrophic failure. Once the number of cycles are computed for a given failure criteria, the following equation can be used to compute the failure rate $\lambda_i$,

$$\lambda_i = \frac{1}{N_f * cycletime} \tag{3}$$

# 2 System Reliability Prediction Modeling

System failure is dependent on the failures of the individual components. The component failure, $\lambda_C$ rate is modeled as [2]

$$\lambda_C = \Pi_{QD} \; x \; \lambda_F + \Pi_{QP} \; x \; (\lambda_{TM} + \lambda_{TH}) + \lambda_{EOS/ESD} \tag{4}$$

where $\Pi_{QD}$ is the die quality factor, depending on the level of line/product qualification (front-end), $\lambda_F$ is the functional failure rate, $\Pi_{QP}$ is the package quality factor, depending on the level of line (back-end) or product qualification, $\lambda_{TM}$ is the thermo-mechanical failure rate, $\lambda_{TH}$ is the temperature/humidity related failure rate and $\lambda_{EOS/ESD}$ is the electrical overstress/electrostatic discharge failure rate.

The functional failure rate is activated by temperature which follows an arrhenius law and power supply voltage which follows an exponential law. If the reference junction temperature is taken as 55 °C (328 °K) and nominal voltage as 10%, the functional failure rate can be modelled as

$$\lambda_F = K_1 \, e^{-\frac{E_a}{k}\left(\frac{1}{Tj} - \frac{1}{328}\right)} \, e^{C(V-1.1V_n)} \tag{5}$$

where $K_1$ is the reference conditions functional failure rate, $E_a$ is the activation energy which depends on the material system and failure mechanism, k is the Boltzmann constant ($8.617385 \times 10^{-5}$ eV/k), Tj is the junction temperature, V is the power supply voltage, $V_n$ is the nominal power supply voltage and C is a constant based on the manufacturers reliability prediction and is between 1 and 2.5.

The thermo-mechanical failure rate can be modelled as

$$\lambda_{TM} = K_2 \, \Pi_{TM} \tag{6}$$

where $K_2$ is package related failure rate in reference conditions value and $\Pi_{TM}$ is a thermo-mechanical environmental factor.

Taking the reference junction temperature, Tj = 55°C and relative humidity, RH = 50%, the temperature and humidity failure rate can be modelled as

$$\lambda_{TH} = K_3 \, e^{\frac{E_a}{k}\left(\frac{1}{Tj} - \frac{1}{328}\right)} \left(\frac{RH}{50}\right)^{2.66} e^{(V-1.1V_n)} \tag{7}$$

where $K_3$ is the temperature and humidity related failure rate in reference condition and temperature and humidity failures have voltage dependence.

The electrical overstress/electrostatic discharge failure rate can be modelled as

$$\lambda_{EOS/ESD} = D\frac{V}{V_{TH}} \tag{8}$$

where D is a proportionality constant (≅ 27.4 x $10^{-9}$ /hours), Vis the average value of electrical field around the compoenent and $V_{TH}$ is the component susceptibility threshold voltage.

Substituting Equations 5 to 8 in Equation 2 the component failure rate, $\lambda_C$ can be calculated and the system failure rate can be calculated from component failure rate depending on how the individual components are connected (series or parallel) in the system. A variation of the simple parallel system is the r out of n system in which at least r out of n elements must function for the MMIC or OIC to function properly. If the n elements are identical, then the risk of failure for such a system is found from a simple binomial distribution.

In a serial architecture which is typical of a multistage power module, the system failure rate $\lambda_s$ can be computed as

$$\lambda_S = \sum_{1}^{n} \lambda_C \tag{9}$$

where n is the total number of devices in a system. For a parallel architecture, which is typical of low noise modules, all devices must fail for the system to fail completely. The system failure rate $\lambda_p$ can be approximately computed as

$$\lambda_P = \frac{1}{\sum_{1}^{n} \frac{1}{\lambda_C}} \tag{10}$$

The prediction of system risk is computed for a given failure mechanism by computing individual component failure rates and then extrapolating the system failure rate. Risk can be expressed as risk to a given failure mechanism or to a specific device failure within a system architecture configuration.

## 2.1 Risk of Complex Electronic Systems

Typical MMIC and OICs in the most common configuration cannot be decomposed into simple series, parallel or r out if n subsystems. Risk to failure analysis of such complex structures can be made using one of the two methods:

**A.** The first method uses the theorem of total probability [3]. This method may be used when key components (MESFETs, photo detections, lasers) that bind the structure are selected, such as a field effect transistor in figure 2.

**B.** The second method for evaluating MMIC or OIC risk to failure is based on a

Boolean truth table [3]. A truth table contains a box for every possible operational state of the MMIC or OIC. In our particular application a state of the system refers to the operational state (working or not working) of all the blocks in the system. Whether the state considered working or not is indicated by a "I" or a "O" in the system column. The state probability is computed for each row that represents a working state. The sum of these state probabilities gives the probability that the overall system will work. These are $2^n$ rows in such a table where n is the number of components in the block diagram. The method therefore becomes impractical for any but small systems.

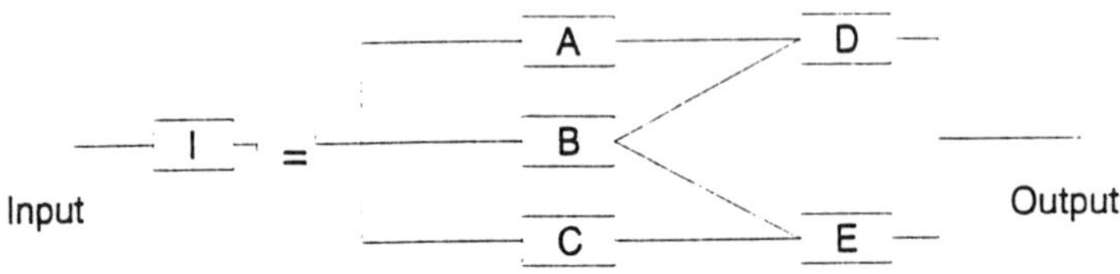


**Figure 2. A complex reliability MMIC block diagram with key FET (B). The key block is B which is also the critical system component.**

# 3 Conclusions

The electronic system risk has been calculated in terms of the failure rate of the individual microwave and optical components. The systems failure rate depends on the architecture and failure rate of individual devices, where the failure rate of individual device is a function of the material system, operating frequencies, operating environment and power dissipation level. So the significant failure mechanisms have been identified and physical models were analyzed to understand the physics of failure. System risk can be estimated by considering series, parallel or complex systems.

## References

1. C. Pusarla, "Design Guidelines of Flip-chip Bonding for Hermetic Microelectronic Packages," Masters Thesis, December 1992, University of Maryland, College Park.
2. Brizoux, M., Deleuze, G., Digout, R., and Nallino, M., "Plastic Encapsulated IC's in Military Equipments Reliability Prediction Modeling", 2nd European Symposium on Reliability of Electronic Devices, Failure Physics and Analysis, Bordeaux, France, October 7-10, 1991, pp. 847-868.
3. W. Feller, "An Introduction to Probability Theory and its Applications, Vol I, 3rd ed 1968, Vol II 2nd Ed, 1970, Wiley, N.Y.

# Error Tolerant Approach Towards Human Error

Kamran Sepanloo[1], Najmedin Meshkati[2]

[1] Nuclear Safety Department
Atomic Energy Organization of Iran
Tehran, Iran

[2] Institute of Safety and Systems Management
University of Southern California
Los Angeles, California 90089-0021, USA

**Abstract**

Human error has been recognized as the main contributor to the occurrence of incidents in large technological systems such as nuclear power plants. The traditional error prevention approach to human error has proved to be insufficient to solve this *dilemma*. Recent researches have concluded that human errors are unavoidable side effects of exploration of acceptable performance during adaptation to the unknown changes in the environment. This is particularly important under difficult unexpected situations where the operator's deteriorated performance may lead to irreversible hazardous processes in the plant. Therefore, it is highly important that the operator's job, which involves effortful and error-prone activities of solving and decision making at the workstation level, be facilitated by proper interface devices and be supported by the needed organizational structures. The innovative error tolerant approach to human error requires more cognitive-friendly designs for interface systems in the control room of large hazardous technological system.

## 1 Introduction

The major share of human errors in the occurrence of accidents in large socio-technological systems has highlighted the importance of reviewing the current approach towards the concept of human error and design of error proof systems. Human error is generally defined as an unacceptable human action which transgresses the boundary of acceptable performance. The boundary is usually defined as preset conditions and relevant procedures which have to be constantly observed. The main goal in pursuit of traditional way of responding to errors is designing less error prone workstations in which the number of human interventions is tried to be reduced to as low as possible.

The current widely used defence-in-depth strategy in large hazardous socio-technological systems helps the adverse impacts of human errors to remain undetected and be forgiven by the system. The redundant, overlapping safety layers prevent the progression of the chain of individual errors and failures from leading to an accident. Frequently, the large safety margins which has been put into the safety layers and systems allow the plant parameters to somewhat exceed the nominal range without serious notification of the operator. The complexity and opacity of the mechanisms in many cases hinder the operators from obtaining a clear understanding of the real changes in the status of the plant.

Similarly, higher level of automation which is usually deemed as the remedy to human reliability deficiency can not eliminate the problem. The issues such as increase of complexity and opacity of the system specially in unprecedented upset conditions, elevated rate of maintenance errors, loss of skill, de-motivation and passiveness of the operators are some of the adverse impacts of more automation. It is highly questionable whether total system safety is always enhanced by allocating functions to automatic devices rather than human operators and there is some reason to believe that control room automation may have already passed its optimum

Learning from the past experience and subsequent modifications have to a great deal improved the operation of hazardous systems, but it seems that following the current "hindsight" approach of error reduction strategies will lead to a point that they will lose their effectiveness. This limitation, however, stems from the implicit static definition of error. In the static image of error, there are certain constant specifications, causes and remedies for the undesired behaviours. Lack of sufficient training, hardware deficiencies, or more recently, structural and management faults are some of the usually cited "root" causes. Generally, there is an attempt towards the "fixation" of the root causes of errors in response to the revelation of human errors in incidents or accidents [1].

However, both human beings and socio-technological systems are not static items. The personal, cultural, technical, economical and political envelope of the system are constantly changing which require the constant adaptation of human and the organization. Therefore, the concept of error must be viewed in the adaptive dimensions as the mismatch between the two changing sides, i.e. human and system. This dynamic mismatch can only be bridged by appropriate human-system interfaces which are designed on the basis of dynamic definition of human error.

# 2 Error Tolerant Approach

Recently, based on extensive research on the role of human element in technological systems, it is known that human error, as the unavoidable side effect of the exploration of degrees of freedom in unknown situations, can not totally be eliminated in modern, flexible, or changing environments.

In order to allow for, and cope with human errors in large technological systems such as nuclear power plants, human errors should be considered as unsuccessful or unacceptable experiments in an unfriendly environment. Therefore, the design of friendly, i.e. error tolerant systems [2, 3, 4] with integrated task and organizational structures should be considered. The interface design should aim at making the boundaries of acceptable performance visible to the operators while the effects of committed errors are observable and reversible. To assist the operators in coping with unforeseen (beyond-procedural) situations, the interface design should provide them with tools (opportunities) to make experiments and test hypotheses without having to carry them directly on potentially irreversible processes [5].

Human error, is considered to occur if the effect of human behaviour exceeds a limit of acceptability. Of course, it is necessary to distinguish clearly between the types of errors induced by inappropriate limits of acceptability, i.e. by the design of the work situation and errors caused by inappropriate human behaviour. Furthermore, in many instances, the working environment can also aggravate the situation. In such unfriendly work environment, once the error is committed, it is not possible for the operator to correct the effects of it before they lead to unacceptable consequences, because the effects of the errors are neither observable nor reversible.

It is known that in traditional work organizations, various task groups must respond to rapid changes which cannot be thoroughly analyzed before implementation of corrective actions. Also, discretionary decisions are made by different people that often interact to produce an unpredictable outcome. Error tolerance is important here, because incompatibility between the solutions chosen by the different groups can have drastic economic and environmental impacts. One solution is an integrated information system that ensures effective horizontal communication that makes the effects of decisions made by team members visible within the work context of each of the other teams. That is, it should be made clearly visible (and hopefully reversible) when decisions made by one group violates the boundary of acceptable design as specified by the other groups.

In this approach, the operator's talent of improvisation is considered as the most reliable factor in directing the plant in unknown upset situations [6]. He can infer the most suitable way through reasonably limited number of trial on error tolerant control system. In doing this, he feels less stress for committing errors, since (in most of the cases) he has the opportunity and enough time to correct his mistakes and recover from their consequences.

# 3 Features of error tolerant systems

An error tolerant system has the characteristics of both human-machine and human-human interfaces. It can be regarded a human machine interface since the operator interacts with the plant through it, and at the same time it is a human-human interface system which informs the other relevant decision makers of the actions taken and also can find the impact of the others' actions on the domain of acceptable behaviour of the operator. In other terms, this interface system provides a means by which the boundaries of safety margins of the plant and the boundaries of allowable actions becomes visible to the operator at any time during the operation of the plant. Error tolerant system can also be considered as a decision support system, since it provides the operator with the actual state of the plant and the consequences of execution of his commands on the plant. The trend of change of area of the space of possibilities (the degrees of freedom) provides him with valuable guidance in directing the plant away from the safety margins borders. The error tolerant systems have the following features:

- simplicity
- transparency
- error detectability
- linearity
- recoverability

The integration of error tolerant systems into the plant should not add to the complexity of the operation. The presentation of the information should be quite simple. The system should facilitate the operator's correct conception about the real status of the plant. This can be achieved by clear presentation of the path of influence of an operator (or combination of actions of operators) on the state of the plant. Error tolerant systems enhance the visibility of human actions in the plant both for the operator himself and the others who monitor his actions. Thus the decisions made by any actor are analyzed and evaluated by a group which greatly enhances the detection of any error. In other words, the group mind checks the correctness of the operators' decisions in view of the instructions and probable outcomes. The linearity of the mechanism of control of the processes by the error tolerant system (i.e. preventing the plant to undergo irreversible changes) enables the operator to take the reverse steps to change the unacceptable error-lead plant status to the initial point of detraction. This feature allows the operator to recover from some committed error through following remedial procedures. The speed of function of error tolerant system must be faster than the rate of deterioration of the plant state due to some erroneous executed command. The time needed for the error tolerant system to reveal the incorrectness of the operator's action should not permit the plant to go through irreversible degradation processes. In other words, the error tolerant system should not expose the plant to any danger of inactiveness from the operator to the rapid dynamics of the safety relevant processes in the plant.

# Conclusion

Since it is impossible to foresee all probable sequence of events leading to accidents, the safety of a large technological system, such as a nuclear power plant is more reliably achieved when the humans could be kept in the main path of control of hazardous processes. To allow the operators to find the proper solution to the unknown situations the control rooms should be equipped with error tolerant systems. In this way, the natural phenomena of occurrence of human error receives more appropriate remedy, avoiding the control of a potentially hazardous system from becoming unsafely hardware-centred which leads to the passivity of operators and degradation of their skills in case of occurrence of an unexpected situation.

Dynamic adaptation to the immediate work environments, both of the individual performance and allocation between individuals, can be combined with a very high reliability only if the errors can be observable and reversible. Error tolerant systems concept is an approach which provides a forgiving cognition environment for the operators to cope with the unforeseen incidents. Error tolerant systems enable the operator to have clear and comprehendible information about the status of the plant and allows the operator to correct his previous errors. The two features of transparency and reversibility provide necessary flexibility for the operator to follow the changes in the plant and timely recover from his error.

## References

1. Sepanloo K., Meshkati N., Kozuh M. The organizational context of error tolerant interface systems. 2nd Regional Meeting in Central Europe, Portoroz, Slovenia, 11-14 September 1995.

2. Rasmussen J. Human error mechanisms in complex work environments. Reliability Engineering and System Safety 1988; 22: 155-167.

3. Rasmussen J. The role of error in organizing behaviour. Ergonomics 1990; 33: 1185-1199.

4. Rasmussen J, Pejtersen A, Goodstein L. Cognitive Systems Engineering. John Wiley & Sons, New York, 1994.

5. Sepanloo K, Meshkati N, Azadeh M, et al. Integration of error tolerant concept into the design of control room of nuclear power pants. Proceeding of 9th Topical Meeting on Problems of Nuclear Safety, Moscow 4-8 September 1995, Vol 1, pp 28-30.

6. Sepanloo K, Meshkati N, Azadeh A, et al. The role of error tolerant interface systems during the emergencies in nuclear power plants. Proceeding of PSA'95 Conference, Seoul November 26-30 1995, Vol 1, pp 335-338.

# Human Error Tolerant Design for Air Traffic Control Systems

Jin Mo, Yves Crouzet
LIS, LAAS-CNRS
Toulouse, France

**Abstract**

In man-machine systems, many system failures are due to operator error. Therefore, human component must be taken into account in the development of safety-critical systems. The work* described in this paper is part of a research project whose goal is to design an operator error tolerant interface for the on-line maintenance of an air traffic control system. Operator error tolerance is aimed at avoiding system failure in spite of the presence of operator error. It can be achieved through three phases: operator error detection, technical system error detection and error recovery. This paper examines each of these phases.

## 1 Introduction

The rapid evolution of technologies in complex systems has caused important changes in the content of human work. The human operator has to perform more and more complex mental tasks. The increase of the complexity of the operator's tasks has led to the fact that human error is a primary cause of 60-90% of all major accidents in aviation, power production and process control [1]. Operator error is one of the main threats to the dependability of complex systems.

This observation has led to various studies that attempt to reduce and eliminate the occurrence of operator error by designing systems which adapt to the characteristics of their users. This approach increases the dependability of man-machine systems by eliminating causes of operator error. Obviously, total elimination of the occurrence of operator error is impossible in any systems. Indeed, the operator frequently copes with delicate situations which need the use of complex knowledge, sometimes under strong temporal constraints. An other approach which is more realistic consists in tolerating operator error, but avoiding its unacceptable consequences. These two approaches are complementary. The first one is operator error prevention, while the second one is operator error tolerance.

The work presented in this paper is part of a research project whose goal is to design an operator error tolerant interface for the on-line maintenance of an air traffic

* This work is supported in part by the French Research Center for Air Traffic Control.

control system. We begin by describing the air traffic control system. Next, we analyze the operator's tasks in the on-line maintenance of this system. Finally, we present different aspects concerning the design of operator error tolerant systems.

# 2 Air Traffic Control System

The air traffic control (ATC) system is a safety-related system. Therefore, a high level of availability and reliability is required. The two main applications of the ATC system are: flight plan processing and radar data processing. The ATC system is a distributed system. It is made up of a set of triple redundant computers, connected to various devices used for air traffic control. The software of each application is replicated on the three computers for fault tolerance purposes. The role of each replica is different. Only a principal replica has an active output, the others act as back up.

# 3 Task Analysis

Our studies have been focused on the interaction between the operator and the system during the on-line maintenance activity of the operator. The operator's mission is to maintain the availability of the system. The main operator's tasks can be classed into: surveillance and management of alarm, system configuration, diagnosis and detection of failed components of the system, restoration of failed components, and database changes.

In order to identify possible operator errors during the on-line maintenance activity, a task analysis was done using procedure scripts which describe the correct sequences of actions required to carry out the operator's tasks. The goal of the task analysis is to obtain the following information:

- potential operator errors in the on-line maintenance activity;
- causes of errors and the severity of their consequences;
- preventive measures to these errors;
- reversibility of erroneous actions and means of error recovery;
- the damage caused by the propagation of errors.

An analysis of system failure modes was also made using failure modes and effects analysis and fault tree analysis to determine the most serious operator errors which directly lead to a system failure. In the present work, a system failure is defined as the discontinuity of system service for air traffic controllers. The errors identified from the above analyses were classified into two categories [2]:

- error of omission: where an action is omitted from the required sequence;
- error of commission: where a required action is incorrectly executed or an extraneous action is inserted in the required sequence.

Error of commission can take on the following forms: choice of a non-required action, repetition of an action, inversion of two actions of a same procedure, and execution of an action outside the allotted time or not in the required conditions.

The information obtained from the above analyses is used for implementing mechanisms of operator error tolerance.

# 4 Operator Error Tolerant Design

The goal of operator error tolerance is to avoid system failure even if an operator error is present. According to the concept of dependability impairments [3], the creation mechanisms of system failure can be expressed by the following chain:

fault → error → failure

An error is that part of the system state which is liable to lead to subsequent failure. A fault is the cause of an error. By propagating through the man-machine interface, an operator error committed during system-user interaction creates technical system errors which can lead to a system failure. To avoid that operator error leads to a system failure, such propagation of operator error must be prevented. This can be achieved by designing a dependable man-machine interface whose role is to place barriers of error confinement between the operator and the technical system. Based on this observation, operator error tolerance can be achieved through three phases:

- **operator error detection**, which allows to identify an operator error before it creates technical system errors;
- **technical system error detection**, which allows to detect an erroneous state of technical system caused by an operator error that has not been detected;
- **error recovery**, where an error-free state of technical system is substituted for the erroneous one.

The last two phases are only used when the first phase is failed. In the remainder of this section, we develop these different aspects.

## 4.1 Operator Error Detection

Operator error detection consists in detecting the erroneous actions of the operator before they are executed by the technical system. Its goal is to confine operator error in the man-machine interface. Operator error detection is particularly important for irreversible actions, because the erroneous state of the technical system caused by this kind of action cannot be replaced by an error-free state through reverse actions.

The operator error detection relies on the identification of the mismatches between the state to be achieved by the current action and acceptable states which can be determined according to different criteria. Such identification can be achieved using several reference models which provide the knowledge about the system and the operator's activities. An erroneous action may be detected by analyzing the pertinence of the states expected from these models [1]. In our present work, a system model and a task model were used for detecting operator error.

The system model is used for determining the evolution of system state according to the operator's actions and other events. It contains the information relating to the system structure, interaction modes between its components, transition conditions of system state, and failure model of each component. This model takes into account both hardware and software components.

The system model of the ATC system was implemented using the STATEMATE tool [4]. This tool offers facilities for modeling and simulating complex systems. The ATC system was modeled using a combination of graphical languages and high-level programming languages. The graphical languages are an extension of state machines. They support the hierarchical modeling and the representation of concurrent states. These possibilities allow to reduce the number of states to be considered when modeled systems are complex. Thanks to the simulation facilities of this tool, the operator's working environment can be simulated by programming expected scenarios which contain different operational procedures and failure events. This possibility offers an experimental environment to test and validate proposed mechanisms of operator error tolerance.

The task model is a representation of the operator's tasks. It describes when and how to carry out different operator's tasks. The structure of the task model is hierarchic. The top level of the model represents operator's tasks. Each task can be decomposed into subtasks and each subtask is defined as a sequence of operator's actions with constraints. A task is characterized by its execution prerequisites and its state. The execution prerequisites of a task specify system states in which the task can be started by the operator. A task can be in an active state or in an inactive state. Initially, all tasks are in an inactive state. A task is in an active state when it has been started by the operator, while it returns to an inactive state when all actions of this task have been performed. Several tasks can be in an active state at the same time. Each action of tasks is associated with several attributes: reversibility, execution conditions, maximum time authorized for its achievement, and the severity of its consequences. The execution conditions of an action specify system states and resources needed for its execution. The information relating to the maximum time authorized for the achievement of each action is used for detecting actions which are not performed in the allotted time.

An action is judged correct if it is performed in the allotted time and meets one of the following conditions:

a) the action is the next step of an active task and its execution conditions are satisfied;
b) the action is the first step of an inactive task and the execution prerequisites of this task are met.

An action which does not meet one of these two conditions is an unforeseen action. It requires further analysis when the man-machine interface cannot determine if it is an operator error or an unpredictable action which has no undesirable consequences. The further analysis can be performed through the cooperation between the operator and the system via the man-machine interface. The man-machine interface has to inform the operator about the consequences of the action, and the operator decides if his action is erroneous or not.

After detecting an operator error, the man-machine interface should inform the operator about its causes and its nature, in order to prevent the operator from doing the same error again and to update the operator's mental representation about the system state.

Operator error can be also detected by the operator. Interface design should improve the capacity of operator's self detection. Such improvement can be obtained by providing the sufficient feedback about the effects of the operator's actions and by allowing the operator to simulate the consequences of his actions.

## 4.2 Technical System Error Detection

Generally, the total detection of operator error cannot be assured. In some circumstances, the time needed to carry out an operator error detection may be too long for urgent actions. If these urgent actions are reversible, the phase of operator error detection for this kind of action may be skipped. Therefore, technical system error detection should be used to identify an erroneous system state due to an undetected operator error or due to an urgent action for which operator error detection has been omitted. Several techniques already exist for detecting technical system error. In general, the application of these techniques is independent of the cause of technical system error. Therefore, they can be also applied to technical system error created by operator error. Some of the most frequently used techniques are [5]:

- timing checks, which are used to verify if the timing constraints of a component in the technical system are being met or not;
- reasonableness checks, which determine if the state of some components in the technical system is acceptable or not;
- diagnostics checks, which use different checks on some components of the technical system to verify if these components are working correctly.

After detecting a technical system error, its causes should be determined. The system should provide the operator with sufficient information so that the operator can determine whether technical system error is due to his erroneous action or due to a technical system fault. In the second case, the operator should perform necessary maintenance tasks to prevent the technical system fault from being activated again.

## 4.3 Error Recovery

Once a technical system error has been detected, it must be removed from the system. Error recovery consists in substituting a normal state of the technical system for an erroneous one before the system fails. Three techniques can be used for error recovery: backward recovery, forward recovery and error compensation [3].

Backward recovery consists of bringing the technical system back to an error-free state already occupied before the occurrence of technical system error. Backward recovery can be carried out automatically by the technical system or manually by the operator. To manually achieve backward recovery, the operator should have means of making the effects of his actions reversible. This can be achieved by providing the operator with the Undo command. This command is used to cancel the effects of an action which has been executed. Implementation of the Undo command requires that the state of technical systems be periodically checkpointed on some storage that is not affected by failure.

Forward recovery consists of finding a new state from which the system can operate. In the ATC system, forward recovery can be done through the reinitialization of failed components.

When error recovery is carried out by the operator, the system should provide the detailed instruction relating to error-recovery strategies.

In the technique of error compensation, the erroneous state contains enough redundancy to enable its transformation into an error-free state. The most frequently employed technique of error compensation is triple modular redundancy (TMR). In TMR, the processing unit is triplicated, and all three units work in parallel. Such a form of redundancy cannot recover the technical system error created by operator error if the operator simultaneously acts on all three units. In fact, operator error affects all three units in the same way (common-mode failure). Let us now consider the following form of redundancy: several units are in active redundancy, but only one unit has an active output at a time. If the operator can independently act on each of them, then operator error can be tolerated. Indeed, if a technical system error created by an operator error is detected in the active unit, then it is disconnected and a backup which is not affected by operator error becomes the active one to support system service. The redundancy implemented in the ATC system belongs to this category. Therefore, it can recover technical system errors created by operator error.

# 5 Conclusion

This paper has presented some concepts for the design of operator error tolerant systems. We began by analyzing the on-line maintenance activity of the operator in an air control system, in order to determine different types of operator errors. A method for tolerating operator error was proposed. It contains three phases: operator error detection, technical system error detection and error recovery. An interface prototype for the on-line maintenance of the air traffic control system is now under development based on the proposed method. This interface prototype will serve as a platform for testing and evaluating the method.

## References

1. Rouse W.B, Morris N.M. Conceptual Design of a Human Error Tolerant Interface for Complex Engineering System. Automatica 1987; vol. 23, no. 2, pp. 231-235
2. Swain A.D. Human factor associated with prescribed action links. Sand 74-0051, Sandia Laboratories, USA, 1974
3. Laprie J.C (Ed). Dependability: Basic Concepts and Terminology in English, French, German, Italian and Japanese, Dependable Computing and Fault Tolerance, vol. 5, Vienne, Autriche, Springer-Verlag, 1992
4. Harel D, Lachover H, Naamad A et al. STATEMATE: A Working Environment for the Development of Complex Reactive Systems. IEEE Transactions on Software Engineering 1990; vol. SE-16, no. 4, pp. 403-414
5. Anderson T, Lee P.A. Fault Tolerance Principles and Practice. Englewood Cliffs, NJ: Prentice Hall, 1981

# Dependable Systems : Error Tolerance and Man-Machine Cooperation

Corinne Mazet, Hubert Guillermain
Technicatome/LIS,
LAAS-CNRS, Toulouse, France

**Abstract**

Technological progress has modified dependability issues by making human errors more visible and critical. To cope with these errors an obvious approach is to insert protections in systems. However, these protections should not be applied locally, as we may run the risk of disrupting cooperation between the components (human, technical) of systems and weakening the overall avoidance of disasters. Within the framework of dependability, this paper tends to present issues on human error tolerant systems and to relate them to man-machine cooperation. Principles of cooperation for error tolerance are outlined.

## 1 Introduction

Advances in the field of hardware and software dependability have made human errors more visible. At present such errors rank first among all causes of aircraft crashes [1] and account for eighty percent of all failures of technical systems [2]. Human errors are also more critical as a result of the paradoxical consequences of technological progress: a) the systems are more efficient but also more complex to handle for the operators whose tasks are increasingly cognitive; b) systems are more dependable and induce a reduction in the perception of technological risks by the operators [3]. Thus, the operators have a tendency to work at the limit of performance with increasingly thin safety margins which reduce the possibilities of action in case of failure. The operators do not make more errors however. In fact, they make less but these errors are more difficult to overcome. Accordingly, operator error may quickly propagate through the system and lead to a catastrophic failure.

Concerning dependability, it seems as if the gains obtained on the technical side were compensated by a system increasingly sensitive to the risk of catastrophic operator errors [2]. Many errors are eliminated by traditional methods (mainly prevention) and traditional protections (interlock system to prevent human error). However, when applied locally these methods tend to disrupt cooperation between the system components, such a cooperation remaining essential however to ensure the overall avoidance of disasters. Consequently, the benefit to be derived from operator intervention reacting positively to unexpected situations would be lost as he would be prevented from performing the corrective actions needed to avoid a technical failure, to limit its consequences or to reestablish the correct functioning [4]. To avoid these dependability impairments, the problems have to be addressed at the level of the whole socio-technical system including the operator(s) and the technical system. This approach advantageously uses the complementarities between the technical and the human components, places the emphasis on the positive role played by the operator with respect to dependability (and not only on his proneness to make errors) and takes into account the impact of cooperation between human and the technical systems to avoid the catastrophic failures. Thus, we can expect significant dependability gains as derived from Error Tolerance (Fault Tolerance in the community from which it is derived, e.g., see [1]).

In what follows the paper identifies the concept of fault tolerance within the framework of dependability, then presents issues on human error tolerant systems related to man-machine cooperation. Finally, practical principles of cooperation for error tolerance are outlined.

## 2. Fault tolerance within the dependability framework

In the field of Human Factors, dependability achievement is mostly limited to error prevention (reduction of the error causes), although it is impossible to eliminate completely human errors and sometimes not desirable. As errors are rare, the operator may tend to build-up the feeling that they cannot happen. Familiarity to error-free situations may distort his error perception and weaken his natural defenses to cope with the unavoidable errors, correct them and minimize their consequences.

Another way to achieve dependability is to cope with the human errors before their undesirable consequences affect people and environment. Consequences of human errors can be minimized, suppressed or avoided by tolerance. According to the unified conceptual framework for dependability [1] the goal of fault tolerance is to avoid that an error leads to a failure, i.e., to break the following causal chain of impairments to dependability:

...failure -> fault -> error -> failure -> fault ->...

According to this scheme: a) the fault is viewed as the cause for an error, prevention aiming at the elimination of fault; b) the failure is the consequence of the error, that tolerance deals with.

Fault tolerance relies on error detection and error recovery. Various techniques can be used for error detection based on redundancy (e.g., definition of error confinement areas, replicated processing or coding of information). Once the error has been detected, recovery can occur according to one of the following three forms: a) backward recovery, in which the system is brought back to an error-free configuration, already occupied prior to error occurrence; b) forward recovery, in which the system is brought to a new state (frequently in a degraded mode) from which it can operate; c) error compensation where the erroneous state contains enough redundancy to enable the delivery of an error-free service from the erroneous (internal) state.

These mechanisms are classically implemented to protect the technical system against its own failures due to internal causes (e.g., programming error) or external ones (physical perturbations, human aggression). Concerning human actions, the application of these mechanisms is confined to actions likely to affect the security or confidentiality of the technical system (unauthorized access). Local defenses are set up but they may turn out to be incompatible with certain useful human actions and with the dependability of the whole system. These limits tend to question the current protection principles in order to make them more flexible to accommodate positive human actions for dependability.

# 3.Man-machine relationship for error tolerance

A simple form of human error tolerant application may be found in the field of office automation. Most text editors contain spelling utilities allowing on-line detection of errors by the system and proposing many suggestions of recovery to be performed by the operator. Most recent programs offer to the operator the opportunities to customize the spelling utilities allowing for an automatic tolerance of his most frequent errors of execution. This kind of logic may be extended to the automatic tolerance of some erroneous actions performed on the interface (mainly erroneous commands) with some precautions on customization. Concerning the efficiency in performing the tasks, benefits are obvious: to cope with human weakness without interfering in his action, no time wasted for the phase of human recovery. Moreover, this solution may be helpful in case of urgent actions which execution can not be delayed. Nevertheless, some risks are associated that may lead to more critical errors. The system substitutes the operator for the correct execution of the action, hence the risk is that the operator keeps making errors while he forgets his proneness to errors. Paradoxically, this promotes opportunities to produce errors with more detrimental consequences when the operator leads to interact with a system less efficient on human error tolerance. For this reason, it may be more adequate to develop human error tolerant systems that keep the operator in the loop of tolerance. This is also a means to account for errrors of intention that cannot be tolerated by means of a customized interface. In the sequel of this section, we first characterize man-machine cooperation for error tolerance and then we introduce some related practical isssues.

## 3.1. Collaboration or Cooperation for Human Error Tolerance

In the field of human error tolerant systems, two approaches are developed; they underline different facets of man-machine relationships, respectively centered on *collaboration* and *cooperation*. These concepts mainly developped in the domain of human interactions may be extented to the analysis of dependable man-machine relationships.

*3.1.1. Collaborative approach.* According to this approach, the technical system is designed to improve the human proficiency in detecting and recovering his own errors [5, 6]. It is meant to assist the operator in realizing correctly his actions and achieving his specific objectives. A situation of collaboration is defined, where the system does not replace the operator to perform his tasks but rather provides some assistance (e.g., via environmental resources) so that the tasks can be carried out correctly (without errors). The operator is fully responsible for tolerating his errors occurring during the execution of the tasks; responsibility is not shared by the system. To define a "kind" environment allowing for error tolerance by the operator (self-detection and self-recovery), the design of the system has to respect the following two principles.

The first one, based on *observability*, requires that the system emphasizes the information feedback about the potential effect of the human action, to make his own error obvious to the operator. This can help the operator in self-detecting discrepancies between the expected effects of his action and its actual results. For example, displays based on redundant information strengthen the noticeable effect of the action. Also, the fact that a command has no effects on the controlled process may be used to identify an illegal action (forbidden by the interface). The aim of the observability is to promote early self-detection (close to the inducing action), mainly to avoid propagation and combination of erroneous actions with more critical consequences. Observability is rather relevant for the self-detection of a slip [5]: the action differing from what was planned, its outcome may be compared with the expected plan. In case of mistakes (error of intention), observability is less useful: the operator executing what he has planned, is not able to noticed that the effects of his actions are incompatible with the current state of the system.

The second principle is based on *reversibility* of actions. It allows the operator to recover from his error (self-recovery) by canceling the effect of the action and by resuming to the previous state of the system. Undo commands that invert the effect of the action, are examples of implementation of this principle, as queuing devices that delay the execution of an action (allowing to cancel the action before its execution). Undo commands require a storage of the system state before the execution of the action. In some cases of personnal computer applications, undo facilities can apply to a sequence of actions, considered as a whole by the operator and the system.

Some supporting function integrated into the system, such as simulation (simulation of the action affecting the current state of the system before execution) meets both the principles of observability and reversibility. The operator can test his planned actions and decide to confirm them (or not) regarding their outcomes.

*3.1.2. Cooperative approach.* According to this approach, the system should be designed to tolerate human errors. Many human errors being complex and subtle to process, it is difficult to provide efficient means for an automatic detection and recovery. Tolerant architectures [7] must rely on man-machine cooperation, where the system and the operator are together involved in performing the common goal of tolerance. Toward this end, tolerance mechanisms may be allocated both to the system and to the operator. Successive loops of control have also to be inserted between them: the operator performs the tasks, their execution is controlled by the tolerance mechanisms of the system, the reliability of these mechanisms is controlled by the operator (sort of on-line testing of the tolerance mechanisms).

In the proposed architectures, error detection is based upon the comparison of the inputs, the actions carried out by the operators on the interfaces, with behavioral models of the systems, the tasks, or the environments [7]. Erroneous actions (and sequences of actions) are detected when discrepancies are identified. Such discrepancies are due to human actions or to erroneous models (faulty detection). Often faulty detection concerns operator actions not predicted by the model. Thus, it is necessary to keep the operator in the loop, so as to cancel the faulty detection and prevent inappropriate error recovery. The operator has to cooperate in the error detection because his own information (for example, resulting from communications with operators) that are neither accessible for the system interfaces (no inputs) nor taken into account by the models, may change the context of execution of the action and make the action correct. The problem of faulty detection is not so crucial in case of models accounting for procedural tasks whose actions may be precisely structured (as some tasks in aeronautics). It is more embarrassing for less formalized tasks which favor human initiative and variability in the execution of actions. In the latter case, as the models might be less reliable, their relevance is questionable. From a practical point of view, it should be better to apply solutions derived from the technical dependability: comparison of information between diverse sources (one of which being the operator), consistency or likelihood check to verify a set of data from the operator against those of the technical system, ...

According to error recovery two main principles may be applied.

1) First, the recovery may consist in automatically replacing the erroneous action by a correct one. Above all, it requires to be sure of the efficiency of the detection. No man-machine cooperation is involved.
2) The recovery may be manual (carried out by the operator) when no correcting action is technically available (for example in unexpected situations that the operator has to face), or when the manual correction is necessary for the operator's activity (for example to understand and control the situation).

Finally, the man-machine cooperation for tolerance is also taken into account by the adaptive aiding device included in the system [7]. Such a device is in charge to address messages, allowing to maintain the operator in error detection to decide whether his action has to be corrected. In case of manual error recovery, the messages concern the error detection and may suggest recovery actions from an integrated explanation or recommendation (which can be obtained by synthesis of warnings).

## 3.2. Some principles of cooperation for error tolerance

Three main situations of man-machine relationships involving or not cooperation can be identified. Choosing one of them is function of the action time available after the error detection, the relative cost of the manual or automatic tolerance, the possibility (or not) of the automatic tolerance, the human activity (observability, reversibility), and the possibility to preserve the operator's self-tolerance.

At the first step of automatic tolerance, no man-machine cooperation is needed. Automatic tolerance may be useful for trivial human error (affecting the efficiency of man-machine system without detrimental consequences on dependability). It can also be used for human error whose automatic recovery is reversible. In case of an urgent action, automatic tolerance depends on the reliability of the detection and on the possibility to substitute the erroneous action for a single action (no dependence on contextual effects). To avoid persistence of a human error that may become critical in a specific (rare) context, it may be helpful to inform the operator with the recovery performed so that he can prevent the error from occurring again.

The second step of tolerance implies cooperation and different loops of control between the operator and the system. More or less demand may be put on the operator regarding the mechanisms of tolerance implanted in the system.

1) First of all, the operator may validate the substitution of his action to be made by the system. The system, stopped upon the suspicion of an error, lets the operator deciding whether the undertaken action is correct. This control loop concerns either faulty error detection or faulty error recovery (error detection being appropriate, but the recovery being incorrect and having to be canceled by the operator before new action).
2) Next, the operator may choose between several suggestions of substitution proposed by the system. This allows to limit the disadvantages of faulty recovery, as mentionned above (extra-delay for the canceling recovery and performing a new action).
3) Finally, the operator may be required to confirm the suspicion of an error and to correct his action. The error detection made by the system is not certain and no recovery is valid.

In this third case, the allocation of tolerance mechanisms between the human and the system let explicitly the operator in the loop of error recovery, and the system in the loop of assistance to the operator. In the other two cases, the role of the operator is essentially devoted to verify the relevance of error detection and error recovery carried out by the system.

An important aspect of the man-machine cooperation lies in the aiding device allowing common means and solutions for error tolerance. The aiding device consists essentially in two explanation functions oriented to the operator. The first explanation function aims at clarifying the actions made by the system, their consequences as well as those of the operator actions. Thus, the operator is able to decide whether he maintains or changes his actions. The second explanation function allows a deeper dialogue with the operator (requests for intention or goals, ...) especially useful when the error detection or error recovery is more complex.

The last step of cooperation for error tolerance occurs when the human erroneous action propagates to the system, i.e., when the mechanisms of human error tolerance and the control on these mechanisms are both failing. Thus, techniques allowing to minimize the consequences of human actions on the system state are necessary. Classic mechanisms of tolerance may be used to protect the system from an internal error. In some cases, man-machine cooperation may be required: the operator aiming at suppressing, minimizing, containing the consequences of his error, e.g., tolerating the error of the system with the help of additional diagnosis devices.

# 4. Conclusion

Within the framework of dependability, fault tolerance aims at minimizing the failure proneness of systems by enabling them to operate despite the presence of errors. After some major progress obtained with respect to hardware and software faults tolerance is now faced with the delicate issue of operator errors. Given the characteristics of human behavior, new challenges emerge to develop an approach devoted to human error tolerance centered on the man-machine cooperation. The human error tolerant systems attach more or less importance to cooperation. We attempted to show the different kinds of cooperation taken into account. We outlined some practical principles of cooperation for error tolerance that imply different levels of human involvement in error tolerance. The operator is devoted to the control of the tolerance mechanisms implemented in the system or involved in error recovery.

# Bibliography


1. Laprie J.C. Dependable Computing: Concepts, Limits, Challenges. FTCS-25 Silver Jubilee, Special Issue, IEEE Computer Society Press, 1995

2. Amalberti, R. Paradoxes on Safety of Large Risk-related Systems Performances Humaines & Techniques, N° 78, 45-54, 1995, (in French).

3. Hollnagel, E. Dependability: Human Reliability Analysis Context and Control, Computers and People Series (B.R. Gaines & A. Monk Eds.), London, England, Academic Press, 1993

4. Guillermain, H. Human Factors in Socio-technical System Dependability. Rapport Technicatome DI/SCE 95/01551, 1995, (in French).

5. Lewis N.G & Norman, D.A. Designing for Error in User Centered System Design: New Perspectives on Human-Computer Interaction, D.A Norman & S.W. Draper Eds. Hillsdale, NJ, USA, Lawrence Erlbaum Ass., 1986.

6. Rasmussen, J. Design for Error Tolerance, Trans. of the American Nuclear Society, vol.45, pp. 358-359, 1983.

7. Rouse W.B. & Morris N.M. Conceptual Design of a Human Error Tolerant Interface for Complex Engineering System, Automatica, vol 23, n°2, pp 231-235, 1987.

# Implementation of Cause-Based Pilot Model for Dynamic Analysis of Approach-to-Landing Procedure: Application of Human Reliability to Civil Aviation

A. Macwan*, J.F.T. Bos#, J.S. Hooijer#

* Delft UT, Faculty of Mechanical Engineering and Marine Technology, Mekelweg 2, 2628 CD Delft, The Netherlands

\# National Aerospace Laboratory NLR, P.O. Box 90502, 1006 BM Amsterdam, The Netherlands

**Abstract**

This paper presents a pilot study about the applicability of Human Reliability Analysis (HRA) in civil aviation. Implementation of a cause-based model of intentional pilot errors in dynamic analysis of Approach-To-Landing procedure is described. The pilot model focuses on errors as deviations from procedures, and was implemented in DYLAM. The results show that the HITLINE methodology is a suitable tool to perform dynamic HRA in the civil aviation domain.

## 1 Introduction

Safety is an important issue in civil aviation. The public experiences risk in civil aviation by the absolute number of accidents rather than by the accident ratio, defined as the number of accidents per flight. The accident ratio seems to have reached a stable level in the last years [1]. The expectation is that the number of aircraft flights will double in the near future, implying that the accident ratio must halve to maintain the same public experience of safety. The current safety enhancement will not be sufficient to achieve the desired level of safety.

About 70% of all accidents occur in the approach, landing, take-off or initial climb phases of flight. In these phases both the aircraft crew and the air traffic controller play an important role. It is therefore essential to have a very clear understanding of the human causes of the failures. There is a need to assess and quantify the human factor in safety analysis. However, yet insufficient knowledge exists on the human factor in the safety assessment in civil aviation.

The purpose of this paper is to describe a pilot study about the practical applicability of HRA in civil aviation. The depth of the investigation is limited to an existing framework of dynamical analysis developed by the Joint Research Centre (JRC) [2]. This paper focuses on the development and implementation of a cause-based model of intentional errors for the dynamic analysis of an Approach-To-Landing procedure.

# 2 Context of the study

The specific ATL procedure used in this study was the same one used in an earlier study by [3]. The first part of the ATL was modelled, starting with the descent from cruise level to the passing of the Novarra VOR at 4000 ft.
The basic steps of the procedure considered here are:
* F1/5/10: extension of flaps to settings 1, 5 or 10 respectively.
* CA: calibration of the altimeter on QNH (actual air pressure at sea level).
* LE: levelling off at 4000 ft.
* CL: reading of the approach checklist, checking the calibration and flap setting.

Use was made of DYLAM as the simulation engine, and of the pilot model developed by JRC, describing the skill-based behaviour, i.e. the actual flying of the aircraft. A deterministic simulation model of the aircraft determined the position and flight conditions of the airplane at all times, with four degrees of freedom.
For details of the context of the study see [3,4].

# 3 Development of the pilot model

The Pilot model from the JRC exercise was extended to include intentional and additional unintentional errors. A CAuse-based Behavioural model for Flight procedures (CAB-Flight) was developed and incorporated in the JRC simulation to simulate these errors. CAB-Flight is a rule-based model describing the pilot's behaviour during the correct and erroneous execution of procedural steps of the ATL. The model is based on the intentional error model of the HITLINE methodology [5,6].

An error in CAB-Flight is defined as a deviation from a procedure. The following error expressions are identified as part of CAB-Flight for this study:
* **unintentional omission** (OM): The pilot unintentionally does not execute an action.
* **unintentional commission** (UC): The pilot unintentionally executes a procedural step in an incorrect way.
* **intentional commission** (IC): The pilot intentionally deviates from the procedure by performing a different action, because of his perception of what should be done in the current situation.
* **procedural action** (PR): Normally the steps of a procedure are linked to specific conditions to perform it. The procedural action occurs when the pilot decides to perform another step of the procedure first.
* **not monitoring** (NM): The pilot fails to monitor the aircraft variables in order to adjust them. This is not a deviation from a procedure but rather a failure at the skill-based level.

HITLINE assumes that not all errors can occur at all times. Thus, for each procedural action, likely errors are identified by considering the procedural instruction within the context of flight. The results for two steps of the procedure are presented in table 1. It should be emphasised that the principle of such mapping is important. Exact mapping may be different from the one presented here. One of the limitations of the dynamic simulation techniques is the very large size of the resulting dynamic event tree. Thus, mapping selected errors for each procedural step helps to reduce the size of the results.

Table 1. Possible errors.

| **procedural step** | **description procedural step** | **possible errors** | **description errors** |
|---|---|---|---|
| CA | calibration of altimeter, setting to local air pressure | OM | not performing calibration |
| | | UC | wrong air pressure |
| | | PR: F1/5/10 | flap setting |
| LE | levelling flight | OM | not levelling flight |
| | | PR:CL,F1 | performing another action |

The skill-based level error of not monitoring can occur at every step. It is regarded as temporarily not addressing the tasks of that level. The only possibility for recovery from errors in the model is successful reading of the checklist. Other possibilities for recovery have been left out for reasons of simplicity.

The causal factors affecting errors are modeled using the so-called performance influencing factors (PIFs). These factors are used both qualitatively (to identify likely errors from the set of possible errors) and quantitatively (to estimate relative likelihood for each error). Three categories of PIF are distinguished, related to aircraft, procedure and pilot. Additionally, two types of PIFs are distinguished:
**scenario-independent**: these PIFs do not change significantly during the development of a scenario. An example is training and experience of the crew.
**scenario-dependent**: these PIFs change during the development of a scenario. Examples include values of parameters and perceived importance of a system.

From an exhaustive list of PIFs, collected from human reliability models such as THERP, evolutionary models, as well as aviation literature and pilot interviews, a small but relevant set of PIFs was selected for implementation in CAB-Flight.

The essential modelling element of CAB-Flight is formed by the relations between PIFs and errors. These relations are represented in mapping tables, which show the generic and specific rules to generate errors as a function of the PIFs. Table 2 shows an example of a mapping table.

Table 2. Mapping table for scenario-dependent PIFs, pilot related.

| Factor | Mechanism | Likely Error |
|---|---|---|
| pilot's diagnosis | confusion when something is wrong | more likely to deviate from procedure |
| perceived importance of certain aspect | fixation on particular aspect / channelling of attention | * unintentional omission<br>* unintentional commission<br>* not monitoring<br>* procedural error |
| interpretation distance to runway | misdiagnosis | * intentional commission<br>* not monitoring |
| mismatch between pilot's expectation and aircraft behaviour | confusion / lack of confidence | more likely to follow procedure |

Relative likelihoods associated with possible pilot actions, including successful execution of procedural steps and possible errors, are estimated as a function of PIFs. For lack of data which is a common problem in human reliability, use is made of subjective judgement to assign numbers to these likelihoods. Parameters used for quantification are estimated independently for different combinations of PIFs. For simplicity, each PIF is treated in a binary fashion, e.g. experience is considered in terms of experienced or novice crew. For a detailed discussion of the quantification scheme, the reader is referred to [4,6].

When CAB-Flight is linked with a dynamic simulation tool, the error generation model provides simulation of errors for each interaction between pilot and aircraft. Additionally, this model carries variables which account for dependency of pilot action at different times. Three cognitive processes are simulated through the use of mapping tables: (1) check behaviour of aircraft variables against expected trends, (2) determine global diagnosis of aircraft status, (3) formulate expectations about behaviour of variables. The results of the cognitive processes are translated into PIF values which are subsequently used to determine possible actions with associated probabilities as mentioned above. This information is then used by the simulation engine to generate possible branches for the dynamic event tree.

# 4 Dynamic analysis results

CAB-Flight was incorporated into the pilot model of the JRC simulator. Additional "failure states" were assigned to simulate the different errors. Associated

probabilities were generated prior to simulation and written in appropriate data files to be used by DYLAM. It should be noted that whereas CAB-Flight is capable of dynamically calculating probabilities, the scheme was not included in the current version of DYLAM. This has been done in a similar, independent application to dynamic study of nuclear power plants [7].
A dynamic event tree is presented in Fig. 1 which includes the following errors: Intentional Commission (IC), more thrust; Procedural action (PR), calibration of altimeter at F5; Unintentional Commission (UC) (flaps to 20 instead of 10) and omission of checklist reading (OM). A hardware failure is also included in the form of flap failure.

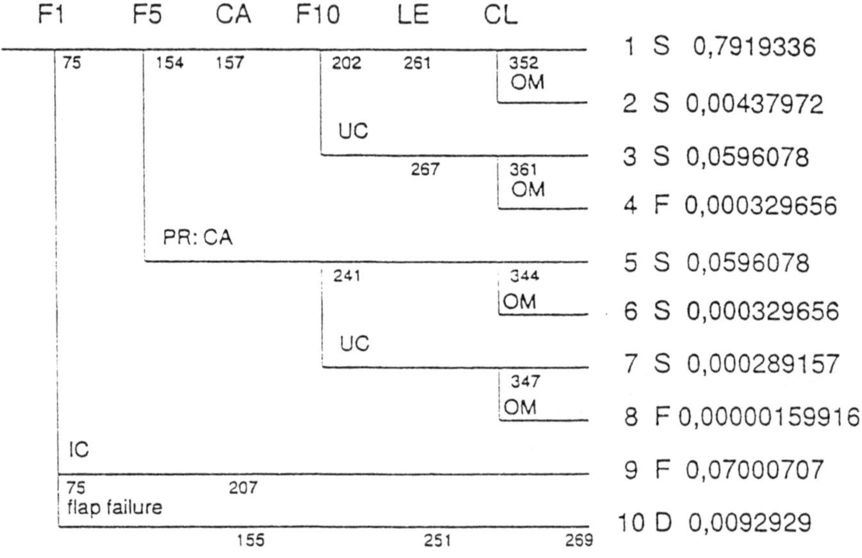


Figure 1 (Dynamic event tree for simulation I)

A comparison with the JRC analysis shows that success probability is 10-40% smaller, which is to be expected since more errors have been introduced to the pilot model of the JRC simulation.

# 6 Conclusions and recommendations

The results show the feasibility of dynamic human reliability study to analyse pilot errors within the context of aircraft procedures. Moreover, the results show that the HITLINE methodology is a suitable tool to perform dynamic HRA in the civil aviation domain. The benefits of a dynamic HRA for civil aviation are to be expected in the qualitative analysis. It will be useful in the design of procedures, training and man-machine interfaces.

Currently, due to lack of useful data, the quantitative side of the analysis is still a weak point. Still the method itself is simple to use. The determination of the error probabilities is traceable for the user. More PIFs can be added without the need to change the calculation method. The advantage of this method is that all probabilities are conditional, depending on the dynamics of the pilot-aircraft system. This makes it possible to take those dynamics into account.

Future research has to consider the operator model development in more detail, like which PIFs needs to be taken into account. Sensitivity analysis must be part of the model development. In addition, model validation is important.

Future research should also help to determine whether *dynamic* analysis for safety assessment will be cost effective in the design of new aircraft.

# Acknowledgement

The authors would like to thank Frederiek Peek for conducting the research with great enthusiasm; Carlo Cacciabue and Giacomo Cojazzi for making it possible for Frederiek to perform part of the research at JRC (Ispra) and providing guidance during her graduation work; and Peter Wieringa for providing valuable remarks and suggestions.

# References

1. Piers et al. The development of a method for the analysis of societal and individual risk due to aircraft accidents in the vicinity of airports. NLR report CR 93372 L, 1993
2. Cacciabue P.C., Cojazzi G. An integrated simulation approach for the analysis of pilot-aeroplane interaction. Control Engineering Practice, vol. 3, no. 2, 1995.
3. Mancini S. Applicazione di modelli per la valutazione dell'affidabilita umana al comportamento di un pilota di un velivo civile. Tesi di Laurea in Ingeneria Aeronautica, Politecno di Milano, Facolta di Ingegneria, 1991.
4. Peek F. Development and implementation of CAB-Flight. NLR memorandum IW-95-013.
5. Macwan A. Methodology for analysis of operator errors during nuclear power plants accidents with application to probabilistic risk assessment. PhD thesis, University of Maryland, 1992.
6. Macwan A., Mosleh A. A methodology for modelling operator errors of commission in probabilistic risk assessment. Reliability Engineering and System safety 45, 1994, pp. 139-157.
7. Groen F. CAB-SIM: a cause based operator model for use in dynamic probabilistic safety assessment for nuclear power plants. Graduation report A-713, Faculty of mechanical engineering and marine technology, Delft UT, Netherlands, 1995

# Dynamic Modelling of Process Equipment in Regularity Analysis

Ivar Skjeldal, M.Sc., Lars H. Katteland, M.Sc
Dovre Safetec AS, P.O.Box 77, N-4001 Stavanger
Norway

**Abstract**

Regularity analyses are currently being used extensively in several stages during offshore oil/gas field developments. As regularity may be translated to profitability, it is vital to predict the regularity as accurately as possible. In gas processing systems, process equipment are often characterised by dynamic behaviour, i.e. there is a non-linearity in the production flow of parallel equipment stages. This paper evaluates solutions of dynamic modelling as a contrast to linear modelling, showing large potential by increasing the accuracy of regularity modelling. Thereby there will also be an increase in the confidence of the decision basis upon which important project decisions may be made.

# 1 Introduction

The extensive use of regularity studies in offshore gas/field projects show the importance of such studies. They form an essential input to the basis for project decisions like production planning and technical considerations.

## 1.1 Measure of production risk

Regularity analyses are used in several stages of offshore projects. During the feasibility phase, such analyses can be used to screen and define field development alternatives, and also further optimisation of the chosen alternative. In the production preparation phase they can be used to provide decision-support for the production planning and optimisation of operational philosophies.

In general, regularity analyses are a form of risk analysis in the sense that they focus on the risk of production loss, i.e. the probability of experiencing production loss and also expected loss of volumes. The risk of production loss is attributed to component failures and repair/maintenance and operational conditions like process shutdown and start-up philosophy. Also the contribution from accidental events (fires, explosions, etc.) is often significant and should be included in these analyses.

MIRIAM is a computerised tool for evaluating operational performance of continuous process plants in terms of equipment availability, production capability and maintenance resource requirements. The program is developed by Statoil/EDS and is under continuous development. MIRIAM is based on a generalised flow network algorithm (multistate statistical theory) [1] and Monte Carlo simulation [2], and may theoretically be used for studies covering a range of complexity levels, from high level facility evaluation to advanced studies including complex maintenance and operational considerations.

## 1.2 Dynamic Behaviour of Process Equipment

In gas processing systems, the process equipment are often characterised by a dynamic behaviour. This is due to the compressibility nature of the gas, i.e. the non-linearity between the pressure and flow rate. This has a significant effect on the process capacity of parallel equipment trains since the process capacity of each train is dependent on the number of operating parallel trains. In other words, if a compression stage consists of say four parallel compressors having a capacity of 4 * 25%, the capacity is typically 3 * 30% in case one train fails.

Previously a linear approach has been used to model the process capacity, mainly due to the fact that there is no direct way to model dynamic behaviour of parallel process equipment within MIRIAM [1]. Linear approach means that loss of parallel trains is associated with linear loss in the process capacity. The effect of such an approach is that the results will be either conservative or optimistic, i.e. lesser/ larger than the "real" value.

## 1.3 Objectives

The objectives with this paper are to show the potential within MIRIAM to facilitate dynamic behaviour of process equipment, and provide solutions to dynamic modelling. Further, the intention is to quantify the difference between linear and dynamic modelling for a specific gas processing system.

## 1.4 Terms

The following terms are applied in this paper:

- *Regularity:*
  The capability of the system to fulfill some given performance targets. May be measured as *production availability* (the ratio of real production to planned production volumes), *demand availability* (the percentage of time that the system produces 100% of the planned production rate) or *on-stream availability* (the percentage of time that production in the system is larger than zero).
- *Monte Carlo simulation:*
  Simulation using the next event technique, see [2] for further description.

- *Multistate models:*
  A component or a system may have a finite number of states and therefore specific algorithms must be used for regularity prediction [3].

## 2 MIRIAM Regularity Model

The model used for the purpose of evaluating the effect of including the dynamic behaviour of process equipment, is a simplified version of a typical gas processing plant. Figure 1 shows the main elements of this model.

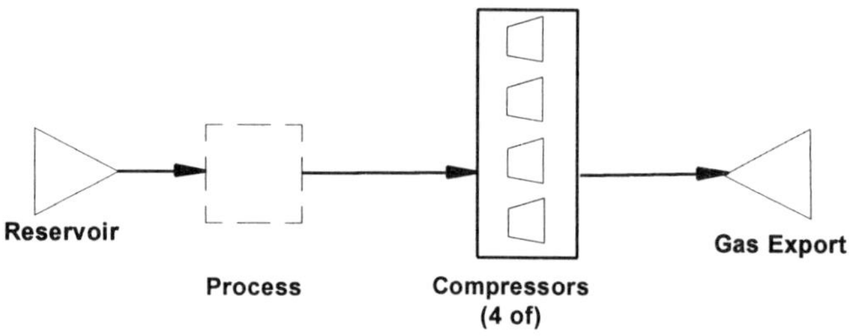


**Figure 1** Main elements of gas processing model used in the simulation

The two boundary points (reservoir and gas export) represent the entry and discharge points of the model. The stage "Process" is a super-component for the remaining process equipment in the system but the compression stage, which consists of four parallel compression trains.

All input data (failure and repair data) are based on typical values used in gas regularity analyses and are assumed to give a realistic picture of a gas processing system. With the data used for compressors, statistical analysis show that the probability of having a state where less than three units are running is less than 0,0008, which corresponds to about 7 hours per year. Therefore, it is considered as relevant in the further work to include only the states of three or four units running.

The test cases developed to illustrate the different dynamic modelling techniques are based on the model as shown in Figure 1. All failure and repair data of the elements are unchanged, however the capacity of each compression train changes. From the design specifications, it is given that the capacity of the compressors are 4 $*$ 26 $MSm^3/d$[1]. If one train fails, the capacity is then 3 $*$ 31 $MSm^3/d$. The gas offtake rate is set to 100 $MSm^3/d$.

In total it is considered four different approaches to the modelling of parallel process equipment. These cases are described in Table 1.

[1] $MSm^3/d$ = million standard cubic metres per day

**Table 1** Description of test cases for modelling of parallel process equipment

| No. | Case Name | Description | Comments |
|---|---|---|---|
| 1a | Linear (con.) | Compression capacity of 4 * 26 | Conservative approach |
| 1b | Linear (opt.) | Compression capacity of 4 * 31 | Optimistic approach |
| 2 | Linear with maximum | Compression capacity of 4 * 31, with maximum process capacity 102 | Correct in case compression stage does not decide maximum capacity. |
| 3a | Dynamic (detailed) | See Figure 2 below, but separate dummy stages for each compressor | Accurate dynamic modelling, increases the model significantly |
| 3b | Dynamic (super) | As Figure 2 below with one super-component for all compressors | Minor increase of model size |
| 4 | Multi-element | Each capacity level modelled by a parallel super-stage, i.e. two stages in parallel consisting of super-stages (one 4 * 26 and one 3 * 31) | Violating stochastic integrity by introducing more equipment than within system design |

The following Figure 2 shows the dynamic modelling approach which is developed. By introducing operational elements and rules, it is possible to model dynamic behaviour of process equipment. The capacities are represented by operational elements (Normal dummy/failure dummy). Flow allocation rules are applied to direct the production flow during normal and failure conditions according to predefined priorities and specified shutdown rules.

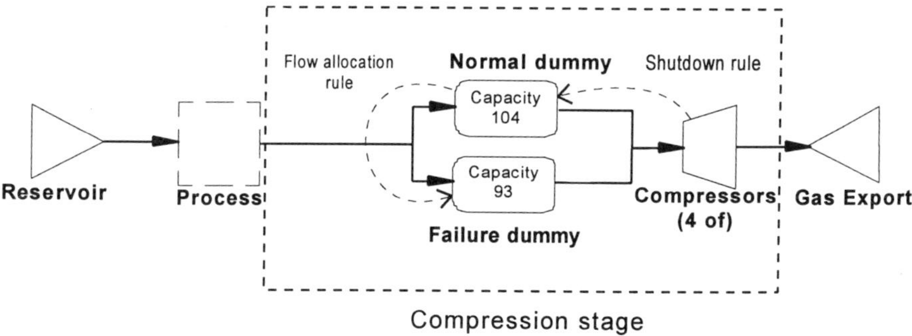


**Figure 2** Dynamic modelling approach of compression stage

During normal operation, the flow is routed through the "Normal dummy". Upon failure in a compression train, the "Normal dummy" is shut down and the flow is routed through "Failure dummy". Thus, the capacity of the compression stage is 4 * 26 = 104 during normal operation, and 3 * 31 = 93 in the state with one failed compressor.

# 3 Simulation Results

For each of the cases as described in Table 1 it has been developed separate model in MIRIAM. The models have been simulated and the results are given in Table 2 below.

**Table 2** Simulation results for the test cases

| No. | Case Name | Production availability (%) | Average production loss ($MSm^3$) |
|---|---|---|---|
| 1a | Linear (conservative) | 97,2 | 1020 |
| 1b | Linear (optimistic) | 97,6 | 890 |
| 2 | Linear with max. | 97,6 | 890 |
| 3a | Dynamic (detailed) | 97,5 | 910 |
| 3b | Dynamic (super) | 97,5 | 910 |
| 4 | Multi-element | 97,4 | 950 |

Measuring the production loss by production availability, the differences between the cases are not very large. However, calculation of average production loss gives an impression of the possible economic loss of the apparently small differences. The results of Case 3a and 3b is to be interpreted as the "real" values.

Case 1a and 1b provide the upper and lower bounds for the results as they are the most conservative and optimistic approaches, as described in Table 1. For a larger model than described in this paper, the difference between the approaches may grow to be very large.

Case 2 shows results equal to Case 1b, which is expected due the equality of models. The limitation of Case 2 is that this approach is only valid in those cases when the system capacity is limited by equipment elsewhere than in the compression stage.

Case 3a and 3b produce equal results, which also is expected. The difference between the two models is the effect on large models. The modelling principle of Case 3a will mean a larger increase of the model than 3b, which subsequently means increased model complexity and running time.

The results of Case 4 shows that this approach provides values between the upper and lower bounds, and may therefore be considered reasonable. However, as mentioned in Table 1, the model is based on more stochastic variables than what is in the system. By averaging the simulation results, the distance to the "real" values is not large, but the model is still inaccurate. Further, it will be very hard to calculate the distribution of production loss in the system, i.e. finding the most critical elements with respect to production loss.

In order to obtain accurate results with Monte Carlo simulation a large number of replications has to be run, especially with components of high reliability. An uncertainty evaluation of the results for Case 3b was performed, showing that a large number of replications (above 300) was necessary to obtain equal distribution of production loss between the compressors. Approximately 50 replications were necessary to stabilise the average of the production availability. For small models as the one for Case 3b it is not very time consuming running several hundred replications. However, for a large model the running time may become too extensive (running time longer than 24 hours) for this amount of replications, and therefore the uncertainty must be given due consideration.

# 4 Conclusions

The results of the simulations show a significant potential of improving the accuracy of regularity modelling within MIRIAM. Linear modelling of parallel process equipment in gas processing systems introduces uncertainty in the model. This may be overcome by applying the relevant features of the MIRIAM program, as described in this paper.

The result of dynamic modelling is, as demonstrated in Table 2, an increased accuracy in the results. Another aspect which also is important is the increased confidence in the regularity model that may be gained from dynamic modelling. This confidence is very important to achieve because the regularity model gives a complete risk picture of the total production chain. It is therefore an unique and perhaps invaluable tool for production risk management and also technical/operational optimisations.

## References

1. EDS Ltd. MIRIAM manuals, user guide and technical reference, EDS Ltd, London, 1990
2. Fishman G S. Concepts and methods in discrete event digital simulation, Wiley 1973
3. Aven T. Reliability evaluation of multistate systems of multistate components, IEEE Trans. Reliability, 1985, R-34:473-479

# A Hybrid (Stochastic and Fuzzy) Methodology for Safety and Risk Assessment*

J. A. Cooper
Sandia National Laboratories
Albuquerque, NM USA 87185-0490

**Abstract**

Probabilistic safety assessments (PSAs) and decision-support risk analyses frequently depend on data that are stochastically variable or subjectively uncertain. Probabilistic calculus approaches have been available to process stochastic variability data. There are now approaches using fuzzy mathematics for processing subjective uncertainty data. The incompatibilities in combining these approaches make hybrid analysis challenging. We have now developed a technique for performing hybrid (stochastic along with subjective) safety and risk analysis.

## 1 Introduction

"Uncertainty " is a generic term for lack of precise data inputs to analyses. The causes of uncertainty can be varied, but are usually described in terms of stochastic variability and subjective uncertainty. Arbitrary mixes of these two types of uncertainty are also possible, but most conventional analytical approaches are tailored to one or the other. Probabilistic calculus (or Monte Carlo/Latin Hypercube Sampling) approaches have been available for many years to process stochastic variability data [1]. There are now approaches using fuzzy mathematics for processing subjective uncertainty data [2]. The incompatibilities in combining these approaches to address real-world data characteristics make hybrid analysis challenging. We have now developed a technique for performing hybrid (stochastic along with subjective) safety and risk analysis.

A somewhat focused view of safety analysis is that the important regime of interest in a spectrum of uncertainty is the extreme that indicates maximum potential loss of safety. The challenge is complicated by an accumulating collection of evidence that safety analyses, even those with a focus on extremes, do not reliably enough portray the chance of safety failure commensurate with the frequency of actual occurrences. There is considerable evidence, e.g., [3,4], that conventional methodology does not provide sufficient warnings in a variety of situations, most significantly when the inputs contain a degree of subjectivity or are dependent. This is the motivation that led to an investigation of meth-

* This work was supported by the United States Department of Energy under Contract DE-AC04-94AL85000.

ods for supplementing conventional PSA techniques with the aims of incorporating potential subjectivity and dependence in input information and of carefully studying the processes leading to extreme values in the resultant variations.

## 2 Fuzzy Mathematics

Pure subjective uncertainty necessarily is based on opinion, and since the results of any analysis are only as good as the analysis inputs, the best possible inputs come from estimators who have the most applicable expertise. Our approach allows subjective inputs and processes them by incorporating fuzzy mathematics applied to "crisp" (conventional Boolean) logic constructs.

In contrast to fuzzy logic, fuzzy mathematics treats operands as fuzzy (subjectively known) numbers with uncertainty along the abscissa, and computes in terms of abscissa values rather than ordinate values. The mathematical basis for combining fuzzy numbers is based on Zadeh's Extension Principle [5]. For addition and multiplication (basic to fault tree and event tree computations), this produces:

$$\mu_{A+B}(z) = \bigvee_{z=x+y} (\mu_A(x) \wedge \mu_B(y)) \tag{1}$$

$$\mu_{A\times B}(z) = \bigvee_{z=x\times y} (\mu_A(x) \wedge \mu_B(y)) \tag{2}$$

These are convolutions basically constructed like those used in probabilistic calculus. For logical consistency (and for mathematical operability), fuzzy numbers must be normal (having a maximum membership value of one) and convex (monotonically increasing with increases in abscissa values from zero to one and then monotonically decreasing from one to zero).

Although relations like those in Eqns. 1 and 2 are meaningful mathematically, there are more efficient ways to implement the operations in software [6]. Under the conditions of greatest applicability to PSA, Eqns. 3 and 4 are mathematically equivalent to Eqns. 1 and 2. The mathematical equivalence of Eqns. 2 and 4 depends on restricting Eqn. 4 to nonnegative numbers. For probability numbers, this condition is met; for the more general situation, a more complex form of Eqn. 4 is required.

$$A_\mu + B_\mu = [a_1(\mu) + b_1(\mu), a_2(\mu) + b_2(\mu)] \tag{3}$$

$$A_\mu \times B_\mu = [a_1(\mu) \times b_1(\mu), a_2(\mu) \times b_2(\mu)] \tag{4}$$

where the subscript 1 indicates the smallest abscissa value for a particular $\mu$, the subscript 2 indicates the largest value for a particular $\mu$, and the computation is done at both extremes for each $\mu$.

# 3 Hybrid Constructs

The hybrid technique described in this paper is derived from classical hybrid numbers. Several years ago, Kaufmann and Gupta [6] suggested that stochastic and subjective uncertainty could be considered together but separately in a pair (f, p), where f is a fuzzy number and p is a probability density function (pdf). These pairs, which they termed "hybrid numbers," can be added together by convolving the respective elements according to normal rules for fuzzy arithmetic and probability theory. The Kaufmann/Gupta formulation of hybrid numbers allows addition and subtraction. Addition is defined as:

$$(f_1,p_1) + (f_2,p_2) = (f_1+f_2,p_1+p_2) \tag{5}$$

where the plus signs on the right side of the equation represent fuzzy max-min convolution and ordinary probabilistic sum-product convolution, respectively. This formulation of hybrid numbers does not directly allow multiplication or a full hybrid arithmetic (e.g., the product of a completely fuzzy number and a completely probabilistic number is undefined). Nevertheless, the construct is useful, when modified slightly, as explained below.

## 3.1 Scale Factors

When the values of input variables are not well known, risk analysts may expect to improve their analyses by incorporating new information that is learned through additional tests, accident assessments, etc. In a Bayesian sense, stochastic information can be improved. Since new input data may only slightly improve the stochastic knowledge about ill-defined situations such as abnormal environment responses, a non-Bayesian hybrid analysis has a useful role. A reasonable approach to this problem that does not assume unavailable stochastic information is to provide for smooth transitioning from subjective (fuzzy) characterization to stochastic characterization as information about inputs is obtained.

### *3.1.1 Uniform Scale Factors*

First, consider the case where the extent of knowledge about a problem is fractionally partitioned between stochastic and subjective portions. An input variable to an analysis whose variation characteristics are known partly stochastically and partly subjectively can be represented by a hybrid number with the relative stochastic/subjective information apportioned according to a scaling fraction:

$$h(x) = a \times p(x) + (1 - a) \times f(x) \tag{6}$$

where p(x) is a cumulative distribution function (cdf), f(x) is a fuzzy number, and a is an estimated scale factor representing the fractional stochasticity of the overall knowledge ($0 \leq a \leq 1$), and where $\times$ and $+$ are operators on x values.

The scale factor is a scalar, which the form of Eqn. 6 suggests can be used to fractionally compress the abscissa (numeric) representation of the probabilistic constituent of variability by a, along with compression of the fuzzy constituent of uncertainty by (1-a). The total variation is then additive along the abscissa, i.e., a scaled sum of the two constituents.

A visual description of the uncertainty represented by a scaled hybrid number is shown in Fig. 1, for an example scale factor of 1/2. The p and f axes represent the variability functions over x due to the constituents of stochastic knowledge and subjective knowledge, respectively. The dashed indication of a fuzzy function has been scaled down by a factor of two along the x axis from the subjective estimate. The dot-dashed indication of a probability function has also been scaled down by a factor of two from the stochastic estimate. The x-axis sum of the scaled variabilities is shown plotted as a three-dimensional hybrid number (solid lines).

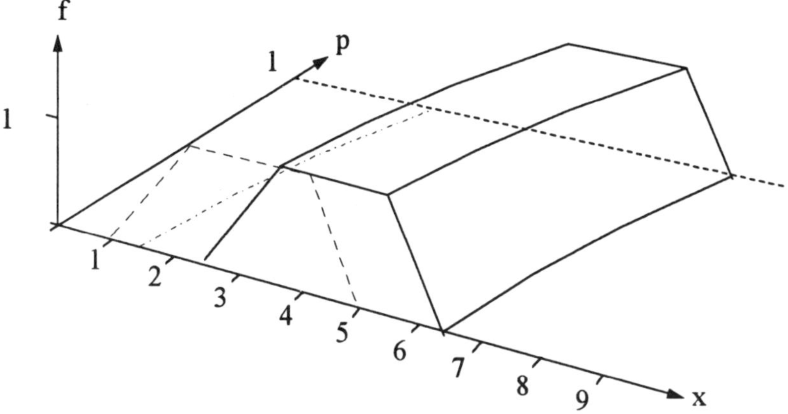


**Figure 1. A Visual Depiction of a Scaled Hybrid Number**

This formulation is understood most clearly if the spread and shape of the fuzzy function and the probability function do not interact with each other, and if separate stochastic and fuzzy mathematics are used.

In the limited case for which the scale factor applies uniformly to all input variables in a mathematical analysis (and therefore also to the output), the conventional mathematical properties (identities, commutative property, and associative property for addition and multiplication, and multiplication distributive over addition and subtraction) hold for scale-factor arithmetic with no further requirements. An important point is that variability for an operand must be entered only one time in equations in which there are multiple occurrences. For example, $A - A \equiv 0$. (We may be uncertain about the value of $A$, but not about the result of subtracting any value of $A$ from itself).

*3.1.2 Individual Scale Factors*

The more general case (allowing the scale factor to be individually chosen for each input operand), is of more practical interest. For this, a subscripted scale factor, $a_i$, will be introduced.

$$h_i(x) = a_i \times p_i(x) + (1 - a_i) \times f_i(x) \quad . \tag{7}$$

As one way to maintain the desired mathematical properties and to also assure that results meet physical expectations, we choose to average operand scale factors to obtain a resultant scale factor for multiplication and to use an average weighted by the relative contribution of the operands for addition. The following binary operations illustrate the concept:

$$h_1 + h_2 = \frac{a_1\hat{p}_1 + a_2\hat{p}_2}{\hat{p}_1 + \hat{p}_2}(p_1(x) + p_2(x)) + \frac{(1-a_1)\hat{f}_1 + (1-a_2)\hat{f}_2}{\hat{f}_1 + \hat{f}_2}(f_1(x) + f_2(x)) \tag{8}$$

$$h_1 \times h_2 = \frac{a_1 + a_2}{2}(p_1(x) \times p_2(x)) + \frac{(1 - a_1 + 1 - a_2)}{2}(f_1(x) \times f_2(x)) \tag{9}$$

where $\hat{p}_i$ and $\hat{f}_i$ represent either point estimates of $p_i(x)$ and $f_i(x)$, respectively, or functions (probabilistic and fuzzy, respectively); the operations on the $p_i(x)$ are ordinary sum-product convolution; and the operations on the $f_i(x)$ are fuzzy max-min convolution. The use of the logic represented by Eqns. 8 and 9 assures that the desired mathematical properties (e.g., associativity, commutativity, distributivity) will be maintained.

Eqns. 8 and 9 can be generalized to n operands (n>2) by extending the number of subscripts in the equations from two to n. Alternatively, n operands can be combined in successive binary operations by retaining information about prior operations. This is analogous to computing the average of the numbers x, y, and z by averaging x and y, multiplying the result by 2 (number of previous operands), adding z and dividing by 3 (total number of operands).

## 3.2 Confidence Factors

Note that if the additive weighting ratios are consistent (e.g., $\frac{\hat{p}_1}{\hat{p}_1 + \hat{p}_2} = \frac{\hat{f}_1}{\hat{f}_1 + \hat{f}_2}$), the scale factors resulting from mathematical operations sum to one. This is an interesting but unnecessary property, e.g., a hybrid number could have a well known fuzzy constituent and a well known stochastic constituent. Since we do not require this relationship, our resultant scale factors are termed "confidence factors."

The attribute of confidence factors is that they provide a metric for the amount of relative knowledge about stochastic information and subjective information, supplementing the indications about the total amount of stochastic variability and subjective uncertainty shown qualitatively.

# 4 Conclusions

For ill-defined inputs, the hybrid mathematics developed in this paper gives an appropriate modeling based on the input data. It therefore could in many cases offer advantages over other approaches. This is basically because the hybrid descriptions and the logical processing required are ideally related to the knowledge base. As stochastic knowledge is gained, the hybrid model mathematically describes the transition from subjective to stochastic information.

## References

1. Breeding R, Helton JC, Gorham ED. Summary description of the methods used in the probabilistic risk assessments for NUREG-1150. Nuclear Engineering and Design 135: 1-27

2. Cooper JA. Fuzzy algebra uncertainty analysis for abnormal-environment safety assessment. Journal of Intelligent and Fuzzy Systems 1994; 2: 337-345

3. Henrion M, Fischhoff B. Assessing uncertainty in physical constants. American Journal of Physics, 54 No. 9: 791-798

4. Morgan MG, Henrion M. Uncertainty. Cambridge University Press, NY 1990

5. Ross T. Fuzzy Logic with Engineering Applications. McGraw-Hill, NY 1995

6. Kaufmann A, Gupta M. Introduction to Fuzzy Arithmetic 1991. Van Nostrand Reinhold

# Markov Models for Quantifying Initiating Event Frequencies in Systems Analysis

Leiming Xing,* Karl N. Fleming,** and Wee Tee Loh*

*PLG, Inc.
Newport Beach, California USA

**Work Accomplished While at PLG;
Now Affiliated with
ERIN Engineering and Research, Inc.
Carlsbad, California USA

## Introduction

Markov models are appropriate whenever the stochastic behavior of the components of a system depends on the state of other components or on the state of the system [1]. There have been interesting publications (e.g., Papazoglou [1], Papazoglou and Gyftopoulos [2], and Blin, et al. [3]) regarding Markov model applications in nuclear power plant PSA. Some researchers [3] were convinced that "it is not possible to use methods such as fault tree analysis to assess the reliability or the availability of time evolutive systems. Stochastic processes have to be used and among them the Markov processes are the most interesting ones." The authors believe that while the time-averaged models have limitations, they could be used as a good estimate under certain circumstances. This paper explores the use of Markov models and fault trees for one class of problems: initiating event frequency estimation through systems analysis.

## Methodology

A two component system is used to illustrate the method of assessing initiating event frequency. Figure 1 shows the Markov diagram of a system with two normally operating components. The independent failure rate, $\lambda_i$ ($hr^{-1}$), and the repair rate, $\mu$ ($hr^{-1}$), of the two components are the same. The two components are also subject to a common cause failure with a rate $\lambda_C$ ($hr^{-1}$). The success criteria of the system is that any one component is sufficient for the system operation.

The probability that the system is at the failed state at any given time t was obtained analytically [4], as follows:

$$\pi_3(t) = 1 + C_{32}e^{\lambda_2 t} + C_{33}e^{\lambda_3 t} \qquad (1)$$

The parameters in Equation (1) are summarized in Table 1.

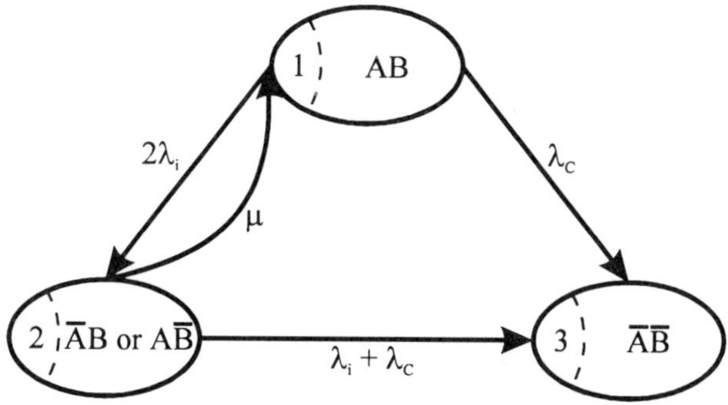


**Figure 1. Simplified Markov Diagram for a Two-Component System**

**Table 1. Solution of a Two-Component System**

| Parameter | Equation |
|---|---|
| $\phi$ | $\sqrt{\mu^2 + \lambda_i^2 + 6\mu\lambda_i}$ |
| $\lambda_2$ | $-\frac{1}{2}[(\mu + 3\lambda_i + 2\lambda_C) - \phi]$ |
| $\lambda_3$ | $-\frac{1}{2}[(\mu + 3\lambda_i + 2\lambda_C) + \phi]$ |
| $C_{32}$ | $-\frac{1}{2\phi}(\mu + 3\lambda_i + \phi)$ |
| $C_{33}$ | $\frac{1}{2\phi}(\mu + 3\lambda_i - \phi)$ |

The system reliability R(t) is the probability that the system is at any state other than the "failed" state; i.e.,

$$R(t) = 1 - \pi_3(t) \tag{2}$$

The hazard rate of the system, $\lambda(t)$, can be obtained:

$$\lambda(t) = -\frac{1}{R(t)}\frac{dR(t)}{dt} \tag{3}$$

If the system failure is treated as an initiating event in a probabilistic safety analysis (PSA) model, then the initiator frequency, or the expected number of system failures, $f_{MK}$, over the mission time $\tau_m$, can be estimated by applying the concept of a nonhomogeneous Poisson process, as follows.

$$f_{MK} = \int_0^{\tau_m} \lambda(t)dt \tag{4}$$

By employing Equation (3), we have:

$$f_{MK} = -\ln[R(\tau_m)] \tag{5}$$

where it is assumed the system is initially in an operable state; i.e., R(0) = 1.

# Comparison of Markov Model and the Fault Tree Approach

## Initiating Events for At-Power PSA

For the two-component system with a typical set of parameter values, $\lambda_i$ = 9.5E-04 ($hr^{-1}$), $\mu$ = 0.1 ($hr^{-1}$), and $\lambda_C$ = 5.0E-05 ($hr^{-1}$), the system hazard rate approaches an asymptotic value within the first 100 hours. For a usual "at-power" PSA in which the mission time is assumed to be 8,760 hours, the system behaves as if it is one component with a failure rate of the asymptotic hazard rate. Thus, a Poisson process with a constant arrival rate is a good approximation for the purpose of initiating event modeling. The asymptotic hazard rate is estimated as [4]:

$$\lambda_{asy} = \lambda_C + 2\lambda_i^2\tau_r \tag{6}$$

where $\tau_r$ is the repair time, and $\mu = 1/\tau_r$.

Alternatively, the fault tree model assumes the system failure occurred if either the common cause failure occurred, or one component failed independently during the mission time, $\tau_m$ (= 8,760 hours), and the other failed either independently or due to common cause failure over the period that the first component is under repair ($\tau_r$). Thus, the fault tree approach estimates the initiating event frequency as follows:

$$f_{FT} = \lambda_C\tau_m + 2\lambda_i\tau_m(\lambda_i + \lambda_C)\tau_r \tag{7}$$

It can be observed that the fault tree is slightly conservative, but since $\lambda_C \ll \lambda_i$, the two approaches essentially get the same results; for the parameter set used herein, the initiator frequency from both approach is 0.59 events per year.

## Shutdown Initiating Event Frequency Modeling

For the shutdown initiator problem, an asymptotic hazard rate can no longer be used because of the short mission time. By assuming the failure rate is much smaller than the repair rate, a correlation was developed [4] to estimate the ratio of initiator frequencies obtained from the Markov model and from the fault tree approach for the two component system, as follows:

$$\frac{f_{MK}}{f_{FT}} = \frac{r + 2x^2\left[1 - \frac{1}{N}(1 - e^{-N})\right]}{r + 2x^2} \tag{8}$$

where $x = \lambda_i/\mu$; $r = \lambda_C/\mu$; and $N = \tau_m\mu$.

Figures 2 and 3 plot the ratio in Equation (8). It is observed that the longer the mission time is, the less important it is to use the Markov model to accurately model the dynamic system behavior. When the common cause failure rate is large, the fault tree achieves a very close estimate with that of the Markov model. However, when the independent failure rate is high and the mission time is short, the effect of repair is important and the system dynamics can no longer be neglected.

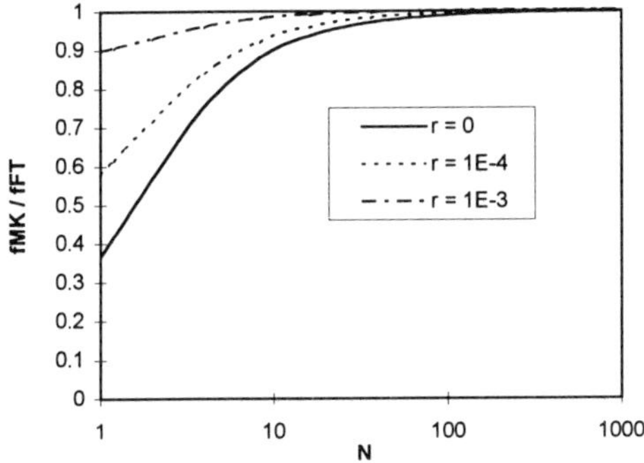


**Figure 2. Effect of Common Cause Failure on Initiator Frequency Ratio (x = 0.01)**

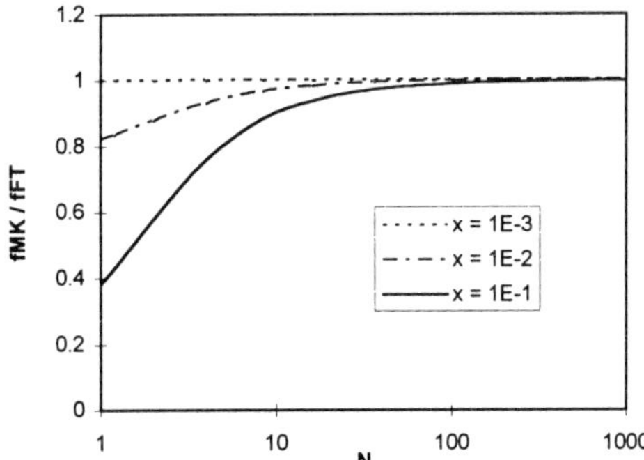


**Figure 3. Effect of Independent Failure on Initiator Frequency Ratio (r = 5E-4)**

A simplified nuclear component cooling water system (NCCW) is used to demonstrate modeling of the shutdown initiating event frequency. As shown in Figure 4, the NCCW consists of two normally operating pump trains (with independent failure rate $\lambda_I$), and one heat exchanger train (with failure rate $\lambda_H$). The repair rate of either pump is $\mu$ and the common cause failure rate is $\lambda_C$. It is observed that the heat exchanger failure has the same impacts to the system as the common cause failure does; i.e., both cause system failure. Thus, the Markov diagram for the simplified NCCW system would be the same as a two-component system if, in Figure 1, the common cause failure rate is replaced by $\lambda_H + \lambda_C$. Table 2 compares ratios of the initiator frequencies, $f_{MK}/f_{FT}$, obtained through Equation (8) and through calculating the initiator frequencies directly via the Markov model, Equation (5), and via the fault tree approach, Equation (7). The independent failure rate of the pump trains is small in the base case, and large in the sensitivity case. The ratio of initiating event frequencies is 1 in the base case and about 0.7 in the sensitivity case.

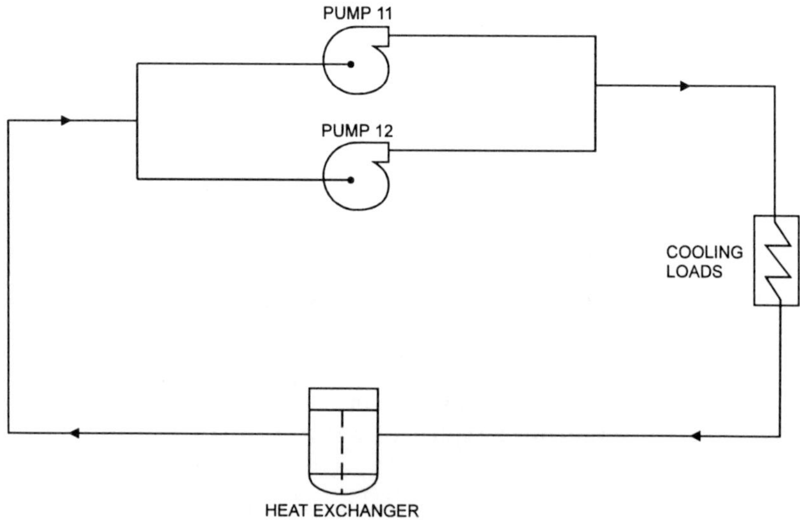


**Figure 4. Simplified System Configuration for NCCW System**

# Conclusions

It is revealed in this study that under most circumstances the fault tree approach would get a very close estimate of the initiating event frequency as in the Markov model. Only when the mission time is short, and the independent failure dominates the failure mode (thus, the effect of repair is more important), the fault tree gives conservative results compared to the Markov model. This may serve as part of the evidence that justifies the practice that the fault tree approach be used for assessing initiator frequencies of some systems. In addition, the fault tree approach is easier to use for the purpose of performing uncertainty analysis. In cases in which the time intervals are rather short, however, there may be some merit in the Markov approach.

**Table 2. Results for a Simplified NCCW System ($\tau_m$ = 7.98E+01 Hours)**

| Variable Name | Base Case | Sensitivity Case |
|---|---|---|
| $\lambda_I$ ($hr^{-1}$) | 7.58E-06 | 2.00E-03 |
| $\lambda_H$ ($hr^{-1}$) | 3.42E-06 | 3.42E-06 |
| $\lambda_C$ ($hr^{-1}$) | 1.47E-07 | 2.00E-06 |
| $\mu$ ($hr^{-1}$) | 5.15E-02 | 5.15E-02 |
| $\lambda_H + \lambda_C$ ($hr^{-1}$) | 3.57E-06 | 5.42E-06 |
| $x = \lambda_I/\mu$ | 1.47E-04 | 3.88E-02 |
| $r = (\lambda_H + \lambda_C)/\mu$ | 6.93E-05 | 1.05E-04 |
| $N = \tau_m\mu$ | 4.11E+00 | 4.11E+00 |
| $f_{MK}/f_{FT}$ (Correlation) | 1.00E+00 | 7.69E-01 |
| Markov Initiator Frequency (per Shutdown) | 2.85E-04 | 9.15E-03 |
| Fault Tree Initiator Frequency (per Shutdown) | 2.85E-04 | 1.28E-02 |
| $f_{MK}/f_{FT}$ (Ratio of above Two Rows) | 1.00E+00 | 7.15E-01 |

# References

1. Papazoglou, IA: Elements of Markovian reliability analysis. In: *Reliability Engineering, Proceedings of the ISPRA-Course held at the Escuela Technica Superior de Ingenieros Navales*, Madrid, Spain, Kluwer Academic Publishers, September 22-26, 1986.

2. Papazoglou, IA, Gyftopoulos, EP. Markovian reliability analysis under uncertainty with an application on the shutdown system of the Clinch River breeder reactor. *Nuclear Science and Engineering* 1980; 73:1-18.

3. Blin, A, Carnino, A, Georgin, JP, Signoret, JP. Use of Markov processes for reliability problems. *Synthesis and Analysis Methods for Safety and Reliability Studies*, Plenum Press, 1980.

4. Xing, L, Fleming, KN, Loh, WT. Comparison of Markov model and fault tree approach in assessing initiating event frequency. Submitted to *Reliability Engineering and System Safety* for publication.

# A Transport Framework for Zero-Variance Monte Carlo Estimation of Markovian Unreliability

J.Devooght,P.E.Labeau *
Service de Métrologie Nucléaire, Université Libre de Bruxelles
Avenue F.D.Roosevelt,50 1050 Bruxelles Belgium

J.L.Delcoux
C.E.N.-S.C.K.
Boerentang,200 2400 Mol Belgium

## 1 Introduction

Markovian modelling is a widespread framework to solve PRA problems [1, 2]. Though this assumption implies that the system is memoryless and thus that a large number of circumstances cannot be modelled in this fashion, addition of supplementary variables [3] or of physical variables [4] allows to handle a much wider range of applications than those satisfying stricto senso the Markovian hypothesis. The drawback of this simple modelling is the large number of states that have to be considered in a realistic application, and thereby the size of the system to solve. Monte Carlo simulation [5, 6] appears to be a practicable way to circumvent this problem.

## 2 Transport-reliability analogy

There exists a conceptual analogy between reliability and particle transport [7] : the free-flight of a particle between two collisions corresponds to the evolution of the system in a state between two transitions, while a collision giving new coordinates to the particle refers to a transition of the system to a new state. A similar role is played by the cross-sections and the transition rates.

In the Markovian assumption, the evolution of $\pi(i,t)$, probability to be in state $i$ at time $t$, is given by :

$$\frac{d\pi(i,t)}{dt} + \lambda_i \pi(i,t) = \sum_{j \neq i} p(j \to i)\pi(j,t) \tag{1}$$

where $\lambda_i$ is the rate of transition out of state $i$ and $p(j \to i)$ is the transition rate from state $j$ to state $i$. Its integral form reads :

$$\pi(i,t) = \pi(i,0)e^{-\lambda_i t} + \int_0^t \sum_j \lambda_j \pi(j,\tau) \frac{p(j \to i)}{\lambda_j} e^{-\lambda_i(t-\tau)} d\tau \tag{2}$$

*Research Assistant (National Fund for Scientific Research, Belgium)

Eq.(2) can be rewritten as an integral stationary transport equation [8] :

$$\Psi(P) = \int T(P',P)Q(P')dP' + \int\int C(P'',P')\Psi(P'')T(P',P)dP'dP'' \quad (3)$$

where $P$ is a point of the phase space of the particle. $Q(P')$ is the source density, $T(P',P)$ and $C(P'',P')$ the transport and collision kernels, respectively. If we assume that absorption and scattering are the only possible results of a collision, $C(P',P'')$ may be decomposed in absorption probability $c_a(P')$, scattering probability $c_s(P')$ and probability density function $C_s(P',P'')$ of the post-scattering coordinates $P''$ following $C(P',P'') = c_a(P')\delta(P''-\bar{P}) + c_s(P')C_s(P',P'')$. $\bar{P}$ is a point outside the phase space region where the simulation takes place, symbolizing the (undefined) post-absorption coordinates.

Setting $P=(i,t)$ , $P'=(j,t')$ and $P''=(k,t'')$, the transport-reliability analogy can be written in the following way :

$$\begin{aligned}
\Psi(P) &= \lambda_i\pi(i,t) && (4)\\
Q(P) &= \pi(i,t)\delta(t) && (5)\\
T(P,P') &= \lambda_i e^{-\lambda_i(t'-t)}H(t'-t)\delta_{ij} \triangleq f(t' \mid t,i)\delta_{ij} && (6)\\
c_a(P') &= \sum_{k\in Y}\frac{p(i\to j)}{\lambda_i} \triangleq \sum_{k\in Y} q(k\mid j) && (7)\\
c_s(P') &= \sum_{k\in X} q(k\mid j) && (8)\\
C_s(P',P'') &= \frac{\delta_{kX}\, q(k\mid j)}{\sum_{k'\in X} q(k'\mid j)}\delta(t''-t') && (9)
\end{aligned}$$

if $X$ denotes the set of operational states and $Y$ the set of failed states.

One is usually interested in estimating functionals of the flux $\Psi(P)$, called reaction rates in transport theory : $R=\int f(P)\Psi(P)\,dP$. Alternatively, we can write them as follows : $R=\int Q(P)M_1(P)\,dP$, where $M_1(P)$ is the expected score due to a starter from $P$. $M_1(P)$ is solution of the adjoint equation of (3).

In this paper, we focus on the unreliability function $U(t_1)$, defined by :

$$U(t_1) = \int_0^{t_1} \sum_{j\in X}\sum_{l\in Y} \lambda_j\pi(j,t')q(l|j)\,dt' \quad (10)$$

if the system is initially in a safe state. The corresponding expected score is $M_1(P) = U_i(t_1-t)$ which is the unreliability at $t_1-t$ if the system was in state $i$ at $t=0$.

In transport theory, any set of estimators $\{f(P,P'), f_a(P'), f_s(P',P'')\}$ (respectively associated with a free-flight from $P$ to $P'$, an absorption at $P'$ and a scattering from $P'$ to $P''$) can be used, provided $M_1(P)$ is conserved. Then, a Monte Carlo simulation is performed by sampling the initial coordinates $P$ from $Q(P)$ and repeating the following operations:

- sample the spatial coordinates $P'$ of the next collision from $T(P,P')$ and add $f(P,P')$ to the score;

- sample the kind of collision from the probabilities $c_a(P')$ and $c_s(P')$; in case of absorption, add $f_a(P')$ to the score and stop, else sample $P''$ from $C_s(P', P'')$and add $f_s(P', P'')$ to the score;

This algorithm can be straightforwardly adapted to our problem, choosing (for instance) the estimators $f_a(P') = H(t_1 - t')$ and $f(P, P') = f_s(P', P'') = 0$.

# 3 Zero-variance method

This formal analogy allows us to study the application of efficient variance reduction techniques developed in Monte Carlo transport theory and based on the moments equations of the score [8, 9]. These methods aim at reducing the second moment while conserving the first one, by correctly biasing the kernels from which next transition times and new states are sampled. The ideal circumstance leads to a zero-variance scheme [8, 9], which is hypothetical, since it assumes an a priori knowledge of the solution. But in practice such a game can be approximated and can lead to an efficient biased simulation. Therefore, it is worth deriving the best of all kernels, in order to look for approximations giving high figures of merit.

A pointwise zero-variance scheme [8], for which the score is identical for any history starting from a given point $P$, leads to the following modified kernels:

$$\hat{Q}(P) = Q(P)\frac{M_1(P)}{R} \tag{11}$$

$$\hat{T}(P, P') = T(P, P')\frac{f(P, P') + m_1(P')}{M_1(P)} \tag{12}$$

$$\hat{c}_a(P') = c_a(P')\frac{f_a(P')}{m_1(P')} \tag{13}$$

$$\hat{c}_s(P') = c_s(P')\frac{\int C_s(P', P'')[f_s(P', P'') + M_1(P'')]dP''}{m_1(P')} \tag{14}$$

$$\hat{C}_s(P', P'') = C_s(P', P'')\frac{f_s(P', P'') + M_1(P'')}{\int C_s(P', P")[f_s(P', P") + M_1(P")]dP"} \tag{15}$$

where

$$m_1(P') = c_a(P')f_a(P') + c_s(P')\int C_s(P', P'')[f_s(P', P'') + M_1(P'')]dP'' \tag{16}$$

According to (6)-(9), these kernels correspond to the following modified laws:

$$\hat{f}(t' \mid t, i) = f(t' \mid t, i)\frac{\sum_j q(j \mid i)U_j(t_1 - t')}{U_i(t_1 - t)} \tag{17}$$

$$\hat{q}(j \mid i, t') = q(j \mid i)\frac{U_j(t_1 - t')}{\sum_{j'} q(j' \mid i)U_{j'}(t_1 - t')} \tag{18}$$

# 4 Reducing variance on a time interval

The Monte-Carlo algorithm given above shows that $U(t_1)$ may be estimated in the same simulation for all values of $t_1$belonging to an interval $[0, t_2]$. But (17) and (18) give a zero-variance estimation of $U(t_1)$ only for the particular value of $t_1$ which appears in them. Moreover the form of (17) makes impossible the selection of times greater than $t_1$. So, if we want to estimate the unreliability function all over $[0, t_2]$, while having a zero-variance for one point of the interval, we have no other choice than making $t_1 = t_2$ in (17) and (18).

With these laws, the curve of the variance ($\sigma^2$) of the estimation of $U(t_1)$ as a function of $t_1$ presents a bell shape : $\sigma^2$ is zero for $t_1 = t_2$ and of course for $t_1 = 0$, and is positive everywhere else. This is certainly not satisfying, for the reason that accuracy is specially needed for unreliability corresponding to the lowest values of $t_1$.

Therefore, we define biased kernels leading to a zero-variance estimation for an averaged value of $U(t_1)$ over $[0, t_2]$, and obtain a variance reduction on the whole time interval by using them to compute $U(t_1)$. These laws are :

$$\hat{f}(t' \mid t, i) = f(t' \mid t, i) \frac{\sum_j q(j \mid i) \int_0^{t_2 - t'} U_j(\tau) d\tau}{\int_0^{t_2 - t} U_i(\tau) d\tau} \tag{19}$$

$$\hat{q}(j \mid i, t') = q(j \mid i) \frac{\int_0^{t_2 - t'} U_j(\tau) d\tau}{\sum_{j'} q(j' \mid i) \int_0^{t_2 - t'} U_{j'}(\tau) d\tau} \tag{20}$$

which could be generalized to any other linear functional of $U(t_1)$.

# 5 Application

We have tested the diverse methods by estimating $U(t_1)$, with $t_1 \in [0, t_2 = 4.8$ h], on a 7-component study case, whose reliability diagram is

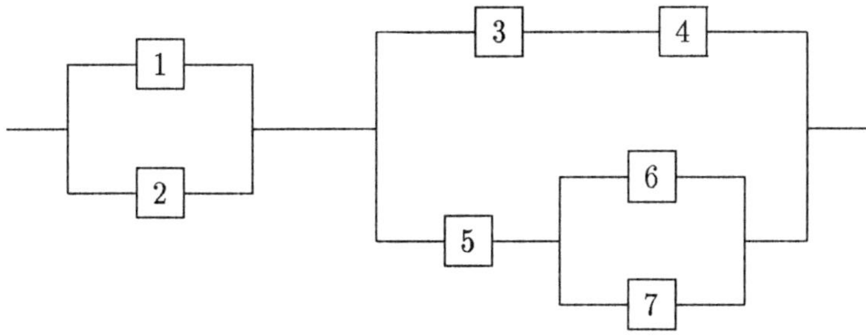


Initially, all the components are working. The following values of the transition rates are considered :

| Component | 1 | 2 | 3 | 4 | 5 | 6 | 7 |
|---|---|---|---|---|---|---|---|
| Failure rate ($h^{-1}$) | 1 | 2 | 7 | 11 | 4 | 17 | 5 |
| Repair rate ($h^{-1}$) | 9700 | 4900 | 8300 | 13200 | 5300 | 10100 | 7700 |

As the still unknown function $U_i(t)$ appears in (17), (18), (19) and (20) we must replace it by an approximation. We have chosen

$$U_i(t) \approx [1 - \exp(-\frac{t}{MTTF_i})] H(t) \tag{21}$$

where $MTTF_i$ is the *Mean Time To Failure* from a starter at $i$. We obtain the $MTTF_i$'s by resolving a system of algebraic equations. Eq.(21) is a crude approximation of $U_i(t)$. For realistic systems, $U_i(t)$ is much better described by a law $U_i(t) = c_i t^{p_i}$, where $p_i$ is the minimal number of transitions necessary to go from $i \in X$ to a state of $Y$.

We also have, in order to simplify the sampling of $t'$, tried to use (18) or (20) along with the unmodified density $f(t'|t,i)$. These different methods

| Type of simulation | Density of $t'$ | Probability of $j$ |
|---|---|---|
| I : analog | $f(t'\|t,i)$ | $q(j\|i)$ |
| II | $f(t'\|t,i)$ | (18) |
| III | (17) | (18) |
| IV | $f(t'\|t,i)$ | (20) |
| V | (19) | (20) |

are compared. All five give unbiased estimations of $U(t_1)$ if correct statistical weights are used. The evolutions of the variance for these methods are shown in fig.1 :

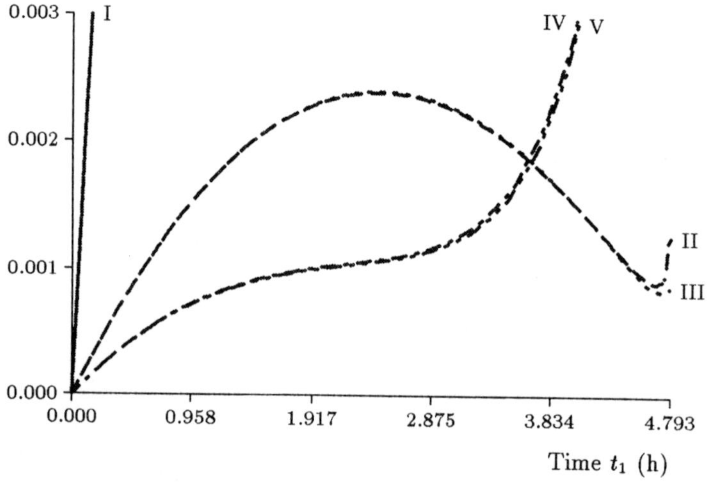


**Fig.1**

Method I gives a quickly growing variance, II and III a bell-shaped curve, and IV and V a plateau lower than the maximum of variance obtained with II or III. So three observations have to be made: 1) the analog method (I) gives the worst accuracy; 2) V gives a variance reduction better distributed over $[0, t_2]$ than III; 3) the modification of $q(j|i)$ has the major impact : II gives the same variance as III and IV the same as V. The third constation is very nice since the modification of $f(t'|t,i)$ brings much more difficulties in the samplings than that of $q(j|i)$. This is confirmed by the comparison of the figures of merit $(= 1/(\sigma^2 \times$ calculation time)) (fig.2)

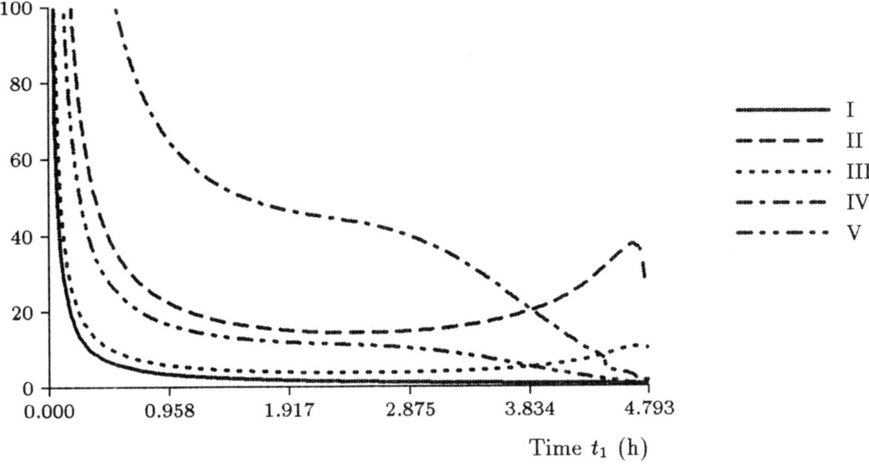


**Fig.2**

which shows that II is quicker than III and IV than V. So the most efficient of the five methods is IV.

# 6 Generalization: probabilistic dynamics

Probabilistic dynamics [4] is a generalization of the classical markovian reliability which takes into account the physical variables $\bar{x} \in \mathcal{R}^n$. The corresponding probability density obeys :

$$\begin{aligned}
\pi(i, \bar{x}, t) &= \int \pi(i, \bar{u}, 0)\delta(\bar{x} - \bar{g}_i(t, \bar{u}))e^{-\int_0^t \lambda_i(\bar{g}_i(s,\bar{u}))ds} d\bar{u} \\
&\quad + \sum_{j \neq i} \int p(j \to i \mid \bar{u}) d\bar{u} \int_0^t \delta(\bar{x} - \bar{g}_i(t - \tau, \bar{u})) \\
&\quad \times e^{-\int_0^{t-\tau} \lambda_i(\bar{g}_i(s,\bar{u}))ds} \pi(j, \bar{u}, \tau) d\tau
\end{aligned} \tag{22}$$

where $\bar{g}_i(t, \bar{u})$ is the solution of the dynamics in state $i$ : $d\bar{x}/dt = \bar{f}_i(\bar{x})$, $\bar{x}(0) = \bar{u}$. (22) is also formally equivalent to the transport equation (3).

We define $U_i(t \mid \bar{x}_0)$ as the unreliability at time $t$ if the system started from state $i$ with variables $\bar{x}_0$ at time 0. We have to specify what we mean by failure in this context. It is not only a transition to a failure state, but it is also the crossing of the border of a safety domain $\mathcal{D}$ in phase space [4]. The equation satisfied by $U_i(t \mid \bar{x})$ takes into account the two contributions to the unreliability:

$$\begin{aligned}
U_i(t \mid \bar{x}_0) &= \int_0^t \sum_{l \in Y} p(i \to l \mid \bar{g}_i(t', \bar{x}_0)) \mathcal{H}_{\mathcal{D}}(\bar{g}_i(t', \bar{x}_0)) e^{-\int_0^{t'} \lambda_i(\bar{g}_i(s,\bar{u}))ds} dt' \\
&\quad + (1 - \mathcal{H}_{\mathcal{D}}(\bar{g}_i(t', \bar{x}_0))) e^{-\int_0^t \lambda_i(\bar{g}_i(s,\bar{u}))ds} \\
&\quad + \sum_{k \in X} \int_0^t p(i \to k \mid \bar{g}_i(t - \tau, \bar{x}_0)) e^{-\int_0^{t-\tau} \lambda_i(\bar{g}_i(s,\bar{x}_0))ds} \\
&\quad \times U_k(t - \tau \mid \bar{g}_i(t - \tau, x_0)) d\tau
\end{aligned} \tag{23}$$

where $\mathcal{H}_{\mathcal{D}}(\bar{x})$ is the characteristic function of $\mathcal{D}$.

# 7 Conclusion

A formal transport analogy has helped us to propose a zero-variance method for Monte Carlo estimation of unreliability. That method has been modified to obtain a variance reduction better distributed over a time interval. A simplification of the biased method turned to be as accurate but much faster. The formal analogy also applies to probabilistic dynamics. Variance reduction methods could be obtained in that domain.

# References


[1] Papazoglou I.A., Gyftopoulos E.P., Markovian processes for reliability analyses of large systems, IEEE Trans.Rel. **R-26**, n$^0$3 (1977), 232-237.

[2] Wells C.E., Bryant J.L., Reliability characteristics of a Markov system with a mission of random duration, IEEE Trans.Rel. **R-34**, n$^0$4 (1985), 393-396.

[3] Cox D.R., The analysis of non-Markovian stochastic processes by the inclusion of supplementary variables, Proc. Cambridge Phil.Soc. **51** (1955), 433-441.

[4] Devooght J., Smidts C., Probabilistic reactor dynamics. I.The theory of continuous event trees, Nucl.Sci.Eng. **111**(1992)229-240.

[5] Lewis E.E., Bohm F., Monte Carlo simulation of Markov unreliability models, Nucl.Eng. Des. **77**(1984)49-62.

[6] Wu Y.F., Lewins J.D., System reliability perturbation studies by a Monte Carlo method, Ann.Nucl.En. **18**, n$^0$3 (1991), 141-146.

[7] Dubi A., et alii, Analysis of non-Markovian systems by a Monte Carlo method, Ann. Nucl. En. **18**, n$^0$3 (1991), 125-130.

[8] Lux I., Koblinger L., Monte Carlo particle transport methods : neutron and photon calculations, CRC Press, Boca Raton (1991).

[9] Gupta H.C., A class of zero-variance biasing schemes for Monte Carlo reaction rate estimators, Nucl.Sc.Eng. **83** (1983), 187-197.

# *"PSA as a Part of a Safety Improvement Program"*

Héctor Kohn

Atucha I NPP, Núcleoeléctrica Argentina S.A.

Lima, Argentina

**Abstract**

## 1 Objectives

This document is intended to show the importance of NPP organisational operations allowing a PSA task force to develop the study in such a way that, its insights and results, become a part of the organisations culture, developing an actual positive impact in attitude of people toward safety issues related with severe accidents. With organisational operations it is meant all those actions such as setting strategies, planning, procuring resources, communicating objectives, managing relationship between functional areas, in order to adjoin a new group, the PSA task force, for which, information the rest of the organisation detain is one of its most critical supplies.

## 2 Introduction

Performing a PSA study is an expensive, time consuming and laborious effort with the risk that its results be forgotten inside its files. Unhappily this is a very frequent fact when the study is performed not regarding credibility and acceptance from the people that could take advantage from applying its results. In other case organisation support is not effective because of the placement of not enough resources. This undesirable danger should be addressed from the very beginning of the planning stage of the study. From an organisational point of view there is a three part relationship that should be managed to achieve success. It involves, the top management, the PSA group and the potential user of the results, that is people of different functional areas at the plant. In the case being presented, there is also an important role of the Regulatory Body as a driving force of the study.

This paper deals with the process of achieving the acceptance and support from people, how important it was to reach the desired goals and which was the impact in the safety attitude of people at the plant.

## 3 Historic summary

On December 1991, the Argentine Nuclear Regulatory Authority (NRA) placed a requirement to perform a Level I PSA to Atucha I NPP. Although it was their requirement, they tried to make easier the way of the utility towards PSA supplying training, PSA documentation, like procedures from other PSA studies and the PSAPACK computer code.

On the utility side, instead of accomplishing plainly that requirement, the organisation fostered its development as an useful safety management tool. At that moment the intention was to strongly support on a quantitative base, all the decision making actions related to safety.

## 4 Our particular case

As any new project, although the strong management support and the significant amount of resources placed, the project faced a subtle resistance from people inside the plant.

In such people there is a tendency to perceive external or new groups as resource draining elements with no practical return. They are suspected of poor knowledge and weak commitment with the plant

A previous attempt to perform a less ambitious study, failed but gave some warnings about the way the PSA should be promoted.

Hence, it was necessary to inform the NPP staff in a clear and efficient way which results could be obtained if the PSA was developed in a proper mode.

This point should be divided into different aspects.

Firstly, Probabilistic approaches were poorly known by the plant people, and performing the PSA was perceived solely as a way to satisfying an external requirement. This fact was confronted by means of a presentation to plant managers that emphasised the potential usefulness of performing the study. The central point of the presentation was the following question, showed in the first slide: **PSA: Regulatory requirement or useful tool?**.

Then, some potential advantages that may arise from the study were enumerated ranging from optimisation of safety-oriented resource allocation to elimination of unnecessary regulatory requirements with the consequent saving of organisational effort.

Another highlighted point was the credibility of results which should be based on sound models and reliable data.

The objectives of underlying this points were:

a) To differentiate ourselves from the out-sourced PSA performers.

b) To get allianced with data owners looking for a common objective.

Another facts were considered in order to plan carefully the following stages.

Some of the difficulties laid on the history and the design conception of this plant, facts that got fixed along twenty years of operation.

Some of this facts were :

- A deterministic DBA design
- Lack of design documentation
- Lack of accident simulation calculation
- One of a kind plant (Heavy water pressure vessel type NPP)
- Isolation from current trends in the nuclear field
- Kingdom Culture in functional areas (Separate and isolated compartment culture)

Any action required a particular approach.

As a consequence a great amount of plant knowledge remained in the mind of experienced personnel. This condition established a first and very important constraint, *the study should by carried out inside the plant* and its people should take an important role performing it. Also suitable information transfer mechanism between involved groups should be implemented. The whole PSA team received a training on the nature of communication inside organisation and how to open communication channels between the plant groups and the PSA group. This knowledge was summarised in the "Ten Commandments for data seeking", that proved to have good results.

As a one of a kind plant, it wont be possible to apply the typical initiating events grouping for either PWR, BWR nor CANDU.

The lack of enough accidents simulation derived in the creation of an area for *determinist calculation* within the probabilistic study.

Once, the group was formed and established some actions were taken in order to put it to work properly. Some internal deep discussion were necessary to clarify the goals and a set of philosophic statements (*mission statement*) was issued.

In parallel a communication action towards the intermediate management was issued, attempting to spread the idea of the potential practical applications of the study and to differentiate its scope from the previous one.

Due to the actual team composition, that is some people with plant experience and some people with PSA knowledge, a cross training course was given. The *program* dealt with the safety systems of the plant and PSA methods. A good result from the mentioned action was a favourable attitude felt from different managers ,i.e: operation, maintenance, as they placed their specialists as instructors.

During the actual execution of the PSA a Technical Adequacy System (technical review plus technical adequacy assurance) was set up and systematically applied to every technical document elaborated and issued by the PSA team.

The fact that such a Technical Adequacy System included in each case the revision of these documents by some specialist of NPP staff, mainly from Operation Department or different Maintenance Department, provided an additional way of spreading the PSA team work within plant staff.

# 5 Results

This strategy did not lasted to give some results, in fact the group became the reference in safety matter. It was dangerous a too early success, for the fact that at that time the bulk part of the model was not developed in sufficient extent to serve the decision making process.

Preliminary results of this study have begun to be used as one of several inputs to make management decisions. The following are examples of applications in which PSA results were considered:

a) Tube leakage in the complementary safe cooling water (UK) system coolers necessitated taking them out of service for repair. Using the results of the PSA study, the station determined that there was very little difference in risk between performing the work at power and shutting down to conduct the repairs. A plant shutdown was thereby avoided.

b) Switchover to emergency cooling on a delta temperature signal between the coolant and moderator is dependent on a signal provided by temperature indicators. Analysis of the system models developed by the PSA study determined the importance of these indicators, and review of operator logs in the control room identified that readings were not recorded on these instruments. To decrease the probability of an undetected failure of these indicators, they were added to the control room operator reading sheet.

c) A modification had been proposed earlier and a package developed to add an automatic shutdown of the reactor on high secondary steam pressure. Results of the PSA study confirmed that implementation of this modification will significantly decrease the overall risk of core damage, and a recommendation has been made to continue with this modification.

d) The initiating event of a stuck open pressuriser spray valve causing low pressure has a relatively high probability of occurrence (once during the life of the plant). A number of countermeasures are available for the operator to deal with this transient, but they are complex and need to be completed within about 20 minutes. Based on these results, it was recommended that a procedure be developed to address this transient.

e) A memorandum was sent to the Training Department asking it to emphasise the importance of certain emergency operating procedure actions determined by the PSA study to be major factors in reducing the risk of core damage.

In addition to the above examples, the station has begun to examine possible changes to optimise periodic testing frequencies. The results of the PSA will be used as input when making decisions to change the frequencies.

A Peer Review mission from WANO on mid 1995 stated the PSA as a NPP Strength and recommended its in-plant execution as a Good Practice.

## 6 Conclusion

Difficulties based on organisational culture and negative experiences were overcome with success. Part of this success relies on effective objective communication.

By different and complementary ways the works and results found by the SPA team were transmitted to the plant staff. In some cases through a participation in technical documents review issued by the SPA, through clarification and eventually simplification of certain maintenance tasks and also through specific training actions.

# Safety and Availability in the Design of Nuclear Power Plants : Conflict or Convergence

BOURGADE E., MAGNE L., Electricité de France,

*1, avenue du Général de Gaulle, 92141 CLAMART CEDEX FRANCE,*

## Abstract

Availability is becoming a major concern in nuclear power plants operation. As efforts are made to improve the availability of future plants during design, questions about compatibility between availability oriented improvements and a satisfactory level of safety may arise. This paper presents the different aspects of the links between availability and safety, and proposes means to ensure a necessary equilibrium between the two concerns during the design process.

## 1. Context

Safety is a key issue in the design and operation of nuclear power plants. The probabilistic aspect of safety, as determined by Probabilistic Safety Assessments (PSA), has become a determining factor in the certification of new units and is becoming increasingly important in operational decisions.

At a time when electricity demand seems to be stabilizing and alternatives to nuclear electricity are beginning to emerge as serious contenders in France, major efforts are being made to improve the availability of nuclear plants [1],[2].

The question naturally arises as to whether efforts aimed at improving plant availability are compatible with those aimed at ensuring plant safety. Of course, one context in which compatibility can and does apply is that of methods and tools. Apart from differences in objectives (safety versus availability) and the stage at which safety and availability considerations are integrated in the design process (this difference doubtless owes more to established working practice than anything else), it would be reasonable to expect a certain degree of synergy at this level. Indeed, initial work on integrating availability into power plant design would seem to bear out this supposition.

However, compatibility must also apply in a second context : to the recommendations produced by safety and availability studies. Even if probabilistic safety studies and availability studies could be totally divorced from one another - and this would entail a serious waste of effort - it would in no way alter the fact that both types of study apply to the same plant. Our major problem therefore concerns the supposedly inevitable conflict between safety and availability recommendations. We can tackle this problem from two angles : theory and practice.

## 2. The Theoretical Angle

In any given system, safety and availability will often appear as antagonistic. For example, in a protection instrumentation and control system, safety considerations will call for good system availability, i.e. the ability to operate when required ; the designer will thus naturally specify a redundant architecture with a 1/n voter (which means that there are n redundant protection signals, and that one is enough to initiate the safety actions, like starting a pump or closing an isolation valve). However, plant availability considerations will call for good reliability, i.e. minimum risk of spurious operation ; here the designer will tend to specify a non-redundant architecture. In this kind of situation, we would generally opt for a k/n architecture (with k greater than 1), as in N4 protection system (2/4). It optimizes both aspects. Only a combined safety/availability approach allows us to reach a mutually satisfactory solution here.

Indeed, even in a general sense it is essential to couple both types of study. Moreover, there is a deeper link ,since the analysis methods will be of the same type, if not identical.

The example illustrated above points out the utility of making joint recommendations. If they are made separately, there might be several iterations before the final solution (2/4) is proposed ; one might object that, in this particular case, the 2/4 solution was found without formal dependability analysis, but it was only reached after 2 or 3 generations of power plants, which seems to be quite a slow optimization process !! More seriously, doubts can be raised about the possibility of reaching a good solution if recommendations are not at least coordinated.

## 3. The Practical Angle

At the design stage, we can address the availability/safety problem by attempting to find answers to two questions : **how to tie in availability studies with safety studies**, and **how to integrate both types of studies in the design process**.

To a very large extent, the specific data input required for dependability studies is in fact common to both availability and safety studies. Dependability analysts are well aware of the difficulties involved in data input, most of which are due to the low availability of designers and the highly specific nature of the data required. It would therefore appear essential to coordinate the data search between the two types of study, in order to minimize the risks of rejection and incoherence. Emerging availability studies during design show that, beyond this general similarity between the two types of input data, some differences exist, especially concerning the definition of failure modes.

Regarding the output from the studies, we must also bear in mind that the results from both types of study are addressed to the same teams, working on the same plant design. Both types of results should therefore be consistent in type and format. The coherency between the two types of studies may be affected because of working organization. Probabilistic safety people often are not part of the design teams, and they are also different from the people who analyze and build availability ; priorities, time schedules and constraints are not the same, and more generally, cultures are often different, creating difficulties in working together.

We conclude that availability and safety studies must be consistent with each other as regards both input and output. Moreover, it will no doubt prove much more efficient to conduct both types of study cooperatively, as we saw in the example situation above.

One difference in the way availability and safety studies are practiced might in fact reveal a more basic underlying difference : safety studies are performed late in the design process, mainly because of the type of input data they require (for example detailed operation procedures, layout, cabling) ; on the other hand, it appears more suitable (and above all more feasible) to run availability studies earlier on in the process. We might therefore legitimately consider that design safety studies should be based primarily on experience feedback, whereas availability studies, starting at this stage, should take form of probabilistic forecasts based on functional modeling. The next part will discuss this statement in more detail.

# 4. When to Perform Safety Studies ?

First of all, as we have stated above, historically, probabilistic safety mainly plays a role at the end of the design process, if not only at the end of the realization and the results of PSA are used to fulfill regulatory requirements, to answer questions of the Safety Authorities, and now in operation, for maintenance management [3] and decision making [4].

The figure 1 illustrates this fact

**Would it be useful**, and **would it be possible**, to analyze new plants from the safety point of view sooner in the design process ?

**Would it be useful** ? The answer to this question is clearly yes. Among the reasons to this answer, we think that two are preeminent : cost efficiency, and profitability.

It would be more cost effective because traditional probabilistic safety studies lead to discussions after nearly all the choices are made ; so, any consequent modification is very costly. For example, let us suppose that probabilistic safety studies , after discussion with safety authorities, lead to state that electrical power supply of a safety system is not reliable enough. It might lead to the addition of some new device, like a diesel generator or a gas turbine, which would be quite expensive because of additional studies and modifications of the already existing plant. The very fact that it is an intruder in an already built plant raises questions about the possibility of ensuring a good level of reliability for the new equipment.

**Would it be possible** ? The answer is more difficult. Actual Spas are quite detailed and take into account very precisely how the plant is operated. One might state that this type of information is mostly known very late in the design. As they have a strong, if not preeminent influence on core melt frequency, the first conclusion seems to be that no PSA can be achieved earlier in the design process. But the situation is a little more complicated : the goal of design probabilistic safety studies is to give indications on design choices from the safety point of view ; it means, that no "absolute" measure is needed, as for regulatory requirements, but only relative measures between different solutions.

Therefore, we think that strong possibilities of such an analysis (a design PSA) exist. Firstly, a Preliminary Risk Analysis can be performed, to identify as soon as possible the main aspects of risk associated with the plant. Secondly, a Functional Analysis with a safety orientation can also be performed, as it was done within the HERMES project.

In addition, we must keep in mind that new designs often mainly modifications, sometimes important, of previous ones. So, in this case, two other possibilities exist. Firstly, there is a lot to learn from experience feedback on previous designs, not only about component failure data, but also about critical parts of the plant, important operation procedures, and probably many other topics. Secondly, some simplified dependability analysis could be realized to make comparisons between several options, concerning for example architecture, components, layout of circuits and components

# 5. Propositions

Despite the inherent difficulties illustrated in this paper, we think, that it would be very valuable to find ways to take into account probabilistic safety during design, earlier than now. It would certainly mean less sophisticated and less accurate evaluations, and studies would be less ambitious.

More than in the case of conventional "regulatory" PSAs, it must again be stressed, that we would only calculate indicators, aimed at helping the designers to make choices, and not providing absolute and accurate measures at all !! Nevertheless, big rewards could be gained, as is illustrated in the third paragraph of this paper. And we think these rewards are likely to increase in the foreseeable future.

Why ? Let us consider a practical example, the safety injection system of a nuclear plant. In recent designs, this system has four pumps, and this choice is based upon experience, previous PSAs results and safety authorities strong recommendations. It might be imagined, that availability analyses performed early in the design of a new plant lead to conclude that three pumps are better for the overall availability of the plant, because of reduced maintenance and fewer applications of technical specifications, let alone the lower investment cost. Only at the final stage, i.e. after realization, would probabilistic safety analyses reveal, maybe, that this new architecture does not allow the plant to reach an acceptable level of safety, i.e. a low enough probability of core melt. In an increasingly environment-oriented and competitive market, it could lead either to very costly modifications, or to the shutdown of the plant (even more costly, as it would never produce electricity !).

More generally, the possibility exists, that efforts to increase availability during design reduce dramatically the existing safety margins, and, unless this reduction is assessed early and correctly handled, licensing problems could be met in the future.

In conclusion, we believe, that consistency must be sought between availability and safety analysis in three fields : **data collection**, **methods and tools**, **results**.

On the last point, availability and safety analyses traditionally do not pursue the same goals, and are performed at different times of the design process, which can be illustrated in figure 1.

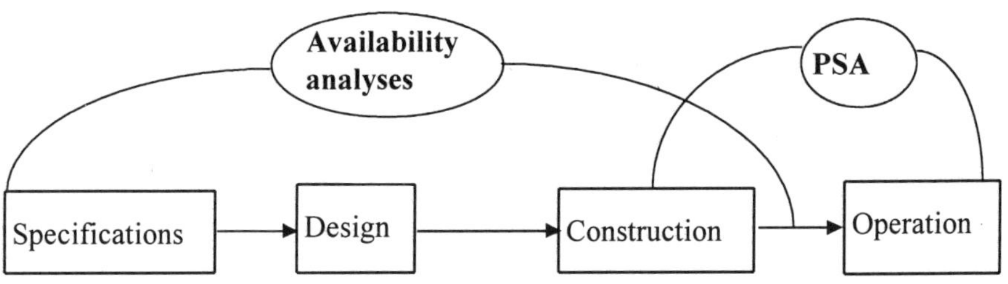


**Figure 1**

We believe that there are obstacles to earlier probabilistic safety analyses, which nevertheless can be overcome, and at least, it is worth trying to progress toward this goal ; the modified design process could be illustrated by figure 2. This process is applied within EPR (European Pressurized Reactor) project ; the designers have begun safety analysis, to help themselves to make choices.

As it is still an emerging concept, the question to answer is : what should be the content of such a design PSA?

As we have tried to demonstrate, we think that, mainly, it should be a functional-oriented PSA, relying heavily in early stages of design on experience feedback of former plants PSAs to provide lacking information. From this point, it will become possible to connect both availability and safety studies, and to correctly integrate them in the design process.

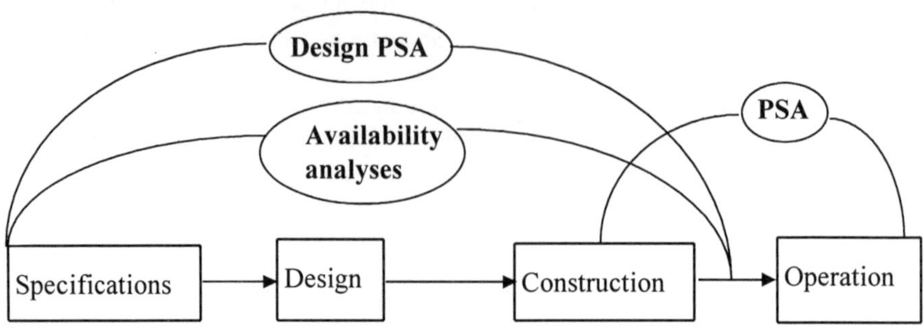


**Figure 2**

# 6. References

[1] C. Degrave, M. Martin-Onraët, "Integrating Availability and Maintenance Objectives in plant design - EDF Approach", Third Joint International Conference of Nuclear Engineering, Vol. 3, pp. 1483-1488, Kyoto, Japan, 1995

[2] E. Bourgade, C. Degrave, A. Lannoy, "Performance Improvements for Electrical Power Plants : Designing-in the Concept of Availability", ESREL'96

[3] F. Ardorino, A. Dubreuil-Chambardel, P. Mauger, "Use of risk importance measures in maintenance prioritization", TCM IAEA, 9-13 October 1995

[4] L. Magne, H. Pesme, "Improvements and Traceability of a PSA to Help Applications and Decision Making", ESREL'96

# Linking Internal and External Safety&Environmental Impact Assessment and Management

Paolo Vestrucci
University of Bologna and Nier Ingegneria, Via del Fossato 5/2
40123 Bologna, I

**Abstract**

Deep changes occurred in the last couple of decades in safety and environmental impact assessment and management in the industrial context, in transportation systems and in general in our societies. This process is very dynamic, regarding all the European Countries, being "robust" to the economical crisis affecting the world during the last five years, and initiating a deep revision of process and organisational aspects.
On one side this is due to the initiatives of the European Union through a series of directives and regulations. On the other side, this reflects a more complex approach of the industries to a market which is more and more integrated.
This paper (a) develops the analysis of the most important acts of the European Union relevant to the Environment, Health and Safety (EHS) pointing out some bottom lines of its policy; (b) shows the links between the EHS and between internal and external sides and (c) with quality; and finally (d) briefly outlinės the objectives of an information system for industrial environment, health and safety management.

## 1. European steps towards EHS management

The European Union steps towards EHS management can be reviewed by the scheme given in Table 1, which yields (through some examples, without pains of completeness) the temporal sequence of directives and regulations with the main objective.
Through this sequence it is possible to recognise that the European policy follows a path such that:

- responsibility,
- transparency,
- voluntariety,
- systematic (and systemic) approach to the problems,

become key-words of a modern development, the so-called sustainable and durable development.

Table 1. Main European Union actions in environment and safety policy.

| # | Year | Document/act | Principal area | Objective |
|---|---|---|---|---|
| 1 | 1980 | Directive 84/360 | Air | Standard definition |
| 2 | 1981 | Directive 82/501 | Major hazards | Prevention |
| 3 | 1984 | Directive 84/360 | Air | Standard definition |
| 4 | 1985 | Directive 85/377 | Env. impact assess. | Prevention |
| 5 | 1986 | Directive 86/280 | Water | Standard definition |
| 6 | 1989 | Directive 89/391 | Safety | Risk elimination/reduction |
| 7 | 1989 | Directive 89/392 | Safety | Risk elimination/reduction |
| 8 | 1990 | Directive 90/313 | Environment | Information access |
| 9 | 1992 | Regulation 880/92 | Ecolabel | Environmental quality (product) |
| 10 | 1993 | Regulation 1836/93 | Eco-audit/management | Environmental quality (site) |

Indeed, the same trend can be grasped from the European V Community Programme of Policy and Action in relation to the Environment and Sustainable Development, in which the transition from a “Command and control”-based system to a more complex dynamic is clearly outlined. This is sketched in Table 2.

Table 2. Some aspects of the evolution in European Union policy.

| FROM | | TO |
|---|---|---|
| Partial approaches & Responsibilities<br>#: 1,3,5 | ⇒ | Global approach & Responsibilities<br>#: 2,4,6,7,9,10 |
| Policy "command & control"<br>#: 1,3,5,6,7 | ⇒ | Market tools & Voluntary mechanisms<br>#: 9,10 |
| "Nobody knows" | ⇒ | Transparency and communication<br>#: 2,4,8,10 |
| Rule observation<br>#:1,2,3,4,5,6,7 | ⇒ | (Complex) system management<br>#:2,4,6,7,8,9,10 |
| Passive/reactive firm:<br>#: 1,3,5 | ⇒ | adaptive/proactive firm<br>#: 6,9,10 |

# 2. Internal & External Environmental, Health and Safety

It is apparent that the dynamic previously outlined corresponds to (in some extent it is the consequence of) an analogous development within the industries.[1] Clearly, a programme like the CEFIC's Responsible Care is a major example of proactive behaviour in the industrial side. On more general basis, it is not difficult to recognise that industrial hygiene and environmental impact are the two faces of the same coin, as major hazards and worker safety. In other words: internal and external aspects need a common approach. This is reflected into the choice of several firms to establish an EHS office or department (which is mandatory as far as safety and hygiene are concerned, after the Directive 89/391). Indeed, it is possible to recognise the global environmental and safety system by adding to the previous considerations the following one: the safety problems usually arise from the degeneration of environmental parameters and this fact indicates a link between environment and safety. This global system can be represented as in Fig.1: the matrix represents the general system, which consists into four elements. Although each element has a precise identity, it belongs to a common system (or matrix), and it interacts with the others in several ways (symbolically, the interactions are put in evidence by the circle in the centre of the matrix).

Figure 1. Global System schematic

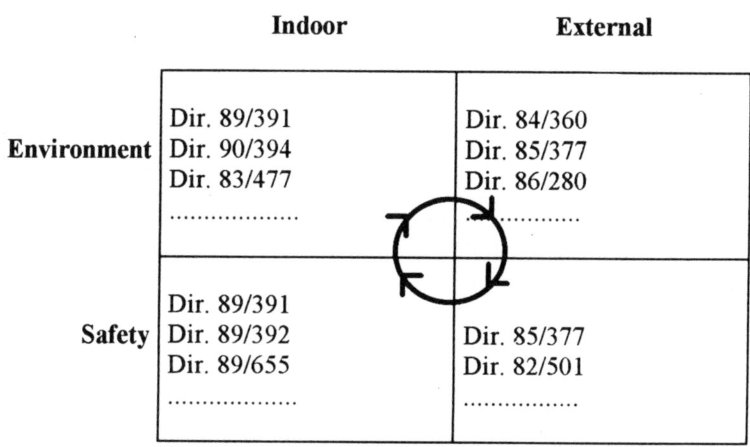


The common approach of internal and external aspects relevant to environment and safety may be recognised also by considering a different point of view: the company stakeholders. Today, in fact, different stakeholder interests influence the company life; stakeholder examples are:

- the shareholders' meeting and investor groups,
- employees,
- clients and consumers,
- decision makers and authorities,
- local community and social organisations or groups (like, but not only, environmentalists),
- banks and insurance companies,
- media,
- .... .

The Figure 1 schematic could be used in order to collocate the stakeholders in a proper matrix element (or elements), once more confirming the systemic structure of our problem.

# 3. EHS and Quality: EHS&Q

The most appropriate context for EHS management system is Total Quality. Indeed, environment, health, and safety can be regarded as components (sometime crucial) of the quality system, in the frame of EN 29004.[2] The EN-29004 contains senses for the firm conduction, for quality and for the firm quality systems. The norm identifies the activities which influence the quality of a product and concern all the phases (from client expectations up to their satisfaction) showed in the quality circle. This representation may be used as the frame for the EHS management system. In fact, although it is stressed that the Total Quality is not a requirement for the EHS management system, it is clear that it may properly regarded as a component of the quality (i.e. company) system itself. In any case, explicit links between quality and EHS are stated in the European regulation,[3] BS 7750[4] and the Draft ISO 14001. For this reason sometime a QEHS (Quality & EHS) structure is defined,[5] in order to have a more integrated and efficient management system.

Of course, EHS management exists independently of quality; nonetheless, different stages of EHS management sophistication can be recognised as the evolution towards EHS management excellence:

regulatory, norm and law requirements compliance ,

EHS management system implementation,

EHS management as one of the principal factors of company management,

integration of EHS into the total quality system.

as sketched in Fig. 2.

Figure 2. The evolution towards EHS management excellence

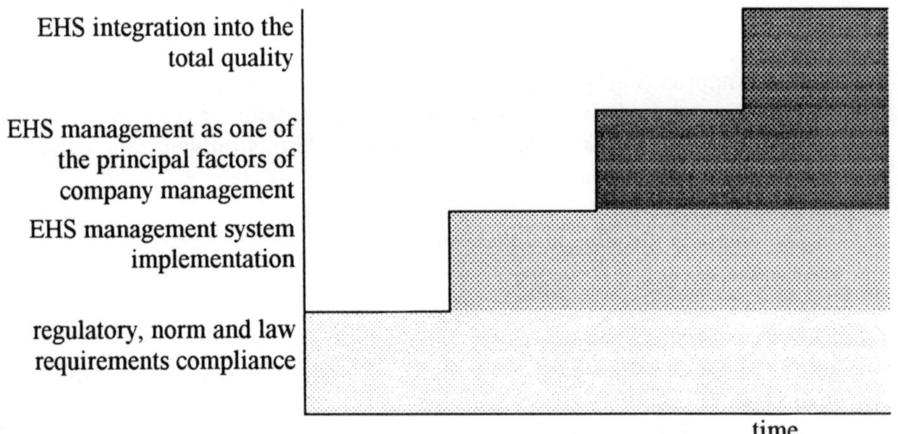


# 4. An information system for EHS management

With many demanding regulations, norms and laws in force today in Europe, the importance of information management to environmental, health and safety

programmes is a strategic issue. As environment, safety, industrial hygiene and medical surveillance requirements increases, so does the need for documentation, reporting and analysis. The regulations, in fact, impose complex procedural, reporting and record keeping requirements on the industry, so that the previously outlined evolution is forced also by this external pressure.
These factors, combined with the need for better internal record management, is leading many organisations to need improving their record keeping systems and their information management systems.[6,7]
As usual, different stakeholders imply different needs for information, as can be grasped in Table 3.

Table 3. Information needs for different stakeholders (examples)

| Stakeholder | type of information required (examples of) | type of reporting (examples of) |
|---|---|---|
| Shareholders' meeting and investors | executive summary of company EHS performance and trend | annual report with main data and meaningful performance indicators |
| EHS personnel | any kind of physical and administrative data relevant to EHS | on-line measurements, balances, technical reports, administrative reports,... |
| Employees | personal exposures, risks, health data; company EHS level and organisation, general information and specific training | Accident reports, HSE trend information, technical manuals and specific data |
| Authorities | demonstration of compliance with laws; evaluation of company environmental impacts | compliance reports; environmental reports |
| Community groups and others | main information about EHS characteristics of the company; trends and benchmarking | annual reports; others specific information tools |

From this short review it is apparent that an efficient EHS management needs an information system as a practical management tool.
Some examples of EHS information systems do exist, but some efforts has to be spent in order to make these efficient and available to medium and small sized companies. The analysis of the problem is out of the scopes of this paper, but the interested reader may refer to Ref.s 6÷9, or in general to researches and the experiences relevant to eco-balance and performance indexes .

## References

1. European Green Table, Environmental Performance Indicators in Industry, Draft handbook, Oslo, 1993.
2. European Standard EN 29000, Quality Management and Quality Assurance Standards Guidelines for Selection and Use.
3. EC Regulation 1836/93, Voluntary Participation by Companies in the Industrial Sector in a Community Eco-management and Audit Scheme, Official Journal OJ L168, 1993.
4. British Standard BS 7750, Specification for Environmental Management Systems, London, 1992.
5. M.Malagoli, P.Vestrucci, E.Camera, A.Biancoli, Design and Implementation of an Integrated Environmental and Safety Quality System, SRA-Europe Fourth Conference, Rome, 1993.
6. Andersen Consulting, Process/1 Reference Manual, 1992.
7. Exxon Company International, OIMS Reference Manual, 1991.
8. Baer, Process MRP helps vinyl manufacture beginner its business, Managing Automation, 1991.
9. Charmas, Software System aids small, large firms, Plastic News, 1991.

# An Approach to Design a Hypertext-based Navigation System for Follow-up of Emergency Operating Procedures

S. W. Cheon, G. O. Park, and J. W. Lee

Korea Atomic Energy Research Institute,
Yu Song P.O. Box 105, Taejon 305-600, Korea
*(Phone) +82-42-868-2941, (Fax) +82-42-868-8357*
e-mail: swcheon@nanum.kaeri.re.kr

## Abstract

*Hypertext is a technique for information navigation, which takes advantages of allowing for the development of large informational spaces, a possibility of considerable merits for the computerisation of operating procedures in nuclear power plants. This paper introduces hypertext-based techniques and our approach of human factors design to develop a hypertext-based navigation system, HyperEOPs, for effectively and efficiently managing emergency operating procedures (EOPs).*

## 1. Introduction

A typical nuclear power plant uses various operating procedures, which are intended to cover all likely plant scenarios. The structure of traditional written emergency operating procedures (EOPs) [1] is normally step and instruction based, i.e., the procedures consist of a step of steps, each step consisting of one or more instructions to an operator about checks to do or actions to take. The written form of procedures typically do not conform to accepted human factors principles and imposes methodological difficulties on representing and using information.

Various attempts [2-7] have been carried out to develop computerised EOPs tracking systems. These systems include EOPTS [2], COMPRO [3], OASYS [4], COPRO [5], and COPMA [6]. SACOM [7] and NORMAT [8] are intended to model the interaction between an operator and EOPs.

This paper presents hypertext-based techniques, and introduces our approach of human factors design to develop a hypertext-based (or hypermedia-based) navigation system, HyperEOPs, for managing EOPs. Navigation is referred as the process of moving through a hypertext information.

Hypertext is a technique for information navigation, which takes advantages of allowing for the development of large informational spaces, a possibility of considerable merits for the computerisation of operating procedures in nuclear power plants. A hypertext system [9,10] is viewed as a set of nodes and links to navigate voluminous information spaces.

Traditional document structure is sequential, i.e., there is a single linear sequence defining the order in which the text is to be accessed. Hypertext is non-sequential, i.e., there is no single order that determines the sequence in which text is to be read.

Comparing with voluminous document-based EOPs, HyperEOPs has the capability to effectively and efficiently follow EOPs in emergency situation.

## 2. Computerisation of Emergency Operating Procedures

EOPs are a well defined framework for emergency operations. The purposes of the EOPs are to assist operators during emergency and to mitigate accidents. EOPs provides diagnostic treatments based on written logic steps, which are voluminous and complicate for effective references under high stress situation.

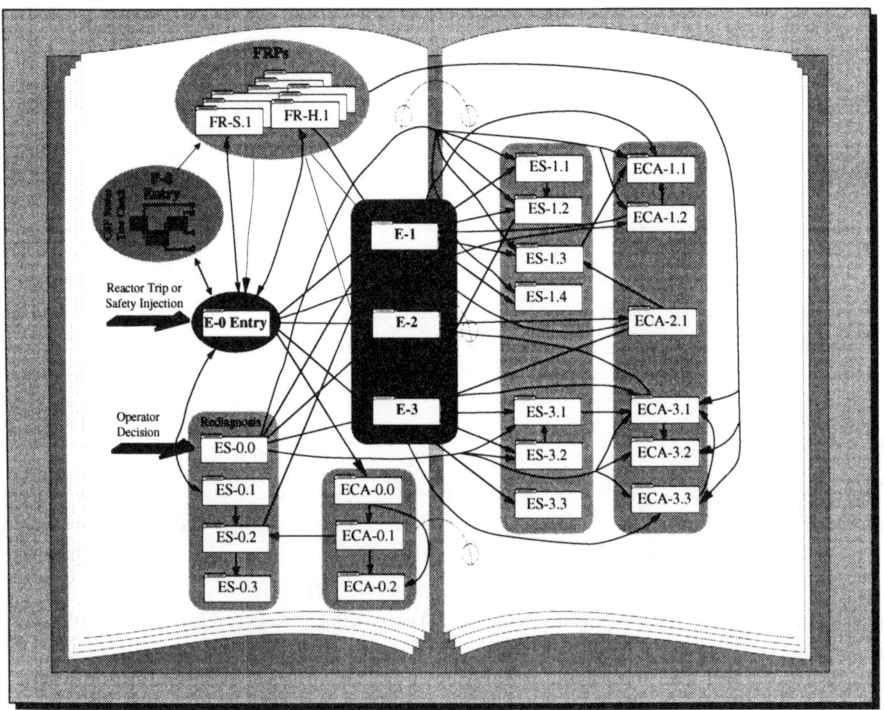


Figure 1. The navigation of guidelines in EOPs

EOPs consists of two main procedures; optimal recovery guidelines (ORGs) and function restoration guidelines (FRGs). Figure 1 shows various guidelines and navigation relationship in EOPs. The ORGs are symptoms based and scenario oriented, and are composed of three types of guidelines; emergency guidelines (E-series), emergency sub-guidelines (ES-series) and emergency contingency action guidelines (ECA-series). FRGs are broadly classified into two classes; critical safety function (CSF) status trees (F-0) and function restoration procedures (FRPs). ORGs are entered upon diagnosis that the reactor protection system limits or the safeguards actuation system limits are exceeded.

The issues for the computerisation of EOPs includes navigation through procedures, formatting and presentation of procedures, help and explanation facilities, and process linking [11,12]. The characteristics of operators in following EOPs include [13]:

- It is important for operators to understand the logic and rationale behind the procedures.

- Operators do not necessarily move linearly through a single procedure path. Operators looked ahead in the procedures, they moved back to earlier steps, and they looked at other procedures in parallel as guidance.

Thus, ease of navigation through procedure networks is likely to be important for facilitating operator's performance in emergency operations. This may be realised by introducing the hypertext techniques.

# 3. Hypertext Techniques

As shown in Figure 2, a hypertext-based system [9] employs a set of *nodes* and *links* to navigate voluminous information spaces. That is, the system employs database and windowing techniques to provide selective access to nodes of information. At the same time, it supports a variety of navigation styles by allowing a user to follow predetermined links from one node to another.

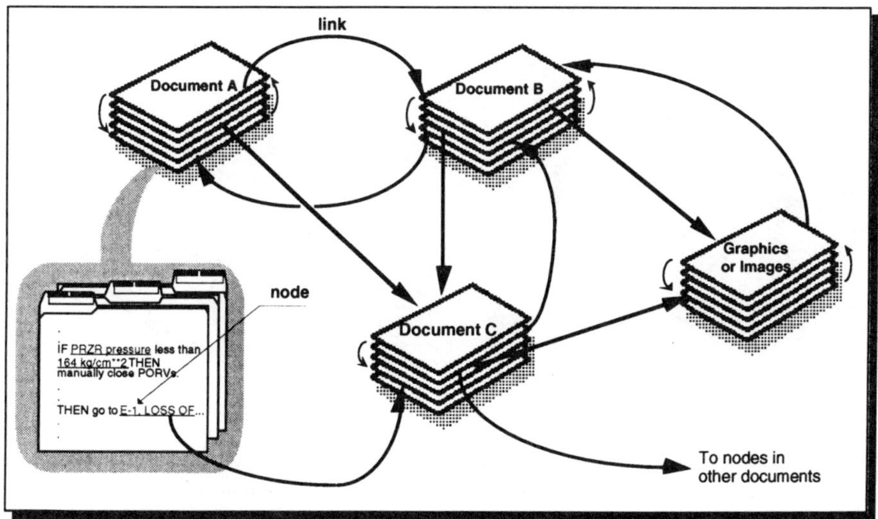


Figure 2. A simple diagram of a hypertext-based system

Various types of documents can be easily implemented in a hypertext-based system by modular writing, i.e., creating a single node and then using the node by creating a link to the contents of the documents.

# 4. Prototype Design of the Navigation System

A prototype navigation system, HyperEOPs, is currently being developed to help the operator verify plant status and take appropriate actions in the emergency situation. The development approach in building the system is to preserve the form and structure of the actual written procedures as much as possible.

Figure 3 shows the structure of the navigation system. The system is implemented on a Power Macintosh 8100 using SuperCard® [14], which is a tool for authoring custom applications (e.g., graphical databases and information systems) and multimedia. This tool can assign a hypertext style to words and phrases and then attach scripts to provide easy nonlinear navigation. English-like script language enables to control over every object and every user action on screens.

As shown in Figure 3, the system consists of two main modules:

- *a navigation handler*, which includes various script codes for the controls of navigation, windows and menus.
- *a hypertext information base*, which contains computerised formats of EOPs and those of EOPs-related documents, such as background EOPs, abnormal operating procedures (AOPs), alarm response documents, and piping and instrumentation diagrams (P&IDs).

For each title step in EOPs, the background EOPs contains its purpose, basic knowledge, required actions, list of instruments/switches, and automatic actions. The alarm response document contains alarm setpoints, automatic actions, probable causes, emergency actions and follow-up actions. These voluminous documents can be efficiently computerised by using a high resolution scanner and image processing and character recognition tools.

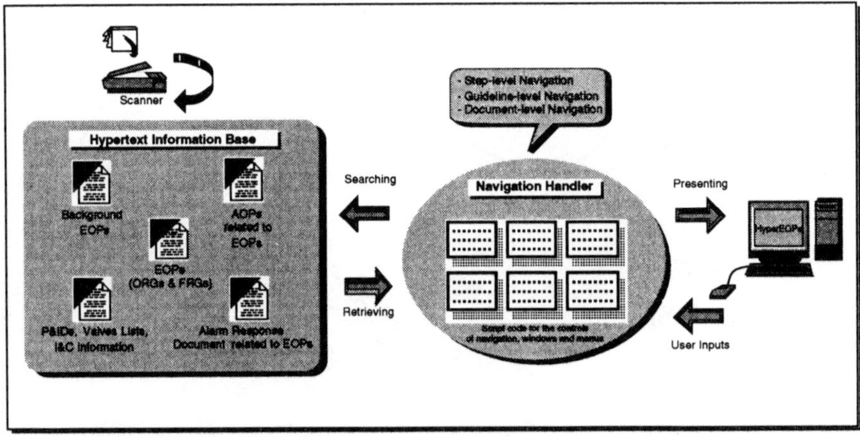


Figure 3. The structure of the navigation system

The system has setup alternatives for the user to explore instead of a single paths of information. HyperEOPs has several different navigation modes based on user's specific information needs to explore instead of a single path of information. The navigation mode include step-level branches in a document, guideline-level branches, and document-level branches. Figure 4 shows the user interface design of the prototype navigation system.

# 5. Discussions and Further Works

The hypertext-base approach may improve operator's understanding of the links between procedure steps and task performance by easy and integrated navigation of multiple documents. Operators can get more information about the required background information in following EOPs. In addition, modular creation of new documents is possible by creating a single node and a link to the contents of the documents. This feature enables to implement and maintain the hypertext-based system efficiently.

The navigation system can also apply to a training system for novice operators. For training purpose, HyperEOPs can generate a set of predefined step flows in EOPs for several emergency scenarios, such as loss of coolant accident (LOCA), steam generator tube rupture (SGTR) and loss of feedwater flow accident (LOFA). Training with the system, operators can familiarise with several emergency situations.

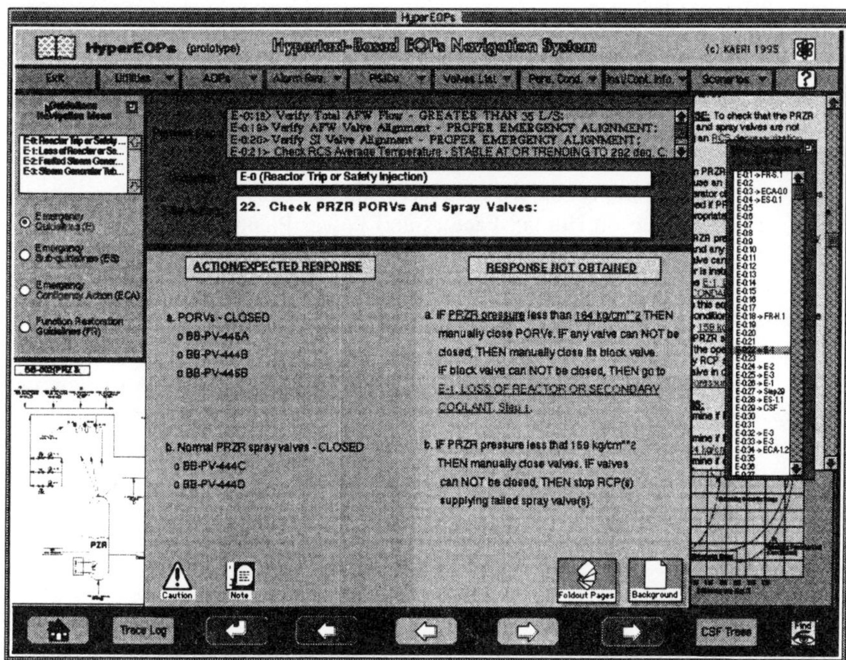


Figure 4. User interface design of the hypertext-based navigation system

To evaluate the performance of the navigation system, the system will be connected with an integrated test facility (ITF), which includes a full-scope plant simulator for human factors experiments. In connection with the ITF, an information module, which can display various plant parameters and plant status, will be implemented on the navigation system.

## Acknowledgment

We would like to acknowledge the financial support of the Ministry of Science and Technology (MOST) for this work.

## References

1. Westinghouse Owner's Group, Emergency Response Guidelines, HP-Rev. 1A, Jul. 1987.
2. William Petrick *et al.*, "A Production System for Computerized Emergency Procedures Tracking," Conf. on Expert System Applications in Power Plants

sponsored by EPRI, Belmont, CA, Dec. 1987.

3. M. H. Lipner and R. G. Orendi, "Issues involved with Computerizing Emergency Operating Procedures," Proc. of AI91 Frontiers in Innovative Computing for the Nuclear Industry, Wyoming, Sept. 15-18, 1991.
4. S. H. Chang *et al.*, "Development of the On-line Operator Aid System (OASYS™) Using Rule Based Expert System and Fuzzy Logic for Nuclear Power Plants," The sixth Int'l. Conf. on Human-Computer Interaction, Yokohama, Japan, July 9-14, 1995.
5. B. Mavko *et al.*, "Computer Managed Tool For Advanced NPP Control Room," Proc. of Topical Meeting on Computer-Based Human Support Systems: Technology, Methods, and Future, Philadelphia, Pennsylvania, June 25-29, 1995, pp. 249-256.
6. S. A. Converse *et al.*, "Computerized Procedures for Nuclear Power Plants: Evaluation of the Computerized Procedures Manual (COPMA-II)," Proc. of the IEEE 5th Conference on Human Factors and Power Plants, Monterey, CA, June 7-11, 1992, pp. 167-172.
7. S. W. Cheon *et al.*, "Development of a Cognitive Task Simulation Analyzer of Operators in Nuclear Power Plants Using Blackboard Techniques," Proc. of the Pacific-Asian Conf. on Expert Systems, Huangshan, China, May 15-18, 1995.
8. S. L. Parrish, W. B. Bobnar and J. F. Kunze, "NORMAT: An Integrated Knowledge-Based Code System for Modeling Emergency Operating Procedures," Proc. of the Topical Meeting on ANS Computer-Based Human Support Systems: Technology, Methods, and Future, Philadelphia, Pennsylvania, June 25-29, 1995, pp. 296-301.
9. J. Conklin, "Hypertext: An Introduction and Survey," IEEE Computer, Vol 20, No. 9, 1987.
10. L. H. Tsoukalas, B. R. Upadhyaya, and N. E. Clapp, "Hypertext-Based Integration for Nuclear Power Plant Maintenance and Operations," Proc. of AI91 Conference on Frontiers in Innovative Computing for the Nuclear Industry, Jackson, Wyoming, Sept. 15-18, 1991.
11. M. Green, E. Hollnagel and Y. Niwa, "Guidelines for the Presentation of Emergency Operating Procedures Using Advanced Information Technology," Proc. of the IAEA Specialists Meeting on Advanced Information Methods and Artificial Intelligence in Nuclear Power Plant Control Rooms, IAEA-12-SP-384.37, Halden, Norway, Sept. 13-15, 1994, pp. 86-93.
12. Y. Niwa and E. Hollnagel, "The Design Of Computerized Procedure Presentation For Nuclear Power Plants," The Sixth Int'l. Conf. on Human-Computer Interaction, Yokohama, Japan, July 9-14, 1995.
13. E. M. Roth, R. J. Mumaw and P. M. Lewis, "Enhancing Crew Performance in Complex Emergencies: What We Can Learn From Simulator Data," Proc. of Topical Meeting on Computer-Based Human Support Systems: Technology, Methods, and Future, Philadelphia, Pennsylvania, June 25-29, 1995, pp. 24-31.
14. Allegiant SuperCard® Macintosh User Guide, Allegiant Technologies, Inc., San Diego, CA, 1994.

# A Probabilistic Environmental Decision Support Framework for Managing Risk and Resources

David P. Gallegos, Erik K. Webb, Paul A. Davis, and Stephen H. Conrad
Sandia National Laboratories
Albuquerque, NM, USA

**Abstract**

The ability to make cost effective, timely decisions associated with waste management and environmental remediation problems has been the subject of considerable debate in recent years. On one hand, environmental decision makers do not have unlimited resources that they can apply to come to resolution on outstanding and uncertain technical issues. On the other hand, because of the possible impending consequences associated with these types of systems, avoiding making a decision is usually not an alternative either. Therefore, a structured, quantitative process is necessary that will facilitate technically defensible decision making in light of both uncertainty and resource constraints. An environmental decision support framework has been developed to provide a logical structure that defines a cost-effective, traceable, and defensible path to closure on decisions regarding compliance and resource allocation. The methodology has been applied effectively to waste disposal problems and is being adapted and implemented in subsurface environmental remediation problems.

## 1.0 Introduction

The environmental decision support framework described herein offers a generalized probabilistic framework for making consistent, defensible, and traceable environmental decisions. These decisions apply to resource management, prioritization of data collection activities, and adequacy of system performance. Most importantly, the framework provides a foundation for negotiation between owner/operators and regulators that facilitates coming to closure on decisions in a timely and cost-effective manner. Risk management, as implemented in this framework, involves the development and application of probabilistic approaches for guiding environmental restoration, risk assessment, and waste management decision making. Resource management, as guided by this framework, is accomplished through up-front, articulated objective setting, and continuous cognizance of those objectives while using sensitivity and data worth analyses to set information collection priorities. An objective of risk management in general, and this framework in particular, has been to synthesize an approach that can be commonly and collectively applied by all parties involved in environmental decisions, including regulators, site operators, policy makers, and other stakeholders.

## 2.0 Existing Approaches and Structures

Traditional approaches to evaluating risk and system performance tend to be straightforward and linear with data collection preceding the actual decision support analyses. Generally, the process begins by constructing a description of the system using existing data. This conceptualization of the system tends to describe system processes and the configuration of boundaries and internal structures with the driving purpose being simply to understand the system. This description is used as the basis for planning site characterization activities. Following this approach, because the system description can be allowed to remain vague, the characterization needs cannot help but be somewhat ambiguous. Decision analysis, in general, involves the process of getting from the data to the decision. In the context of environmental risk management, such analyses can range from making direct inferences from the data to conducting some (typically deterministic) calculation or modeling of system behavior. If data collection has been either insufficient or superfluous, it is commonly not discovered until the end of the process. For the linear approaches, the connection between data collection and the decision to be made tends to be more implicit than explicit because the feedback loops from analysis back to data collection tend to be weak (data collection largely precedes analysis). Because traditional approaches have been recognized to be time-consuming and inefficient, improved approaches have been proposed.

The Environmental Protection Agency (EPA) has developed the Data Quality Objectives (DQOs) approach with the goal of explicitly linking data collection to the decision [1]. It formalizes the process of planning data collection by unambiguously stating the objective of the study (i.e. the decision to be made), and specifying tolerable limits on decision errors (i.e. the degree of certainty required before a decision can be made). The presumption is that reaching consensus on these issues during the planning process will allow for improved identification of the type, quantity, and quality of data that will need to be collected to support decision making. The DQO approach uses classical statistics and hypothesis testing to help make determinations on sampling design. Because the DQO approach relies on statistical inference, it is not particularly amenable to incorporating knowledge of contaminant transport processes.

The Streamlined Approach for Environmental Restoration (SAFER) combines elements of the DQO process with an approach for contingency planning to facilitate the management of uncertainty [2]. The SAFER approach provides a framework for planning and conducting remediation in a more efficient manner than more traditional approaches. It extends the DQO process by tracking implementation of the decision, looking for deviations, and having contingency plans ready. While these approaches provide improvements over more traditional, linear approaches, they are still limited by their forward-looking approach that attempts to anticipate the data required for decision making. Iterative approaches provide much stronger feedback loops, allowing for a much more explicit link from data collection to the decision.

## 3.0 Description of the Decision Support Framework

The environmental decision support framework described here utilizes an approach that incorporates process modeling, performance assessment and decision analyses, and will eventually incorporate cost/benefit analysis, and site sampling optimization techniques. Quantitative estimates of risk to human health, risk to the environment, dose to humans, or other appropriate performance measures are based on a probabilistic assessment of the site conditions. This type of analysis is desirable because it explicitly incorporates the uncertainty in information about the natural system and in turn, provides a representation of the uncertainty associated with the performance of the system; consequently, it provides a complete set of information necessary to make robust decisions regarding site safety, to direct additional site characterization, or to define site remediation schemes. The results also provide a consistent framework for comparing alternative system conceptualizations or comparing alternative engineered designs.

In developing and applying this methodology, an iterative approach has been stressed in which the definition of performance objectives sets the context of all subsequent analyses and decisions. Consequence analyses are conducted early on, and through sensitivity and data worth analyses, are used to guide the collection of site characterization data. In turn, the resulting new site characterization data are incorporated into each successive iteration of the analysis. The advantages of this approach is that decisions are focused around agreed-to performance measures, uncertainty in models and parameters are clearly articulated, and means are provided for tracking the reduction of the uncertainties associated with the technical analysis as more data are collected. The iterations in the process are repeated until either the uncertainty has been reduced to the point that a clear decision can be made or we find that continued data collection necessary for further reduction in uncertainty is prohibitively expensive. Ultimately, the decision makers want to minimize the probability of making a wrong decision while at the same time minimize the cost and time that it takes to make and support the decision. To address these constraints, this approach offers a logical and integrated framework for guiding technically-based decisions and for facilitating negotiations around items that are truly critical to the decisions.

## 3.1 Detailed Description of Technical Components

The decision framework is an iterative process consisting of nine steps. The integrated framework is shown in Figure 1 and the components are described in detail below.

### *3.1.1 Definition of Performance Objectives*

Performance objectives are defined in the initial step in the process because they act as the driver for all subsequent steps. Defining performance measures requires an analyst to state a question or type of analysis, the alternative answers to the question, and the specific criteria that will be used to make a decision. In addition, a threshold value for the criteria (e.g., 5 ppm for a maximum concentration limit or $10^{-6}$ excess cancer risk) and probability of meeting this threshold (e.g. 95% probability of ...) must be specified. Obviously, a wide range of thresholds, metrics, and questions can be phrased in this way. Each permutation establishes a need for a slightly different analysis. Nevertheless, the generalized framework is still applicable.

### *3.1.2 Data and Information Assimilation and Management*

The first iteration of the steps in the framework is based on existing data, information or knowledge. To accomplish this task the data may need to be assembled, organized and established in some form of data base. Additionally, the quality and source of information should be evaluated and maintained. Then the analyst should attempt to assimilate the existing information in order to make a site interpretation as part of the next step.

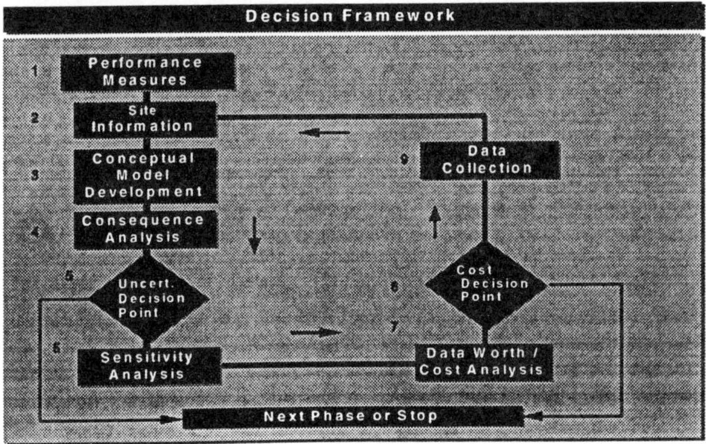


**Figure 1** Environmental Decision Support Framework

### 3.1.3 Conceptual Model Development

Conceptual model development requires an interpretation of data to develop a set of assumptions about the physical system. The uncertainty about pathways and processes (conceptual model uncertainty) as well as their controlling factors (parameter uncertainty) must also be defined. Thus, if more than one conceptual model is plausible or presented by an interested party, then each must be defined and evaluated through all steps of the decision framework to determine its validity and significance. Once these models have been developed, and documented, a probabilistic numerical simulation technique must be selected that represents the suite of specified assumptions. Then, the necessary input parameters and their associated uncertainties to perform each analysis must be defined. To do this, each parameter is defined as a distribution from which individual values are extracted for each probabilistic simulation.

### 3.1.4 Consequence Analysis

The core of this analysis is performing probabilistic physical process simulation. Generally this can be performed using Monte Carlo techniques. Then, once the simulation step is complete, the results are compiled in the form of a statistical distribution. The amount of effort and computational power necessary to accomplish this task is determined from the complexity of the conceptual model and the corresponding simulation tools.

### 3.1.5 Performance Evaluation and Decision Making

The uncertainty decision point is an evaluation of the simulation results as compared to the explicitly defined performance measures. This is generally accomplished graphically using either histograms, cumulative distribution functions, or probability maps for two and three-dimensional analysis. If the estimated distribution of concentration/risk/dose values either grossly exceeds or falls below the threshold and specified probability, an unambiguous decision can be made and defended. If the output distribution falls across the threshold and contains wide ranges of uncertainty, the user can decide to continue the analysis to define what and how much additional data collection would be of value in reducing uncertainty enough to make a decision.

### 3.1.6 Sensitivity Analysis

Sensitivity analysis is defined as a comparison of variations in input parameters with variations in the resulting output values over the previously defined distributions for each input. Techniques for performing this step include rank correlation or regression analysis on normalized input distributions. The purpose of this process is to identify those parameters which cause the greatest variance in estimates of the output based on current estimates of the input parameter distributions. The results of this analysis may help identify parameters for which collecting new data should be evaluated in the data worth step.

### 3.1.7 Data Worth

Data worth involves specifying the options one might take relative to the original question (e.g., safe or not safe, selection of remedial options) and what data one might collect to further refine the confidence in a decision. Quantitative approaches to data worth primarily involve establishing a decision tree and assigning to the component branches of the tree, probabilities of that set of conditions occurring (either from the consequence analysis or a similar computation) and the cost that would accrue if that happens. In addition, if data collection or experimental work is part of the option, the likelihood that the field or laboratory work will yield useful information is a factor in the probability of conditions occurring. By compiling this information, one can determine the least cost, highest probability of success, least time consuming, or some combination of the above.

### *3.1.8 Cost Evaluation and Decision Making*

This step involves determining which of the various options is most cost effective. If none of the additional data collection options is cost effective, the decision maker must accept the default decision resulting from the uncertainty decision point. If one option is most cost effective, then that option (remedy, or data collection) is pursued.

### *3.1.9 Data Collection*

In this step of the process, new data is collected, checked, and added to the existing data compilation. Given that the information gathered was defined through a data needs analysis, it should alter the analyst's understanding of the system and cause some change in the conceptual model, selection of numerical model, or parameter distributions. Additionally, this new data may help eliminate one or more of the alternative conceptual models. The altered conceptual model is then used in the succeeding consequence analysis and so forth.

# 4.0 Application and Development of the Decision Framework

## 4.1 Low-Level Waste Disposal and Decontamination and Decommissioning

One of the initial driving forces for the development of a software implementation of the framework (called the Sandia Environmental Decision Support System or SEDSS) was the need to provide a user-friendly platform for the Nuclear Regulatory Commission (NRC) methodology that is used in assessing the long-term performance of low-level radioactive waste (LLW) disposal sites. This user-friendly platform was also intended to supply the NRC, the states responsible for LLW disposal regulation, and site operators with a consistent technical basis for regulatory analysis of proposed sites. The LLW performance assessment methodology consists of the structured approach to decision making employed by SEDSS, models used to simulate release and transport of radionuclides from a LLW disposal facility to humans, and methods of treating model and data uncertainty. Since then, the focus of the NRC decision-related work at Sandia has been in decontamination and decommissioning (D&D) of nuclear facilities licensed by the NRC. Here the tools developed for LLW disposal are being evaluated for their applicability to D&D problems. Once evaluated, a new capability will be added to SEDSS to allow for the analysis of D&D sites.

## 4.2 Uranium Mill Tailings Remediation

The Uranium Mill Tailings Remedial Action (UMTRA) program is part of the U.S. Department of Energy's effort to restore uranium mining, processing, and enriching facilities. Specifically UMTRA is responsible for 22 sites of abandoned uranium mills. As UMTRA turns its attention to groundwater remediation, there are several possible approaches that can be used for each facility. Among these is an option to perform no additional remediation and allow the natural movement of groundwater to flush contaminants to the surface drainage where they are diluted to below harmful levels. This is termed the "no-further-action" alternative. To assess whether this alternative is protective of human health, the UMTRA program is planning to apply the decision framework. This is a safety assessment with two alternative answers, "do something" or "do nothing." The determination will occur based on the sites ability to meet a specified set of concentration limits everywhere over the site at a regulatory defined 100 year compliance timeframe with a probability specified by DOE analysts. This probability has not yet been set with the regulatory authority, the NRC. If the "do something" alternative is chosen then a more detailed analysis, again using the framework, can be used to determine which of the proposed alternative remedies should be used and refine alternative engineering designs. Actual field implementation is planned after beta release of the SEDSS during the summer of 1996.

### 4.3 Superfund

Beginning in 1992, Sandia has provided technical assistance to EPA's Superfund program in developing improved strategies for site characterization for designing remediation and groundwater monitoring systems, and obtaining data for performing risk assessment involving the groundwater pathway. Prior to that time, the Superfund program had been criticized for mandating the implementation of ineffectual and costly site characterization strategies in their regulations and in their guidance. Implementation of these strategies led to superfluous or insufficient data collection, and inconsistent decision making. To address this problem, improved site characterization strategies for the groundwater pathway were developed, including an examination of the links between site characterization and each of the following: risk assessment, remediation selection, remedial progress tracking, and monitoring to assure contaminant containment. The synthesis of these improved strategies into a comprehensive approach resulted in the formation of the environmental decision support framework that we have presented in this paper. Since that time, the EPA Superfund program has promoted development of the SEDSS software system to automate the approaches advocated in the framework.

## 5.0 Summary and Status

The risk-based decision support framework discussed in this document will be embodied in a software tool called the Sandia Environmental Decision Support System (SEDSS). The framework and the tool are being developed to provide for more cost-effective and transparent decisions regarding environmental restoration and waste management problems. The decision framework and its embodiment in SEDSS is designed to streamline the decision making process by: (1) providing a platform for negotiation between site operators and regulators, (2) quantifying the uncertainty associated with the decision making process, and (3) extending the decision making process from the level of setting clean up criteria and acceptance levels to the level of making decisions about whether to proceed with no-further-action alternatives, to continue characterization, or to choose a remedial alternative. The methodology employed is probabilistic to provide for a direct quantitative connection between data collection and decision making, and to provide an explicit mechanism for calculating risk. This methodology relies on contaminant fate and transport modeling to allow for assessing risk level and setting clean up goals and to provide added knowledge about the extent and nature of contamination. The inclusion of cost/benefit and data optimization routines is designed to aid decision makers in choosing efficient and effective data collection and remediation strategies. Additionally, implementing the methodology into a user-friendly software system will allow for wide use and application of the methodology. And finally, the participation of the regulatory bodies, EPA and NRC, should enhance the likelihood of successful applications of the SEDSS methodology and software.

### References

1. United States Environmental Protection Agency, Guidance for the data quality objectives process. EPA QA/G-4. 68 p., 1994.
2. Blacker, S.M. , Cost-effective environmental restoration and corrective action using the data quality objective process. proceedings of HMC/Superfund '92, December 1-3, 1992, Washington, D.C., pp. 1022-1025, 1992.

# A Risk-based Methodology for Addressing Environmental Risk from a Wide Variety of Contaminated Sites at the Idaho National Engineering Laboratory[a]

Robert L. Nitschke
Lockheed Martin Idaho Technologies
Idaho Falls, USA

The Idaho National Engineering Laboratory (INEL) is a multidisciplinary applied-engineering laboratory located in Southeastern Idaho, covering approximately 1430 $km^2$ of high semi-arid desert on the eastern Snake River Plain. The laboratory has a rich history of a wide variety of activities spanning over 45 years. The primary functions have been related to nuclear reactor research and development, and radioactive waste management treatment, storage, and disposal. The Navy was the first to use the site, as a gunnery range for testing 41-cm-diameter battleship guns; another area was used by the Army as an aerial gunnery range. During its first forty years, the INEL was home to fifty-two reactors, most of them first-of-a-kind. In addition to the testing of these reactors, fuel testing and reprocessing of Naval fuel was performed. As a result of many of these past operations, the INEL was placed on the Environmental Protection Agency's (EPA's) Superfund National Priorities List (NPL) in November of 1989. This action set into motion the requirements for performance of investigations to "determine fully the nature and extent of any threat to the public and the environment caused by . . . hazardous substances . . .". This listing and the resultant actions identified over 400 potential hazardous waste sites that needed to be addressed. These sites ranged in complexity from a rubble pile with construction debris, to radioactively contaminated pond sediments, to a 0.4-$km^2$ burial ground consisting of large quantities of transuranic, low-level, and mixed transuranic and low-level wastes. It was quite obvious that the degree of attention and information that would be needed to make informed decisions about what remedial actions if any these sites

---

a. Work supported by the U.S. Department of Energy, under DOE Idaho Operations Contract DE-AC07-94ID13223.

might require should vary greatly. This paper describes the development and successful implementation of a graded and streamlined approach being used to provide the necessary and appropriate information needed by the decision makers to make sound and cost-effective cleanup decisions.

The graded and streamlined risk-based approach consists of four different levels of analysis and documentation. *Graded*, in this context, refers to the computational rigor applied, the quality of data used, and the treatment of uncertainty. *Streamlined* refers to the process used to get consensus on the entire process. This includes not only the overall approach, the methodology, and key parameter values, but also how the risk information would be used. It was recognized early, that no matter how good the approach and method selected, unless there were agreement and understanding with the end-users on the information being provided, the effort would be in vain. This approach allows for the judicious allocation of limited resources (both money and manpower) and still meets the statutory time constraints under the Superfund law. The methodology and select parameter values were formalized in a set of guidance documents to ensure consistency when appropriate, standardize conservative assumptions, and provide a uniform presentation of the risk information.

The simplest evaluation is referred to as a Track 1 risk assessment. This assessment is appropriate for what are termed *low probability hazard sites*. Low probability hazard sites are those that are suspected to have low or no quantities of residual contamination and would pose no unacceptable risk. Even the existence of some of these sites was based on extremely limited information, such as someone's recollection of past practices. The methodology, in conjunction with the organized collection of historic data and other pertinent information, is used to develop a qualitative risk assessment. This methodology develops risk-based soil screening concentrations to evaluate potential hazardous and radioactive contaminants in a consistent, logical, timely, cost-effective, and defensible manner. For the purposes of this screening, humans are regarded as a sensitive indicator species for the ecosystem as a whole. The methodology focuses on the major environmental media (air, soil, and groundwater), pathways (ingestion, inhalation, and, in the case of radioactive contaminants, direct exposure), receptors (adults and children), and exposure scenarios (industrial and residential) to identify the significant risk drivers if any. As a result of conservative algorithms, site-specific parameter values, and exposure parameters, this screening is sufficiently conservative to justify no further action at sites with very little risk as well as further consideration of sites and contaminants that may pose a significant risk.

As an example, based on historical information, limited sampling information, and process knowledge, sediments in an abandoned leach pond contained an estimated 10 mg/kg of cadmium. Upon conduct of a qualitative risk assessment, the risk-based soil screening criteria for a Hazard Quotient of 1 and for a residential exposure scenario is 135 mg/kg. Based on this information, the probability of adverse health effects from cadmium can be considered negligible and a no-further-action decision could be selected. Forty-two percent of the original list of potential hazardous sites were categorized and addressed using this Track 1 process. This allowed more time to be spent on the more hazardous sites and saved a tremendous amount of money.

The second level of analysis is called a Track 2 risk assessment. Track 2 sites are low-probability hazard sites where insufficient data are available to make a decision regarding the risk level or to select or design a remedy. The goal of the Track 2 process is to use existing qualitative and quantitative data (such as the Track 1 process) to minimize the collection of new environmental data. A structured format consisting of a series of questions and tables is used to generate a reproducible and defensible method. The questions are as follows:

- What are the waste generation processes, locations, and dates of operation associated with the site?
- What are the disposal processes, locations, and dates of operation associated with this site?
- Is there empirical, circumstantial, or other evidence that sources exist at this site? If so, list the sources and describe the evidence.
- Does site operating or disposal historical information allow estimation of the pattern of potential contamination? Discuss the estimated patterns of potential contamination over time.
- Estimate the length, width, and depth of the contaminated region as it is today. What is the known or estimated volume of each source as it is today?
- What is the known or estimated concentration or quantity of each hazardous substance/constituent at each source? If quantities are estimates, explain carefully how each estimate was derived.
- Is there empirical, circumstantial, or other evidence of a release and/or contamination in a pathway? Discuss the evidence. Address each potential pathway.
- Is there evidence that hazardous substances or constituents are present at any of the potential sources today? If so, describe the evidence. For each pathway, discuss fate and transport.

The qualitative risk assessment then proceeds, generating both backward calculations of risk-based screening soil concentrations as well as forward calculations of risk. As in the Track 1 process, conservative fate and transport algorithms, site specific parameter values, and exposure parameters are used to give an overall conservative estimate of the potential risk posed by the site being evaluated. As a result, the risk information can justify one of three outcomes:

1. There is no unacceptable risk; therefore no further action is required.
2. The site poses an unacceptable risk and sufficient information exists to select a remedy. An interim action will be initiated.
3. A more detailed risk analysis or remedy selection is required, the site will proceed to the formal CERCLA Remedial Investigation/Feasibility Study (RI/FS) process.

An acid pit that was used to dispose of liquid organic and inorganic wastes, some of which were radioactively contaminated, is an example of a site where sampling data were needed and collected and the Track 2 process used to evaluate the potential environmental risk. While there was much historical knowledge, the process knowledge was sketchy and assurance of a reasonably complete inventory of contaminants was lacking. Results from the additional Track 2 sampling effort were inconclusive, and a decision was made to add the acid pit to a surrounding site that is being addressed by the next level of analyses. For the INEL as a whole, 31% of the sites were addressed using the Track 2 process. Again, by applying more appropriate levels of rigor and data needs, much time and money were saved.

The third level of analyses in this approach is called a Baseline Risk Assessment (BRA) and follows the standard EPA risk assessment guidance. See the EPA report entitled "Risk Assessment Guidance for Superfund (RAGS)" for complete details on the process. The BRA is a deterministic presentation of the risk using upper 95th percentile data and provides a qualitative evaluation of the uncertainty. Four percent of the sites are being addressed using this full-blown approach.

An example of a site for which the traditional EPA BRA approach was used is a disposal site for containerized radioactive waste contaminated with less than 10 nCi/g of transuranic (TRU) nuclides. The waste was predominately evaporator salts from a nuclear weapons assembly plant. At the time of closure, there were over 18,000 metal drums and over 2,000 plywood boxes, for a total waste volume of over 10,000 cubic meters. Closure was performed by placing plywood or polyethylene over some of the containers, followed by a soil cover 0.9 to 1.8-m thick. The risk evaluation allowed for 100 years of institutional control and proceeded 1000 years into the future. Both an occupational and a hypothetical residential scenario were considered with five exposure routes and three locations being addressed. The risk characterization indicates that the carcinogenic risk for current and future hypothetical scenarios were below or within the National Contingency Plan (NCP) acceptable risk range of $10^{-4}$ to $10^{-6}$. The only unacceptable adverse health effect was from nitrate contamination of the groundwater to an infant about 200 years into the future. Key factors in this conclusion were the condition of the soil cover and the infiltration rate. Based on these results, the decision-makers decided on a limited action. This action consisted of a contouring operation on the soil cover and a monitoring program to determine the water infiltration rate and the erosion rate to confirm modeling assumptions.

The fourth and most robust evaluation takes the standard BRA approach and uses probabilistic risk analysis (PRA) techniques to provide a distribution of possible risk results and a quantification of the associated uncertainties. Of all the sites on the INEL list, only one is being considered for this very rigorous approach. The site in question is a 0.4-km$^2$ burial ground containing large quantities of radioactive and hazardous substances. The site was established in 1952 for disposal of solid low-level radioactive waste generated by INEL operations. Beginning in 1954 and continuing for the next 16 years, transuranic waste was also buried. The waste was buried in pits, trenches, and soil vaults. The acid pit discussed above is also located within the disposal site boundaries. The source volume is about 300,000 cubic meters. These wastes contain several thousand kg of TRU, several hundred thousand liters of volatile organic compounds such as carbon tetrachloride, large quantities of fission and activation products such as Cs-137 and Co-60, respectively, and other radioactive and hazardous materials. The environmental setting for this site can be briefly described as a high desert plain located about 190 meters above the water table. The vadose zone consists of approximately 7 to 10 meters of surficial sediments with 180 meters of fractured basalt interspersed with three major sedimentary interbeds. Mountains to the west and north purge passing air masses of available moisture, resulting in an arid to semiarid climate. The climate is characteristically warm and dry in the summer and cold in the winter. Average annual precipitation is about 20 cm, snowmelt being a major contributor. Vegetation is dominated by sagebrush, wheatgrass, and rabbit brush. Burrowing rodents and lagomorphs are abundant, the deer mouse being the most abundant. Owing to the complexity of both the environmental setting and the waste constituents and forms, as well as the potential cost of full treatment (over $2 billion), the need for more refined risk information and a better understanding of the uncertainties associated with the risk assessment is warranted. This effort, while time-consuming (two years) and costly (a couple of million dollars), nonetheless has the potential to save hundreds of millions of dollars and will ensure that the dollars that are spent will be spent wisely.

This tiered approach for evaluating environmental risk at the INEL has played an integral and critical role in the successful environmental restoration program. The application of these graded levels of analytical rigor to the different sites has resulted in cost savings/avoidance of over $100 Million for the American taxpayer.

## References

1. Track 1 Sites: Guidance for Assessing Low Probability Hazard Sites at the INEL, DOE/ID-10340(92), July 1992, Revision 1

2. Track 2 Sites: Guidance for Assessing Low Probability Hazard Sites at the INEL, DOE/ID-10389, January 1994, Revision 6

3. Risk Assessment Guidance for Superfund, Volume 1, Human Health Evaluation Manual (Part A), Interim Final, EPA/540/1-89/002

# Ecological Vulnerability Analysis: Towards a New Paradigm for Industrial Development

D.A. Sarigiannis and G. Volta

European Commission
Joint Research Centre
Institute for Systems, Informatics and Safety
21020 Ispra (VA), Italy

**Abstract**

Current day risk analysis practices traditionally include plant safety assessment and the related effects on the health of the workers and the public. This kind of analysis, however, usually fails to describe and take into account the direct and indirect effects of novel technology development on the natural and human ecosystem and its sustainability potential. This work presents the concept of ecological vulnerability analysis as an effective tool for a rigorous assessment of the interaction between industrial technology and the environment. The key issues associated with the effective implementation of this approach into specific process and energy industries are discussed and the relevant environmental risk indicators are described. Moreover, the steps required to integrate the results of this analysis to the design of novel technological systems under increased uncertainty are delineated. This approach to integrated environmental risk assessment may lead to the development of a new paradigm for industrial development.

## 1 Introduction

Traditional PRA techniques used for accident analysis (e.g. WASH 1400 [1]) include a limited array of hazard indicators like fatalities and property damage; usually, however, they do not include the consequent environmental damage and they are limited to operational analyses of the plant in question.

Studies on the fuel cycles of energy technologies and the production systems of other industrial products essentially treat only deaths as the major expression of technological hazard of interest to human analysts. This fundamentally anthropocentric approach to risk and hazard analyses has in part been due to the lack of effective implementation of ecological economics concepts.

Ecological economics concepts do attempt to capture in an integrated description the whole array of interactions between man-made and natural world. They remain, however, obscure to most engineers and industry decision-makers as they depict a very macroscopic picture of the real world without much relevance to actual industrial decision-making.

An intermediary approach that will facilitate decision- and policy-makers and technology developers to deal with real-life problems while enabling them to have a comprehensive picture of the effects their decisions have on the human and natural ecosystem is required. Such an integrated technology assessment can only be based on sustainability indicators because in this way the system is viewed in an external observer perspective.

# 2 Ecological compatibility and vulnerability

The risk-integrated technology assessment and development methodology suggested in this work aims to describe industrial systems as entities interacting with their surrounding human and natural environment. The societal and ecological systems in which novel technologies are introduced are characterized by their capacity to absorb the negative side-effects of technology. This property includes all fundamental functions and services of an ecosystem. The integrated assessment of technological risk for public health and the natural environment may thus be viewed as an assessment of the compatibility of the induced perturbation with the absorbing capacity of the ecosystem. This could be seen as a global perspective for evaluation of the different kinds of risk related to industrial innovation.

Methodologically, the assessment entails the identification and quantification of the fluxes and transformations of material and energy from the beginning (the primary resources) to the end of the production and utilization cycle (cradle-to-grave approach). The critical element in this kind of analysis is the comprehensive and realistic estimation of ecosystem vulnerability, and, hence, the sine qua non conditions of sustainability.

The most recent methodological tools that have resulted from this perspective include *life-cycle analysis* [2], *resource intensity analysis* [3], *resource productivity and material input per unit service (MIPS) accounting* [4], *material input-output accounting* [5]. These methods have their origins in the energy analysis methods developed in the 70s, which paid attention to the energy flows throughout the life cycle of a product. In essence, vulnerability and sustainability are viewed as properties related to material transformation. Finally, the impact on the ecosystem is determined as the potential for "waste" generation by the particular technology.

The assessment of ecological compatibility requires the widening of the perspective under which technological systems are viewed to include the natural environment with which they interact. Use of industrial ecology approaches to the analyses of technological processes facilitates the study of the driving factors influencing the specification of flows of key materials and energy between different stages of the technological system and its external environment. The implementation of such an approach for industrial development leads to design-for-environment methodologies [6],[7]; they integrate social value judgment with multisectoral information into tools and systems that enable the industrial implementation of environmentally preferable technologies and practices and the internalization of environmental objectives and constraints into organizations and operations. Such applications include manufacturing [8] and process design [9], material and technology choice [10],[11], product takeback and recycling to the market [12].

A major methodological problem for this kind of global perspective is the determination of the appropriate degree of aggregation of different vulnerability indicators used in the analysis. Here, four groups of indicators will be assumed:

i. depletion of natural resources;
ii. disposal of toxic/radioactive and non-recyclable or bio-degradable material as waste;
iii. energy loads disposed on environmental reservoirs;
iv. probabilistic risk values calculated for the public and/or operators of industrial plants.

Each group may be analyzed to more process-specific indicators in order to provide an informative picture of the environmental pressures exercised by particular technological choices. Such a disaggregation, however, cannot be done a priori. It should satisfy the requirements for comprehensive representation of the technology in question and of its impacts on the natural and human environment by being of variable complexity. The topological representation of the values of this set of indicators in the $n$-dimensional space (where $n$ is the overall dimension of the indicators set) for the examined technology gives an overall picture of the impact of the technology with the sustainability requirements of the relevant ecosystem. For the four indicators groups described above this graphical representation is given in figure 1.

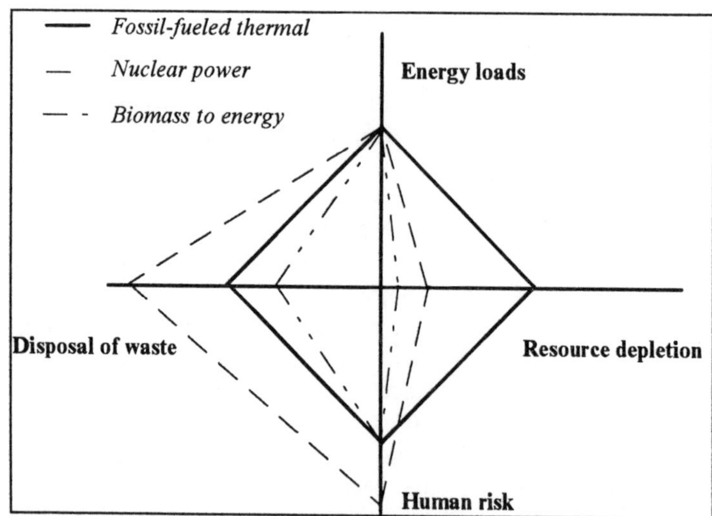


Figure 1: Integrated comparison of selected energy technologies with regard to their ecological compatibility

A direct measure of the ecological compatibility of industrial technologies is given by the surface delineated by the rectangles shown in the 4-dimensional space. If the eco-compatibility indicators are further disaggregated to $n$ the corresponding measure would be the volume of the corresponding sphere in the $n$-dimensional space.

Through normalization of the $n$-dimensional ecological compatibility space comprehensive intertechnology comparisons can be realized without sacrificing the possibility to perform detailed comparisons of particular aspects of technology. This measure can be a useful tool for the identification of the best technological solutions in order to provide goods and services needed by society while putting the minimum pressure on human and environmental health.

# 3 Technological system representation

The principal bottleneck for the development of the technology assessment method described above is the coherence and comprehensiveness of the representation of the technological system and its interactions with the ecosystem. The technology representation and assessment model suggested herein is based on a multi-level object-oriented representation [13] of industrial processes as follows:

I. *The Structural (Physical) Dimension.*

The structural dimension of a technology representation includes the physical and topological description of the technology and its sub-division in components. This model is organized in a hierarchical structure made of structural levels; each level contains a set of interconnected components. Moving down this hierarchical tree the amount of detail in the physical description of the components increases. Depending on the use of the model and the experiential information available, different levels of detail can be introduced.

This model can be coupled with the energy and life-cycle analyses of the materials used for the fabrication and scheduled maintenance of the technology components.

II. *The Functional Dimension.*

The functional representation describes the various functions and objectives fulfilled by the technology in question and its components. These functions converge towards the overall objective function of the given technology which in general consists of the transformation of materials and/or energy to useful products / services. The functional representation defines also the interface between the examined technology and other relevant technologies (e.g. management, control, and service systems).

III. *The Operational / Non-functional Dimension.*

This representation elucidates how different operational configurations of a technological system are related to functional and non-functional requirements (safety, accidental release of hazardous materials, etc.). The behavior of components according to their structural and operational attributes are described using formalisms such as Failure Mode and Effect Analyses (FMEA), transfer functions or production rules.

The information generated by the development of the multi-level model allows the construction of a set of indicators suitable for multicriteria assessment qualification and selection of the technology. The indicators relate to:

(a) materials flow and degradation;
(b) economics (cost, rate of return, capital recovery factors, etc.);
(c) environmental pressures and impacts;
(d) social pressures (e.g. major hazards, risks to public health).

The output of the model includes the dynamic evolution of the technology in time or, in other terms, the life cycle of the process in question. The assessment can therefore include the impact of the new technology on the overall sustainability of the ecosystem to which it will be introduced.

The efficiency and practicality of this multi-level representation of technological systems as a tool for integrated technology assessment is, however, conditioned by the following requirements:

(a) *Temporal coherence* of the different representations. This is a non-trivial problem related to the late effects of technology on the natural environment and the public health. A typical example is given by the environmental and health considerations of nuclear power plants decommissioning; in such a case, the adverse effects associated with the normal operation of the reactor extend much beyond its operational lifetime. Integrated analyses of the safety and environmental characteristics relevant to the development of new reactor types should however account for such retarded effects and incorporate them into the overall life-cycle assessment of the technology.

(b) Development of a *normal surface in the sustainability phase space* using the indicators chosen for each type of representation. This is very important for the consistent and comprehensive comparative assessment of different technologies with regard to the degree of perturbation implicit in their introduction in a societal system. What is however the appropriate norm according to which the different aspects of ecological vulnerability can be measured ? Such measures should account for the differences in the cultural and environmental context to which the technologies in question are to be introduced. In the perspective presented in this work, absolute limits to, say, air pollution from a given technology do not exist, since, for instance, the actual needs for environmental and civil protection differ between urban and rural areas. Such differences pose, though, significant problems in the development of a consistent representation of ecological vulnerability potential in the phase space of the chosen sustainability indicators.

(c) Definition of the *limiting surface* (threshold to catastrophe) for each type of representation. The integrated representation of the pressures put on the social and ecological system to which industrial technologies are introduced allows the identification of limits beyond which fundamental functions and services of the ecosystem cannot be sustained. Each system dimension as defined in the model representation above poses, however, different limits resulting in difficulties in the determination of the overall limiting surface in the system phase space.

# 4 Conclusions

Normalization processes in international and national authorities currently push toward integrated assessment of the environmental and safety consequences of industrial and energy technologies. The need for a holistic approach to the development of new technologies by taking into account the sustainability of the natural ecosystem creates the impetus for a shift in the industrial development management paradigm. This shift will require the development of new methodologies for the assessment of the adverse side-effects of technological development.

The analytical approach suggested in this work aims to provide a tool that will facilitate the integrated assessment of the diverse impacts of technology on the human and natural environment, and its sustainability potential. The concept of ecological vulnerability analysis albeit general enough to allow the holistic evaluation of the adverse impacts of particular technologies can be readily applied on actual industrial installations.

To date, however, the bottleneck for methodology development has been the coherence of the multiple system representation that serves as basis for ecological vulnerability analysis. Future work will focus on the development of:

(a) coherence conditions of the different model configurations, and

(b) the detailed mathematical description of the ecological vulnerability limiting conditions in order to integrate them in technology optimization algorithms.

The ultimate goal of this work is the use of ecological compatibility and vulnerability analyses for (i) the integrated assessment and (ii) development of novel industrial technologies .

# References

1. U.S. Nuclear Regulatory Commission. Reactor safety study. NUREG 75/014, ex-WASH 1400, 1975

2. Society of Environmental Toxicology and Chemistry. A technical framework for life-cycle assessment, 1991

3. Stahel W.R. Long-life system utilization: new strategies of product design, marketing and risk management for waste prevention, Proc. intern. con. on innovation, industrial progress and environment. FEANI, 1991, pp. 25-33

4. Schmidt-Bleek F. Wieviel Umwelt braucht der Mensch ? Backhaus, Wuppertal, Germany, 1994.

5. Nebbia G. Proposta di una rappresentazione input/output dei flussi di materia nella biosfera e nella tecnosfera. AIS meeting on statistics for environmental Management. Roma, 1996.

6. Allenby B. Industrial ecology gets down to Earth. IEEE Circuits and Devices Magazine 1994; Vol. 10, 1:24-28

7. Sarigiannis D.A. Computer-aided design for environment in the process industries. Comp. Chem. Eng. 1996 (in press)

8. Graedel T.E., Allenby B.R., Linhart P.B. Implementing industrial ecology. IEEE Technology and Society Magazine 1993; Vol. 12, 1:18-26

9. Sarigiannis D.A. On the incorporation of safety and environmental considerations in computer-aided design of novel process synthesis. In: Avouris N., Page B. (eds.) Environmental informatics---methodology and applications of environmental information processing. Kluwer Academic, Dordrecht, 1995, pp. 409-428

10. Ehrenfeld J.R. Industrial ecology---a technological approach to sustainability. Hazardous Waste and Hazardous Materials 1992; Vol. 9, 3: 209-211

11. Allen D.T. Using waste as raw material---opportunities to create an industrial ecology. Hazardous Waste and Hazardous Material 1993; Vol. 10, 3:273-277

12. Stalmans M. Lifecycle analysis, environmental quality and consumer products. Proc. intern. con. on innovation, industrial progress and environment, 1991, FEANI, pp. 11-20

13. Nordvik J.P., Carpignano A. Poucet A. Computer-based system modelling for reliability and safety analysis. Proc. topical mtg. on computer-based human support systems: technology, methods and future, ANS, Philadelphia, PA, 1995, pp. 211-217

# Comparison of Life Cycle Risks in Various Power Generation Systems

Leiming Xing, David H. Johnson, James C. Lin, Harold F. Perla
PLG, Inc.
Newport Beach, California USA

## Introduction

The quantitative comparison of risks associated with different power generation technologies has been a topic of increasing interest since the 1970s. Among the methodology advancements, the concept of a multiple dimensional framework for comparison has emerged. Haddad and Gheorghe [1] characterized the different dimensions of health risks and presented them using four risk attributes:

1. Source: Routine or Accidents
2. People at Risk: Workers or Public
3. Exposure: Short, Medium, or Long Term
4. Effects: Fatal or Nonfatal.

Methodology considered both the fatality risk as well as the environmental impacts.

This paper presents an application of the multidimensional framework for risk comparisons of six power technologies: Light Water Reactor (LWR); Coal; Oil; Liquefied Natural Gas (LNG); Liquid Metal Fast Breeder Reactor (LMFBR); and Solar Photovoltaic (PV). The LMFBR technology considerations include mixed oxide fuel (MOX) cycle and the Advanced Liquid Metal Reactor (ALMR) cycle. In each case, the full cycle is considered from the extraction of the fuel to the disposition of the waste and decommissioning of the plant.

## Methodology

The comparison of health impacts is achieved through development of a risk database. Risk matrices were constructed for each technology. One dimension of the risk matrix is each fuel cycle element of a power technology. The other dimension lists all combinations of the risk attributes described in the Introduction. The measure of risks is fatalities or injuries per annual generation of 1 GW of electricity. The data is scaled to account for a 100% capacity factor in order to compare different power options in a common basis; i.e., amount of energy generated. For the environmental impacts, the measures are quantities (metric tons) of air pollutants and greenhouse gas emitted from each cycle element and land use ($km^2$) per annual generation of 1 GW of electricity.

Over a hundred reports and papers were reviewed to collect information for the risk matrix development. Multiple entries were put into the same matrix element with the literature source cited when different risk ranges were reported. Synthesis of data was performed to assess a range of impact measure for each risk matrix entry.

# Insights of Health Risk Perspectives for the Selected Power Cycles

## Fossil Fuel Cycle

The power generation portion of the fuel cycle dominates the public fatality risks from normal operation of a coal-fired power plant.. An upper bound risk of 1.3E+02 fatalities per GWe-y was reported by Hamilton [2]. The data was estimated based on a coal power plant without sulfur removal technology (FGD). If sulfur removal with 90% efficiency is employed, the estimated risk would be an order of magnitude lower. However, the resulting estimate is still much larger than that for the LWR option. According to Boffetta, et al. [3], sulfur and metals emitted from a coal power plant capable of promoting the oxidation of $SO_2$ to $H_2SO_4$ in the presence of water and hydrogen peroxide are concentrated in the ultrafine fraction of coal ash. Occupational exposure to sulfuric acid has been associated with increased risk of laryngeal and other respiratory cancers in a number of epidemiological studies. In addition, coal mining and transport also contribute to the total cycle risks.

The oil fuel cycle also emits large quantities of $SO_x$. An upper bound risk of 100 public deaths per GWe-y due to air emissions from normal operation of a oil power plant was given by Comar and Sagan [4].

The available risk data indicate that the LNG power technology is a "cleaner" and "safer" alternative than the coal and oil fuel cycles. However, LNG storage and transportation may pose threats of catastrophic accidents.

## Nuclear Options

For the LWR power cycle, more data are available from the literature; some representative publications include Hamilton [2], Comar and Sagan [4], and Gotchy [5]. Milling is a dominant contributor to fatality risk arising during normal operation both to the public and workers, due to release of radon gas from the process of crushing and grinding the ore. Mining is another LWR fuel cycle element that contributes significantly to several risk dimensions.

Most research and power fast reactors in the world use a mixture of $UO_2$ and $PuO_2$ as fuel. These MOX reactors can be classified into two groups: pool design or loop design. Examples of pool type reactors include EBR-II (U.S.), PFR (U.K.), Phenix (France), BN-600 (Russia), and Super Phenix (France). Reactors that employed loop design include Joyo (Japan), FFTF (U.S.), Monju (Japan), SNR 300 (Germany), and BN-350 (Russia).

Most information regarding health risks from the MOX cycle was taken from Sherwood [6]. Occupation exposure in breeder power plant dominates the occupational fatality risk during normal operation. The public fatality risk due to release of Kr-85, C-14, and H-3 from normal operation of the reprocessing facility is also an important risk contributor.

The ALMR refers to the PRISM design by General Electric [7]. The PRISM takes into account worldwide fast reactor development experience and has a number of unique characteristics; e.g., modular design, passive safety, and metallic fuels. PRISM belongs to the "pool" type LMFBR.

Most risk data for the ALMR cycle was estimated based on information from Michaels [8]. Both LMFBR cycles (MOX and ALMR) presents similar risk perspective, and both have lower risks than the LWR cycle because exclusion of milling and mining of uranium.

## PV Power Cycle

The development of solar PV energy production has not progressed to the design and operation of a central station with a large-scale plant capacity, because of the high cost of silicon solar cells and the low energy density from solar irradiation. The capacities of the world's two largest single installations of PV plants, both located in the U.S., are 6.5 and 1 MWe, respectively [9].

Information for the PV power cycle was mostly taken from Moskowitz, et al. [9], and Inhaber [10]. The risk data were scaled to a per GWe-y basis. PV device fabrication and raw material preparation dominate the public fatality risk during normal operation. The data for PV device fabrication has large uncertainty because different fabrication technologies emit different kinds and quantities of air pollutants. The upper bound risk for PV device fabrication corresponds to the CdS process where sulfate exposure contributes 6.9E+01 fatalities per GWe-y and cadmium exposure contributes 1.4E+01 fatalities per GWe-y.

## Comparison of Total Cycle Risks

Table 1 summarizes fatality risks from the total fuel cycle of each power technology. Only four dominant risk dimensions were listed in the summary table. The total cycle risk is a summation of all the risk dimensions in the risk matrix.

**Table 1. Fatality Risk Data (Unit: Fatalities per GWe-y)**

| Impact Group | | Public | | Worker | | Total |
|---|---|---|---|---|---|---|
| Operation Mode | | Routine | Accident | Routine | Accident | |
| Duration | | Medium | Short | Medium | Short | |
| LWR | Max | 2.3E+00 | 1.7E-01 | 6.7E-01 | 7.5E-01 | 4.7E+00 |
| | Min | 5.3E-03 | 4.0E-05 | 3.9E-02 | 2.5E-02 | 1.6E-01 |
| Coal | Max | 1.5E+02 | 2.0E+00 | 8.8E+00 | 4.1E+00 | 1.6E+02 |
| | Min | 1.0E+00 | 1.0E-01 | 8.8E-02 | 3.7E-03 | 1.2E+00 |
| Oil | Max | 1.0E+02 | 1.0E-01 | -- | 1.4E+00 | 1.0E+02 |
| | Min | 1.0E+00 | 1.0E-03 | -- | 1.2E-01 | 1.1E+00 |
| LNG | Max | 1.9E-01 | 2.1E-01 | -- | 1.0E+00 | 1.4E+00 |
| | Min | 4.1E-03 | 4.5E-04 | -- | 3.8E-02 | 4.3E-02 |
| MOX | Max | 1.7E-01 | 3.1E-02 | 1.0E+00 | 4.5E-01 | 1.7E+00 |
| | Min | 1.9E-02 | 1.9E-03 | 9.9E-03 | 4.8E-02 | 7.8E-02 |
| ALMR | Max | 6.1E-02 | 3.1E-02 | 8.8E-01 | 4.5E-01 | 1.4E+00 |
| | Min | 6.6E-03 | 1.9E-03 | 8.0E-03 | 4.8E-02 | 6.4E-02 |
| PV | Max | 1.1E+02 | 3.8E-01 | 3.2E-01 | 3.9E+00 | 1.2E+02 |
| | Min | 7.6E-01 | 1.6E-01 | 4.0E-03 | 8.9E-01 | 1.8E+00 |

The coal, oil, and PV cycles have larger public fatality risks during normal operation because of emission of the air pollutants. The coal power cycle has the largest public accident risk due to coal transportation.

The coal cycle leads the occupational fatality risks during normal operation because of diseases associated with coal mining; e.g., black lung or coal workers pneumoconiosis. The coal and the PV cycles have larger risks for occupational accident risks. This is due to coal mining and transport accidents for the coal cycle, and power plant maintenance, transport, and construction accidents for the PV cycle.

The LMFBR (including ALMR and MOX cycles) and LNG power cycles have very low total fatality risks. Total fatality risk for the LWR cycle is higher than the LMFBR and LNG cycles but it is more than one order of magnitude lower than coal, oil, and PV cycles.

## Comparisons of Environmental Impacts

Table 2 tabulates the environmental impact data of the seven power generation options. The coal and oil cycles emit the largest quantities of $SO_x$. The upper bound values for coal and oil cycles correspond to $SO_x$ emissions without abatement. The MOX and ALMR cycles emit the least amount of $SO_x$. For the $NO_x$ emissions, the fossil fuel cycles have much larger emissions than the other cycles. All of the fossil fuel cycles emit large amounts of $CO_2$, which is a greenhouse gas. The coal and the

**Table 2. Environmental Impact Data**

| Power Option | | Total Fuel Cycle Impacts | | | |
|---|---|---|---|---|---|
| | | $SO_x$ (Tons per GWe-y) | $NO_x$ (Tons per GWe-y) | $CO_2$ (Tons per GWe-y) | Land ($km^2$ per GWe-y) |
| LWR | Max | 4.8E+03 | 1.1E+03 | 6.9E+04 | 5.3E+00 |
| | Min | 3.2E+02 | 4.7E+02 | 6.9E+04 | 4.3E+00 |
| Coal | Max | 1.6E+05 | 3.6E+04 | 9.3E+06 | 1.2E+02 |
| | Min | 2.3E+04 | 2.3E+04 | 7.8E+06 | 9.1E+01 |
| Oil | Max | 5.8E+04 | 4.3E+04 | 6.6E+06 | 9.5E+00 |
| | Min | 2.8E+04 | 1.2E+04 | 6.0E+06 | 8.5E+00 |
| LNG | Max | 9.2E+02 | 2.0E+04 | 5.6E+06 | 2.0E+01 |
| | Min | 2.7E+01 | 2.8E+03 | 4.0E+06 | 1.8E+01 |
| MOX | Max | 2.0E+02 | 2.5E+02 | -- | 2.9E+00 |
| | Min | 2.0E+02 | 2.4E+02 | -- | 1.3E+00 |
| ALMR | Max | 2.0E+02 | 2.5E+02 | -- | 2.9E+00 |
| | Min | 2.0E+02 | 2.4E+02 | -- | 1.3E+00 |
| PV | Max | 2.9E+03 | -- | -- | 1.5E+02 |
| | Min | 1.1E+03 | -- | -- | 3.4E+01 |

PV cycles have the largest land requirement. The MOX and ALMR cycles have the least land requirement.

# Conclusions

The following are general observations from this study:

- The environmental impacts associated with the fossil fuel cycles, especially the coal and the oil fuel cycles are generally large due to emissions of air pollutants (e.g., $SO_x$ and $NO_x$) and the greenhouse gas ($CO_2$).

- The PV power cycle requires large land inventory.

- The coal, oil, and the PV power cycles have larger health risk impacts than the other cycles.

- The nuclear options, especially the LMFBR options, are most favorable in terms of health and environmental impacts.

The current work is based on a survey of available literature. Areas of further research could include an in-depth study of the dominant risk contributors to the power cycles and a study of long-term risks to future generations.

# References


1. Haddad, S, Gheorghe, A. Issues in comparative risk assessment of different energy sources. *International Journal of Global Energy Issues* 1992; 5:174-187.

2. Hamilton, LD. Comparative risks from different energy systems: evolution of the methods of studies. *Nuclear Safety* April-March 1983; 24:155-172.

3. International Atomic Energy Agency: Electricity and the environment. In: *Proceedings of the Senior Expert Symposium*. Helsinki, Finland, May 13-17, 1991.

4. Comar, CL, Sagan, LA. Health effects of energy production and conversion. *Ann Rev Eng* 1976; 1:581-600.

5. Gotchy, RL. Health effects contributable to coal and nuclear fuel cycle alternatives. U.S. Regulatory Commission, NUREG/CR-0332, August 1977.

6. Sherwood, G: Liquid metal fast breeder reactor (LMFBR) risk assessment. In: *Proceedings of the Society for Risk Analysis International Workshop on Uncertainty in Risk Assessment, Risk Management, and Decision Making*, Knoxville, Tennessee, September 30-October 3, 1984.

7. General Electric. PRISM preliminary safety information document. GEFR-00793, UC-87Ta, Appendix A, December 1987.

8. Michaels, GE. Impact of actinide recycle on nuclear fuel cycle health risks. Oak Ridge National Laboratory, ORNL/M-1947, UC-802, June 1992.

9. Moskowitz, PD, et al. Health and environmental effects document for photovoltaic energy systems. BNL 51676, UC-63 (Photovoltaic Conversion TIC-4500), September 1983.

10. Inhaber, H. Energy risk assessment. Gordon and Breach, Science Publishers, Inc., 1982.

# Preliminary Requirements for a Knowledge Engineering Approach to Expert Judgment Elicitation in Probabilistic Safety Assessment

G. Guida, P. Baroni, G. Cojazzi*, L. Pinola*, R. Sardella°

Università degli Studi di Brescia, DEA, Brescia, Italy
*European Commission, Joint Research Centre, ISIS, Ispra, Italy
°Università degli Studi di Bologna, DIEM, Bologna, Italy

**Abstract**

In this paper the problem of expert judgment (EJ) in Probabilistic Safety Assessment is recognised as a true knowledge problem. A critical analysis of traditional EJ methodologies leads to highlight some common weak points of theirs. The suggestion is made to reformulate the expert opinion issue in a Knowledge Engineering setting in order to reach a sound basis for comparison, integration and justification of expert knowledge and relevant uncertainty.
The proposed innovative approach is described in its goals, and requirements for its development are detailed.

# 1 Introduction

Among the primary objectives of Probabilistic Safety Assessment (PSA) of nuclear plants there is the representation of the overall risks related to their existence and operation. Due to the complexity of the systems under analysis, a large amount of information is inevitably involved. Different types of knowledge sources must be considered and integrated to obtain global results, and the inherent uncertainty affecting them must be correctly quantified and propagated. Expert judgment is generally needed to identify, interpret and process available information and the related uncertainty. As a matter of fact, structured expert judgment, substantiated by adequate rationales, is accepted as major source of information when:

- no experimental data or validated computer codes are available for predicting the occurrence and the evolution of the phenomena of interest;
- the interpretation of available data or the applicability or validity of current models and codes are questioned.

In the nuclear domain, structured expert judgment has been widely used, and various applications can be found, spanning from PSA of power stations [1] to waste repository studies [2, 3]. The most extensive application of expert judgment in PSA is described in NUREG-1150 [1]. The approach adopted there is based on a highly structured and formal methodology for the whole expert opinion acquisition process.

The problem of eliciting, modeling and representing expert knowledge - and the related uncertainty - has been extensively tackled in past years in the field of Knowledge Engineering [4, 5]. A variety of approaches, methods, and tools have been developed and successfully applied in several practical applications. This paper presents a first attempt to exploit a knowledge engineering approach in the PSA task. More specifically, in section 2 some critical considerations about classical EJ methodologies are presented. This discussion is hence used as a starting point in section 3, to formalise a set of high-level requirements a knowledge-based approach to expert judgement should comply with. Finally, section 4 illustrates the practical conditions for the applicability of the just framed knowledge-based methodology, and a phase structure for its development.

# 2 A Knowledge Engineering Viewpoint

From a knowledge engineering point of view, some limitations and weak points of the expert judgment elicitation methodology employed in NUREG-1150 can be identified.

First of all, even if one of the principles that guided the development of this methodology was the use of state-of-the-art techniques, in NUREG-1150 the proposed elicitation methods appear rather informal and primitive in comparison with those currently applied in the knowledge engineering field [6, 7]. No specific interview techniques are taken into account for single-expert elicitation, nor any specific discussion techniques are applied for group-expert elicitation.

Another weak point can be found in the absence of careful distinction between elicited knowledge and elicited uncertainty about that knowledge. The origin of knowledge and especially of the relevant uncertainty (that is, of the evidence in favour of an acquired piece of knowledge) is not explicited and represented, unless in a very informal way. Moreover, as far as the representation of uncertainty is concerned, only uncertainty about specific facts is considered, while uncertainty about the relations between such facts or about complex aggregations of facts is not taken into account.

In NUREG-1150 expert rationales and motivations are collected mostly in a textual form and are considered at a merely descriptive level. In this way, the reasoning process followed by an expert to generate a given assessment is not explicitly represented, and a formal model of the knowledge the expert uses is almost impossible to obtain.

Furthermore, the integration of different sources of evidence is performed only at a numeric level, without considering such fundamental aspects as the representativeness of the sources considered, their possible inter-relations, the reasoning paths the experts exploit to produce their results, the motivations that underlie the generation of different - sometimes even contrasting - assessments, etc.. As a consequence, a real integration of the judgments derived from different experts is not done: only mechanical aggregations are allowed, carrying inherently weak semantics.
Finally, NUREG-1150 expert judgment elicitation methodology represents uncertainty exclusively by means of probabilities and this can turn out to be not definitely satisfactory. Depending on context, it is well known that for representing and quantifying uncertainty and for reasoning with and about uncertainty, a variety of different methods could more conveniently be applied [5, 8-10].

Methodologies that do not provide an explicit representation of expert knowledge and reasoning, are generally unable to model knowledge and reasoning that are behind expert assessment. Such methods are opaque, since they hide the mental models and the deep reasons on which the results of expert judgment rely. If experts are simply asked about the results of their reasoning, these issues - in our opinion of primary importance - remain beyond the scope of expert judgment elicitation. Moreover, improvements are scarcely foreseeable: asking different experts or asking the same expert over and over can hardly contribute to a better understanding of expert knowledge and reasoning.

# 3 Requirements for an Innovative Knowledge-Based Approach

The limitations discussed above for current approaches to expert judgment elicitation significantly affect the potential use of the acquired judgments. In fact, the resulting shallow representation entails the following drawbacks:

- it does not allow to carry out meaningful comparisons of the judgments expressed by different experts;
- it does not support significant integration of expert judgments derived from different experts;
- it is unable to provide deep and detailed justifications for the conclusions reached.

It seems therefore reasonable to undertake the construction of an innovative approach to expert judgment elicitation whose main goal is to overcome such limitations. In order to accomplish this goal, the new methodology should meet some high-level requirements, which are summarised in the following three main points.

- modeling: the methodology should provide criteria and techniques to model both expert knowledge in a specific domain of interest and the various forms of uncertainty affecting it;
- reasoning: the methodology should provide criteria and techniques to capture, model and reproduce reasoning paths used by experts when facing problems in the specific domain of interest, with particular attention to the methods used for assigning meaningful uncertainty quantification to specific propositions of interest (namely, the issues of safety assessment);
- acquisition: the methodology should provide specific techniques and tools to assure that knowledge, uncertainty, and reasoning acquisition from experts are carried out effectively and properly, without introducing disturbing factors which could spoil elicitation results.

From the above stated high-level requirements it follows that a knowledge-based approach [6] might be appropriate. In fact, the issues of knowledge representation, uncertainty modeling, reasoning formalization, and knowledge acquisition have been tackled and explored for a long time by the knowledge engineering community. Therefore, what is suggested here is to profitably reuse the results achieved in this field for expert judgment elicitation in the domain of safety assessment. A methodology satisfying the above stated high-level requirements should be able to overcome the practical limitations of current approaches: in fact, only knowledge and reasoning underlying judgments are the sound basis for comparison, for integration, and for justification.

In the following we present more detailed requirements for this new knowledge-based methodology, organised according to some general properties: scope, results, structure, transparency and traceability, applicability and usability, quality orientation.

**1. Scope**

1.1 The methodology should include specific techniques to represent knowledge in order to assure:
   - expressive power: the representation of all concepts of interest at the desired level of detail must be possible;
   - naturalness: the representation must be natural and easy to understand for domain experts;
   - economy: the representation must be concise and simple to articulate.

1.2 The methodology should include specific techniques to represent uncertainty in order to assure:
   - soundness: the definition of uncertainty must be consistent, with clear and explicitly stated semantics;
   - cognitive plausibility: uncertainty attributes must be easily understood by domain experts;
   - concreteness: it must be actually possible to acquire and validate all the elements needed to represent uncertainty with reasonable confidence

about their validity (domain experts should not be overstressed, or asked information they can not be supposed to possess);
- meta-information: there should be an explicit representation of the characteristics of different uncertainty sources, to be used in the comparison and combination of relevant uncertain judgments.

1.3 The methodology should include specific techniques to reason under uncertainty that may ensure:
- clear semantic: any function used to propagate and summarise uncertainty should have clear semantics and produce cognitively plausible results, according to general common-sense principles recognised by the expert community;
- goal and context sensitivity: the global goals of the reasoning process and the main feature of the actual application context should be explicitly represented, so that it should be possible to select, among a set of uncertainty propagation and aggregation functions, the most coherent with stated goals and current context.

1.4 The methodology should include specific techniques to reason about uncertainty such that:
- individual assessment of knowledge sources must be possible (validity, level of confidence, potential, etc.);
- cross assessment of knowledge sources must be possible (ranking);
- global assessment of the set of available knowledge sources must be possible (coverage, redundancy, support, etc.).

1.5 The methodology should specify methods for eliciting, modeling, representing, integrating, and verifying knowledge and for managing the whole knowledge acquisition process.

1.6 The methodology should include techniques for the analysis of the specific issues to deal with and for issue-oriented processing of knowledge and uncertainty.

## 2. Results

2.1 The methodology, when applied, is expected to provide results that:
- are technically correct;
- are easily understood and reasonably accepted as valid by a community of recognised experts;
- are cognitively plausible;
- can be justified by means of credible and valid arguments.

## 3. Structure

3.1 The methodology should be clearly structured in a hierarchical way so as to enforce disciplined application.

3.2 The methodology should be modular in order to allow its economic application to a variety of cases, where only some of the issues considered in its design are really relevant, and to facilitate its reuse.

3.3 The methodology should be defined in very general terms, including conditions for applicability and quality assurance.

**4. Transparency and traceabilty**

4.1 The methodology should include explicit assumptions about what aspects of the domain are included in the model (coverage), which is the level of detail of the representation (granularity), which are the types of uncertainty considered.

4.2 The methodology should be clear and transparent, allowing total visibility on each individual phase, task, or activity.

4.3 The methodology should allow easy documentation of all activities carried out.

**5. Applicability and usability**

5.1 The methodology should apply to a sufficiently large variety of cases.

5.2 The methodology should be easily applicable in concrete cases with the support and under the responsibility of dedicated specialists (knowledge engineers).

5.3 The methodology should require a reasonable effort (learning, cognitive load, time, etc.) from domain experts.

5.4 The time and cost for applying the methodology should not exceed the expectations of potential users.

**6. Quality orientation**

6.1 The methodology should include the definition of a set of characteristics selected for quality assurance.

6.2 The methodology should include the statement of the quality level required for each characteristic.

6.3 The methodology should include the definition of metrics, rating criteria and global assessment procedures for quality assurance.

6.4 The methodology should explicitly support the implementation of assurance procedures methods.

# 4 Towards a Practical Implementation

The methodology specified in the previous section should be applicable if the following conditions are met:

- recognised domain experts actually exist and are available to analyse and articulate their knowledge and reasoning;
- expert judgment is the result of a knowledge-intensive task (possibly including also non knowledge-intensive steps, like numerical computations);
- expert judgment is the result of the mental processes occurring in the minds of domain experts and relying on the exploitation of a large and complex corpus of knowledge.

In order to develop a practical implementation of the knowledge-based methodology specified above, a suitable working group should be established first of all. It might include, for example, a couple of knowledge engineers, a couple of domain experts with previous experiences in safety assessment tasks, a specialist in uncertainty representation, and a group leader. Later, a set of already faced and well-documented safety assessments problems can be selected in order to constitute a (not exclusive) reference for assessment and validation during methodology development.

The development for the methodology might then be organised incrementally and might develop through four main phases.

Phase 1

The working group should first focus on the issue of methodology scope and work on the identification of the representation and reasoning techniques related to requirements 1.1 to 1.4. The technical soundness of the proposed techniques and their ability to cover a sufficiently large set of safety assessment problems should be then preliminarily tested and possibly refined and revised considering the selected problems and with the cooperation of domain experts.

Phase 2

Requirements 1.5 and 1.6 should then be faced, and proper methods for knowledge processing (from elicitation to verification) defined and tested.

Phase 3

A structure for the methodology is later defined (according to requirements 3.1 to 3.3) and a preliminary formalization is undertaken. The working group should assure that the requirements concerning transparency and traceability (4.1 to 4.3) and quality orientation (6.1 to 6.4) are carefully met. Such issues should be the subject of specific verification activities after the first formalization has been completed.

Phase 4

The formalised methodology is eventually extensively tested, experimenting with complete safety assessment cases. The requirements concerning applicability and usability (5.1 to 5.4) are taken into account and the characteristics of the results produced are assessed (2.1).

Through these phases, a draft version of the methodology is produced. If the developed methodology seems to satisfy all the requirements stated above, a large-scale validation and refinement campaign should be eventually planned and implemented.

The main benefits expected from the exploitation of knowledge-based technology include:

- the achievement of a deeper understanding and of a formal representation of the motivations underlying expert judgment;
- the possibility to exploit this understanding in order to critically and rationally compare and integrate different expert judgments;
- the use of advanced knowledge elicitation techniques in order to avoid improper influences on expert (such as overstressing) and to filter out possible distorted behaviours of the experts (such as overconfidence or underestimation).

A preliminary experiment in applying a knowledge-based approach to safety assessment is presently being carried out at the Joint Research Centre, European Commission, Ispra (Italy), in the frame of the ongoing benchmark exercise on expert judgment in nuclear PSA.

# References


1. USNRC. Severe Accident Risks: An Assessment for Five U.S. Nuclear Power Plants. United States Nuclear Regulatory Commission, NUREG-1150, Vol 1, 1990
2. USNRC. Elicitation and Use of Expert Judgment in Performance Assessment for High-Level Radioactive Waste Repositories. SNL, prepared for United States Nuclear Regulatory Commission, NUREG/CR 5411, 1990
3. Thorne M.C. The use of expert opinion in formulating conceptual models of underground disposal systems and the treatment of associated bias. Reliability Engineering and System Safety 1993; 42: 161-180
4. Kidd A. (ed.). Knowledge Acquisition for Expert Systems, A Practical Handbook. Plenum Press, New York, 1987
5. Shafer G., Pearl J. (eds.). Readings in Uncertain Reasoning. Morgan Kaufmann, San Mateo, CA, 1990
6. Greenwell M. Knowledge Engineering for Expert Systems. Ellis Horwood, Chichester, UK, 1988
7. Firley M., Hellens D. Knowledge Elicitation - A Practical Handbook, Prentice-Hall, London, UK, 1991
8. Clark D.A. Numerical and symbolic approaches to uncertainty management in AI. Artificial Intelligence Review 1990; 4: 109-146
9. Ng K.C., Abramson B. Uncertainty management in expert systems. IEEE Expert, 1990; April: 29-47
10. Léa Sombé Group. Reasoning under incomplete information in artificial intelligence: A comparison of formalisms using a single example. International Journal of Intelligent Systems 1990; 5(4): 323-472

# Catastrophe Risk Analysis in Technical Systems Using Bayesian Statistics

*Elizabeth Saers Bigun*

**Center for Safety Research, KTH, Royal Institute of Technology, and Department of Statistics, Stockholm University Stockholm, Sweden**

**Abstract**

In this paper we present risk analysis models which are built on expert assessments of future risks. These models calibrate and aggregate judgements of the experts in order to predict the future risks. The models are constructed by applying Bayesian statistics.

# 1 Introduction

The global needs for complicated technical systems are increasing. To possess a holistic perspective of these systems in risk management is important. In this context it is essential to analyse catastrophe risks. Hence the use of expert judgements of the concerned risks are unavoidable, although there are several problems connected with the use of judgements which at first glance appear biased and subjective. In this paper we present risk analysis models which are built on expert assessments of future risks. These models calibrate and aggregate judgements of the experts in order to predict the future risks. Bayesian statistics offers powerful tools to combine both subjective assessments and observed data which we used it in this work, Bernardo & Smith (1994).

A catastrophe is defined as an event which occurs with small probabilities, is concentrated in time and space and have extensive negative consequences for human life, health and environment. Consequently there are not enough data in order to build risk analysis models on "classical" statistical methods. Therefore expert judgements of the future risks constitute valuable source of information.

There are several papers which treat the use of expert judgements and reconciliation of expert judgements; Lindley (1979, 1982, 1985 and 1986), French (1980,1985), Apostolakis (1988), Morris (1974, 1977, 1983), Pulkkinen (1993), West (1988) and Winkler (1981). There are some relevant Bayesian approaches in the reliability analysis whose applications are nuclear plants, Pörn (1990) or oil platforms, Huseby (1988), Gåsemyr & Natvigs (1991, 1992). There are several papers on the human assessment, expert judgements and the human risk perception abilities, and they can be find in Kahneman & Slovic & Tversky (1982). More references are available in Bigün (1994A, 1995B).

# 2 The Theoretical Models

In this chapter we suggest models which calibrate and aggregate the future risk assessments of the experts. A more extensive presentation can be found in Bigün (1994B, 1995B).

The experts are asked to give their personal opinions on both past and future events (see appendix ). The questions about the past are asked in order to calibrate the risk assessments of the experts. For simplicity we assume that the experts and their answers are independent.

We use the following notations related to the expert assessment of an unknown variable e.g. the number of major accidents in 1984-1988: $X_{ij}$ denotes the i-th expert's assessment of the most probable value of the unknown variable, for the j-th time period; $i = 1, 2, ..., k$ and $j = 1, 2, ..., n, n+1$ (where 1 to n are indices for the past time periods and n+1 is the index for the future time period). $Y_j$ is the true value of the unknown variable, for the j-th time period (where $Y_j$ is already observed for $j = 1, 2, ..., n$ but not for $j = n+1$ ). Finally $Z_{ij}$ denotes the assessment error of expert i, $Z_{ij} = Y_j - X_{ij}$. We denote the expected value of $Z_{ij}$ by $b_i$ . Assume that the $Z_{ij}$ given $b_i$ are normally distributed;

$$(Z_{ij}|b_i) \in N(b_i, \sigma_{ij}^2). \tag{2.1}$$

We assume first that the variances $\sigma_{ij}^2$ are known and that $Z_{ij}$ are independent for all $i$ and $j$.

We suppose that $b_i$ has a non-informative prior distribution; $b_i \in N(0,\infty)$. Then the posterior distribution of $b_i$ is normal with mean and variance

$$M_i = \sum_{j=1}^{n} (z_{ij} / \sigma_{ij}^2) / \sum_{j=1}^{n} (1/\sigma_{ij}^2) \text{ and } V_i = 1 / \sum_{j=1}^{n} (1/\sigma_{ij}^2), \tag{2.2}$$

respectively.

Since $(Z_{i,n+1}|b_i)$, according to (2.1), is normally distributed, the predictive distribution of $Z_{i,n+1}$ given $z_{i,1}, z_{i,2}, ..., z_{i,n}$ is also normal with mean $M_i$ and variance $V_i + \sigma_{i,n+1}^2$ . If we know $X_{i,n+1}$ the predictive distribution of $Y_{n+1} = (Z_{i,n+1} + x_{i,n+1})$ is also normal;

$$(Y_{n+1}|z_{i,1}, z_{i,2}, ..., z_{i,n}, x_{i,n+1}) \in N(M_i', V_i'), \tag{2.3}$$

where $M_i' = x_{i,n+1} + M_i$ and $V_i' = V_i + \sigma_{i,n+1}^2$.

According to (2.3) the i:th expert's assessments about the future risks will be calibrated by means of the expected value in (2.2).

Assume that the $k$ independent experts are asked to give their assessments of the unknown variable in the past and in the future. The posterior distribution of $Y_{n+1}$ given all opinions of experts, is normal; , (2.4)

$(Y_{n+1}|z_{1,1}, z_{1,2},\ldots, z_{1,n}, x_{1,n+1},\ldots, z_{k,1}, z_{k,2},\ldots, z_{k,n}, x_{k,n+1}) \in N(M'', V'')$ where

$M'' = (\sum_{i=1}^{k} M_i' / V_i') / \sum_{i=1}^{k}(1/V_i')$ and $V'' = (\sum_{i=1}^{k} 1/V_i')^{-1}$

This is the posterior distribution of $Y_{n+1}$ after having observed the assessment errors of the $k$ experts. The prior distributions of $b_i$ and $Y_{n+1}$ are non-informative.

Since the variances $\sigma_{ij}^2$ are not known we must estimate them. We suppose that the experts give the precisions correctly except for an individual proportionality constant;

$$s_{ij} = a_i \sigma_{ij}^2. \qquad (2.5)$$

Assume, a priori, the following non-informative joint distribution of $a_i$ and $b_i$ :

$$f(a_i, b_i) = (1/a_i). \qquad (2.6)$$

Assume also that $(Z_{ij}|a_i, b_i, s_{ij}) \in N(b_i, \sigma_{ij}^2)$ are independent for all $i$ and $j$. The posterior distribution of $a_i$ may now be computed as the marginal distribution of $f(a_i, b_i|(z_{i1}, s_{i1}), (z_{i2}, s_{i2}),\ldots,(z_{in}, s_{in}))$:

$$f(a_i|(z_{i1}, s_{i1}), (z_{i2}, s_{i2}),\ldots,(z_{in}, s_{in})) = \qquad (2.7)$$

$$= \beta^{(n-1)/2}(a_i)^{(n-3)/2}\exp(-a_i\beta) / \Gamma((n-1)/2),$$

which is a Gamma distribution; $\Gamma((n-1)/2, \beta)$ with $\beta = (1/2)(G_i - D_i)$;

$D_i = (\sum_{j=1}^{n}(z_{ij}/s_{ij}))^2(\sum_{j=1}^{n}(1/s_{ij}))^{-1}$ and $G_i = \sum_{j=1}^{n}(z_{ij}^2/s_{ij})$. Since $(\sigma_{ij}^2|s_{ij}) = s_{ij}/a_i$ we have (2.8)

$$E(\sigma_{ij}^2|s_{ij}, (z_{i1}, s_{i1}), (z_{i2}, s_{i2}),\ldots,(z_{in}, s_{in})) = s_{ij}E(1/a_i) = (s_{ij}(G_i - D_i))/n-3, \; n>3.$$

Assume that the experts are able to assess $s_{ij}$ in at least two different ways. In the analysis below these are called method 1 and method 2. Method 1 is built on questions about the probabilities:

**Method 1**: $s_{ij} = 1/P_{ij}^2$, (2.9)

where $p_{ij}$ denotes the i-th expert's probability assessment that $X_{ij} = Y_j$ (with a given precision e.g. $\pm$ 10%) i.e. a measure on the i:th expert's certainty on her assessment $X_{ij}$.

Method 2 utilises the assessments of the minimum and maximum values of the unknown variable:

**Method 2**: $s_{ij} = (U_{ij} - L_{ij})^2$ (2.10)

where $L_{ij}$ denotes the i-th expert's assessment of the **minimum** value of the unknown variable, for the j-th time period, $i = 1, 2, ..., k$ and $j = 1, 2, ..., n, n+1$ (where 1 to n are indices for the past time periods and n+1 is index for the future time period) and $U_{ij}$ denotes the i-th expert's assessment of the **maximum** value of the unknown variable, for the j-th time period, $i = 1, 2, ..., k$ and $j = 1, 2, ..., n, n+1$ (where 1 to n are as in $L_{ij}$). The appendix contains the actual wording which may be used in order to obtain $p_{ij}$, $L_{ij}$ and $U_{ij}$ from the experts.

In this context we deal with positive variables which only take positive values. Therefore, a multiplicative assessment error model, see (2.11), instead of an additive one is more suitable. Define $X_{ij}$ and $Y_j'$ to be the logarithm of the assessed and the true values, $X_{ij}'$ and $Y_j'$. Let $Z_{ij}$ denote the difference between $Y_j$ and $X_{ij}$;

$$Z_{ij} = (Y_j - X_{ij}) = \ln(Y_j') - \ln(X_{ij}') = \ln(Y_j' / X_{ij}') \quad (2.11)$$

Then for a given $X_{ij}$ the $Y_{n+1}'$ in (2.3) and (2.4) will be log-normally distributed.

In this chapter we assumed that both within and between expert assessments are independent. This assumption is of course not realistic. Bigun (1996) treats models where some of these dependencies are taken into account.

# 3 Some results from an empirical study

The models, presented in Chapter 2 are evaluated by means of the results of pilot survey which was performed during the fall 1994 on risk analysis of major civil aircraft accidents in Europe, among 21 Swedish experts in aviation safety. The survey was built on a questionnaire where most of the questions were about the number of accidents and the number of fatalities due to the major accidents. The time periods for the experts' assessments are 1984-1988, 1989-1993, 1994-1998 and 1999-2003. The accident locations (occurred or presumed to occur) are all in Europe. There were separate questions about western (as it looked like during the existence of the Soviet Union) and whole of Europe. The concerned aircraft are (or were) registered in Europe with jet-engines and with weight over 5700 kg. Since not all accidents are of interest, the experts were forced to use their own judgements when they answered questions about the past risks; the statistics for the occurred accidents were known to us but were not directly available for the experts because of they concerned a small subset of the accidents.

The main results of this evaluation study are: (1) The prediction models in Chapter 2 seem to work satisfactorily. They take very well care of experts who are too certain or too uncertain in their assessments causing under- or overestimation of risks. The aggregated models are mostly effected by experts who have less assessment errors having the highest confidence. (2) Calibration of the experts' assessed risks are needed. Most of our experts' assessments are calibrated with respect to their overestimation of risks and to their overconfidence in their own assessment ability. (3) The introduced methods in (2.9) and (2.10) do not lead to remarkably different results. (4) All individual distributions are positively skewed. The aggregated distributions are less skewed.

A more extensive presentation of the results of the pilot survey can be found in Bigun (1995A, 1995B).

# References


1. Apostolakis G. Expert Judgements in Probabilistic Safety assessment. Accelerated Life Testing and Experts' Opinions in Reliability. Proceedings of the International School ofPhysics "Enrico Fermi", 28 July-1 August 1986. Amsterdam, North-Holland, 116-131. Edited by Clarotti C. A. & Lindley D.V..
2. Bernardo M. J. & SMITH M. F.A. Bayesian Theory. Wiley & Sons, Chichester, 1994.
3. Bigun, S. E. Risk Analysis of Catastrophes Using Bayes' Methods. Research Report Department of Statistics, RRDS 1994A:2, Stockholm University.
4. Bigun, S. E. Risk Analysis of Catastrophes Using Bayes' Methods I: Models which build on experts' judgements. Research Report Department of Statistics, RRDS 1994B:10, Stockholm University.
5. Bigun, S. E. Risk Analysis of Catastrophes Using Bayes' Methods II: Results from the empirical studies. Research Report Department of Statistics, RRDS 1995A:1, Stockholm University.
6. Bigun, S. E. Risk analysis of major civil aircraft accidents in Europe. European Journal of Operational Research. 20th anniversary, 1995B, special issue.
7. Bigun, S. E.. Bayesian Prediction Based on Few and Dependent Data. In progress.
8. French S. Updating of belief in the light of some else's opinion. J. R. Statist. Soc. A 1980; 143:43-48.
9. French S. Group consensus probability distributions: A critical survey. Bayesian statistics 2 1985; 183-202.
10. Gåsemyr J. & Natvig B. Using expert opinions in Bayesian estimation of component l lifetimes in a shock model - a general predictive approach. Statistical Research Report, university of Oslo, No. 4, 1991.
11. Gåsemyr J. & Natvig B. Expert opinions in Bayesian estimation of system reliability in a shock model - the MTP connection. Statistical Research Report, university of Oslo, No. 2, 1992.
12. Huseby A.B. Combining opinions in a predictive case. Bayesian statistics 1988; 3:641-651.

13. Kahneman D. & Slovic P. & Tversky A. Judgement under uncertainty: Heuristics and biases. Cambridge University press, Cambridge, 1982.
14. Lindley V. D & Tversky A. & Brown V. R. On the reconciliation of probability assessments. J. R. Statist. Soc. A, part 2, 1979; 142:146-180.
15. Lindley V. D. The improvement of probability judgements. J. R. Statist. Soc. A, part 1, 1982; 145:117-126.
16. Lindley V.D. Reconciliation of discrete probability distributions. Bayesian statistics 1985; 2:375-390.
17. Lindley V. D & Singpurwalla. Reliability (and faultree analysis using expert opinions), ASAS 1986; 87-90.
18. Morris A. P. Decision analysis expert use. Management science 1974; 20:1233-1241.
19. Morris A. P. Combining expert judgements: a Bayesian approach. Management science 1977; 22:679-693.
20. Morris A. P. An axiomatic approach to expert resolution. Management science 1983; 29:24-32.
21. Pörn K. On empirical Bayesian inference applied to poisson probability models. PhD thesis, Linköping University, No. 234, Linköping, 1990.
22. Pulkkinen U. Methods for combination of expert judgements. Reliability Engineering and system Safety 1993; 40:111-118.
23. West M. Modelling expert opinion. Bayesian statistics 1988; 3: 493-508
24. Winkler R.L. Combining probability distributions from dependent information sources. Management Science 1981; 27: 479-488.

# Appendix

## Below is question extracted from the questionnaire which consisted of 19 questions.

**(A) Estimate [1] the total number of accidents for the following time periods in Western Europe and the probability that the given estimation is (or will be) equal to the true values with a $\pm$10% marginal failure of the true value.**
**(B) Estimate a lower and an upper boundary for the total number of accidents in Western Europe so that You are quite sure[2] that the true[3] value is between these boundaries.**

| Time period | (A) Total number of accidents in WESTERN EUROPE | (A) The probability that the estimation is equal to the true value $\pm$10% marginal failure (Express as a number between 0 and 100) | (B) Lower boundary of the total number of accidents in WESTERN EUROPE | (B) Upper boundary of the total number of accidents in WESTERN EUROPE |
|---|---|---|---|---|
| 1984 -88 | | | | |
| 1989-93 | | | | |
| 1994-98 | | | | |
| 1999-03 | | | | |

**1)Give the most probable value according to Your belief. 2) You may interpret "quite sure" as 75% probability. 3) The true current or future occurrences.**

# An Exercise on Bayesian Combination of Expert Judgment for Climatic Predictions at the Yucca Mountain Site.

R. Bolado*
A. Lantarón**
J.A. Moya*

(*) Cátedra de Tecnología Nuclear, Universidad Politécnica de Madrid
(**) Consejo de Seguridad Nuclear
Madrid, Spain

**Abstract**

## 1.- Introduction.

Knowledge uncertainty is of major concern when a Performance Assessment of any High Level Nuclear Waste Repository is under development. Its nature and the scarcity of data or the impossibility to obtain them are the main reasons to use expert judgment to characterize that uncertainty. At present, there is a clear idea of what an expert elicitation process should be like, its phases and the techniques to be applied to debias judgments. Nevertheless, some controversy exists about the way to combine judgments from several experts to produce a common judgment, and if they should be combined or not [1]. The authors of this paper think that the proper way to tackle this problem is through a bayesian information updating process.

## 2.- The Case Study.

A case study [1] has been selected in which predictions for future climate in the vicinity of the Yucca Mountain site from five experts are elicited. In that study experts provided several quantile values for the magnitudes of interest, and there was no reference to any specific distribution. In the documentation associated to the process each expert provided a short informal paper reporting the basis of their judgments. A resume of the elicitation session held with each expert was also reported. The present study has been focused on the experts' opinions about the change in the mean annual temperature one hundred years in the future, about 2100 AD (that average is defined in an area of 50 Km around the site in a period of one hundred years centered in the aforementioned date).

# 3.- The Methodology.

The methodology followed in this paper is the usual one in the bayesian framework, though the specific approach or modelling of the problem may vary slightly from some authors to others [2], [3]. In this approach three elements are considered in Bayes' formula: The prior knowledge of the decision maker (DM) about the quantity ($\theta$) to be assessed ($\pi_0(\theta)$), the likelihood of the estimates given by the experts for the same quantity (L(expert judgment/$\theta$)), and the posterior distribution obtained by the DM after combining the prior knowledge and the expert provided information ($\pi(\theta)$). In the case studied it has been considered that the normal hypothesis for the likelihood could be appropriate. Moreover, the distributional information provided by the experts is transformed into two measures; a centralization measure, the mean or median, and an spread measure, usually the standard deviation. So, the distributions provided by the five experts are transformed into two vectors, the first one with the five means of the distributions, ($m_1,\ldots,m_5$), and the second one with the five standard deviations, ($s_1,\ldots,s_5$). The application of Bayes' formula implies

$$\pi(\theta/\vec{m},\vec{s}) \;\alpha\; L(\vec{m},\vec{s}/\theta) \cdot \pi_0(\theta) \; . \tag{1}$$

Under these conditions Lindley [2] suggests two hypotheses. The first of them is to suppose that most of the information is contained in the evaluations of the means or medians, except in the cases of very large values of the standard deviations stated by the experts, since in those cases neither the means nor the standard deviations provide any valuable information. This hypothesis simplifies the mathematical problem. The second hypothesis is about the modelling of expert biases, precisions and correlations. Biases and correlations among experts may be modelled assuming that the likelihood is a five-dimensional (in the case study) multivariate normal distribution with means $\alpha_i + \beta_i\,\theta$, standard deviations $\sigma_{ii} = \gamma_i s_i$ and covariances $\sigma_{ij} = \rho_{ij}\gamma_i s_i \gamma_j s_j$; i,j= 1,...,5. Under these hypotheses, and with a non - informative normal prior, the DM will assign to the parameter an a posteriori normal distribution with mean and variance

$$m' = \left(\vec{\beta}^T\, \Sigma^{-1}(\vec{m}-\vec{\alpha})\right)\cdot(\sigma')^2 \quad , \; and \quad (\sigma')^2 = \left(\vec{\beta}^T\, \Sigma^{-1}\vec{\beta}\;\right)^{-1} , \tag{2}$$

where $\Sigma$ is the variance and covariance matrix. Usually, the inverse of the variances that appear in the process are called precisions.

This modelling of experts' opinions is interpreted as follows. If the DM assigns values $\alpha_i=1$ and $\beta_i=1.1$ to expert i it means he thinks expert i overpredicts the value of $\theta$ in 1 plus a 10% of $\theta$. If the DM states $\gamma_i=0.85$ it means he thinks expert i overestimates his own uncertainty about $\theta$ approximately in a 15%. The elicitation process is expected to incorporate debiasing tools that allow the DM to assign 0 values to all the $\alpha_i$ and values 1 to all the $\beta_i$ and $\gamma_i$. Correlation among experts is a natural event associated to all the knowledge held in common by them and the way in which they use it to generate their opinions. Focusing the attention in the latter aspect of correlation, Chhibber [3] proposes the evaluation of concordance probabilities in thought experiments to estimate subjective correlation coefficients.

Let us suppose that no debiasing term is needed to model experts' means, so that all the $\alpha_i$ are null and all the $\beta_i$ are equal to one. Let us suppose that not all the experts are equally credible, in general each one of them is characterized by a different $s_i$ and has been assigned a $\gamma_i$ ($\sigma_i = \gamma_i s_i$), and that each pair of them has a common correlation coefficient $\rho$. Then, the posterior distribution mean is a weighted average of the experts' means, where the weight associated to expert i is

$$(\vec{e}^{\,t}\Sigma^{-1})_i = \frac{[1+(k-1)\varrho]\ \sigma_i^{-2} - \varrho\ (\sum_{j=1}^{k} \sigma_j^{-1})^2}{(1-\varrho)\ [1+(k-1)\varrho]} \quad , \tag{3}$$

where the transposed vector multiplying $\Sigma$ has all its components equal to one. Clemen [4] shows that the precision of the posterior distribution becomes

$$(\sigma')^{-2} = \frac{[1+(k-1)\varrho] \sum_{j=1}^{k} \sigma_j^{-2} - \varrho\ (\sum_{j=1}^{k} \sigma_j^{-1})^2}{(1-\varrho)\ [1+(k-1)\varrho]} \quad , \tag{4}$$

that is an addition of individual precisions modified by a function of the common correlation coefficient and corrected by terms depending on the crossproducts of the inverse of experts' standard deviations.

Two surprising phenomena arise at this point, the first one is that negative weights appear [4] if the common correlation coefficient verifies the inequality

$$\varrho > [\ 1 - k + \sigma_i \sum_{j=1}^{k} \sigma_j^{-1}]^{-1} \quad , \tag{5}$$

though this is not a serious problem. The second phenomenon is related to the precision of the posterior distribution. It should be expected that the precision of the posterior decreases as the common correlation coefficient increases. Nevertheless, the analysis of equation 4 shows that precision is not monotonically decreasing with $\rho$, but it shows vertical asymptotes at $\rho=1$ and $\rho=-1/(k-1)$, and a minimum between those values. None of these phenomena appear when all the experts are equally credible (common variance), and it is somewhat more complex when there are different correlations between each pair of experts. In the general case it is only necessary to take into account that weights are applied on experts' debiased means.

## 4.- An Application of the Methodology.

As it was said in point 2 in this paper, the case studied is one in which five experts are asked about the likely change in the annual mean temperature from now to 2100 AD, as defined in that point. The problem that the DM tackles is the modelling of the five experts he is working with. This means the evaluation of the experts' biases and precisions, and the correlation coefficient for each pair of them. The assessment of experts' bias in the mean is a difficult task that needs a deep DM's expertise, however their effect on the posterior mean is easy to be predicted; these are two reasons for not considering those biases in this work.

Information supplied by the experts has to be analyzed in order to obtain a better estimation of the experts' precisions and correlations. For the reference case all the experts considered that the most important physical phenomenon affecting the climate and specially the temperature in the next 100 years is the greenhouse warming. All of them pointed out GCM's (Global Climate Models) as the appropriate tool for estimating the variables under study. Three of them run several GCM codes translating the outputs explicitly in the provided results, the other two give the estimations based on these models but it is not explicitly stated that they run the codes. This induces correlations between the experts, nevertheless the hypotheses considered were different depending on the expert, for example, when evaluating the greenhouse warming expert A considers as unique $CO_2$ emission scenario the one that expert B considered as the worst case. This leads to conclude that the correlations between experts could be lower than those expected from the first reading. Expert precision could be assessed from two different hypotheses. The first one is based on considering that the spread given by the expert for the demanded variable is a good estimate of his own uncertainty about the variable. The second one consists in taking into account that experts are usually overconfident, the largest spread provided by the experts could be used as the best estimate of the common uncertainty. The latter hypothesis is probably the most appropriate in this case, since expert A produced a larger spread when taking into account more phenomena affecting the climate than the rest of the experts (e.g. the effect of sulphates, not included in the GCM's), although the former could also be considered for sensitivity analysis. For modelling experts the previous rationale has been considered.

The data provided by the experts are shown in table 1. A normal distribution has been fitted to the data given by each expert. Each normal fitted distribution has the same median provided by the expert, while the standard deviation has been estimated as the maximum of the set $\{(x_{50}-x_5)/1.645,(x_{95}-x_{50})/1.645\}$, where $x_5$, $x_{50}$ and $x_{95}$ are respectively the 5, 50 and 95 percentiles provided by the expert.

| | PERCENTILE | | | | | Std. dev. |
|---|---|---|---|---|---|---|
| Expert | 0.05 | 0.25 | 0.5 | 0.75 | 0.95 | |
| A | -2.0 | 0.4 | 3.0 | 6.0 | 8.0 | 3.04 |
| B | 1.0 | 1.5 | 2.0 | 2.6 | 4.0 | 1.22 |
| C | -0.3 | 1.7 | 2.8 | 3.6 | 4.2 | 1.86 |
| D | -1.0 | 0.0 | 1.0 | 1.5 | 2.0 | 1.22 |
| E | -0.5 | 0.5 | 1.0 | 1.4 | 2.2 | 0.91 |

Table 1 Data provided by the experts and fitted normal distribution parameters.

Four bayesian aggregates have been considered, and they are compared with the mechanical aggregate with equal weights reported in reference [1]. Results are summarized in table 2 and figure 2. The base case is the first bayesian combination (bayesian 1) which hypotheses are independence, $\rho=0$, and common precision for all the experts, that of expert A. Bayesian 2 differs from bayesian 1 in increasing the common correlation coefficient to 0.3. Bayesian 3 is a sensitivity case in which independence is assumed, like in bayesian 1, but the variance (precision$^{-1}$) assigned to each expert is the same he has estimated for the uncertain variable. Bayesian 4 differs from bayesian 3 in increasing the common correlation coefficient to 0.3.

| | PERCENTILE | | | | | Std. dev. |
|---|---|---|---|---|---|---|
| Aggregate | 0.05 | 0.25 | 0.50 | 0.75 | 0.95 | |
| Mechanical | -0.56 | 0.82 | 1.96 | 3.02 | 4.08 | 1.53 |
| Bayesian 1 | -0.28 | 1.04 | 1.96 | 2.88 | 4.20 | 1.36 |
| Bayesian 2 | -1.35 | 0.63 | 1.96 | 3.32 | 5.27 | 2.01 |
| Bayesian 3 | 0.53 | 1.09 | 1.48 | 1.87 | 2.43 | 0.58 |
| Bayesian 4 | -0.10 | 0.65 | 1.17 | 1.69 | 2.44 | 0.77 |

Table 2 Mechanical and bayesian aggregates of experts' distributions.

The first two bayesian combinations produce a posterior distribution with the same mean that the mechanical aggregate, as expected when applying equation 3 to experts with the same precision (all the weights are 0.2). Bayesian 1 produce the most similar result to the mechanical aggregate. The increase of $\rho$ from 0 to 0.3 produces a noticeable increase in the posterior spread. Dramatically different results are obtained when the expert estimate of the uncertainty about the variable is assigned to his own opinion. In this cases the weights are nevermore equal, acquiring larger weights those experts with the larger precisions. In the third bayesian case experts A,B,C,D and E are assigned respectively weights 0.036, 0.228, 0.098, 0.228 and 0.41. This is the reason for the convergence of the posterior mean to those given by experts B, D and E. This effect is even stronger in the fourth bayesian case, in which $\rho=0.3$ and the condition in inequality 6 is verified (its right hand side has a value 0.14 in this case). Now, experts A,B,C,D and E are assigned respectively weights -0.046, 0.229, 0.021, 0.229 and 0.567, and the mean resembles strongly the one provided by experts D andE. It is clear that assigning precisions in this way forces the convergence of the posterior to the opinion of the most precise experts, and this phenomenon becomes predominant when correlation is considered. Again, considering correlation induces precision decrease. Figure 1 shows the evolution of precision as a function of $\rho$ in the cases of common expert variance (lower curve) and different expert variance (upper curve). The minimum in the upper curve in this case is at $\rho=0.47$. That region has not been reached in this application.

## 5.- Conclusions.

Bayes' formula provides a powerful and structured procedure to combine expert opinions about uncertain variables. The capability to model expert biases, correlation and precisions allows the DM to build a posterior distribution consistent with his own view of experts' expertise and the rules of probability. Modelling those biases, correlations and precisions is not an easy task and is risky in the sense of affecting dramatically the posterior result.

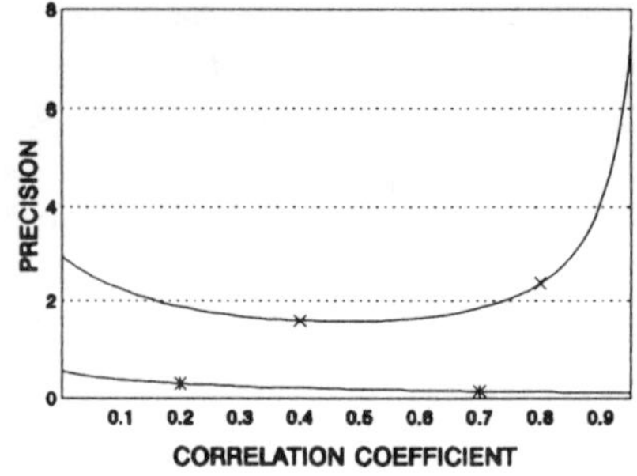


Figure 1 Behaviour of precision versus $\rho$ for common ($*$) and different ($\times$) expert variances.

Although it has been reached a mature estate in expert elicitation processes, they lack specific tools to estimate in a proper way expert correlations and precisions. The importance of getting good estimates for this parameter has arisen as a key point in this expert opinion combination process as can be deduced from the application of the methodology performed in this paper, since high precisions and correlations may produce fast convergence to opinions of the most precise experts. This effect is undesirable if overconfidence bias has not been properly corrected.

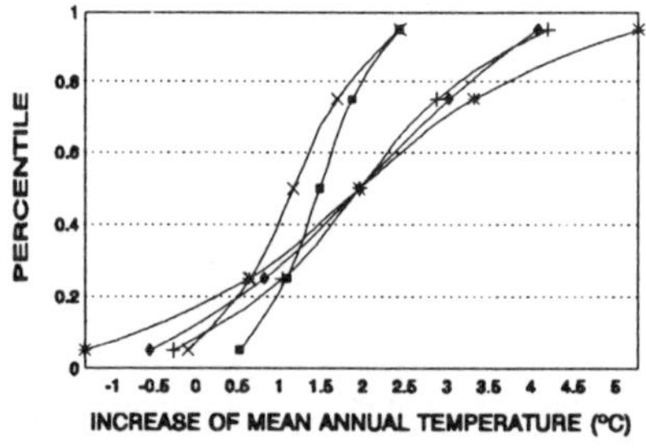


Figure 2 Mechanical (◆), bayesian 1 (+), bayesian 2 ($*$), bayesian 3 (▪) and bayesian 4 ($\times$) aggregates.

## References.

1. DeWispelare AR, et al. Expert Elicitation of Future Climate in the Yucca Mountain Vicinity. Iterative Performance Assessment Phase 2.5. CNMWRA 93-016. San Antonio. 1993
2. Lindley DV. The Use of Probability Statements. In: Accelerated Life Testing and Expert Opinions on Reliability. North Holland, Amsterdam, 1988, pp 25-57 (Proceedings of the CII Course of the International School of Physics "Enrico Fermi")
3. Chhibber S, Apostolakis G, Okrent D. On the use of expert judgments to estimate the pressure increment in the Sequoyah containment at vessel breach. Nuc Tech 1994; 105:89-103
4. Clemen RT, Winkler RL. Limits for the precision and value of information from dependent sources. Oper Res 1985; 33 No 2:427-442

# Formal Bases for the Construction of Possibilistic Expert Judgement Systems

Sandra A. Sandri
Brazilian National Institute for Space Research
São José dos Campos, Brazil

**Abstract**

We discuss here the formal bases for the elicitation, assessment and pooling of expert opinion in the possibilistic framework. We also bring a short discussion on the concept of probabilistic calibration and of one of its implementation in continuous domains.

## 1 Introduction

The knowledge of experts is very useful to evaluate unknown parameters in the field of reliability and safety analysis of newly designed installations, since statistical data are not always available, especially regarding rare or destructive events, or for devices whose novelty implies a scarcity of experimental data at the time when the safety analysis must be carried out. In order to get useful information from the experts, several problems must be solved. The first one is a proper modeling of the pieces of data supplied by a single expert about a given parameter. This type of data is almost never precise and reliable because the expert only possesses a rough idea of the value of quantitative parameters, due to limited precision of human assessments and to the variability of such values. The second task to be solved is to assess the quality of the expert, namely his accuracy, and the precision of his response. Lastly when several expert responses are available, they must be combined so as to yield a unique, hopefully better response.

In most systems the uncertainty model adopted is related to the probabilistic framework (see [1] for a survey); in this work we discuss possibilistic approach to expert judgement (see [2] for details). We also discuss the notion of *calibration* of expert opinions [2] [3] and compare it to a notion of *accuracy*. Section 2 brings an overview of possibilistic elicitation of information; section 3 discusses the notion of calibration and of possibilistic assessment; section 4 deals with the pooling of uncertain pieces of information and section 5 brings the conclusion.

## 2 Elicitation

Elicitation of expert opinion is traditionally carried out in the probabilistic framework, usually through the use of quantiles or of parametrized families of functions. However, a pure probabilistic model of expert knowledge is not so satisfactory, since probabilistic information looks too rich to be currently supplied by individuals. Moreover, information supplied by individuals is not so much tainted with randomness, but rather incomplete and/or imprecise.

Possibility theory [4][5] is a framework capable of capturing the concept of imprecision. A possibility distribution $\pi$ on a variable x in X is a mapping from X to [0,1], such that there exists at least one value $x_0$ in X such that $\pi(x_0) = 1$.

$\pi(x_i) = 1$ (resp. $\pi(x_i) = 0$) means that it is completely possible (resp. impossible) that$x_0$ is the real value of x. It is interesting to note that a possibility distribution can be obtained from a probability distribution and vice-versa (see [8] for details).

Let v be a variable defined on a given domain X. The simplest form of a possibility distribution [4] for v is given by $\pi(x) = 1$ if $v \in [s_l, s_u]$, 0 otherwise, where $[s_l, s_u]$ is an interval of X. This type of possibility distribution is naturally obtained from experts claiming that "the value of v lies between $s_l$ and $s_u$". This way of expressing knowledge is more natural than giving a point-value $x_0$ for v right away, since it allows for some imprecision, but it is still rather uninformative.

A natural way of eliciting information from experts consists in asking them to supply several intervals $A_1,\ldots,A_m$ directly, together with levels of confidence $\lambda_1,\ldots, \lambda_m$. The level of confidence $\lambda_i$ can be conveniently interpreted as the least probability that the true value of v hits $A_i$ (e.g. from the point of view of the experts, the proportion of cases he has observed where the realization of v lies in $A_i$). Having obtained the set of intervals $A_i$ and levels of confidence $\lambda_i$ it is then easy to construct a possibility distribution [6]. In practice, a limited number of intervals is already capable of providing a rich amount of information, e.g. the 3 intervals using $\lambda_1 = 0.05$, $\lambda_2 = 0.5$, and $\lambda_3 = 0.95$.

# 3 Assessment

An expert can be deficient mainly in respect to 2 aspects: he can be innacurate, giving values that are inconsistent with the actual information; and/or imprecise, furnishing non-informative assessments.

Most expert judgement systems try to detect and treat these defficiencies; either the analyst knows the experts' defficiencies and is able to furnish coefficients which will modify the experts' estimates, or the experts are submitted to a battery of tests, and evaluated thereupon.

In the first case, which occurs notably on systems adopting the Bayesian model, the analyst has to be in a way a super "expert" himself, in order to account for the deficiencies of the experts. In the second case, in order to assess the quality of experts, the latter are asked questions whose answers are known, and are rated on the basis of these results. The questions pertain to the true values of a series $v_1, v_2,\ldots, v_n$ of "seed" (test) variables; the values of these parameters are either known by the analyst and not known by the experts, or more often can be determined afterwards by means of physical experiments, or other means. In this framework, experts are usually assigned a global "quality" index, that takes into account the precision and accuracy of his assessments.

Before entering the subject of possibilistic assessment we discuss the notion of calibration in the probabilistic framework, and of one of its implementations.

## 3.1 Discussion on Winkler's Calibration Concept

Winkler describes in [3] his notion of probabilistic *calibration*. In his view, a weather forecaster can be viewed as perfectly well-calibrated if it rains on 10% of the days for which he says that the probability of rain is .1, on 20% of the days for which he says that the probability of rain is .2, and so on. In our point of view, a more useful notion is that of *accuracy*: the weather forecaster can be considered perfectly accurate if it rains on the days he is sure that it will rain, and it does not rain on the days he is sure it won't rain. A perfectly calibrated expert would therefore not be considered as perfectly accurate. The contrary is true however and, therefore, the notion of accuracy is intuitively more severe than that of calibration.

Moreover, a measure based on the notion of calibration will be necessarily global because it is impossible to assess the calibration of an expert in relation to each individual variable. This is not ideal because experts are expected to be both calibrated and precise; and if these aspects are measured separately it is in principle possible to have an expert qualified as "good" when he is imprecise whenever his opinions hit the variable realization, and precise otherwise. This drawback does not occur with the notion of accuracy, which can be measured individually.

Finally, the implementation of the notion of calibration is not straightforward when the universe of discourse of a variable is not dichotomic, and even more so when it is not discrete at all. Cooke [1] presents an ingenious implementation of this notion for continuous domains using probability quantiles. It presents however some other problems besides those discussed above.

Let $x_i$ represent the value of a share $x_i$, i=1,10, with $x_i$ taking values in the interval [0,10] unities. Let us suppose we have asked 3 economists to assess how much these 10 shares will be worth in the following month. Each economist $E_j$ furnishes a triple $q = (q_{5\%}, q_{50\%}, q_{95\%})$ for each share $x_i$, where $q_{k\%}$ is the k% quantile, ie k% of the probability mass assigned by the expert lies in the interval $[0, q_{k\%}]$. Let us suppose that a month later the values of shares $x_1$ to $x_5$ are found to be 3.5, and the values of shares $x_6$ to $x_{10}$ are found to be 6.5 (see Fig. 1).

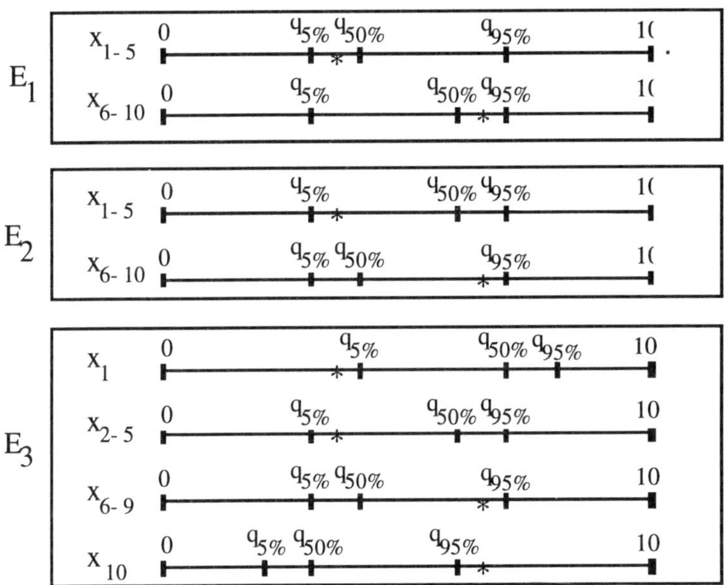


Figure 1: Assessments of economists $E_1$, $E_2$ and $E_3$, for variables $x_1$ to $x_{10}$. Symbol "*" indicates the realization of each variable.

Let us suppose economist $E_1$ has furnished the triple (3, 4, 7) for variables $x_1$ to $x_5$, and (3, 6, 7) for variables $x_6$ to $x_{10}$ (see Fig. 1). In the same way, $E_2$ has furnished (3, 6, 7) for variables $x_1$ to $x_5$, and (3, 4, 7) for variables $x_6$ to $x_{10}$. If we consider that the probability mass is uniformely distributed inside each interval interquantile, then intuitively $E_1$ is more accurate then $E_2$. Indeed, the realization of each variable occurs on the interval that $E_1$ considers the most plausible one, to the contrary of $E_2$. However, the calibration measure as implemented in [1] is not capable of making this distinction, even though it is somewhat compatible with the notion of calibration as stated by Winkler.

Let us suppose that $E_3$ has furnished the triple (4, 7, 8) for $x_1$, (3, 6, 7) for $x_2$ to $x_5$, (3, 4, 7) for $x_6$ to $x_9$, and (3, 4, 7) for $x_{10}$. $E_3$ gives thus the same values as $E_2$ for variables $x_2$ to $x_9$, and in relation to these variables, he can be considered as accurate as $E_2$. The realization of variables $x_1$ and $x_{10}$ occurs in the intervals that $E_3$ considered the less plausible ones and therefore, in relation to these variables, he is less accurate than $E_2$, and in consequence he is clearly less accurate than $E_1$. However, using the calibration measure implemented in [1] $E_3$ will considered to be more calibrated than both $E_2$ and $E_1$. This result is somehow coherent with the notion of calibration as stated by Winkler, since the variable realizations are more distributed in the intervals given by $E_3$.

We can thus see that the calibration notion as implemented in [1] can lead to erroneous results when compared to a more restrictive notion such as accuracy.

## 3.2 Possibilistic Assessment

Let v in X be a seed variable whose value x* is precisely known. Let $\pi_E$ be the assessment supplied by an expert E, to describe his knowledge about v. It is easy to see that the greater $\pi_E(x^*)$, the more accurate is the expert. Indeed if $\pi_E(x^*) = 0$, the expert's assessment totally misses x* while if $\pi_E(x^*) = 1$, x* is acknowledged as a usual value of v. Hence a natural measure of accuracy is given by

$$A(E, v) = \pi_E(x^*)$$

The larger is $\pi_E$, the more imprecise (hence under-confident) the expert. A reasonable precision index is then

$$Sp(E,v) = (|\pi_U| - |\pi_E|) / |\pi_U|$$

where $|\pi_E| = \int_X \pi_E(v)\, dv$, and $\pi_U = 1, \forall x \in X$.

On the whole, the overall rating of the expert regarding a single seed variable can be defined as

$$Q(e,v) = A(e,v) \cdot Sp(e,v)$$

that requires him to be both calibrated and informative in order to score high.

Using simple arithmetic mean over the individual scores of the seed variables a global measure of quality can be obtained for an expert. Overall measures of accuracy A(E), precision Sp(E) can be obtained in the same way, but it is important to note that that generally $Q(E) \neq A(E) \cdot Sp(E)$. In other words, it is important to derive Q(E) from individual indices Q(E,v) in order not to assign a good quality index to an expert who is precise only when he is inaccurate, and vice-versa.

# 4 Pooling

The basic principles of the possibilistic approach to the pooling of expert judgments are first that there is no unique combination mode, and second that the choice of the combination mode depends on assumptions formulated by the analyst about the reliability of experts and the dependence of their assessments. Based on these assumptions several formal modes are supported by possibility theory, as shall be seen below. These methods also form a reasonable set of alternatives for the analyst to test when he has been able to assess the experts' quality, as seen above.

Basically, we can divide the combination modes in two classes: one in which the experts are viewed as a set of sources in parallel to be combined in a symmetric way, and another in which the opinion of an expert may be taken more into account in the determination of the result. Let $\pi_i$ be the possibility distribution supplied by expert i, for $i \in K$. A small set of combination modes is:

*Conjunctive Mode*

If the experts are considered to be reliable then the response of the group of experts is given by a distribution $\pi_c$ which can be defined by $\pi_{min}(x) = \min_{i \in K} \pi_i(x)$. This mode makes sense if all the $\pi_i$ significantly overlap; if the degrees in $\pi_c(x)$ are very low this mode of combination makes no sense. Generally, agreement between experts is due to common background, and the idempotence of min deals with such a kind of redundancy.

*Disjunctive Mode*

The most careful optimistic assumption about a group of experts is that one of them is right, without specifying which one. This assumption corresponds to obtaining a distribution $\pi_d$ as response, which can be defined by $\pi_{max}(x) = \max_{i \in K} \pi_i(x)$. This is a very conservative pooling mode that allows for contradiction between experts but may lead to a very poorly informative result, although not always a vacuous one.

*Trade-off Mode*

One type of consistency based trade-off between the conjunctive and disjunctive modes of pooling is to use a measure $\kappa$ of conflict between the experts and define $\pi_{trade}(x) = \kappa_{12} \cdot \max(\pi_1(x), \pi_2(x)) + (1 - \kappa_{12}) \cdot \min(\pi_1(x), \pi_2(x))$. This index gives the conjunctive (resp. disjunctive) mode if $\kappa = 0$ (resp. $\kappa = 1$). Index $\kappa$ can be for instance defined as $\kappa_h = 1 - h(\pi_{min})$, with $h(\pi) = \sup_x \pi(x)$, or as the Jacquard index J defined by a quotient of fuzzy cardinalities yielding $\kappa_J = 1 - J(\pi_1, \pi_2)$, where $J(\pi_1, \pi_2) = |F_1 \cap F_2| \,/\, |F_1 \cup F_2|$, and $\mu_{Fi} = \pi_i$.

An interesting class of trade-off methods is that of the symmetric sums [4]. An operator of this class is such that it gives high membership degrees to values on which both experts have high confidence, and low membership degrees to those on which both experts have small confidence. An idempotent parametrized class of this operators can be defined by

$$\pi_{sum\text{-}\alpha}(x) = \begin{cases} \pi_{max}(x), \text{ if } \pi_{min}(x) \geq \alpha \\ \pi_{min}(x), \text{ if } \pi_{max}(x) \leq 1 - \alpha \\ \pi_m(x), \text{ otherwise} \end{cases}$$

where $\alpha \in [0,1]$, and $\pi_m$ stands for the arithmetic mean on the $\pi_i(x)$.

*Discounting Experts*

Let us suppose that the degree of certainty that a given expert $e_i$ is reliable is known, say $w_i$. Then it is possible to account for this information by changing $\pi_i$ into $\pi'_i = \max(\pi_i, 1 - w_i)$ (as suggested in [4]). When $w_i = 1$ (reliable expert),

$\pi'_i = \pi_i$ and when $w_i = 0$ (unreliable expert), then $\pi'_i = 1$. Note that $w_i = 0$ does not mean that the expert lies, but that it is impossible to know whether his advice is good or not. Once discounted, expert opinions can be combined conjunctively.

*Priority Aggregation of Expert Opinions*
Let us suppose that we have determined a partition on the set K of experts into classes $K_1, K_2, \ldots, K_q$ of equally reliable ones, where $K_j$ corresponds to a higher reliability level than $K_{j+1}$, for j = 1,q. Then the above symmetric aggregation schemes can be applied inside each class $K_j$. The combination between results obtained from the $K_j$'s can be performed upon the following principle: the response of $K_2$ is used to refine the response of $K_1$ insofar as it is consistent with it. When $\pi_1$ is obtained from $K_1$ and $\pi_2$ from $K_2$, then a reasonable combination rule is [7] $\pi_{1-2} = \min(\pi_1, \max(\pi_2, 1 - \text{cons}(\pi_1,\pi_2)))$, where $\text{cons}(\pi_1,\pi_2) = 1 - \kappa_{12}$. Note that when $\text{cons}(\pi_1,\pi_2) = 0$, $K_2$ contradicts $K_1$ and only the opinion of $K_1$ is retained ($\pi_{1-2} = \pi_1$), while if $\text{cons}(\pi_1,\pi_2) = 1$ then $\pi_{1-2} = \min(\pi_1,\pi_2)$.

# 5 Conclusion

We presented here the formal basis for the construction of expert judgements systems in the possibilistic framework. The elicitation of information in the possibilistic framework is more user-friendly than that in the probabilistic one.

In what concerns pooling, the possibilistic approach is much richer than the probabilistic one, and presents less problems in relation to an eventual dependance between the sources. In practice, with only a small set of methods we can find satisfying ways of pooling experts opinions.

The possibilistic evaluation does not overload the system analyst, can be easily checked by the experts themselves, and does not lead to incoherences. One of the main contributions of this work is to discuss the notion of calibration, used in the probabilistic framework, in relation to the more strict notion of accuracy, used in the possibilistic framework. We have shown that the notion of calibration intrisically leads to erroneous results if compared to accuracy, and that implementations of it in continuous domains lead to some extra drawbacks. These differences in a conceptual level have been also confirmed in practice, using the possibilistic expert judgement system PEAPS (see [2]).

# References


[1] Cooke R.M. *Experts in uncertainty*, Dept of Mathematics, TU Delft, The Netherlands, 1989.

[2] Sandri S., Dubois D., Kalfsbeek H., "Elicitation, assessment and pooling of expert judgements using possibility theory", *IEEE Tr. on Fuzzy Systems*, 3, pp 313-335, 1995.

[3] Winkler R.L. "On "Good" Probability Appraisers", in *Bayesian Inference and Techniques* (P. Goel, Zellner A., eds), Elsevier, 1986.

[4] Dubois D., Prade H., *Possibility Theory*, Plenum Press, 1988.

[5] Zadeh L.A., "Fuzzy sets as a basis for a theory of possibility", *Fuzzy Sets an Systems*, 1, pp. 3-28, 1978.

[6] Dubois D., Prade H., Sandri S. "On Possibility/Probability transformations". *Proc. IFSA'91*, Brussels, Belgium, 1991.

[7] Dubois D., Prade H. "Default reasoning and possibility theory," *Artificial Intelligence*, 35, pp 243-257, 1988.

# Reliability Analysis of Operating Procedure in Team Performance

Takehisa Kohda*, Takayoshi Tanaka*, Yoshihiko Nojiri**, and Koichi Inoue*
* Department of Aeronautics and Astronautics, Kyoto University
Kyoto 606-01, Japan
** Marine Technical College, Ministry of Transport
12-24, Nishikura-machi, Ashiya, Hyogo 659, Japan

**Abstract**

To prevent the effect of human errors in a team of multiple members who execute a given task, how they take actions to achieve it must be clarified. This paper proposes a framework for qualitative analysis of a team performance based on a human behavior model. Each member goes around a hierarchical problem-solving structure to achieve his current task goal. The effectiveness of a team performance depends on a kind of collaboration, communication. A communication is assumed to occur when a member needs help from others to achieve his current goal. An illustrative example of a tank filling task show the representation of a team performance model and its simulation result.

# 1 Introduction

In operation of complex man-machine systems, the role of humans have been replaced by those of machines, which decreases the number of operators. As there still exists some roles for human operators, their collaboration is more important than before. Considering the team performance composed of several members in man-machine systems, their interactions thorough communications and collaboration have an important effect on their achievement of a given task. The team performance model must consider these factors.

Several works have been studied on team performance: a conceptual framework for classifying the collaboration operations [1], analysis of teamwork behavior using simulator experiments [2,3], and studies of team collaboration forms using simulation models [4,5]. This paper gives a framework for evaluating team performance qualitatively based on a human behavior model expressed in Petri net models[6]. Qualitative analysis shows how members work together to achieve the goal through their interaction. A team model is derived based on assumptions on human behaviors. An illustrative example shows the representation of a team performance model and summary of its simulation results.

# 2 Team Performance Model

## 2.1 Assumptions on Human Behavior

Consider a case where a team of multiple members try to achieve a given task goal. Team performance is described in terms of individual members' behaviors. Based on his knowledge, a given mission, and input data from the environment, each member is assumed to work independently or interacting with the others as follows.

[Assumption 1] A goal which motivates a member to take an action is given by himself or orders from the other members.

[Assumption 2] A member is in a stand-by position where he takes no action, if his current goal is empty.

[Assumption 3] To achieve his current goal, a member takes the following hierarchical problem-solving action.

(3-1) If he knows an action to achieve his current goal based on his experience or training, he immediately does it.

(3-2) If he knows no actions but a plan to achieve his current goal, he modifies his current goal based on it.

(3-3) If he knows neither actions nor plans to achieve his current goal, he tries to modify his current goal based on knowledge or information on the system.

(3-4) If he cannot achieve his current goal by himself, he tries to obtain help from the other members or support systems.

## 2.2 Meaning of Assumptions

In assumption 1, the detection of an abnormal state in the plant operation corresponds to the case where a member sets his goal by himself. A stimulus from the external world can give a motivation to an operator. Orders from his boss can be also a motivation for his action. Assumption 2 shows that a member does nothing if he has no motivation to take an action.

The problem-solving action in assumption 3 is similar to Rasmussen's SRK model [7]. Cases (3-1) and (3-2) corresponds to the situation which a member has experienced, while case (3-3) and (3-4) are unexperienced situations. In case (3-2), he changes his current goal into one of its subgoal according to his plan. Sequentially dividing his current goal into its subgoals, the requirement to achieve the goal can be represented in terms of basic actions. Basic actions are minimal in the sense that it need not be divided into its subgoals. If its achievement conditions are satisfied, the member take the basic action. Basic actions correspond to case (3-1). Case (3-3) corresponds to the case where a member replaces his goal with its equivalent goal. In case (3-4), the function of human interfaces is important for support systems, while human relations can affect the effectiveness of communication.

## 2.3 Human Behavior Model

Figure 1 shows the overall structure of a member's problem-solving action. The terms in Figure 1 are defined as follows:

Current goal — a goal for a member to achieve immediately; his focus

Task goal — an original goal for a member to be achieve thorough the sequence of actions; he does nothing unless the task goal is given

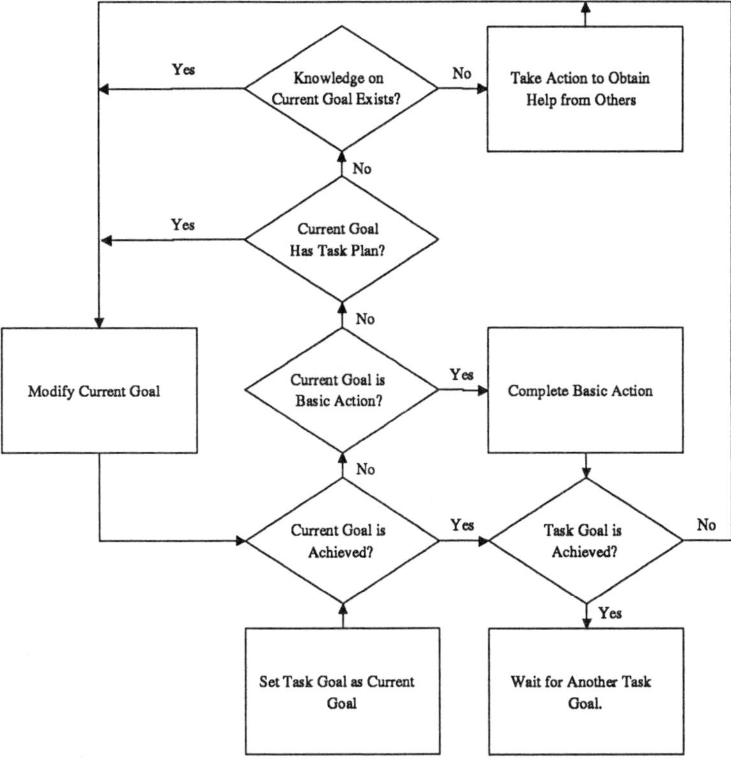


Figure 1 Human Behavior Model

| | |
|---|---|
| Basic action | a goal which cannot be divided into its subgoals |
| Task plan | a sequence or combination of its subgoals to achieve a goal |

## 2.4 Representation of Human Behavior Model

To achieve his current goal, each member utilizes such information as a relation between a goal and its subgoals (i.e., task plans), knowledge on system behaviors and status of other members and the system. Task Analysis [8] can clarify what kind of information is necessary to take an action, what actions are necessary to achieve a goal, and what condition must be satisfied to take an action. Based on task analysis result, the above knowledge is represented as follows.

### *2.4.1 Task Plan*

A task plan, a relation between a goal and its subgoals, can be obtained by analyzing how it can be achieved by accomplishing its subgoals. The relation between a goal and its subgoals are represented as a production rule, whose condition part represents a logical combination of its subgoals necessary to obtain a goal. Based on a task plan, the current goal can be modified as a logical combination of its subgoals. Sequentially resolving a goal by its task plan, the entire task plan for a task goal can be obtained in terms of basic actions. When a task goal is given, he takes actions based on the hierarchical task plan whose top is the given task goal.

*2.4.2 Basic Action*

Basic actions are located at the branch of a hierarchical task plan, which represents an action to achieve its upper goal. For a person to take a basic action or operational action, its achievement conditions must be met. An operational action changes system and environmental states, or his mental model of them. Thus, a basic action can be represented by a pair of precondition and postcondition. which are expressed in the same manner as transitions in Petri net models [6].

*2.4.3 Knowledge on System Behavior and Teams*

Generally, the condition of a system can be specified by the system states. A member judges the system condition based on observed state variables. Unobserved system state variables can be estimated from observed state variables based on their relations. Normal condition of a system is specified by an input-output relation, whose deviation can indicate its malfunction. The effect of a state variable deviation on another one can be also represented by a cause-effect relation of state variables.

The information on team members is represented as a pair of an attribute and its value in the same way as the system state. The search of an appropriate member for a task can be represented as the identification of a member who has the corresponding attribute lists.

## 2.5 Human Errors

Human errors whose effect can be considered in the above framework are 1) operational errors due to misconception of system states and team members, and 2) misunderstanding due to incorrect knowledge or time pressure. A member's misconception is represented as some difference between his mental model and the true state, which leads to omission errors or wrong actions. Incorrect knowledge is represented as an incorrect relation of state variables or wrong task procedure. The effect of behavioral human errors can be also represented as different output states from specified states for normal one. The effect of time pressure is represented as incomplete search of rules or knowledge, which leads to the misuse of rules. Thus, the effect of such human errors are represented as some difference between state variables and mental models.

## 2.6 Qualitative Analysis

The proposed framework can obtain event sequences in a team to achieve the mission task goal, where each member takes actions based on the same problem-solving behavior model using his knowledge and given information. A collaboration or communication depends on the knowledge and task distribution among members and the hierarchical structure of members. Event sequences obtained by modifying the contents of the team performance model show their effect on the team performance.

# 3 Example of Tank-filling Operation

## 3.1 Description

Consider a team operation of tank-filling task. The team consists of a supervisor and two operators. The supervisor cannot perform plant operations by himself and only gives orders or checks operator actions based on only the reports from operators.

Operators perform plant operations and inform the supervisor of their actions. The system is composed of a pump, a valve and a tank connected in series by pipes. An operator cannot deal with {pump and valve} and {the tank} simultaneously. To fill the tank, the pump must function after the valve is opened. After completing the task, the pump must be stopped and the valve is closed.

## 3.2 Task Knowledge Representation

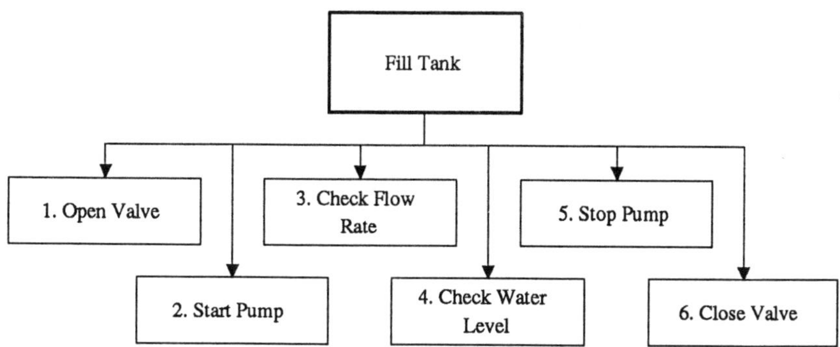


Figure 2 Tank-filling Task

Figure 2 shows the entire task plan in terms of 6 basic actions. From the viewpoint of the supervisor, these basic actions are divided into three orders: a) "run the pump", b) "check the water level" and c) "stop the pump". If the supervisor is told to fill the tank, he just gives three orders sequentially. An operator who gets orders must do two basic operations for orders a), b) or c). Basic actions are performed by two operators depending on the task allocation. An operator who finishes a basic operation must report his result to the supervisor.

Basic actions are represented by a pair of precondition and postcondition. For example, for opening the valve, the precondition is: [the valve is closed](mental model and true state) and the postcondition is: [the valve is opened](mental model and true states). Similarly, behavioral human errors can be represented by a pair of precondition and postcondition. An omission error can be represented as: the postcondition is changed into [the valve is opened](mental model)AND[the valve is closed](true state), while the precondition is the same.

## 3.3 System Representation

The system is represented in terms of pump, valve and tank state as follows:
1) the pump has two states: running or stopping, 2) the valve has two states: opened or closed, and 3) the tank state is represented by water level and flow rate. The water level is below, within or above the desired level, while the flow rate is negative, zero or positive. Normal system conditions are specified by functional relations of these system variables: positive correlation of water level and flow rate, and consistent relation of {valve and pump states} and {flow rate}. Based on these relations, the system states change after an operational action, system changes due to failure, or time passage. Each member's mental model of system states are the same as the

system state representation.

### 3.4 Qualitative Analysis

Changing the distribution of knowledge on basic actions among two operators (or changing the task allocation), its effect on the event sequence in a tank-filling task is examined. The obtained result shows that the frequency of unnecessary communications such as a question on unknown procedures depends on the supervisor's understanding of operators' capability as well as task allocation among operators. With regard to the prevention of operators' errors, the result shows that knowledge on the system behavior with operators' reports of their actions gives the supervisor a chance of correcting their errors. These results can be accepted generally.

## 4 Conclusions

This paper proposes a framework for qualitative analysis of a team performance, which utilizes both the task analysis result and a human behavior model. Interactions with other members, collaboration and communication, are defined to occur when a member needs help from others to achieve his current goal. The method can obtain event sequences for a simple tank-filling task whose result is generally accepted. The prototype program based on the proposed method is now being developed using an expert system shell G2, which needs further improvement as well as verification and validation through more practical examples.

## Acknowledgement

This paper includes a result related to the research program executed through Special Coordination Funds of the Science and Technology Agency of the Japanese Government.

## References

1. Schmidt K. Cooperative work: A conceptual framework, Position paper at workshop in New Technology, Distributed Decision Making, and Responsibility, Bad Homburg, May 1988, FCI Publication #89-1, FCI Soborg Denmark, 1988
2. Rouse W B, Cannon-Bowers J A, Salas E, The role of mental models in team performance in complex systems, IEEE Trans Sys Man Cybernet 1992; SMC-22-6: 1296-1308
3. Ujita H, Kubota R, Fuji-ie M, Experimental analysis of plant operator crew communication, Ningen-Kougaku, 1993: 29-4: 239-248 (in Japanese)
4. Furuta K, Kondo S, Group reliability analysis, Reliab Engng System Safety 1992; 35: 159-167
5. Yoshimura S, Takano K, Sasou K, Team behavior simulation model - Proposal of the conception-, Journal of Scientific & Industrial Research, 1994: 53; 574-578
7. Rasmussen J. Skills, rules, and knowledge; Signals, signs, and symbols and other distinctions in human performance models, IEEE Trans Sys Man Cybernet 1983; SMC-13-3: 257-266
6. Peterson J L. Petri net theory and modeling of systems, Prentice-Hall, 1981
8. Diaper D(ed), Task analysis for human-computer interaction, Ellis Horwood Ltd Chichester, 1989

# A DEFINITION AND MODELING OF TEAM ERRORS

Kunihide Sasou, Ken'ichi Takano, Seiichi Yoshimura
Human Factors Research Center, CRIEPI
2-11-1, Iwato-kita, Komae-shi, Tokyo 201, Japan

## 1 Introduction

Rapid progress in computer technology, such as neural networks, fuzzy logic theory, artificial intelligence, etc. has brought us several support systems for use in decision-making. One example used around the world is the support systems used for the stock exchange. Not every decision-making process, however, completely relies on such support systems, people make decisions by taking into consideration the information gained from the support systems. Moreover, a decision is generally made by a group of people rather than by individuals.

There seems to be a lot of merits in having a team, the most important of which is being able to help each other. One member can help another when the other person is busy, is about to mishandle an operation, or when a bad decision is about to be made. These merits do not always work well however, especially in stressful situations. For example, it was a stressful situation that prevented the pilot and co-pilot from recognizing a certain flight condition in the China air accident at Nagoya airport on April 26, 1994. Both pilots misunderstood a system function of the plane related to landing [1]. Consequently, a catastrophic accident occurred as a result of their misunderstood.

Therefore, the authors conducted manned experiments and observed mistakes in teams' decision-making processes in order to discuss "Team Error", and discuss the factors affecting team errors. Modeling and simulation of team errors are also described in this paper.

## 2 Experiments

*Facility:* Manned experiments were conducted in order to observe mistakes in teams' decision-making processes. The control panel is 4 meters wide, 1.8 meters high, with 55 alarms, 40 indicators, 6 chart recorders and 29 switches to operate a simplified plant, the behavior of which is calculated in a workstation. A series of information of the plant operating conditions is presented on the indicators and so forth. Subjects (operators) can watch them and turn switches to change the plant status. The workstation also records much data such as sounded warnings, operators' controls, transition of plant parameters, etc. The behavior of operators and conversations are also recorded on laser discs.

*Subjects:* Four teams of 3 male students listed in Table 1 were formed, having different backgrounds and skill levels in plant operations. The subjects in teams A & B received 50 hours education and training, in order to operate the simplified plant. After the experiments, two subjects were placed in other teams to form teams of members with different levels of experience. They received 20-100 hours of education and training, depending on their roles in their own team.

Table 1 Subjects, their backgrounds and education/training hours

| Team | Role | Subject | Age | Background | Education/ Training hrs. |
|---|---|---|---|---|---|
| | Leader | S.M* | 25 | Economics | 50 |
| A | Reactor | O.I | 23 | Literature | 50 |
| | Turbine | A.B+ | 19 | Social Science | 50 |
| | Leader | A.K | 24 | Economics | 50 |
| B | Reactor | A.H | 22 | Agricultural Science | 50 |
| | Turbine | T.K | 20 | Psychology | 50 |
| | Leader | S.M* | 25 | Economics | 100 |
| C | Reactor | K.F | 25 | Biotechnology | 50 |
| | Turbine | H.F | 20 | Law | 20 |
| | Leader | A.B+ | 19 | Social Science | 100 |
| D | Reactor | T.I | 23 | Computer Science | 50 |
| | Turbine | K.I | 26 | Business Economics | 20 |

Note : * and + attached to the subjects' initials means they are the same person.

*Events:* Each team faced several abnormal events listed in Table 2. The operators were unaware of the events. In each condition, 1 or 2 abnormal events occurred and they became aware of something happening when the first warning sounded. Under certain conditions, additional failures occur; these are critical because troubles may occur in warnings which sounds at first when an abnormal event happens and troubles may occur in equipment required to counteract abnormal events.

Table 2 Abnormal events for the manned experiments

| No. | Abnormal Event | Additional Failure |
|---|---|---|
| 1 | Deterioration of the function of Ejector A | Ejector B trouble |
| 2 | Accidental open of Condensate water withdrawal valve & Leakage at Condensate water pump A | |
| 3 | Deterioration of the function of Ejector A & Accidental open of Feed water control valve | |
| 4 | Accidental stop of Feed water pump A | No warnings related to the pump failure |
| 5 | Accidental close of Feed water control valve | No warning related to the valve failure |
| 6 | Leakage at Condensate water pump A & Accidental open of Condensate water withdrawal valve | Condensate water pump A & Condensate water supply valve troubles |

*Results:* Through observations and analyses of 24 trials (4 teams x 6 conditions), several mistakes were found as listed in Table 3. Most of them involved false decision-making in response to the accidents.

Table 3 Mistakes observed in the experiments

Example 1......Deviation from the operational rule (Team A, Event No. 5)

The reactor operator was manually controlling Feed water control valve because its automatic control circuit had failed. Then the reactor water level decreased rapidly due to his poor manipulation. Therefore, the reactor operator proposed inserting the control rod to keep the reactor water level. However, the proposal was inadequate in this situation. He should have decreased the primary recirculation pump flow. According to the interview with them just after the trial, the leader said that he had wondered they should have decreased the flow (this was correct idea) when the reactor operator made his proposal. Despite this, the leader accepted his proposal and permitted him to do it.

Example 2......False selection of the object (Team B, Event No. 2)

The turbine operator identified a leakage point after a warning sounded, and had to start Condensate water pump C before stopping the failed pump A. After starting up the pump C, the turbine operator by mistake stopped the pump B which was operating normally. He told the leader to stop the pump B just before the operation. However, no one noticed that he had made a mistake. The record of their communication shows that the turbine operator correctly identified the pump A as the leakage point and told the others. However, by mistake he stopped the pump B. In the interview with them just after this trial, the turbine operator said he recognized that the leakage had occurred in the pump B. He was confident to stop the pump B and his confidence stopped the others doubting his judgment.

Example 3......Not starting a pump (Team B, Event No. 4)

The operators started to identify the cause of the event. The leader noticed that Feed water pump A had stopped through checking the lamps on the control panel. Then, the turbine operator shouted that they should decrease the primary recirculation pump flow. The others accepted his proposal and the reactor operator did it. This is an unbelievable mistake. This event requires only to start the pump C. Of course they know the procedure to start the auxiliary pump when a pump fails, they didn't. In the interview the reactor operator said he had started to decrease the primary recirculation pump flow because it had been proposed by another operator.

Example 4......Deviation from operational rule (Team C, Event No. 3)

This shows another example of a mistake. They started up Ejector B and were manually controlling Feed water control valve. To maintain the condenser vacuum, the leader instructed to stop Feed water pump A and decrease the primary recirculation pump flow and the reactor operator did it. When the primary recirculation pump flow reached the minimum level, the leader instructed to insert Control rod. According to the operational rules, they should stop one of two Condensate water pumps before inserting the rod in this situation, but they did not do it. In addition, the condenser vacuum, which was the parameter they wanted to keep, was recovering, so they did not have to insert Control rod. In the interview just after the trial, the leader confessed to the interviewer that he had been tense when instructing to insert the rod.

Example 5......Wrong instruction compounding the problem (Team D, Event No. 6)

The reactor operator was manually controlling Feed water control valve to keep the reactor water level constant, but the reactor water level decreased and the leader instructed the reactor operator to withdraw Control rod. In the interview just after this trial, the reactor operator said he wondered that the withdrawal of the rod would further decrease the reactor water level but he followed the leader's instruction. That is, in spite of realizing that the leader's instruction was wrong, the reactor operator didn't point it out and followed it.

Example 6......Narrow escape from a mistake (Team A, Event No. 4)

This is not an example of a mistake, but a narrow escape from a mistake. After recognizing the cause of an event, the leader instructed the reactor operator to decrease the primary recirculation pump flow to keep the condenser vacuum. However, the vacuum continued to deteriorate and the turbine operator called out that the vacuum had reached 670. Vacuum "670" indicates they have to manually shut down the reactor. In spite of this, the leader instructed the reactor operator to manually stop Feed water pump A and to decrease the primary recirculation pump flow. Then, the turbine operator suggested to manually shut down the reactor. Therefore, the leader instructed the reactor operator to shut down the reactor. This was a narrow escape from a misjudgment.

Before conducting the experiments, an examination was conducted to evaluate what knowledge the operators got during the education and training period. The results indicate that the operators involved in making incorrect decision had the knowledge which was necessary to make correct decisions. That is, they could not use the knowledge.

# 3 Team Error and Influential Factors

## 3.1 Definition of Team Error

There are many merits in working in a team, cooperation and communication improve productivity and make the working conditions safer. However, moments of high stress can destroy a team's ability to function coherently to make proper decisions and choose the correct course of actions. Examples 1 to 5 demonstrate misjudgments made by teams. However, it is important to remember that even though a final decision on a particular course of actions is agreed upon by the members of a team, the possible ideas are nevertheless generated by individuals. Even when a team makes a mistake, not every operator always makes a misjudgment. The interviews with operators just after each trial revealed that they sometimes felt the proposals or instructions inadequate (see example 5). The decision-making process of a team starts with one individual proposing an idea, and if the other team members can not find fault with the idea, then the idea is accepted. Therefore, the entire process of decision-making hinges on the ability of group members to identify weaknesses or faults in each other's ideas. In other words, a team error made by a team can be defined ***as inability to find a fault in a wrong decision, to point it out or the inability to choose a correct decision.***

In accordance with this definition, team errors in examples 2, 3 & 4 were caused by the inability to find faults in wrong decisions and team errors in examples 1 & 5 were caused by the inability to point them out. Although team errors caused by the inability to choose the correct decision were not observed in the manned experiments, the potential for this kind of team error can not be neglected. Example 6 shows the case where the ability to find a fault, to point it out and to choose the correct decision worked well.

## 3.2 Factors affecting team errors

Whether or not an operator notices a fault in a wrong decision depends on individual ability. Meanwhile, the inabilities to point errors out and to choose a correct decision seem to depend on the relationships between members rather than on individual ability. For example, in example 5 the reactor operator wondered that the leader's instruction was incorrect but said nothing to the leader. The team leader was probably recognized as being more senior with more experience and therefore under a moment of high stress, the reactor operator hesitated to point out the problem, even though a better course of actions was known. Referring to social psychology [2,3,4,5], the authors therefore took several factors into consideration and made a model of the relations among them. The authors assumed that whether a listener accepts a speaker's idea depends on the speaker's persuasion, listener's arousal level and the credibility. Persuasion is also assumed to consist of ascendancy, confidence and the speaker's arousal level. Credibility is assumed to depend on rank, experience and reliability. Though these factors do not include all aspects of decision-making, the authors believe that these are the most critical.

# 4 Simulation of team's decision-making process

The authors defined the relations between the factors and developed the disagreement solution model [6] to describe the team's decision-making process. This section shows how the disagreement solution model describes the process where a team makes a team error. Instances of contradiction or opposing opinions between members of the operating team in the experiments were closely analyzed and the requirements for activating the disagreement solution model were evaluated or estimated in order to trace the calculation in the mechanism.

Table 4 Simulation result for the conflict situation in example 1

| | Simulation | | Experiment | |
|---|---|---|---|---|
| | Team Leader | Operator | Team Leader | Operator |
| Idea | Decrease of the primary recirculation pump flow | Insertion of Control rod | Decrease of the primary recirculation pump flow | Insertion of Control rod |
| Conversation | | | | Oh, the reactor level is reaching to the low-low warning level. |
| | | May I insert Control rod now? | | I'll insert Control rod. |
| | Agreed. Do it now | | Agreed. | |
| | | Agreed. | | |
| Result | Although his idea was different from the operator's, the leader accepted the operator's idea which was not suitable. The correct response here is to decrease the primary recirculation pump flow. | | | |

Table 4 shows the simulation result for the conflict situation in example 1. In this simulation, the first speaker was the reactor operator. The leader and the operator had different ideas and they wanted to execute them immediately. The leader became confused in this situation and was not so confident, wondering whether they should have decreased the primary recirculation pump flow. As a result of the estimations and assumptions on the requirements for activating the disagreement solution model, the model generated the conversation shown in this table, which shows a similar chronology to the conversation in the manned experiment.

As shown in this table, the disagreement solution model can successfully generate a result identical to that in the manned experiment. In terms of the chronology of conversation, small differences are observed: For example, while the simulation result includes the operator's reply indicating that he accepted the team leader's instruction in Table 4, a similar result did not occur in the manned experiment. However, the differences found here are inconsequential since they do not affect the decision-making process. The simulation result also seems to indicate that factors such as persuasion, arousal level, etc. are significant in influencing the decision-making of a team facing abnormal events in a plant.

# 5 Conclusion

Various mistakes occurring in the decision-making process of a team were observed in the experiments with four teams of three male students. Most of them involved incorrect decision-making in order to deal with the accidents. It is important to remember that even though a final decision on a particular course of actions is agreed upon by the members of a team, the possible ideas are nevertheless generated by individuals. The decision-making process of a team starts with one individual proposing an idea, and if the other team members can not find fault with the idea, then the idea is accepted. Therefore, the entire process of decision-making hinges on the ability of group members to identify weaknesses or faults in each other's ideas. This paper proposed that a team error made by a team can be defined as *the inability to find a fault in a wrong decision, to point it out, or the inability to choose a correct decision.* In addition, the disagreement solution model to simulate the team's decision-making process was developed. The model considers several factors affecting team errors such as ascendancy, arousal level, etc. Instances of contradiction or opposing opinions between members of the operating team were closely analyzed, and the authors evaluated or estimated several requirements for activating the model through the analysis to trace the calculation in the mechanism of the model. As a result, although there are small differences in the chronology of conversation, the disagreement solution model can successfully generate a result identical to that in manned experiments.

## References

1. Gero, D., 1993. Aviation Disasters, Patrick Stephens Ltd.
2. Hovland, C.I. et al., 1951. The influence of resource credibility on communication effectiveness, Public Opinion Quarterly, 15, pp.635-650
3. Perry, J.B. et al., 1978. Collective behavior: Response to social stress, St.Paul: West Publishing Co.
4. Janis, I.L., 1972. Victims of group think, Boston: Houghton Miffin
5. Stoner, J.A.F., 1961. A comparison of individual and group decisions including risk, Unpublished master's thesis, School of Industrial Management, MIT
6. Sasou, K., et al., 1995. Modeling and simulation of operator team behavior in nuclear power plants, Symbiosis of Human and Artifact, Y. Anzai et al. (Eds.), Elsevier Science

## Acknowledgement

The authors wish thank professor Dr. Haraoka of Kurume University and Professor Dr. Kitamura for their insights and helpful words of advice. We would also like to thank Mr. Hada of Nihon Unisys Ltd. who was in charge of programming.

# An Efficient Human Reliability Assessment Using Soft Data and Bayesian Updates*

J. L. Auflick, Ph. D.

Los Alamos National Laboratory**

S. A. Eide

Idaho National Engineering Laboratory***

J. A. Morzinski

Los Alamos National Laboratory**

F. K. Houghton

Los Alamos National Laboratory**

## 1 Introduction

In post-incident investigations of catastrophic events like Three Mile Island and Chernobyl, human error has been identified as a principal root cause in 50% to 90% of all such accidents [1]. In recognition of how human error significantly affects overall levels of risk in complex systems, the U. S. Department of Energy (DOE) specifically requires that all Final Safety Analysis Reports (FSARs) " include the application of methods such as deterministic safety analysis, risk assessment, reliability engineering, common cause failure analysis, human reliability analysis (HRA), and human factors safety analysis techniques as appropriate in the identification, investigation, elimination, mitigation, or control of vulnerabilities of the facility to accidents and accidental releases [2]."

In an effort to upgrade an existing FSAR at a critical materials facility at the Los Alamos National Laboratory (LANL), 16 accident scenarios were selected from an initial pool of 832 (derived from a pre-existing hazard assessment) and were used in a refined frequency and consequence analysis. Initially, each of the 832 scenarios was placed in one of four risk matrices based on primary consequence categories: public health, co-located worker health, worker health, and environmental damage. Within each of the four matrices, individual scenarios were assigned (1) a frequency interval of 1 through 5 (where interval 1 was 1.0 E-1/yr and interval 5 was 1.0 E-6/yr.) and (2) a consequence level ranging from "A", loss of life or significant facility damage or contamination, to "D", no permanent health effects and minor or no facility contamination. Finally, those scenarios with a risk ranking of 3 or 4 and a consequence level of C or D were excluded from further consideration.

The analysts then selected a subset of representative scenarios (the highest risk ranking scenarios within each of the four primary consequence categories) with the

* This work is published as Los Alamos National Laboratory Report LA-UR-95-2161. Los Alamos National Laboratory is operated by the University of California for the United States Department of Energy under contract W-7405-ENG-36. Opinions expressed in this report are those of the authors and do not necessarily reflect the opinions of the Department of Energy or the Los Alamos National Laboratory.

** P. O. Box 1663, MS F-684, Los Alamos, NM, USA 87545

*** P. O. Box 1625, Idaho Falls, ID, USA 83415-3850

caveat that at least one scenario was chosen for each consequence category-scenario type combination. This resulted in a final list of 16 scenarios representing criticalities (CRT), explosions involving radionuclides (EXR), fires (FRR), chemical (MRC) and radiological (MRR) releases, and seismic events. Each of these 16 scenarios was modeled by a simplistic fault tree to display the dominant combinations of events that could lead to selected outcomes. Fault trees helped clarify various assumptions and model parameters based on the physics of the relevant processes, the behavior of the systems under consideration, and the potential for human error.

# 2 Human Reliability Analysis and Bayesian Process

These fault trees contained 26 human errors that had to be quantified in a cost-effective, plant-specific HRA meeting DOE requirements. Numerous HRA techniques were available, ranging from screening techniques (providing relatively "gross", conservative estimates for human failures, without the expenditure of large amounts of time and effort) to detailed approaches (requiring the collection of extensive task analysis information for *each* human error action, an expensive, labor intensive activity). Because of the FSAR's limited budget and time constraints, two screening techniques, the Accident Sequence Evaluation Program (ASEP) [3] and Human Cognitive Reliability (HCR) [4], were selected to model and quantify the human error actions.

To collect HRA data, the analysts first conducted several walkdowns of specific operations to familiarize themselves with the respective processes, operator actions, and accident scenarios. Next, the analysts conducted several rounds of interviews with subject matter experts (SMEs) who were most familiar with the modeled processes. All walkdowns and interviews also used a detailed human factors checklist to capture operational data and information about the human-machine interface. This information was correlated with a review of a computerized database looking for site-specific unusual occurrence reports (UORs) for incidents related to the scenarios.

The ASEP Screening HRA for Pre-Accident Tasks was the primary method used to quantify (assign human error probabilities, HEPs, and error factors, EFs) for each human error action. ASEP decomposes individual human error actions into errors of omission (EOMs) and errors of commission (ECOMs). For ECOMs, ASEP assigns a basic HEP of 0.01 which is (1) modified (based on the presence or absence of four specified recovery factors) and (2) categorized by placing the action in one of nine ASEP cases. ASEP also models EOMs, but in the accident scenarios in this analysis, the omission of required procedural steps actually provided protective functions to the serial, step-by-step processes involved in the scenarios. For example, in scenario FRR 26 the related EOM, omitting the procedural step to remove the crucible, prevents the error action of spilling the crucible of liquid Pu. Four actions did not fit into the ASEP paradigm and were quantified by using the HCR screening model which assigned basic HEP point estimates based on whether a human error was defined as being either skill-, rule-, or knowledge- based. However, it should be noted that both ASEP and HCR do not provide as thorough an analysis of the operational environment, compared to the detailed HRA techniques. As a result, they tend to produce conservative results, i.e. high estimates of HEPs, which may not reflect actual operating experience.

Output from the quantification phase provided screening HEPs (assumed median values lognormally distributed) and associated EFs for each human error. To deal with the conservatism in the HEPs, expert elicitation was used to capture plant-specific data which was factored into a Bayesian process that modified the screening

HEPs. SMEs reviewed both the scenarios and HEPs at a later date, to provide a reality check of the human error modeling and quantification. At this time, SMEs were also asked to estimate the number of errors and the number of opportunities for errors in specific scenarios during operations over the past few years. Data collection was most efficient during these sessions because the human errors were clearly characterized and the most knowledgeable plant personnel were present. Ideally, this kind of plant-specific operational data should be found in detailed record logs, but in this case these records did not exist. Since the SMEs were not accustomed to thinking about their processes in probabilistic terms, the analysts utilized a structured approach during these interviews to help the SMEs provide their data. The resulting plant-specific information was called "soft" data because of the reliance upon operator estimates rather than detailed plant records. This data collection effort produced soft data for 13 of the 26 modeled human errors. In the remaining cases, the Bayesian process was not able to be used because no such failure opportunity had occurred or the SMEs were unable to reach a consensus about how many opportunities for failure (demands) there had been, thereby preventing an accurate calculation of the denominators in equations (3.5 and 3.6).

An approach based on Bayes' Theorem integrated the soft data with the human error models. Bayes' Theorem uses deductive reasoning to weight prior information with empirical evidence allowing analysts to calculate probabilities of causes based on the observed effects. It is an effective and consistent link between the inductive logic (i.e., indirect probabilities) required for scientific reasoning and the deductive logic (direct probabilities) that most people know how to use [5]. Specifically, in simple terms, Bayes' Theorem states that the conditional probability of the hypothesis "A" (being true) given the soft data "B", is proportional to the conditional probability of B given A, times the probability of A, e.g.,

$$\Pr(A|B) \propto \Pr(B|A) \times \Pr(A) \text{ [eq. 3.1].}$$

As Sivia [6] points out, "the full implications of Bayes' Theorem do not become apparent until we discover that the theorem applies equally well to cases in which A and B are any arbitrary propositions and the probabilities assigned to them represent merely our belief in the truths (or otherwise) of the propositions." Cox [7] reportedly proved this generalization while he was considering the rules necessary for logical and consistent reasoning. Bayes' Theorem then, is a paradigm where "expert opinions can be combined with experimental results and statistical observations to produce quantitative measures of risk in a given system [8]."

The Bayesian process in this paper was identical to that often used for mechanical component failure rate estimation [9] and was similar to the method used as part of the Savannah River Site Human Error Database for Nonreactor Nuclear Facilities [10]. Each of the screening median HEPs was first converted to a mean value (lognormally distributed) by:

$$mean\,/\,median = \exp\{0.5[(\ln EF)\,/\,1.645]^2\} \text{ [eq. 3.2]}$$

where EF = error factor (95$^{th}$ percentile/50$^{th}$ percentile) and ln = natural logarithm. Then, for the 13 instances where plant-specific soft data were available, mean HEP values and associated EFs were used as the prior distributions for the Bayesian update. Individual prior distributions were made to fit to beta distributions by the following equations:

$$\alpha + \beta + 1 = (1 - Mn_1)\,/\,\{Mn_1[\exp((\ln(EF_1)\,/\,1.645)^2) - 1]\}$$

[eq. 3.3]; and $\alpha = Mn_1[\alpha + \beta + 1) - 1]$ [eq. 3.4], where

$Mn_1$ = mean of prior lognormal distribution and $EF_1$ = error factor of prior lognormal distribution. Next, for each of the human errors, the number of events (n) for a given process and the number of demands (D) were factored into equations determining the beta distribution for the posterior:

$$MN_2 = (\alpha + n) / [(\alpha + \beta + 1) + D - 1] \text{ [eq.3.5] and}$$

$$EF_2 = \exp\{1.645[(\ln(1 + (1 - Mn_2) / (Mn_2(\alpha + \beta + 1 + D))))^{0.5}]\}$$

[eq.3.6], where $Mn_2$ = mean of the posterior beta distribution and $EF_2$ = error factor of posterior beta distribution. Results were converted to a lognormal posterior distribution by: $Mn_3$ = $Mn_2$ and $EF_3$ = $EF_2$; where $Mn_3$ = mean of the posterior lognormal distribution and $EF_3$ = the correlated error factor.

For example, in scenario CRT 85 (Table 1) a criticality event was postulated because there was too much special nuclear material (SNM) in a container as a result of an ECOM where the technician made an analysis or mislabeling error. Walkdowns and task analyses allowed the analysts to classify this ECOM as ASEP Case 2, with a median HEP of 0.0001(EF=16). A review of the UORs found one significant accident precursor resembling the hypothesized accident scenario. Expert elicitation verified this and ascertained that there had been approximately 15,000 transactions (demands) where this type of error could have occurred. The operators basically estimated how many containers were processed, on the average, on a daily basis and results were extrapolated to estimate the number of yearly demands. The basic median HEP was then converted to a mean value using equation 3.2. Then, the resulting mean HEP (0.00041), now used as the prior distribution, was made to fit to a beta distribution using equations 3.3 and 3.4. Finally, equations 3.5 and 3.6 calculated the beta distribution for the posterior, i. e., the posterior mean value (0.00007) using the evidence, or soft data (where n = 1 and D = 15,000). A similar process was used for all scenarios that had soft data.

# 4 Results

HRA results for the Bayesian process for this FSAR are shown in Table 1. Columns in this table identify the scenario, median HEPs and EFs, prior mean HEP, events (*N*), demands (*D*), and the posterior mean HEP.

| Scenario and Description | HEP (EF) | Prior Mean | *N* | *D* | Post. Mean |
|---|---|---|---|---|---|
| **CRT 85**: Extra SNM in container from analysis error, mislabeling, etc. | .0001 (16) | .00041 | 1 | 15,000 | .00007 |
| **CRT 85**: More mislabeled SNM enters room - analysis error, mislabeling, etc. | .0001 (16) | .00041 | 1 | 15,000 | .00007 |
| **CRT 85**: SNM containers placed too close to each other. | .001 (10) | .0027 | 1 | 15,000 | .00008 |
| **EXR 14**: Fail to establish glovebox vacuum. | .001 (10) | .0027 | 0 | 70 | .0012 |
| **FRR 26**: Oxygen line mistakenly connected to outside Argon manifold. | .001 (10) | .00266 | 0 | 2 | .0026 |
| **FRR 26**: Operator opens both doors to air-filled drop box. | .0001 (16) | .00041 | 0 | 1,000 | .00005 |
| **FRR 28**: Operator prematurely removes | .0001 | .00041 | 0 | 1,600 | .00004 |

| Scenario and Description | HEP (EF) | Prior Mean | *N* | *D* | Post. Mean |
|---|---|---|---|---|---|
| crucible - spills molten Pu. | (16) | | | | |
| **FRR 88**: Forklift operator drops-punctures waste container. | .0005 (10) | .0013 | 2 | 5,000 | .00042 |
| **MRC 18**: Operator upsets liquid nitrogen dewar in freight elevator (doors closed). | .001 (10) | .0027 | 0 | 2,184 | .00007 |
| **MRC 27**: Operator error during connection of halogen gas tank to manifold. | .0001 (10) | .00027 | 0 | 3 | .00027 |
| **MRC 27**: Operator drops halogen gas cylinder. | .01 (5) | .016 | 0 | 3 | .017 |
| **MRR 12**: Poor worker response to glove box breach causes significant exposure. | .05 (5) | .081 | 0 | 20 | .019 |
| **MRR 75**: SNM container lost or misplaced in plant. | .0001 (16) | .0004 | 1 | 2,500 | .0004 |

Table 1. HRA Results (NOTE: SNM = special nuclear materials).

In general, where at least one human error had occurred, the Bayesian update process resulted in a posterior human error probability lying between the model result and the result based solely on the soft data. In cases where no plant failures occurred, the Bayesian process lowered the model result. For the 13 human errors with soft data, the Bayesian process lowered the model probabilities by an average reduction of a factor of seven and significantly lowered the scenario frequency estimates.

This augmented approach to HRA generated several benefits. First, the use of screening techniques allowed the analysts to conduct the HRA in a timely and cost effective manner. Next, by collecting soft failure data and using the Bayesian process, the analysts were able to reduce the conservatisms in the screening techniques. In addition, combining plant-specific data with estimates for human error rates incorporated actual plant operating experience into the analysis, resulting in more accurate results and a more realistic risk profile for the facility. Finally, this process met or exceeded all of the DOE's stringent requirements for HRAs. Results from this study are now being incorporated into a current LANL laboratory directed research and development (LDRD) project that is compiling a database of HEPs from different operational environments and is developing an expert system to access various HRA techniques as well as the HEP data.

## References

1. Hollnagel, E., Human Reliability Analysis Context and Control. Academic Press, London, 1992.
2. U. S. Department of Energy, Order 5480.23, Nuclear Safety Analysis Reports. U. S. Department of Energy, Washington, 1992.
3. Swain, A., Accident Sequence Evaluation Program Human Reliability Analysis Procedure, NUREG/CR-4772. U. S. Nuclear Regulatory Commission, Washington, 1987.
4. Hannaman, G. W. et al., Human Cognitive Reliability Model for PRA Analysis, NUS-4531. Electric Power Research Institute, Palo Alto, CA, 1984.

5. Sivia, D., Bayesian inductive inference, maximum entropy, and neutron scattering. Los Alamos Science, Los Alamos Natl. Lab., Los Alamos, NM, 1990.
6. ibid., #5
7. Cox, R. T., Probability, frequency, and reasonable expectation. American Journal of Physics, 1946; 14: 1-13.
8. Apostolakis, G., The concept of probability in safety assessments of technological systems. Science, 1990; 250: 1359-1364.
9. U. S. Nuclear Regulatory Commission, Probabilistic Risk Assessment Procedures Guide, NUREG/CR-2300, Section 5. U. S. Nuclear Regulatory Commission, Washington, 1983.
10. Benhardt, H., Eide, S., Held, J., et al., Savannah River Site Human Error Data Base Development for Nonreactor Nuclear Facilities, WSRC-TR-93-581. Westinghouse Savannah River Company, Aiken, SC, 1994.

# Development of Dynamic Human Reliability Analysis Method Incorporating Human-Machine Interaction

**Ryuji KUBOTA & Kouji IKEDA**
Hitachi Works, Hitachi Ltd.
Saiwai 3-1-1, Hitachi, Ibaraki, 317, Japan

**Tomihiko FURUTA & Akira HASEGAWA**
Institute of Human Factors, Nuclear Power Engineering Corporation
Fujita Kanko Toranomon BLDG., 6F Toranomon 3-17-1,
Minato-ku, Tokyo, 105, Japan

## 1. Introduction

In recent years, in order to collect data relating to human factors, the performance of the operator crew under abnormal events is acquired by VTR, etc., using full-scale operation training simulators or research simulators, and the response time, communication, etc. are analyzed[1]. In the experiments using these simulators, operator responses are found to vary between the case where the safety limit of the plant is reached quickly even though the action is taken as a countermeasure after the scram, and the case where the safety limit of the plant takes a long time to reach and the operator waits for the next operation task or action since the behavior of the plant changes slowly and there is plenty of time to response required action. Also, when a simulated event which is beyond the operational procedure is generated, the operators take appropriate actions based on training such as OJT (On the Job Training) and simulator training, though the response time becomes longer. Thus, the Human Reliability Analysis Method which handles interactions with the plant has been developed to simulate the cognition process of the operator corresponding to the plant behavior and the degree of abnormal event. The method was evaluated scenarios using in existing simulator experiment to verify that the human characteristic parameters applied as inputs of this method could be collected from data of experiments using the simulator.

## 2. Human reliability analysis method considering interaction with the plant

The authors previously developed a dynamic human reliability analysis method that considered the interaction with the plant, to compare and evaluate the response time between the case where the safety limit of the plant is quickly reached and the case where it is not[2]. However, as the modeling method in this old analysis used a "process waiting time" between successive operation tasks, the analysis could be carried out only when the plant behavior was slow. This method was not suitable for situations requiring quick action, such as plural operation tasks in succession, parallel tasks, and so on.

Accordingly, the interaction with the plant has been improved to enable adequate modeling even when the safety limit of the plant is reached quickly.

### 2.1 Cognition mechanism

The cognition mechanism of the operator is modeled by the Monte Carlo calculation using the probabilistic network method, as well as by applying Rasmussen's decision-making model as shown in Fig. 1 to the information processing phase and by establishing the detection phase and the operation phase before and after it, respectively.

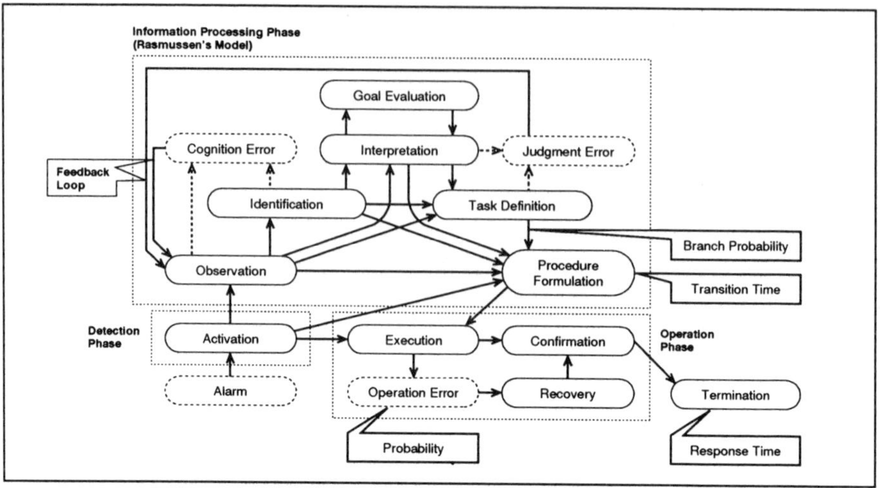


**Fig. 1 Basic Model of a Human Reliability Analysis Including Rasmussen's Model**

## 2.2 Interaction with the plant

In the existing model[2], a node of "process waiting time" between successive operation tasks was established, and the time interval when the plant behavior, which is the object of the analysis, was beforehand analyzed in another program using thermodynamic equations was given to this node as the time untill generation of the next trigger signal. However, the response time in this model was simply added, since an interval of the plant behavior was assumed to be constant even though the Monte Carlo calculation was used in this model.

Thus, improvements were made to incorporate dynamic characteristic equations expressing the thermodynamic behavior of the plant and to include plant behavior which varied for each execution of the Monte Carlo calculation. By these improvements, different response time to the next operation task can be simulated as differences in plant behavior, when the response time is delayed due to cognition or judgment errors though the event is the same, and when the response time is shortened due to correct operation.

# 3. Preliminary analysis for data collection

In this research, simulator experiments using multiple engineering work stations (EWSs) will be conducted. A preliminary analysis has already been carried out to confirm that the data of this simulator experiment can be applied to this model.

## 3.1 Study of data collection method and data survey method with simulator experiment

### *3.1.1 Data collection method*

The conversation data which becomes the base of the human characteristic parameters required to verify this model, and the response time must be collected. The response time is the time from alarm generation to completion of the diagnosis conversation, or to completion of operation execution. The conversation data is collected using the VTR by simultaneously recording the voice of the wireless microphone attached to the operators.

*3.1.2 Data Survey method*

(1) Transition time

Each node of the cognition process of this model is given a transition time (time spent in each node) as an input. This transition time is an inherent value at each node of the cognition process and can be expressed as a constant or as a distributional curve. This distributional curve was assumed to be a normal distribution expressed by a mean value, standard deviation, maximum value and minimum value.

To induce the distribution of the transition time from the simulator experiment, the conversation data obtained from the simulator experiment, etc. is analyzed by the method shown in Fig. 2. First, each node of cognition process is assigned to the conversation data, and then the time interval between these cognition nodes is calculated. Summing up the time intervals of these cognition nodes, the mean value, the standard deviation, the maximum value and the minimum value are calculated. The normal distribution determined by these calculated values is input as the transition time inherent in each node of the cognition process.

| Time | Content of Alarm and Conversation | Classification of Cognition Node |
|---|---|---|
| T1 | Alarm "Flow rate rise of dry well sump" | (Alarm) |
| T2 | Operator A "Occurrence of flow rate rise of dry well sump." | "Activation" |
| T3 | Operator A "Dry well sump is $\Delta\Delta m^3$." | "Observation" |
| T4 | Operator A "Rise of dry well sump from normal." | "Identification" |
| T5 | Shift supervisor "Confirm dry well sump pressure." | "Confirmation" |
| : | | |

Transition time: T2 - T1 = Transition time of activation
T3 - T2 = Transition time of observation
T4 - T3 = Transition time of identification
T5 - T4 = Transition time of confirmation

If the branch is defined as branching from origin to branch destination, then,
Branch from (alarm) to "activation"
Branch from "detection" to "observation"
Branch from "observation" to "identification"
Branch from "identification" to "confirmation".

If the number of branch destinations is defined as the sum of individual branches, then,

$$\text{Branch Probability} = \frac{\text{Number of branch origins}}{\text{Total number of branch destinations}}$$

**Fig. 2 Calculation Procedure of Transition Time and Branch Probability**

(2) Branch probability

Since there are several kinds of branches (arcs) from a given node of the cognition process in this model to the next cognition node, the branch probability is required for each branch. Although the branch probability depends on the node configuration modelled for each operation task, there is usually a main branch. In other words, except when the branch probability in the same arc is different between the decision-making task by the rule base and that by the knowledge base, the same branch probability can be applied in most cases.

To induce the branch probability from the simulator experiment, the conversation data obtained by the simulator experiment, etc. is analyzed by the method shown in Fig. 2, like the transition time. First, the nodes of the cognition processes are assigned to the conversation data, and then those branched node are counted for each cognition node. The branch probability can then be calculated by dividing the individually summed number of branches by the total number of branches.

## 3.2 Preliminary analysis

### *3.2.1 Selection of human characteristic parameters*

To demonstrate that the human characteristic parameters of this model can be obtained by the simulator experiment of this research, these parameters were induced from the data of previous simulator experiments and the results were as follows. For the transition time, the data published by ORNL (Oak Ridge National Laboratory)[3] can be used without correction. However, since there are some differences between the basic model of this research and that of ORNL, the data published by the Japanese BWR Joint Research[4] was used for those parts where data was not available from ORNL source. The provisional values of the branch probability were determined using the data of SNL (Sandia National Laboratory)[5]as a reference.

The transition times which were thus set are shown in Table 1 and the branch probabilities are shown in Table 2.

**Table 1 Transition Time of Basic Model Used in This Research**

| Node name | Distribution type | Mean value | Standard deviation | Minimum value | Maximum value | Source |
|---|---|---|---|---|---|---|
| Activation | Constant value | 5.0 | – | – | – | (1) |
| Observation | Normal distribution | 13.4 | 12.3 | 1.2 | 45.0 | (2) |
| Identification | | 3.8 | 3.8 | 0.1 | 12.0 | (2) |
| Interpretation | | 26.6 | 13.9 | 5.0 | 55.0 | (2) |
| Goal evaluation | | 50.0 | 50.0 | 0.0 | 100.0 | (1) |
| Task definition | | 11.9 | 2.0 | 0.3 | 67.0 | (2) |
| Procedure formulation | | 32.7 | 15.3 | 8.0 | 60.0 | (2) |
| Execution | | 27.5 | 12.5 | 15.0 | 40.0 | (1) |
| Recovery | | 27.5 | 12.5 | 15.0 | 40.0 | (3) |
| Confirmation | | 50.0 | 50.0 | 0.0 | 100.0 | (1) |

[Source] (1) Provisional value summarized from published values of experiment data of BWR Utilities Joint Research of Japan
(2) Published numeric values of ORNL experiments
(3) Same transition time as execution is assumed.

**Table 2 Branch Probability of Basic Model Used in This Research**

| Node name | Node name of branch destination | | | | | | | | | | | | |
|---|---|---|---|---|---|---|---|---|---|---|---|---|---|
| | Activation | Observation | Identification | Interpretation | Goal evaluation | Task definition | Procedure formulation | Execution | Cognition error | Judgment error | Operation error | Recovery | Confirmation |
| Activation | | 0.8 | | | | | 0.1 | 0.1 | | | | | |
| Observation | | | 0.699 | 0.1 | | 0.1 | 0.1 | | 0.001 | | | | |
| Identification | | | | 0.2 | | 0.399 | 0.1 | | 0.001 | | | | 0.3 |
| Interpretation | | | | | 0.1 | 0.799 | 0.1 | | | 0.001 | | | |
| Goal evaluation | | | | 0.3 | | 0.7 | | | | | | | |
| Task definition | | | | | | | 0.3 | 0.699 | | 0.001 | | | |
| Procedure formulation | | | | | | | | 1.0 | | 0.001 | | | |
| Execution | | | | | | | | | | | 0.062 | | 0.938 |
| Operation error | | | | | | | | | | | | 1.0 | |
| Recovery | | 0.1 | | | | | | | | | | | 0.9 |
| Confirmation | | 0.4 | | | | 0.4 | | 0.2 | | | | | |

### *3.2.2 Preliminary analysis*

(1) Comparison with conventional model

The comparison between the model using the conventional "process waiting time" and the model of this research was analyzed for the response time until manual scram in a small LOCA event. For the response until the manual scram in this small LOCA, two successive operation tasks are required: reduction of reactor power from leakage detection of small LOCA event and execution of manual scram. In this new model the response time is calculated according to the time calculated with the thermodynamic equations of the plant behavior.

The result is shown in case 1, 2, and 3 of Fig. 3., the time until generation ofthe trigger signal of the second operational task is not added a constant time like the conventional method but varies for each random number of the Monte Carlo calculation.

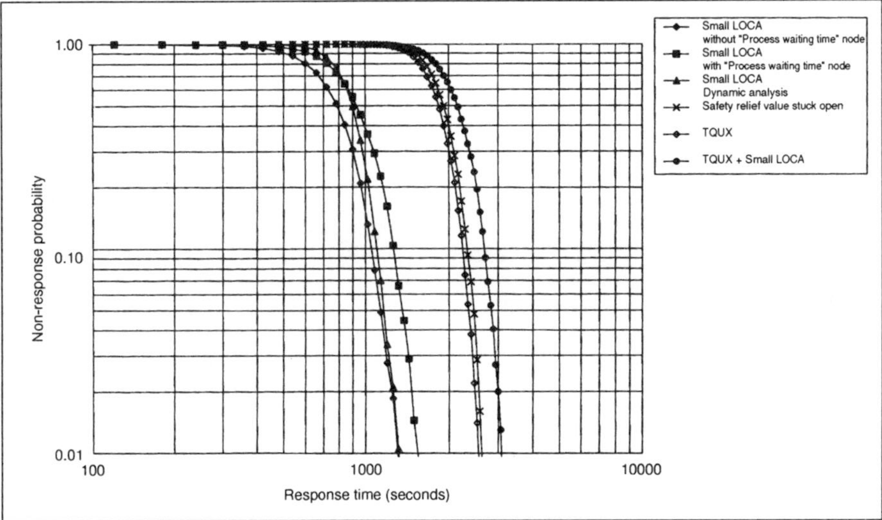


**Fig. 3 Results of Analysis**

(2) Comparison between short time and long time until safety limit

The evaluation was carried out for both a short time and a long time until the safety limit. In the case of a BWR plant, the safety limit becomes for longer time beyond the model of simulator because of the inherent safety of the plant. Therefore, in this research, the safety limit is assumed to be the limit of the countermeasures which the operator can take to ensure safety of the plant, in order that the operator may take action in the simulator experiment and data may be collected. The before-mentioned small LOCA event is an example where the safety limit quickly reached, while the safety relief valve stuck open event is an example where there is much time. Case 3 and case 4 in Fig. 3 show the result of evaluating response time until execution of manual scram for these two events.

(3) Case of event beyond operation procedure

In the simulator experiment, the operation crew took appropriate actions even for events beyond the symptom based EPG (Emergency Procedure Guidline), though the response time becomes longer. This is because the operators judge such events based on their OJT experience, simulator training, etc.

Accordingly, it is assumed that this effect can be well simulated by making the transition time of the "Goal Evaluation" node longer in this model. Therefore, this research evaluated the response time until reducing the reactor pressure, for the loss of all feedwater event (TQUX event), which is an event within the scope of the operation procedure and for the loss of all feedwater + small LOCA + high pressure water system failure event, which is an event beyond the operation procedure. This evaluation result is shown in case 5 and case 6 of Fig. 3.

# 4. Considerations

The non-response probability used in the PSA (Probabilistic Safety Assessment) is not the probability of failure to respond but the probability of delay to respond. In the same way as when the safety limit of the plant is not reached quickly, this non-response probability can be treated by modeling that, even if a severe event occurs and the safety limit of the plant quickly reached, the error is corrected according to the plant behavior and the response time is longer due to the time required for the error correction. In the conventional method, the two cases where the safety limit of the plant is quickly reached or not, could not be treated using the same method. However, this new model could simulate the interactions between the operators and the plant in such cases.

# 5. Conclusions

Interactions with the plant have been incorporated, multiple operators can be treated, and a model which can correspond flexibly to the individual differences of operators and different tasks has been realized. A new dynamic human reliability analysis method has been thus developed in place of THERP (Technique for Human Error Rate Prediction) and TRC (Time Reliability Correlation) methods.

In the future, human characteristic parameters provisionally used in this research will be collected from simulator experiments, the validity of the model will be verified experimentally, and the model will be modified to a genera purpose human reliability analysis method.

## Acknowledgments

This research was conducted as a study entrusted by the Ministry of International Trade and Industry.

## References

1. R. Kawano, R. Kubota, et al., "Plant Operator's Behavior in Emergency Situations by Using Training Simulators," 11th Congress of the International Ergonomics Association, Paris, 1991.
2. R. Kubota, T. Furuta, et al., "Development of Dynamic Human Reliability Analysis Method," PSA '95 International Conference on Probabilistic Safety Assessment Methodology and Applications, Seoul, Korea, Nov. 1995.
3. D. B. Barks, et al., "Nuclear Power Plant Control Room Task Analysis: Pilot Study for Boiling Water Reactors," NUREG/CR-3415, 1983.
4. H. Ujita, et al., "Behavior Evaluation of Plant Operators using Cognition Model," Proceedings of the Forth Human Interface Symposium, 1988. [In Japanese]
5. A. N. Beare, et al., "A Simulator-Based Study of Human Errors in Nuclear Power Plant Control Room Tasks," NUREG/CR-3309 (SAND83-7095), 1984.

# Probabilistic Safety Analysis of a Plant for the Production of Nitroglycol including Start-up and Shut-down

Ulrich Hauptmanns
Otto-von-Guericke-Universität Magdeburg
D-39016 Magdeburg, Germany

Jaime Rodríguez
Gesellschaft für Anlagen- und Reaktorsicherheit (GRS) mbH
D-50667 Köln, Germany

## 1 Introduction

Probabilistic safety analyses for process plants generally only address accidents occurring during production operation (cf.[1],[2]). This is in contrast with the numerous start-ups and shut-downs required in most chemical plants. For this reason an investigation is presented which covers all the three operational phases mentioned. It was performed for a plant producing nitroglycol, a very powerful explosive. The presentation is limited to the production part, although other plant areas as, for example, storage were investigated as well [3].

## 2 Brief Description of the Plant and its Safety System

Figure 1 gives the P&I diagram of the continuous production of nitroglycol using an injector. The reaction is between a mixture mainly of nitric (26%) and sulphuric acid (69%) and glycol. The acid mixture at 0 °C is driven by air pressure from tank (1) through the injector (2) where it sucks in glycol at 30 °C from tank (3). The exothermal reaction producing nitroglycol is virtually instantaneous. The quantity of glycol introduced and hence the reaction temperature is regulated by the control loop T03 leading to an injector outlet-temperature between 46 and 48 °C.

The process described has the advantage of containing only about 4 kg of explosible nitroglycol as compared with the reactor-based procedure where 1000 kg may be present.

The safe operation of the plant hinges on the fulfilment of the following conditions:

(1) excess of nitrating acid according to the mass flow of glycol
(2) upper temperature limit for the process of 52 °C
(3) correct composition of the nitrating acid mixture

In order to assure that condition (1) is satisfied the following equipment is provided:

- interlock of acid pump P1 impeding the start-up of the plant before the pump is in operation

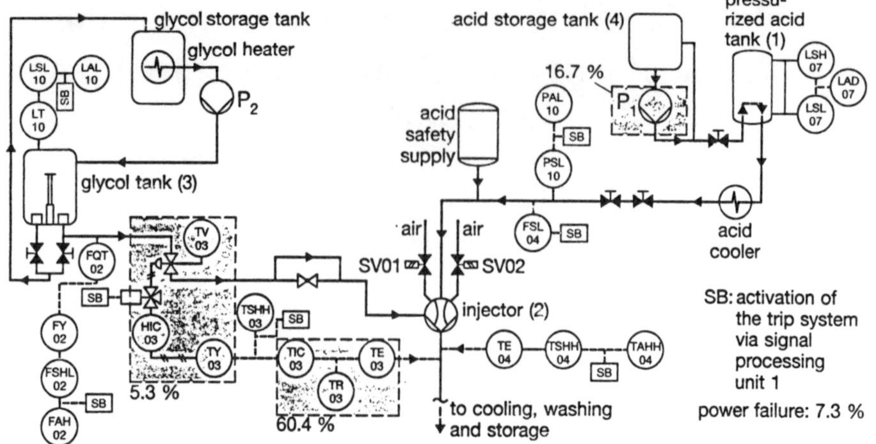


Figure 1:P&I diagram of the injector process for the production of nitroglycol (shaded areas mark main contributions to an explosion)

- pressure control PISL 10/1 in front of the injector; it activates a trip on low pressure thus making sure that the pressure in the injector is sufficiently high
- flow control in front of the injector by flow-meter FSL 04 which causes a trip in case of insufficient flow
- control of the level in tank (1) by the switch LSHL 07 which activates an alarm upon which the operator is supposed to shut down the plant

The control of the glycol supply is monitored as follows:

- flow control FQT02 causing a trip in case the flow is too high or too low
- level control L10 in tank (3) leading to a trip should the level be too high or too low

Reaction temperature is maintained within a permissible range around its setpoint by regulating the glycol supply via control valve TV03 activated by the measuring loop T03. The corresponding safety measures are as follows:

- temperature switch TSH03, which forms part of the control loop T03 and causes a trip in case the temperature of reaction were too high
- temperature measuring loop T04 which redundantly to T03 activates a trip

In addition, the measuring loops supply a number of alarms which keep the operator informed about the state of the plant and in some cases require him to shut it down.

For tripping the plant the vacuum in the injector is broken redundantly by opening valves SV01 and SV02. Additionally, control valve TV03 is fully opened thus allowing the glycol supply line to be emptied into tank (3) and avoiding the possibility of a contact with nitrating acid (or its vapour), nitroglycol formation and its possible explosion in dry state being the consequence.

# 3 Safety Analysis

## 3.1 General Procedure

In order to prepare quantification a qualitative safety analysis of the plant was performed using the HAZOP methodology based on the application of guide words to the process. This was supported by the evaluation of records of accidents in similar installations. It was found that safety measures against the undesired event, an explosion, are provided in the plant for all initiating events thus conceived. These are assigned to three categories:

- initiating events during start-up liable to cause an explosion during start-up itself
- initiating events during shut-down liable to cause an explosion during shut-down itself
- events during start-up, shut-down or both with potential repercussions during the operational phase such as e.g. weakening of safety systems by leaving valves in a wrong position during start-up

Whilst events from categories (1) and (2) were analysed mainly using the THERP method [4]. The investigation for (3) used fault trees and the THERP method for the human acts involved. Quantification was based on

- plant-specific reliability data for independent failures of technical components evaluated in a previous investigation [5]
- estimated β-factors for common-cause failures
- probabilities for human error from the Handbook [4]
- operational data like times between functional tests (mostly 8h or 80h) and the number of start-ups and shut-downs

The reliability data for components were evaluated with the Bayesian procedure and non-informative prior distributions (cf. [6]). In the case of common-cause failures estimatimation became necessary since the investigation [5] produced no evidence of the occurrence of such failures. The uncertainty factors were increased multiplying the original distribution with a log-normal distribution of expected value of 1 and

- a factor of $K_{95} = 3$ for technical components in order to account for the fact that there is a degree of uncertainty involved in assessing the "future" with data from the past
- a factor of $K_{95} = 5$ for the basic human error rates from [4] in order to accomodate the uncertainty involved in applying this data to a situation different in many respects, e.g. cultural background, training etc., from that of its origin.

## 3.2 Specific example

By way of an example the case of the failure of the trip system during the shut-down procedure will now be discussed.

The plant is shut down by opening a hand-operated valve to the injector in order to break the vacuum. After that the automatic trip is activated opening the two valves SV01 and SV02 and the control valve TV03 in the glycol supply line (cf. section 2).

This action is performed about 600 times per year. Should it be forgotten to open the hand-valve or should the automatic trip be activated in the first place contrary to stipulations in the operating manual and were both valves SV01 and SV02 to fail closed, the vacuum in the injector would be preserved. The opening of valve TV03 - a safety-prone action - would then cause an explosion because there would be an excessive supply of glycol. This means that a procedure devised for safety reasons is counterproductive in this case. The corresponding event tree is shown in Figure 2 along with the data used for ist quantification. Its evaluation leads to the expected frequency of occurrence $H = h \cdot [p+(1-p) \cdot p] \cdot u = 3.1 \cdot 10^{-5}\ a^{-1}$ $/K_{95} = 13.2$

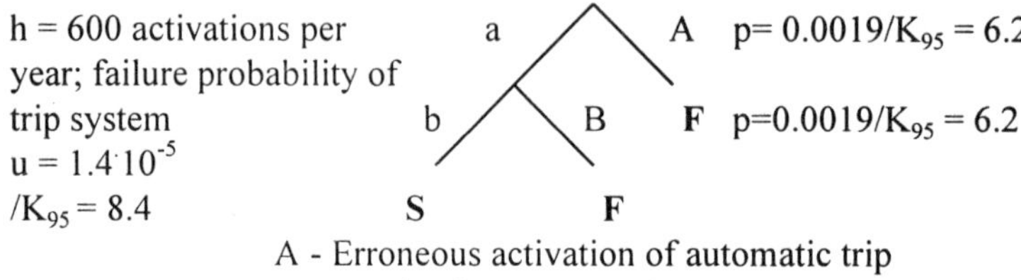


A - Erroneous activation of automatic trip
B - Hand-valve not opend

Figure 2: Event tree for the evaluation of trip-system failure during shut-down (all probabilities are mean values)

# 4 Results of the Analysis

## 4.1 Existing design

The contributions of the different initiating events to an explosion during production are given in Figure 3.

The results may be interpreted as follows:

- human error during start-up contributes 4.2% to the expected frequency of explosion during production,
- if human error leading to an explosion already during start-up and shut-down is included, the contribution rises to 8.3%,
- if, additionally, human error committed during start-up and shut-down, which paves the way for an explosion during production is included and it is supposed that all human acts are performed with perfection the expected frequency of explosion drops to 86%.

The following failures contribute most to an explosion, as also indicated in Figure 1

- failure of the common part of the measurement T03 (60.4%)
- operational failure of the nitrating acid pump P1 (16.7%)
- failure of electric power supply (7.3%)
- temperature controller T03 erroneously opens valve TV03 (5.3%)

| Initiating event | | Safety function | Undesired event |
|---|---|---|---|
| Denomination | Expected frequency $a^{-1}$ | Unavailability | Expected frequency $a^{-1}$ |
| power supply fails | $4.6 \cdot 10^{0}$ | $1.5 \cdot 10^{-4}$ | $6.9 \cdot 10^{-4}$ |
| common part of TE03 and TIC03 fails | $2.7 \cdot 10^{0}$ | $2.1 \cdot 10^{-3}$ | $5.7 \cdot 10^{-3}$ |
| T03 wrongly opens TV03 | $2.4 \cdot 10^{-1}$ | $2.1 \cdot 10^{-3}$ | $5.0 \cdot 10^{-4}$ |
| pump P1 does not start | $2.2 \cdot 10^{-1}$ | $6.4 \cdot 10^{-5}$ | $1.4 \cdot 10^{-5}$ |
| pump P1 does not run | $2.2 \cdot 10^{1}$ | $7.3 \cdot 10^{-5}$ | $1.6 \cdot 10^{-3}$ |
| connecting line (1)-(4) not open | $8.8 \cdot 10^{-1}$ | $9.1 \cdot 10^{-6}$ | $7.9 \cdot 10^{-6}$ |
| control not switched to „automatic"during start-up | $8.3 \cdot 10^{-2}$ | $1.5 \cdot 10^{-4}$ | $1.2 \cdot 10^{-5}$ |
| valves in cooling circuit not opened during start-up | $4.8 \cdot 10^{0}$ | $5.1 \cdot 10^{-5}$ | $2.4 \cdot 10^{-4}$ |
| pressurized air set to< 120 Mpa | $2.5 \cdot 10^{-1}$ | $2.0 \cdot 10^{-4}$ | $5.0 \cdot 10^{-5}$ |
| soda solution contains too little sodium carbonate | $1.1 \cdot 10^{0}$ | $1.8 \cdot 10^{-5}$ | $2.0 \cdot 10^{-5}$ |
| start-up | | | $3.9 \cdot 10^{-4}$ |
| shut-down | | | $3.1 \cdot 10^{-5}$ |
| Sum (including initiating events not mentioned here) | | | $1.0 \cdot 10^{-2}$ |

Figure 3: Contributions of different initiating events during production, start-up and shut-down to the expected frequency of explosion of the plant

## 4.2 Possible improvements and comparative results

Possible improvements addressing the weak areas of the design shown in Figure 1 are given in Figure 4.

| Measure | Reduction of the expected annual frequency of explosion |
|---|---|
| (a)Reduction of the interval between functional tests from 80 h to 8 h for flow- measurement F02 | $1.0 \cdot 10^{-2}$ to $4.3 \cdot 10^{-3}$ |
| (b)redundant design of flow measurement F02 | $1.0 \cdot 10^{-2}$ to $4.3 \cdot 10^{-3}$ |
| either measure (a) or (b) and additionally installation of a flow-meter between acid tank (4) and pressurized acid tank (1) with alarm for low mass flow and instruction to shut-down | $4.3 \cdot 10^{-3}$ to $2.8 \cdot 10^{-3}$ |

Figure 4: Possible improvements of the design of the plant and their effects

A number of further possible improvements were identified. However, their effect was not quantified, since the safety level attained after implementing the quantified improvements was deemed to be sufficient. This is especially true because harm to the public cannot be caused by the plant given its remoteness from housing areas.

Since not only the expected values were calculated but also the effect of data uncertainties was propagated through the analysis, centiles are stated along with expected values for both the original and improved designs in Figure 5.

| | 5%-Centile | Expected value | 95%-Centile |
|---|---|---|---|
| **Original design** | $1.2 \cdot 10^{-3}$ $a^{-1}$ | $1.0 \cdot 10^{-2}$ $a^{-1}$ | $3.1 \cdot 10^{-2}$ $a^{-1}$ |
| **Modified design** | $3.4 \cdot 10^{-4}$ $a^{-1}$ | $2.8 \cdot 10^{-3}$ $a^{-1}$ | $8.6 \cdot 10^{-3}$ $a^{-1}$ |

Figure 5: Characteristic values of the distributions for explosion frequencies

# 5 Conclusions

- the methods employed, HAZOP and fault tree analysis along with THERP for human error proved adequate for the investigation
- the plant has a balanced design with respect to safety (no under or overdesign)
- the relatively small contribution of start-up and shut-down to explosions of 14% can be explained by the fact that the reaction in this type of process only starts at the end of the start-up phase and ends immediately after shut-down
- no special safety devices are required for start-up and shut-down
- the reliability of human acts is very important (the plant model showed sensitivity to modifications of human error probabilities)
- a number of easily implemented proposals of improvement were made which lead to a reduction of the expected frequency of explosion by a factor of 4
- additional potential for improvement was identified, but not quantified because the design was considered adequate after implementing the quantified improvements

It should be emphasized that the proposals for improvement stemmed mainly from the quantitative part of the analysis with the weak areas identified and their relative importance serving as a starting-point.

## Reference List

1. Risk Analysis of Six Potentially Hazardous Industrial Objects in the Rijnmond Area- A Pilot Study. D.Reidel Publishing Company, Dordrecht 1982
2. Hauptmanns U;Sastre H. Safety Analysis of a Plant for the Production of Vinyl Acetate. J.Chem.Eng.Jpn.,1984;17(2):165-173
3. Hauptmanns U;Rodríguez J. Untersuchungen zum Arbeitsschutz bei An- und Abfahrvorgängen von Chemieanlagen. Bundesanstalt für Arbeitsschutz Fb 709, Dortmund 1994
4. Swain AD; Guttmann HE. Handbook of Human reliability Analysis with Emphasis on Nuclear Power Plant Applications. NUREG/CR-1278, October 1988
5. Doberstein H.;Hauptmanns U;Hoemke P;Verstegen C;Yllera J. Ermittlung von Zuverlässigkeitskenngrößen für Chemieanlagen. GRS-A-1500, Köln, Oktober 1988
6. Martz HF;Waller RA. Bayesian Reliability Analysis. Krieger Publishing Company, Malabar, Florida 1991

# Risk Based Approach to Estimating Tank Waste Volumes

J. Young
S. A. Driggers
Science Applications International Corporation
Richland, Washington, USA

G. A. Coles
Westinghouse Hanford Company
Richland, Washington, USA

## 1 Introduction

A project was proposed at the Hanford site to provide additional waste storage tanks. This project would construct new waste tanks in two different tank farm areas, and a related project would construct a new cross-site line between the areas. These projects were intended to ensure sufficient space and flexibility for continued tank farm operations, including tank waste remediation and management of unforeseen contingencies. The objective of the tank waste volume assessment was to support determination of the adequacy of the free-volume capacity provided by these projects and to determine related impacts.

The existing waste volume projections were based on point estimates for three cases: best, optimum, and worst. Furthermore, the current estimates addressed only a very limited number of potential combination of events which could impact volumes. Therefore, a risk based approach was taken to identify the possible free-volume requirements based on combinations of potential operational decisions and upset events and their associated likelihoods and uncertainties.

The scope of this assessment was limited to one tank farm area within the Hanford Site. It was assumed, for simplicity in this study, that no limitation on transfer of waste to other tank farm areas exists; even though free-volume limitations in other areas could affect successful operation. Two different time periods (1995 through 1998 and 1999 to 2005) were analyzed because a new cross-site tie line improving the probability of successfully performing transfers to other tank farm areas and the proposed tanks could not be available until 1999.

## 2 Assessment Approach

This risk based approach identified the possible paths that future operations could take and quantified the likelihoods and free-volume impacts of such paths. Development of the possible paths was performed in two parts: first, identification of the operational volume and waste handling issues; and second, the development of possible combinations of handling system configurations and failures. The operational volume and waste handling issues and failures addressed in this risk assessment are given in Table 1. The combinations of waste handling issues and failures (operational scenarios) were developed, combined, and quantified using event tree analysis.

The event tree analysis consisted of developing three separate event trees.

1) An event tree (Figure 1) addressing potential waste volumes that could be generated depending on which tank farm operational decisions might be made;

2) A second tree (Figure 2) identifying possible combinations of facility upsets (waste generator) and tank leaks which also impact the waste volume that must be handled; and

3) An event tree identifying possible combinations of tank farm configuration and operability states that impact the ability to handle waste volumes.

These three event trees were linked together in the following fashion. The first two trees were linked together to determine the potential volumes of waste which must be handled, the conditions under which they would be generated, and the likelihood of achieving a potential volume. The sequences (paths) developed by combining these two event trees were grouped by similar characteristics (e.g., potential for similar free-volume generation). Each of these groups became the "initiating event" for the final event tree that addressed waste handling issues. Dependencies between the first two event trees and the waste handling tree were addressed explicitly in the postulation of event sequences for each initiating group and implicitly through the conditional likelihood assigned to the events in the sequences postulated in the third event tree. The results of the combination of the three trees were binned into five volume bins for each of the two time phases addressed in the assessment.

Final determination of the volume associated with each bin was determined by uncertainty analysis. This was accomplished by modeling the variability of the volume associated with each sequence based on the volume prediction. Both triangular and normal distributions were used depending on the data available. The uncertainty analysis used the Crystal Ball PC software. Distributions of the volume of waste in each bin were developed from the Crystal Ball simulations. These distributions were used to identify a mean free-volume prediction and 75% and 90% confidence limits.

The likelihood of each bin was developed from the likelihood for the sequences in each bin. Event likelihoods were based on data or expert opinion, depending primarily on the nature of the event. Failure event (e.g., tank leaks) and operational upset likelihoods were derived from available data. The likelihood of events which were driven by programmatic decisions was based on a semi-quantitative scale which assigned a numerical value to the degree of certainty the decision would be made. For example, an event that managers were sure would occur was assigned a likelihood value of 0.8 and event that they doubt would occur 0.2.

Because of the nature of the input data (e.g., based on expert opinion or derived from conditions which have changed) and the controversy associated with some programmatic issues, sensitivity analyses were performed on the results to gain insight into certain dependencies and the effect of certain input assumptions.

# 3 Summary of Results

For the 1995 through 1998 time frame, the most likely situations (79% chance) will result in free-volume requirements in excess of 5,000 kgal. There is a 75% confidence that the free-volume requirements will be less than approximately 8,600 kgal and a 90% confidence that it will be less than 9,200 kgal. The major contributors to the total waste volume in the 1995 through 1998 time period are normal facility-generated waste, salt well pumping of the single-shell tanks (SST), SST leaks, and facility upsets.

For the 1999 to 2005 time period, the most likely situations (65% chance) will result in mean free volume requirements between 4,200 and 5,800 kgal. There is a 75% confidence that the generated waste will not exceed 5,500 kgal and a 90% confidence that the free volume will not exceed 5,800 kgal. The major contributors to these totals in the 1999 to 2005 time frame are normal facility-generated waste, salt well pumping of the remaining SSTs, waste generated by unplanned facility activities, the retrieval of the TX-107 and TX-118 solids, SST leaks, and facility upsets.

Table 1. Operational Volume and Handling Issues.

| Issues | |
|---|---|
| Facility waste | Facility-generated waste volumes |
| Dilution volume | Volumes from dilution of tanks |
| Salt well pumping | Salt well pumping volumes |
| Flexibility | Volumes from unplanned activities |
| Retrieval | Volumes from retrieval of single-shell tanks |
| Leaks and Upsets (failures) | |
| Single-shell tank leak | Single-shell tank leaks |
| Double-shell tank leak | Double-shell tank leaks |
| Facility upset | Facility upsets |
| Waste volume handling | |
| SY Farm | SY Farm tank, line, and pump availability |
| Transport | Transfer by tanker car or truck versus pumping |
| Cross-site | Operability of cross-site transfer line |

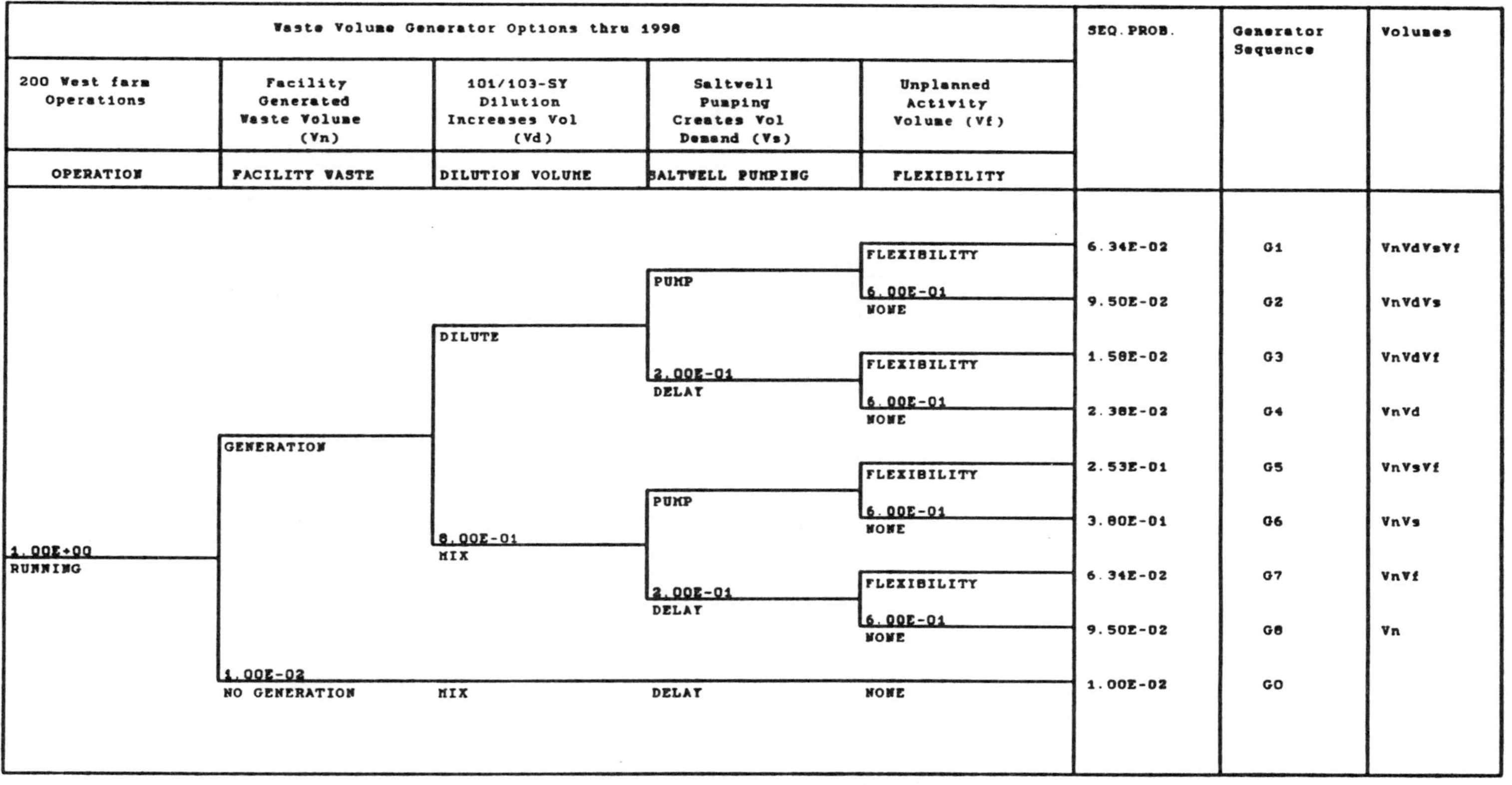


Figure 1 Waste Volume Generator Options, 1995 through 1998.

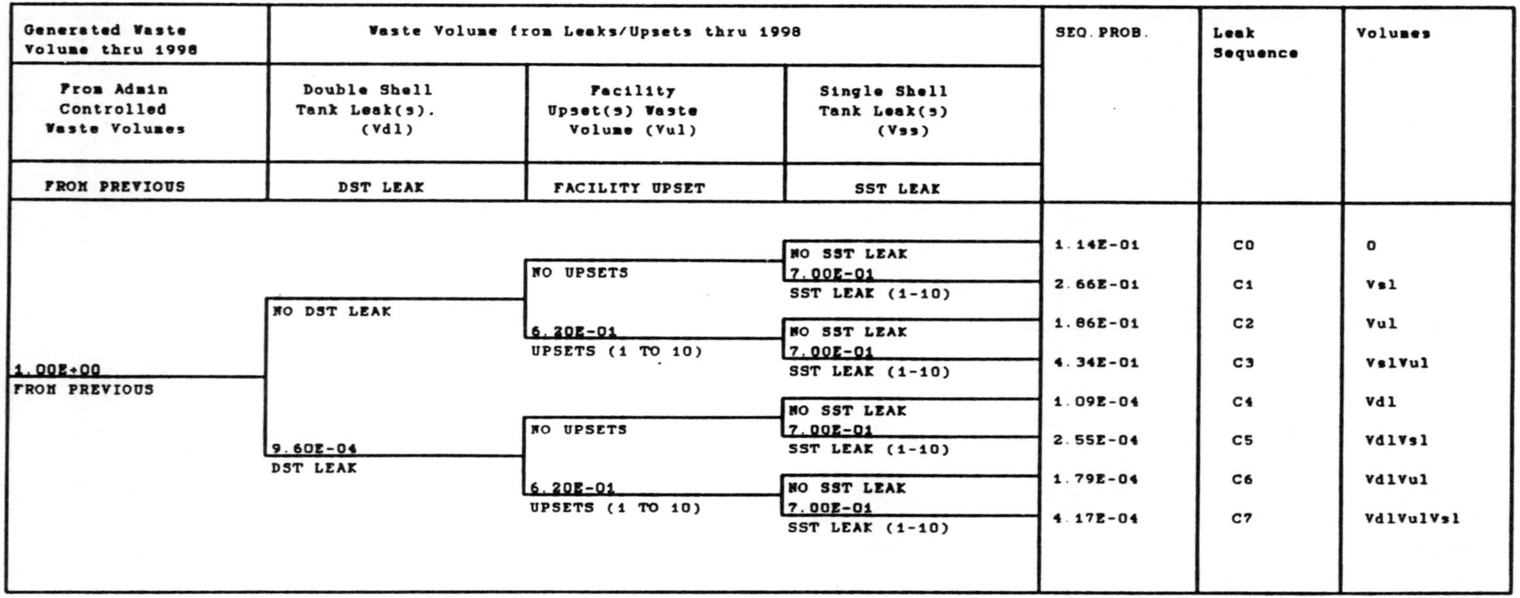

| Generated Waste Volume thru 1998 | Waste Volume from Leaks/Upsets thru 1998 | | | SEQ. PROB. | Leak Sequence | Volumes |
|---|---|---|---|---|---|---|
| From Admin Controlled Waste Volumes | Double Shell Tank Leak(s). (Vdl) | Facility Upset(s) Waste Volume (Vul) | Single Shell Tank Leak(s) (Vss) | | | |
| FROM PREVIOUS | DST LEAK | FACILITY UPSET | SST LEAK | | | |
| 1.00E+00 FROM PREVIOUS | NO DST LEAK | NO UPSETS | NO SST LEAK | 1.14E-01 | C0 | 0 |
| | | | 7.00E-01 SST LEAK (1-10) | 2.66E-01 | C1 | Vsl |
| | | 6.20E-01 UPSETS (1 TO 10) | NO SST LEAK | 1.86E-01 | C2 | Vul |
| | | | 7.00E-01 SST LEAK (1-10) | 4.34E-01 | C3 | VslVul |
| | 9.60E-04 DST LEAK | NO UPSETS | NO SST LEAK | 1.09E-04 | C4 | Vdl |
| | | | 7.00E-01 SST LEAK (1-10) | 2.55E-04 | C5 | VdlVsl |
| | | 6.20E-01 UPSETS (1 TO 10) | NO SST LEAK | 1.79E-04 | C6 | VdlVul |
| | | | 7.00E-01 SST LEAK (1-10) | 4.17E-04 | C7 | VdlVulVsl |

Contingency Waste Volumes MWTFLK2.TRE 2-19-95

Figure 2 Waste Volume from Leaks/Upsets, 1995 through 1998.

# A Level III PSA for the Inherently Safe CAREM-25 Nuclear Power Station[1]

J. Barón, J. Núñez, S. Rivera
C.E.D.I.A.C., Engineering Faculty
Cuyo National University
Mendoza, Argentina

**Abstract**

A Level III PSA has been performed for the inherently safe CAREM-25 nuclear power station, as a requirement for licensing according to Argentinean regulations.
The CAREM-25 project is still at a detailed design stage, therefore only internal events have been considered, and a representative site has been assumed for dose estimations.
Several conservative hypothesis have been formulated, but even so an overall core melt frequency of 2.3E-5 per reactor year has been obtained.
The risk estimations comply with the regulations.
The risk values obtained are compared to the 700MWe nuclear power plant Atucha II PSA results, showing an effective risk reduction not only in the severe accident probability but also in the consequence component of the risk estimation.

## 1 Introduction

Argentinean regulations [1] impose the elaboration of a level III PSA as a requirement for nuclear power plant licensing. Both the annual probabilities and the radiological consequences to the public must be assessed and the resulting representative severe accident sequences must satisfy the acceptance criterion indicated in figure 1, as the thick line. Any point above the line is unacceptable and requires further development to reduce its probability, its consequence, or both.

## 2 The CAREM-25 Project

The inherently safe CAREM-25 Nuclear Power Plant is a low power (25MWe) integrated reactor concept, designed for electricity production in remote areas, and therefore, designed to operate independently of external power supplies.

---

[1] Work sponsored by Invap S.E.

It has a PWR-like design with twelve once-through steam generators inside a pressure vessel. The top of the pressure vessel constitutes the pressurizer, and the primary coolant circulates by natural circulation from the core (at bottom) through a chimney to the upper part, then downwards through the steam generators. This natural circulation concept avoids the use of pumps and large diameter pipes. The only penetrations through the pressure vessel are those for the secondary system and those for safety systems, auxiliary systems and instrumentation purposes (small diameter pipes).
The core is fueled with enriched uranium (4%) and operates in the epithermal range to provide for large negative reactivity coefficients for easy operation. It has burnable poison (gadolinium) and a relatively low power density, compared to commercial PWRs.

The safety functions make full use of inherently safe processes and they do not need any electric supply. The plant has two independent shutdown systems (gravity-driven hydraulic rods and boron injection by natural convection), a two-stage gas-driven emergency injection system (providing a massive reflooding and a flow-controlled injection), a passive gravity-driven two-phase emergency cooling system, and a pressure-suppression steel containment that can evacuate accident heat to the external air by natural convection. At least two redundancies are provided for each safety function.
The reactor protection system is software-based and uses a distributed logic approach. Besides the logical redundancies, all safety components are designed fail-safe and do not need any external power supply for actuation.

At present, the CAREM-25 project is in the detailed design stage, and several experimental facilities are set up to demonstrate the basic operating principles, including a full size critical facility.

# 3 PSA Procedure

A classic PSA procedure has been used, with special provisions to consider the peculiarities of the CAREM design. These provisions included specific methods developed for the definition of initiating events, and for the analysis of the containment response. The analysis extended not only to the operating state of the plant, but also to the refueling state.

The data for component failures was obtained from IAEA [2]. For human behavior, models and data from [3] were used.

## 3.1 Initiating Events Definition and Selection

As the CAREM-25 NPP has a unique design, a specific three-step procedure was developed to obtain a list of initiating events. This procedure starts with the identification of all the radiation sources at the plant and what are the barriers that separate them from the public. The second step identifies the possible mechanisms for

the failure of these barriers, in order to obtain a *complete* list of initiating events. The third step is a simplification process, based on common attributes, to obtain a manageable group of initiating events to be developed in event trees.
As long as a definite site does not exist at the moment, only internal initiating events were considered in the present work. In any case, the CAREM-25 NPP is designed for seismic, wind and other external loads, and it is expected that the contribution from external events to the overall risk will be smaller than or comparable to that from internal ones.

## 3.2 Event Trees and Fault Trees

Each initiating event was developed into an event tree, and fault trees were built for each of the safety systems or actions postulated. Special consideration was made for the actuation of the safety systems during power loss, as long as no signal processing is needed in that case.
The fault trees and event trees were programmed and solved with PSAPack [4].
One of the quantification results is that the signal failure (which is based in a software voting logic) is dominant for all the safety systems, as their design is very simple and redundancy is provided. For the software itself, a failure probability of E-3 per demand has been assumed in the present work. This is considered to be very conservative, based on the software development process and quality assurance program implemented.
The event tree sequences were grouped into plant damage states, with common attributes. The probability of each plant damage state was assigned to the sum of its contributors, and the worst sequence in the group was selected as representative for containment response and source term estimation. This procedure is conservative.

## 3.3 Containment Response

The response of the CAREM containment was studied from the phenomenological point of view. This containment is composed by an upper wetwell and a lower drywell, where the pressure vessel is located.
A full revision of containment failure modes was performed, and their possibility for the particular CAREM design was examined in detail. After qualitative estimation, expert opinion was used to quantify each containment failure mode for each plant damage state, according to [5].
Containment event trees were developed and quantified, and combinations of plant damage state - containment failure mode were groped into eight release categories, from which six correspond to core melt sequences.

## 3.4 Source Term and Dose Estimation

The initial radioactive inventory of the core was obtained with the ORIGEN code. Then, for each release category, the release fraction was obtained by analogy with other PSA studies [6],[7]. This is very conservative, as no credit is taken for any of the CAREM-25

peculiarities that effectively will mitigate the releases (i.e., low power density, large water mass in the vessel, etc.), both in magnitude and in timing.
With these releases, dose estimations on a critical group located two kilometers from the plant were obtained, considering cloud immersion, inhalation (for lungs and thyroid) and ground exposure, with no provision for countermeasures. The atmospheric parameters of the Atucha site were used. On this basis, a comparison to Atucha II PSA results [8] is meaningful.

# 4 PSA Results

The results obtained provide an overall core melt frequency of 2.3E-5 per reactor year, from which 62% correspond to the blackout scenario, 37% to LOCA scenarios, and less than 1% for other transients and reactivity accidents. The high contribution from the blackout scenario comes from the fact that a yearly probability of 1 is assumed for loss of external power, which is assumed to produce a total blackout. No credit is taken for any recovery action. These highly conservative assumptions are presently under revision because they distort the overall results of the PSA, even when the licensing requirements are fulfilled.

On a system bases, the passive long-term cooling system failure is responsible for 70% of the overall core melt frequency, while 29% correspond to safety injection failures and less than 1% for shutdown system failures. The contribution of the software signal processing system is dominant in all the sequences except blackout.

Regarding risk estimations, the results for the six release categories (CL) involving core melt, are presented in figure 1. For each CL six different estimations are provided, which correspond to the six atmospheric conditions (a...f) from the Pasquill-Gifford dispersion model, representative of the Atucha site.

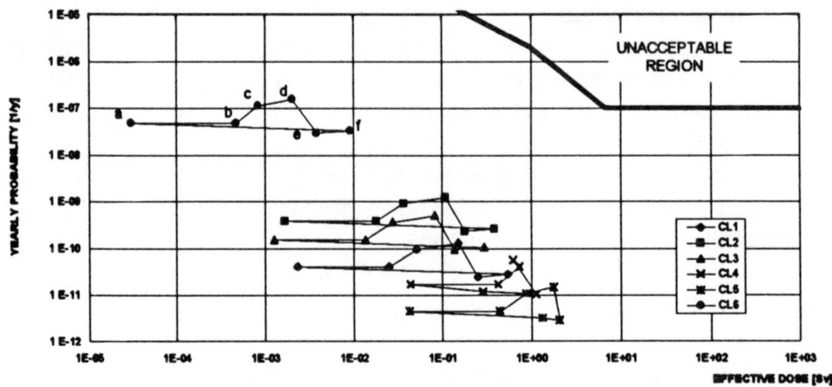


Figure 1. Risk estimations and acceptability for CAREM-25

It is observed that all the estimations lay well below the acceptability level, both on probability and consequence. The release category in the upper left (CL6) corresponds to the blackout scenario with late containment failure, and it is expected to be substantially reduced in their probability if more realistic hypothesis are used in its calculation.

# 5 Comparison to Atucha II PSA Results

A comparison was performed of the results obtained, with the results presented in [8] for Atucha II. In order to make them comparable, the probabilities for the CAREM-25 estimations were multiplied by 28, which gives an overall power of 700 MWe that is consistent with the Atucha II power. The results are plotted in figure 2. Thick lines correspond to Atucha II [8], and thin lines to 28 CAREM-25 stations.

From the 28 CAREMs group, the upper left release category may be substantially reduced in this case, as it correspond to a blackout scenario with no recovery possible, and this will not be the case if several CAREM plants are installed at the same location. From the probability point of view, it must be noted that the use of inherently safe systems provide for a significant reduction in the yearly probability of severe accidents, ranging from one to two orders of magnitude.

From the consequence point of view, the reduction is even more drastic, ranging from one to three orders of magnitude, for the same energy output. This is due to the reduced power of each CAREM unit, which makes the consequences of any single severe accident much reduced. This fact is particularly important because in the CAREM case, the maximum expected doses are of about 1 Sv, and therefore the appearance of early fatalities is almost precluded.

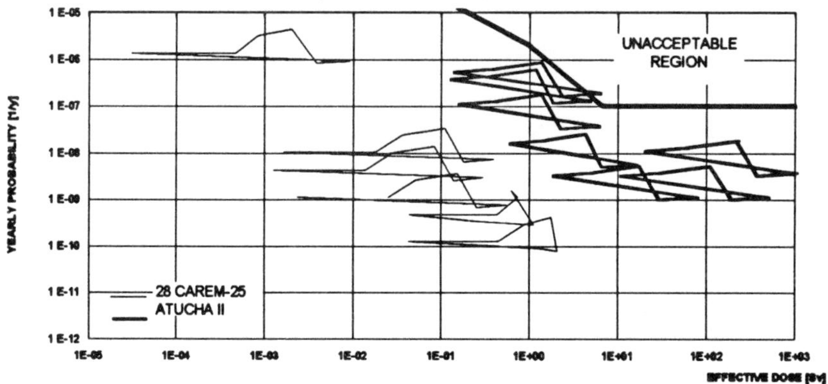


Figure 2. Risk estimation comparisons among Atucha II and 28 CAREM-25 NPPs

# 6 Conclusions

The severe accident probabilities and consequences were estimated for the CAREM-25 NPP, and the results obtained fulfill the requirements of Argentinean regulations with a large margin.

Several conservative hypothesis were assumed on the calculations, and these will be reviewed as the detailed design of the plant proceeds. Also, some other improvements are expected after the inclusion of several design modifications presently under discussion (among these are ex-vessel cooling, containment inertization and software independent review).

Only internal events were considered in the present work, and the study will be extended to external events when a definite site is selected.

The comparison of the risk estimation to the 700 MWe Atucha II PSA results (properly scaled) show two important facts: the first one is the reduction in severe accident probabilities due to the inherently safe concept of CAREM. The second one is the effective reduction on the expected doses in the public, due to the low power of each CAREM unit, which may shift the risk from catastrophic consequences to very limited effects, eventually eliminating the possibility for early fatalities.
These effective risk reduction for smaller, inherently safe NPPs, shows a trend to follow towards a safer use of nuclear energy.

## References


1. Ente Nacional Regulador Nuclear. Norma Básica de Seguridad Radiológica AR10.1.1. rev.1, Buenos Aires, 1995.

2. IAEA-TECDOC-478. Component Reliability Data for use in Probabilistic Safety Assessment, Vienna, 1988.

3. NUREG/CR-1278. Handbook on Human Reliability Analysis with Emphasis on Nuclear Power Plants Applications, Sandia National Laboratory, 1983.

4. Bojiadiev, A., Vallerga, H. PSAPack v4.2 computer program. IAEA, Vienna, 1993.

5. NUREG/CR-4551. Evaluation of Severe Accident Risks, 1990.

6. NUREG-75/014 (Wash-1400). Reactor Safety Study. An Assessment of Accident Risks in U.S. Commercial Nuclear Plants, 1975.

7. German Risk Study - Main Report. EPRI translation EPRI NP-1804-SR. 1981.

8. CNEA. Preliminary Safety Analysis Report - NPP Atucha II. Appendix B, 1981.

# A comparison study of qualitative and quantitative analysis techniques for the assessment of safety in industry.

Ir. J.L. Rouvroye, W.M. Goble MSc., Prof.Dr.Ir. A.C. Brombacher
Eindhoven University of Technology
Eindhoven, The Netherlands

Ing. R.Th.E. Spiker
GTI Industrial Automation
Apeldoorn, the Netherlands

**Abstract**

The primary function of safety systems is avoiding personal injuries or death and environmental pollution. Major problem is that, at this moment, there is no standardised method for the assessment for safety in the process industry. Many companies and institutes use qualitative techniques for safety analysis while other companies and institutes use quantitative techniques. The authors of this paper will compare different quantitative and qualitative techniques and will show that different analysis techniques show widely different results. It will also be shown that the qualitative techniques are highly dependent on human judgement while the quantitative techniques suffer from highly uncertain underlying data. Therefore in the second part of the paper a new quantitative technique will be presented that can be used for the assessment of safety, availability and triprate in the process industry. This technique can take into account effects of uncertain data and identify parameters that influence safety and/or availability dominantly. Finally the paper presents an example of the new technique applied to a practical system used in Dutch petrochemical industry.

## Introduction

Petrochemical processes and oil & gas production sites can harm people and environment when running out of control. To prevent this dangerous situation all these processes have to be equipped with mechanical and instrumental protection systems. The systems are referred to Emergency Shut-Down (ESD) systems or Safety Instrumented Systems (SIS). These systems, together with the relevant parts of the process system itself, are commonly referred to as Safety Related Process Equipment.

Emergency Shut-Down systems keep industrial processes within safe operation conditions or shut the process down in a predetermined way in the case of exceeding specified, unsafe, process conditions. If these events can not be handled by the protection systems or the protection systems do not operate, Fire & Gas systems

mitigate the consequences. The primary function of both types of safety systems is avoiding personal injuries or death and environmental pollution.
Protection and Fire & Gas systems are expensive both as initial investments as well as maintenance. The 'safety quality', essential system size and validation techniques of especially the instrumented protection systems are today the subject of discussion and investigation.
In the first place the social communities do not accept risks for life and environment from industrial activities. They know there will always be a certain risk, but it has to be as low as possible. The latest is the source of a lot of problems and uncertainty, because what will be an acceptable risk and who will make that decision? In the second place the industries do not like to invest heavily in protection systems. The costs are considered to be not productive and disturbing the balance of competitiveness, when not all the industrial players have to meet the same plant safety requirements.
A concerning aspect in the process industry is the lack of unity in applying Quantitative Risk Analysis (QRA) for the whole involved process. Before an adequate ESD system can be installed one has to complete a QRA. The probability for a dangerous event to happen of a process without protecting systems must be estimated. This risk must be compared with government guidelines concerning the acceptable minimum risk or company standards. The difference in both probabilities can be eliminated by protection systems in combination with other measures (see Figure 1).

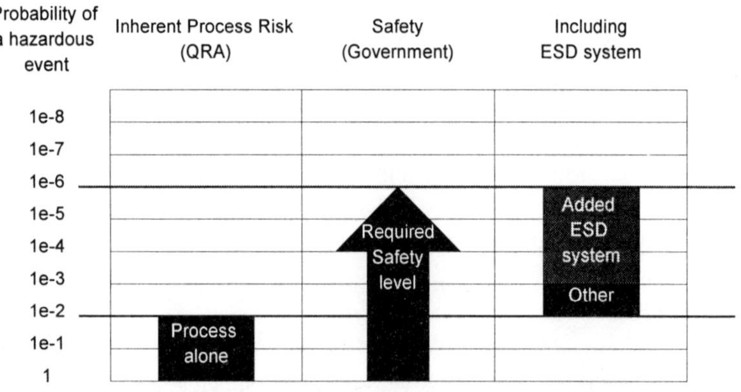


Figure 1. Quantitative demands on process and ESD systems

An other concern especially for the management in process industry is the introduced unavailability by adding an ESD system. Because these systems can shut down (part of) the production process a safe error (an error resulting in a nuisance trip) in these systems can have very big financial consequences. One part of these costs can be related direct to the process equipment, for example personnel costs and material used for the repair. Other costs can be described as indirect costs, for example production loss (no or less (good) product), loss of clients and goodwill, extra costs because of environmental pollution. Because of the total costs related to the availability of the ESD system, an estimation of the availability and number of trips in a certain period of time can help in selecting or optimising the ESD system to be used.

# Techniques currently in use for safety/availability assessment of ESD systems

The authors have compared different techniques in use for safety assessment [1]. They used the following comparison aspects:
Ability for the technique to:

- rank and compare safety
- calculate the probability of unsafe failure
- incorporate effects of test and repair
- include time dependent aspects
- include the effect of uncertain data

In this paper two extra comparison aspects are introduced (see Table 1):

- the ability to predict the availability
- the ability to predict the triprate

Table 1. Comparison different analysis techniques

| | Safety ranking comparison | Availability prediction | Probability of Unsafe Failure | Effects of test & repair | Triprate prediction | Time dependent effects | Effects of uncertain data |
|---|---|---|---|---|---|---|---|
| Expert analysis | ✓ | ✓ | | | | | |
| FMEA | ✓ | | | | | | |
| FTA | ✓ | ✓ | ✓ | ? | ✓ | | |
| Reliability Block Diagram | ✓ | ✓ | ✓ | | | | |
| Parts Count Analysis | ✓ | ✓ ? | ✓ | | | | |
| Markov Analysis | ✓ | ✓ | ✓ | ✓ | ✓ | ✓ | |

This leads to the following conclusions:

- A large number of different techniques and tools is used to analyse the safety of industrial processes
- All techniques can be used to compare and to rank safety of different systems
- Only the quantitative methods provide a means to calculate the probability of an unsafe failure
- The different quantitative methods take different aspects into account; therefore it is not unlikely that different methods will lead to different results in terms of safety
- The quantitative techniques base themselves on mathematical analysis using certain models. The data underlying these models can be highly uncertain

For the aspects availability and triprate the following conclusions can be made:

- All quantitative techniques can predict the availability
- Predicting the triprate is only possible by means of Failure Tree Analysis (FTA) or Markov analysis
- The data used in the models for the availability and triprate prediction can be highly uncertain

# The new technique: Enhanced Markov analysis

In order to overcome the problems, mentioned above a new technique was developed. This technique consists of the following steps:

- Markov analysis to analyse different (safe and unsafe) failure modes in a system
- Uncertainty analysis to analyse the effect of uncertainty in data
- Statistical sensitivity analysis to derive dominant parameters influencing safety and/or availability

Additional aspects, introduced in this paper, are

- Implementation of (im)perfect testing
- Availability and triprate calculation.

The technique will be illustrated using a practical case.

# Practical case

The practical usage of the combination of the enhanced Markov analysis technique can be demonstrated by the following case study. In this study, the safety and availability of a High Integrity Pressure Protection System (HIPPS) is evaluated. The evaluated pressure protection system is used in the Dutch gas industry to protect the down-stream equipment. It is an independent last line of defence in the case of over-pressure (pressure higher than design pressure of the downstream equipment).

## Configuration of the process

The pressure protection system is a de-energise to trip system. For the layout of the system see Figure 2. Because of the chosen lay-out both shutoff valves close when one or more sensors signal a pressure becoming too high. The system is tested functionally every third month. Once every year the shutoff valves are inspected as well. After five years the system is completely overhauled. Because the shutoff valves are relatively large, the repair rate (1 / Mean Time To Repair) for the shutoff valves differs from all other component repair rates, which are assumed to be the same.

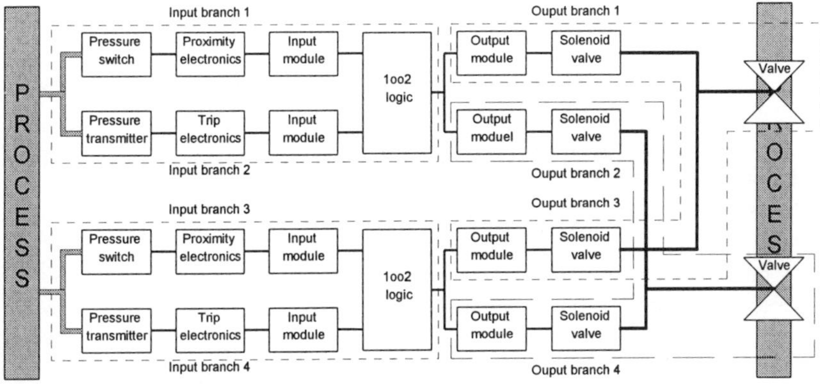


Figure 2. Layout pressure protection system

## Effects of testing the system

One of the problems in safety is the issue of undetected failures. Basically, if no special precautions are taken, there is only one way to detect undetected failures: when in case of a process hazard the safeguarding system fails to respond. Although the likelihood of undetected failures is often quite low, because of the fact that failures of this category might exist for years in a system the ultimate probability of (long-term) existence of these failures can become quite high. See Figure 3.

Figure 3. Cumulating probability of undetected failures

One of the solutions, common applied to solve this problem is the use of extended test routines, that are performed on a regular basis, in order to avoid that the probability of unsafe failures becomes unacceptable. However, although the introduction of testing might help avoiding/removing undetected failures also the use of testing has its limitations. Quite often the assumption is made that after a test the system can be considered new. In Markov terms this implies that the probability of having an operational system is assumed 1 and the probabilities of other states are assumed 0.

This situation, however, assumes perfect testing of a complete system. In many situations this might not be the case. In practical situations there are different kinds of tests applied to different parts of the system. Every test category will cover a different cross-section of the system. The different cross sections in a plant with respect to testing are:

- Sections that are tested during normal operation of a system. (On line tested sections)
- Sections that are tested during a (scheduled or unscheduled) stop of (a section of-) a plant. (Off line tested sections, minor tests)
- Sections that are tested during major overhaul of the system (Off line tested sections, major tests)
- On top of this there can be sections in a plant that are never tested. (Never tested sections)

Usually the first three of these test types are not mutually exclusive. Cross sections that are tested during on-line tests are usually covered in off-line tests as well. A common (but not necessarily correct) approach is to assume that a higher level test includes all the parts of a system covered in a lower level test as well. See the figure below.

The following paragraphs will show the results of enhanced Markov analysis, including effects of uncertain data, combined with effects of imperfect testing on the test case described above. A more detailed description of the analysis is given in [1,2]. The generic approach on Markov modelling is given in [3].

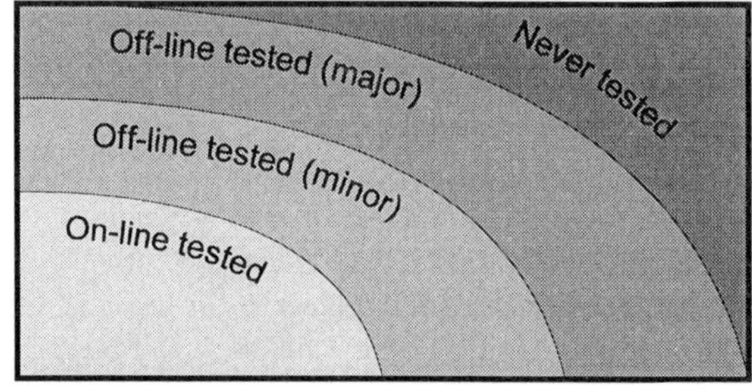


Figure 4. Coverage of different tests

## Analysis results

Calculations using the Markov analysis result in the following information:

1. The Probability of Failure on Demand (Pfd) for the system as function of time (see Figure 5). In this figure the effect of periodical testing can be seen clearly

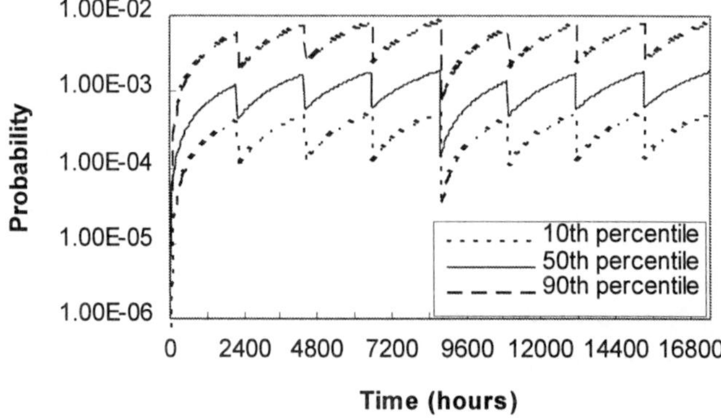


Figure 5. Probability of failure on demand versus time

(the dips at each multiple of 2200 hours = 3 months). The testing is implemented as non-perfect testing (minor test off-line). This can be seen in the figure because after testing the Pfd is not starting at the same starting value as at 0 hours. The test once every year is a more thorough off-line test resulting in more repaired undetected failures and thus in a lower Pfd after this test. This can be seen in Figure 5 in the starting point of the dip at 8800 hours, which is deeper than the one for the dips at 2200, 4400 and 6600 hours. Because of uncertainty in data the results are not represented by a single nominal curve but as a region.

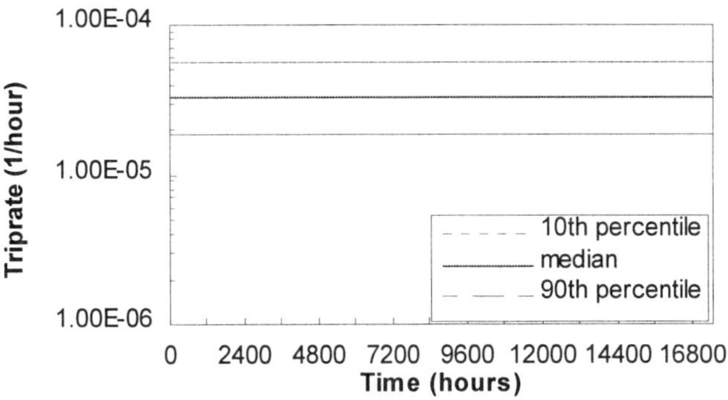


Figure 6. Triprate versus time

2. Triprate. In Figure 6 the triprate is presented. Because the effects of periodic testing only affect the probability to be in states with undetected failures these tests hardly have any influence on the triprate.
3. Plot of a pareto diagram of the statistical sensitivities (see Figure 8). In this plot the variables are arranged in order of their so-called statistical sensitivity coefficient to the triprate at the end of the simulation time. The statistical sensitivity analysis technique for uncertainty analysis is developed by Elias and Spence [4]. The statistical sensitivity coefficient can vary between 0 (parameter not important), and 2 (parameter has dominant influence). This plot gives an indication of the parameters that dominantly influence the systems triprate. This information can be used to improve the system. In this case study the most important parameters influencing the triprate are the failure rates for the valves and the

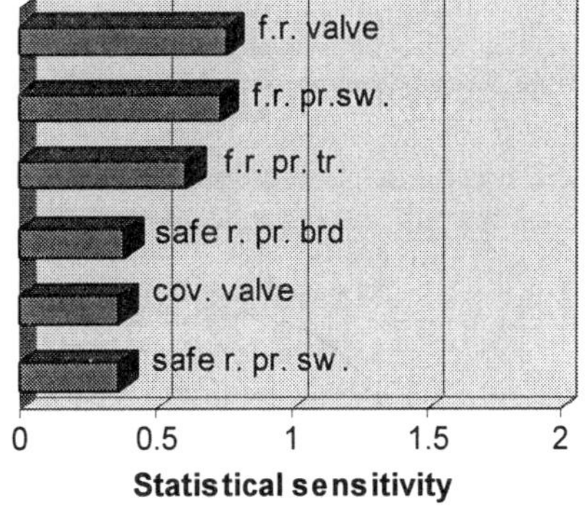


Figure 7. Pareto diagram of statistical sensitivities of parameters for the triprate

pressure sensors. Although not presented here this technique can be used of course also for the results of the probability of failure on demand (Pfd) and the probability to be in the failed safe state (Pfs).

## Conclusions

This leads to the following conclusions:

- There are strong demands to analyse the safety and availability of process equipment
- There is no standard method to analyse the safety and availability of process equipment
- Different analysing methods lead to widely different results
- Common used data used in quantitative methods has a high degree of uncertainty
- Markov analysis covers, at this moment, the largest number of safety/availability aspects but still has the problem of uncertain data
- Combination of Markov and Monte Carlo analysis gives not only an indication of the safety/availability of a system but also of the uncertainty in safety/availability
- Statistical sensitivity techniques allow identification (and improvement) of parameters dominantly influencing the (uncertainty in-) safety/availability
- Including effects of imperfect testing on different levels covers the effects of large system overhaul tests on the safeguarding system, also in non-ideal cases.
- The presented techniques can be applied successfully on practical industrial systems

## References


1. Rouvroye, J.L., et al, Uncertainty in safety, New techniques for the assessment and optimisation of safety in process industry. SERA-Vol. 4, Safety Engineering and Risk Analsysis, ASME 1995
2. Brombacher A.C., Rouvroye J.L., Safety Analysis NAM HIPPS systems using Statistical Sensitivity techniques, Eindhoven University internal report, October 1995
3. Electrical (E) / Electronic (E) / Programmable Electronic Systems (PES) for Use in Safety Applications - Safety Integrity Evaluation Techniques, ISA Technical Report Draft 4, June 1995
4. Spence R., Singh Soin R., Tolerance Design of Electronic Circuits, Addison Wesley, Wokingham UK), 1988
5. Brombacher A.C., Reliability by Design, John Wiley & Sons, Chichester, 1992

# Cost-Benefit-Risk Analysis Spreadsheets for Probabilistic Safety Assessment Applications

James K. Liming, Donald J. Wakefield

PLG, Inc.
Newport Beach, California USA

## Introduction

As background to the technical approach and analytical tools described in this paper, it is appropriate to present an overview of current issues and developments in the commercial nuclear power industry that provide strong motivation for application of probabilistic safety assessment (PSA) to trade-off studies involving costs, benefits, safety, and associated risks at nuclear facilities. There have been several ongoing and recent developments within the U.S. Nuclear Regulatory Commission (NRC) related to the formulation and application of policy on the use of PSA and risk-based regulation in general. Since the landmark development and review of the original "Reactor Safety Study" (WASH-1400) in the late 1970s, the issue of PSA applications has been a topic of great interest and controversy in the commercial nuclear power industry and at the NRC. The issue of PSA applications to reactor plant regulation has gained momentum, particularly since the issuance and implementation of NRC Generic Letter 88-20 and its supplements that mandated the Individual Plant Examinations (IPE) and subsequent Individual Plant Examinations for External Events (IPEEE). Virtually all of the commercial nuclear power plants in the United States chose to perform a plant-specific PSA to address these requirements. In an effort to improve the state of the art of plant-specific PSAs, the NRC sponsored a study that resulted in a second landmark report on PSA called, "Severe Accident Risks: An Assessment of Five U.S. Nuclear Power Plants," (NUREG-1150) in 1991. Several key reports, most notably NUREG/CR-4550 and NUREG/CR-4551, have been issued as supporting documentation for NUREG-1150. In August 1993, the NRC issued a draft "Regulatory Analysis Technical Evaluation Handbook," (NUREG/BR-0184) which was designed to promote high-quality applications of PSA technology to regulatory issues. In November 1993, the NRC Probabilistic Risk Assessment (PRA) Working Group issued a draft report called, "A Review of NRC Staff Uses of Probabilistic Risk Assessment," (NUREG-1489) which documented detailed recommendations on the application of PSA technology in the areas of licensing and other regulatory functions. In August 1994, the NRC issued two key internal policy documents, "Proposed Policy Statement on the Use of Probabilistic Risk Assessment Methods in Nuclear Regulatory Activities" (SECY-94-218) and "Proposed Agency-Wide Implementation Plan for Probabilistic Risk Assessment" (SECY-94-219). Finally, in December 1994, the

NRC issued in Volume 59, Number 235 of the *Federal Register* its "Use of Probabilistic Risk Assessment Methods in Nuclear Regulatory Activities; Proposed Policy Statement," which more firmly and formally established the NRC's permanent move toward risk-based regulation and acceptance of PSA applications.

The NRC has issued a number of rules and policies in recent years, in addition to those presented in the previous paragraph, which promote or support the application of PSA technology. One important example is the issuance in 1991 of the "Requirements for Monitoring the Effectiveness of Maintenance at Nuclear Power Plants" (10 CFR 50.65) widely known as the Maintenance Rule. Another example, which forms one of the primary motivations for the technical tool described in this paper, is the NRC's formation of a cost-beneficial licensing action (CBLA) program in April 1993. This program was initiated specifically to provide nuclear utilities the opportunity to have a direct-path program at the NRC designed to accept, review, and approve burden-relief licensing amendment requests based on cost-benefit-risk arguments. The key to successful PSA applications is to develop the applications at a high enough quality level to satisfy strict licensing decision-making requirements while keeping the methods, tools, and associated case studies accessible to operating plant staffs at reasonable and acceptable cost.

# Cost-Benefit-Risk Analysis Approach

A general decision process flow for nuclear utility implementation of plant-specific operations, maintenance, and design enhancements, which includes support from cost-benefit-risk analyses, is presented in Figure 1.

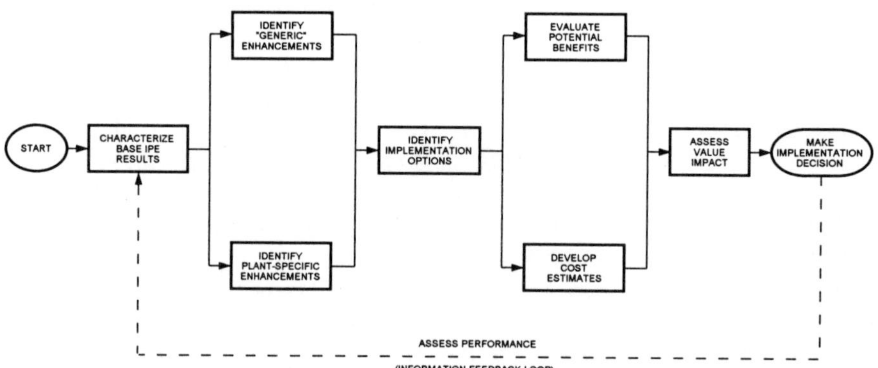


**Figure 1. Process for Identification and Evaluation of Potential Plant Enhancements**

This paper focuses on an analytical tool that consists of linked spreadsheets that contain encoded baseline information from a plant PSA and generic information from regulatory guidance documents. This information is used to form a set of five transformation matrices: one from initiating events to plant damage states, the second from plant damage states to key release categories, the third from key release categories to accident progression bins, the fourth from accident progression bins to averted population dose, and the fifth from averted population dose to cost-benefit figures of merit. These transformation matrices are encoded into computer spreadsheets, using common Windows or DOS based "off-the-shelf" software such as Excel, Lotus 123, or QuattroPro, to calculate the risk and benefit tradeoffs associated with proposed plant enhancements. The process used in calculating cost-benefit-risk figures of merit associated with potential plant enhancement decision-making is summarized in Figure 2. The primary figures of merit applied in this approach are plant damage state (PDS) frequency, key release category (KRC) frequency, accident progression bin (APB) frequency, projected population radiation dose (in person-rem exposure), net benefit (in dollars), and benefit-to-cost ratio.

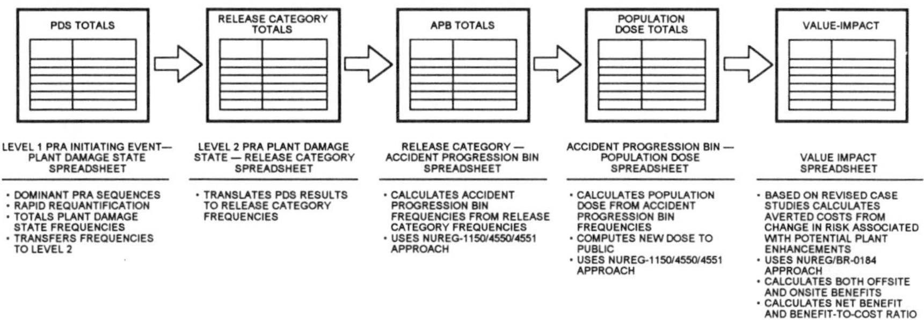


**Figure 2. Process for Calculating Potential Benefit of Enhancements**

# Development of the Cost-Benefit-Risk Spreadsheets

The spreadsheets developed for this tool have been developed such that the first column of the spreadsheet contains the input variables and the last row of the spreadsheet contains the resultant or output variables. The internal cells of the spreadsheets contain the products of the transformation factors for input variable to output variable calculation and the associated input variables. The separate spreadsheets defined here may be treated as separate individual spreadsheet files linked through the software, or may be formed as separate worksheets in a spreadsheet workbook or notebook. In the case studies performed for this paper, the authors found that treating the spreadsheets as separate (but linked) files proved to be best, because of the possibility of exceeding current spreadsheet software array size and file size limitations and because of the slower run times experienced with exceptionally large spreadsheet workbooks. However, with the rapid improvement of spreadsheet software technology and personal computer hardware capabilities, these limitations may soon be eliminated or significantly mitigated by future software versions and newer faster personal computers.

Information from the first two of the five linked cost-benefit-risk spreadsheets (the initiating event (IE) to PDS spreadsheet and the PDS to KRC spreadsheet) can generally be extracted directly from the Level 1 and Level 2 PSAs performed for the IPE and IPEEE requirements. In fact, some PSA software packages have a feature which enables the user to form two $\phi$-M matrices containing the necessary parameters of IE-PDS and PDS-KRC spreadsheets. If this feature is not available, the analyst must group all Level 1 event sequences or cutsets that map from a specific IE to a specific PDS and calculate the transformation factor by dividing the PDS frequency associated with the specific IE by the IE frequency. This enables the analyst to form the IE-PDS transformation matrix. Likewise, the analyst must group all Level 2 event sequences or cutsets that map from a specific PDS to a specific KRC and calculate the transformation factor by dividing the KRC associated with the specific PDS by the PDS frequency. This enables the analyst to form the PDS-KRC transformation matrix.

To continue with the process of developing the five basic transformation matrices, the analyst must select a peer unit from the five NUREG-1150 plants. The KRC-APB transformation matrix is formed by the analyst simply assigning each KRC to one of the ten APBs defined in NUREG/CR-4551 for their peer unit. Therefore, the transformation factors in the KRC-APB spreadsheet have a value of 1.0 or 0.0. In general, many more than one KRC may be associated with a single APB. The characteristics that define an APB are based on the primary accident progression attributes that influence a source term, e.g., the timing and failure mode of the reactor pressure vessel and containment. The analyst must assign this categorization based on the physical impact of each KRC on the plant. Thorough knowledge of this physical impact is generally available in the PSA documentation for specific KRCs, but should be repeated in any case study documentation on the use of this tool in cost-benefit-risk studies. The resultant of the spreadsheet is the APB frequency in events per reactor-year.

Next, the analyst must form the APB to population radiation dose (PD) spreadsheet. The goal is to calculate the projected population dose within a 50 mile radius of the plant associated with core damage events. The person-rem dose is the product of the radiation level experienced in a given area and the exposed population in that area. A simplified formula for calculating offsite dose is:

Dose = (Constant) * ($\chi$/Q) * (Sum of activity for Various Isotopes) * (Sum of energy released from the Decay of Each Isotope)

The analyst must evaluate the detailed terms which make up this equation for the peer plant, and then develop conversion factors to account for plant-specific characteristics of the plant of interest in his case study(ies). The key factors generally are source term, population distribution, and site-specific atmospheric dispersion characteristics. Many of these factors can be converted using a simple ratio. For example, for the source term conversion, it is generally acceptable to use the ratio of the thermal power rating of the plant of interest over the thermal power rating of the peer plant. The analyst must scale these factors to convert from the NUREG-1150 peer plant to the plant of interest. The analyst must carefully document his evaluation of these factors in his case study report. Then using the more general version (contained in

NUREG/CR-4551) of the simplified formula presented above, the analyst may form the APB-PD transformation matrix. This spreadsheet simply takes the APB frequency values for each of the ten APBs, converts them into their respective PD rate values, and sums the PD rate values over all ten APBs. At this point the PD rate is multiplied by the expected remaining life of the plant to calculate PD. This spreadsheet, as well as the other presented in this paper, can be modified to include both a base case spreadsheet and "revised" case spreadsheet presentation on one printout. This facilitates showing the basis for calculation of "averted" PD, which is the result of subtracting the revised case PD from the base case PD. If the revised case shows an increased PD, then the "averted" PD will be negative.

Finally, the analyst performs a cost-benefit-risk or "value-impact" (VI) analysis on the difference between a base case and a revised case. He can calculate the "value" of the averted PD by multiplying the PD from the APB-PD matrix by $1,000.00 per person-rem (the recommended averted dose value criterion established by the NRC). Then, using cost estimates for the cost of the revision, and using guidance provided in NUREG/BR-0184, the analyst can perform a detailed value-impact assessment including both onsite and offsite cost impacts. While space considerations prohibit the authors from presenting all five of the cost-benefit-risk spreadsheets, a simplified example of a point estimate PD-VI spreadsheet is presented in Table 1.

**Table 1. Value-Impact Assessment Matrix**

| APB ID | AVERTED POPULATION DOSE (Man-REM) | BASE CASE DOSE COST ($) | REVISED DOSE COST ($) | AVERTED POPULATION DOSE VALUE ($) | COST-AVERTED DOSE RATIO ($/Man-REM) |
|---|---|---|---|---|---|
| 1 | 0.00E+00 | $0.00 | $0.00 | $0.00 | |
| 2 | 3.88E-01 | $3,176.56 | $2,788.50 | $388.06 | |
| 3 | 8.39E+00 | $18,473.04 | $10,080.30 | $8,392.74 | |
| 4 | 1.79E-01 | $711.85 | $533.04 | $178.81 | |
| 5 | 1.00E+01 | $40,884.89 | $30,884.46 | $10,000.43 | |
| 6 | 4.92E-02 | $2,672.38 | $2,623.15 | $49.23 | |
| 7 | 1.45E+01 | $137,502.18 | $123,012.89 | $14,489.30 | |
| 8 | 3.96E-02 | $349.21 | $309.58 | $39.63 | |
| 9 | 0.00E+00 | $0.00 | $0.00 | $0.00 | |
| 10 | 9.84E-05 | $38.58 | $38.48 | $0.10 | |
| TOTAL | 3.35E+01 | $203,808.69 | $170,270.40 | $33,538.29 | $885.55 |

| USER-PROVIDED DATA | |
|---|---|
| DISCOUNT RATE (%) | 7 |
| UNDISCOUNTED COST FOR CLEANUP AND DECONTAMINATION ($) | $1,545,000,000.00 |
| CLEANUP TIME (YEARS) | 10 |
| EXPECTED REMAINING PLANT LIFE (YEARS) | 40 |
| ENHANCEMENT COST ESTIMATE ($) | $29,700.00 |
| ESTIMATED AVERTED DOSE VALUE ($/Man-REM) | $1,000.00 |

| BASE CASE | |
|---|---|
| EXPECTED DECONTAMINATION COST ($) | $1,539,939.94 |
| EXPECTED REPLACEMENT POWER COST ($) | $2,027,347.39 |
| EXPECTED TOTAL ON-SITE COST ($) | $3,567,287.33 |

| REVISED CASE | |
|---|---|
| EXPECTED DECONTAMINATION COST ($) | $1,342,700.64 |
| EXPECTED REPLACEMENT POWER COST ($) | $1,767,679.75 |
| EXPECTED TOTAL ON-SITE COST ($) | $3,110,380.40 |

| COST-BENEFIT RESULTS | |
|---|---|
| ENHANCEMENT NET BENEFIT ($) | $460,745.23 |
| ENHANCEMENT BENEFIT-COST RATIO ($/$) | 16.51 |

As shown in this table, factors such as discount rate, decontamination costs, cleanup time, and proposed enhancement implementation and maintenance cost values are considered. The cost-benefit results are presented in terms of dollars per person-rem averted, enhancement net benefit, and enhancement benefit-to-cost ratio. Also, pay-back period may be added if desired.

These five linked spreadsheets may be reconstructed to yield "reverse" spreadsheet results based on a case study of any selected input/output parameter in a particular transformation matrix by simply reversing the columns and rows of the original "forward" spreadsheets and taking the inverses of all the transformation factors which appear in the internal cell formulae. This provides a useful tool for sensitivity case studies.

Uncertainty calculations are promulgated through the spreadsheet calculations by defining each input variable as a probability distribution, and then by applying standard formulae if one assumes all input and output cells maintain a lognormal distribution format, or, more rigorously, by applying standard probability distribution software designed to be used in conjunction with standard computer spreadsheets. Examples of this type of companion software are @Risk and Crystal Ball. Both probability density distributions and cumulative probability distributions for key risk and cost factors can be displayed and applied in the decision-making process. Only when uncertainty analysis is fully integrated with these calculations can the analyst fully define the risk associated with various competing decision factors.

# Conclusions and Recommendations

The authors of this paper feel that proper application of this tool to support decision-making associated with operation and maintenance of nuclear power plants could provide a strong, technically defensible basis for maintenance of acceptable levels of safety and continuous improvement in cost-effectiveness and cost-competitiveness of the commercial nuclear industry in the world energy mix. The authors recommend that the NRC, EPRI, and individual nuclear utility companies develop and apply tools like the linked cost-benefit-risk analysis spreadsheets to identify quantitatively prioritized lists of options for safety and cost-effectiveness improvements at nuclear power plants, and that they encourage use of these tools to help implement and track the impacts of these improvement options. The future health of the world nuclear power industry may well depend on the progressive and prudent application of these types of decision support tools and techniques.

# Design Improvements Based on External Events PSA for Reference Plants

Beom-Hee Jeong, Beom-Su Lee, Sun-Koo Kang
Korea Power Engineering Co., Inc.
Seoul, Korea

**Abstract**

Nuclear power plant construction in Korea has continued steadily since the 1970s. The pattern of these construction activities has been to build 2 identical units at the same site referencing the latest two previous units. An external events Probabilistic Safety Assessment (PSA) has been performed for nuclear power plants under construction. Based on the PSA study, two types of design improvements have been identified. The first one is those associated with the software (design and engineering). The second type of improvements is those related to hardware and is only applicable to the new plants due to the high costs of replacement. The sensitivity study showed that the hardware change can significantly reduce the risk induced by the external events. This approach will gradually enhance the safety of the future plants from external events.

## 1. Introduction

Nuclear power plant construction in Korea has continued steadily since the 1970s. The pattern of these construction activities has been to build 2 identical units at the same site referencing the latest two previous units.

In Korea, a Probabilistic Safety Assessment (PSA) including external events analysis has been performed on plants under construction. The study showed that the risk induced by internal events is well balanced and significantly reduced compared to that of existing plants by applying design improvements, but the risk due to external events is not so significantly changed . The reason has been determined to be due to the large uncertainties inherent in the external events PSA.

Despite the large uncertainties, several design improvements were identified from the results of the study. The improvement items could be grouped into two types. One is those mainly associated with the software (design and engineering), and the other is those associated with the hardware. However design improvements requiring hardware changes is difficult to implement especially to the plants under construction due to large cost impacts. Therefore, these design improvements requiring hardware change would be best suited for future plants.

This paper will describe the results of the external events PSA of a reference plant, design changes to new plants, and the impacts of the design change to plant safety. Fire events were the most significant contributors for adversely affecting plant safety based upon the PSA results. Therefore most design changes applied to new plants have been fire related and review and analysis of fire event is a central focus of this study.

# 2. External Events PSA for the Reference Plants

The overall Core Damage Frequency (CDF) of the reference plants due to external events was estimated to be 3.1E-5/yr. The CDF due to internal flooding was found to be negligible while the CDF due to seismic event and internal fire were 1.4E-5/yr and 1.7E-5/yr respectively. These estimates was high compared to the CDF due to other internal events, which was estimated to be approximately 8.5E-6/yr.

## 2.1 Internal Flooding

No flooding scenarios were found to be significant in terms of core damage frequency. All flooding areas were screened out during the screening process and thus no detailed analysis was performed. The design features of the reference plants in terms of flooding induced risk are follows:

- The sea water, which is an unlimited flooding source, is not allowed in the primary auxiliary building where almost all safety-related equipment is located.
- The emergency drain paths and emergency drain sumps have the capacity to accommodate the maximum possible flooding inside the plant.
- The concept of safety-related divisional separation prohibits the flood propagation from one division to the other division.

## 2.2 Seismic Events

The total core damage frequency due to the seismic events was estimated to be 1.4E-5/yr, and this estimate was high compared to that of internal events. However, this estimate did not include the CDF due to seismic/fire interaction problems described below. The significant seismic-induced events were the seismic-induced failures of Diesel Generators (DGs), Condensate Storage Tanks (CSTs), and electrical cabinets which have relatively high seismic capacities. The value of High Confidence of Low Probability of Failure (HCLPF) is 0.41g or higher. The sensitivity study showed that the uncertainties in the seismic hazard curves used in the PSA have dominant impacts on the CDF.

## 2.3 Seismic-Fire Interaction

A seismic-fire interaction problem was found during the seismic analysis. There is an automatic $CO_2$ fire suppression system, which is seismic category II, in emergency DG rooms. Thus there is some possibility that the $CO_2$ suppression system can spuriously flood the DG room due to seismic-induced relay chatter. Since the DG is designed to take combustion air from inside the room, the $CO_2$ flooding would make the DG unavailable. The relay chatter can also inadvertently trip the fans and close fire dampers of DG rooms, which can cause the DG unavailable due to loss of room cooling. The CDF induced by the seismic-fire interaction scenario was estimated to be 1.5E-4/yr, and due to the high CDF, it was decided to implement the necessary design changes even during the construction.

### 2.4 Internal Fire

The CDF due to fire was estimated to be 1.7E-5/yr. The significant fire events were found to be a fire originating in the MCR and a fire originating in, or propagating into either division A or B switchgear (SWGR) rooms. The CDF due to a fire in the MCR and switchgear room was estimated to be 1.4E-5/yr, which is approximately 80% of the total CDF due to fire. This is primarily due to the fact that the divisional separation concept cannot be applied in the fullest extent in these areas.

The most significant fire induced event was turned out to be the Loss Of Off-site Power (LOOP). The analysis also showed that fire induced Loss Of Coolant Accident (LOCA) could occur due to an electrical short-circuit on the control cables of the High-Low pressure boundary isolation valves.

## 3. Design Improvements

As described above, there were no significant flooding scenarios, and thus no design improvement items were identified. Considering the relatively higher seismic capacities and large uncertainties inherent in the seismic hazard curves, the seismic induced CDF was judged to be acceptable compared to other seismic PSA results. The efforts to reduce the seismic induced CDF is postponed till more reliable seismic hazard data could be obtained.

Therefore the efforts to identify design improvement items were focused on the seismic-fire interaction and fire events. Based upon the results of the study, the following items were identified to reduce the CDF.

### 3.1 Software Related Items

a. The doors between each DG room and the respective switchgear room should be regularly inspected to ensure that they are kept closed at all times, including during testing and maintenance.

b. Plant procedures for MCR evacuation should specify necessary operator actions, to prevent fire induced LOCA and to reestablish operational status of those components which are not monitored or controlled from the Remote Shutdown

Panels (RSPs).

c. Operators should be properly trained in manual fire suppression techniques and procedures.

d. Establish the procedure for transferring the source of feedwater from the CST to the Raw Water Tank (RWT) to maintain a secondary cooling supply and train operators for the procedure.

## 3.2 Hardware Related Items

a. To resolve the seismic-fire interaction problem, the relays that can cause $CO_2$ flooding if chattered should be replaced by the non-chattering relays and the intake for the DG combustion air should be changed to outside of the building.

b. To prevent the fire induced LOOP in each switchgear room, the design should be changed for the offsite power supply to the safety-related switchgears. Each startup transformer should supply off-site power only to a switchgear located in a associated division.

c. Two remote shutdown panels need to be electrically independent from the MCR. In the reference plants, only one of two RSPs are independent from the MCR.

d. To prevent fire induced LOCA, electrical power supply to isolation valves and cable routing to the valves should be changed so that a fire cannot induce hot-shorts for both isolation valves.

e. To prevent fire propagation from DG room to switchgear room, the doors between the rooms should be removed.

## 3.3 Application of Design Improvements

As described above, the CDF arising from external events is judged to be acceptable compared to that of previous PSAs[1], except the seismic-fire interaction problem. Thus for the plants under assessment, only the design improvements related to the software were implemented with the exception of item 3.2.a and 3.2.e resolved the seismic-fire interaction problem. Item 3.2.e was included since the change did not incur additional cost or delay of the construction schedule. Other hardware related design improvement items were judged that additional cost or construction schedule delay outweighed benefit or improvement.

For new plants, however, all design improvement items, regardless of the type, would be incorporated during initial design stage with little cost impact.

# 4. Impacts of the Design Changes to Future Plants

To identify the impacts of design improvements to the safety of new plants, some sensitivity analysis was performed. By applying item 3.2.a, the new plants do not have the seismic-fire interaction problem described above. Other design improvements mainly affect the CDF from fire induced effects. Following table summarizes the safety improvement comparison between the reference plants and new plants due to fire in specific areas.

| Fire In Areas | Core Damage Frequency (/yr) | |
|---|---|---|
| | Refer. Plants | New Plants |
| MCR | 5.99E-6 | 3.27E-6 |
| SWGR Rooms | 7.54E-6 | 4.40E-7 |
| Aux. EERs | 1.91E-6 | 9.73E-8 |
| CSRs | 6.09E-7 | Negligible |
| EPAs | 1.11E-6 | Negligible |
| Total | 1.72E-5 | 3.81E-6 |

Based upon the above results, it is conclude that the safety could be improved considerably with the implementation of the design change at the early phase. The major reasons of lower CDF comes from the facts that there are no possible LOCA sequence in the MCR or the Auxiliary Electrical Equipment Rooms (Aux. EERs), and that there are no LOOP sequence in the switchgear rooms. The independence of the RSPs from the MCR also contributed in lowering the CDF.

## 5. Conclusions

The external events PSA performed on the reference plants identified two types of design improvement items. One was related to the plant administrative activities such as procedures, etc. while the other was related to hardware. It was found that cost of implementing hardware related improvements to the plants under assessment was prohibitive. Thus, only the items that could significantly affect the safety of the plants or that can be done without additional costs, were implemented to the plants under assessment. In case of new plants, the hardware related items could be applied with little cost impact and it was determined that the design improvements could significantly improve the safety of the plants.

The external events PSA will be performed for new plants also to identify design improvement items. These PSA results will be used as an input for the design of the next generation of new plants. This approach will definitely enhance the safety of the future plants.

## References


1. J.A. Lambright, S.P. Nowlen, V.F. Nicolette, and M.P. Bohn. Fire Risk Scoping Study: Investigation of Nuclear Power Plant Fire Risk Including Previously Unaddressed Issues. (NUREG/CR-5088). U.S. Nuclear Regulatory Commission, Washington, D.C. 1988

2. U.S. Nuclear Regulatory Commission, Procedural and Submittal Guidance for the Individual Plant Examination of External Events (IPEEE) for Severe Accident Vulnerabilities. (NUREG-1407). Washington, D.C. 1991

3. Electric Power Research Institute, Fire Induced Vulnerability Evaluation (FIVE). (EPRI TR-100370). Palo Alto, California, 1992

4. W. Parkinson, G. Solorzano, B. Najafi, M. Marteeny, and K. Bateman. Fire Events Database for US Nuclear Power Plants. (NSAC/178L). Electric Power Research Institute, Palo Alto, California, 1992

# Use of PSA Results for Improving French PWR's Safety Examples and Insights

J.M. LANORE - F. PICHEREAU
IPSN/DES
Fontenay-aux-Roses - France

## Abstract

Although the safety of French PWR's relies on deterministic principles, probabilistic assessments have been used as a complement for improving safety.

The paper will describe on real examples the design or operation improvements based on PSA results. Insights can be drawn from this experience relating to the role of PSA in the decision making process.

## 1 Introduction

Although the safety of French PWRs relies on deterministic principles, the probabilistic approach was considered since the 1970s as an important complement for safety analysis.

In 1977, the Safety Authorities defined probabilistic safety objectives, but these objectives were only considered as orientation values and generally the probabilistic aspect is not the only basis for making a decision. However, in some cases, the probabilistic results gave a sufficiently high importance to a problem for considering that the decision relies mainly on PSA.

The objective of this paper is to give an overview of the important safety improvements and plant modifications implemented on French PWRs for which PSA results played a dominant role. From this experience some insights can be drawn relating to the place of PSA in a decision making process.

# 2 Probabilistic Safety Objectives

In 1977, during the examination of the major technical options for the 1300 MWe reactors, the Ministry of Industry fixed an overall probabilistic objective expressed as follows :

« In general terms, the design of a plant which includes a pressurized water nuclear reactor should be such that the overall probability that the plant could be the source of unacceptable consequences should not exceed $10^{-6}$ per year.

This implies that, whenever a probabilistic approach is used to assess whether a family of events must be taken into account in the design of the reactor, the family must effectively be taken into account if the probability of its leading to unacceptable consequences exceeds $10^{-7}$ per year. This value should not be exceeded unless it can be clearly demonstrated that the probability calculations used are sufficiently pessimistic.

Moreover, it would appear desirable for Electricité de France to continue to make all possible efforts to extend the use of the probabilistic approach to the largest number of events possible as quickly as possible.

By virtue of the above, Electricité de France shall... examine, case by case, whether simultaneous failures on redundant trains of vital safety systems should be taken into account in designing plants which include a pressurized water nuclear reactor... « Realistic » calculation methods and assumptions may be used for these studies. »

This text gives rise to a number of remarks :

a) The overall objective is stipulated in terms of « unacceptable consequences ». These « unacceptable consequences » are not specified by legislation or regulations.

b) The $10^{-6}$ value is an « objective » for a PWR plant, but Electricité de France is not required to demonstrate that this objective has been achieved.
A deterministic approach and not an overall probabilistic analysis is still used to justify the design features selected to avoid any unacceptable risk.

c) The $10^{-7}$ per year value is more practical for operational use and is used in the approach to determine the risks generated by external events ; for example, the value is applied to several families or aircraft crash events.

d) Unlike the deterministic approach, where the assumptions and calculations must be pessimistic, the probabilistic approach must be based on the most realistic values.

# 3 Probabilistic Studies Related to the loss of Redundant Safety Systems

## 3.1 « H » procedures

Following the letter of 1977, partial probabilistic studies were carried out by EDF for investigating the probabilities and consequences of the loss of redundant safety systems.

These studies showed the need for complementary measures to achieve a satisfactory safety level. Specific procedures, called « H » procedures (H for « hors dimensionnement », i.e. beyond design basis), if necessary implementing supplementary equipment, have been established :

- H1 procedure « total failure of the heat sink » ;
- H2 procedure « total failure of the steam generator water supply » (includes primary « Feed and Bleed ») ;
- H3 procedure « total loss of electrical supply » (led to the implementation of a gas turbine on the sites. Moreover a turbo-alternator was added with the aim of providing supply for water injection to the primary pump seals and to instrumentation and control) ;
- H4 procedure « total loss of LPSI or CSS during long term post-LOCA situations » (including interconnection between LPSI and CSS for mutual back up).

The H procedures are implemented on all the French PWRs (900 MWe - 1300 MWe - 1400 MWe).

## 3.2 Reactor protection system of the 1400 MWe PWR

For the latest series of French PWRs (1400 MWe-N4), in order to fulfill the safety objectives in case of total loss of reactor scram (ATWS), and due to the difficulties for quantifying software reliability, a diversified scram system was implemented.

The diversified system known as 'ATWS' trip system, actions as follows :

- trip the breakers supplying the control rod hold clutches,
- start up the auxiliary feed water to the steam generators,
- trip the turbine,
- isolate the steam generator purge.

The ATWS trip system is implemented with hardware and software totally different to that used on the protection system. The use of separate measurement points, instrumentation and independent rod trip breakers furthers the diversification.

# 4 Global PSAs - Scope and Results

After the partial probabilistic calculations, and although it was not a regulatory requirement, two French PWR Probabilistic Safety Assessment (PSA) studies were terminated in 1990.

The first of these studies (PSA 900) concerns a standard reactor of the 900 MWe series, and was carried out by the « Institut de Protection et de Sûreté Nucléaire » (Institute of Health Physics and Nuclear Safety). It was financed by the « Service Central de Sûreté des Installations Nucléaires » (Central Service for the Safety of Nuclear Installations - Safety Authority).

The second study (PSA 1300) was carried out by the utility « Electricité de France » on Unit 3 (1300 MWe) of the Paluel nuclear power plant.

A mutual external review was carried out, and a common data base was used for both studies.

Both PSA 900 and PSA 1300 are level 1 PSAs. In their present form the studies do not cover internal and external hazards such as fires, floods, and earthquakes.

The total frequency of core damage obtained from the study of the 900 MWe units is $5 \times 10^{-5}$/reactor-year and the total frequency of core damage obtained from the study of the 1300 MWe Paluel unit is $10^{-5}$/reactor-year.

One interesting finding from these PSA studies is the considerable contribution to core melt probabilities arising from situations connected to the shutdown states, which account for 32 % of the total core melt probability for PSA 900 and for 56 % in the case of PSA 1300. These high values may result from the fact that there are generally no automatic systems to counter accident situations in shutdown conditions and that human action is therefore necessary.

The core melt frequency is particularly high in case of loss of Residual Heat Removal System (RHRS) during mid-loop operation because only a short time is available for the operator to take any action, due to the low primary coolant inventory.

Other particular sequences initiated by a spurious boron dilution of the primary coolant were also identified.

Although the PSA's completion was not a regulatory requirement, after the presentation of the results, the Safety Authorities required from the utility to propose plant modifications in order to reduce the frequency of these sequences, due to their high contribution to core melt frequency.

These sequences were similar for 900 MWe, 1300 MWe and 1400 MWe plants, and led to modifications of all series.

## 4.1 Loss of RHRS during mid-loop operation

The initiating event can occur when the primary level must be lowered to mid-loop operation level. An extra lowering of the RCS level of a few cm may cause loss of RHRS pumps (over draining events). Experience feedback shows that RHRS pumps have been temporalily lost in this way several times in the world, and in particular on French units (5 events during 500 year.plant).

A total loss of RHRS pumps leads to core uncovery if no water is supplied by the operator in a relatively short delay. This delay before core uncovery depends on the residual power and the position of the RCS openings and could be as short as 15 minutes, in particular circumstances.

The first probabilistic assessment of this particular sequence led to a corresponding core melt frequency (CMF) of about $5.10^{-5}$ per year.

Considering this relatively high probability and the fact that during shutdown the containment integrity was not guaranteed, the Safety Authorities required from the utility (Electricité de France) in 1990 immediate dispositions to deal with this particular sequence and, over the medium term, definitive measures enabling the risk of core damage to be reduced to some $10^{-6}$/plant.year at mid-loop operation.

It can be noted that some immediate measures were decided before the completion of the PSAs, and for this reason the preliminary assessment of the sequence was not included in the overall results.

### *4.1.1 Preliminary measures*

Immediate measures were taken by the utility with the main objective of helping the operator in the prevention and mitigation of the accident :

- technical specifications are modified to ensure a sufficient delay for operator action ;
- a mobile ultrasonic level sensor is implemented on a hot leg ;
- special measures are taken concerning containment management.

The probabilistic assessment of the sequence, taking into account the preliminary measures, was about $5.10^{-6}$ with a large uncertainty. The uncertainty is due to the difficulties related to human reliability evaluations, especially to assess a priori the effect of improved detection means and of longer time windows.

### *4.1.2 Definitive measures*

In order to reach the above mentioned probabilistic objective, the utility proposed a new set of improvements concerning human and machine aspects :

- a permanent monitoring of the reactor cavity and the RCS level from the control room ;
- an early warning system for vortex phenomena detection ;
- an increased operating margin on the RHRS pumps relative to vortex formation ;
- improved incidental and accidental procedures ;
- improved operator training.

Moreover, an important hardware modification was defined :

- an automatic RCS water supply, actuated on a criterion representative of a loss of RHRS pumps.

With all these improvements the assessed CMF corresponding to the total loss of RHRS is about $3.10^{-7}$ per year.

As mentioned above, it is difficult to evaluate quantitatively the benefit of the improvement measures on human reliability before any experience feedback, and the uncertainty is important. The benefit due to the automatic system is easier to demonstrate.

The Safety Authorities have considered that the assessed CMF was significantly reduced by these measures which should be rapidly implemented on all the plants.

## 4.2 Rapid boron dilution

The main accident sequence of rapid boron dilution identified in the PSA is the following :

- initially the plant is at hot shutdown, a normal boron dilution is in progress before start-up ;
- a loss of off-site power supply causes the trip of the primary coolant pumps. Since the CVCS pumps are backed-up by auxiliary power, the dilution goes on ;
- if the flow rate is insufficient for boron mixing, and if the operator does not stop the dilution neither restart a primary pump rapidly, a pure water slug is created in a loop ;
- when after some delay a primary pump is restarted, the pure water reaches the core. Due to the rapid dilution, a reactivity accident can occur (energy higher than 300 cal/g).

The probability of this sequence was about $10^{-4}$/reactor x year.

Despite pessimistic assumptions related to the physical phenomena, due to the high probability and to the potential severe consequences, an immediate measure was implemented (before the completion of the PSA) : in case of RCP trip an automatic isolation of dilution and suction of CVCS pumps from borated RWST.

This immediate preliminary modification allowed to reduce the dominant sequence probability by a factor of about 100. However, other rapid dilution sequences were identified with non negligible probabilities but high physical uncertainties. In order to define a final modification, several actions are in progress :

- physical studies (neutronic - thermalhydraulic),
- experimental studies (mixing experiments on a 1/5 scale facility),
- more complete identification of all the boron dilution scenarios.

The results of these studies and the final modification proposed by the utility are presently discussed by the Safety Authorities.

## 4.3 Other PSA based modifications

- For the completion of the PSAs, an important effort was devoted to human factor analysis, especially by help of simulator experiments. This analysis and the high contribution of human factor in the overall results led to several modifications and improvements of operating and emergency procedures. PSA results are also used for operator training.

- The results of the 900 MWe PSA were used for the safety reassessment of the 900 MWe series, by analysing all the dominant sequences. Following this analysis it was decided to implement two automatisms (already existing on the 1300 MWe series) : automatic suction of the CVCS pumps from the borated RWST in case of loss of heat sink (high temperature in let-down line) and in case of slow boron dilution (high flux alarm).

# 5 Insights

Some interesting insights can be drawn from these examples :

- Generally the decision does not rely totally on probabilistic assessment. Other elements (experience feedback, safety studies...) are also considered.

- Decisions are sometimes taken before completion of the PSA : functional qualitative analysis and preliminary probabilistic assessment can be sufficient to indicate the importance of a problem.

- Quantitative safety objectives are useful guidance, but are not used as limits. Relative considerations (dominant sequences, comparison of plant states, choice of the assumptions,...) are very important.

- Difficulties arise when the credibility of PSA results is not sufficient, due to possible (known or unknown) uncertainties. In order to improve this credibility, some ways seem particularly important :

* quality of data : the question of basic data (reliability data, common cause failures, human factor) is always an important factor for the credibility of a PSA. In France the relatively important experience feedback on an homogeneous series of plants is a favourable situation. However in several cases the quality of data (as explained in particular in 4.1) is still a source of uncertainty ;
* sensitivity studies, taking into account the possible ranges of variation for some hypothesis have to be presented and discussed ;
* quality assurance and external review : a detailed review by independent experts is necessary in order to precisely assess the exhaustiveness of the PSA as well as the validity of the various assumptions made. The extensive mutual external review of the French PSAs by EDF and IPSN was very helpful ;
* international exchange : a large international exchange on the experience in the field of PSA use is particularly necessary. Reference to an international state-of-the-art is often a useful indication.

## References

1. EPS 900 - A Probabilistic Safety Assessment of the Standard French 900 MWe pressurized Water Reactor - IPSN - Main Report - April 1990

2. EPS 1300 - Probabilistic Safety Assessment of Reactor unit 3 in the Paluel Nuclear Power Centre (1300 Mwe) - EDF - Overall Report - May 31, 1990

# Risk Assessment and Life Prediction of Complex Engineering Systems

Michael D. Garcia and Ravi Varma
Los Alamos National Laboratory
Los Alamos, New Mexico, U.S.A.

A. Sharif Heger
The University of New Mexico
Department of Chemical and Nuclear Engineering
Albuquerque, New Mexico, U.S.A.

**Abstract**

Many complex engineering systems will exceed their design life expectancy within the next 10 to 15 years. It is also expected that these systems must be maintained and operated beyond their design life. This paper presents a integrated approach for managing the risks associated with aging effects and predicting the residual-life expectancy these systems. The approach unifies risk assessment, enhanced surveillance and testing, and robust computational models to assess the risk, predict age, and develop a life-extension management procedure. It also relies on the state of the art in life-extension and risk assessment methods from the nuclear power industry. Borrowing from the developments in decision analysis, this approach should systematically identify the options available for managing the existing aging systems beyond their intended design life.

## 1 Introduction

This paper establishes specific technical issues and challenges to assess a complex system in terms of aging effects, life prediction, prediction of failure modes, and providing cost-effective retrofits for aging, complex engineering systems. In this paper we define complex in terms of the basic requirement to integrate the risk assessment to a fundamental understanding of the engineering materials used in the system as they deteriorate with age, material synergistic effects and compatibility issues, and the evaluation of the resulting change in total system reliability. Risk assessment involves identifying and characterizing most common materials failure modes of aging, in-service components. For example, corrosion with or without stresses may lead to "fracture" which constitute major materials failure modes. Radiation damage in materials (e.g., beryllium, polymers) that results in swelling and decreased tensile strength becomes prone to mechanical fracture. Hydriding of zirconium alloys, for example, will degrade the mechanical strength of the element, and this may lead to fracture.

Life prediction of aging, complex engineering systems will involve the development of robust computational models that utilize, a real-time experimental data base. While a physical foundation is essential to achieve generality and some measure of confidence in extrapolations, phenomenological constraints is equally crucial for achieving reliability and predictive value in descriptions of macroscopic behavior, despite the enormous complexity of underlying physics. There are lessons to be learned from the research programs [1, 2, 3] on structural aging and aging of nuclear power plants. Much of this work can provide a basis and relevant objectives for developing an integrated plan for life-prediction of aging complex systems. Other aging research programs such as those related to conventional weapons and aircraft must be researched for relevance.

In this approach, it is also critical that existing system surveillance units (if they exist) provide information and data to support computational model development and predict anticipated failure modes. For example, we need to detect and measure the initiation of material degradation which may involve corrosion, the progression stage which may involve microcracks developing during corrosion particularly in presence of stress, and the ultimate materials failure involving mechanical fracture of the metal/alloy specimen. Certainly measurement tools are needed; some of which may be readily available while others may have to be developed for the particular application. Additionally, surveillance must be enhanced to recover a refined set of data and information that directly supports the computational models, such as obtaining time-dependent material properties. Where there is no data, we must rely on accelerated-aging testing to simulate the aging process and extrapolate our models.

Of particular interest is the development of robust computational models to predict failures or sequences of failures (coupled failures) that have not been observed in existing surveillance or shelve life units, due to synergistic effects or existing units that are not old enough to have initiated or detectable failures. Common-mode failures are of particular interest since they may affect a significant number of common systems.

The four critical competency development areas are :

1) Probabilistic risk assessment (PRA) and reliability,
2) Enhanced surveillance - destructive and nondestructive evaluation (NDE),
3) Constitutive model development and aged material properties, and
4) Deterministic computational code modeling.

# 2 Critical Capabilities and Associated Technical Issues

## 2.1 Probabilistic Risk Assessment (PRA) and Reliability

It is important to organize and systematically evaluate system operation and system reliability. With PRA system models, failures can be analyzed in terms of how they might be initiated, how failures progress, how they might be mitigated, and how human actions might affect their progression. Failure modes for each complex system will differ because each system has different common materials (for example, different types of explosives, different special materials, different types of foams, etc.), different ages, different fabrication techniques, and different assembly procedures. Each complex system also has a unique set of mission environments that it must endure. The same components in the system also have different reliabilities due to statistical differences in the manner in which they are machined, formed, welded, etc. and they have different material properties. In view of these differences, it is appropriate to a perform failure modes and effects analyses (FMEA) that identify the unique set of potential failure modes for each system.

Generally, failures are defined as an incident or condition and sequence of events that causes the complex system to operate unsafely, unreliably, and not cost-effectively. Failure modes that may be considered are fracture, stress corrosion, pitting, cracking, radiation induced embrittlement, radiation damage, hydriding, wear, etc. or a combination of any these failure modes. Scientific data accumulated from a comprehensive aging research program together with careful documentation of surveillance experience make this tool more effective as a practical tool for managing and assuring system safety and reliability, for example, in weighting the relative merits of inspection, surveillance, testing, and maintenance as measures to assure acceptable reliability. This can be accomplished with the general PRA framework.

## 2.2 Enhanced Surveillance - Destructive and Nondestructive Evaluation (NDE) Assessment

Current techniques in support of model development includes metallography, tensile tests, radiography, photography, physical inspection, surface characterization, weight testing, gas testing, chemistry, and acoustic resonance. In general, however, the emphasis associated with these tests was to ensure conformance to manufacturing specifications. The requirements from the aging and life prediction standpoint is to define an enhanced and robust set of surveillance techniques and corresponding data that directly supports the development of accurate failure and aging computational models.

### *2.2.1 Residual-Stress Measurements*

A complementary area for model development is the characterization of residual stresses induced into as-built components resulting from welding, high-temperature forming, machining, or fabrication. Redistribution of these stresses in as-built configurations will occur due to cracking, volume expansion due to radioactive decay daughter-product production, or by other means. Residual stresses can be measured using various methods, including X-ray, and strain gauge techniques. The enhancement of old techniques or the development of new techniques such as using neutron diffraction techniques should be explored and developed. Furthermore, improved fabrication techniques should be explored to relieve stresses (stress-improvement techniques) such as induction heating and mechanical techniques, especially since potentially different fabrication techniques may be implemented in the future.

### *2.2.2 Crack Detection Measurements*

In general, conventional NDE methods fail to detect incipient damage which can lead to crack initiation, followed by failure. It is also unlikely these methods will determine the nature or location of these cracks or damage within the component or system. Several improved NDE techniques have been recently developed for estimating life consumption. These include strain monitoring techniques, microstructural techniques, hardness-based techniques, and ultrasonic testing. Acoustic Emission (AE) are result of stress waves in stressed materials due to microstructural events leading to cracks. An important feature of AE is that it is non-destructive and non-intrusive and that with suitable sensors, the entire device or component can be evaluated for flaw or cracks in one measurement session. Because of known stress-strain relationships, accurate measurement of residual stress levels in a structure can be made by measuring strain using highly sensitive resistance strain gauges. Fiber optic, heterodyne, dual probe interferometer for laser ultrasonic crack detection has been refined to an extent such as that flaws or crack depth of 100 micron sizes may now be measured.

## 2.3 Constitutive Model Development and Aged Material Properties

The purpose of this area is the development of constitutive material models for aging system materials. The development of this aged-material data base is a basic requirement for the development of robust, computational models. This will require the measurement of critical material properties as a function of time. The data base will be used to assist in the prediction of long-term deterioration of critical system components. The bulk macroscopic mechanical properties of interest are typically derived from stress-strain measurements such as the elastic modulus, yield stress, hardness, creep strength, and ultimate stress.

The difficulty associated with development and measurement of the degraded or aged material properties is the fact that it is difficult, if not impossible, to simulate the natural-aging processes that occur within nuclear materials. For example, it is well-known that radiation damage to organic materials may depend not only on the absorbed dose but also on the irradiation time and dose rate. Furthermore the presence of external gas species such as oxygen that may present for diffusion into the organics may strongly influence the amount of permanent damage to the material. Other limitations are the fact that naturally-aged materials are not available for the ages of interest, such as ages greater than 20-years. Additionally, most of the data on relevant reactor material properties are at high dose-rates and high temperatures, that typically occur in nuclear reactors. Virtually all existing information on the aging effects of temperature and radiation on the properties of organic materials comes from artificial-age conditions.

## 2.4 Deterministic Computational Code Modeling

In general, the computational models (e.g. finite element models) must be developed in conjunction with the aged-material data base and constitutive models for predicting the residual-life of complex systems. Furthermore, the models must be verified against accurate experimental data or extrapolations that are theoretically sound. With aging systems there is greater emphasis on predicting the type of damage or state-of-stress that may develop in a system as it ages and the initial state of stress as it is delivered for operation. For example, these computational models will be used to simulate residual stresses present in the components due to welding or fabrication and then modeling and superimposing the mission environments. This level of damage or initial state-of-stress (such as residual stresses, cracking, corrosion, etc.) then serves as initial conditions for subsequent mission environment calculations and the assessment of how the system performs. These initial conditions vary depending upon the age of the system and system type.

Accurate computational modeling of welding must model the weld cooling rate, temperature gradient of the weld pool, turbulence, multiphase flow, surface tension flow, transient heat transfer, weld solidification microstructure, convection of nuclei and grains, and evaporation. Of particular interest is the requirement for stress-strain relationships that are a function of microstructure. Existing models also ignore high-temperature strain-rate dependence and creep. Refinement of the existing models are required to better understand the welding process and how it affects the residual-life of system components.

In some cases existing flaws or microcracks on a component or part (e.g., aircraft wing panels) under applied load or residual stress may suddenly and rapidly propagate. Crack-growth-rate characteristics for components are important in predicting residual-life of system components. Cracks of less than a critical-size will likely increase in size during mission environments. The key issues here are will these cracks be initiated, how they propagate if initiated, what are critical sizes, how fast these cracks grow in size in mission environments, and do the cracks affect system operation. The key concept of the model is the coupling between the internal and the external environments involving all the variables that have been observed to have a significant role in crack growth.

Corrosion is characterized by spatially separate anodic and cathodic areas and is accompanied by a flow of electric current for perceptible distances through the metal surfaces. The localized corrosion modes, namely pitting, crevice corrosion, and stress corrosion is more destructive than general corrosion (e.g., corrosion of a steel coupon in sea water , ~ .13 mm per year). Pitting corrosion is usually attended by the breakdown of passivating film on the metal surfaces. Such areas are adjacent to weld zones in a metallic structure. The use of empirical and deterministic models for calculating crack growth rates and for predicting damage functions for pitting corrosion in stainless steel-403 has been demonstrated to be very successful in their applications to nuclear power reactor plants, particularly for the heat transport circuits. For the first time the extent of surface damage by localized corrosion can be predicted.

# 3 Conclusion

This paper has established the framework, issues, limitations, tools, and competencies required to assess an aging, complex, engineering system in terms life prediction and assessment. It addresses the importance of :

- integrating the information derived from surveillance units with the system assessment or establishing an enhanced surveillance management program,
- integrating fundamental information on material behavior to the risk assessment,
- improving measurement techniques on surveillance systems and integrating this information, such as aged-material properties, to the deterministic computational models,
- performing refined calculations of the mission environments that incorporate defects, welding or fabrication residual stress effects, or aged-material properties, and

## References

1. Morris, and Vora, Nuclear Plant Aging Research (NPAR) Program Plan, NUREG-1144 (Rev. 1), Division of Engineering Technology, Office of the Nuclear Regulatory Research, U.S. Nuclear Regulatory Commission, Washington, DC, July 1985

2. Edited by Sinnappan, Meligi, Narayanan, Bond, Power Plant Systems/Components Aging Management and Life Extension, 1991, PVP-Vol. 208, ISBN 0-7918-0802-5

3. Edited by Shah, Macdonald, Aging and Life Extension of Major Light Water Reactor Components, 1993, ISBN 0-444-89448-9

# PSA Activities by the Principal Working Group 5 (PWG5) of the Committee for Safety of Nuclear Installation (CSNI) of the OECD

B. Kaufer
OECD/NEA
Paris, France

R. K. Virolainen
Finnish Centre for Radiation and Nuclear Safety
Helsinki, Finland

## 1 Introduction

In conjunction with restructuring and orientation of the whole OECD/NEA programme the Committee of Safety of Nuclear Installations (CSNI) decided that the activities on water reactor safety should be managed by five Principal Working Groups. Consequently the PWG5 on Risk Assessment was set up in 1983.

In general terms the mandate given to the PWG 5 by the CSNI is to deal with practices and methods of PSA, to exchange information on national efforts to develope safety goals and to assess the role of safety goals in licensing and to exchange information on national PSA programmes and current research and assist other CSNI groups on questions relevant to PSA.

A real fact today is that in the course of increasing number of PSAs (~ 200 PSAs completed), the scope of applying PSA for safety issues resolution is extended much. Accordingly, more than 20 countries are running or are aiming to run so called Living PSA application. The extending use of PSA for resolutions of practical issues has raised justified questions of the consistency of PSA methods in use. In order to respond to this challenge, PWG 5 has made comparisons and reviews of PSA practices, and methods in use and under development.

The group has critically reviewed the results and practices of more than 20 PSA studies in context of different projects.

Irrespective of fairly sound and mature technology, level 1 PSA methods are still partly questioned. Methods in use for analysing CCFs and data dependences, human factor, external events and modelling uncertainties still need further refinements. Hence, the group has thoroughly reviewed and compared the specific methods questioned [1]. Also, Living PSA tools and practices need still further development.

Some risk analysis methods are still under development, it is, not well established. Accordingly, analysis methods on low power and shutdown (LPS) mode are insufficient for high quality assessment. PWG 5 has compared LPS risk analyses from eight countries in general terms. More thorough review is still needed for assessing the methods in detail.

An extensive review of reliability data collection and processing systems in OECD and non-OECD countries has been completed and a supporting workshop was organized in Toronto in the spring 1995.

# 2 Overview of Few Completed Tasks

## 2.1 Data Acquicition and Evaluation

An extensive review on data acquisition and processing systems in no less than 44 organizations in 26 countries was made based on responses received to the questionnaire [2]. Thanks to IAEA also several non-OECD countries participated in the study. It appeared that no less than half of the organizations have completed one or more Probabilistic Risk Assessments Studies (PSAs) and only three organizations do not have and do not plan to implement data collection system. In addition over half of organizations informed that they plan to maintain a Living PSA, which shows that the concept of Living PSA is spreading quickly. As to the coverage of the data collection, about one third of the organizations collect data on all plant components. In case of incomplete data collection coverage, instrumentation and electronic components are typically excluded.

Contrary to the good coverage of data on mechanical components, only few organizations collect plant specific data on human interactions and CCFs. In general human error data is scarce and is planned to be provided mainly by simulator training sessions in the future.

An important issue the responses raised was that a strong Quality Assurance program on data evaluation is necessary.

## 2.2 Low Power NPP Operation and Shutdown Conditions (LPS)

Low power and shutdown (LPS) risk analyses [3] from Belgium, Finland, France, Germany, Netherlands, Sweden, United Kingdom and USA have been surveyed.

A few LPS studies have been completed so far and the studies show that LPS risks are not negligible. In some studies LPS dominates core damage frequency. Large uncertainties are an inherent part in LPS PSAs, because traditional PSA methods are partly insufficient. Consequently LPS PSA does not yet have the same degree of confidence as PSA for full power operation.

There are several special features in LPS assessments which makes it different from a PSA for operational states including

- plant configuration changes with time in the course of refuelling outage
- automatic safety systems are to a large extent unavailable
- maintenance actions are performed in complex environment
- less strict Tech-Specs requirements are used during shutdown
- less EOPs available for LPS states, because traditionally LPS states are regarded as very safe
- lack of specific LPS data (components as well as human factors)
- lower decay heat, more time available for recovery
- some barriers are partly unavailable (eg. containment is open).

Data from normal operation state may not always be valid for LPSA conditions.

However human reliability analysis (HRA) is the weakest part of LPS safety analyses. It appears that the traditional analytical HRA methods used in full-power PSA are insufficient for LPS analyses.

Traditional PSA methods have been combined with special approaches such as interview techniques, group discussions, barrier analysis. Instead, further development is needed in order to reach proven systematic body of LPS methods. Also fire and flooding risks need more attention in LPS PSA. Task will proceed providing new information for the second stage of the task.

## 2.3 Risk Based Monitoring of Safety System's Availability

The reasons for status monitor development can be seen as a combination of several factors, including, increases in numbers of safety systems and technical specifications, increased flexibility in applying technical specifications, increased use of on-line maintenance and repair and use of PSA for safety justification.

The practicability of on-line risk-based monitoring of safety systems' availability status has been demonstrated by over 5 years of experience at each of four nuclear power reactors in the UK [4]. Within this experience many thousands of status changes have been monitored and evaluated.

Status monitors have obvious safety advantages, in that they effectively prevent the existence, for any significant time, of high risk safety system outage combinations. This cannot be achieved only by applying traditional rules to each safety system individually.

In a regime of increasing Technical Specifications complexity, status monitors can provide rapid and accurate advice on responses to unplanned system outages. A rapid and accurate response is otherwise difficult when many Technical Specifications must be consulted.

The Heysham application is an example of direct use of a PSA model. It is recalculated in response to any changes of status, with results leading to three different action categories - normal, urgent and immediate. This approach requires an agreement between operator and regulator giving PSA a very significant role in day to day operational control.

The Torness application is an example of a use of a comprehensive list of 'Allowed States', agreed in advance by the regulator. Two different action categories - normal and urgent belong to the allowed states but all other states require immediate action.

# 3 Ongoing Studies

## 3.1 Human Interactions: Critical Operator Actions and Data Issues

Human factor has been mentioned as one of high priority studies in all safety research plans in context of CSNI. The focus of the task is on the dynamic operator actions which represent the most serious modeling challenge due to scarceness of data. According to most PSAs they also have the highest safety significance in relative terms. Interactions which occur prior to the initiating event or which lead to a plant transient are not ignored within this task but have a lower priority. Based on a number of available PSAs critical operator actions during accident conditions are identified, and probabilities assigned to selected interactions and the approaches used for generating these data are reviewed. Potential for improvements of the current data situation through wider use of already established approaches as well as novel approaches are considered.

Through a detailed questionnaire developed for this task, 10 countries are sharing information on HRA methods in practice and on the human interactions identified as important. Detailed responses have been completed and evaluated for four BWRs, eight PWRs and five advanced reactors.

The responses provide descriptions of the quantification process as implemented in the studies, and list the HRA-based improvements of design and procedures. In some cases the analyses were supported by simulator experiments.

From the point of view of identification of critical operator actions the expectation was that the responses would provide information which partially could be of generic nature. In fact, some actions identified as common important ones for BWRs and PWRs, respectively, are currently subject to detailed treatment within the task. These include manual depressurization and liquid control system actuation for BWRs, and "feed and bleed", alignment for recirculation and loss of RHR for PWRs. Commonalities and differences in the approaches used to model the selected actions are being examined on the basis of "traceable" descriptions that can be followed step by step, for example from Performance Shaping Factors (PSF) rating to the Human Error Probability (HEP) value.

The report will address special issues in HRA, associated with particular modeling difficulties. This includes operator actions under special conditions (e.g. external events, low power and shutdown operational modes), errors of commission, dependencies and recoveries, transferability of simulator-based data, and impact of organization and management.

## 3.2 PSA Level 2 Methodology and Severe Accident Management

The objective of the work is to review current Level-2-PSA (AM) methodology and to investigate how Level-2-PSA can support accident management programmes, i.e. the development, implementation, training and optimisation of accident management strategies and measures. New severe accident research results are to be taken into account and investigated how level 2 PSA has to be structured and conducted in a best way to support AM.

The work is quite far based on NEA/CSNI work made by Principal Working Groups and other special groups. More than 20 OECD/CSNI publications are referenced. Extensive review is made of recent research on severe accident phenomena including related integral and single phenomena code calculations.
Current approaches towards accident management involving human interactions and human reliability analysis which are issue areas in severe accident management, are reviewed. Uncertainty analysis and expert judgment which are vital parts of level 2

analysis, are emphasized.

The results of no less than 15 recent level 2 PSAs are reviewed. Based on the studies the influence of major design differences on severe accident behaviour is examined through quantities such as total in-core zirconium (related to hydrogen generation), power/containment volume ratio and fuel mass/containment volume ratio. Accident management provisions modelled in the studiese e.g. sprays, hydrogen control, water injection and depressurization and level 1 and 2 methodology used are reviewed.

The key phenomena that need to be modelled due to their significance for accident progression are containment by-pass or failure modes, magnitudes and timing for releases and the assessment of AM-measures.

The important issues of level 2 methodology to be considered in the study are, interface between level 1/level 2 PSA including binning of core hazard/damage states to plant damage states and influence of failed preventive AM-measures, phenomena and modelling of systems behaviour under severe accident conditions. Further, the modeling of SAM-measures, treatment of uncertainties due to differing results of code calculations, and uncertain initial and boundary conditions and expert judgment are of great importance.

## 3.3 PSA Based Plant Modifications and Backfits

In the work several representative examples have been provided of safety improvements made by PSA. The examples give information of type of plant modifications, main insights received (positive aspects and difficulties encountered).

The report gives detailed information of plant changes made at nuclear power plants in Finland, France, Netherlands, Sweden, Switzerland, United Kingdom and USA.

Technical insights received from PSAs show that almost all PSAs led to several plant modifications and backfits. The most frequent modifications are related to the EOPs and other procedures, and technical specifications. Several of the modifications concern instrumentation and control system. Few examples of the important modifications are:

- 3rd feedwater system and a diverse short term reactivity control system are installed at Magnox station in UK
- redesign of the whole ventilation system of electrical and control equipment room at Loviisa NPP in Finland
- a new AFWS is installed at Oskarshamn NPP in Sweden
- an additional make-up water system is installed at Ringhals NPP in Sweden
- installation of fire resistant wall, a water resistant door and assuring the anchorage of electrical cabinets at Dodeward in the Netherlands.

The general insights confirm that the strength of PSA in making decisions for plant and procedure modifications lays in the ability to deal with complex interactions between system CCFs and human factors. Accordingly PSA is widely used for justifying the plant modifications.

# 4 Long Term Outlook of PSA Activities

The current use of PSA in safety regulation and in support of plant operation covers mainly level 1 PSA (internal events, fire flooding, seismic) but a number of level 2 PSAs are in progress in several countries [5]. In addition the number of low power and shutdown PSAs is increasing steadily. Typical PSA uses are the identification of weaknesses, backfitting and design. This includes the identification and comparison of alternative designs and procedures, incident analysis or precursor studies, AM Planning, Optimization of AM procedures and derivation of AM measures.

Use of PSA to support plant operation includes exemptions, improvements and optimization of Limiting Conditions of Operations (LCOs), improvements of operation training programmes, aging analysis and development of probabilistic safety indicators.

The use of PSA to support regulation includes prioritization of inspection tasks, periodic safety reviews, support during emergencies and emergency planning, quality assurance programme optimization, changes to regulations, safety decision making and cost-benefit analysis.

The future use of PSA in safety regulation is much reflected by safety management activities and improvements in technology.

A great part of the future PSA applications suppose that well established PSA methods and sophisticated PSA codes, even real time risk management tools, are available. The enhanced requirements for PSA methodology and PSA tools have given incentives for further development of PSA technology. This is well argued due to the fact that, the more specific is the issue to be resolved the more strict are the requirements for the consistency of methods and tools. Accordingly the methods and implementation used for ex. in real time safety monitoring applications still deserve careful reviews and comparisons and possibly refinements.

The major trend in regulatory activities and plant safety management is a move towards more risk based approaches which in turn very much emphasizes the needs for more consistent PSA methods and data and user friendly Living PSA tools with capabilities of real time or almost real time applications.

## References

1. State of the Art of Level 1 PSA Methodology [NEA/CSNI/R(92)18]
2. Reliability Data Collection and Analysis to Support Probabilistic Safety Analysis (PSA) [NEA/CSNI/R(94)25]
3. Shutdown and Low Power Safety Assessment - A Status Report [NEA/CSNI/R(93)13]
4. Risk Based Management of Safety System Availability, NEA/CSNI/R(94)12
5. Regulatory Approaches to PSA, Report on the Survey of the National Practices [NEA/CNRA/R(95)2]

# An Evaluation of the Effectiveness of the US Department of Energy Integrated Safety Process (SS-21) for Nuclear Explosive Operations Using Quantitative Hazard Analysis

by


S. R. Fischer, H. Konkel, T. Bott, S. Eisenhawer, J. Auflick, K. Houghton, K. Maloney, L. DeYoung, and M. Wilson

Los Alamos National Laboratory
Probabilistic Risk and Hazard Analysis Group

Sandia National Laboratory
Assessment Technologies Department



**Abstract**

This paper evaluates the effectiveness of the US Department of Energy Integrated Safety Process or "Seamless Safety (SS-21)" program for reducing risk associated with nuclear explosive operations. A key element in the Integrated Safety Process is the use of hazard assessment techniques to evaluate process design changes in parallel or concurrently with process design and development. This concurrent hazard assessment method recently was employed for the B61-0, 2 & 5 and W69 nuclear explosive dismantlement activities. This paper reviews the SS-21 hazard assessment process and summarizes the results of the concurrent hazard assessments performed for the B61 and W69 dismantlement programs. Comparisons of quantitative hazard assessment results before and after implementation of the SS-21 design process shed light on the effectiveness of the SS-21 program for achieving risk reduction.


## 1 Introduction

In response to criticism from the Defense Nuclear Facilities Safety Board (DNFSB) regarding its Nuclear Explosive Safety Study process and concern over the continuing practice of "auditing" safety into nuclear explosive processes and in recognition of the large uncertainty in the response of high explosives (HE) to various insults, the US Department of Energy (DOE)/Albuquerque initiated the Seamless Safety "SS-21" program to design safety into a nuclear explosive process. In December 1993, the DOE formally initiated the SS-21 program, and a demonstration project was started based on the B61-0 Center Case Disassembly, to demonstrate, in part, the viability of conducting hazard assessments concurrent with process design and development to facilitate risk reduction. The goal of the SS-21 program or

Integrated Safety Process is to "integrate established, recognized, verifiable safety criteria into the process at the design stage rather than continuing the reliance on reviews, evaluations and audits." The entire SS-21 design process is verified by a hazard assessment (HA) that is conducted concurrently with the Seamless Safety process development and implementation activities.

For the B61 and W69 HAs, the "final" or SS-21 process was compared with the HA performed on the initial or baseline process to provide an indication of overall risk reduction. A comparison of the risks associated with the initial dismantlement process and the SS-21 process provides a qualitative estimate of the effectiveness of the SS-21 process to achieve risk reduction.

## 2 Purpose and Scope of the Hazard Assessment

The scope of the HA is defined in DOE/AL Order 5610.11A and draft DOE-STD-XXXX-95, "Preparation Guide for the U.S. Department of Energy Hazard Analysis Reports for Nuclear Explosive Operations," and specifies that the HA must address all aspects of worker and public safety and environmental protection. The HA, which is based on traditional HA techniques,[1,2] must address all nuclear explosive operations and associated activities and must identify all hazards using a step-by-step review of the entire operation. Human reliability and human factors analyses are performed and are used to help determine accident-sequence likelihoods. For accident sequences resulting in high consequence (i.e., HE detonation, HE deflagration, nuclear detonation, and fire), a thorough and detailed analysis of accident sequences is performed. For these high-consequence accident sequences, sufficient analytic detail, including uncertainty analyses, is included in the HA. The HA also identifies and categorizes safety-significant or safety-class systems, structures, and components as well as identifies operational safety controls.

The HA provides the basis for the Hazard Analysis Report (HAR) and the Nuclear Explosive Hazard Assessment (NEHA). The NEHA focuses on the high-consequence accident sequences and is provided to the Nuclear Explosive Safety Study Group during their review and scrutiny of the proposed process. The HAR identifies the dominant accident sequences for a wide range of accident types. The identified accident sequences can be ranked in terms of importance and reviewed by design teams to identify opportunities for risk reduction. The HAR establishes the bases for administrative controls and for the identification of positive measures. The HA focuses risk-reduction efforts on the higher risk operations. The HAR, of which the NEHA is a subset, provides a thoroughly documented safety basis for the specific nuclear explosive operation that can be maintained as a "living document" to support future change control and risk-reduction measures.

## 3 Conduct of the B61 and W69 Hazard Assessments

A comprehensive discussion of the HA process used to conduct hazard assessments for nuclear explosive operations was provided in Refs. 3 and 4. During the preparation phase of the HA, an extensive data gathering effort was undertaken to evaluate weapon response in the disassembly accident environments and to obtain information on past operating incidents. Both the baseline and SS-21 process HAs were performed using a multi-step process involving observations of disassembly activities using a trainer, hands-on experimenting with the trainer, and extensive reviews of process videos. This accident-sequence identification phase was iterative in that interim HAs were conducted identifying high-risk candidate activities for which process redesign efforts were undertaken. During reviews of procedures and videos, accident sequences were developed by the HA Team using guide words, historical events, member experience and training, and "What If?" questions.

The developed accident scenarios and the scenario consequences were reviewed in detail by the HA team, which then developed an estimate of the likelihood or probability of the accident sequence as well as the conditional probability or likelihood of the consequence given that the accident sequence occurred. The overall likelihood of the accident occurring with the stated consequence was the product of the accident sequence and consequence likelihood probabilities. This product was used to determine the overall likelihood bin for the risk-ranking process as summarized in Table 1.

Likelihoods were assigned to the discrete events in the accident sequences through review of occurrence reports and data relating to equipment failure probabilities and human reliability. Accident consequences were estimated based on a review of historical and experimental data and models related to accident phenomena, such as evaluation of HE tests conducted at Pantex and at Los Alamos National Laboratory to determine the probability of detonation if an HE component were dropped. Table 2 summarizes the consequence severity categories used for the HA. Using likelihood and consequence categories determined from Tables 1 and 2, respectively, risk severity levels were assigned to each accident sequence in accordance with the

**Table 1. Probability or Frequency Categories**

| Category | Definition |
|---|---|
| I—Normal | Events that are planned or expected to occur with a frequency between once per weapon dismantled to once in every 1000 weapons dismantled. These events would be expected to occur one or more times per dismantlement campaign (nominally 1000 weapons). ($1 >$ Probability per dismantlement $\geq 10^{-3}$) |
| II—Anticipated | Incidents or events of moderate frequency that may occur once or more during the dismantlement campaign (nominally 1000 weapons). ($10^{-3} >$ Probability per dismantlement $\geq 10^{-5}$) |
| III—Unlikely | Incidents or events not expected, but that may occur during the dismantlement campaign (nominally 1000 weapons). ($10^{-5} >$ Probability per dismantlement $\geq 10^{-7}$) |
| IV—Very Unlikely | Incidents or events that are credible but not expected to occur during the dismantlement campaign (nominally 1000 weapons). ($10^{-7} >$ Probability per dismantlement $\geq 10^{-9}$) |
| V—Improbable | Exceedingly low probability incidents or events that cannot be judged incredible over the life of the facility but are extremely unlikely to occur during the dismantlement campaign (nominally 1000 weapons). ($10^{-9} >$ Probability per dismantlement) |

**Table 2. Consequence Severity Categories**

| Category | Definition (Bounding Consequences) | |
|---|---|---|
| | Worker | Facility |
| A<br>Catastrophic | **Loss of Life** as a result of chemical, physical (e.g., explosion), or nuclear-related hazard.<br>• Lethal chemical >> ERPG-3 | **Significant Facility Damage** or contamination resulting in loss of facility for future use. |
| B<br>High | **Severe Injury/Permanent Disability**<br>• Exceed lifetime occupational radiation limits<br>• Physical injury resulting in permanent disability<br>• Chemical exposure > ERPG-3 | **Moderate to Significant Facility Contamination and Damage**<br>• Repair and cleanup possible but quite expensive |
| C<br>Moderate | **Lost Time Accident but No Disability**<br>• Chemical exposure < ERPG-3<br>• Exceed annual/quarterly worker radiation dose limits<br>• OSHA reportable injury | **Facility Contamination Minor Facility Damage**<br>• Repair and cleanup possible at moderate expense |
| D<br>Low | **No Significant Impact: Minor or No Injury**<br>• Minor recordable injury<br>• Chemical exposure < ERPG-1 | **Minor or No Facility Contamination**<br>• Minor facility damage |

matrix in Table 3. These risk severity levels then were used to identify those postulated accidents that represented unacceptable risk and to rank recommendations for risk reduction.

# 4 Comparison of the Baseline and SS-21 Disassembly Process Risk for the B61 and W69

Table 4 presents a comparison of the Baseline and SS-21 disassembly processes for the B61 from the overall risk rank perspective. A reduction in the overall number of accident sequences for the SS-21 process was achieved primarily through tooling and procedure modifications. When viewed from the overall risk rank standpoint, the SS-21 process achieved a 67% reduction in risk rank 2 sequences. Risk rank 3 scenarios are about the same for both processes. The SS-21 process resulted in about a 50% reduction in risk rank 4 scenarios.

In Table 5, HE detonation (HED) event likelihoods are summarized for both the W69 and B61 Baseline and SS-21 processes. As presented in Table 5, the SS-21 process resulted in a significant reduction in the likelihood of HEDs when compared

**Table 3. Risk Matrix**

| Severity of Consequence | Likelihood of Postulated Accident | | | | |
|---|---|---|---|---|---|
| | I | II | III | IV | V |
| A | 1 | 1 | 2 | 3 | 3 |
| B | 1 | 2 | 3 | 3 | 4 |
| C | 2 | 3 | 3 | 4 | 4 |
| D | 3 | 4 | 4 | 4 | 4 |
| E | NH | NH | NH | NH | NH |

**Table 4. Overall Risk Rank Comparison—B61**

| Risk Rank | Baseline Process | SS-21 Process |
|---|---|---|
| 1 | 0 | 0 |
| 2 | 18 | 7 |
| 3 | 54 | 50 |
| 4 | 33 | 17 |
| NH | 2 | 1 |

**Table 5. Comparison of HED Event Likelihoods**

| Event Likelihood | B61 Baseline Process | B61 SS-21 Process | W69 Baseline Process | W69 SS-21 Process |
|---|---|---|---|---|
| I | 0 | 0 | 0 | 0 |
| II | 0 | 0 | 0 | 0 |
| III | 15 | 1 | 5 | 0 |
| IV | 12 | 12 | 32 | 15 |
| V | 14 | 26 | 14 | 23 |

with the Baseline process for both the W69 and B61. Similarly, Table 6 presents a comparison of worker injury and radiation dose events (excluding HEDs) for the Baseline and SS-21 processes for the W69 and B61. Again, both the B61 and W69 SS-21 processes achieved a significant reduction in worker risk.

# 5 Conclusions

In summary, from a qualitative perspective, both the SS-21 designed B61-0 and W69 disassembly processes resulted in a significant reduction of HED, worker injury, accidental worker radiation exposure, and facility contamination risk as compared with the Baseline process. This significant reduction in risk for both the B61 and W69 dismantlement programs is the result of major tooling and procedural enhancements made as part of the SS-21 Integrated Safety Process.

The qualitative results show that using the SS-21 design process can result in a significant reduction in risk for nuclear explosive operations. Note that there are inherent risks associated with work on nuclear explosives that cannot be eliminated entirely with any process design. However, the SS-21, or Integrated Safety Process, can be used to effect an overall reduction in risk and an overall improvement in safety for nuclear explosive activities.

**Table 6. Comparison of Industrial Accident/Worker Injury/Radiation Dose Events**

| Risk Rank | B61—Baseline Process | B61—SS-21 Process | W69—Baseline Process | W69—SS-21 |
|---|---|---|---|---|
| 1 | 0 | 0 | 0 | 0 |
| 2 | 3 | 3 | 0 | 0 |
| 3 | 21 | 10 | 30 | 4 |
| 4 | 31 | 17 | 53 | 39 |

## References


1. Center for Chemical Process Safety (CCPS), Guidelines for chemical process quantitative risk analysis, 1989.
2. Center for Chemical Process Safety (CCPS), Guidelines for hazard evaluation procedures, 2nd ed., 1992.
3. S. R. Fischer, H. Konkel, et. al., Use of hazard assessments to achieve risk reduction in the USDOE stockpile stewardship (SS-21) program. Los Alamos National Laboratory document LA-UR-95-1670, April 1995.
4. T. F. Bott and S. W. Eisenhawer, A hazard analysis of a nuclear explosives dismantlement. Los Alamos National Laboratory document LA-UR-95-1774, May 1995.

# PSA Application for the Safety Assessment of Nuclear Power Plants in Germany

H.P. Berg, R. Görtz, T. Schaefer, H. Schott
Federal Office for Radiation Protection (BfS)
Salzgitter, Germany

## 1 Introduction

Nuclear power plant licensees and regulators use a broad variety of methods and tools to ensure that facilities are designed, constructed, and operated at a high level of safety. Probabilistic methods have added important attributes that enhance and expand the safety assessment and safety management processes.

One attribute is a detailed, realistic model of the plant, systems, equipment, and human responses to a broad spectrum of possible initiating events. The model is a valuable platform for examination and understanding of complex plant-specific dependencies that arise from physical, functional, environmental, and human interactions that are often difficult to identify using other methods.

The main objective and benefit of PSA is to provide qualitative and quantitative information and insights into plant design, performance, and environmental impact, including a systematic and structured identification of the contributors to risk.

For the regulatory body which has primary responsibility to review and audit all aspects of design, construction, and operation to ensure that an acceptable level of safety is maintained throughout the plant life, PSA is an important element in this process because it facilitates consistent understanding and communication between the operator and regulator. The probabilistic framework and the PSA models provide a common basis for examination of safety issues, operational events, and regulatory concerns and for determining their plant-specific safety significance.

Therefore, the regulatory body in Germany is supporting the application of PSA in a twofold way. Firstly, guidelines are developed for the performance of PSA and its review. Secondly, a comprehensive investigation programme is conducted aimed at completing and improving PSA methods, data and analytical tools.

## 2 PSA as an Integrated Part of Periodic Safety Reviews

The most important application of PSA in Germany is presently the periodic safety review (PSR) of operating nuclear power plants, which has been recommended in 1988 by the German Reactor Safety Commission [1] amended by further recommendations on the performance of such a review [2]. It should be pointed out that in Germany the PSR is performed by the licensee on voluntary basis. There is no legal obligation for this task.

The goal of PSR is to describe the safety-related plant state as well as its mode of operation in a comprehensive and systematic manner. Moreover, the safety-related plant state as well as the mode of operation must be assessed with regard to a sufficient damage precaution.

This requires, in particular, a comprehensive identification to what extent the development of safety-related requirements and experiences with the safety-related measures in the respective NPP including operational experiences have been taken into account. The PSR results in the comparison of the effectiveness of available safety-related measures with the requirements according to the current regulations, identification of possible differences in that comparison as well as their classification according to safety-relevance, and - if necessary - determination of measures to improve the safety state as well as their urgency.

PSR mainly consists of the parts "analysis of the safety state of the plant and its operation" (deterministic analysis) and "probabilistic safety analysis" according to the state of the art of science and technology. For both types of analyses, guidelines are elaborated by BfS on the basis of discussion with the Federal Ministry, State Ministries, utilities and technical support organizations.

The PSA guideline deals with conducting a level 1+ PSA (i.e. level 1 plus active containment systems but without consideration of core degradation).

In principle, accident management measures should not be taken into account. However, the effects of accident management, as far as they are contained in the operating and emergency manuals, may be analyzed separately for evaluation of certain plant vulnerabilities.

Realistic (best estimate) boundary conditions are to be used for the analysis. In principle, plant-specific data should be used in the analyses to achieve optimal modelling of the plant. For those components whose independent failures contribute substantially to system non-availability a plant-specific assessment is required. The data characterising the failure behaviour have to be determined from operational records. For a transition time, however, a generic data base for typical components and media can be used.

Large fault trees and small event trees are recommended. Common cause failure methodology is discussed in chapter 4.

As to the event and fault tree analysis, lists of initiating events are given for all types of NPP in operation in the Federal Republic of Germany; current discussions are concentrated on the inclusion of fire events, external events and non-full power states.

The results from probabilistic safety assessments are to be taken into account for the identification and assessment of the safety relevance of insights gained in the deterministic analysis and, thus, supports the assessment of the plant's safety state.

The summation value of core damage frequencies serves for the supplementing judgement of the present plant safety level, whereas the comparison of the single frequencies serves for the judgement of the safety-related balancedness of the plant's safety concept. If event sequences show an increased single frequency, a low period of time between the initiating event and occurrence of damage or estimations show increased radiological consequences for the environment of the plant this is considered to be an indicator for the necessity of remedial measures. Extent and urgency of these measures can be derived from the analytical results of PSA.

From the regulators' point of view, PSA results support but not determine regulatory decision making, taking into account considerable uncertainties of PSA results reflecting the existing PSA limitations. Therefore, probabilistic safety limits are not emphasised.

On the other hand, the experts which review the qualitative and quantitative PSA results have to evaluate the plant's safety state taking into account existing national and international PSA results and evaluation schemes.

At the present state of discussions with different expert groups concerning the evaluation of the safety state of the plant under PSR, probabilistic safety figures might be used as additional indicators to support the decision regarding the necessity and priority of possible backfitting measures.

# 3 Regulatory Activities to Support PSA Application and Development

The PSA review guideline is still in the conceptual phase. It has to cover the objectives of PSA review, the management of the review process and related organizational questions. The definition of initiating events, event sequence analysis, PSA computer codes, data collection, uncertainty and sensitivity analyses, human reliability, common cause failures and documentation including quality assurance must be covered adequately in the review process.

Major effort within the investigation programme is concentrated, among others, on a practicable approach of PSA for external events and on the evaluation of the significance of low power and shut-down conditions for PSA results for German nuclear power plants.

Moreover, activities are focussed on the issues common cause failures and human actions in order to achieve a consent on the models, methods and data which shall be included as recommendations in the PSA guidance.

# 4 Common-cause Failure Analysis

In [3], the following kinds of dependent failures are distinguished:

- Failure to function of several redundant components or partial systems, occurring as the consequence of one single failure. These are so-called "causal failures".

- Failure to function of several redundant components or partial systems, resulting from functional interdependencies, i.e. from the system structure. There can, for example, exist functional dependences on one common support system, on one common drive mechanism or on human maloperations.

- Failure to function of several redundant components or partial systems of the same or similar design due to common, though undetected, causes.

The first two dependences are treated by the fault-tree analysis.

The third type of function failure from common cause is internationally known as "common-cause failure" (CCF). Failures of this kind are modelled separately and taken into account in the analysis using reliability data.

The results for system reliability and the integral safety level can easily be dominated by CCFs; highly reduncant systems, such as are generally the case in KWU plants, are particularly sensitive to this.

A large number of models based on theories of varying complexity for evaluating CCFs have been developed. In Germany, for example, the GRS applied a modified Binomial Failure Rate (BFR) model [3], and KWU the Stochastic Reliability Analysis (SRA) model [4].

The approach developed by GRS [3] is based upon the observation that non-lethal shocks can occur with significantly different values of the coupling parameter. The usual BFR-model, therefore, is modified by defining separate coupling parameters for each observed CCF-event for a given type of component and mode of failure. One key point in selecting the approach was not to underestimate the failure rates involving large numbers of components on highly redundant systems. The data base used consists of licensee event reports (data bank BEVOR at BfS), events from the international Incident Report System and further operational experiences.

The Stochastic Reliability Analysis (SRA) model describes dependent and independant failures in an integrated way [4]. CCF-events are modelled by two different states occurring with probabilities $a_0$ and $a_1$, where $a_0$ corresponds to lethal shocks. Independent failure occur with probabilitiy $1\text{-}a_1\text{-}a_0$. The distribution of the failure rate has three maxima, contrary to a homogenous population.

To reduce conservatism with respect to common cause failure analysis, it is recommended in [5] to perform a qualitative analysis of the components of interest with respect to the level of defence against common failure causes. The operational aspects and the technical characteristics of the individual components must be analyzed base on known common cause failure mechanisms that have been observed in the past in order to establish the individual component vulnerability. Non-applicable failure mechanisms can be ruled out altogether and the remaining failure candidates can be classified as shown in Table 1. When mechanisms leading to lethal cause effects to be considered justify the assumption of a decreasing failure rate with increasing group size, the application of the binomial failure rate model is supported.

Furthermore, the result of this analysis can be used to identify defensive measures against failure candidates.

| CCF mechanisms | |
|---|---|
| Lethal | Accelerated stochastic process |
| Test failure<br>Maintenance error<br>Calibration error<br>Design deficiency<br>Deviating operating data | failure processes such as<br>corrosion<br>erosion<br>vibration<br>deformation<br>growth of cracks<br><br>due to changes of lubricant,<br>changes if spare parts,<br>changes of media,<br>intrusion of dirt, etc. |

Table 1: Classification of causes for common failures [5]

# 5 Concluding Remarks

PSA in the regulatory process is used supplementary to the deterministic approach in licensing, supervision of operating plants and in the assessment of the design of future reactors. In the past, PSA results have partially supported decisions on backfitting. This holds especially for the German Risk Studies for PWR and BWR.

A major driving force in this process in the periodic safety review comprising PSAs for all operating NPPs in Germany to be performed in time intervalls of ten years. It is intended to have level 1+ PSAs for all NPPs completed until the end of this decade. At present, PSAs have been elaborated for more than half of the operating NPPs.

The regulatory body in Germany is supporting the application of PSA in different ways. Guidelines are developed for the performance of PSA and its review. A PSA guideline has been drafted and substantial experience has been gained with its application leading to revision and amendments to the draft. A PSA review guideline is still in the conceptual phase. A concept for systematic precursor evaluation has been developed and is under discussion with experts; a case by case analysis of precursors is already performed. Moreover, a comprehensive investigation programme is still ongoing aimed at completing and improving PSA methods, data and analytical tools, in particular concerning common cause failures, human factors, external events as well as operational states others than full power.

# References


[1] Reaktor-Sicherheitskommission:
Abschlußbericht über die Ergebnisse der Sicherheitsüberprüfung der Kernkraftwerke in der Bundesrepublik Deutschland durch die RSK.
Bundesanzeiger Nr. 47a vom 08. März 1989

[2] Reaktor-Sicherheitskommission:
Periodische Sicherheitsüberprüfung (PSÜ) für deutsche Kernkraftwerke. Durchführung der PSÜ.
Bundesanzeiger Nr. 158 v. 28. August 1995

[3] Kersting, E.; von Linden, J.; Müller-Ecker, D.; Werner, W.:
Safety Analysis for Boiling Water Reacotrs, A Summary.
GRS-98, Köln, July 1993

[4] Dörre, P.:
Dependant failure - a multiple-state stochastic description. Reliability Engineering and System Safety 1992, No. 35

[5] Breiling, G.; Oehmgen, T.:
Determination of input data for probabilistic safety Assessment. Kerntechnik 1995; 60 No. 2-3: 105-109

# The Role of PSA to Improve Safety of Nuclear Power Plants in Eastern Europe and in Countries of the Former Soviet Union

L. Lederman, J. Höhn
International Atomic Energy Agency
Vienna, Austria

## 1. Introduction

There are currently 59 nuclear power plants (NPPs) of WWER and RBMK types in operation and some 15 under construction in countries of Eastern Europe and the former Soviet Union. The safety of these plants, particularly the WWERs and RBMKs of the first generation, has been a matter of concern.

In the past several years national programmes for safety re-assessment have been strengthened in the countries constructing and operating these plants and several bilateral and multilateral assistance projects have been initiated to improve their safety. These safety assessments were mostly based on deterministic reviews and engineering judgement. The results obtained have formed the basis for the programmes of safety upgradings (PSUs) and the priority actions established in the countries operating WWER and RBMK NPPs.

More recently, probabilistic safety assessments (PSAs) have been initiated for these NPPs. To date, most WWERs and some of the RBMKs have PSA studies at various levels of completion.

The IAEA has been providing assistance for the review of the completeness and adequacy of PSUs and for the peer review of PSA studies. Some 250 generic safety issues have been identified by the IAEA for the WWER NPPs. These issues have been further ranked in four groups (categories I-IV), based on their perceived impact on the plants' defence in depth [1,2,3].

The scope and the technical quality of PSAs reviewed by the IAEA [4] vary considerably. The IAEA is also reviewing the impact of level 1 PSA results on the PSUs of WWER NPPs [5]. Preliminary insights from this review are discussed next.

## 2. WWER-440/230 NPPs

PSA results and PSU measures have been reviewed for Bohunice Unit 1 and Novovoronezh Units 3,4 NPPs. Substantial differences were noted in the scope and results of the PSAs for these two plants, and hence the applicabilitiy of the methodology.

**Bohunice Unit 1**

The Bohunice PSA was performed by Electrowatt UK in the frame of the PHARE Programme. It includes internal initiating events including internal fires and floods. Seismic events were not modelled. A basic model was developed based on the status of Unit 1 in July 1992 and subsequently changed to reflect the plant modifications introduced during the period of the "Small Reconstruction" until 1993. As a result of implementation of these modifications, the core damage frequency per year was reduced from 1.7 E-3 to 8.9 E-4.

**Novovoronezh Units 3,4**

A PSA for Novovoronezh 3,4 was initiated based on a request by Rosenergoatom to support the NPP modernization programme. The work was managed by Atomenergoproject, Moscow. A set of preliminary results was obtained for some initiating events. The reported core damage frequency per year was 1.8 E-3. In 1990 a Special Operation Regime was approved by the regulatory body in Russia concerning Novovoronezh 3,4 NPP. The concept of safety upgrading for Novovoronezh 3,4 was developed in 1992 based entirely on results of deterministic analysis.

A new PSA study is under way in the frame of the TACIS programme but the results were not available in 1995.

**Insights from PSAs**

There are significant differences in the insights between the Bohunice and Novovoronezh PSAs. For Bohunice the dominant contributions to core damage are from sequences involving the trip of the 2 turbine generators with subsequent failure to shutdown the reactor; pressurizer LOCA and LOCAs (7-32 mm) with subsequent failure of the high pressure injection system. For Novovoronezh the dominant contributors to core damage arise from loss of long-term heat removal through the secondary side with the failure of technological condensers; loss of offsite power; primary to secondary leaks and LOCAs (32-100 mm).

The differences in results stem mostly from plant specific design differences and modelling assumptions. Relevant design differences include the number of technological condensers, the water inventory and redundancy in the emergency feedwater system and the number of diesel generators. Frequencies and grouping of some initiating events are also substantially different. Generic values used for LOCA frequencies vary by one order of magnitude and some important transient initiators (e.g. steam line breaks) also have significant differences.

Some PSU measures have shown to have a negligible contribution to the total frequency of core damage. At Bohunice these include: installation of a new fully redundant essential service water system, installation of two additional super emergency feedwater pumps, replacement of the old neutron flux measurement system and installation of a new steam dump station to the atmosphere on each main steam line. For Novovoronezh these include the replacement of auxiliary feedwater pumps.

It was also noted that a large number of PSU measures are not modelled in the PSA. Some measures are not modelled because of the limitations in the scope of the PSAs while others are not possible to include directly in PSA modelling. The first group includes: new ventilation system in electrical cabinets, upgrades for seismic hazards, internal floods and fires (Novovoronezh) and instrumentation and control (I&C) improvements. The latter group includes measures related to reactor pressure vessel integrity (embrittlement), quality assurance and enhancing safety culture.

# 3. WWER-440/213 NPPs

PSA results and PSU measures have been reviewed for Bohunice Unit 3, Dukovany Unit 1 and Paks Unit 3 NPPs.

**Bohunice Unit 3**
The Bohunice 3 PSA was performed by VUJE Inc. and RELKO Ltd. and sponsored by the Slovak Power Enterprise. The scope of the study includes internal initiating events including internal floods and fires.

An analysis of human actions was performed as part of the PSA study. There were several human actions considered in the scope of this study for which there are no operating procedures and only limited operator training. These human actions include: feed-and-bleed, draining of the bubbler tower into the confinement sump, and rapid primary depressurization using relief valves in the secondary circuit.

**Dukovany Unit 1**
A level 1 PSA was performed for Dukovany NPP by the Nuclear Power Research Institute (UJV) Řež and financed by the Dukovany NPP. It includes internal initiating events and floods. The analysis of internal fires is due to be completed in 1995. The core damage frequency reported was 1.8 E-4 per year.

Based on insights derived from earlier application of PSA to Dukovany in 1992, new operating procedures were developed for several important operator actions that had previously not been proceduralized. These actions include: feed-and-bleed, draining of the bubbler tower into the confinement sump, and rapid primary depressurization using relief valves in the secondary circuit.

**Paks Unit 3**
A level 1 PSA for Paks 3 was performed by VEIKI and Paks NPP and sponsored by

the latter in the frame of the AGNES (Advanced General and New Evaluation of Safety) project. The calculated CDF of the current plant state was 5.0 E-4 per year.

A detailed analysis of human actions was performed as part of the PSA study. Feed-and-bleed cooling was not considered in the base case. This resulted in a very high contribution of certain sequences to CDF when compared to the PSA studies for other WWER-440/213 plants which considered feed-and-bleed.

The CDF is dominated by sequences involving loss or reduction in feedwater (FW) flow and loss of steam flow. One sequence which contributes nearly 50% of the CDF is the rupture of the FW collector with consequential failure of the main, auxiliary and emergency FW systems. Other important sequences involve rupture of FW line, steam line or steam collector with consequential failures of FW systems.

**Insights from the PSAs**

Differences in PSA results are mainly from modelling assumptions, frequencies of initiating events and the scope of the human reliability assessment for each PSA.

There are orders of magnitude differences in the frequency and differences in grouping of some initiating events between the three plants (e.g. secondary pipe breaks, very small LOCAs, pressurizer LOCA and loss of offsite power). Modelling assumptions which most significantly impact the PSA results include: not considering feed-and-bleed cooling in Paks, no credit for recovering the external electric power grid within four hours in Bohunice and the fire in the turbine hall with consequential loss of secondary cooling and operator failure to initiate feed-and-bleed, which is included only in the Bohunice PSA. The accident sequence involving small LOCAs (< 40mm) with operator failure to realign the low pressure emergency core cooling system (ECCS) pump suction to prevent overflow of low pressure ECCS tanks and subsequent core damage was a major contributor to the total core damage frequency at Bohunice and Dukovany and was not modelled in Paks.

The loss of reactor coolant pump (RCP) seal cooling has a significant effect on the PSA results. At Dukovany, loss of cooling for two hours has been assumed to result in seal failure and a non-recoverable loss of primary inventory to the RCP motor room. At Paks and Bohunice the loss of RCP seal cooling was not considered. The design of the RCP seals is virtually the same in these plants.

For Dukovany, 4 groups of PSU measures have been defined. Group 1 measures are those required to decrease the CDF to less than 1.0E-4. All issues ranked in the IAEA's category III and IV (defined in ref. 1) are included in this group.

Other important PSU measures not modelled in the PSA are related to the means used to control the reactor pressure vessel embrittlement, improvements in the control room, fire safety, seismic qualification of the decay heat removal system and equipment in the secondary circuit, water level measurement in the steam generator rooms and hydrogen removal.

Paks PSA results have been used to rank PSU measures into three groups. Therefore the relocation of the superemergency feedwater system was ranked as a measure in the extremely high priority group due to its extremely high contribution to the total core damage frequency.

A real-time risk monitoring system is being developed for Bohunice by SAIC in cooperation with RELKO. Similar work was completed for Dukovany by SAIC and UJV Řež and is being used since January 1995.

# 4. WWER-1000/320 NPPS

PSA results and PSU measures have been reviewed for Balakovo WWER-1000/320 Unit 4, Kozloduy WWER-1000/320 Units 5,6 and Temelin 1 NPP. Major differences were also noted in the scope of PSAs for these three plants.

**Balakovo Unit 4**

A level 1 PSA study for Balakovo Unit 4 was started based on a request by the Russian Ministry for Atomic Energy. It was managed by AEP Moscow and performed in co-operation with OKB Gidropress and the RRC Kurchatov Institute. The PSA includes internal initiating events. The estimated core damage frequency per year was 1.9 E-5. A level 3 PSA, sponsored as part of the EC TACIS programme, should be completed in 1996.

For Balakovo more than 75% of the total core damage frequency results from 2 sequences initiated by loss of offsite power (LOOP). The first involves LOOP followed by diesel generators' failure and in the second LOOP is followed by depressurization system failure with additional water supply failure to emergency feedwater tanks.

**Kozloduy Units 5,6**

The Kozloduy level 1 PSA was performed by Risk Engineering Ltd, Bulgaria. It includes internal events, internal fire and seismic initiators. The IAEA Generic Reliability Database and the computer code PSAPACK Vers.4.0 were used in the study. The core damage frequency reported was 3.7 E-4 per year. An IAEA IPERS mission was carried out at the end of 1994.

The main contribution to core damage at Kozloduy stems also from an accident sequence initiated by loss of offsite power and followed by failure of the emergency feedwater system. A single steam generator (SG) tube rupture with subsequent failure of steam dump to the atmosphere and failure of the low pressure injection system is the second largest contributor to core damage.

**Temelin 1 NPP**

A level 2 PSA for the Temelin NPP was completed under the overall direction of the NUS Corporation. It was sponsored by Řež/Temelin NPP. The PSA includes internal and external events (fire, floods, seismic events, others) and a living PSA as safety monitor. Due to the mixture of eastern and western equipment suppliers, several sources were used for the assessment of reliability data. The level 1 PSA which also includes the shutdown mode resulted in a core damage frequency of 7.6 E-5 per year for internal initiating events only.

For Temelin the dominant contribution to core damage is a SG collector cover failure followed by operator failure to depressurize and cool down before the containment sump is depleted. SG failure caused either by collector cover failure or tube rupture lead to accident sequences which contribute nearly 80% to the total core damage frequency.

**Insights from PSAs**
The PSA results provide insights into plant safety which differ considerably from plant to plant.

A comparison of accident sequence results reveals significant differences in the modeling assumptions and initiating event frequencies used at the various plants. Two of the top four sequences at Kozloduy, for example, do not result in core damage at Temelin. In addition, a large number of dominant accident sequences at Temelin (including several LOCA and SGTR sequences) were not modelled at Balakovo and Kozloduy. This indicates that there were also major differences in success criteria assumptions and in the modelling of human actions.

The reason why SGTR and SG collector rupture initiators have a much higher contribution to core damage frequency at Temelin is due to a frequency of SG collector cover failure that is two orders of magnitude larger than that used at Kozloduy. Two additional diesel generators to be installed at Temelin and the different initiating event frequencies estimated for Balakovo and Kozloduy are the main reasons for the differences in the contributions of accident sequences initiated by loss of offsite power among the three plants.

PSU measures not modelled or assessed as "not important" to safety by the PSA for Temelin include: reactor vessel head leak monitoring system and reactor pressure vessel embrittlement and monitoring, and the capacity of the emergency batteries.

# 5. Conclusions

In general, PSA results are being used both to define new programmes of safety upgradings and to complement existing ones. However, the differences in the scope and technical quality of PSAs performed to date limit a wider sharing of insights among plants, even those of the same type. PSU measures and priorities defined on the basis of some PSA insights need careful consideration before they can be adopted.

Major differences can be generally related to modelling assumptions, scope of human reliability analysis, data (particularly grouping and generic frequencies of initiating events which in some cases differ up to two orders of magnitude) and design differences. It is also noted that there are some quite conservative assumptions in some PSAs, and PSA results are often dominated by these assumptions. Furthermore, failure to model operator actions in the WWER PSAs have a severe impact on the risk profiles. Therefore the priority of PSU measures defined on the basis of these PSA insights need careful consideration.

PSA results have also identified several PSU measures with a negligible impact to reduce the frequency of core damage. Another group of PSU measures is not modelled in PSAs either due to the limited scope of some PSAs or because they cannot be qualitatively considered in PSAs. Priorities of those PSU measures can only be assessed based on deterministic considerations and engineering judgement.

The results of this review indicate that there is a need to harmonize models and data used in PSAs of WWERs before their insights can be used to enhance PSU measures. The IAEA will continue to assist in the peer reviews of PSAs and to serve as a forum for exchange of information between PSA practitioners, regulators and NPP operators.

## References

1. INTERNATIONAL ATOMIC ENERGY AGENCY. Ranking of Safety Issues for WWER-440 Model 230 Nuclear Power Plants. IAEA-TECDOC-640, Vienna, February 1992

2. INTERNATIONAL ATOMIC ENERGY AGENCY. Safety Issues and their Ranking for WWER-440 Model 213 Nuclear Power Plants. IAEA-EBP-WWER-03 (in publication)

3. INTERNATIONAL ATOMIC ENERGY AGENCY. Safety Issues and their Ranking for WWER-1000 Model 320 Nuclear Power Plants. IAEA-EBP-WWER-05 (in publication)

4. Gubler R. Probabilistic Safety Assessment for the Nuclear Power Plants of the WWER Reactor Line, Proceedings of the International Conference on Probabilistic Safety Assessment Methodology and Applications, PSA'95, November 26-30, 1995, Seoul, Korea

5. INTERNATIONAL ATOMIC ENERGY AGENCY. Impact of PSA Studies on the Programmes of Safety Upgradings of WWER NPPs, WWER, WWER-SC-152, Vienna, October 1995

**Note:** The authors want to thank W. Puglia from SAIC for his contribution in the preparation of ref. [5].

# The Joint EC/USNRC project on Uncertainty Analysis of Probabilistic Accident Consequence Codes: Overall Objectives and New Developments in the Use of Expert Judgement

L.H.J. Goossens[a], F.T. Harper[b], G.N. Kelly[c], C.H. Lui[d]

[a]Delft University of Technology, Delft, the Netherlands

[b]Sandia National Laboratories / Albuquerque, NM, USA

[c]European Commission / DG-F-6, Brussels, Belgium

[d]United States Nuclear Regulatory Commission, Washington, DC, USA

**Abstract**

This paper describes the results of an ongoing EC/USNRC joint project on the uncertainty analysis of probabilistic accident consequence codes for nuclear applications. The uncertainty distributions are obtained using a formal expert elicitation procedure based on methods from Delft University of Technology and NUREG-1150 experience in the US.

## 1. Introduction

The development of two new probabilistic accident consequence codes - MACCS [1] in the US and COSYMA [2] in the European Union (EU) - was completed in 1990, and both codes have been distributed to a large number of potential users. These codes have been developed primarily, but not solely, to enable estimates to be made of the risks presented by nuclear installations, based on the postulated frequencies and magnitudes of potential accidents.

Fairly comprehensive assessments of the uncertainties in the estimates of the radiological consequences of postulated accidental releases of radioactive material have already been made, both in the US and EU, using predecessors of both codes. Fundamental to these assessments were estimates of uncertainty (or more explicitly, probability distributions of values) for each of the more important model parameters. In each case these estimates were largely done by those who developed the accident consequence codes, as opposed to experts in each of the many different scientific disciplines featured within an accident consequence code.

## 2. Formal Use of Expert Judgement

The formal use of expert judgement has the potential to circumvent this. Although the use of expert judgement is common in the resolution of complex problems, it is most often used informally and rarely made explicit. The use of a formal expert judgement process has the considerable benefits of an improved expression of uncertainty, greater clarity and consistency of judgements, and an analysis that is more open to scrutiny.

In 1991, both the European Commission (EC) and the United States Nuclear Regulatory Commission (USNRC) were giving consideration to initiating independent studies to better quantify and obtain more valid estimates of the uncertainties associated with the predictions of accident consequence codes.

## 3. Joint Project Objectives

The broad objectives of both the USNRC and EC in undertaking the joint USNRC/EC Consequence Code Uncertainty Study are:

1. to formulate a generic, state-of-the-art methodology for uncertainty estimation which is capable of finding broad acceptance;
2. to apply the methodology to estimate uncertainties associated with the predictions of probabilistic accident consequence codes (COSYMA and MACCS) designed for assessing the consequences of commercial nuclear power plant accidents;
3. to contribute to the regulatory process, for instance, to decision making on risk acceptability
4. to provide an input to the identification of research priorities.

Within these broad objectives, small differences in emphasis exist between the two organizations about how the results of the project may subsequently be used. The emphasis within the EC is primarily on methodological development and its generic application, whereas the USNRC, while sharing this interest, is also interested in the potential use of the methods and results as an input to the regulatory process. In particular, this would complement the work completed within the USNRC-sponsored NUREG-1150 study, where the detailed analysis of the uncertainty in risk estimates was confined to consideration of the contributions arising from uncertainties in the probability, magnitude, and composition of potential accidental releases.

The ultimate objective of the USNRC/EC joint effort is to systematically develop credible and traceable uncertainty distributions for the respective code input variables using a formal expert judgement elicitation process. Each organization will then propagate and quantify the uncertainty in the predictions produced by their respective codes.

# 4. Accident Consequence Codes Expert Panels

These ACA codes consist of several modules, with numerous input parameters. A relatively small number of parameters have been selected as key input paramters on the basis of importance to final outcome and perceived contribution to uncertainty. These key input parameters were selected based on previous sensitivity analyses of the ACA codes or modules. The important phenomenological areas for which the key parameter uncertainties are assessed with expert judgement are:

* atmospheric dispersion and deposition
* behaviour of deposited materials and external exposure of people
* transfer through food chains
* the kinetics and metabolism of radioactive material taken into the body (internal dosimetry)
* early or deterministic health effects from radiation exposure
* late or somatic health effects from radiation exposure.

The current joint project is undertaken with the aim of quantifying the uncertainty on key parameters in these areas. For instance, in the area of atmospheric dispersion and deposition uncertainty distributions were estimated for the following parameters:

1. a library of uncertainty distributions, assessed by the experts, over **quantities of interest to the model** that are potentially observable in nature (elicitation variables) such as concentrations in the plume, standard deviation of lateral plume spread, dry deposition velocities and fractions washed out by rain, and
2. a library of uncertainty distributions for the parameters in the models, processed from the experts' assessments of the elicitation variables, such as the power law coefficients of lateral and vertical plume spread and the wash out coefficient parameters, which are the parameters used in both accident consequence codes.

The expert judgement methodology applied in this project is a combination of methods from previous US and EC studies: the NUREG-1150 method [3,4] and methods developed in Europe at Delft University of Technology [5,6]. The formal use of expert judgement elicitation and evaluation follows a procedure set out in a protocol [7].

In all panels experts were carefully selected based on the following nomination criteria: reputation and experience in the field of interest; number and quality of publications; familiarity with uncertainty concepts; diversity in background; awards; balance of views; interest in the project, and availability for the project.

In the joint project a programmatic decision was made by USNRC to assign all experts equal weight, i.e. all experts on each respective panel were treated as equally credible. It was also decided to explore the use of performance based weighting schemes (see [7]).

## 5. Atmospheric dispersion and deposition panels

The results of the Atmospheric disperion and deposition panels carried out in 1992-1994 are reported in [8,9].

For the elicitations in total 16 experts were selected, for the dispersion panel and the deposition panel (8 experts each) and particpated in the elicitations.

The aggregated elicited uncertainty distributions on dispersion and deposition represent state-of-the-art knowledge in these areas. These elicitation distributions concern physically measurable quantities, conditional on the case structures provided to the experts for non-complex terrains. They were developed by the experts based on a variety of information sources. For dry deposition the distributions capture uncertainty in dry deposition velocities of particles of different sizes, iodine and methyl iodide over different surfaces. For wet deposition the distributions capture the uncertainty on the fraction of particles, iodine and methyl iodide removed by rain. The dispersion distributions capture uncertainty in plume centerline concentrations, plume spread, and off-centerline concentration ratios at several downwind distances for four weather conditions. The distributions were all processed into distributions for the corresponding code input parameters: dry deposition velocities, washout coefficient constants and dispersion power law coefficients.

## 6. Food chain Panels

The results of the two food chain panels carried out in 1994-1995 are published in [10]. 16 experts were selected for the two panels. The diversity of interest was large and it was recognised that not all experts had the same broad view on the whole field.

The experts on the food chain panels provided uncertainty assessments in the following areas:

- soil migration and root uptake into plants
- interception, retention and translocation in plants
- resuspension onto plants
- transfer from feedstuffs to milk and meat
- retention in the gut of animals
- biological half-life in meat of animals

- animal diets.

The quantities of interest to the FARMLAND model [11] (used for COSYMA) and the COMIDA model [12] (used for MACCS) will also be useful for many applications in the evaluation of food chain transfer. The parameter distributions are used for uncertainty analyses of the FARMLAND and COMIDA models.

# 7. External Exposure Panel

The results of the deposited materials and related doses uncertainty assessments carried out in 1994-1995 are published in [13]. The experts provided assessments for the external doses to individuals, (both indoors and outdoors), for activity deposited on the ground and other surfaces in urban and rural areas. The experts also provided assessments on population behaviour.

# 8. Health Effects Panels and Internal Dosimetry Panel

The remaining three panels on Early or deterministic health effects, Internal dosimetry and Late or somatic health effects have commenced in 1995 and will be published late 1996 [14,15,16]. 31 experts in total were selected for these panels.

The experts were asked to estimate the probability of occurrence of several deterministic health effects, given patterns of exposure that might be encountered following a nuclear accident. Consideration was limited to deterministic effects following exposure of the bone marrow, the gastrointestinal tract, the lung and the skin. These results will be processed for use by the MACCS and COSYMA models. The model for deterministic health effects implemented in MACCS and COSYMA uses a Weibull function and a threshold to calculate the probability of a deterministic health effect.

The somatic health effects panel was limited to an assessment of cancer. The aim is to elicit uncertainties in the risks of various types of radiation induced cancer, taking account extrapolation to low doses and low dose rates, as well as the projection of risks over a lifetime and across populations with different baseline cancer rates.

The internal dosimetry elicitation questions address uncertainties in the biokinetic parameters used in the models for inhalation and ingestion. For inhalation, the parameters addressed include particle deposition in different regions of the respiratory tract, movement of particles out of the respiratory tract to the gut and dissolution and transfer to blood. For ingestion, movement between regions and absorption to blood are considered.

# 9. New Developments

Uncertainties in COSYMA and MACCS will be estimated separately by the respective EU and US teams. In each case the overall uncertainties in the predictions of the codes will be estimated together with those in individual modules. These separate assessments will however be made using an agreed methodology developed jointly by the two teams. Account will be taken of uncertainties in all code parameters; more 'convential' approaches will be used for those quantities not subjected to formal expert judgement.

## References


1. Chanin D.I. et al. MELCOR Accident Consequence Code System (MACCS): User's Guide. Report NUREG/CR-4691, Albuquerque/NM/USA, 1990
2. Kelly G.N. et al. COSYMA: A new programme package for accident consequence assessment. Report EUR 13028, Luxembourg, 1991
3. USNRC. Reactor risk reference document. Report NUREG-1150, 1987
4. Hora S., Iman R. Expert opinion in risk analysis: the NUREG-1150 methodology. Nucl Sci and Engng 1989; 102:323-331
5. Goossens L.H.J., Cooke R.M., van Steen J.F.J. Expert opinion in safety studies. Report to Dutch Ministry of Housing, Physical Planning and Environment. TUDelft/TNO Apeldoorn, 1989
6. Cooke R.M. Experts in uncertainty. Oxford University Press, Oxford, 1991
7. Goossens L.H.J., Cooke R.M. Procedures Guide for the Use of Expert Judgement in Uncertainty Analyses. PSAM III Conference, Crete, June 1996
8. Harper F.T., Goossens L.H.J., Cooke R.M., et al. Probabilistic accident consequence uncertainty analysis: Dispersion and deposition uncertainty assessment. Report NUREG/CR-6244, EUR 15855, Washington,DC/USA, and Brussels-Luxembourg, 1995
9. Cooke R.M., Goossens L.H.J., Kraan B.C.P. Methods for CEC\USNRC accident consequence uncertainty analysis of dispersion and deposition - Performance based aggregating of expert judgements and PARFUM method for capturing modelling uncertainty. Report EUR 15856, 1995
10. Goossens L.H.J., Harper F.T., Brown J. et al. Probabilistic accident consequence uncertainty analysis: Food chain uncertainty assessment, Report for EC and USNRC, to be published 1996
11. Brown J., Simmonds J.R. FARMLAND: A dynamic model for the transfer of radionuclides through terrestrial foodchains, Report NRPB-R273, HMSO, London, 1995

12. Abbott M.L., Rood A.S. COMIDA: A radionuclide food chain model for acute fallout deposition, Report EGG-GEO-10367, EG&G, Idaho, USA, 1993
13. Goossens L.H.J., Harper F.T., Boardman J. et al. Probabilistic accident consequence uncertainty analysis: External exposure from deposited material uncertainty assessment, Report for EC and USNRC, to be published 1996
14. Harper F.T., Goossens L.H.J., Haskin F.E. et al. Probabilistic accident consequence uncertainty analysis: Early (deterministic) health effects uncertainty assessment, Report for EC and USNRC, to be published 1996
15. Goossens L.H.J., Harper F.T., Little M. et al. Probabilistic accident consequence uncertainty analysis: Late (somatic) health effects uncertainty assessment, Report for EC and USNRC, to be published 1996
16. Goossens L.H.J., Harper F.T., Harrison J.D. et al. Probabilistic accident consequence uncertainty analysis: Internal dosimetry uncertainty assessment, Report for EC and USNRC, to be published 1996

# Dealing with Dependencies in Uncertainty Analysis

Roger Cooke Bernd Kraan
Delft University of Technology*

**Keywords**: dependence, rank correlation, minimum information, uncertainty analysis, expert judgement, cobweb plots.

## 1 Introduction

It has long been known that significant errors in uncertainty analysis can be caused by ignoring dependencies between uncertainties [1]. New techniques for estimating and analyzing dependencies in uncertainty analysis have been developed in the course of the joint CEC/USNRC accident consequence uncertainty analysis and in a recent analysis of uncertainty in failure probabilities for underground gas pipelines. These are described here. For the mathematics of dependence modeling see [2] and the references therein. Post-processing [4] also represents a way of assessing dependencies, but this is not discussed here. We discuss how to elicit dependencies from experts, how to combine them, and how to analyse dependencies in the output of an uncertainty analysis.

## 2 Lumpy and smooth elicitation strategies

The best source of information about dependencies is often the experts themselves. The most thorough approach would be to elicit directly the experts' joint distributions. The practical drawbacks to this approach have forced analysts to look for other dependence elicitation strategies. One obvious strategy is to ask experts directly to assess a (rank) correlation coefficient. Even trained statisticians have difficulty with this type of assessment task [5]. Two approaches have been found to work satisfactorily in practice. The choice between these depends on whether the dependence is *lumpy* or *smooth*. Consider uncertain quantities $X$ and $Y$. If $Y$ has the effect of switching various processes on or off which influence $X$, then the dependence of $X$ on $Y$ is called *lumpy*. In this case the best strategy is to elicit conditional distributions for $X$ given the switching values of $Y$, and to elicit the probabilities for $Y$. This might arise,

*Delft University of Technology, Faculty of Applied Mathematics and Informatics, Department of Statistics, Stochastics and Operations Research, Mekelweg 4, 2628 CD Delft, The Netherlands.

for example, if corrosion rates for underground pipes are known to depend on soil type, where the soil type itself is uncertain. In other cases the dependence may be *smooth*. For example, uncertainties in the biological half lifes of cesium in dairy cows and beef cattle are likely to be smoothly dependent.

Within the joint CEC/USNRC study a new strategy has been employed for eliciting smooth dependencies from experts. When the analyst suspects a potential smooth dependence between (continuous) variables $X$ and $Y$, experts first assess their marginal distributions for $X$ and $Y$. They are then asked:

> *Suppose $Y$ were observed in a given case and its value were found to lie above the median value for $Y$; what is your probability that, in this same case, $X$ would also lie above its median value?*

Experts quickly became comfortable with this assessment technique and provided answers which were meaningful to them and to the project staff. If $F_X$ and $F_Y$ are the (continuous invertible) cdf's of $X$ and $Y$ respectively, the experts thus assess $\pi_{\frac{1}{2}}(X,Y) = P(F_X(X) > \frac{1}{2} \mid F_Y(Y) > \frac{1}{2})$.

Consider all joint distributions for $(X,Y)$ having marginals $F_X$, $F_Y$, having minimum information relative to the distribution with independent marginals $F_X$, $F_Y$, and having rank correlation $\tau(X,Y)$, $\tau \in [-1,1]$. For each $\tau \in [-1,1]$, there is a unique value for $\pi_{\frac{1}{2}}(X,Y) \in [0,1]$. Hence, we may consider the $\tau(X,Y)$ characterizing the minimal information distribution as a function of $\pi_{\frac{1}{2}}(X,Y)$. More generally, we consider for each $r_1 \in (0,1)$ and $r_2 \in (0,1)$, $\tau$ as a function of $\pi_{r_1,r_2}$, where

$$\pi_{r_1,r_2}(X,Y) = P(F_X(X) > r_1 \mid F_Y(Y) > r_2).$$

To illustrate the procedure we choose $r_1 = r_2$; it is not difficult to generalize the procedure when $r_1 \neq r_2$. The functions $\tau(\pi_r)$ have been computed with the uncertainty analysis package UNICORN [2]. The results for $r = 0.05, \ldots, 0.95$ are shown in Figure 1.

When there is a single expert having assessed $\pi_{\frac{1}{2}}(X,Y)$, then we simply use the minimum information joint distribution with rank correlation $\tau(\pi_{\frac{1}{2}})$ found from Figure 1. When several experts are combined via linear pooling, a complication arises. Since the medians for $X$ and $Y$ will not be the same for all experts, the conditional probabilities $\pi_{\frac{1}{2}}(X,Y)$ cannot be combined via the linear pooling. However, the marginal distributions can be pooled, resulting in cdf's $F_{X,DM}$ and $F_{Y,DM}$ for $X$ and $Y$ for the Decision Maker (DM). Let $x_{DM,50}$ and $y_{DM,50}$ denote the medians for DM's distribution for $X$ and $Y$. With each expert we associate a minimum information joint distribution; for each such distribution we can compute the conditional probabilities $P(X > x_{DM,50} \mid Y > y_{DM,50})$. Since these conditional probabilities are defined over the same events for all experts, they can be combined via the linear pool. This yields a value for $\pi_{\frac{1}{2}}$ for DM, for which we can find the corresponding $\tau$.

A simple graphical method for doing this can be given if for each experts the numbers $F_X(x_{DM,50})$ and $F_Y(y_{DM,50})$ are not too different. In this case we associate with each expert a number $r = (F_X(x_{DM,50}) + F_Y(y_{DM,50}))/2$.

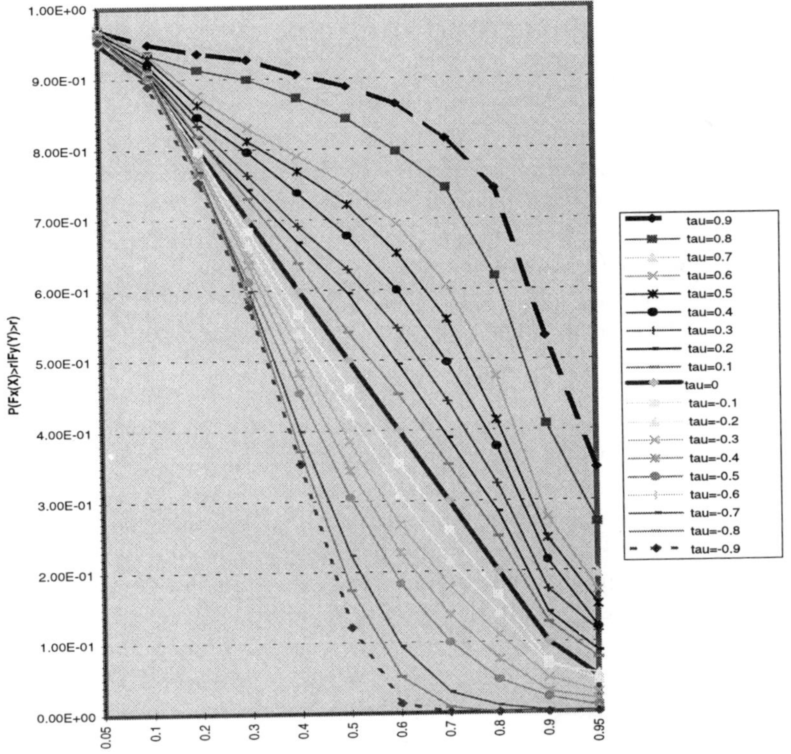


Figure 1: $\pi_r$ *for minimal information distributions with rank correlation* $\tau$

We then use this value of $r$ together with the value of $\tau$, determined from the experts response $\pi_{\frac{1}{2}}$, to read the value of $P(X > x_{DM,50} \mid Y > y_{DM,50})$ from Figure 1. The steps are summarized as follows.

1. For each expert query $\pi_{\frac{1}{2}}(X,Y) = P(F_X(X) > \frac{1}{2} \mid F_Y(Y) > \frac{1}{2})$
2. For each expert $e$, find $\tau_e$ which passes through $(\frac{1}{2}, \pi_{\frac{1}{2}})$ in Figure 1.
3. Take linear pooling of experts' marginals, find medians $x_{DM,50}$, $y_{DM,50}$.
4. For each expert find $r = (F_Y(y_{DM,50}) + F_X(x_{DM,50}))/2$, and $\tau_e(r)$ from Figure 1.
5. Take linear pooling of the $\tau_e(r)$ to find

$$\pi_{DM} = P_{DM}(F_{X,DM(X)} > \frac{1}{2} \mid F_{Y,DM(Y)} > \frac{1}{2})$$

6. Find $\tau$ from Figure 1 which passes through $(\frac{1}{2}, \pi_{DM})$, this is $\tau_{DM}$.

# 3 Analyzing dependence in output with cobweb plots

The problem of understanding and communicating information about dependencies in the results of an uncertainty analysis is inverse, as it were, to the problem of eliciting dependencies. If the dependencies are smooth then familiar regression, rank regression and partial rank correlation measures can convey useful information about how the uncertainty in output variables depends on uncertainty in input variables. Unfortunately, such measures may easily miss important information. Using only such measures, one never knows whether something is being missed.

For this reason a graphical tool, called *cobweb plotting* has been developed to study dependence. To form a cobweb plot, selected variables are represented as parallel vertical lines on which the variables' percentiles have been marked. Each sample from an uncertainty analysis is mapped as a line intersecting the vertical lines at the percentile points realized in that sample. The result of plotting a few hundred samples suggest a cobweb, and contains all the information in the empirical joint distribution of the variables quantile functions. To study and extract information from a cobweb plot, the user can filter all lines passing through selected intervals on the vertical lines. This is equivalent to conditionalizing the joint distribution on percentile intervals of the variables.

Figures 2 and 3 show cobweb plots for frequency of leak per km. yr. due to corrosion (corlk) for a given type of underground pipeline, taken from [3]. The variables which influence this frequency are the free corrosion rate (crf), the corrosion rate for partial cathodic protection (crp) the probability of small and large unrepaired damage to the pipe from third party intervention (ps and pl respectively), the probability of damage to the coating from third parties (pc3) and the probability of damage to the coating from the environment (pcen). All of these quantities are uncertain and the uncertainties have been assured with structured expert judgement. In Figure 2 we have conditionalized on the upper 20 percentiles of the uncertainty distribution for corlk. A strong association is discovered with pcen. In Figure 3 we have conditionalized on the lower 20 percentiles of the distribution for corlk. Now there is no strong association with pcen, but a strong association with values of crp between the 40th and 60th percentiles. In both cases, the pattern of dependence could not be foreseen from the structure of the model, but once it is revealed, it can be understood in terms of the model. The pipe in this example is so old that new leaks from corrosion would come predominantly from scenarios in which the environment was very aggressive, causing small damages to coating which could lead to failure in a period of time comparable to the age of this pipe. If the free corrosion rate were very high, then it is likely that the pipe should already have failed from another cause, unless the probability of more severe damage was low. The relations between the frequency of corrosion leak, the

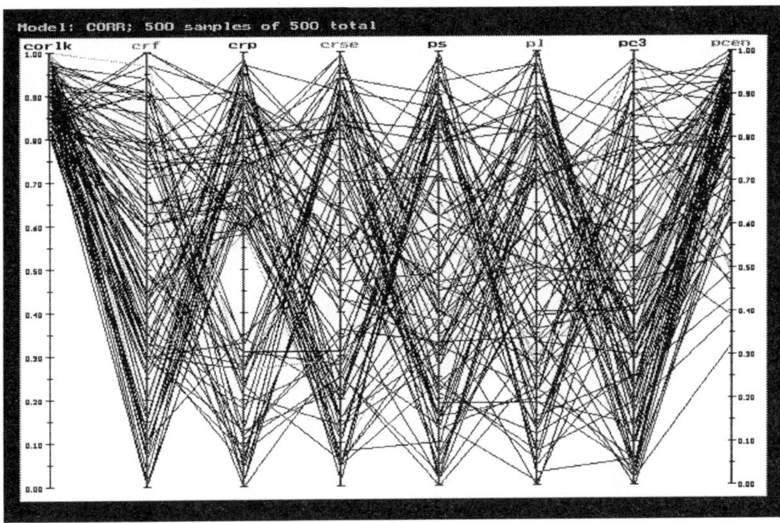


Figure 2: *Cobweb plot conditionalized on corlk above 80% quantile*

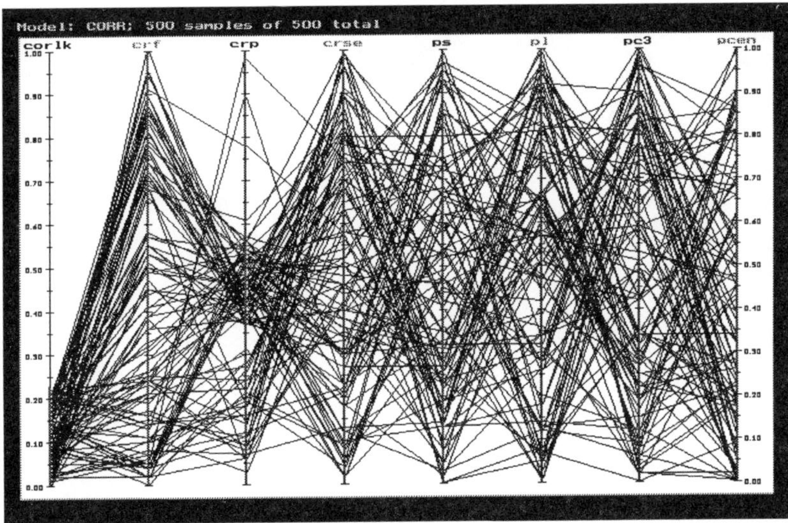


Figure 3: *Cobweb plot conditionalized on corlk below 20% quantile*

partial corrosion rate and the probability of damage due to the enviromnent would not be revealed by rank correlations ($\tau$(corlk,pcen) = 0.63, $\tau$(corlk, crp) = 0.16).

# References

[1] Apostolakis G. and Kaplan S. Pitfalls in risk calculations. Reliability Engineering 1981:2, 135-145.

[2] Cooke, R.M. UNICORN; Methods and Code for Uncertainty Analysis, AEA Technology, Warrington, Cheshire, UK, 1995.

[3] Cooke, R. and Jager, E. and Geervliet, S. Failure frequency of underground gas pipelines. Proceedings, ESREL'96-PSAM III, Crete, 1996.

[4] Cooke, R.M., Kraan, B.C.P. Processing of Expert Judgment Assessments into Paramter uncertainties. Proceedings, ESREL'96-PSAM III, Crete, 1996.

[5] Gokhale, D. and Press, S. "Assessment of a prior distribution for the correlation coefficient in a bivariate normal distribution" J.R. Stat. Soc. A 1982: vol. 145 P.2, 237-249.

# Two Approaches to Model Uncertainty Quantification: a Case Study

E. Zio, G.E. Apostolakis
Department of Nuclear Engineering
Massachusetts Institute of Technology
Room 24-221, Cambridge
MA 02139-4307, U.S.A.

**Abstract**

The evaluation of the performance of high-level radioactive waste repositories implies the use of simulation models for predicting the system evolution. The complexity of the system together with the large spatial and temporal scales imposed by the normative regulations are such to introduce broad uncertainties in the analysis. In this paper we address the problem of quantifying the uncertainty associated to the model representation of groundwater flow and contaminant transport in unsaturated, fractured tuff. Two approaches to the problem of aggregating the results coming from multiple models are here considered. The first one focusses on the plausibility of the various models, i.e. on their likelihood to provide a correct representation of the real system. The second one concentrates on the accuracy of the predictions of the models, which are considered as devices providing information helpful for the estimation of the quantity of interest.

## 1 Introduction

Model uncertainty is considered a primary source of uncertainty for the study of the performance of mined geologic repositories proposed for the disposal of nuclear wastes. When modeling a natural system, the information one would like to have available is of a substantially higher order than what can be inferred from observations. The consequences of this lack of identifiability are usually apparent in the absence of a unique 'best' combination of modeling hypotheses and parameter values that fit the data (many combinations are 'equally good') and in parameter estimates with high error variances and covariances.

In the practice of performance assessments, the quantitative evaluation of model uncertainty is strongly affected by the complexity of the system and by the large spatial and temporal scales required by the analysis. The predictive capabilities of the models used cannot be observed over the time frames and spatial scales at which they are required to apply. As a consequence, the analyst cannot obtain empirical confirmation of the validity of a model from observations. The evaluation of the models must then rely on the subjective interpretation of the information available at the time of the analysis. This leads to the conclusion that any attempt to address the issue of model uncertainty in a quantitative manner will rely on the use of expert judgment. Within a risk assessment, the way in which model uncertainty is characterized and expert opinion is elicited and used are complex and controversial issues which need to be clarified and possibly framed in a robust procedural structure.

In this paper we discuss the application of two mathematical formulations, previously presented in the literature [1-3], to a case study concerning the uncertainty due to alternate modeling assumptions for the description of a hypothetical high-level nuclear waste repository in unsaturated, fractured tuff formations. In the next section we briefly outline the basic concepts of the two formulations employed. Section 3 then contains a description of the case study and the results of the quantitative assessment of the uncertainty associated with the models. Finally, in section 4 we draw some conclusions on the matter discuss in the paper.

## 2 Two mathematical formulations for the treatment of model uncertainty

The quantitative assessment of model uncertainty can be addressed from two distinct directions, corresponding to two different theoretical formulations.

The first approach consists of considering a set of $N_S$ models $\{M_i\}$, where each model $M_i=(S_i,\underline{\theta}_i)$ consists of an alternate structure $S_i$ based on *alternate hypotheses* which are plausible in light of the existing information, and the associated parameter vector $\underline{\theta}_i$. Each model in the set provides a different description of the unknown quantity $y$ which we assume is in the form of aleatory distributions

$$F_i(y) \equiv F(y| M_i) = F(y|S_i, \underline{\theta}_i), \qquad i= 1, 2, \ldots, N_S, \tag{1}$$

which are conditional on the structure of the model as well as on the values of the parameters. Note that we use the term *aleatory* to indicate uncertainty due to natural variability of phenomena which cannot be modeled by deterministic expressions.
The family of distributions $\{F_i(y)\}$, which properly represents the uncertainty in the unknown $y$ due to uncertainty in the models' structure and parameters' values, can be probabilistically combined in a summary measure by means of a standard Bayesian approach[1]. The joint density function $\underline{\psi}(S_i,\underline{\vartheta}_i)$ which expresses the analyst's beliefs regarding the numerical values of the parameters and the physical validity of the model hypotheses can be expressed as

$$\underline{\psi}( S_i,\underline{\vartheta}_i) = \underline{\pi}_i(\underline{\vartheta}_i|S_i)p(S_i) \tag{2}$$

where $\underline{\pi}_i(\underline{\vartheta}_i| S_i)$ is the epistemic probability distribution function (pdf) of the vector of model parameters $\underline{\vartheta}_i$, conditional upon the choice of model structure $S_i$, and $p(S_i)$ is the epistemic probability which expresses the analyst's confidence in the set of assumptions underpinning the model. Here the term *epistemic* is used to characterize the uncertainty due to incomplete state-of-knowledge.
By applying the theorem of total probability, we can assess the unconditional aleatory distribution F(y) in the form of the standard Bayesian estimator

$$F(y)=\sum_{i=1}^{N_S}\int F_i(y)\pi_i\left(\underline{\vartheta}_i\middle|S_i\right)d\underline{\vartheta}_i\,p(S_i) \tag{3}$$

This average value has often encountered objections when employed in a decision-making environment. The argument is that the decision maker should be aware of the full epistemic uncertainties that $\underline{\pi}_i$ and $p$ represent. The average value can lead to erroneous decisions, particularly when the epistemic uncertainty is very large, since the average can be greatly affected by high values of the variable even though they may be very unlikely. In general, we agree that the entire distribution of the uncertainties should be presented to the decision-maker who may then choose his own summary measure upon which to base the decisions.

The second approach, directs its attention to the accuracy of the predictions of the models. A simple formulation of this approach, previously employed[1], is the so-called *adjustment-factor* approach which focuses on one single model of the process under study and aims at appropriately modifying its output so as to allow for the uncertainty in the model description.

Suppose that one single model M* is selected as reference to describe the process. Even though this model may provide the best representation of the system, there still are significant uncertainties associated with it. To formalize this situation, we introduce a factor E*, which may be additive ($E^*_a$) or multiplicative ($E^*_m$), so that our assessed value for the quantity y can be written as

$$y = y^* + E_a^* \qquad (additive) \tag{4a}$$

$$y = y^* E_m^* \qquad (multiplicative) \tag{4b}$$

The factor E* is in general unknown and the uncertainty associated with it can be represented in the form of a distribution g(E*). The resulting distribution F(y), for example in the additive scheme, is given by

$$F(y)=\iint z(y-E_a^*,E_a^*;\underline{\vartheta})\,w(\underline{\vartheta})\,dE_a^*\,d\underline{\vartheta} \tag{5}$$

where $z(y^*,E_a^*\,;\,\underline{\vartheta})$ is the joint aleatory distribution of y* and E* with parameters $\underline{\vartheta}$ and $w(\underline{\vartheta})$ is the epistemic distribution representing the state-of-knowledge uncertainty in the parameters. A similar expression can be derived for the multiplicative version of the approach.

## 3 A case study: alternate models of groundwater flow and contaminant transport in unsaturated, fractured tuff

In this section we re-examine six alternate models of groundwater flow and contaminant transport in unsaturated, fractured tuff, previously presented by Gallegos, Pohl and Updegraff[4], with the objective of quantitatively assessing the uncertainty associated to the flow and transport predictions, within the theoretical frameworks discussed in the previous section. The six models considered are based on a set of common assumptions and simplifications but differ by some fundamental hypotheses on the flow and transport mechanisms in the system. For further details on the specific models, the reader should consult reference 4.

Following the alternate-hypotheses approach, we can subjectively assign values to the probabilities of the alternate hypotheses (Table 1). The application of these probabilities as weights in the Bayesian estimator of eq. (3) leads to a measure of cumulative release as indicated in Table 2 where it is compared to the values obtained by using only $S_4^{ft}$, chosen as reference. Again we argue that the estimates obtained by means of eq. (3) give a more accurate representation of the uncertainty inherent in the model structure. Notice how the estimates are quite close in all cases, as the mean is driven by models $S_3^{ft}$ and $S_4^{ft}$ which happen to have similar probabilities and predictions.

Suppose now that we wish to represent the uncertainty in the output predictions regarding the cumulative release of I-129 given by the six models, by expanding directly the prediction of a selected best model $S^*$ through a properly assessed adjustment factor $E^*$. To be consistent with the earlier analysis we choose $S^* = S_4^{ft}$. Indeed, one could very well think of a two-step process which combines the two approaches proposed in section 2: first, expand the model set and select the best model $S^*$; then modify the predictions of $S^*$ so as to account for its inherent uncertainty as indicated by the available information, including that provided by the identified alternate models. This process has the potential of leading to a very efficient and explicit method for evaluating and propagating model uncertainty. In our simple example, we simply use the information given by the alternate models to build a distribution for the modification factor $E^*$, which then represents model-to-model uncertainty. In most practical cases, other information combined with expert judgments will also be used by the analyst to arrive at the final distribution for $E^*$. Notwithstanding the paucity of the data, for the purpose of this example, the residuals $(y_i - y^*)$ of the five alternate models can be equally weighed to produce a mean value $\overline{E}^* = -21.37$ and a standard deviation $\sigma_{E^*} = 616.66$, which indeed reflects the large uncertainty present in the predictions given by the models. The solid line in Figure 1 represents the distribution of $E^*$, under a normality assumption. Taking into account the judgments on the credibility of the various alternatives, expressed in terms of probabilities $p(S_i)$, we can modify the distribution just obtained by using the probabilities of Table 1, to estimate a Bayesian predictive distribution according to eq. (3) (dotted line in Figure 1). In this case, this leads to a slight shift of the curve towards more negative values and a reduction in the uncertainty, the new mean and standard deviation values being -4.48 and 262.32, respectively.

## 4. Conclusions

This paper addresses the problem of quantifying model uncertainty. Two mathematical frameworks are presented which, in principle, account for the uncertainty associated with models. In the first approach, the idea is to systematically construct a set of models based on alternate hypotheses which, in light of the available information, provide plausible descriptions of the system under analysis. We have referred to this method as the *alternate-hypotheses* approach.

For decision-making purposes, it is often necessary to combine the results of the uncertainty assessment into an aggregated measure. The alternate-hypotheses approach provides a sound theoretical framework for the evaluation of a predictive distribution to be used in formal decision analysis.

The other mathematical formulation, the *adjustment-factor* approach, which can be used to account for model uncertainty amounts to identifying a 'best' model and then appropriately modifying its predictions by means of an adjustment factor. This factor is in general unknown and the uncertainty associated with it is represented in the form of a distribution which is intended to reflect the uncertainty in the models.

The application of these approaches to a case study regarding six alternate conceptual models for the groundwater flow and contaminant transport in unsaturated, fractured tuff has shown that the two formulations can lead to different results.

The proprer use of these approaches in practice is strongly connected to, and dependent on, the elicitation of expert judgments to estimate the relevant quantities.

**ACKNOWLEDGMENT:** This work is based on research supported by the U.S. Nuclear Regulatory Commission under grant NRC-04-93-093. Although this paper is based on research funded in part by the U.S. Nuclear Regulatory Commission, it presents the opinions of the authors, and does not necessarily reflect the regulatory requirements or policies of the USNRC.

## References

1. G. E. Apostolakis, *A Commentary on Model Uncertainty*, Proceedings of Workshop I on Advanced Topics in Risk and Reliability Analysis, A. Mosleh, N. Siu, C. Smidts and C. Lui Eds., Annapolis, Maryland, Oct. 20-22, 1993, NUREG/CP-0138, US Nuclear Regulatory Commission, Washington, DC.

2. D. Veneziano, *Uncertainty and Expert Opinion in Geologic Hazards*, Symposium in Honor of R. V. Whitman, MIT, October 7-8, 1994.

3. E. Zio, G. Apostolakis, D. Okrent, *Towards a Quantitative Treatment of Model Uncertainty in the Performance Assessment of High-Level Radioactive Waste Repositories*, Proceedings of WM'95, Tucson, Arizona, Feb. 26-Mar. 2, 1995.

4. D. Gallegos, P. I. Pohl and C. D. Updegraff, *Preliminary Assessment of the Impact of Conceptual Model Uncertainty on Site Performance*, High Level Radioactive Waste Management, Proceedings of the Second International Conference, Las Vegas, Nevada, April 29-May 3, 1991.

*Table 1. Probability values for the groundwater flow and contaminant transport models*

| $S_i^{ft}$ | $p(S_i^{ft})$ |
|---|---|
| $S_1^{ft}$ | 0.01 |
| $S_2^{ft}$ | 0.04 |
| $S_3^{ft}$ | 0.1 |
| $S_4^{ft}$ | 0.45 |
| $S_5^{ft}$ | 0.3 |
| $S_6^{ft}$ | 0.1 |

*Table 2. Comparison of cumulative releases to water table (Ci)*

| | Tc-99 | I-129 | Cs135 | Np-237 |
|---|---|---|---|---|
| $S_4^{ft}$ | 5.3e4 | 515 | 0.0 | 0.0 |
| Bayes | 6.9e4 | 470.5 | 672 | 1.7e-3 |

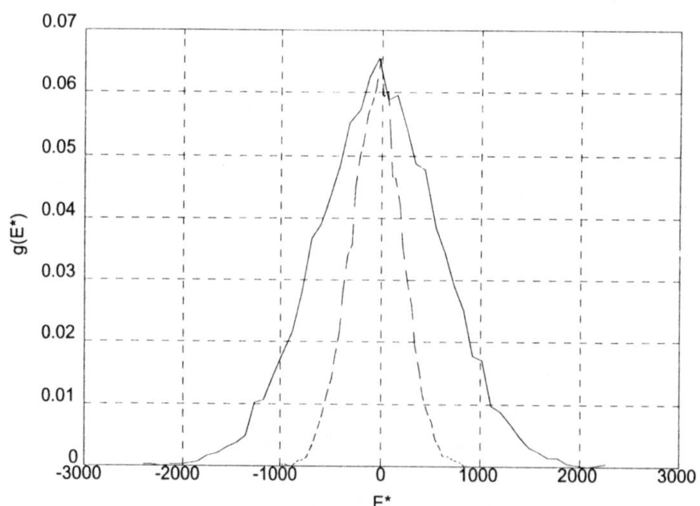


*Fig. 1. Comparison of the distributions of E* obtained by simple(solid) and weighed averaging (dashed)*

# A glance to uncertainty from an epistemological point of view

L. Pinola, R. Sardella*

European Commission, Joint Research Centre
Institute for Systems Engineering and Informatics
21020 Ispra (VA), Italy

*Università degli Studi di Bologna
DIEM, Viale Risorgimento 2
40136 Bologna, Italy

**Abstract**

A PSA is a methodological frame in which different types of knowledge are melted with their relevant uncertainties. In PSA literature, various uncertainty taxonomies have been proposed. This paper proposes an epistemological ground as the most suitable base for discussion on such questions. Starting from an evolutionary analysis of life of theories, the conclusion is drawn of adopting a knowledge engineering approach to reconduct uncertainties to their knowledge counterparts.

## 1. Introduction

The goal of a Probabilistic Safety Analysis (PSA) of a nuclear power plant is the identification of all the accident scenarios leading to reactor damage states. The risk profile is the result of the assessment of the occurrence frequencies of such scenarios, the relevant radioactive releases, and the consequences on public health and on the environment. This process requires the homogenisation at different levels of different types of knowledge, such as theories, models, expert judgment, and so on. The complexity and variety of the involved knowledge have been tackled in PSA by means of a methodological frame. On the one hand, this approach carries along the advantages of manageability, modularity, as all the typical features of a non-rigid, wide-band structure; on the other hand, this point of view is forced to face the integration without an overall theoretical skeleton.

As regards the types of knowledge intervening in the analysis, some are derived from well-established physical theories, either deterministic or not, some belong to the field of empirical laws, and some even to the domain of scientific heuristics.

Accordingly, each knowledge item calls for a different and distinct representation scheme.
Moreover, knowledge is pervasively affected by "uncertainty" to various extent: even those theories largely accepted because of their dated validity, cannot be expected to be completely accurate, nor definitive. The temporary nature of all human knowledge must be recognised, and inherent "uncertainty" must be coped with.
In such a perspective, the PSA methodology does not constitute an exception: it clearly states among its objectives that of describing the risk (associated with the existence and operation of nuclear power plants) and the relevant uncertainties. For this purpose, various taxonomy are proposed in PSA literature [1] dealing with different kinds of uncertainty. Remarkably, the IAEA Procedures Guide [2, 3] suggests a classification following the criterion of differentiating the sources of uncertainty among:

1. Lack of completeness. It would never be possible to assure that all accident scenarios have been taken into account.
2. Modeling adequacy. Incomplete understanding of the involved physical phenomena and questionable applicability of models (in its generic meaning) prevent a totally accurate description of accident scenarios.
3. Input parameters uncertainty. Because of scarcity of data, the quantification of most parameters should resort to engineering judgment.

A slightly different taxonomy is the one proposed in NUREG-1150 [4], where a first distinction is put between stochastic and subjective uncertainty. Stochastic uncertainty refers to those systems that under the same environmental conditions may behave in many different way; it is the expression of the intrinsic variability of those systems, and, as such, is pertinent to "reality". Nonetheless, in some cases, it could also reflect our limited resolution power on the systems [5]. On the contrary, subjective uncertainty is deemed to be a property of the analysts performing the study since its roots are to be found in the current incomplete knowledge, peculiar of human understanding. From a more applicative point of view, as uncertainties in a PSA should in any case be imposed on the parameters of the codes used for the computations, subjective uncertainty is separated into:

1. Modeling uncertainty, that refers to the gap between the so-called "high level" summary parameters actually expressed in the codes and the fundamental physical parameters, i.e., essentially a lack of understanding.
2. Data uncertainty, that refers to the imprecise knowledge one can possibly have even on physical parameters.

As these classifications show, it is difficult to keep a clear-cut distinction among their items. Furthermore, NUREG-1150 itself claims that "no consistent effort was made to differentiate between the effects of the two types of uncertainties" (modeling and data uncertainties), meaning that these uncertainties very often appear intertwined in the analysis and hardly could be processed in separate manner.

## 2. Two Stages of Life for Theories

Our considerations start suggesting that a mandatory distinction should be made between two phases of life for theories (theories meant in a broad sense): a first stage of infancy before their computed realisation, and a second stage of maturity when they overcome plenty of counter-experiments, either theoretical or empirical. The transition from the first to the second stage may be questioned on the ground that, strictly speaking, there is no such thing as a definitive verification of a theory, as well as there is not even a definitive falsification by means of empirical observations [6]. But, as already said, our point here is not the ontological matter; on the contrary, on an epistemological base, we recognise theories fundamentally as hypotheses that could be dismissed as a result of some methodological decision. So, we purposefully go on speaking about the second stage of the life of theories, after they have passed severe tests and they are increasingly corroborated having gained a high degree of intersubjective acceptance.

What intimately characterises the first stage is the dominance of the psychological component in the activity of theory construction. This creative process is absolutely subjective in nature, and does not follow any identifiable procedural form.

The constitution of a theory is urged by the recognition of unsatisfactory traits of the available knowledge, and aims at organising facts, explaining mechanisms, joining various theoretical systems, etc.. Some pre-existing knowledge material and the analysis of special observational cases match intuition in a speculative activity when fresh ideas arise and new knowledge is formed.

During such a development, the employed *logics* seem to be restored to their most basic function of prescribing admissible inferences, so they approach, at most, a minimalist, or weaker [7], logical formulation. Indeed, the choice of a formal inferential paradigm should be such as to admit local contradictions without necessarily implying the inconsistency of the overall paradigm [8].

The transition from infancy to maturity leads the theory to a posterior restructuring, typically an axiomatized version of the theory is issued, that consists in its rearrangement into a hypothetic-deductive scheme. The main trust of this process is to express the theory in a format susceptible to falsification, plainly regarding falsification from a purely logic point of view. A label of "scientificity" may be hence attached to the theory, and its status recognised as that of communicable knowledge, essentially a non-circular system constituted of: a set of basic objects (axioms), the links (interrelationships) among these objects, plus a suitable set of strategies/mechanisms for inferring new objects and relations according to an intended goal (representational, interpretative, predictive, etc.).

The requirement of structuring knowledge in a communicable form should be interpreted as the only way to reach the possibility of exchanging information in critical discussion. This postulates also that each supporter of a theory makes explicit the conditions in the presence of which he is willing to take the theory falsified.

It is worth noting that, in the attempt to achieve a sort of pure, objective knowledge purged of any subjective feature, the reformulation of theories into some axiomatized form of theirs is always in progress.

# 3. Unifying Theories

With the same objective of pushing aside subjectivities, the process has been enlarged to the ensemble of scientific theories [8], trying to settle them in a unique formal system. But, if single formal systems could be given a linear shape, the set of mathematical theories, not to mention the set of all scientific theories, have not yet succeeded to be framed in those same terms, that is as a unique formal system with no circularities.
Well-known limitative theorems (Gödel's theorems, 1931) about the deduction power of coherent axiomatizable formal systems (at least as potent as arithmetic) prevent a sensible structuring of a global linear frame. In other words, some epistemological requirements for formal systems are incompatible between themselves, namely: coherence, axiomatizability, syntactic completeness. It should be finally pointed out that the mentioned limitations are not only inherent to formal systems; it is not their own structure to be responsible for these problems, indeed strong antinomies like the liar paradox appear also at an intuitive level of reasoning.

Summarising, two considerations may be drawn from the foregoing discussion. First, since even the integration of "easy" (without any physical referent) mathematical theories in a unique formal frame seems to be impossible because of unavoidable circularities, it is apparent how useless it is to pursue that same aim of formal integration for the whole cluster of theories (most of which, "hard" physical theories both at the 1st and 2nd stage) playing a role in the PSA analysis. Secondly, in a PSA it would be safer to distinguish and keep separate theories currently in their infancy and theories having reached their maturity because of their different ontological hypotheses, reasoning logics and formalization degree.

# 4. Prediction

The structuring of theories in a hypothetic/deductive format is introduced not only in response to a methodological requirement, but also for satisfying a more empirical demand: at least from the engineering point of view, the value of scientific knowledge can be directly linked to his predictive power, since the absence of this capability prevents any planning of actions.
Both for falsification and prediction purposes, theories, expressed as universal statements, need to be applied: only their conjunction with some particular statement representing boundary/initial conditions, yields the deduction of new particular statements that can be eventually checked against empirical observations.
In this sense, theories can be thought of as generators of models, and are indeed these latter that undergo testing procedure. The debate around model falsification has engaged epistemologists and philosophers for long time, but we would like to accept a rather practical stand of saying that a model could be considered falsified when it is no longer able to answer the problem it was called to.

Actually, the intervened inadequacy for a model is the outcome of a methodological judgment practised by the community of scientists reaching intersubjective agreement [9].

As concerns prediction, three basic ingredients enter the picture: the forecaster should be knowledgeable of the system boundary/initial conditions, of the laws ruling its behaviour, and, furthermore, their validity should be assumed for all the time of the prediction. Hence, some intrinsic uncertainties can be singled out. From a strictly technical point of view, uncertainties can be ascribed to the unavoidable approximation affecting the knowledge of boundary (and initial) conditions of a system, or even to the very impossibility of making crucial experiments for their determination. From an epistemological point of view, uncertainties stem from the assumption of having completely circumscribed the system, with the identification of all those factors (internal and external) on which it depends and on which it does not depend. Finally, from a methodological point of view, it should be mentioned that uncertainties may originate from an unbalance between the limited logical capacity of the forecaster and the degree of complexity of the relationships describing the system evolution. For deterministic systems, two approaches for the analysis are broadly adopted. The first one consists in performing a whole set of analyses, each relevant to a different set of variables kept as constant. As outcomes, a spread of scenarios is obtained, and these are meant to cover all the most significant features of the process. The second approach is a statistical analysis in terms of aggregate variables; the objective of the analysis is shifted toward the prediction of global system behaviour by means of lumped parameters.
And what about indeterministic systems? In this respect, we begin to refer the question to proper ontological limits.
It has always been disputed on what may be referred to as “stochastic uncertainty”, apparently inherent in nature, and what is called “subjective uncertainty”, expressing our state of imperfect partial knowledge [10].
From the ontological perspective, natural stochasticity characterises systems whose present state points to (may evolve toward) sets of future states. For these systems, past and future are not equivalent, since the past is made of a chain of fired possibilities, whereas the future appears to be intrinsically undetermined. The point is that, if such systems exist, the concept of possibility would be ontological as well as logical: a smoothed degree of truth would be given ontological sense. On the contrary, if the non-uniqueness in the prevision is caused by imperfect and incomplete knowledge about a deterministic system, then the truth (false) value of propositions about future states cannot be decided in advance, and the concept of possibility would make sense only in a logical context.
Unfortunately, metaphysical decisions underpinning ontological statements are hardly judgeable, and there is often no room enough for critical debate. Thus, for the purpose of settling the present discussion, our point is to skip any ontological question and to focus only on the epistemological traits of the problem. Within this viewpoint, "stochastic" is only an epistemological attribute for identifying system modeling frame, regardless either of their own inherent randomness or of a coarse resolution adopted by the analyst.

## 5. Uncertainty Sources

Tackling the uncertainty issue from an epistemological viewpoint, it is apparent that uncertainty itself is but a sort of knowledge and it should be represented as such. This conceptual attitude requires every form of uncertainty to be brought back to its root causes, namely pieces of knowledge not satisfactorily mastered. Knowledge engineering techniques are readily available to build proper models where explicit representation of those unsatisfactory traits is extensively carried out [11].
When the proper domain for handling knowledge has been identified, we suggest a sketchy list of items to aid in the procedure of tracking and making explicit the different kinds of uncertainty sources. The suggested list follows the evolution lines of theories: uncertainties are looked for in parallel with the course of life of theories, i.e., beginning with their initial, intuitive formulations until the attainment of more formalised communicable knowledge. At the same time, the focus of attention never leaves out of consideration the flaws (uncertainties) due to imperfect knowledge considering the intended use of that knowledge, primarily that of making predictions.

*1st stage - infancy*

1. Natural language

   Common-sense meaning of terms
2. Tacit knowledge

   Not directly and consciously accessible knowledge [11]
3. Reasoning mechanisms [12]

   Revisable, common-sense reasoning

   Intuition, analogy, metaphor
4. Explanation of unsatisfactory traits of existing knowledge

   Theoretical-progressive theories: increasing the set of predictable facts

   Empirical-progressive theories: increase the set of predicted facts that comply with empirical observations [9]

*2nd stage - maturity*

5. Axiomatization procedure

   Theories deal with idealised referents, not with physical objects supposed to inhabit reality [13]
6. Intended use of knowledge

   Representational, interpretative, predictive [7]
7. Formal language

   Specialised definition of symbols
8. Basic propositions

   Axioms
9. Logics

   Formalised inferential rules
10. Epistemological requirements

    Coherence, syntactic completeness
11. Relationships with other related theories

    Background knowledge

Ad-hoc hypotheses

12. Realisation of the formal system

    Establishing a correspondence between the symbols of the theory and the universe of the discourse

13. Truth concept

    How to compute the truth value of a compound statement as a function of the truth values of its components [8]

*Building models from theories: prediction*

14. Definition of (spatio/temporal) boundary conditions (technical viewpoint)
15. Explicit assumptions on the permanence of validity of boundary conditions and laws for the time-lapse of prediction
16. Explicit description of conditions for accepting a model as falsified
17. Explicit identification of factors on which the system depends and factors on which it does not depend (epistemological viewpoint)
18. Explicit decision about the resolution power of the model (methodological viewpoint)

    Deterministic vs. stochastic models

# 6. Conclusions

This paper has the main objective of highlighting what is, in our opinion, a convenient ground for screening uncertainties with respect to their different roots. In this way, it could be conceivable both to use different representations of uncertainty and to cope with the need of distinguishing among the relevant effects.

Recognising uncertainties as true knowledge items, immediately indicates knowledge engineering as the reference setting. One dictate in this field of artificial intelligence states that the perspective of building a comprehensive theory of knowledge is utopic. On the other hand, knowledge engineering has demonstrated that effective results may turn out when knowledge problems are referred to proper (goal-oriented) contours and are tackled with tailored techniques [14].

Performing a PSA results in merging more formal theories with rather heuristic knowledge. Expert judgment, formalised to various extents, appears pervasively in the analysis; in particular, resorting to expert opinion is unavoidable when human experts are the only possible knowledge processors. In this sense, it could be argued that experts actually construct new knowledge in the form of theories in their first stage of life. An effective exploitation of this knowledge should come after its extensive acquisition and modeling.

From a knowledge engineering perspective, the suggested list would constitute a guiding pattern for prompting the analyst's attention on uncertainties in knowledge.

## References

1. USNRC. Model Uncertainty: Its Characterisation and Quantification. In: Proc. of Workshop I in Advanced Topics in Risk and Reliability Analysis, October 20-22, 1993, Annapolis, Maryland. United States Nuclear Regulatory Commission NUREG/CP-0138, 1994
2. IAEA. Procedures for Conducting Probabilistic Safety Assessments of Nuclear Power Plants, (Level 1). A Publication Within the NUSS Programme, Safety Series No. 50-P-4, International Atomic Energy Agency, Vienna, 1992
3. IAEA. Procedures for Conducting Probabilistic Safety Assessments of Nuclear Power Plants, (Level 2), Accident Progression containment analysis and estimation of accident source terms. A Publication Within the NUSS Programme, Safety Series No. 50-P-8, International Atomic Energy Agency, Vienna, 1994
4. USNRC. Severe Accident Risks: An Assessment for Five U.S. Nuclear Power Plants. United States Nuclear Regulatory Commission, NUREG-1150 vol 1, 1990
5. USNRC. Evaluation of Severe Accident Risks: Methodology. SNL, prepared for United States Nuclear Regulatory Commission, NUREG/CR 4551 vol. 1, rev. 1, 1993
6. Popper KR. Realism and the Aim of Science. In: Bartley III WW (ed) Postscript to the Logic of Scientific Discovery. Routledge, London, 1992
7. Barr A, Feigenbaum EA. Representation of Knowledge. In: The Handbook of Artificial Intelligence. Addison-Wesley Publishing Company, 1981, pp 141-222
8. Dalla Chiara Scabia ML. La logica. Arnoldo Mondadori Editore S.p.A., Milano, 1979
9. Lakatos I. Proof and Refutations. The Logic of Mathematical Discovery. Cambridge University Press, Cambridge, 1976
10. Kaplan S. Formalisms for Handling Phenomenological Uncertainties: the Concepts of Probability, Frequency, Variability, and Probability of Frequency. Nuclear Technology 1993; 102: 137-142
11. Guida G, Tasso T. Design and development of knowledge-based systems. John Wiley and Sons, Chichester, 1994
12. Léa Sombé Group. Reasoning under incomplete information in artificial intelligence: A comparison of formalisms using a single example. International Journal of Intelligent Systems 1990; 5(4): 323-472
13. Bunge M. Foundations of Physics. Springer-Verlag Berlin Heidelberg, 1967 (Springer Tracts in Natural Philosophy vol 10)
14. Kidd A (ed.). Knowledge Acquisition for Expert Systems, A Practical Handbook. Plenum Press, New York, 1987

# Evaluation of New Developments in Cognitive Error Modeling and Quantification: Time Reliability Correlation

Bernhard Reer
Forschungszentrum Jülich GmbH (KFA), D-52425 Jülich
Oliver Sträter
Gesellschaft für Anlagen- und Reaktorsicherheit mbH (GRS), D-85748 Garching

**Abstract**

A subset of methods for the probabilistic analysis of cognitive errors of control room operators is investigated. The subset concerns the time reliability correlation approach. New methods (EdF, HCR/ORE) exhibit partial improvements compared to older ones (THERP, HCR), especially concerning the incorporation of simulator data. However, with respect to modern findings in cognition of human beings, there is still a need for further developments of probabilistic methods.

## 1 Introduction

In probabilistic risk assessment (PSA), it is necessary to consider human-related events as elements of the overall model of system safety. Human reliability analysis (HRA) deals with the assessment of the corresponding human error probabilities (HEPs). Cognitive errors of control room operators are known as important factors in accident causation. Their quantification in HRA is criticized by many experts, e.g. [1]. In view of this criticism, this paper evaluates some advances in HRA with emphasis on cognitive error analysis.

## 2 The Problem of Cognitive Error Analysis

Webster's dictionary [2] defines cognition as the act or process of *knowing*. Rasmussen [3] distinguishes three levels of human behavior, (1) skill-based, (2) rule-based, and (3) knowledge-based. Therefore, it seems to be obvious to categorize cognitive errors under the heading of knowledge-based behavior. However, the relation of behavior levels to real objects is not a 1:1 representation and categorizing into Rasmussen's behavior levels is more difficult than expected, because human beings are able to accommodate themselves to different situations. They learn behavior and become skilled if they do an action several times. This enables human beings to reduce information load by performing some cognitive activity on the subconscious level. To a wide extend, this is independent from the observable complexity of the action by 'objective' measures (e.g., the number of task-elements, number of information-units, use of procedures, available time window) and is making HRA of cognitive errors difficult. Basis for this accommodative performance is the recognize-act cycle of human information processing. If the recognition of a situational pattern does not imply an abnormality (e.g., new situation), then the human will act on subconscious or low conscious level (e.g., walking until we stumble). If the recognition does imply an abnormality, then the conscious processing will be involved for this abnormality. This so-called 'cognitive dissonance' is a mismatch of learned or getting used behavior and actual situation (cf. [4]).

If these capabilities of human beeings are considered together with the objective measures, a more realistic interpretation and classification of observed error types may be provided. An application of a bad a procedure may be classified as a knowledge-based decision error as well as a skill-based error of habit, for instance. This depends on the situation (e.g., due to a lacking right procedure) as well as on the habit of the operator (e.g., due to lacking dissonance of the operator leading to overconfidence). If both factors (external cognitive stress and internal cognitive strain) are modeled within one approach, nine primary cognitive error types may be distinguished in a first attempt (see [5]).

Therefore, in long term, this aspect of the operator's perception of his level of cognitive behavior is to be included into the current HRA methods for cognitive error assessment. Current HRA methods offer simplified approaches to deal with such complexities of human cognition. Two of them are evaluated by the authors in [5], (1) time reliability correlation (TRC), and (2) expert judgment by performance shaping factors (PSFs). The next section presents a brief summary of [5] according to the former approach, TRC.

# 3 Time Reliability Correlation

## 3.1 Derived Human Error Probabilities and their Limitations

The TRC approach is usually applied for analyzing the diagnosis of an abnormal event in a nuclear power plant. In this context, a TRC-derived HEP is the probability pr(T>t) that the required time T for diagnosis is greater than the available (or allowable) time t for diagnosis. Table 1 summarizes some TRCs of the HRA methods THERP (Technique of Human Error Rate Prediction [6]), EdF'sPHRA (Electricité de France's Probabilistic HRA [7]), HCR (Human Cognitive Reliability model [8]), and HCR/ORE (HCR/ (based on) Operator Reliability Experiments [9]).

The THERP TRCs are based on expert judgment. Nevertheless, for $t < 30$ min (see Table 1), they are closed to EdF's TRCs which are based on French simulator data (see [10] and [11] for details).

The HCR TRCs are derived from several simulator studies, see [12] and [13] for details. They are presented in [8] as curves for a time scale which is normalized by the median time $T_{0.5}$ required to perform the respective cognitive task. Therefore, HCR is only to a limited extent comparable with THERP and EdF. For a comparable case, $T_{0.5} = 2$ min, Table 1 shows that even the most adverse HCR curve (knowledge) is more optimistic than the nominal curve of THERP.

However, such comparisons must be seen in the light of additional features of the methods. One feature concerns the limitation of the TRC approach for the extreme case that the available time t is very large. EdF [7] covers such a case by a so-called "residual" HEP which remains constant beyond a certain time limit, see Table 1. HCR considers "faulty detection or cognitive processing" as an additional, time-independent diagnosis error; for analysis the confusion matrix approach is recommended in [8]. In HCR/ORE [9] such time-independent error is denoted as "failure to formulate correct response", and a specific model (based on eight error mechanisms) for quantification is outlined. THERP [6] gives no explicit time-independent or residual HEP for diagnosis. However, de facto, the HEP of 0.0001 at t = 60 min could be interpreted as residual, because beyond 60 min the HEP decreases with a negligible rate of one order of magnitude per day, see Table 1. Nevertheless, compared to 0.005 (EdF, curve 1 or 1'), 0.0001 is a very optimistic residual HEP.

| HEP | **0.5** | **0.1** | **0.01** | **0.001** | **0.0001** | **0.00001** |
|---|---|---|---|---|---|---|
| THERP, nominal curve | 2 min $= T_{0.5}$ | 10 min $= 5\ T_{0.5}$ | 20 min $= 10\ T_{0.5}$ | 30 min $= 15\ T_{0.5}$ | 60 min $= 30\ T_{0.5}$ | 1500 min $= 750\ T_{0.5}$ |
| EdF, curve 1 | $\approx$ 3 min $= T_{0.5}$ | $\approx$ 9 min $= 3\ T_{0.5}$ | $\approx$ 21 min $= 7\ T_{0.5}$ | Not applicable, because for t > 30 min both curves remain at a constant residual HEP of 0.005 | | |
| EdF, curve 1' | $\approx$ 9 min $= T_{0.5}$ | $\approx$ 12 min $= 1.33\ T_{0.5}$ | $\approx$ 21 min $= 2.33\ T_{0.5}$ | | | |
| HCR, knowledge | $T_{0.5}$ | 2.74 $T_{0.5}$ | 5.84 $T_{0.5}$ | 9.34 $T_{0.5}$ | 13.19 $T_{0.5}$ | 17.27 $T_{0.5}$ |
| HCR, rule | $T_{0.5}$ | 2.12 $T_{0.5}$ | 3.88 $T_{0.5}$ | 5.75 $T_{0.5}$ | 7.68 $T_{0.5}$ | 9.68 $T_{0.5}$ |
| HCR, skill | $T_{0.5}$ | 1.52 $T_{0.5}$ | 2.15 $T_{0.5}$ | 2.74 $T_{0.5}$ | 3.29 $T_{0.5}$ | 3.82 $T_{0.5}$ |
| HCR/ORE | Six curves for a normalized time scale (see [9]); the concrete percentiles are property of the Electrical Power Research Institute (EPRI) | | | | | |

Table 1. Comparative Evaluation of HRA Methods According to Percentiles of the Time Required for the Diagnosis of an Abnormal Event

## 3.2 Related Performance Shaping Factors

Additional features (concerning TRC curve selection and HEP modification) of the TRC-related methods (introduced in 3.1) are outlined in Table 2. These features could be interpreted as PSFs in the widest sense. On the first view, it seems that the consideration of PSFs is very inhomogeneous among the several methods. However, a more detailed view discovers some common aspects. For example, EdF's PSF "complexity" could be interpreted as superposing for PSFs like "training/experience" (THERP, HCR) and "additional abnormal event" (THERP).

| Method | TRC-Related PSF | Effect |
|---|---|---|
| THERP | *Training* (exercises) with respect to the abnormal event | Choosing among three TRC curves, the nominal one (see Table 1), its lower bound (LB), or its upper bound (UB); see [6] for guidelines |
| | *Additional abnormal event* annunciated closely in time | Diagnosis will take 10 min longer per additional abnormal event |
| | Annunciated display (*alarm*) as diagnosis support | Recovery factor for diagnosis HEP; special model in [6] |
| EdF'sPHRA | *Complexity* of the situation | Choosing among four curves, 1, 1', 2, or 3; curves 2 and 3 refer to extremely difficult diagnosis; they are based on pessimistic THERP curves |
| | Safety engineer | Recovery factor for diagnosis HEP; special model in [7] |
| HCR | *Cognitive level* of the diagnosis task, skill-based, rule-based, or knowledge-based; see [8] for guidelines | Choosing among three normalized curves; see Table 1 |
| | Nominal median *time* $T_{0.5}$ *required* for diagnosis | Scale (absolute) of the curve |
| | Operator *experience*<br>*Stress* level<br>Quality of *man-machine interface* | Modifying the nominal $T_{0.5}$; special model in [8] |
| HCR/ORE | *Reactor type*, boiling (BWR) or pressurized (PWR) water reactor<br>Response type dictated by the dynamic of the accident | Choosing among six normalized curves; see [9] for details |
| | Median *time* $T_{0.5}$ *required* for diagnosis | Scale (absolute) of the curve |
| | Specific PSFs and recovery factors for each (out of a set of eight) relevant error mechanism which contributes to the failure to formulate correct response | Additional, time-independent contribution to the diagnosis HEP; special model outlined in [9] |

Table 2. Performance Shaping Factors Related to Time Reliability Correlation

Furthermore, Table 2 shows that, in contrast to models for a absolute time scale (THERP, EdF), normalized time scale models need a direct numerical estimation of the median diagnosis time $T_{0.5}$. On the one hand, this is an advantage because such an option for estimation results in a high flexibility. However, on the other hand, such option produces two elements of uncertainty, firstly, the estimation of $T_{0.5}$

itself, especially if no data are available. Secondly, due to the procedure of normalizing (dividing by $T_{0.5}$), the absolute time reserve is not considered adequately, especially if a value of $T_{0.5}$ is estimated which is not covered by the underlying TRC data. A resulting pitfall may be illustrated by Table 1 ($t < 30$ min): for *knowledge-based behavior* and $T_{0.5} = 3$ min, HCR is not so far from EdF's curve 1 (*easy diagnosis*), however, for *skill-based behavior* and $T_{0.5} = 9$ min, HCR gives a good approximation of EdF's curve 1' (*more difficult diagnosis*).

In HRA practice, another problem of HCR is that the behavior levels (knowledge, rule, skill) are not sufficiently equitable to real levels of cognitive behavior, see Section 2. However, according to this point, the ORE program resulted in HCR/ORE [9] – a modified version of HCR.

Moreover, Table 2 points at two critical features of THERP. Firstly, the guideline for quantifying recovery of diagnosis failure is too optimistic, see case studies in [6] where response to a certain alarm is modeled as an independent recovery factor. Psychological findings and operating experience indicate that common causes could result in both, failing of diagnosis and ignoring of the relevant alarm (which is designed as recovery factor). According to this point (diagnosis failure recovery), EdF and HCR/ORE are more realistic than THERP. Secondly, the option (in THERP) to use the lower bound (LB) of the diagnosis HEP curve could tend to over-optimistic results, because the nominal curve of THERP is already slightly more optimistic than the most optimistic data-based curve of EdF, see Table 1. Therefore, if at all, THERP's LB option should be used with care.

# 4 Conclusions

Time reliability correlation (TRC) offers an attractive approach in cognitive error analysis. This approach circumvents sophisticated error identification for a cognitive task. Instead, TRC produces a holistic HEP of the total cognitive task. However, the TRC approach has limitations.

As criticized by Reason [1], it seems to be unlikely that psychologically different processes (like slow processing of information or failure to detect the onset of the emergency) could be described by the same TRC curve. To some extent, current HRA methods diminish this weak point by introducing a residual or time-independent HEP for cognitive tasks. In this context, it should be noted that a time-independent error cannot always be clearly differentiated from a time-dependent error. A time-independent error like misreading could, in principle, also contribute only towards a delay without causing complete failure.

A further weakness of current TRC methods is the neglect of the dependence between available time t and required time T. The time perceived as available by the crew could influence the time of performance in such a way that the crew works particularly slowly or particularly fast. Comparable compensation effects can be corroborated by ergonomic findings [14].

HRA in practice should take credit of advances in TRC. Unfortunately, published information about new developments (EdF, HCR/ORE) is sparse. Nevertheless, if a HRA practitioner relies on well-documented methods (like THERP or

HCR), he should apply them with care by considering their weak points. Context sensitive assessment [15] including cognitive habits and TRC may be one solution.

## References

1. Reason J. Human Error. Cambridge University Press, Cambridge, 1990
2. Webster's New Collegiate Dictionary. G. & C. Merrian Company, Springfield, MA,1975
3. Rasmussen J. Information Processing and Human-Machine Interaction. North-Holland, Amsterdam, 1986
4. Wickens C.D. Engineering Psychology and Human Performance. C.E. Merrill Publishing Company, Columbus, Toronto, 1984
5. Reer B., Sträter O., Mertens J. Evaluation of Human Reliability Analysis Methods Addressing Cognitive Error Modelling and Quantification. Forschungszentrum Jülich GmbH (KFA), Jülich, 1996 (in press)
6. Swain A.D., Guttmann H.E. Handbook of Human Reliability Analysis with Emphasis on Nuclear Power Plant Applications, Final Report. NUREG/CR-1278, Washington, DC, 1983
7. Mosneron Dupin F., Villemeur A., Moroni J.M. Paluel Nuclear Power Plant PSA: Methodology for Assessing Human Reliability. 7th International Conference on Reliability and Maintainability, Brest, France, 1990
8. Hannaman G.W., Spurgin A.J., Lukic Y. A Model for Assessing Human Cognitive Reliability in PRA Studies. IEEE Third Conference on Human Facors in Nuclear Power Plants, Monterey, CA, June 23-27, 1985, Institute of Electronic and Electrical Engineers, New York, 1985
9. Moieni P., Spurgin J., Singh A. Advances in Human Reliability Analysis Methodology, Part I: Frameworks, Models and Data. Reliability Engineering & System Safety 1994; 44: 27-55
10. Villemeur A., Moroni J.M., Mosneron Dupin F., Meslin T. A Simulator-Based Evaluation of Operators' Behaviour by Electricité de France. Proceedings of the International Topical Meeting on Advances in Human Factors in Nuclear Power Systems, April 21-24, 1986, American Nuclear Society, La Grange Park, IL, 1986
11. Mosneron Dupin F. Is Probabilistic Human Reliability Analysis Possible?. EdF International Seminar on PSA and HRA, Paris, November 21-23, 1994
12. Hall R.E., Fragola J., Wreathall J. Post Event Human Decision Errors: Operator Action Tree / Time Reliability Correlation. NUREG/CR-3010, Washington, DC, 1982
13. Hannaman G.W., Worledge D.H. Some Developments in Human Reliability Analysis Approaches and Tools. Reliability Engineering & System Safety 1988; 22: 235-257
14. Hacker W. Arbeitspsychologie. Huber, Bern, 1986
15. Dougherty E. Context and Human Reliability Analysis. Reliability Engineering & System Safety 1993; 41: 25-47

# Expert Judgment in Human Reliability Analysis

Lasse Reiman
Finnish Centre for Radiation and Nuclear Safety
Helsinki, Finland

## 1 Introduction

The use of expert judgment in a probabilistic safety assessment (PSA) is often inevitable. One reason for this is the fact that experimental or statistical information, on the basis of which assessments could be made, is not easily or not at all available. Using expert judgment is not extraordinary in science. Almost every scientific endeavor calls for human judgments to be made. Often these judgments are implicitly understood and thereby not explicitly stated. Analysts typically make decisions in defining problems, establishing boundary conditions, collecting and analysing data and selecting analysis methods. Each of these decisions is based on expert judgment by the analyst.

How do experts assess the probability of an uncertain event or the value of an uncertain quantity? Tversky and Kahneman have shown that people rely on a limited number of heuristic principles which reduce the complex tasks of assessing probabilities and predicting values to simpler judgmental operations [1]. These heuristics are quite useful, but sometimes they lead to severe and systematic errors.

If the biases in an expert judgment are to be determined, then the expert's reasons for his beliefs must be found. Expressing these reasons permit a critical comparison of various experts' assumptions and possibly lead to modifications in their way of thinking by allowing the exchange of ideas. The biases and dependence amongst experts can be better understood if the analyst knows the reasons for judgments. This information can also be used to rank experts on the basis of the sophistication of their analysis.

An essential part of a human reliability analysis is a thorough qualitative analysis of the human actions that are being assessed. In the case studies described in ch. 2 and reported in details by Reiman [2] the qualitative analysis was done and methods for the quantitative analysis were developed by the author and experts from different organizations took part in some phases of the quantitative analysis.

## 2 Case Studies

### 2.1 Analysis of Operator Error Probabilities

A modified SLIM-based approach was used in the analysis of cognitive operator actions at the TVO NPP. A group interaction method that followed the principles of the Nominal Group Technique (NGT) was adopted for the first SLIM session. The session was conducted at the power plant and the operator actions to be evaluated were first demonstrated to the experts in the main control room of the plant. When the first estimates of weights and ratings were done each expert briefly went through his assessments giving some arguments for them. Thereafter the experts had a possibility to make questions to each other. The discussion was controlled by the chairman of the session. Finally the assessments were revised without any subsequent discussion.

Based on the experiences of the first SLIM session new methods to evaluate weights were searched to improve the weight assessment procedure. Different methods to assess weights were reviewed. Consequently, the Analytic Hierarchy Process (AHP) of Saaty was chosen as the method to assess weights in the second SLIM session.

The TRC-SI -method was used in the analysis of cognitive operator actions at the Loviisa NPP [2]. As results of simulator experiments were not available, expert judgment had to be used in the assessment of TRCs. Depending on the familiarity of the situation, the experts gave their assessments in the form of time or in the form of a value of a critical parameter. The expert judgments were aggregated using a procedure based on the Bayes Theorem.

In the evaluation of the TRCs the analyst first described the accident sequence without operator actions and the expert evaluated how the crew would act. The interview took place in the form of a dialog. Also the progression of the most important parameters and the use of EOPs were gone through. After the whole sequence of events has been discussed once, the analyst asked the expert to evaluate the diagnosis time for the best crew and for the worst crew they can think in the situation.

The method is based on the hypothesis, that the assessments of the experts are mainly based on their experience of accident and disturbance situations at the plant-specific simulator. Therefore, the combined TRC of the experts has to be corrected to take into account the influence of factors that are present at the plant but not at the simulator. A stress index (SI) was defined and used to correct the TRCs formed directly from the assessments of the experts. The stress index was assessed by the analyst.

## 2.2 Analysis of Test and Maintenance Personnel Errors

When the behaviour and error possibilities of nuclear power plant personnel are analysed, main attention has often been focused on the operating crew working in the main control room of the plant. This is justified, taking into account their essential role in accident management. Their chances of succeeding might be, however, worsened if there exist latent equipment failures at the plant as a consequence of test and maintenance activities.

The aim of the second case study was to develop and test methods intended to analyse human errors, which may take place in connection with scheduled test or maintenance activities outside the main control room. Errors that were studied were of a type that have a possibility to stay unnoticed (latent) at least until the next regular test or the next refuelling shutdown. The dependence of errors between tasks performed in redundant subsystems of a safety system is the most important issue when the safety significance of test and maintenance errors at NPPs is considered.

Error possibilities related with some test and maintenance activities at the TVO and Loviisa NPPs were assessed using the Paired Comparisons and Ranking methods. The methods were used also to assess the dependence of errors related with these activities. The dependence factor k was used as a measure of dependence. The assessment of the dependence factor k has to be based on information concerning the different factors that contribute to dependence. To provide this information, a detailed qualitative analysis of human actions is necessary.

Three tasks were given to the experts. Task A was an assessment of the human error probabilities of the 12 cases using the Paired Comparisons method. Task B was an assessment of the dependence of human errors of the same cases, also using the Paired Comparisons method. Task C was an assessment of the dependence of human errors using the Ranking method. The Paired Comparisons method was modified in task B so that the experts were asked to explain the reasons for their judgments. The purpose of this modification was to make sure that a careful consideration of every comparison is made and to make it possible to study afterwards the experts' reasonings. Three experts from VTT and three experts from STUK made the assessments.

# 3 Summary of Results

## Study 1: Analysis of Operator Errors

For the analysis of cognitive operator actions a modified version of the SLIM and the TRC-SI method were used. Also other methods were used for comparison. In the TRC-SI method a stress index (SI) is used to correct the TRC, formed from the assessments of the experts, to take into account differences between behaviour at a simulator and at a real plant. The stress index was determined using a model that evaluates the stress level of the operators based on the characteristics of an incident. Sensitivity analyses showed that the method is not very sensitive to the hypotheses that it is based on. A method was developed for the weighting of the judgments of experts. In this study the weighting of experts had only a minor effect on the results.

It is evident that great uncertainties are related with the quantification of operator errors. However, when the rankings produced by different methods were compared, it was found that the correlation was statistically significant.

A strive for quantitative results necessitates a detailed qualitative analysis, which often is the most important part. A great amount of valuable qualitative information was produced in the study.

## Study 2: Analysis of Test and Maintenance Personnel Errors

The human reliability and dependence of errors in some test and maintenance activities were assessed using the Paired Comparisons and Ranking methods. One of the most difficult problems in the use of these methods is that they require a large number of judgments in order to produce reliable estimates. Therefore the possible limitations because of the number of experts required was analysed first.

The Paired Comparisons method was modified so that the experts were asked to explain the reasons for their judgments when they assessed the dependence between errors. This made it possible to attain also qualitatively useful results in the study.

The dependence factors for typical omission type errors in test and maintenance activities were found to lie between 0.15 and 0.25. The estimation of the dependence factor based on an analysis of failure reports of one Finnish utility supported the assumptions used in the quantitative analysis.

# 4 Conclusions

First of all it is emphasized that a thorough qualitative analysis is essential to achieve credible reliability estimates. The qualitative analysis should include interviews of the personnel, who perform the tasks, and walk-throughs of the tasks.

In the quantification of control room operator error probabilities plant and sequence specific methods are preferable. The TRC-SI method developed in the study is such a method. The use of the SLIM can be recommended based on the experiences of the study.

Concerning the methods used in the analysis of test and maintenance tasks, the Paired Comparisons was found to be a useful method. The modification of the method, where the experts were asked to give arguments for every judgment, made it possible to obtain information which was also qualitatively useful. Because of resource requirements, the Paired Comparisons can be applied only in limited scope studies. For practical PSA work, a simple dependence model is needed. The model should be validated in studies with data from operating experiences.

In both case studies the experts were asked to express reasons for their judgments. This made it possible to achieve important qualitative information. They can also be used to evaluate the biases in judgments at least qualitatively. Usually the internal consistency of the judgments was rather good in this study, whereas between-expert agreement was not as good but still satisfactory. An important finding was made in the analysis of test and maintenance activities. It was found out that the cultural factors of the organization may have an effect on the judgments of experts and may thus be a source of correlation between experts.

A structured, formal method should be used in the elicitation of expert judgment. Structuring can include using a predesigned set of questions, allowing only particular kinds of communication between the experts and requiring a certain response mode to be used. It also includes a documentation of the method and of how the experts arrived at their final opinions. Such a documentation allows the formal method to be reviewed by peers. Improvements are needed in the current PSA practices in this respect.

A structured group interaction method (The Nominal Group Technique) was used in the SLIM analysis. The process improved the consistency of the assessments. The qualitative usefulness of the method was also high. Most experts expressed satisfaction with the method. The NGT can be recommended if behavioural aggregation is used.

An important question is, how could the reliance on expert judgment in human reliability analysis be reduced. As concerns the behaviour of control room operators, data should be collected regularly at the training simulators in transient and accident situations. This should not disturb the training of operators because to be able to assess TRCs it is sufficient only to measure the time taken for a diagnosis of the situation during a normal retraining exercise. Another, more useful approach would be to perform tests especially for research purposes, as has been done in France. In these test also qualitative data has been gathered (for example departures from strict observance of procedures). This has made it possible to model extraneous operator actions in PSA, which is quite unusual.

As regards the test and maintenance errors, more attention should be paid at the plants to the documentation of errors and their causes and this information should be used to analyse errors in depth. Testing and maintenance activities should be developed so that dependent human errors would be avoided to the extent possible. The efficiency of different inspection practices as error detection mechanisms should be examined in further studies. To reduce uncertainties of PSA, by revealing possible latent organizational problems, analyses of organizational behaviour should also be made.

# References

1. Tversky A, Kahneman D. Judgment under uncertainty: Heuristics and biases. Science 1974; 185:1124-1131
2. Reiman L. Expert judgment in analysis of human and organizational behaviour at nuclear power plants. PhD Thesis, Technical University of Tampere, 1994 (STUK-A118)

# International Survey of PSA-Identified Critical Operator Actions

S. Hirschberg and V. N. Dang
Paul Scherrer Institute
Villigen, Switzerland

L. Reiman
STUK
Helsinki, Finland

**Abstract**

This paper describes some of the central activities within a task of OECD Nuclear Energy Agency Principal Working Group No. 5 on Risk Assessment. The task is entitled "Critical Operator Interactions and Data Issues" and aims at surveying both Human Reliability Analysis (HRA) methodology and results of Probabilistic Safety Assessments (PSAs) related to operator actions. Some insights concerning methods used and characteristics of the analyses of operator actions identified as common for BWRs and PWRs, respectively, are provided and illustrated by specific examples.

## 1 Background

The treatment of human interactions is considered one of the major limitations in the context of PSA. While many PSAs have been successful at identifying critical operator actions, large uncertainties are normally associated with the quantitative estimates of these contributors. Furthermore, the present approaches as applied in PSAs are limited in scope (e.g. errors of commissions are either ignored or treated superficially). New, dynamic methods, primarily aiming at the resolution of the issues of cognitive errors including errors of commission are emerging but their full scope applications within the PSA framework belong to the future. As a recognition of the importance of human interactions and of the need to exchange experiences from their treatment, a relevant task was initiated within PWG5 in 1994. This paper provides examples of some preliminary insights.

## 2 Scope and Objectives

The focus of the task is on the dynamic operator actions which represent the most serious modeling challenge due to the scarceness of data. According to most PSAs, among the different categories of human interactions, they normally also have the highest safety significance. Interactions that occur prior to the initiating event or that lead to a plant transient are not ignored within this task but have a lower priority. Moreover, the concentration is on the actions related to accident prevention; accident mitigation may be included in the future but will probably be considered for a subsequent task after the completion of the present one.

The specific objectives are:

1. Based on a number of available PSAs, to identify critical operator actions during accident conditions, with particular emphasis on interactions of potentially generic interest.
2. To review probabilities assigned to selected interactions and the approaches used for generating these data.
3. To consider the potential for improvements of the current data situation through wider use of already established approaches as well as novel approaches.
4. To identify examples of major modifications in procedures and design resulting from PSA findings that relate to operator actions.

In order to meet these objectives the activity was divided into two parts, i.e. a survey of PSA-identified critical operator actions and a state-of-the-art review including development trends and needs. The present paper addresses only the survey.

# 3 Survey Approach

Through a detailed questionnaire developed for this task, Finland, France, Germany, Italy, Japan, the Netherlands, Spain, Switzerland, the U.K. and the U.S. are sharing information on Human Reliability Analysis (HRA) methods in practice and on the human interactions found to be important. The information was assembled in three phases:

1. **First survey.** This covered the general facts about plants and PSAs, list of important operator actions with the emphasis on the dynamic ones, discussion of the treatment, results of sensitivity analysis, and HRA-based plant modifications (if any). Responses were completed and evaluated for four BWRs, nine PWRs and five advanced reactors. Table I provides the overview of some characteristics of the PSAs and the HRAs carried out.
2. **Detailed treatment.** Some actions identified from the responses as common important ones for BWRs and PWRs, respectively, were subject to a detailed treatment. For BWRs, these include manual depressurization and standby liquid control system actuation; for PWRs, "feed and bleed", and either alignment for recirculation or loss of Residual Heat Removal (RHR). Commonalities and differences in the approaches used to model these actions were examined on the basis of "traceable" descriptions that could be followed step by step, for example, from Performance Shaping Factor (PSF) ratings to the Human Error Probability (HEP) value. Whenever needed, additional details were requested to establish a comparable set of facts about each study.

## Table I Some characteristics of surveyed PSAs and HRAs

| Task Member | CH | CH | D | D | F | F | I | I | I |
|---|---|---|---|---|---|---|---|---|---|
| BWR | B | | | | | | SBWR | | |
| PWR | | P | | P | P | P | | AP600 | |
| Other | | | GCR | | | | | | PIUS |
| PSA Charact.[1] | 2 / I,E FP | 2 / I,E FP | 3 / I FP | 1 / I FP | 1 / I A | 1 / I A | 3 / I,E A | 3 / I,E A | 1 / I,E A |
| HRA Methods Used For Cat. C Operator Actions | FLIM | FLIM | THERP extended with features of HCR, SLIM, HEART, APJ | SHARP both ASEP and THERP | SHARP simulator data simulator -based TRCs ASEP | SHARP simulator data simulator -based TRCs ASEP | SHARP1 ORE/HCR THERP | SHARP1 THERP | SHARP1 THERP |

| Task Member | J | J | J | NL | NL | SF | SF | SP | UK |
|---|---|---|---|---|---|---|---|---|---|
| BWR | B | | | B | | B | | | |
| PWR | | P | | | P | | P | P | P |
| Other | | | LMFBR | | | | | | |
| PSA Charact.[1] | 2 / I FP | 2 / I FP | 1 / I,E + 2 / I FP | 3 / I,E A | 3 / I,E A | 1 / I,E A | 1 / I,E FP | 1 / I +FIRES FP | 3 / I,E A |
| HRA Methods Used For Cat. C Operator Actions | THERP OAT/TRC expert judgment | THERP OAT/TRC expert judgment | SHARP OAT/TRC THERP | OAT/TRC and THERP | SHARP ORE/HCR and THERP | SHARP –[2] | simulator -based TRCs or THERP diagnosis curves expert judgment | SHARP HCR and THERP | HEART / THERP |

[1] PSA Level / I = Internal Events; E = External Events
Modes FP = full power; A = all modes

[2] SF- Probabilities from THERP diagnosis curves used as priors in a Bayesian update with simulator data or plant-specific judgment factors.

# 4 Comments on Methods Used

A wide spectrum of HRA methods were represented in the responses. This includes[1]: (a) Decomposition or Database Techniques (THERP/ASEP, HEART); (b) Time Dependent Methods (OAT/TRCs, HCR); (c) Expert Judgment Based Techniques (APJ, SLIM/FLIM).

In some cases the analyses were supported by simulator experiments. Only a few studies use a single method; in most cases several techniques were combined. The responses provide a number of insights concerning mixing and matching different methods, criteria in the choice of HRA overall approach, and views of reviewers and PSA users. Some of the observations on methods follow below:

- The choice of the most suitable approach may depend on the application. Thus, screening (conservative) approaches may be fully adequate when analyzing a plant under construction or in the conceptual stage. On the other hand, for operating plants the implementation of more refined best estimate approaches is desirable and much more feasible. While the HEART method has not been widely used in nuclear PSAs, the experiences from a full scope application for a plant under construction are encouraging. An "in-house" extension of HEART was implemented in order to account for within-system dependencies. Focus provided by HEART on ways of error reduction is beneficial; the main difficulty lies in the sensitive judgments concerning the appropriate fraction of the maximum effect for a given error producing condition.
- Simulator-based models have definite merits but their applicability is still restricted, which makes it necessary to combine them with engineering/expert judgment. Problem areas include[2]: failure to formulate the correct response, rare situations, time windows exceeding 30 minutes, aggregation of test samples, accounting for factors other than time, difficulties to simulate some situations (particularly during shutdown), and transferability of simulator data to real accident situations.
- Techniques based on expert judgment are extensively applied in PSAs[3]. For the SLIM family of methods significant differences exist between SLIM/SLIM-MAUD and FLIM. Gradual refinements of FLIM are for example reflected in implementations in the Swiss PSAs. There are some key characteristics (dimensions) of SLIM/FLIM implementations, which contribute significantly to

---

[1] A recent review of the different methods and current issues in HRA may be found in [1].

[2] Reference [2] provides a detailed summary of French experiences from application of simulator-based approaches, with emphasis on problem areas.

[3] Discussion of the applications of expert judgement in HRA and its use for analysis of organisational behaviour, as well as case studies are provided in [3].

the validity and consistency of the results, independently of the theoretical basis of SLIM. They include [1]: expertise, elicitation (group process, PSF weighting and rating processes), index formulation, and calibration (data sources, grouping of actions).

# 5 Detailed examination of common operator actions

An example showing the characteristic features of the analyses of manual depressurization, identified as one common important action for BWRs, is given in Table II. There is no intention to directly compare the HEP estimates since the plants represent a variety of designs and differences exist with respect to the degree of automation, procedures and context of the actions. However, the factors that drive the numerical values can be identified.

For manual depressurization the time windows used in the PSAs vary between 20 and 45 minutes; most (but not all) analyses consider stress associated with this action as high and consequently the most significant error producing condition among the PSFs. The estimated HEPs are predominantly of the order of $10^{-2}$, although a much lower value (determined by the short median response time) applies to the Dutch plant. The expectedly non-dominant execution part has either been modeled using THERP, neglected or is considered to be implicitly covered by the integral evaluation (FLIM). In the base case of the Finnish study 41 minutes are available for the decision part (4 minutes within the 45-minute time window were allocated to the manual actions to execute depressurization). The 95th percentile curve of the ASEP nominal diagnosis model was used to establish a prior ($4.9 \times 10^{-3}$) which was then updated using the results of simulator exercises to obtain the final estimate of $1.3 \times 10^{-2}$. In the Japanese case the time window is shorter but the full time available (i.e. 30 minutes) was allocated to the diagnosis period. The same TRC as in the Finnish analysis was used but since no simulator data were available the final estimate is lower.

Other important common actions for BWRs and PWRs have been examined in a similar manner. The task report is in progress. Apart from the results of the survey, it discusses current methods and their limitations, special issues in HRA, as well as revolutionary approaches and priorities for future developments.

## References

1. Dang V. N. and Hirschberg S. Human Reliability Analysis in Probabilistic Safety Assessments: Current issues, the Swiss studies and options for research. Prepared by Paul Scherrer Institute, Würenlingen and Villigen, for the Swiss Federal Nuclear Safety Inspectorate, HSK-AN-2887, December 1995.
2. Mosneron-Dupin F. Is Probabilistic Human Reliability Assessment possible? Presented at EDF International Seminar on PSA and HRA, Paris, 21 - 23 November 1994.
3. Reiman L. Expert judgment in analysis of human and organizational behavior at nuclear power plants. Ph.D. Thesis, Finnish Centre for Radiation and Nuclear Safety, Helsinki, STUK-A118, December 1994.

## Table II Treatment of manual depressurization in different PSAs

| Country-Study | CH - Müh | I - SBWR | J - B1100 | NL - Dod | SF - TVO | |
|---|---|---|---|---|---|---|
| Initiating event | General transient (non-ATWS) | sLOCA | General transient (non-ATWS) | General transient (non-ATWS) and small LOCAs | Transient resulting in loss of all main feed pumps | |
| Sequence(s) | Feedwater and Reactor Core Isolation Cooling (HP Injection) failed | Failure of automatic start of HP Injection and depress. systems | Stuck open safety relief valve<br>HP Core Spray failure | (Multiple sequences) | HP Injection failed | Stuck open safety relief valve<br>HP Injection failed |
| Fussell-Vesely | 1.2 E-1 | 1.2 E-1 | 1.2 E-1 | 7.73 E-3 | 2.4 E-2 | 2.9 E-3 |
| R. Ach. W. | – | – | – | 78.4 | – | – |
| Seq.frequency /yr | 8.3 E-7 | 3.1 E-8 | 1.2 E-7 | 4.6 E-7 | 3.4 E-7 | 4.1 E-8 |
| HEP | 1.30 E-2 | 8.8 E-2 | 2.9 E-3 | 1.0 E-4 | 1.3 E-2 | 3.6 E-2 |
| Time available | 25 min. | 35 min. | 30 min. | 40 min. | 45 min. | 20 |
| Median response time | n/a | 15 min. | n/a | 2 min. | n/a | n/a |
| Important PSFs | seven equally weighted PSFs<br>'Preceding and concurrent actions' and Stress rated worst | moderate dependence RO-SRO<br>Stress level: 'grave emergency' (factor=5) | moderately high stress, good interface (for calculation of execution part) | "no burden" assumed | Stress related to decision-making burden<br>Procedures and training good | |
| DDD | integral evaluation | 8.69 E-2 | 2.7 E-3 | 1.0 E-4 | 1.3 E-2 | 3.6 E-2 |
| Execution | | 9.72 E-4 | 2. E-4 | neglected | considered neglibible | |
| Recovery | implicit | credited (exec.) | credited (exec.) | implicit | implicit | implicit |
| Methods used to assess this action | FLIM | HCR/ORE for DDD<br>THERP for execution part | THERP:<br>-Nominal diagnosis median curve<br>-Execution[1] | TRC | ASEP nominal updated with simulator data | |

[1] Execution part of HEP includes recovery factors and hardware failures (HRA event tree).

# Analyzing the Loss of RHR Using a Dynamic Operator-Plant Model

V.N. Dang[a], N. O. Siu[b]
Nuclear Engineering Department
Massachusetts Institute of Technology
Cambridge, MA 02139, U.S.A.

**Abstract**

OPSIM is a dynamic operator-plant simulation model intended for the analysis of operator-plant interactions, including the cognitive behavior of the operators. This paper describes the model and the application of OPSIM to the analysis of a pressurized water reactor (PWR) scenario, the loss of Residual Heat Removal (RHR) during mid-loop operation.

## 1. Background

To address 'errors of commission' and cognitive error issues, a number of simulation models that treat the dynamic evolution of the operator-plant system have been developed, e.g. [1-4]; some current efforts are [4],[5],[6]. These consider the interactions between the response of the plant (the evolution of the accident sequence) and that of the operators and thus the dynamic context of operator actions more broadly and explicitly than has been generally the case in the current PSA framework of event trees and fault trees.

This paper describes an application of OPSIM to treat the response of an operator to a loss of RHR during shutdown. The cognitive model of the operator in OPSIM uses a blackboard architecture; the overall simulation is implemented in C++ within a discrete event simulation framework [7]. In contrast to to other models, OPSIM emphasizes operator behavior based on schemas, "large" rules described below, in combination with procedure-following. The model is applied to the analysis of the response of a single operator in loss of RHR scenarios; the purpose of the runs described in this paper is to demonstrate the schema-based operator model and its application to the analysis of cognitive errors.

### 1.1. The Loss of Residual Heat Removal (RHR)

The loss of RHR scenario analyzed in this work occurs during a 'mid-loop' plant state in a PWR. In this state, the reactor coolant system (RCS) level is lowered to the nominal mid-level of the primary coolant loops in order to allow some

---

[a] To whom correspondence should be addressed. Current address: Paul Scherrer Institute, CH-5232 Villigen PSI, Switzerland

[b] Idaho National Engineering Laboratory, P.O.Box 1625, Idaho Falls, ID 83415-3855, USA

maintenance operations, such as work on steam generator requiring their draining. Under these conditions, the RCS is typically vented and at atmospheric pressure and the reactor pressure vessel head is installed. Loss of RHR in this plant state is of interest from a risk standpoint because of the relatively short time before boiling conditions are reached in the RCS.

The RHR system removes decay heat by circulating RCS coolant through its heat exchangers. Thus, loss of RHR events can be divided into two groups according to their causes: those related to flow, e.g. RHR flow is lost, and those related to heat removal, e.g. the heat exchangers fail to remove the decay heat. In mid-loop conditions, the low RCS coolant level (which leads to the short time to boiling) is also of concern because it increases the potential for vortexing and cavitation in the RHR pumps, which would cause them to shut off.

### 1.2. The Scenario Simulated and Nominal Response of the Operator

In the specific scenario simulated for the analyses discussed here, the plant is in mid-loop conditions with a single active RHR train; the second RHR train is in standby. The scenario starts with a spurious closure of a component cooling water (CCW) valve, isolating CCW flow to the heat exchanger of the active RHR train.

An abnormal operating procedure (AOP) guides the operators in loss of RHR events during "reduced inventory" conditions. This single procedure is intended to deal with the various causes that may lead to the loss of decay heat removal, i.e. insufficient RCS level, vortexing and cavitation of RHR pumps, loss of the heat exchanger, etc. In the nominal response to the simulated scenario, the operator should enter the procedure after detecting increasing RCS temperature. The procedure instructs the operator to verify RCS level, flow through the RHR, conditions for RHR pump operation, and heat sink for the RHR heat exchanger. If RHR operation cannot be restored in a timely manner, the procedure instructs the operator in establishing alternate means of decay heat removal.

## 2. The Operator Model in OPSIM

### 2.1. A Schema-Based Model

The model of the individual operator is based on the schema theory, a descriptive model of how people use schemas, a type of mental model, to understand and respond to familiar and unfamiliar situations. In the schemas, cues from the plant, such as indications and alarms, are associated with plant states (i.e. a given event), system or equipment states (e.g. cavitation in a pump). Secondly, schemas associate responses, the actions that the operator should take in a given situation, with these states. Schemas are activated when their cues are identified, leading to a current assessment of a situation; the active schemas are then used to identify the response appropriate for the context. Because expectations are also associated with a situation in the schemas, active schemas also affect subsequent interpretations. A more detailed description of schema theory is provided in [8].

From a modeling standpoint, schemas are approximately equivalent to the frames or scripts used in some artificial intelligence applications to model dynamic knowledge and behavior in context-rich problems. Schema theory may explain a range of human behaviors in complex problem-solving contexts such as nuclear power plant operations, in well-trained situations as well as in unfamiliar situations in which the cognitive load is large or the time too limited for knowledge-based reasoning at the level of "basic principles". Ref. [9] indicates that the schema or frame appears to describe airplane pilot diagnostic behavior, which shares features of problem-solving in the nuclear plant context. In the terms of the skill-/rule-/knowledge-based behavior framework [10], schema theory deals with the rule-based and part of the knowledge-based modes of behavior.

The schematic knowledge in OPSIM concerns mainly plant operations and scenarios. The operator model also contains generic knowledge for following and interpreting written procedures as well as other procedural knowledge. In addition, the OPSIM operator model is able to resolve tasks referred to in procedures and to insert other tasks (e.g. a response associated with a state).

### 2.2. Errors

OPSIM can treat a variety of errors and their effects. Errors can be modeled at the level of the knowledge base (long-term memory) to reflect erroneous mental models, for instance, incorrect knowledge of plant layout, inadequate rules for situation assessment from plant indications, or inappropriate responses for given states of the system. In addition, errors in cognitive processing, e.g. recall of knowledge from long-term memory or of plant indications, may also be treated.

All forms of procedural knowledge, that is, written and memorized procedures and procedures associated with schemas, can be easily varied because all are represented in a canonical text form and not part of compiled code. As a result, both variations and errors in procedural knowledge can be modeled; these might be the result of simplifications of a written procedure, of errors of memory, or of confusion with other similar task procedures or with task procedures for different situations. Besides cognitive errors, the appropriateness of written procedures in combination with 'typical' operator responses and expectations can therefore be examined in variations of a scenario (in time response or in contributing causes).

The runs discussed in this paper address mainly two issues: first, potential consequences of a slip, which is recovered, compared to those for a mistake; second, expectations and assumptions and their effects on diagnosis and response.

## 3. Analysis Results

### 3.1. Operator Responses Treated

Two sets of runs are described in this paper. In the first set, the nominal response as described in Section 1.2 is treated along with several responses in which the operator applies schematic knowledge to decide whether the written procedure should be applied (i.e. the procedure 'entry conditions' are fulfilled); in some cases, the operator is also postulated to address the scenario cause directly, without entering the procedure.

In the second set of runs, the analysis considers a slip and a mistake. In the scenario, the operator unintentionally reads indications for the RHR (inactive) train B while checking the heat sink (component cooling water (CCW)) for the RHR train A heat exchanger. In the case of the slip error, the operator recovers; on the other hand, in the case of the mistake, the train A/train B switch of the CCW instrumentation in the operator's knowledge base persists throughout the simulation.

Figure 1 shows the temperature evolution for the nominal response to the operator from the first set of runs and its corresponding timeline (first from top), which

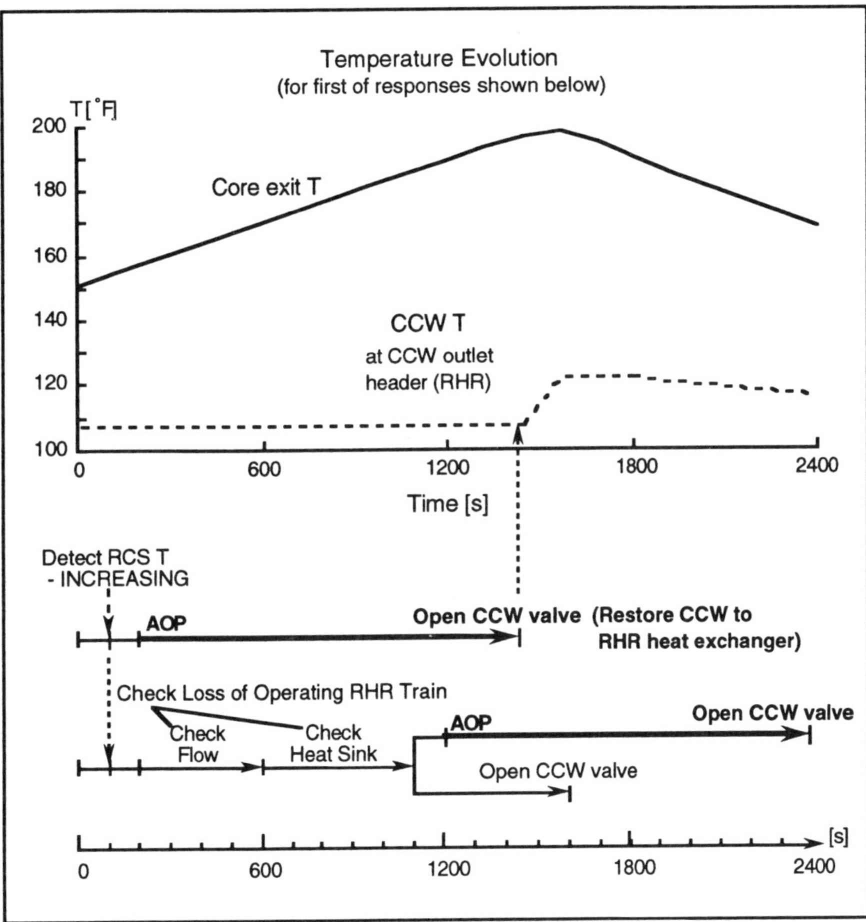


Figure 1. Simplified timelines for error-free responses and temperature evolution for one case.

indicates the major phases of the response through to the restoration of CCW to the RHR heat exchanger. Two other responses simulated are sketched in the timelines shown at bottom. In these postulated schema-based responses to the same scenario, the operator spends some time trying to determine the cause of the increasing RCS temperature; he checks for flow problems in the RHR system and for loss of RHR heat sink, both of which lead to a rise in RCS temperature. In the first of these, the operator then enters the procedure; in the second, the operator restores CCW (without entering the procedure) when he determines the problem with RHR heat sink and its cause. In all cases, the plant model coupled in the simulation model shows that CCW temperature initially increases sharply when CCW is restored; with some delay, the core exit temperature trend 'turns' down.

Quantitative response data such as the times for key elements of the operator response, the behavior of important plant parameters, e.g. the temperature at the core exit, and the margins for these are obtained for each response. With the currently assumed values of the durations of actions and of cognitive processes, the analysis shows, for example, that following the written procedure is slower than the postulated schema-based responses in some cases and faster in others.

### 3.2. Slips and Mistakes

Table 1 summarizes the outcomes of the runs with the slip and the mistake. In the reference case (without errors) and the slip (1A) case, the operator correctly identifies the problem. In the slip (1B) case and in the mistake case, the operator incorrectly concludes a control valve failure and no RHR flow to the RHR heat exchanger. Figure 2 shows the timeline of the slip (1B) case leading to this erroneous

Table 1. Outcomes of simulation runs with slips and mistakes.

| | Case | | | |
|---|---|---|---|---|
| 37.5 mins. after CCW isolation | Reference case (no errors) | Slip (1A) | Slip (1B) | Mistake |
| RHR Train A assessment | unavailable: CCW isolation valve - failed closed (correct) | | unavailable: no RHR flow to RHR heat exchanger due to control valve failure | |
| RHR Train B assessment (is actually in service) | in service | in service | in service | unavailable: CCW isolation valve failed to open |
| Final state | Verifying RCS Temperature - Stable or Decreasing | | | Checking RCS level and initiating alternate decay heat removal |
| RCS Temperature | ~ 175 °F, Decreasing | | RCS T ~ 186 °F, Decreasing | |

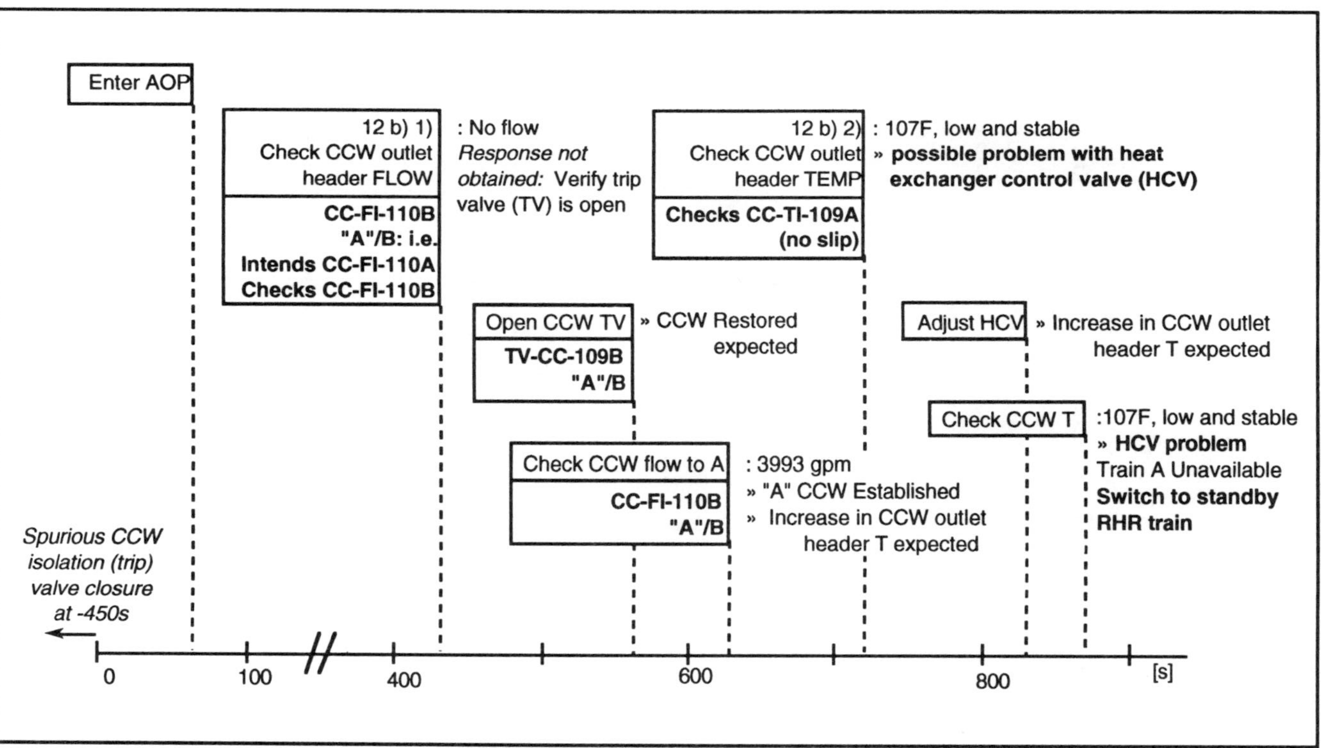


Figure 2. Part of timeline for a simulated response. Case with slip error (1B).

conclusion. The boxes show the actions taken by the operator; the lower part of the box shows the action actually taken when it differs from the intent (shown in quotes). To the right of the boxes are the indications read by the operator, and the conclusions and expectations these lead to.

Note that regardless of the diagnosis, the operator switches the RHR operating train to train B in all cases. However, in contrast to the slip (1B) case shown in the figure, the operator is predicted in the mistake case to conclude that train B is also unavailable after placing RHR train B in operation. This cognitive error is based on his expectation that the CCW temperature will increase as the train is placed in service, which is not met because he has erroneously checked the train A temperature trend. Consequently, he decides, as directed by the procedure, to implement an alternate means of decay heat removal that is not necessary in this context.

# 4. Conclusion

A schema-based model of operator behavior has been developed for modeling misdiagnosis and other cognitive behaviors and errors, resulting in the implementation of an operator-plant simulation, OPSIM. Compared to other models, OPSIM places more emphasis on operator cognitive behavior and potential errors during procedure-following. This allows interactions between the plant operating procedures and the rules, expectations, and assumptions used by the operator during the evolution of a scenario to be analyzed. The causes for situation assessment or response selection errors, for procedure misinterpretation, and for deviations from the procedures can thus be considered. In addition, the schema-based model has potential advantages for modeling the operators' adaptation of procedures and responses in unusual situations. For the validation and further development of OPSIM in progress, better characterizations of operator behaviors and knowledge are most important.

# References

1. Woods DD, Pople HE Jr, Roth EM. The Cognitive Environment Simulation as a tool for modeling human performance and reliability. NUREG/CR-5213, US NRC, DC, 1989
2. Cacciabue PC, Mancini G, Bersini U. A model of operator behavior for man-machine system simulation. Automatica 1990; 26(6):1025-1034
3. Dang VN, Huang Y, Siu N, Carroll JS. Analyzing cognitive errors using a dynamic crew-simulation model. In: Proceedings of the 1992 IEEE Fifth Conference on Human Factors and Power Plants, Monterey, CA, 7-11 June 1992
4. Tanabe F. Human factors research in JAERI. Halden-Japan Joint Symposium on Fuel and Material and Man-Machine Systems Research, Tokyo, 22-23 May, 1995
5. Dang VN, Siu N. Simulating operator cognition for risk analysis: current models and CREWSIM. In: Proceedings of PSAM-II, San Diego, CA, 20-25 Mar, 1994, pp. 066/7-13
6. Shen SH, Smidts C, Mosleh A. IDA: A cognitive model of nuclear power plant operator response during abnormal conditions. UMNE-94-006, U of Maryland College Park, 1994
7. Dang VN. Modeling cognition for accident sequence analysis: development of a dynamic operator-plant simulation. PhD thesis, Massachusetts Institute of Technology, Cambridge, to be published, 1996
8. Posner MI (ed). Foundations of cognitive science. The MIT Press, Cambridge MA, 1989
9. Smith PJ, Giffin WC, Rockwell TH, Thomas M. Modeling Fault Diagnosis as the Activation and Use of a Frame System. Human Factors 1986; 28(6):703-716
10. Rasmussen J. Human Reliability in Risk Analysis. (Sec. 1.6.1.) In: Green AE (ed). High Risk Safety Technology. John Wiley & Sons, Ldt, 1982, pp 143-170

# Derivation of Fatality Probability Functions for Occupants of Buildings Subject to Blast Loads

R M Jeffries, L Gould and D Anastasiou
WS Atkins Consultants, Epsom, UK
A P Franks
Health and Safety Executive, Merseyside, UK

## Abstract

Risk assessments for proposed developments in the vicinity of major hazard plant include consideration of the effect of possible blast loads from explosions. Blast can cause injury to people either through direct effects, secondary effects such as the impact of fragments of glazing or cladding, or collapse of the building. Some commonly used existing methods for assessing fatalities due to debris or building collapse do not take into account the characteristics of the building, and consequently may not give sufficient information to assess occupant vulnerability with confidence. A procedure has been derived for use in risk assessments where explosions are considered. The procedure is used to determine the failure sequence of a building subject to increasing blast loads and investigate the effect on people within the building of either debris or glazing fragments generated by the blast load or partial/ total collapse of the load-bearing structure. In this way the comparative vulnerability of occupants of different building types to blast loads can be assessed based on the building's primary constructional characteristics. This paper concentrates on the initial stage of the work in which the procedure for predicting vulnerabilities was developed.

## 1. Introduction

In the UK, the Health and Safety Executive (HSE) provide advice to local planning authorities on the safety implications of proposed developments in the vicinity of major hazard plant. In order to do this, a quantified risk assessment approach which produces estimates for both individual and societal risk may be used. This QRA includes consideration of the effect of possible blast loads from explosions which can cause injury to people either through direct effects, secondary effects such as the impact of fragments or the total/partial collapse of buildings. At present many of the methods used to assess fatalities due to building collapse do not fully take into account the building characteristics. For example, within the computerised analysis tool RISKAT, fatalities due to building collapse caused by overpressure are assessed from a probit derived from World War II flying bomb data.

This paper describes an ongoing project to develop a procedure for assessing the vulnerability of occupants of different buildings subject to overpressures produced from vapour cloud explosions. The failure sequence of a building subject to a blast load has been determined and the effect on people within the building of either debris generated by the blast load striking them or partial/total collapse of the load bearing structure has been assessed. The aim is to produce a consistent methodology for assessing the comparative vulnerability of occupants of differing types of buildings to blast loads, based on the building's primary constructional characteristics.

## 2. Human Vulnerability

The overall purpose of the study is to derive a methodology for determining the vulnerability of the occupants of buildings to blast overpressures. This is dependent on the level of overpressure, and the type and construction of the building. The probability of fatality is dependent on the probability of acquiring a specific injury and the probability of that injury being fatal.

In general, three categories of blast induced injury can be identified [1]:

**Primary injury** is due to the direct effects of blast overpressure on people. The location of most severe injuries is where the density differences between adjacent body tissues are greatest, i.e. the lungs, ears, abdominal cavity, larynx and trachea.

**Secondary injury** is due to impact by missiles generated by the explosion, and to building collapse. This can give rise to laceration, penetration and blunt trauma.

**Tertiary injury** is due to displacement of the entire body followed by high decelerative impact loading which is when broken or fractured limbs can occur.

In this study, we are primarily interested in causes of fatality for building occupants, and hence the most relevant injury type is secondary injury caused by missiles, either from failed glazing or structural components, or from building collapse. Primary and tertiary injuries are less important at the ranges and overpressure levels considered, although impairment of hearing or lung damage may affect the ability of people to escape from collapsed buildings.

## 3. Structural Assessment

The capacities of buildings subject to blast loading have been assessed, as an input into the calculation of the vulnerability of the building occupants. Buildings of interest for this study are those existing in urban areas, typical of sites to be considered in the planning of a major on-shore hazardous installation. This includes residential housing, together with offices, retail developments, hospitals, schools and leisure centres, and covers the whole range of building types and geometries. In estimating building capacities, the pressure loads, structural response and structural capacities have been investigated for generic building types, as outlined below.

### 3.1 Pressure Loads

For a vapour cloud explosion (VCE) of known combustion energy, the TNO multi energy method [2] can be used to estimate peak incident overpressures, pulse shapes and durations at a range of distances from the source. Curve 7 is typically used for VCE combustion energies. This method has been used in order to define the ranges of pressure and impulse of interest, but it should be noted that the methodology for predicting vulnerability is independent of the blast model selected.

For a VCE, at locations close to the source, the pulse shape is close to that of a pressure pulse, but as the wave travels away from the source it 'shocks' up, and in the far field the pulse resembles a shock pulse. Quantifying the 'shocking up' effect is difficult to do: Puttock [3] presents a method for predicting pulse shape at a distance from the source, but his method cannot quantify the partial shock at intermediate distances. This may be important in estimating structural loads.

## 3.2 Structural Loads

The load experienced by a structure as a result of an incident overpressure is a complex function of the incident pulse, the building length and breadth and the vented areas. Externally, the horizontal loading can be calculated from:

- **Reflected Pressure**: at the front face, the primary loading comes from the reflected pressure, which can have a value considerably higher than the peak incident overpressure, dependent on the building orientation with respect to the blast wave. The interaction of the building with the incident wave results in a rarefaction wave progressing from the edges of the front face towards the centre, which causes the initial reflected value to reduce in time, and eventually, the reflected pressure reduces to the incident overpressure plus the dynamic pressure described below. It is suggested [2] that for a pressure pulse, the rarefaction wave will effectively counteract the reflection such that the resulting pressure will at no time be higher than the incident pressure plus the dynamic pressure. This dependency of the level of reflection on the pulse shape has implications for intermediate shaped pulses, where the pulse shape cannot at present be quantified. Assuming that the pulse shape is a shock wave is generally the most conservative approach.
- **Dynamic Pressure / Drag**: the secondary loading comes from a combination of the incident overpressure and the building drag as air is displaced around the building.
- **Rear Face Loading**: when the blast wave reaches the rear of the building, it may exert a force on the rear face that opposes the front face load and reduces the overall translational load. This is dependent on the time taken for the wave to reach the rear face, i.e. the building length and the wave speed, plus the rise time of the pulse.

In addition to the external loads, the building may be subject to an internal load if any of the external glazing or cladding fails. The average internal pressure rise can be estimated from the pressure differential across the opening, the area of the opening and the volume of the building [4]. Calculation of an average pressure within the building does not predict the time difference between pressure increase at the front of the building and pressure increase at the rear. Using this approach is conservative, however, as it underestimates the internal pressure at the front face and thus overestimates the overall load to which the building is subjected.

The external pressures acting on the front and rear faces are converted into loads by multiplying them by the total available area of the respective building face. Thus the internal pressure reduces the overall load, primarily by reducing the front face area available for external loading. This is particularly important for framed structures, where pressure relief due to cladding failure may be high enough to ensure that overall building collapse does not occur.

## 3.3 Structural Response and Failure Sequence

The dynamic response of a building is dependent on its natural frequency and the elastic limit of the structural components. Using a non-linear single degree of freedom elasto-plastic model, the maximum displacements, velocities and accelerations in the structure can be derived under a prescribed pressure load [1]. If the displacement of the structure is known, the maximum allowable overpressure for a particular pulse duration can be predicted. The method for assessing dynamic response is applied to specific structural components in order to assess their maximum overpressure ranges. Having done this, the failure sequence for the building can be determined. For the majority of buildings, the first structural component to fail under blast loading will be the glazing, followed by light cladding, walls and finally the building frame.

The effect on the overall building response of the different load components, and their relative importance to the overall vulnerability prediction has been assessed [5]. It has been shown that the building response is dependent on the amount of pressure relief afforded by front face glazing failure, although this is more noticeable for a shock wave than for a pressure wave, primarily due to the reflection of the shock wave but also to the finite rise time inherent in a pressure wave which results in an internal pressure similar to the external pressure. The internal pressure is also very important for determining pressure differentials across the external building walls, and hence the probability of wall failure. Rear face pressure relief can play an important part in reducing overall response, particularly for buildings which are relatively short with respect to their width i.e. for which the travel time to the rear face is short.

## 4. Vulnerability Assessment

The general methodology for assessing building occupant vulnerability is based both on a knowledge of the behaviour of the structure and the failure capacities of the various components, and the individual vulnerabilities to the different hazards, as shown schematically in Figure 1. The most widely used information concerning the fatality probability for different types of injury is presented in Baker et al [1].

### 4.1 Glazing / Cladding Failure

Injuries from glazing can be either due to penetration or blunt trauma. The most onerous vulnerability criterion identified is that which relates to the probability of skull fracture, which is based on in vivo experiments on sheep and dogs performed by Fletcher et al [6]. Injuries from debris are also primarily due to blunt trauma i.e. skull or other bone fracture.

In order to use these criteria, it is necessary to predict the possible fragment mass and velocity arising from the failure of a glazing/cladding panel under dynamic loading. For glazing, [1] provides equations for mean mass and velocity, based on reference [6]. The velocity equation appears to give a reasonable agreement with the limited amount of experimental data available, but the mass equation may under-predict the observed mean mass considerably at low pressures [7].

For debris, no such simple equations are available, and it is necessary to estimate a suitable size of fragment. Having done this, it is possible to estimate its velocity from the strain energy required to initiate failure and the calculated velocity of an unconstrained fragment for the incident blast wave. Larger fragments have been shown to be more injurious than smaller ones, and hence the largest realistic size has been used to estimate debris vulnerabilities.

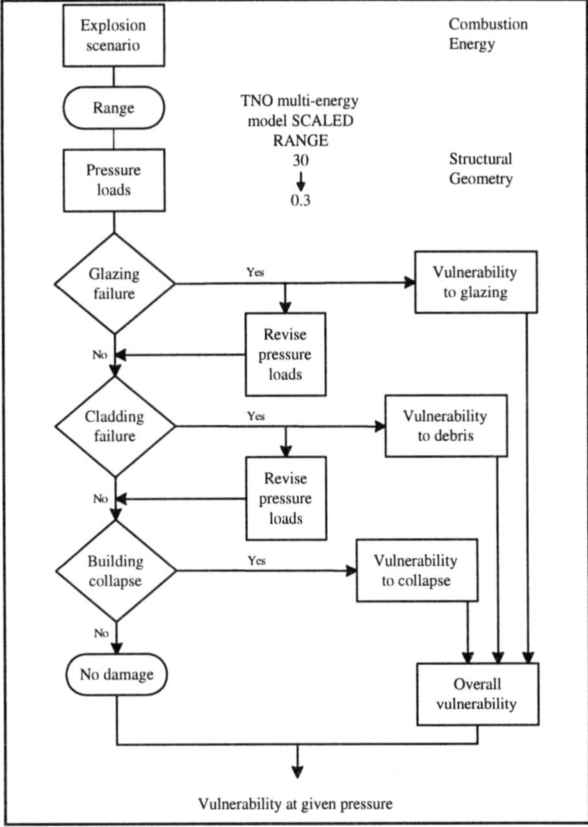


**Figure 1** Vulnerability Prediction

## 4.2 Building Collapse

In addition to vulnerability to glazing and fragments of debris, the fatality probability in the case of complete building collapse needs to be considered. This is particularly important for brick buildings, such as residential housing, where the brick walls form the loadbearing frame, and wall failure leads to collapse. Information has been obtained from references detailing the number of fatalities due to building collapse during earthquakes. This is generally relevant, with the following provisos. Firstly, the time taken for a building to collapse in an earthquake may be much longer than in an explosion, in which case, occupants have more time to leave the building, and they are less likely to be trapped in the collapsed structure. In addition, the fatalities can depend on the type of building: it is estimated that in a masonry structure, 20% of the occupants trapped will receive fatal injuries on collapse, whereas in a reinforced concrete structure, 40% of the injuries received will be fatal [8]. However, the directional nature of the loading for an air blast on the side of a building may result in localised collapse, whereas movement of the foundations in an earthquake may cause global structural collapse. Thus in an explosion, it may be more appropriate to factor the percentages by the proportion of the building which has collapsed. Post collapse, people trapped in the rubble will die if they are not rescued. In an earthquake, the fatalities post-collapse are largely dependent on the speed at which emergency rescue services can be facilitated. This is less so for an explosion where the damage may be localised and rescue efficiency is expected to be high.

There appears to be little further information available in the literature relating to experimental work on human response, although descriptions of the results of terrorist bombs and accidental or military explosions may contain some data. Descriptions of the effects of terrorist bombs, [9, 10] focus mainly on the extent of damage and the numbers of casualties/fatalities, but little information is given concerning the causes of injury or death or the locations of the fatalities.

## 4.3 Overall Vulnerability

The overall vulnerability of building occupants as a result of damage due to blast loads is based on a summation of the effects of glazing, wall debris and collapse, taking care to prevent 'double-counting' of fatalities. Uniform building occupancy has been assumed.

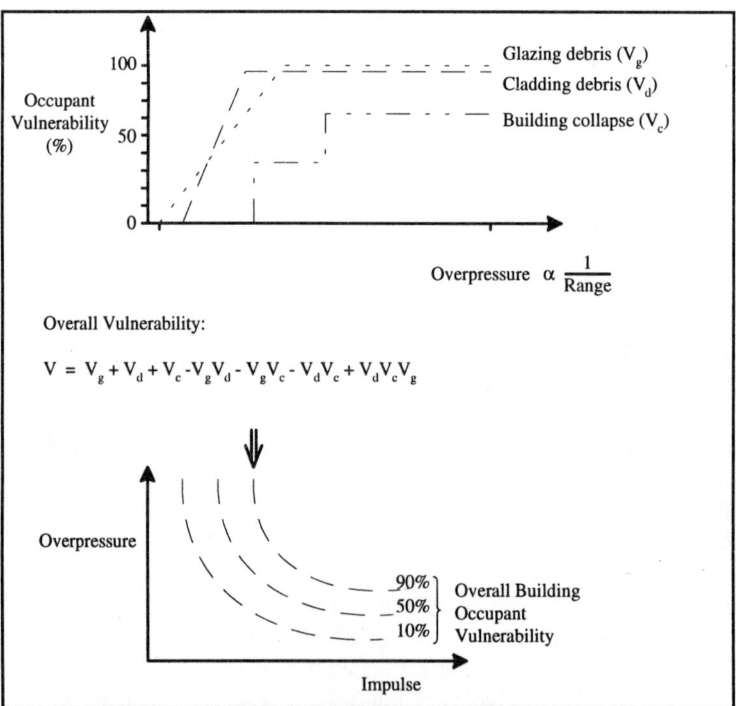


**Figure 2** Schematic Vulnerability Curves

At fairly low overpressures, the failure pressure of the glazing in the building is likely to be the dominant factor. As the pressure increases, the vulnerability is assessed by considering the speed at which fragments are travelling, the maximum range of the fragments, and the area over which the fragments are generated. For a low proportion of glazing in the building, the corresponding vulnerability will be low. Vulnerability due to debris will generally start to occur at higher pressures, dependent on the failure pressure of the cladding or walls and the differential pressure across those walls. Again, the value is dependent on the maximum range, the velocity and the potentially vulnerable area. Building collapse may occur at higher pressures dependent on the extent to which the pressure is relieved through glazing and/or cladding failure. This is shown schematically in Figure 2.

## 4. Conclusions

A methodology has been derived for assessing the probability of fatality for occupants of a building subjected to a blast load, based on the construction characteristics of the building. This methodology is to be applied to various generic building types and the results compared against available historical and experimental data.

## References

1. Baker, W.E., Cox, P.A., Westine, P.S., Kulesz, J.J. and Strehlow, R.A. 'Explosion Hazards and Evaluation'. Elsevier Scientific Publishing Company, Oxford, 1983.
2. 'Methods for the Determination of Possible Damage', CPR 16E, TNO, 1992.
3. Puttock, J.S. 'Fuel Gas Explosion Guidelines - the Congestion Assessment Method', ICHEME Symposium Series No 139, 1995.
4. 'Fundamentals of protective design for conventional weapons' Technical manual 5-855-1 Headquarters, Department of the Army, 3rd November 1986
5. 'Derivation of Fatality Probability Functions for Occupants of Buildings Subject to Blast Loads.', WS Atkins Science and Technology, To be published.
6. Fletcher, E.R., Richmond, D.R. and Yelverton, J.T. 'Glass Fragment Hazard from Windows Broken by Airblast'. Lovelace Biomedical and Environmental Research Institute, 1980.
7. Nowee, J.'Dynamic Failure Pressure and Fragmentation of Thermally Hardened Window Panes', TNO Report No. PML 1985 - C - 103, 1985.
8. Coburn, A.W., Spence R.J.S. and Pomonis, A. 'Factors Determining Human Casualty Levels in Earthquakes. Mortality predictions in building collapse.' The Martin Centre for Architectural and Urban Studies, University of Cambridge, UK.
9. Jenkins, L.D. 'The St. Mary Axe Bomb: Coping with the result of a Large Explosion in a City Centre', AME Annual Conf., Torquay, 28 June - 1 July, 1993
10. Brismar, B, and Bergenwald, L 'The terrorist bomb explosion in Bologna, Italy, 1980. An analysis of the effects and injuries sustained', Journal of Trauma 22 (3), 1982

# Development of Real-time Dose Assessment System for Korean Nuclear Power Plant


Kuk Ki Kim and Kun Jai Lee
Dept of Nucl. Eng., KAIST
Kusong-dong, Yusong-gu
Taejon, 305-701, Korea
(82-42) 869-3818

Won jong Park
KINS
Kusong-dong, Yusong-gu
Taejon, 305-338, Korea
(82-42) 868-0283



**Abstract**

The dispersion of the released radionuclides and corresponding exposed doses are calculated using Gaussian-Puff model for the emergency response The RADCON (RADiological CONsequence analysis) version 3.0, newly developed, is capable of simulating emergency situation by considering continuous washout phenomena, and provide a function of effective emergency planning.

The official regulatory institute, KINS (Korea Institute of Nuclear Safety), has been authorized and developing Computerized technical Advisory system for the Radiological Emergency preparedness (CARE).

In this paper, in line with the CARE system, we presented the result of a modularized intermediate-level emergency dose asessment computer code which is operatable on PC. The source files are coded by using C language in order to increase the compatibility with the other computer systems. The developed code is modularized to adjust the functions and characteristics of each module for easy understanding and further modification of the code.


# 1. Introduction

In a NPP(nuclear power plant) severe accident, radionuclides are dispersed into the air by the mechanism of dispersion. Then external and internal exposure to the public will be followed. The estimated radiation dose to the public near the site must be estimated accurately and will provide very useful and important information for the decision-makers of the emergency situation.

After a few NPP severe accidents occurred in other countries, the importance of developing the emergency dose assessment computer codes has been realized in Korea also. Most of the Korean NPP's are located on the coastal regions, and the surrounding terrain around the sites are usually very complex and hilly. So recently much attention has been focused to develop a very complex and large scale computer code for the effective emergency preparedness.

However, in a NPP severe accident, it is also very important to predict how much time it will take until the legal institutes and local government respond and take action to protect the public. By a large scale computer code, it usually takes a very long time to respond to the NPP severe accident. For a real-time dose assessment, sometimes it is as much necessary to develop an intermediate-level emergency dose assessment computer code as the larger one. The official regulatory institute, KINS, has been authorized and developing CARE system. In this paper, in line with the CARE system, the result of a modularized intermediate-level emergency dose asessment computer code which is operatable on PC is presented.

# 2. Theoretical Background

## 2.1 Atmospheric Dispersion

In order to simulate the NPP emergency situation in mesoscale range, Gaussian puff model is assumed to calculate atmospheric dispersion. The derivation of the equation is discussed in many places including the works of Pasquill (1974)[1], Gifford(1968)[2] and Csanady(1973)[3]. This model has three basic assumptions[4]

1. Continuous plume can be approximated by a finite number of puffs released in succession.
2. Along-wind and cross-wind diffusions occur to the same extent.
3. The height of the puff center above ground level is constant.

For the washout rate calculation, Brenk and Vogt's[5] model is utilized and Sehmel's model[6] for a dry deposition velocity (0.01 m/s) for all radionuclides was adopted. In this work, the processes such as weathering and resuspension are ignored, so it is assumed that the only depletion process for the radionuclide deposited on the ground surface is a radioactive decay.

Brigg's formulation[7] and the NRC method[8] were adopted to RADCON 3.0 to calculate the horizontal and vertical diffusion coefficients. Puff release rate can be specified by user, however, the default value is set to 6 puffs per hour(i.e. each puff releases every 10 minutes).

## 2.2 Dose Calculation

The exposures considered in this code are shown in Table 1. Cloud shine is the major source for an external dose from a series of puffs. Its calculation is very time-consuming and contains lots of uncertainties compared with the other dose calculation procedures. It is due to the discrete gamma energies and the difficulty of considering the real distribution of the radionuclides in a puff. When the sigmas describing the distribution of the concentratioins in a puff are large enough compared to the mean free paths of gammas emited by radionuclides in the puff, the puff can be treated as if it composed of a cloud with infinite dimensions. So the semi-infinite cloud model calculates the cloud shine.

Total whole body dose is the sum of external dose and internal whole body dose. Total whole body dose and thyroid dose are important in the emergency response and used as main criteria for the decision-making.

| External Exposure | Internal Exposure |
|---|---|
| External Dose From Puff (Cloud shine) | Internal Whole Body Dose |
| External Dose From Contaminated Ground (Ground Shine) | Internal Lung Dose |
| | Age-Dependent Thyroid Dose |

Table 1. Exposures considered in this code

# 3. Code Features

In our previous work[9], the source files had been originally coded by using C language to be operatable on SUN-4 SPARC workstation. We have further modified and extended this code as it can be operatable on PC(486 or 586) whose operating system is LINUX. LINUX is operatable in the protected mode. Basically RADCON 3.0 has been developed to be operatable on the X-window system. We used hanterm to enable own Korean language input and output. We developed RADCON 3.0 with two versions : First one is for X-window system (with hanterm), and the other is WWW (World Wide Web) type which is operatable at Netscape(TM). Anyone who gets WWW browsers such as Netscape(TM) and Mosaic can access this code. It performed best at Netscape(TM) version 2.0.

RADCON 3.0 has provided the following improved advantages.

- it would be easy, portable, and still compatible with the network system.
- offer much improved user friendliness and compatibility with the CARE system.

## 3.1 Modularization

RADCON 3.0 is modularized to adjust the functions and characteristics of each module for easy understanding and further modification of the code. Each module consist of several subprograms. The function of each module is summarized in Table 2. The modularization was accomplished to make it easy for users to analyze and to modify the code in the future.

## 3.2 Area Specification

A 40X40 square grid has been adopted into RADCON 3.0 as a default, however, the grid size and resolution are able to be adjusted by users. RADCON 3.0 is capable of simulating the heights of a series of moving puffs by adjusting topography around the release point. The topographic data put into as a input file for each site. Three-dimensional topographic data file for Kori site of Korean Nuclear Power Plant is developed and exercised , and RADCON 3.0 has been customized to the site with the topographic data.

| Module | Function of Module |
|---|---|
| RADCON | Main Module |
| INIPUT | Initialization and Input Module |
| METOPO | Meteorological and Topographic Data Processing Module |
| SOTERM | Source Term Calculation Module |
| TRADIF | Transport and Diffusion Module |
| DOSCAL | Dose Calculation Module |
| CHECK | Dose Checking Module |
| WRITER | Output-Writer Module |
| FOODCH | Food Chain Module |
| SUPPLE | Supplementary Module |
| ISODOSE | Iso-dose line plotting Data Output Module |
| GRAPHIC | Graphic Tool Calling Module |

Table 2. Function of Each Module

## 3.3 Graphic Tool

The GNUPLOT software is applied to plot the results of simulation and the GHOSTVIEW code is used to see the actual graphs. These software is available freely through ftp. The lists of graph modules are showed in Table 3. Click the words to see the graphs of the following data. Each graphic file is produced with the form of postscript file.

| Word | Graph |
|---|---|
| AIRNDP | Ground-level Air Concentration (Without Depletion) |
| AIRDPL | Ground-level Air Concentration (With Depletion) |
| DEPOST | Surface-deposited Concentration |
| INHLNG | Inhalation Lung Dose |
| THYINF | Infant Thyroid Dose |
| INHWBD | Internal Whole Body Dose |
| EXTSIC | External Whole Body Dose from Semi-infinite Cloud |
| EXTGRD | External Whole Body Dose from Ground |
| EXTPLM | External Whole Body Dose from Puff |
| TOTWBD | Total Whole Body Dose |
| THYCHD | Child Thyroid Dose |
| THYADL | Adult Thyroid Dose |

Table 3. The Lists of graph modules

# 4. Relationship with CARE

In Korea, a project to establish computerized technical advisory system for the radiological emergency preparedness, such as ARAC(Atmospheric Release Advisory Capability) and SPEEDI(System for Prediction of Environmental Dose Information) has been developed by KINS. It is a nation-wide NPP emergency planning system and the objective of work is to establish a real-time dose assessment system for nuclear emergency. This emergency planning system is called CARE.

The real-time system under development to be installed in the emergency preparedness system CARE is composed of several models such as wind field atmospheric diffusion, and radiation exposure. A random-walk method has been adopted in atmospheric diffusion. In CARE system, FADAS which is developed by KAERI(Korea Atomic Energy Research Institute) is a real-time dose assessment system for nuclear emergencies. And RADCON 3.0 is connected with CARE system through network to receive a meteorological input data and source terms. If the main utilized network system has some problem and CARE system is not operating properly, RADCON 3.0 can be independently as an intermediate level for a rough calculation.

# 5. Results and Discussion

In order to describe the actual severe accident, six hypothetical accident scenarios have been assumed and sensitivity analysis for the potentially significant eight input parameters has been performed. We have analyzed the effects of source term, atmospheric stability, mixing layer lid-height, rainfall rate, wind speed and the height of release point. The effects of source term and meteorological data(wind speed and stability) is more dominant than the other parameters. It is also proved that the cloud shine and internal whole body exposure are important pathways for the short-term scale, while the ground shine become more important for the long-term scale in the emergency situation of the assumed NPP accident.

The simulated results were compared with those of the Radiological Assessment System for Consequence Analysis(RASCAL), version 2.1[10]. RASCAL is developed for NRC personnel by ORNL(Oak Ridge National laboratory) and utilized in 1989 as a federal emergency response tool. However it has adopted Gaussian plume model for the atmospheric dispersion

The comparison of RASCAL 2.1 and RADCON 3.0 is shown in Fig 1.

# 6. Conclusion

RADCON 3.0 has been developed as an intermediate level dose assessment computer code for the NPP severe accident. And it is connected with CARE system through network to receive a meteorological input data and source terms. If the network has some problem and CARE system is not operating properly, RADCON 3.0 can be operatable independently to simulate the emergency situation.

When we compared with RASCAL version 2.1 the same input data, RADCON 3.0 was turned out to give fairly good predictions. The dose-monitoring and decision-making capability, and the graphic tool added in the code is expected to make it easier to prepare an emergency situation effectively. It is expected that RADCON 3.0 can be adopted as a part of nation-wide emergency planning system, in order to support the main code system and obtain quite reasonable dose information in a relatively short time.

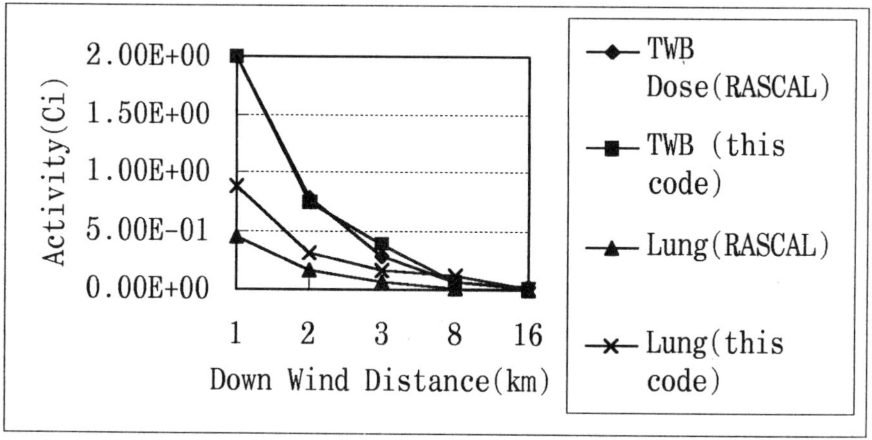


## References

1. Pasquill, F. Atmospheric Diffusion, 2nd Ed, Halstead Press, New York, 1974

2. Gifford, F.A. An Outline of Theories of Diffusion in the Lower Layers of the Atmosphere. Meteorology and Atomic Energy, U.S. Atomic Energy Commission. 1968: 65-116

3. Csanady, G.T. Turbulent Diffusion in the Environment. D. Reidel, Boston, 1973

4. John E. Till and H. Robert Meyer, Radiological Assessment, NUREG/CR-3332, ORNL-5968, U.S. NRC, 1983

5. Brenk H.D. and Vogt, K.J., The Calculation of Wet Deposition from Radioactive Plumes, Nuclear Safety, Vol.22, 3, 1981

6. Sehmel, G.A., Particle and Gas Dry Deposition : A Review , Atmos. Environ, Vol.14, 1980

7. Briggs, G.A., Diffusion Estimation for Small Emissions, ATDL Contribution File No. 79. Atmospheric Turbulence and Diffusion Laboratory, NOAA, Oak Ridge, Tennessee, 1973

8. Eimutis, E.C. and Konicek, M.G., Derivations of Continuous Functions for the Lateral and Vertical Atmospheric Dispersion

9. Jae Hak Cheong, Kun Jai Lee, et al., A Study on the Development of a computer Code and Establishment of Its Application Plan for the Evaluation of Environmental Impact from an NPP Severe Accident, KINS/HR-042, Korea Institute of Nuclear Safety, 1992

10. G.F. Athey, A.L. Sjoreen, T.J. McKenna, RASCAL Version 2.1 User's Guide, NUREG/CR-5247, ORNL-6820/V1/R2 Vol. 1, Rev. 2, 1994

# Internal Control of Safety, Health and Environment (SHE) in Industry: an Effective Alternative to Direct Regulation and Control by Authorities?

Jan Hovden
Norwegian University of Science and Technology (NTNU)
Trondheim, Norway

**Abstract**

The Internal control (IC) concept is still vague and open ended. This gives, however, an opportunity for analyzing IC systems of enterprises in development from the minimum documentation of administrative procedures and deviation control, required by the regulations, to more dynamic and adaptive systems and activities.

## 1. Introduction

The Norwegian experiment with the so called internal control reform was partly an attempt to develop new approaches and means to cope with new challenges of misfits between technology and regulation. In order to match regulatory means to regulatory risk problems, the traditional functions and responsibilities for assessment and detailed control were delegated to the enterprises themselves. Each company should be free to adjust their SHE management systems to own needs and special risk problems. Based on functional acceptance criteria the authorities use systems audits supplemented by on-site verifications to monitor the compliance with laws and regulations.

Internal control (IC) of safety was first introduced as the main principle for controlling safety in Norwegian offshore oil activities in the eighties. In 1992 the reform was extended by new regulations mandatory for *all* private and public enterprises. In addition the scope of IC was extended from occupational safety and major hazard control to encompass product safety, working environment, emergency planning and protection of the external environment against pollution and better treatment of waste. The IC regulations imply a close collaboration and integration between all the SHE responsible regulatory authorities. The short-word "SHE", demonstrating the integration of the fields, came into common use in 1990.

Compared to the traditional control regime, the IC reform gives more attention to and requirements on:

- the obligations of the responsible person of the enterprise
- systematic and documented actions based on principles for written SHE objectives and systems for deviation control similar to the quality assurance area.

- system audits as a tool for control within the enterprise, and for the authorities in their monitoring of enterprises.

The IC concept for SHE management is rather similar to European trends in SHE management as revealed in EU directives, requirements on certification, auditing, etc, though there are some differences. Therefore, the experiences of the "IC experiment" should have some relevance and interest in an international context.

# 2. Review of Research and Development on IC

The first empirical basis for evaluating IC systems was in the Norwegian offshore petroleum activity. The high risks of the activity, the hostile environment of the North Sea and the technological complexity and rapid changing technology reveal both major hazards problems and a high risk working environment with a need for a new control philosophy resulting in the Norwegian Petroleum Directorate's IC Regulations. It is described as controlling procedures, organizational functions, documentation and the installations' physical condition, and claims that IC of safety and work environment will be fulfilled in a Quality Assurance (QA) system which includes all the activities of a company. The systems control performed by the authorities is regarded as an additional control to the control performed by the companies, not as replacing it. According to Quale et al (1989) IC should be a concrete and living activity in the organization, with the intention of contributing to SHE management in a systematic manner. A crucial factor for success was whether the line management really acted as though they felt responsible for the IC system.

A main issue, which got a political solution by the 1992 IC regulations for on-shore industry, was about the inclusion or exclusion of small and medium sized enterprises in the IC reform. The result was to include *all* enterprises. Those who opposed the inclusion of enterprises smaller than 50 employees was right according to evaluation studies (Hovden & Skage, 1994) in that the focus on documentation and formal procedures did not fit the traditions, competence and needs of very small companies. The abstract systems thinking language of the regulations was also difficult to understand for people with low formal education. As a result very few of these enterprises succeeded in implementing IC the first two years after the reform was put into force. On the other hand, the failure of IC of SHE in small enterprises has resulted in special resources, research and information tailored specially for them for improving SHE conditions also in these enterprises accepting less strict formal requirements of procedures and with more focus on actions and activities in practice.

Based on the documentation from the IC research activities and governmental documents on the IC reform, Flagstad (1995) developed an intentional model of the desired results of the IC reform in the enterprises, see figure 1.

The model is a compromise between different views, but emphasizes the contents of the final IC regulations. Flagstad extracts three basic pillars: 1) the participative aspect with the safety representative system and the ideology of the Working Environment Act of 1977, 2) demands for a systematic approach, documentation and dynamics, and 3) the responsibility for the IC system, both legally and practically, within the enterprise. The model suggests that there are certain connections and consequences that are important to obtain the desired results from the three basic pillars.

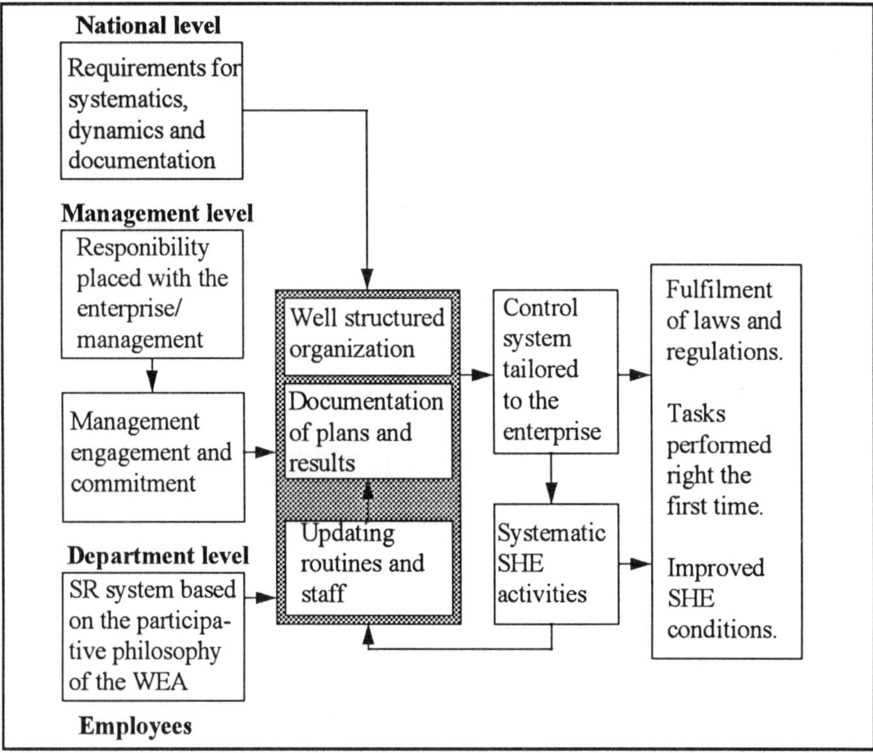


*Figure 1* *The intentional model for IC presented as a logic chain of inference (from Flagstad, 1995).*

# 3. IC Evaluation Studies

The IC evaluation study starting in 1990 and finished in 1994 is mainly contract research for the regulatory authorities and the parties in industry (Hovden & Skage, 1994). This gave the priorities for the aims of the study as follows: (1) monitoring and feedback on the diffusion, the status and trends in implementing the IC reform in industry (descriptive approach), and (2) investigating pitfalls and success factors in the implementation of IC systems (exploratory approach). The data sources were annual questionnaires to pioneer enterprises, a big representative interview survey, and more than 20 in-depht qualitative case studies.

## 3.1. Indicators for the Status of the IC Implementation

In the special sample of pioneer enterprises 56% assessed that they had succeeded to develop a more or less complete IC system. On the other hand, in the representative sample in 1993 the figure was only 13%. In big companies classified as high hazard industry by authorities more than 90% had established IC systems according to the regulations. Another way of expressing this is that the enterprises with developed IC systems (13%) represent over half of all employees in Norway if we generalize this finding to the target population (Saksvik & Nytrø, 1995). The above results refer to measuring the status of implementing IC by a subjective direct measure (i.e. the enterprise representatives' own opinions of the progress on a 4 point scale from "not started" to "finished").

There has been a slow but steady dispersion of the reform to new branches and small enterprises according to the authorities in 1995. However, the patterns are the same: the small enterprises and the service sector are far behind the big high hazard industries in complying with the intentions of the reform. When the IC regulations entered into force in 1992 the Government made a strategic goal stating that IC will be implemented in all enterprises before 1997. The goal is now regarded as unrealistic by all parties.

## 3.2. Changes due to the Implementation of IC

A survey of the pioneer sample tells about the relationship between the reported development of IC and assessments of improvements of SHE conditions and related performance criteria in enterprises. This is showed in table 1.

A general trend in table 1 is that there is a positive correlation between improvements in SHE conditions and status of IC implementation. For improvements in conditions like occupational accidents, absence more than 4 days, continuous pollution, and operation regularity, the correlation with status of IC implementation is statistically significant. Most of the other variables for SHE conditions indicate a similar but weaker trend.

The representative sample of 1993 revealed a number changes in SHE practices in enterprises as a result of the introduction of IC. The most frequently reported changes attributed to the IC regulations were:

- New strategic plans (42%)
- Better documentation (56%)
- Integration of IC and QM (53%)
- More / better risk assessment (46%)
- More clear lines of responibility (58%)
- Increased SHE awareness (69%)

In the surveys of the pioneer enterprises from 1990 to 1992 no clear and significant relationships between IC status and factual statistics on SHE improvements could be documented. These results are a contradiction to results in table 1 using subjective judgements on SHE improvements in the enterprises. There are however many methodological fallacies in the interpretation of the results. The representative survey of more than 2000 enterprises gives a partly different picture of the effects (Saksvik & Nytrø, 1995). A regression analysis of IC status and background variables shows that IC status correlates significantly with absenteeism ($p < .01$), but the main explanatory factor for the improvements in factual SHE performance was bad SHE results (base-line) before the reform.

*Table 1. Improvement of the SHE conditions, etc., related to the status of IC implementation: Percentage reporting improvements in SHE conditions for each category of IC status. (Compare the percentages horizontally.)*

| Improvement of SHE conditions, etc. | Status for IC development (in %) Not started | Under-way | (Almost) complete | Total | (n) |
|---|---|---|---|---|---|
| Occ. accidents*) | 23 | 33 | 66 | 41 | (228) |
| Strain injuries | 14 | 27 | 35 | 27 | (202) |
| Absence, > 4 days*) | 15 | 31 | 44 | 33 | (206) |
| Absence 0-3 days | 14 | 38 | 30 | 32 | (206) |
| Disability pension/ early retirement | 0 | 9 | 7 | 7 | (185) |
| Cooperation climate | 27 | 32 | 32 | 31 | (214) |
| Fire risks | 33 | 51 | 51 | 49 | (208) |
| Continous pollution*) | 30 | 51 | 62 | 52 | (175) |
| Sudden releases | 32 | 46 | 55 | 47 | (166) |
| Waste treatment | 52 | 60 | 63 | 60 | (213) |
| Complaints/penalties from authoritories | 29 | 24 | 38 | 28 | (204) |
| Customer complaints | 28 | 28 | 43 | 32 | (188) |
| Productivity | 38 | 57 | 60 | 56 | (207) |
| Operation regularity*) | 12 | 41 | 47 | 41 | (185) |
| Manning: - increase | 29 | 19 | 30 | 23 | (205) |
| - lay-off | 36 | 63 | 54 | 57 | (205) |

*) Means siginificant correlation for $p \leq 0.05$ (chi square test, DF=2, improvement vs. stable and worsening together).

## 3.3. The Relationship Between Control Authorities and Industry

Quite a lot complains about the role of the authorities were revealed in the questionnaires and in the case studies. They had expected higher competence, better advisory services and qualified system audits from the authorities. An educational programme for inspectors started in 1992 when the reform entered into force. Little "home work" and coordination had been done before, and the result was confusion and disappointment about the new role and functioning of the regulatory bodies. In fact, this was a major threat for the success of the reform.

The IC regulations give little specific guidelines on what is meant by the IC concept and what the regulatory bodies will require in an audit if their system. The guidelines to § 11 say "The measures shall be adapted to each enterprises, depending on its size and the scope of its activities. Thus the individual enterprise itself decide how the internal control system will be designed and what measures will be implemented". Small enterprises asking an inspector for advices, and getting the answer cited above, get angry and feel helpless, while proactive enterprises with competence and resources use it as an opportunity for tailoring effective safety management systems. A paradox for the IC reform so far is that it has contributed more to making enterprises with good SHE records better, than in improving the SHE conditions in ignorant enterprises.

The auditing methods used by the authorities and within enterprises are mainly based on the principles in quality assurance. The methods fit best to big, stable bureaucratic organizations, and give lower rating to flexible dynamic organizations, small enterprises with few formal procedures, etc. regardless of the factual SHE conditions. Both in their information to the enterprises and in search for alternative auditing concepts the regulatory bodies are now focusing more on the processes, the actions and activities and less on formal procedures and documentation compared to the situation 3-4 years ago.

# 4. Discussions

The studies have shown that the meaning and interpretations of the IC concept have changed during the reform period. In general I will argue that this *ambiguity* is positive for future development of management of SHE systems in Norway, despite that in the short run many, especially small enterprises, got numbed in their IC implementation due to lack of clear and detailed guidelines on how to do it, - or they were cheated by smart consultants selling elegant standard packages of IC systems.

The current focus in IC development is more on change processes, adaptation, actions and activities. Based on these trends, it can be concluded that so far the IC reform has partly fulfilled the intentions of being dynamic and adaptive.

The scope of IC systems developed from major accident risk management to include occupational accidents and working environment in general. From 1990 product safety and pollution control were included in the development of the reform. The advantage of integration seems to be: (1) More effective identification of common root causes to the problems, - and consequently more focus on general problems related to top management factors. (2) Experience transfer: the different problem areas had traditionally worked isolated from each other, and the reform inspired exchange of experiences, knowledge and ideas.

The evaluation studies have also revealed some difficulties related to the integration of the problem areas: (1) The regulatory authorities have tried, but not really succeeded in harmonizing their control strategies and practice according to the requirement of integrated IC of SHE. (2) The different groups of experts in the SHE areas have felt professional positions threatened by this integration and the new requirements of knowledge in management and system analysis instead of the discipline oriented phenomena problem analysis. (3) It has been a tendency to ignore the fact that some problems and their solutions are genuinely safety related, health related, quality related, etc, and do not fit into the general scope of SHE and quality.

The IC concept at the time the regulations entered into force, was mainly focusing on deviation control. However, the ambiguity of the IC concept, and the reflections on modifications indicating a dynamic development of the reform, are in accordance with trends and challenges for the future as described by Rasmussen (1994). Of cause, the

IC approach to SHE-management does not have answers to all new threats. However, in some aspects the reform represents experiences and trends to be considered in general and international discussions of successful approaches to safety management in the future.

# 5. Conclusions

As demonstrated by the research refered in this paper, the IC reform has not been a failure, but the goals set for the implementation of the reform and the expected effects in improved SHE conditions have not been achieved. Lots of success cases can be demonstrated, but also many examples of ritual formal systems and documentation at higher levels in the organization with no effects on the execution level and practice at work-places. At least a Hawthorn effect can be documented: SHE problems have got more attention in enterprises due to the IC reform.

In my view, the most prosperous aspect of the IC reform is the debate and reflections it has stimulated among the significant actors and parties, including the fact that the contents of the IC concept are still developing, and therefore promising for learning and adaption to new SHE problems.

## References

Flagstad K.E. The Functioning of the Internal Control Reform. Ph.D. thesis, University of Trondheim, NTH, 1995.

Hale A.R. et al. Extension of the Model of Behaviour in the control of Danger. Vol. 3, Industrial Ergonomics Group, Birmingham Univ./Safety Science Group, TU Delft, 1994.

Hovden J. & Tinmannsvik R.K. Internal Control: A Strategy for Occupational Safety and Health. Experiences from Norway. Journal of Occupational Accidents, 12:21-30, Elsevier, 1990.

Hovden J. & Skage B. Internkontroll av HMS: Evaluering og erfaringsoverføring. SINTEF report no STF75 A94017, Trondheim, 1994.

Quale T.U. et al Under et internkontrollregime. Report 4/89, Work Research Institute, Oslo, 1989.

Rasmussen J. Risk Management, Adaptation, and Design for Safety. in Sahlin N.E. & Brehmer B. (eds.) Future Risks and Risk management. Kluwer, Dortrecht, 1994

Saksvik P.Ø. & Nytrø K. Implementation of Internal Control (IC) of Health, Environment and Safety (HES) in Norwegian Enterprises. An Evaluation and a Model for Implementation. 7th European Congress on Work & Organizational Psychology. Györ, Hungary, 1995.

# Societal Risk - An Operator's View

D J Hewkin[1]
Leader ESTC Risk Assessment Study
Ministry of Defence
London WC1X 8RY

## 1 Introduction

The Ministry of Defence in the United Kingdom (MOD(UK)), is well aware of the increasing demand for public accountability, and the need to explain its activities to interested parties (stakeholders). Comparison is required between hazardous activities performed by MOD(UK), and more familiar hazards elsewhere. Procedures are reasonably well developed for the estimation of risks to an individual, both for the workforce and members of the general public, but we are less content with arrangements for estimating and managing risks to exposed groups of people. This paper is intended to publish some recent work by MOD(UK) on societal risk from its explosives storage activities, and to comment on the application of the results to risk management.

## 2 Background

MOD(UK) has a well defined management procedure for the safe control of explosives storage and handling. This calls for the operator of a facility to identify people at risk and to satisfy him- (or her-) self that risks are both tolerable and as low as reasonably practicable. In many cases, this can be achieved by following prescriptive rules which have been shown (for example by benchmark tests) to ensure an acceptably safe condition. In other situations a site specific risk assessment is necessary.

We have previously published [1,2] details of the procedure we have developed for assessing risks from explosives storage activities. A management protocol has been developed which combines the advantages of prescriptive rules (for routine and low risk situations) with risk assessment procedures (more appropriate for higher risks). The authority and advice of senior management is an essential input to the procedure where local powers are likely to be exceeded. Safe working arrangements are based upon the estimation of risks to representative exposed individuals.

---

1

Dr D J Hewkin, TL-RAST(ESTC) Room 610 Lacon House,
Theobalds Road, London WC1X 8RY
Telephone (UK) (0) 171-305-6379 fax (UK) (0)171-305-6022

However, situations do arise when estimates of risks to individuals do not give adequate weight to a particular situation. For example, when large numbers of people are exposed to a low risk, each member of the group will bear a negligible individual risk - especially if the duration of any exposure is short. However, if the very unlikely hazard were to occur while large numbers of people happened to be in the area, then the number of fatalities could be unacceptable. MOD(UK) is in the process of producing guidance to its line managers on the management and control of societal risk.

MOD(UK) has used an accepted [3] procedure for estimating societal risk. This is based upon a Complementary Cumulative Frequency Distribution of the expected number of fatalities[2] from credible accidents and is usually called an F-N curve. Results are compared with the limited information available (eg [3]), and this paper will discuss various opportunities for risk management which arise from this procedure.

## 3 Estimation of Societal Risk Within MOD(UK)

The assessment of individual risks from MOD(UK)'s explosives storage facilities to people at exposed sites is carried out using a specially developed software package [4].

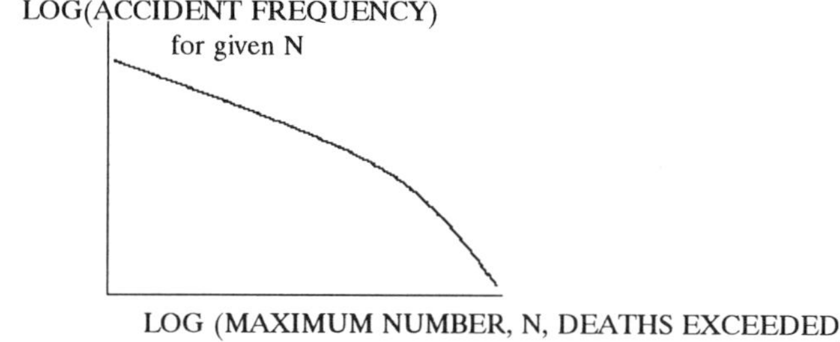


Figure 1 F-N Plot for Single Site

This offers a methodology with which to build up societal risk estimates, and the procedure can be adapted to take account of situations which are intermittent, such as use of nearby transport routes. An illustrative example is available from the author. The societal risk output consists of a log - log plot of the complementary cumulative frequency, F, of more than N expected deaths, and is generally concave downwards in shape (figure 1). This reflects an imposed

---

2 The use by MOD(UK) of the risk of death to estimate societal risk does not imply that accidents which result in many deaths are probable as the result of our explosives storage activities. It is simply that, of the parameters which could be used to estimate societal risk, injury, monetary cost, public response, etc. the risk of death is the most readily quantified value, and it is one of the least contentious to estimate. It is frequently possible to make allowance for other variables (eg material damage costs or strategic value) at a specific location if required.

reduction of cumulative frequency, as N increases, and the observation that credible accidents from an explosives storage facility can only afflict a finite population in a specific land area.

## Societal Risk From F-n Data

**Table 1**

| SITE | | ANNUAL FREQUENCY FOR ACCIDENTS WITH FATALITIES IN RANGE: | | | | | |
|---|---|---|---|---|---|---|---|
| | MAX | 400 | 100 | 50 | 20 | 10 | 5 |
| | MIN | 101 | 51 | 21 | 11 | 6 | 1 |
| 01 | | | | | | | 6.7E-4 |
| 02 | | | | 1.5E-4 | 3.5E-4 | 1.0E-3 | 2.3E-3 |
| 03 | | | | 4.0E-4 | 6.0E-4 | 8.0E-4 | 5.0E-4 |
| 04 | | | | | | 8.6E-8 | 1.4E-8 |
| 05 | | | | 3.2E-5 | 3.3E-4 | 3.8E-4 | 1.5E-2 |
| 06 | | | | 9.7E-4 | | 3.3E-4 | 2.5E-3 |
| 07 | | | | | | 9.6E-5 | 3.3E-2 |
| 08 | | | | | | | 1.0E-4 |
| 09 | | | | | | 1.1E-4 | 2.9E-4 |
| 10 | | | | | 1.3E-4 | 5.3E-3 | 3.8E-2 |
| 11 | | 1.2E-6 | 3.0E-6 | 1.1E-6 | 2.2E-6 | 1.6E-4 | 1.8E-3 |
| 12 | | | 1.0E-8 | 6.0E-5 | | 7.4E-4 | 1.8E-3 |
| 13 | | | | | | | 7.9E-4 |
| 14 | | | | 3.1E-5 | 2.0E-5 | 2.6E-5 | 9.0E-3 |
| 15 | | 1.2E-5 | 2.6E-4 | 1.1E-4 | 9.2E-4 | 8.0E-4 | 6.0E-4 |
| 16 | | | | 1.4E-3 | 6.0E-4 | 1.8E-3 | 5.8E-3 |
| 17 | | 3.0E-4 | 9.7E-3 | 5.0E-3 | | 8.0E-3 | |
| SUM | | 3.1E-4 | 1.0E-2 | 8.0E-3 | 2.6E-3 | 1.9E-2 | 1.1E-1 |
| | | --------------- EXPECTED ANNUAL FATALITIES FOR EACH RANGE ------------ | | | | | |
| | MAX | 0.13 | 1.03 | 0.40 | 0.05 | 0.19 | 0.55 |
| | MIN | 0.031 | 0.51 | 0.16 | 0.03 | 0.09 | 0.11 |
| | | --------------- ANNUAL TOTAL EXPECTATION BETWEEN 2.3 AND 0.9 ----------- | | | | | |

**Table 2**

**Cumulative Frequency(per Year)of Accidents with More Than N Deaths**

(For 17 Major Mod Explosives Storage Sites)

| >100 | >50 | >20 | >10 | >5 | >1 |
|---|---|---|---|---|---|
| 3.1E-4 | 1.1E-2 | 1.9E-2 | 2.1E-2 | 4.0E-2 | 1.5E-1 |

The procedure was carried out using records of the contents at 17 explosives storage sites and these results are summarised in Tables 1 & 2. The results cover the majority of large scale non operational MOD(UK) explosives activity in UK. An average of less than 3 fatalities per year is predicted from all of this activity, and the implicit assumptions within the procedure make this a worst case estimate.

## 4 Management of Societal Risk Within MOD(UK)

MOD(UK) has chosen not to impose threshold criteria for tolerable societal risk. This is because we consider that the tolerability of a societal risk for a site will depend on many site specific variables. We also believe that the line manager who operates the site is best placed to know the sources of hazard and the people who will be at risk. Centrally imposed thresholds will be too low in some situations (and hence be ineffective) but will be too high in others (and hence waste resources). MOD(UK) has chosen instead to carry out regular audits of its safety arrangements which can review the steps taken by local management to identify risks and to manage their undertakings to ensure the safety of those who may be affected.

## 5 Is Societal Risk Redundant?

Although, it is clear that MOD(UK) can meet its duty to control risks to groups of people, it is not easy to ensure that risk reduction plans achieve a benefit appropriate to the resources required.

In the UK, the Health and Safety at Work Act requires operators to ensure, so far as is reasonably practicable, the health safety and welfare of those who may be affected by their undertakings. It is clearly recognised that the resource costs to produce any reduction in risk must be taken into account in any decision on whether it is necessary to reduce risks from an already tolerable level in order to achieve "As low as reasonably practicable" within the meaning of the regulations under the Act.

The currently accepted view in the UK seems to be that regulations are necessary not only to limit individual risk, but also to reduce the likelihood of accidents where many deaths can be expected but where the risk to any individual might be defended as tolerable.

However, if ALARP is interpreted on purely financial terms, the benefits in terms of numbers of deaths avoided would not help management to decide between an improvement which prevented an estimated 200 single fatality accidents per year, from one which prevented a single accident per year with an estimated 200 deaths. Experience suggests that our society is usually much less concerned about frequent single fatality accidents (eg road deaths) than occasional catastrophic events (eg aircraft crashes) even though the estimated annual loss of life on the roads is much greater.

We suggest that society is entitled to take a more complex view of the comparison. It is widely held that the level of risk which an individual considers tolerable can be rationalised in terms of the degree of control which he/she has over the possibility of an accident, and the extent to which the outcome of the accident is dreaded. It is reasonable we suggest, for these considerations by an individual to be transferred to society at large and used to consider the risks run by others. Thus society is prepared to accept relatively familiar road accidents in which the participants have a perceived degree of control even though the expected annual loss of life may be much higher than that for a single major accident. The expected response to the major accident and many innocent victims would be national and longlasting.

It is also relevant to consider the types of people who make up the group of people at risk. In many cases those exposed can be perceived to:

be unaware of the risk they were exposed to
trust the risk generator to protect them
lack any significant control over the risk
comprise vulnerable members of society
etc.

Since laws and regulations in a democratic society must meet the wishes of the majority of those who are to be regulated, it follows that there is likely to be support for regulations which demand consideration of societal risk - particularly when the members of the exposed group could be seen as non-involved and/or vulnerable - even when individual risk bourn by members of a group can be defended as tolerable.

Groups such as those listed above are now qualifying for sympathy and help from society and the regulator. As risk generators, we need to ensure that we have taken sufficient care to protect exposed groups before reaching our decision to continue production, or to resource risk reduction measures.

## 6 Regulation of Societal Risk

Given that we have argued a case for assessing societal risk and encouraging local line-management to demonstrate ALARP and tolerability in terms of societal risk, to what extent can regulations specify acceptable conditions which the operator should meet?

Evans [5] has pointed out that utility theory can be applied to detriments (or disutility) such as accident data, and that there are illogicalities (incoherences) associated with the use of a linear risk threshold criterion on an F - N curve. It is sometimes impossible to identify a preferred solution from a set of alternatives. A procedure which Evans prefers is to estimate the annual expected (ie average) number of fatalities. It is only marginally more complicated to calculate a risk averse societal risk where the value of lives lost increases as a power of the number of fatalities, but this is not pursued here.

This procedure may be more defensible mathematically, but it reduces societal risk to a single number which is a function of the average expected number of deaths per year. It gives the operator no information about worst credible accidents nor about which groups of people are at risk from different events.

We consider that whilst societal risk may be difficult to control by regulation, the use of F - N plots is of considerable value to a site operator as a decision aid management tool. It enables him/her to compare alternative risk reduction proposals and to support a specific course of action or priority allocation on the basis of his appreciation of local conditions (for example by comparing F - N curves for different potential solutions as in figure 2).

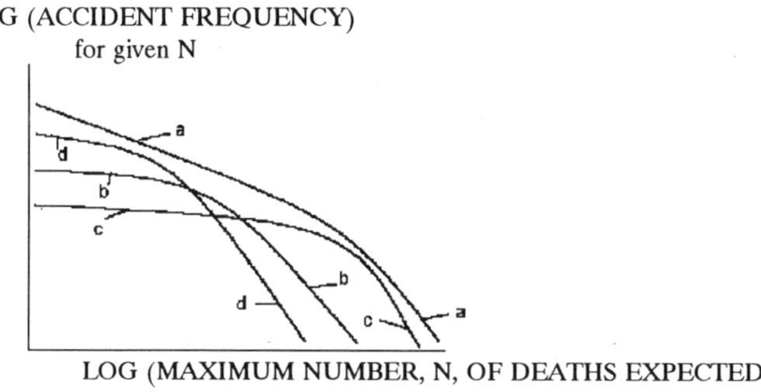


Figure 2 Family of F-N Curves for different options at the same site

This approach seems to us to be consistent with the "Goal Orientated" Philosophy of our Health and Safety Regulators. The F-N curves also remind site operators of the numbers and locations of people likely to be involved in reasonably foreseeable accidents. This information is essential for constructing the contingency plans which are required for the control of major accident hazards.

## 7 Conclusions

Societal risk estimates from MOD(UK) Explosives Storage facilities, do compare reasonably favourably with the limited data available from elsewhere. The site operators can defend their operation as tolerable, but they must then justify their view that risks they generate are as low as reasonably practicable.

The CCDF or F-N curve is of value as a management decision aid, but the imposition of linear acceptable risk criteria on such curves is not desirable.

## 8 Acknowledgements

The data used in this paper were provided by Mr R A Drake, Mr V J Gill, Sqn Ldr G B Jones and Col H A Hanning. Thanks are particularly due to Dr D W Phillips of AEA(Technology), Mr Gill and Mr Drake for their contributions and helpful comments during drafting.

## References

1. J Connor and D Hewkin, Control of Risk from Explosives Stores within MOD(UK), PSAM(II), San Diego, March 1994.

2. D Hewkin, V J Gill et al, Risk Management of Explosives Stores, presented at US DoD Explosives Safety Seminar Miami August 1994.

3. Health and Safety Executive, The Tolerability of Risk from Nuclear Power Stations - HMSO London 1992 ISBN 0 11 886 368 1.

4. D W Phillips and R G J Robinson, RISKEX - A PC based Software Suite for Quantified Risk Assessment of Ammunition and Explosives Depots, presented at 1st Australasian Explosive Ordnance Symposium, PAPARI-93, Canberra, 26-9 Oct 93.

5. A W Evans and N Q Verlander, What is wrong with Criterion FN lines for judging the tolerability of risk? presented at Conference on Risk Analysis and Assessment, Institute of Mathematics and its Applications, University of Edinburgh, April 1994.

# Risk-Based Improvement of Nuclear Power Plant Inservice Test and Inspection Programs

James K. Liming, Wee Tee Loh, Wade M. Lardner
PLG, Inc.
Newport Beach, California USA

Allen C. Moldenhauer, C. Richard Grantom
Houston Lighting & Power Company
Wadsworth, Texas USA

## Introduction

Nuclear power plant inservice testing (IST) programs were developed to ensure the reliable operation of safety-related pumps and valves. Codes, standards, and guidelines for these tests have been developed by the American Society of Mechanical Engineers (ASME) Operations and Maintenance Committee at the request of the U.S. Nuclear Regulatory Commission (NRC). The essential regulation governing this process of testing is 10CFR50.55, which has been implemented via ASME Section XI, both for passive component examination (welding, studs, etc.) and for active component testing (pumps and valves).

For the past several years, both the nuclear industry and the NRC have devoted significant attention and resources to improve the performance of pumps and valves. Industry concerns have been raised associated with the restrictive nature and basis for ASME IST requirements and their impact on plant operation. Overly restrictive requirements can complicate plant operation, cause unwarranted operating costs, and most importantly, degrade plant safety through needless component testing and undue burden during plant outages.

Developments in the industry demonstrate an acceptance of the use of risk-based approaches for support in justification of modifications to nuclear plant procedures, policies, and design. These approaches use an individual plant examination (IPE) to identify prescriptive regulations that have marginal or no safety benefits.

The objectives of this project were: (1) apply risk-based technologies to IST components to determine their risk significance; (2) apply a combination of deterministic and risk-based methods to determine appropriate cases where testing requirements can/should be eliminated or significantly relaxed; (3) support submittal of code relief request(s) to the NRC using STPEGS analysis results; and (4) ensure the results and insights from this project are available for use by appropriate staff.

The scope of this project included application of plant-specific methods for identifying procedures to reduce IST-related regulatory requirements and commitments that require significant resources to comply with and/or implement, but contribute little to safe or reliable operation of STPEGS.

# Methodology Overview

The basic approach and associated methodology applied in this project is presented in detail in the lead plant Implementation Guidelines for Optimizing Inservice Testing Program Using IPE Results document [1]. Additional details on the methodology are presented in NUMARC 93-01 [2] and 93-05 [3]. The RISKMAN® computer program was used to quantify the PSA models and calculate the basic event importance measures for this project. Details on the RISKMAN PSA methodology and software use are presented in the RISKMAN user manual [4]. The most current version of the RISKMAN software that existed during the quantification stages of this project was applied. This software was RISKMAN Version 6.0. Other software used to support this project included Microsoft Access, Microsoft Excel, and Mircosoft Word.

The RISKMAN PSA and other software methodology discussions are presented in detail in the RISKMAN user manual [4], PLG-0675 [5], and other associated software users guides. These methodology discussions are not repeated herein. However, it is appropriate to document how component importance measures were derived in this project and some specific issues regarding component safety importance determination from basic event importance measures calculated within the STPEGS RISKMAN PSA models.

In the STPEGS PSA, the Fussel-Vesely importance (FV), the Risk Achievement Worth (RAW), and several additional basic event importance measures are calculated directly within RISKMAN for the basic events modeled in the PSA. The "component" importance measures for pumps and valves required in this analysis must be calculated outside RISKMAN, using the RISKMAN basic event importance measures as fundamental elements, because each component in a model can include multiple basic events which represent different failure modes or causes for the component of interest. The two primary figures of merit for calculating importance for the 10 pilot plants in this program are component FV and RAW.
The component FV values were calculated as follows:

$$FV_C = \sum_{n=1}^{N} (FV_N)$$

where $FV_C$ is the total FV for component C (the component of interest) and $FV_n$ is the FV for basic event n associated with the component C. The value N is the total number of basic events associated with component C (including initiating event model basic events and all common cause events). The individual values $FV_n$ are obtained from the basic event importance files contained in the RISKMAN basic event importance runs performed for this project.

The component RAW was calculated as follows:

$$RAW_C = 1 + \sum_{n=1}^{N} (RAW_n - 1)$$

where $RAW_C$ is the total RAW for component C (the component of interest) and $RAW_n$ is the RAW for basic event n associated with the component C. The value N is the total number of basic events associated with component C (again, including initiating event model basic events and all common cause events). The individual values of $RAW_n$ are obtained from the basic event importance files contained in the RISKMAN basic event importance runs. The resulting value of $RAW_C$ calculated via this equation is a conservative upper bound on the actual component RAW, but this calculation process was chosen as a practical alternative in this project to eliminate the need for multiple RISKMAN quantification runs.

These two equations are applied under the assumption that component importance measures can be derived from a linear combination of the associated basic event importance values for each component of interest. These component importance measure calculations were performed within a Microsoft Access database.

For IST components modeled in the PSA, and not truncated in the risk quantification process, the pilot project team established risk ranking criteria based on the FV and RAW importance measures derived from PSA calculations. Based on the combination of FV and RAW values, each specific component was placed in one of four risk quadrants (A, B, C, or D) based on the established risk ranking criteria. The risk quadrant system developed by the RB-IST pilot project team is presented in Figure 1. General conclusions about the components classified in each of the risk quadrants are presented in the notes to Figure 1.

Following analysis and classification of IST components through application of the PSA, an expert panel was convened to review the results of the component classification. The purpose of this review was to integrate deterministic analyses and qualitative insights into the IST component safety-significance classification process. The basic disciplines represented on the expert panel included System Engineering, Probabilistic Safety Assessment, Operations (Senior Reactor Operator), Design Engineering, Maintenance Engineering, Maintenance Rule Implementation Engineering, and Test (IST) Engineering. Procedures were developed for the expert panel review process. A quorum of staff members/disciplines was required to constitute valid meetings of the expert panel.

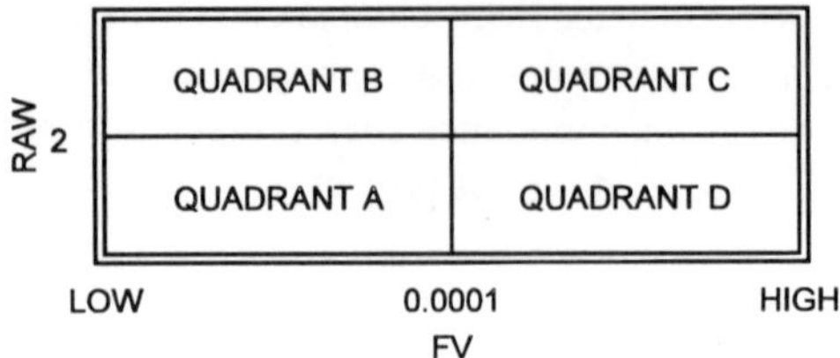


- **QUADRANT A:** Components with low FV and low RAW have very little impact on the plant when they fail; thus, these are candidates for either no IST requirements or very low cost IST requirements (run to failure in a preventive maintenance program).
- **QUADRANT B:** Although the expected impact of the component is very low risk, when the failure rate increases significantly there is a substantial risk impact; thus, inclusion in an IST program is warranted (same conclusion for preventive maintenance program).
- **QUADRANT C:** These components with relatively high impact on overall plant risk and substantial additional risk impact when the failure rates increase significantly, are candidates for a highly focused and effective IST program (safety considerations are perhaps more important than cost considerations); these are the same components that rate a very comprehensive predictive maintenance program.
- **QUADRANT D:** Although these components have a relatively high impact on overall plant risk (perhaps due to the consequences of the component failure), the increased failure rate of the component has a very little extra impact on plant risk; thus, inclusion in an IST program (or preventive maintenance program) is not effective. The application of this component in this system configuration is perhaps a poor or inadequate design application.

**Figure 1. FV-RAW Quadrants of Risk**

The expert panel reviewed the results of the PSA risk significance classification of IST components, and, based on additional qualitative considerations, reclassifies each component into one of two safety categories (*more* or *less* safety significant). For this project, the focus was on the IST components classified as risk-insignificant from the PSA, and on IST components truncated from or not modeled in the PSA. These are the components that are the leading candidates for relaxation of IST testing requirements. The qualitative considerations forming the bases for the final safety significance classification by the expert panel are documented in a component review spreadsheet developed for the expert panel review. This documentation process provides for consistency and future repeatability.

# Results

There are 597 components (pumps and valves) in the IST program at STPEGS Unit 1. Of these, a total of 284 are modeled in the Level 1 PSA, and 309 are modeled in the Level 2 PSA (which includes all Level 1 components as a subset).

Tables 1 and 2 present summaries of results at the IST component type level of indenture for the Level 1 IPEEE data set. The component breakdown for Tables 1

**Table 1. Level 1 IPEEE Basic Event Importance — Fussel-Vesely Importance Summary**

| Type | FV > 1E-2 | 1E-2 > FV > 5E-3 | 5E-3 > FV > 1E-3 | 1E-3 > FV > 1E-4 | 1E-4 > FV > 0 | Trun-cated | Total IST and IPE Components | IST Components Not Modeled in PRA | Total IST Components |
|---|---|---|---|---|---|---|---|---|---|
| AOV | | 2 | 11 | 3 | 6 | 5 | 27 | 73 | 100 |
| CV | | 4 | 21 | 24 | 16 | 12 | 77 | 94 | 171 |
| HOV | 4 | | | | | | 4 | 6 | 10 |
| MOV | 2 | 5 | 35 | 33 | 24 | 25 | 124 | 34 | 158 |
| MV | | | | | | | | 43 | 43 |
| PORV | 1 | 1 | | | | | 2 | | 2 |
| PUMP | 12 | 5 | 6 | 1 | | 3 | 27 | 9 | 36 |
| SOV | | | | | | | | 54 | 54 |
| SRV | 20 | | | | 3 | | 23 | | 23 |
| **Total** | **39** | **17** | **73** | **61** | **49** | **45** | **284** | **313** | **597** |

**Table 2. Level 1 IPEEE Basic Event Importance — Risk Achievement Worth**

| Type | RAW > 2 | 2 > RAW > 0 | Truncated | Total IST and IPE Components | IST Components Not Modeled in PRA | Total IST Components |
|---|---|---|---|---|---|---|
| AOV | 13 | 9 | 5 | 27 | 73 | 100 |
| CV | 52 | 16 | 9 | 77 | 94 | 171 |
| HOV | 4 | | | 4 | 6 | 10 |
| MOV | 60 | 54 | 10 | 124 | 34 | 158 |
| MV | | | | | 43 | 43 |
| PORV | 2 | | | 2 | | 2 |
| PUMP | 24 | | 3 | 27 | 9 | 36 |
| SOV | | | | | 54 | 54 |
| SRV | | 23 | | 23 | | 23 |
| **Total** | **155** | **102** | **27** | **284** | **313** | **597** |

and 2 includes the following IST component types: pumps, air-operated valves (AOV), check valves (CV), hydraulically-operated valves (HOV), motor-operated valves (MOV), manual valves (MV), pressurizer power-operated relief valves (PORV), solenoid operated valves (SOV), and safety relief valves (SRV).

# Conclusions

The safety benefit of testing depends on the balance between its value for improving equipment reliability and its adverse impact on safety; e.g., component wearout, operator distractions, etc. The inherent safety value of a low-risk component is already low. Further, if the test is redundant to other tests or programs that affect reliability, its value for improving equipment reliability can be even smaller. Finally, there can be negative safety impacts due to the test, such as degrading component performance, which necessitates corrective maintenance or components being left in the wrong position subsequent to a test. If negative safety impacts are found, the net value of a test may be neutral or even negative.

Of the 597 IST components in the IST program at STPEGS, 417 valves are classified as *less* safety significant based on the criteria established for this project. This report and the backup PSA models, data bases, and expert panel survey worksheets provide a strong technically-defensible basis for requesting significant relaxation of IST test requirements for these 417 valves. It is the conclusion of this project analysis that the test frequency for each of these valves can justifiably be extended to once every two refueling cycles not to exceed four years between tests. Pump test frequency will be addressed in a separate submittal.

# References

1. TU Electric Company. Comanche Peak Steam Electric Station implementation guidelines for optimizing inservice testing program using IPE results. Rev. 0, June 1994.

2. NUMARC 93-01. Industry guidelines for monitoring the effectiveness of maintenance at nuclear power plants. 1993.

3. NUMARC 93-05. Guidelines for optimizing safety benefits in assuring the performance of motor-operated valves. December 1993.

4. PLG, Inc. RISKMAN® PSA workstation software. Release 6.0, January 26, 1995.

5. PLG, Inc. South Texas Project probabilistic safety assessment. Prepared for Houston Lighting & Power Company, PLG-0675, May 1989.

# Test Policy Optimization for a Complex System : An Application for the Differential Model for Equivalent Parameters (DMEP)

D. VASSEUR [(1)] - M. EID [(2)]

(1) Electricité de France - Studies and Research Division - Surveillance Diagnostic Maintenance Department - Chatou - France

(2) Atomic Energy Commission - Nuclear Reactors Directorate - DMT/SERMA - Saclay - France

**Abstract**

One of EDF's current priorities is the optimisation of the preventive maintenance in all French nuclear power stations. This optimisation involves a rationalisation of the choice of equipments to be maintained and maintenance tasks to be carried out, as well as a judicious choice of intervals between these tasks. This work is being carried out in cooperation between EDF and the CEA (Atomic Energy Commission), and suggests a procedure to provide assistance in optimising intervals between maintenance tasks respecting a global unavailability target. This work is based on the DMEP.

## 1 Introduction

In 1990, the maintenance of French nuclear power stations represented 2.5% of the construction cost per year, and was continuously increasing.

This is why EDF started in 1990 a maintenance optimisation through a Reliability Centered Maintenance procedure (OMF : Optimisation de la Maintenance par la Fiabilité) [1] in order to control maintenance costs, while maintaining or improving the safety and availability of units.

This procedure is based on an analysis of the functional consequences and the criticality of equipment failures on the unit. A preventive maintenance program has subsequently been proposed in which tasks have been chosen rationally to counter the critical failure modes found.

These preventive maintenance tasks must be carried out at periodic intervals that depend firstly on changes in the equipment failure rate with time, and secondly on global criteria to be respected such as the maximum allowable equipment unavailability.

Work carried out as part of a common EDF-CEA Research and Development program has resulted in a procedure for choosing maintenance task intervals satisfying these criteria. This procedure was developed using the Differential Model for Equivalent Parameters [2] and was tested on data derived from experience feedback on French nuclear power stations.

# 2 Sample Application : the choice of a periodic test policy for emergency diesel generators in nuclear power units

## 2.1 Summary of the problem

Emergency diesel generators for nuclear units carry very rarely out their designed function, therefore periodic tests are required to determine any failures that occur while on standby. During these tests diesel generators are started up and then operated for about an hour. These periodic tests which restore some of the conditions necessary for correct operation of generators (oil film on engine block liners) also cause ageing of the engine block, particularly on the pistons, due to the speed of the start ups.

The problem in which we are interested is the following :
Taking account of the maximum allowable unavailability of the emergency diesel generators, reliability and maintainability data for engine block components, and the systematic preventive maintenance policy applied to engine block components, at what intervals should tests be carried out ?

## 2.2 Data about the chosen system

The first item we considered was the emergency diesel generators engine block, which is the functional sub-assembly most affected by problems induced by periodic tests.

The data necessary to handle this problem were extracted from analyses carried out on emergency diesel generators in 1992, as part of the OMF procedure. The functional analysis used in this method states that the chosen "engine block" sub-assembly (S) is composed of three functional entities in series, which may themselves be composed of mechanical components also in series. These entities and the associated components are as follows:

the combustion chamber ($S_1$) consisting of the piston ($C_{11}$), the liner ($C_{12}$) and the piston ring ($C_{13}$),
the distribution device ($S_2$),
the crankshaft ($S_3$) consisting of journal bearings at small end of connecting rod and link ($C_{31}$ and $C_{32}$), journal bearings at big end of connecting rod and link ($C_{33}$ and $C_{34}$), and the crankshaft).

Failure rates during operation were assigned to each elementary component by analysing data obtained from experience feedback collected from French nuclear power stations as part of the OMF procedure, as follows:

| | |
|---|---|
| liner | $\lambda = 950.10^{-6}/h$ |
| piston | $\lambda = 0.01 \left(\frac{t}{225}\right)^{0.3}$ |
| piston ring | $\lambda = 0.0018 \left(\frac{t}{632}\right)^{0.16}$ |
| distribution device | $\lambda = 10^{-6}/h$ |
| connecting rod small end journal bearing | $\lambda = 10^{-20}/h$ |
| link small end journal bearing | $\lambda = 10^{-20}/h$ |
| connecting rod big end journal bearing | $\lambda = 165.10^{-6}/h$ |
| link big end journal bearing | $\lambda = 10^{-20}/h$ |
| crankshaft | $\lambda = 10^{-20}/h$ |

These components are not repairable.

Furthermore the result of this analysis of information feedback was an estimate of the average failure rate in standby ($\lambda_a(t)=7.6.10^{-7}/h$) and of the average failure rate in operation of the engine block ($\lambda_f(t)=4.8.10^{-6}/h$).

The current test policy consists of carrying out monthly tests lasting about one hour. Furthermore, the components of the combustion chamber are replaced every five years regardless of their condition, as part of a systematic preventive maintenance program.

## 2.3 Modelling the problem

One of the difficulties with this problem is how to take account of a failure rate, when some of its basic components are variable with time. The Differential Model for Equivalent Parameters has made this solution possible.

### 2.3.1 *The Differential Model for Equivalent Parameters*

The Differential Model for Equivalent Parameters (DMEP) is the main tool used to carry on this study. It has been developed in the CEA since 1986. The main interest of the DMEP is that it allows taking into account : i) the time dependence of the basic component's failure rate in a complex system, ii) the possible different patterns of failure rate (in operation, in standby, ...), iii)the instantaneous corrective actions such as standard exchange of components. An exhaustive presentation of the DMEP is beyond the scope of this paper. Shortly, the model is based on the following basic idea that a system is fully determined by :

- a logical expression describing its failure or its success as a function of the failure and/or the success of its elementary components,
- a system of differential equations describing the kinetics of the system which could be presented as follows

$$\frac{d}{dt}P = -\lambda P + \mu Q$$

$$\frac{d}{dt}R = -\lambda R$$

$$\frac{d}{dt}S = -\mu S$$

where P and Q are respectively the availability and the unavailability of the system, $\lambda$ and $\mu$ are its failure and repair rates, R and S are respectively its reliability and non -reparability.

### 2.3.2 *Modelling the problem*

In order to model sequences of engine block waiting times ($T_a$ = 719 hours) and operating times ($T_f$= 1 hour), we write:

$$S = M_a - M_f$$

where $M_a$ and $M_f$ are the engine block failure conditions during shutdown and operation respectively, defined by:

| | | |
|---|---|---|
| $M_a$ | $\lambda = \lambda_a(t), p = p(t)$ | $0 < t < T_a$ |
| | $\lambda = 0, p=1$ | $T_a < t < T_a + T_f$ |
| $M_f$ | $\lambda = 0, p=1$ | $0 < t < T_a$ |
| | $\lambda = \lambda_f(t), p = p(t)$ | $T_a < t < T_a + T_f$ |

$\lambda_f(t)$ is calculated from data concerning basic components. The failure rate of the engine block functional sub-assembly during operation is modelled by the logical sum of the failure rates of the elementary mechanical components, due to its series structure.

$$S = S_1 + S_2 + S_3$$

where $S_1 = C_{11} + C_{12} + C_{13}$

$S_3 = C_{31} + C_{32} + C_{33} + C_{34} + C_{35}$

### *2.3.3 Results*

The Differential Model for Equivalent Parameters produced the variation of the unavailability of the engine block with time (Figure 1).

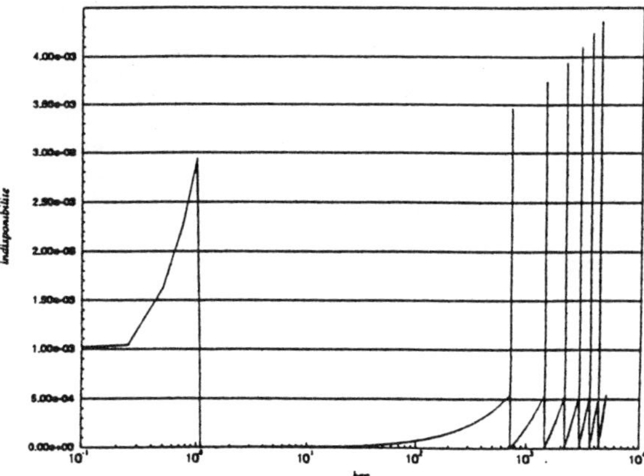


Figure 1: Variation of engine block unavailability with time

This figure clearly shows that the engine block unavailability increases with time.

This series of calculations was also used to calculate an average failure rate of the engine block at different moments in time. The average failure rate obtained during 55 tests, which is the number of tests carried out during 5 years, after which a standard replacement of the combustion chamber is systematically made, was $4.7.10^{-3}$/h which agrees well with the average value calculated using experience feedback data and which partly validates the model.

Finally the model was used to evaluate different test policies. These simulations were used to plot a family of curves showing the unavailability of the engine block at the end of each test campaign, as a function of the number of tests carried out for various test intervals (Figure 2).

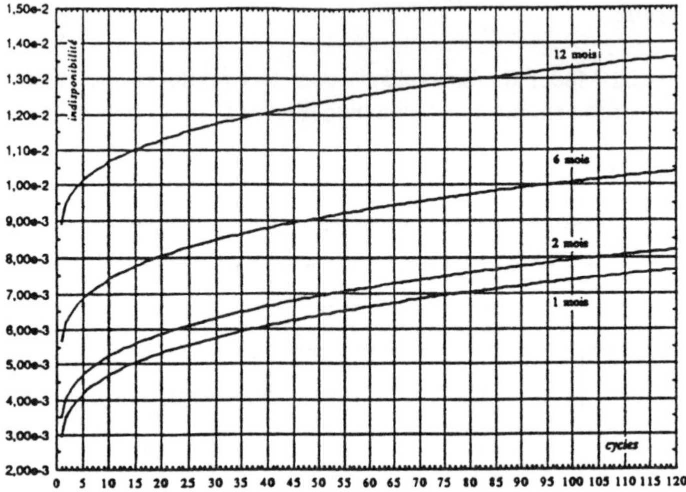


Figure 2: Unavailability of the engine block as a function of the number of tests for various test policies

The preventive maintenance policy applied to the engine block at the present time (1 test per month and a standard replacement every 5 years) gives an engine block unavailability of $6.6.10^{-3}$. The above representation demonstrates that the same unavailability level can be reached with lesser number of test campaign (1 test each two months) and longer period between standard replacement (every 6.6 years).

# 3 Conclusion

Data derived from the OMF approach and from experience feedback of the French nuclear units used through the Differential Model for Equivalent Parameters have allowed to evaluate different periodic test policies of a given equipment, and to measure their impact on systematic preventive maintenance carried out on this equipment respecting a target unavailability defined beforehand.

**References**

1. JACQUOT J.P., LEGAUD P., ZWINGELSTEIN G.: Development of reliability centered maintenance (RCM) methodology for *Electricité de France* nuclear plants: a pilot application to the CVCS system. LAEA-TECDOC-658, Vienne 1992.

2. BID M., DUCHEMIN B.: A new method for the determination of failure and repair rates of complex systems. Reliability 89, Brighton 1989, 4C/5.

# Electrical Substation Performability and Reliability Indicators Modelled by Non-Homogeneous Markov Chains

Agapios PLATIS[1,2], Nikolaos LIMNIOS[1], Marc LE DU[2]
(1)Université de Technologie de Compiègne, Compiègne, France
(2) Electricité de France, Clamart, France

## 1 Introduction

In the scope of operating safety in electrical substations, EDF is brought to perform reliability and availability computations. These computations declined on the substation performance are used to estimate non-quality indicators. However, these different computations use models which rely on simplifying assumptions of functioning and dysfunctioning probabilistic data. Generally, the main assumption retained in order to evaluate reliability and availability parameters is that failure and repair rates of the components are time-constant. This is the case of time-homogeneous Markov systems. The above assumption allows to use traditional models to evaluate reliability and availability parameters, such as Markovian graphs or fault trees. Therefore, in these models, we use mean values for the failure rates and the repair rates. In the same way, certain parameters involving the system performance are computed with mean values (mean value of an electrical substation load). However, for electrical power systems, many parameters may vary with time, especially on the hour of the day and the period of the year. As a consequence, if we take into account hazard rate time variation, the traditional models are no longer valid for this kind of problems. This dependence of the hazard rate function with time leads us, naturally, to investigate non-homogeneous Markov models. In this paper, we first define and give basic developments of transient analysis of non-homogeneous Markov chains then we provide a new formulation for reliability and performability indicators. This new formulation naturally includes state probabilities, reliability, availability, maintainability but also some useful performance indicators that are more dedicated to power systems such as expected energy not supplied on a time interval. Finally a numerical application is solved to illustrate this method. In this example, we highlight the differences in the results obtained by classical and new modelling but also the new information we can access.

## 2 Non Homogeneous Markov Chains Transient Analysis

### 2.1 Definitions and Transition Functions

Let $E=\{1,2,\ldots,s\}$ be a set of points that represent the system state space and $X=\{X_n;\ n\in \mathbb{N}\}$ an $E$-valued stochastic process on a probability space, whose probability measure is $\mathbb{P}$. $X$ is a Markov chain if, for all $j\in E$ and all $n>0$, we have

$$\mathbb{P}(X_{n+1}=j \mid X_0, X_1,\ldots, X_n)=\mathbb{P}(X_{n+1}=j \mid X_n)$$

We define for each $i, j\in E$, $p_n(i,j)= \mathbb{P}(X_{n+1}=j \mid X_n=i)$, the transition probability from state $i$ to state $j$ in the time interval $[n, n+1]$. The probability $(p_n(i,j);\ n\in\mathbb{N},\ i,j\in E)$ is called *transition function* of chain $X$. It is a one step transition probability. We can also define multiple step transition probabilities : $p_{n,m}(i,j)=\mathbb{P}(X_m=j|X_n=i)$ for $m>n\geq 0$ $p_n(i,j)=p_{n,n+1}(i,j)$. If $p_n(i,j)$ does not depend on $n$, the chain is time-homogeneous.

### 2.2 State Probability Vector

Let $\alpha$ be a probability distribution on $E$, considered as the initial distribution of the chain, that is $\alpha(i)= \mathbb{P}(X_0=i)$, we define $\alpha$ as the corresponding vector and for each $n=1,2,\ldots$ , let $\mathbf{p}_n=(p_n(i,j);\ i,j\in E)$ be the transition matrix. The Markov chain $X$ is completely defined by its initial distribution $\alpha$ and the transition function $(p_n; n\in\mathbb{N})$. We have, for all $n\leq 1$ $\mathbf{p}_{0,n}= \mathbf{p}_0\, \mathbf{p}_1 \ldots \mathbf{p}_{n-1}$. Hence

$$P_j\,(n)=\mathbb{P}(X_n=j)=(\alpha \prod_{k=0}^{n-1} \mathbf{p}_k\,)(j), \text{ and } \mathbf{P}(n)=\alpha\, \mathbf{p}_{0,n}$$

$P_j(n)$ is called state probability at time n and the vector $\mathbf{P}(n)=(P_1(n), \ldots, P_s(n))$ is called state probability vector at time $n$ [2].

## 3 Reliability and Performabilty Analysis

### 3.1 Transition Matrix Decomposition

Let U be the set of Up states and D the set of Down states, with $v=|U|$, and $U\cup D=E$, $U\neq\varnothing$, and $U\neq E$. The transition matrix can be written $\mathbf{p}_n = \begin{bmatrix} \mathbf{p}_n^U & \mathbf{p}_n^{UD} \\ \mathbf{p}_n^{DU} & \mathbf{p}_n^D \end{bmatrix}\begin{matrix} U \\ D \end{matrix}$ with columns indexed by $U$, $D$.

### 3.2 Reliability and MTTF

The *Reliability* at time n, is given by the relation

$$R(n)=\mathbb{P}(\forall i\leq n, X_i\in U)=\alpha_1\,(\prod_{k=0}^{n-1}\mathbf{p}_k^U\,\mathbf{1}_v\,),\ (n\geq 1) \text{ and } R(0)=\sum_{i\in U}\alpha(i)$$

with $\mathbf{1}_v=(1, \ldots, 1)^T$ a $v$-dimensional vector whose elements are ones, $\mathbf{p}_n^U$, the sub-matrix of $\mathbf{p}_n$ obtained by eliminating the rows and the columns of $\mathbf{p}_n$ not belonging to U and $\alpha_1$ the sub-vector of $\alpha$ corresponding to the Up states subset U.

The *MTTF* will be given by $MTTF=\alpha_1\left(\mathbf{I}+\sum_{k=0}^{\infty}\prod_{r=0}^{k}\mathbf{p}_r^U\right)\mathbf{1}_v$

### 3.3 Maintainability and MTTR

The *Maintainability* at time n, is given by the relation

$$M(n)=1-\mathbb{P}(\forall i\leq n, X_i\in D)=1-\alpha_2\,(\prod_{k=0}^{n-1}\mathbf{p}_k^D\,\mathbf{1}_{s-v}\,),\ (n\geq 1) \text{ and } M(0)=1-\sum_{i\in D}\alpha(i)$$

with $\alpha_2$ and $\mathbf{p}_n^D$, the sub-vector and sub-matrix corresponding to the subset D.

The *MTTR* will be given by $MTTR=\alpha_2\left(\mathbf{I}+\sum_{k=0}^{\infty}\prod_{r=0}^{k}\mathbf{p}_r^D\right)\mathbf{1}_{s-v}$

### 3.4 Availability

The *instantaneous Availability* at time n, is given by the following relation

$$A(n)=\mathbb{P}(X_n\in U)=\alpha(\prod_{k=0}^{n-1}\mathbf{p}_k\,\mathbf{1}_{s,v}\,),\ (n\geq 1) \text{ and } A(0)=\sum_{i\in U}\alpha(i)$$

with $\mathbf{1}_{s,v}=(1,\ldots,1,0,\ldots,0)^T$ : $s$-dimensional vector whose $v$ first elements are ones and the other are zeros.

## 3.5 Frequency of Entering in State i at Time n

We define the frequency of entering in state $i$ at time $n$, by

$$f_i(n)=\sum_{l\in B-\{i\}} P_l(n-1)\,\mathbf{p}_n(l,i)/h \quad \text{with } h \text{ the digitalization step}$$

## 3.6 Expected Energy Not Supplied on a Time Interval

The Expected Energy Not Supplied on a Time Interval (EENSTI) is defined as the mean value of the energy not supplied in a fixed time interval [t, t+θ] due to one or more interruptions in that interval, where L(u) is the demanded load at time $u$ and $g_i(u)$ the supplied load in state $i$ at time $u$ (L and g are bounded functions).

$$EENSTI(t,\theta)=\mathbb{E}\sum_{i\in B}\int_t^{t+\vartheta}\left(L(u)-g_i(u)\right)\mathbf{1}_{\{X_u=i\}}\,du=\sum_{i\in B}\int_t^{t+\vartheta}\left(L(u)-g_i(u)\right)\mathbb{P}_i(u)du$$

# 4 Numerical Application

This is the case of an electrical substation which is supplied by two lines L1 and L2 (coupled with circuit-breakers D1 and D2). In this configuration only line L1 supplies the substation while L2 stands by. In case a fault takes place on L1 that causes the opening of circuit-breaker D1 (of L1), a restoration system closes circuit breaker D2 (of L2). This restoration, however, goes along with a brief supply interruption of the customers.

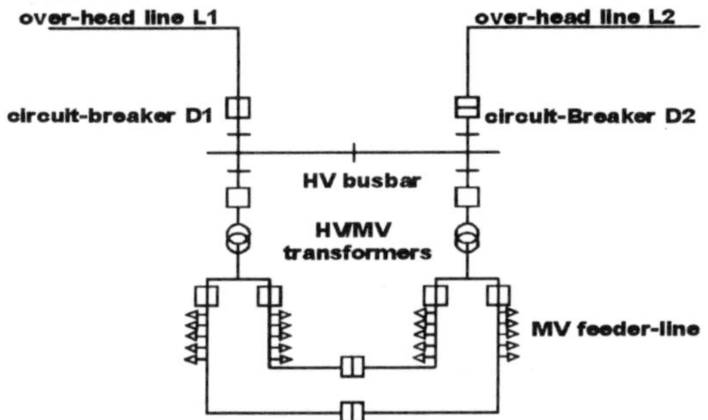


*Figure 1 :* Electrical Substation Diagram

We have modelled this system by an eight-state Markov chain whose transition diagram is given in the following figure (figure 2).

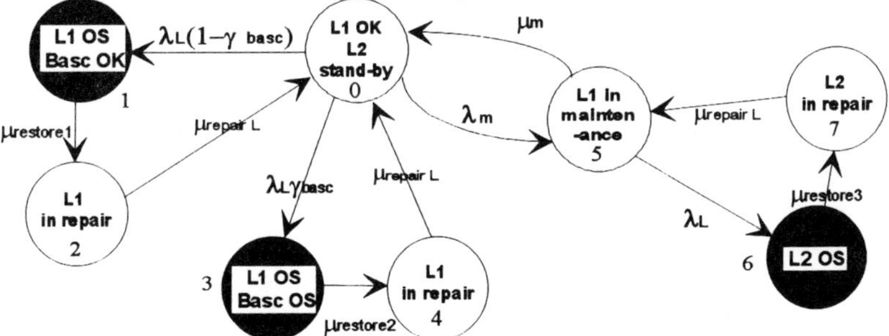


*Figure 2 :* State Transition Diagram

State 0 corresponds to the operational state, L1 is available, L2 stands by. State 1 corresponds to a customer supply interruption due to a fault on line L1. State 2 corresponds to the repair of L1 while L2 supplies the substation and the restoration system has succeeded. State 3 corresponds to a customer supply interruption due to a fault on L1 while the restoration system has failed. State 4 corresponds to the repair of L1 while L2 supplies the substation and the restoration system has failed. State 5 corresponds to the electrical isolation of L1 for maintenance. State 6 corresponds to a customer supply interruption due to a fault on L2. State 7 corresponds to the repair of L2 while the customers are supplied by another source.

## 4.1 Failure Rates, Repair Rates and Reward Rates

The line failure rate has two peaks. One corresponding to the higher amount of lightning in the summer while the second corresponds to the hard climatic conditions of the winter. The repair-rates correspond to the restoration delay and to the repair rates. We suppose that when an operator is sent to repair in the winter, this delay is higher due to the climatic conditions in that season. This repair rate evolves during the day. Operator's reaction is faster in the working hours rather than in the night. Load varies according to the season but it also depends on the hour of the day.

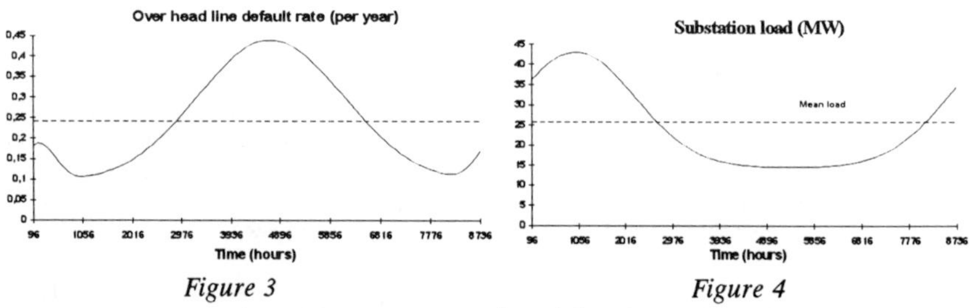


*Figure 3* *Figure 4*

The transition matrix is obtained by $p_n(i,j)=\begin{cases} a_{ij}(nh)h & \text{if } i\neq j \\ 1-a_{ij}(nh)h & \text{if } i=j \end{cases} \; i,j\in E$ where $A(t)=[a_{ij}(t)]_{i,j\in E}$ is the transition rates matrix and $h>0$ the unit of time for the Markov chain.

## 4.2 Reliability and Performability Evaluation in Classical Modelling

In this section, we will give some of the basic indicators, modelled with homogeneous Markov chains, used by EDF.

### *4.2.1 Interruption Frequency*

It is the annual average frequency of being interrupted [3]

$$Freq_i=\Pi_i\Big(\sum_{1\le j\le n\ j\neq i} a_{ij}\Big) \text{ with } \Pi_i \text{ the stationnary distribution}$$

### *4.2.2 Energy Not Supplied*

It is the annual average load multiplied by the average interruption delay [1]

$$ENS=Load \times 1/\sum_{1\le j\le n\ j\neq i} a_{ij}$$

### *4.2.3 Worth of the Annual Energy Not Supplied*

The reward of the ENS depends on the total amount of ENS of the interruption and is given by semi-linear scale : a constant cost for less than 30 MWh events, a linear increasing cost for events between 30 and 100 MWh, and again a constant cost for greater events.

### *4.2.4 Sensibility of the Entry Data*

If we increase the load by 30%, the performance indicators will be increased. However, if load is increased in the winter or in the summer, what is the impact on the indicators?
If we decide to change the working periods of the operators, that is make them more present, when the interruptions take place, while keeping the same total working hours, what will be the impact on the final results? If we decide to reduce the working hours of the operators while changing presence time, will the ENS necessarily increase?
Today, our maintenance policy is to make more electrical isolations in the summer because of the load that is lower in that season, but what would be the impact if we decide to change that policy and make more electrical isolation in the spring or in the autumn? To all these questions, we cannot answer by using classical models, we need a sharper model that takes into account the time variation of the hazard rates. Moreover the assumption that the hazard rates are constant is no longer valid.

## 4.3 Numerical Results

| | Interruption frequency | | Annual ENS (MWh) | | Worth of the ENS (FF) | |
|---|---|---|---|---|---|---|
| | *Classical* | *Non-homog* | *Classical* | *Non-homog* | *Classical* | *Non-homog* |
| Initial case | 0,24 | 0,24 | 0,39 | 0,52<br>33% | 22,7 | 16,3<br>-28% |
| Increasing load by 30% | | | | | | |
| in the summer | 0,24 | 0,24 | 0,51 | 0,68<br>33% | 39,9 | 40,7<br>2% |
| in the winter | 0,24 | 0,24 | 0,51 | 0,68<br>33% | 39,9 | 19,5<br>-51% |
| Changing the working hours of the operators | 0,24 | 0,24 | 0,39 | 0,49<br>25% | 22,7 | 15,6<br>-31% |
| Changing the electrical isolation periods | 0,24 | 0,24 | 0,39 | 0,47<br>20% | 22,7 | 18,1<br>-20% |

*Table 1 : Results Comparison Table*

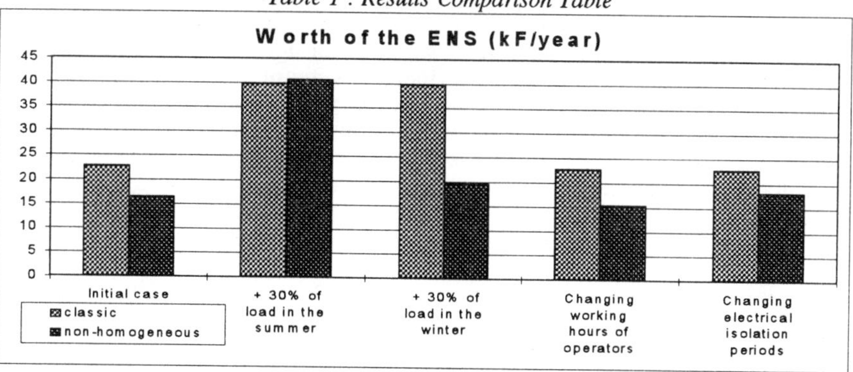


*Figure 5*

## 4.4 Interest of this Modelling

The former questions can be answered now.

### 4.4.1 *Initial Case*

From the initial case, and modelling with non-homogeneous Markov chains, we can see differences in the results of the ENS (+33 %) and in the reward of this ENS (-28 %).

### 4.4.2 *Increasing Load by 30 %*

We can see in the non-homogeneous case that increasing load in the summer will have an important impact on the reward of the ENS, while an increment in the winter will have a smaller impact on this reward of the ENS. We can explain that by the fact that an electrical substation is designed to resist to a high amount of load in the winter but not in the summer where electrical isolation is more frequent. The difference of the reward indicator in the summer and in the winter reaches 50 %.

### 4.4.3 *Changing the Working Hours of the Operators*

If the operators are in front of their control panels when the load is maximum, the ENS will drop by 6 %. With the classical modelling, no difference can be remarked since we have the same average manual restoration rate. The difference in the reward of the ENS reaches 30 %.

### 4.4.4 *Changing the Electrical Isolation Periods*

Electrical isolation for maintenance is generally done in the summer because of weak load. If this isolation is more spread over the year, the ENS will decrease by 10 % but the reward of this ENS will increase by 11 % while having a difference with the traditional modelling of 20 %.

### 4.4.5 *More Information*

We can see that the results obtained for the frequency are exactly the same for all cases, however an instantaneous indicator is more interesting than having a constant value : we can orientate the improvements. An other information that cannot be obtained by the classical modelling is that we have an access to distributions, for instance the distribution of the ENS per incident.

## 5 Conclusion

In this article, we have treated a particular type of Markov systems : the case of time non-homogeneous Markov systems in discrete time. In order to have measures adapted to this kind of system, we have formulated some reliability and performability measures including new indicators more dedicated to electrical systems. The real interest of this method, that is taking into account hazard rate time variation, is to get more accurate and more instructive indicators, such as the Worth of the ENS where the difference in the results has reached 50 % in our numerical application. But it is also interesting to see that this method allows us to access new performability indicators that cannot be obtained by classical methods such as the distribution of the ENS per incident. Finally we have to bear in mind that this modelling is powerful when hazard rates are time varying, but it is not necessary to use it on any case especially when the rates (reward included) do not vary much because of the amount of computation needed.

## References

1. Allan R.N., Billinton R., Reliability evaluation of power systems. Plenum press, 1990.
2. Limnios N., Processus Stochastiques et Fiabilité. CS-61 DEA course UTC, 1995.
3. Pagès A., Gondran M., Fiabilité des systèmes. Eyrolles, 1980.

# A Multi User Information System concerning Safety and Maintenance.

Jette Lundtang Paulsen, Risø National Laboratory, Denmark
Joan Dorrepaal, Delft Technical University, The Netherlands

Nuclear equipment, e.g. safety systems, are complex, high technology systems that must operate for long periods of time without serious failure and with a very long total life. A great amount of redundancy and diversity is used in nuclear facilities to ensure the safety of the plant. Repairs, inspections and overhaul of equipment are usually done at specific time intervals, when the plant is down for refuelling. This process generally follows a pattern of increasing complexity, depending on the operating times accumulated by the system.
The physical environment in which the equipment operates is very severe and can have serious detrimental effect on the complex mechanical and the electronic components of the equipment. High temperatures, high vibration, high humidity and the presence of radiation take their toll.
This means that throughout its operating phase, the reliability characteristics may start deviating from those predicted by the supplier.
The nuclear power plants in the Nordic countries have established a common data base to where all repairs on components are reported. 14 nuclear power blocks exist today in Sweden and Finland. The licensee event data base is also establshed as a common data base. The purpose of these data bases are safety related and used as the data bases from where the reliability of the safety critical components are calculated.
The aim of the present work is to use the data base for aging and maintenance assessment of the components too. Good maintenance is a requirement for good safety.
The personnal at the power plants have access to the data bases from terminals and this gives them the opportunity to use the data for their own purposes.
The investigation of different known methods and developement of new statistical methods to support the different users with information for use in their daily work is the main part of the work. The users of the final tool are primarily meant to be

- Maintenance planners
- Safety departments
- Authorities
- Craftsmen
- Managers
- Utilities
- Operators

The users need for information can be divided into the following main topics.

- Maintenance
- Aging and condition
- Safety
- Economy

To extract the right information some questions have to be discussed.

- What is good maintenance
- How do we measure good maintenance
- Which data are available
- How to present the results for the users

One simple answers on good maintenance

- Minimize maintenance cost but keep safety high

To minimize maintenance cost but still keep safety high, is a combination of

- having a good quality of the components,
- good quality of the repair work and
- do the preventive repairs at the right time.
-

So the following is a list of possible indicators to measure or calculate.

- The repair rate is comparable with other identical components from other manufacturers.
- The repair rate is comparable with identical components on other plants.
- The amount of early failures after repair is low.
- The amount of unplanned repairs is low.
- The relation between preventive and corrective repairs are reasonable
- The failure rate is constant or not increasing
- The used time to maintenance is not increasing
- The unavailability due to repairs is low

One question is to discuss what the users want to measure, an other question is the possibility to get the information from the data available.
The failure reports contain the following information.

- name of the plant
- system number
- component identification
- the time the failure was detected.
- the time the component went out of duty.
- the time the repair started
- the time the component was available
- the manhours of the repairs

The report also contains a coding system with letters which give information of

- how the failure was detected
- the failure mode
- what has been done during repair
- the failure type

The reports also contain a text based description of the failure event, how it was detected and the repair of the failure.

The information is used in the program, which do all the necessary statistical work. The interface of the program is shown on fig.1.

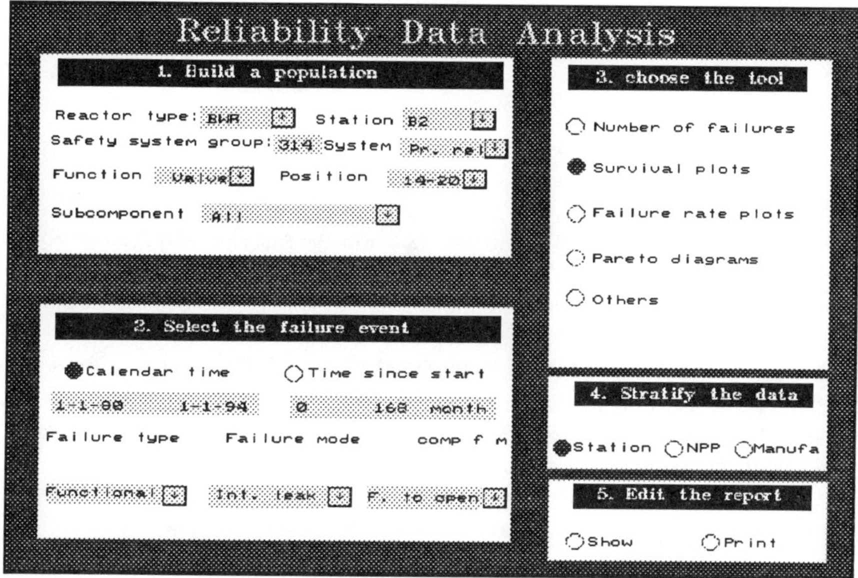


Fig.1

The main idea behind the design of the program is the possibility for the user to pool data from selected plants, from selected components, selected failure modes and selected time interval. This facility gives the user the possibility to select the plant in question, pool data from several plants and compare the results. A great amount of graphs can be retrieved to be able to elicit a problem. The system of graphs can be used as a hierarchical system where the user starts with a kind of overview graphs and then dive deeper and deeper down into an indicated problem.

A strategy is to start at system level, continue at component level, then failure mode level etc.

All levels can be shown for one plant or the data can be pooled for several plants.

In the following are listed some graphs which can be extracted from the data.

1. Number of repairs/year
2. Manhours/year
3. Repair hours/year
4. Unavailability due to ...
    - Repairs (all)
    - Repairs of critical failures
    - etc.
5. Early failures after repair
6. Subsurvival functions
7. TTT-plot
8. Repairs as func. of Time between repairs (TBR)
9. $U_{prev}$ to $U_{corr}$
10. Mean number of repairs per. component (MT-plot)
11. Failure rate
12. Repair rate
13. Mean Time Between repairs
14. Pareto plots

Three graphs are shown here just to illustrate how the same problem can be observed in different modes.
The problem are the early failures after repair, which is an indication of quality of the repair work.
The graph on fig.2 shows for one plant the number of failures per year for a system of three pumps. It also shows the amount of these failures which appear less than 3 month after last repair ( the upper part of the bars). The interval for detection of early failures can be selected too.

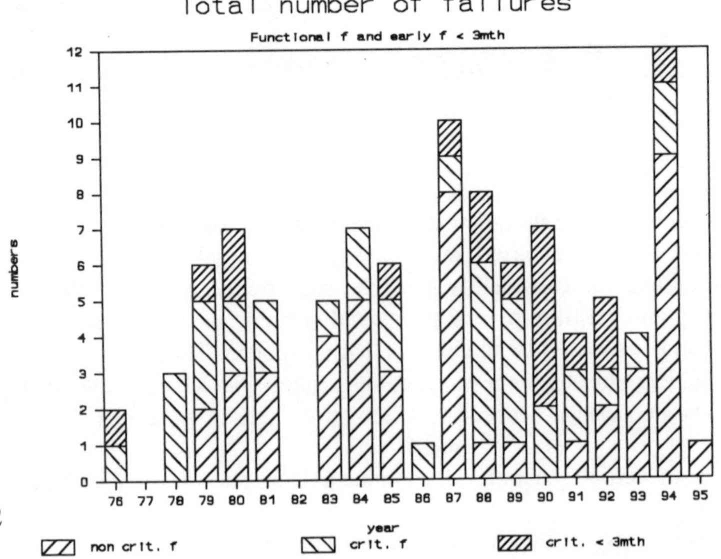


Fig.2

The graph on fig.3, subsurvival functions, use the mathematics for competing failure modes but look at detection modes instead. The detection modes are in this study classified as the failure detected during preventive maintenance or detected during corrective maintenance.
The graphs show the probability for the next repair to be detected either as a corrective repair or a preventive repair. The steepness in the beginning of the curves tell about the amount of early failures after repair.
In this pump system the corrective repairs are dominating, The probability that the next repair will be a corrective repair is highest.

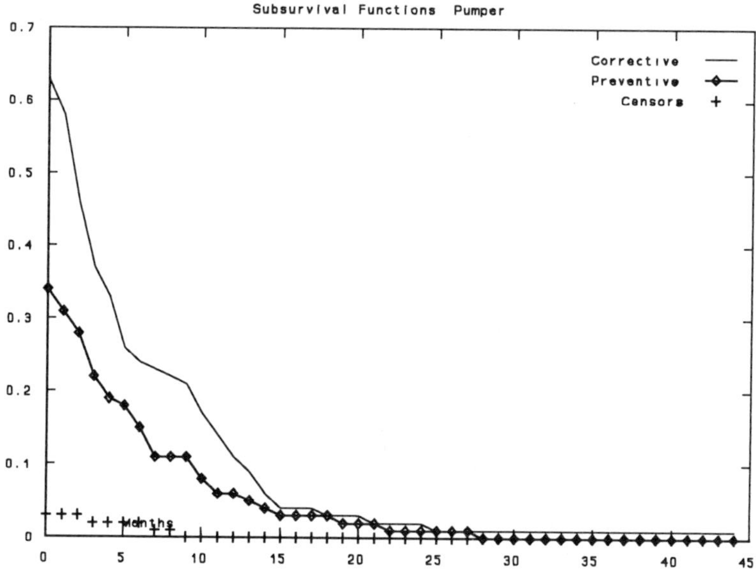


Fig.3

These two first plots was on system level, 3 pumps, the next step is to make a plot on an individual level. The plot choosen seen on Fig.4 is the accumulated number of failures for each pump. The x- coordinate is chosen as monthe to be able to detect early failures. It is easy from this plot to see that there is a difference in the amount of failures from the one pump to the others, and it is rather easy to detect periods with early failures.

The final interpretation of the results and the question about changing components or changing some plans require a knowledge of the systems that the users are expected to have.

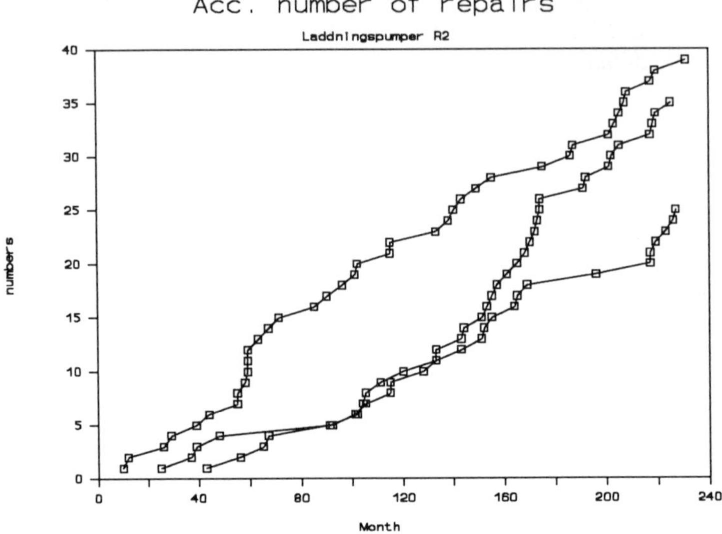


Fig.4

The work is a part of the Nordic Nuclear Safety research program NKS and the work has been done in collaboration with Sydkraft, Barsebäck, Swedish Nuclear Inspectorate (SKI) , The data base office (TUD) and Delft Technical University.

**References**

(1) Bergman,B. and Klefsjö,B. (1984) The total time on test concept and its use in reliability theory. Operation Research,32,596-606.

(2) Dorrepaal, J.W, Reliability analysis tool.
Thesis. TUDelft, Dec. 1995. SKI xxxx report.

(3) Cooke, R., Bedford, T., Meijlson, I. and Meester, L. Design of Reliability Data Bases for Aerospace Applications, Report to the European Space Agency Technical Centre, Department of Mathematics, report nr 93/110, Delft.

(4) Paulsen,J.L., Cooke, R., Concepts for measuring maintenance performance and methods for analysing competing failure modes. ESREL'94, 9th International Conference Reliability Maintability. La Baule, France

(5) Paulsen J.L., Clementz,M. Pilot Study on Maintenance indicators. Risø-I-667, NKS/SIK1 (92) 18 May 92

# Risk Based Regulation Using Mathematical Risk Models

Vincent Brannigan J.D.* Carol Smidts Ph.D. **

* Dept. of Fire Protection Engineering, UMCP vb15@umail.umd.edu
**Reliability Engineering Program, Dept. of Materials and Nuclear Engineering, University of Maryland at College Park. College Park Md. 20742 USA

**Abstract**

Technological regulation involves predicting and anticipating technological failures. Mathematical risk models can be used to develop risk based regulation. However, regulation is a process to determine compliance with social norms, and the appropriate standard for mathematical risk models is unclear. Minimum regulatory standards must be developed for the use of mathematical risk models.

## 1 Introduction

Safety regulation using mathematical risk models has been developed in the nuclear power industry. Nuclear power plants are so complex that they forced the development of both probabilistic safety assessments and risk based regulation. But predicting nuclear power plant system failures has not been very successful. Recently it was predicted that the Maine Yankee nuclear power plant would have only a few hundred cracked tubes in the steam generator. The variables affecting performance were thought to be well known, and prediction was expected to be easy. But more than 6000 cracked tubes were found. [1] Cracking is simply not well understood. As another example, Thermolag was used for fire protection of control system wiring, but at least some of the tests used to support the use of Thermolag were fraudulent. Mathematical risk models are proposed to show that the fire hazard is so low that the problems with the material can be ignored.[2] But progress has been made in risk modelling. New variables have been developed for the risk models. For example, until the ThermoLag problem was revealed, there was no consideration of the risk of possible fraud in product approvals. The Nuclear Regulatory Commission is moving steadily towards risk based regulation, but acknowledges that it must move very carefully given the unknown nature of many of the risks.

## 2 Undefined Systems and Model Uncertainty

The most important problem for risk modeling is whether the underlying problem is well understood or not. Most engineering analysis is focused on solving the problems of "**defined**" systems. Systems which are well defined tend to involve well

understood physical laws, have small values in any uncertainty term, tend to not involve substantial uncontrolled human variables and are relatively simple. Real world safety problems are often "**undefined**" in that physical laws are poorly understood, there are high uncertainties in any risk term, the systems is complex, there is little control over the range of potential variables and human actions such as public response, sabotage, and weather conditions tend to dominate the problem.
"Undefined" problems can be described as occupying a multidimensional space where defined problems are at the origin. The more undefined components in a problem, the less certainty we can have in a purely "technical" solution to a problem.[3]

Proponents of mathematical risk models often describe well defined environments where the system stress and all critical variables are simple and known or knowable. For example, in structural engineering gravity is a constant and even wind is well bounded. Further the assumptions are easily tested in wind tunnels or by loading the structure. This builds confidence that the problem is well defined. However, there is no guarantee that undefined portions of the system are actually understood, as was shown in the recent KOBE earthquake. If the mechanism of injury is not known, uncertainty in predicting performance cannot be corrected with a safety factor.

# 3 Regulation as Prediction

Societies respond to technical uncertainties with regulation, All technological regulation inherently involves predictive activity. The regulatory system attempts to predict, prevent or mitigate injuries. Probabilistic risk assessment (PRA) has been developed to give a quantitative structure to risk analyses which had previously only been qualitative. The use of PRA is an attempt to formalize and validate a portion of the risk based regulatory approach. However the creation of formal risk assessment methodologies cannot substitute technical analysis for the political and legal judgmental components of the regulatory process. The level of risk to be imposed on unwilling members of the public is not merely a decision for technical professions and practitioners. The public decides what risks it will take and public policy decisions must not be subsumed under the technological analysis. One of the most important public decisions is how to deal with uncertainty. For the purpose of this article the term **Risk** is a statement of the probability of an occurrence and **Uncertainty** is a statement of confidence in the risk estimate. Society has to decide how to deal with high uncertainty in the risk estimate. As just one example PRAs often have a safety level where the conditions become "untenable". In the model, as long as the tenability threshold is not violated the path is considered tenable. But just how good is the threshold assumption? Is it supported by overwhelming technical evidence, or even an informed consensus? The belief that a tenability threshold actually exists is a public policy decision not a technical consultant's decision. The decision reflects the type of error which the society prefers, I.e. would it prefer to err on the side of precaution or costs savings?

In traditional legislative regulatory activities technological decisions are made as part of the **Political** process, with the acceptance by the political process of the inherent limitations of knowledge. The legislature's decision is not expected to be scientifically valid. The technical criteria embodied in the statute are simply the society's best political guess as to how to remedy the problem. **The ability to enact regulatory predictions politically means we do not have to accept them scientifically.** While political acceptance of a risk analysis methodology by a regulator in no way "validates" the underlying technological determinations, it is an alternative method of satisfying society. The problem comes when political decision makers are not aware of the unknowns in the technical process. Long feedback loops between technical decisions and disasters, fundamental uncertainties in the evaluation of new technology and the inability to completely understand extremely complex systems make regulation a combination of political and technical decision making. Even when the regulator has technical training, the regulator is still making fundamentally political decisions.

# 4 Political and Technical Decisions in Modelling Risk

It is critical to track the policy and technical components of decision making. **Technical** decisions are those which deal with scientific or technological phenomenon that are the subject of well defined scientific or technical decision processes. **Policy** decisions are those which involve weighing of competing social, legal, cultural, technical and other judgmental factors in the regulatory process. Traditional regulatory systems normally evaluated the technical issues as "facts" and then presented technical conclusions as an input to responsible policy officials. The officials then combined the facts of the matter with the delegation of policy judgement in the statutes to regulate the risk. In traditional regulation, the domain of the technical and the policy decision maker were easily separated. Questions were clearly either technical or policy oriented, and the decision maker could easily sort them out. This regulatory model worked well in environments in which technical disputes were minor and easily understood. Risk based regulation using mathematical models presents an entirely different problem. PRAs can contain substantial policy judgments presented as technical statements of fact. PRAs can implicitly reflect the risk analyst's judgement on what are arguably political or policy decisions. Such implicit policy decisions might include:

**Estimates of probabilities which involve social actions** e.g. the probability of terrorist attacks, and the probability that the regulatory agency will continue to act consistently in the future.

**Estimates of technological events for which insufficient data exists for definitive technical understanding.** e.g. earthquakes, some human responses to accident conditions, low dose radiation. It often includes the use of expert judgement. [4]

**Choices among different schools of technical thought** Resolving disputes is difficult if there is no common core of beliefs. For example, there are distinct differences between Bayesian and Non Bayesian analysis of probabilities.

**High uncertainty in any risk term** Since risks are expressed as numbers, they can be manipulated using simple mathematical tools. However, if there is uncertainty in the risk estimate it cannot be reduced by manipulation.

# 5 Regulatory Use of Mathematical Models

Effective regulation in this type of complex environment requires creating an interactive regulatory system which effectively connects both the political and technical domains of the decision making.[5] The vital link is to create an iterative process which identifies decisions and makes preliminary assignments to either the technical or political domains. As the problem becomes more defined, it is possible to continue to refine the assignment to political and technical domains of the decision. However, it is vital not to overstate the ability of the risk model to predict accidents. Many catastrophic accidents are rare events and are the product of complex causal chains. Mathematical models which were developed for **Explanation** of accident phenomena (such as discovery of root causes) may not be suitable for regulatory **Prediction** of future accidents. Accidents are the product of a complex web of events, all which can be described as a "cause" of the accident. Perception of a specific "cause" for an accident may be a function of the fact that a specific accident actually occurred, and the methods used for investigating that accident. Human error at the time of an accident is often caused by latent defects in the system design. Even assembling the known root causes and extrapolating them will not necessarily result in a complete predictive model.

This is especially true if proof of causation is required. From a scientific point of view causation is difficult to prove, but correlation is a very useful substitute. Many risk models do not clearly identify the standards used for inferring causality from correlative models and use events which are not known to be causally related. But regulators require "causation" even when causal statements have high levels of uncertainty. The issue of causation is critical because regulators expect to be able to alter a "causative" event and have a real effect on the ultimate rate of injury. If the data element is merely correlated with the risk, regulators may be led to incorrect conclusions.

# 6 Conclusion

The regulatory use of Mathematical risk models requires addressing key issues:

1) The political or technical components of the risk assessment must be expressly defined.
2) There must be a clear distinction between political and technical decisions

3) Detailed analysis of the sources of uncertainty in the risk assessment must be presented.
4) Policy oriented rules should be developed for inferring causality from correlations
5) Predictive uncertainty normally increases as the system becomes more complex and undefined.
6) In undefined systems, retrospective analysis may not be able to give us adequate information on potential disaster paths or the root cause of accidents.
7) Predictions of future levels of human actions may have high levels of inherent uncertainty.

## References

1. FEDERAL REGISTER NRC Issuance of Director's Decision Under 10 CFR 2.206 60 FR 63737 December 12, 1995

2. Stellfox, David "Jackson to stay the course on thermo-lag; but resolution still way off" Inside N.R.C. 11/27/1995 17:24

3. Shlyakther, Alexander "Statistics of Past Errors as a Source of Safety Factors for Current models", Proceedings of Workshop I in Advanced Topics in Risk and Reliability Analysis, Model Uncertainty : its Characterization and Quantification, October 20-22, 1993, NUREG/CP-0138

4. Cooke R. Experts in Uncertainty: Opinion and Subjective Probability in Science, Environmental Ethics and Science Policy Series, Oxford U. Press, New York, 1991

5. Brannigan V., Meeks C.,"Computerized Fire Risk Assessment Models: A Regulatory Effectiveness Analysis", Proceedings of Workshop I in Advanced Topics in Risk and Reliability Analysis, Model Uncertainty : its Characterization and Quantification, October 20-22, 1993, NUREG/CP-0138

# An Integrated Probabilistic and Fuzzy Multi-objective Method for Hazardous Waste Risk Management

Timothy J. Ross
Department of Civil Engineering
University of New Mexico
Albuquerque, New Mexico 87131, USA

**Abstract**

Multi-objective decision making involves the selection of alternatives where the simultaneous satisfaction of two-or-more objectives dictates the decision outcome. Using this idea, risk assessment and risk management, although separate issues, can be combined in a single framework where decisions about remediation to reduce risks can be made based on an integrated approach where probabilistic methods and fuzzy methods are used to characterize both random and non-random forms of uncertainty. The ability of fuzzy logic to integrate numeric data like costs and risk, with linguistic knowledge such as political and economic sustainability is shown to be a powerful paradigm in the risk management area. A case study based on a US government facility involving a hazardous waste is used to illustrate features of the method.

# 1 Introduction

In the United States there are more than one-thousand hazardous waste sites on the national priorities list (NPL), and creating a policy to determine the order of clean-up of these sites is a controversial and difficult task. Risk assessment is now widely used to characterize sites for placement on the NPL. In making decisions, not only are risk values important, but factors such as social, economic, and political sustainability are also having more impact on decision-making. Although risks are usually quantified, their determination is based on many subjective assumptions.

Risk is defined as the likelihood of an adverse health effect, usually a carcinogenic death, due to exposure to an environmental hazard. This hazard is typically toxic or radioactive waste. Conventional risk analysis uses the concept of classical Boolean logic which is based on crisp numbers to estimate the risks. However, due to limitations of time and money on the collection of data, crisp numbers seldom represent the actual estimate. In the absence of complete data subjective estimates form the basis for determining the behavior of all the components of the risk assessment process. Subjective estimates are qualitative in nature; the estimates are generally best described by vague linguistic terms which cannot be appropriately represented by crisp numbers.

Fuzzy sets, which inherently consider the vagueness in data and which uses a set of values instead of a single value, are being widely used in assessing the non-random form of qualitative data. In recent years a number of papers have been written on the application of fuzzy logic to the process of risk assessment and risk management [1, 2]. Ross, et. al, [3] recently applied fuzzy Bayesian decision making to the selection of an appropriate waste remediation technology in risk management.

In this paper the use of fuzzy logic in the decision making process of risk management is combined with the traditional probabilistic approach of assessing likelihoods of death due to exposure to a toxicant. Fuzzy numbers can be used to assign numerical values to vague linguistic terms, such as low risks and high concentrations, for example, whereas probabilistic estimates are used to quantify numerical entities like the average incidence of cancer over a large population exposed to a given toxicant. Such an approach enables fuzzy logic and probability theory together to make more realistic estimates of risks, given the enormous uncertainty involved. Such realistic estimates can be used in making decisions about the reduction of these risks.

# 2 Multi-Objective Decision Making

Risk management of environmental contaminants involves the need to consider a number of options to reach a decision. In the past, decisions were made with respect to the satisfaction of one economic objective, which usually was the minimization of losses or the maximization of profits. With the increasing awareness of degrading environmental quality and the discovery of a plethora of contaminated sites, there is an urgent need to make prudent decisions that consider all the factors within the constraints of limited financial and temporal resources. Such decisions, therefore, are functions of many variables, which, in the decision-making process represent the numerous objectives that need to be satisfied for an efficient and appropriate decision.

In conventional decision analysis, objectives to be satisfied are usually reduced to a crisp number, and the decision making process consists of the optimization of a function of these objectives. However, the uncertainty associated with the satisfaction of many subjective objectives results in an uncertainty that is best represented by a range, or set, of numbers rather than by one crisp number.

What follows is a description of an approach to combine probabilistic information and fuzzy information is a single framework for conducting a risk assessment. This assessment then represents a component in a multi-objective decision making method for risk management.

## 2.1 Fuzzy Bayesian Analysis

The combined probabilistic and fuzzy method illustrated here makes use of a Bayesian analysis modified to accommodate fuzzy information, fuzzy states of nature and fuzzy actions (decision alternatives) [3]. Such an approach combining probabilistic and fuzzy analysis of uncertain information is developed in many places in the literature; readers not familiar with this method could review, for example, Terano, et. al. [4]. For contaminated waste sites the states of nature in the Bayesian scheme could be the risk posed to human health, and new information could be associated with tests conducted on site to better determine risk–or the extent of contamination.

The definition of the true states of nature is subjective and depends on the decision maker. For example, five crisp (non-fuzzy) states of nature based on specific health risks to a given toxicant could be considered in an analysis: $s_1=risk_1=10^{-8}$; $s_2=risk_2=10^{-6}$; $s_3=risk_3=10^{-5}$; $s_4=risk_4=10^{-4}$; and $s_5=risk_5=10^{-3}$ (e.g., a risk of $10^{-2}$ means that there is an expected cancer death of 1-in-100 people exposed to the carcinogen). We define these states as crisp because they are defined on deterministic values of risk, whose values vary according to the chemical and the corresponding maximum contaminant level (MCLs) considered for the site. Prior probabilities assigned to each of these states, expressing the existing opinion regarding the likelihood of the various risk levels, for example, could be:

$$p(s_1) = 0.1,\ p(s_2) = 0.2,\ p(s_3) = 0.3,\ p(s_4) = 0.3,\ p(s_5) = 0.1.$$

In classical (non-fuzzy) Bayesian decision analysis, these would be the states that would be used in the analysis.

In practice, due to large variations in the concentrations of the chemicals in the environmental media and to the uncertainties associated with testing (determining contaminant concentrations and geographic extent) at the sites, we can rarely assign a definitive risk value to each defined state. As with most problems in life dealing with state definitions using ambiguous natural language, there is overlap in the meaning between two or more states, i.e., the state of a site can be "approximately benign" or "around moderately hazardous" simultaneously. Usually it is more natural for humans to define a site by such linguistic terms. Fuzzy decision making is based on assigning membership functions to these linguistic terms, and then using them to determine the optimum alternative.

## 2.2 Fuzzy Sets and Membership

In fuzzy sets an ambiguous variable is usually represented by a function called a membership function. For example, the risk of a carcinogenic death due to exposure of a given toxicant might only be known approximately, based on animal studies. Suppose the animal studies show that the extrapolated risk to the average human exposed to this toxicant is *moderate*. Moreover, this risk is estimated to be centered around the numerical value of $10^{-5}$. A mathematical distribution on the domain of numbers in the set “approximately $10^{-5}$”, as shown in Fig. 1, is then assigned to

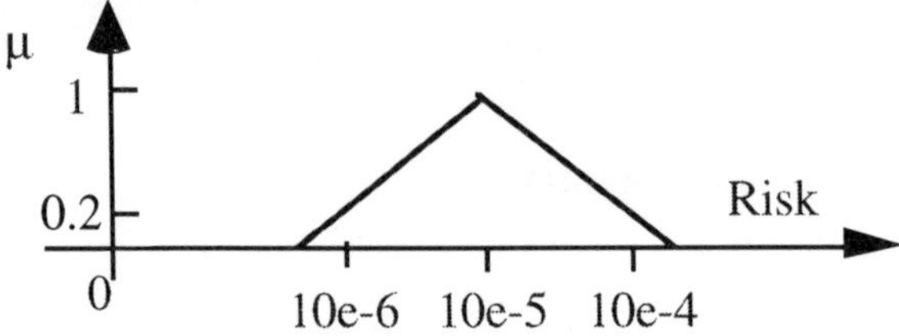


Figure 1. Fuzzy membership function for the variable "moderate risk”

this linguistic variable. We could assign a linguistic label to this approximate value as well, for example the label *moderate risk*. In the figure a membership of 1 is assigned to the risk number $10^{-5}$, i.e., $10^{-5}$ belongs completely to the set “moderate risk”, a value of 0.2 is assigned to the risk numbers $10^{-4}$ and $10^{-6}$, indicating these numbers have much lower membership in the set “approximately $10^{-5}$”, or in the set "moderate risk".

The membership values are assigned based on the accuracy and the degree of confidence in the available data. When used in numerical analysis this function appears as,

$$\text{“moderate risk”} = \text{“risk is} \sim 10^{-5}\text{”} = \frac{0.2}{10e-6} + \frac{1.0}{10e-5} + \frac{0.2}{10e-4}$$

## 2.3 A Fuzzy Multi-objective Decision Making Method

In the multi-objective development, we begin by defining a set of n alternatives, A={$a_1$, $a_2$, ..., $a_n$} and a set of r objectives, O={$O_1$, $O_2$, ...,$O_r$}. Let $O_i$ indicate the ith objective. Then the degree of membership of alternative a in objective $O_j$, denoted, $\mu_{Oj}(a)$, is the degree to which alternative a satisfies the criteria specified for this objective. We seek a decision function which simultaneously satisfies all of the decision objectives; the decision function, D, is given by the intersection of all the objective sets,

$$D = O_1 \cap O_2 \cap \ .... \ \cap O_r \tag{1}$$

Therefore, the grade of membership that the decision function, D, has for each alternative a is given by,

$$\mu_D(a) = \min[\mu_{O1}(a), \mu_{O2}(a), \ ..., \ \mu_{Or}(a)] \tag{2}$$

The optimal decision, a*, will then be the alternative which satisfies,

$$\mu_D(a^*) = \max_{a \in A}(\mu_D(a)) \tag{3}$$

We now define a set of preferences, {P}, which we will constrain to being linear and ordinal. Elements of this preference set can be linguistic values such as none, low, medium, high, absolute, or perfect, or they could be values on the interval [0, 1], or other scales [5]. These preferences will be attached to each of the objectives to quantify the decision-maker's feelings about the influence that each objective should have on the chosen alternative. Let the parameter, $b_j$, be contained on the set of preferences, {P}, where i = 1, 2, .., r. Hence, we have for each objective a measure of how important it is to the decision maker for a given decision. The decision function, D, now takes on a more general form when each objective is associated with a weight expressing its importance to the decision maker. This function is represented as the intersection of r -tuples, denoted as a decision measure, M($O_i$, $b_i$), involving objectives and preferences,

$$D = M(O_1, b_1) \cap M(O_{21}, b_2) \cap \ ... \ \cap M(O_r, b_r) \tag{4}$$

A key question at this point in the development of this calculus is what operation should relate each objective, $O_i$, and its importance, $b_i$, that preserves the linear ordering required of the preference set, and at the same time relates the two quantities in a logical way where negation is also accommodated? It turns out that the classical implication operator satisfies all of these requirements, although others have been proposed [5]. Hence, the decision measure for a particular alternative, *a*, can be replaced with a classical implication of the form,

$$M(O_i(a), b_i) = b_i \to O_i(a) = \overline{b}_i \vee O_i(a) \tag{5}$$

Therefore, a reasonable model will be the joint intersection of r decision measures,

$$D = \bigcap_{i=1}^{r} (\overline{b}_i \cup O_i) \tag{6}$$

and the optimal solution, $a^*$, is the alternative that maximizes D. If we define,

$$C_i = \overline{b}_i \cup O_i, \text{ hence, } \mu_{Ci}(a) = \max\left[\mu_{\overline{b}_i}(a), \mu_{Oi}(a)\right] \tag{7}$$

then, the optimal solution, expressed in membership form, is given by,

$$\mu_D(a^*) = \max_{a \in A}\left[\min\{\mu_{C1}(a), \mu_{C2}(a), \ \ldots, \ \mu_{Cr}(a)\}\right] \tag{8}$$

## 2.4 Case Study

Suppose the US Army needs to select a remediation technology to reduce the risk at a contaminated trichloroethylene (TCE) site at one of its bases. TCE was a common industrial and commercial solvent and degreaser used for maintenance activities at Army posts until the mid-1970's, and is now a known carcinogen. Its usual pathway to human exposure is through groundwater, particularly drinking-water wells. Among the many alternative technologies available to the Army, three methods to treat the contaminated soil and water are considered in this example: (i) pump and treat with aerobic stripping (Aerob), (ii) pump and treat with ultra-violate radiation (UV). and (iii) aerobic stripping with bioremediation (Bio). The Army (the decision maker) has defined five objectives which impact their decision: (i) the risk of the contaminant (Risk) to human health, ii) ecological risks (ER), (iii) the life-cycle cost of the remediation technology (Cost), (iv) the political sustainability (PS), and (v) the economic sustainability (ES) of the proposed treatment. Moreover, the Army also decides to rank their preferences linguistically on the unit interval, based on participation by the public. Hence, the Army sets up their problem as follows,

$$A = \{\text{Aerob, UV, Bio}\} = \{a_1, a_2, a_3\}$$
$$O = \{\text{Risk, ER, Cost, PS, ES}\} = \{O_1, O_2, O_3, O_4, O_5\}$$
$$P = \{b_1, b_2, b_3, b_4, b_5\} \rightarrow [0,1]$$

The five objectives carry different units. Risk values are probabilistic values which give a projection of the excess risk to human health and ecology due to exposure to TCE; i.e., the probability of the number of excess cancer deaths per 100,000 population or a decrease in the native species population; costs are expressed in dollars calculated for a particular base-year; political and economic sustainability would generally be expressed linguistically since these are very subjective variables. Because we need to address the problem from the same units, all the variables will be reduced to the unit interval using membership functions.

From this scenario of objectives, the Army first expresses, in standard fuzzy notation, the remediation alternatives with respect to the objectives,

$$\underset{\sim}{O}_1 = \left\{\frac{0.4}{Aerob} + \frac{1}{UV} + \frac{0.1}{Bio}\right\} \quad \underset{\sim}{O}_2 = \left\{\frac{0.5}{Aerob} + \frac{0.8}{UV} + \frac{0.3}{Bio}\right\}$$

$$\underset{\sim}{O}_3 = \left\{\frac{0.7}{Aerob} + \frac{0.8}{UV} + \frac{0.4}{Bio}\right\} \quad \underset{\sim}{O}_4 = \left\{\frac{0.2}{Aerob} + \frac{0.4}{UV} + \frac{1}{Bio}\right\}$$

$$\underset{\sim}{O}_5 = \left\{\frac{1}{Aerob} + \frac{0.5}{UV} + \frac{0.5}{Bio}\right\}$$

For example, the aerobic remediation technology alternative has a membership of 0.4 in the first objective, $O_1$, the UV remediation technology alternative has a membership of 1.0 in the first objective, $O_1$, and the bioremediation technology (Bio) alternative has a membership of 0.1 in the first objective, $O_1$. Since the first objective is human health risk reduction, this simply means that the UV technology is more likely to reduce health risks than the aerobic method, and the aerobic method is more likely than the bioremediation method to reduce risks. Now, the army ranks the importance of the objectives and develops a preference vector. The preferences assigned are, $b_1 = 0.8$, $b_2 = 0.6$, $b_3 = 0.9$, $b_4 = 0.7$, $b_5 = 0.5$. Therefore, from Eq. 6, for the aerobic method,

$$D(a_1) = D(\text{Aerob})$$

$$= (0.2 \vee 0.4) \wedge (0.4 \vee 0.5) \wedge (0.1 \vee 0.7) \wedge (0.3 \vee 0.2) \wedge (0.5 \vee 1) = 0.3$$

Similarly, for the two other remediation alternatives,

$$D(UV) = 0.4 \text{ and } D(Bio) = 0.2.$$

Now, using Eq. 8 to determine the optimal solution,

$$D^* = \max \{0.3, 0.4, 0.2\} = 0.4.$$

Therefore the Army chooses the second alternative: it makes the decision to remediate the site using ultra-violet radiation.

# 3 Conclusions

Most of the decisions we face involve the satisfaction of a number of objectives. Many of these objectives are more subjective than objective in nature. Fuzzy logic is a promising method that is gaining popularity in terms of modeling uncertain and vague information. Probabilistic methods for assessing human health risk are now well accepted in the literature because they are based on numerical information from laboratory studies or simulations involving pharmacokinetic models for cell mutations in the development and growth of cancers. Unfortunately, because these studies either involve extrapolations from one species to the next, or they involve uncertainties in the true mechanisms of cancer, the uncertainties plaguing the probabilistic estimates are usually of the non-random kind. The case study in this paper exemplifies the importance and the utility of fuzzy logic.

## References

1. Anandalingam, G. and Westfall, M. Selection of Hazardous Waste Disposal Alternative using Multi-Attribute Utility Theory and Fuzzy Set Analysis, J. Environmental Systems 1989; 81 (1), 69-85.
2. Lee, Y. W., Dahab, M. F. and Bogardi, I. Fuzzy Decision Making in Groundwater Nitrate Risk Management, Water Resources Bulletin 1994; 30 (1), 135-148.
3. Ross, T., Trumm, J. and Donald, S. Decision-Making Paradigm for Risk Assessment of Hazardous Wastes using Fuzzy Sets, Proc. ASCE First Congress in Computing, Washington, DC, ASCE Publications 1993; 1, 200-207.
4. Terano, T., Asai, K. and Sugeno, M. Fuzzy Systems Theory, Academic Press, 1992.
5. Yager, R. A New Methodology for Ordinal Multi-objective Decisions Based on Fuzzy Sets, Decision Sciences 1981; 12, 589-600.

# Evaluating the Control of Safety in Maintenance Management in Major Hazard Plants: a Research Model

Heming, B., Hale, A.R., Smit, K.. van Leeuwen, D., Rodenburg, F.
Delft University of Technology & Rijnconsult
The Netherlands

## 1 Introduction

Maintenance in the process industry is connected with a significant proportion of the serious accidents occurring in the industry [1,2,3]. These findings led the Department of the Dutch Social Affairs Ministry responsible for the assessment of company safety reports for major hazards plant, to commission a study of the management of safety in maintenance in the total life cycle of a plant to assess the potential of improvement [4,5].

In order to investigate the quality of management of safety related to maintenance a model was developed of an ideal management system covering the full life cycle of the plant or installation from design and construction to normal operations and the maintenance phase itself. The model was validated by peer review in five companies with high reputations for the control of safety in maintenance. An audit instrument and a questionnaire were developed from the model. The audit formed the basis for the studies in eight major hazard companies. The questionnaire was used to obtain results from 47 major hazard companies. This paper describes the model and the thoughts behind it and discusses the usability of the model as shown by the results of the surveys. The paper of Rodenburg [6] gives in detail the audit and survey results.

## 2 Modelling the Management of Safety in Maintenance

### 2.1 Life cycle of a plant

The primary purpose of maintenance is to prevent deviation in plant functioning, which can threaten not only production but also safety (preventive maintenance) and to return a plant to full functioning after a breakdown or disturbance (corrective maintenance). The frequency with which maintenance needs to be carried out and the ease and safety with which it can be done are factors which are determined by decisions made in the design and construction phases of the life cycle, such as the choice of the equipment and instrumentation, the design for shutting down parts of the plant, or the accessibility of items of plant for maintenance work. Maintenance work is also frequently associated with modifications to the plant, which will alter the plant operations and hence its safety and require plant operators to learn some new operating routines. Any modelling of the management of safety related to maintenance must therefore take account of the whole plant life cycle. There must be both feed-forward loops of information and decisions from earlier to later phases of the life cycle and feedback loops to communicate lessons learned, or changes made to earlier life cycle phases for the existing or for new plants.

## 2.2 Models of Maintenance Management

Models of maintenance management [7] emphasise the development of a maintenance concept or plan during the detailed engineering phase of a plant design, which specifies when and how each significant part of the plant will be maintained. In particular the maintenance concept specifies whether preventive maintenance will be carried out, and if so whether this will be condition-dependent or at fixed intervals, or whether the plant will only be maintained after breakdown. The maintenance management model distinguishes three levels:

- *a management level* at which objectives, policies and standards are determined and adjusted on the basis of periodic evaluation;
- *a control level* (planning and procedures level) at which the maintenance procedures are kept within the prescribed standards; this has two sub-levels, the determination and development of the required resources (spare parts, trained personnel, methods, facilities and documentation) and the scheduling and work planning to match the work load to the resources;
- *an execution level* at which the activities are performed, from plant isolation and handover to completion of the tasks and return of the plant to the operations personnel; within the execution level the life cycle phases of the plant are visualised.

## 2.3 The Deviation Concept

The concern with a life cycle view of control and the division into three levels of management are also to be found in the models of safety management developed in Delft [8]. These models are based upon a conception of safety management as a series of interlocking problem-solving cycles focused on the prevention and correction of deviations which can lead to harm (injury, damage, ill-health). The deviation concept can also be found as a central feature in maintenance management. The types of deviations from full functioning of a plant or system, the ease of predicting or detecting them and the dynamics of their development play a fundamental role in deciding on the type and frequency of maintenance. Figure 1 illustrates the relationship of the two deviation concepts.

## 2.4 The Research Model

These essential similarities between the models of safety management and maintenance management enables a combined model to be formulated illustrating the

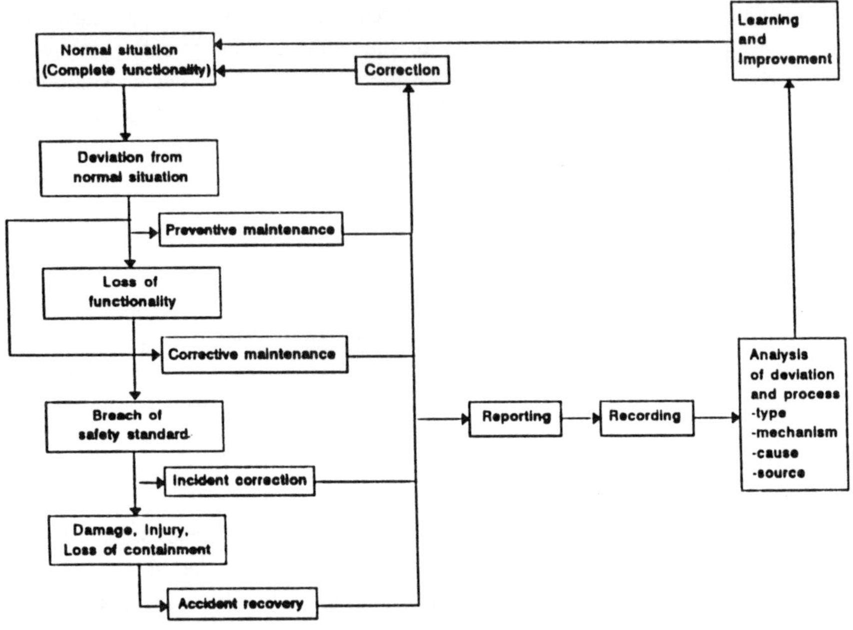


Figure 1
The Relationship between Safety and Maintenance in a Deviation Concept

steps and information flows which should be found in an ideal management system to take proper account of safety in relation to maintenance. The model is shown in figure 2 on the next page. The activities of dealing with breakdowns or plant failure (corrective maintenance) and the execution of preventive maintenance are directed by the maintenance concept. In addition the maintenance function deals with plant modifications for improving functional capabilities, legal compliance, reliability and maintainability and the reduction of lifetime costs. The same service may be responsible also for plant design and engineering, but must in any case be involved in these phases inputting to design reviews aspects such as reliability and maintainability. In this phase also the maintenance concept should be developed making use of feedback from maintenance experience and data. The whole activity is directed and steered by maintenance policy, whilst each step in each process needs to incorporate safety criteria, derived from the safety policy (shown in figure 2 as "s" feeding in to each box) to ensure the safe completion of each step as well as the safety of the output of the step. The box at the right hand side of the figure represents the learning processes which feed into the continuous cycle for improving safety in all the processes.

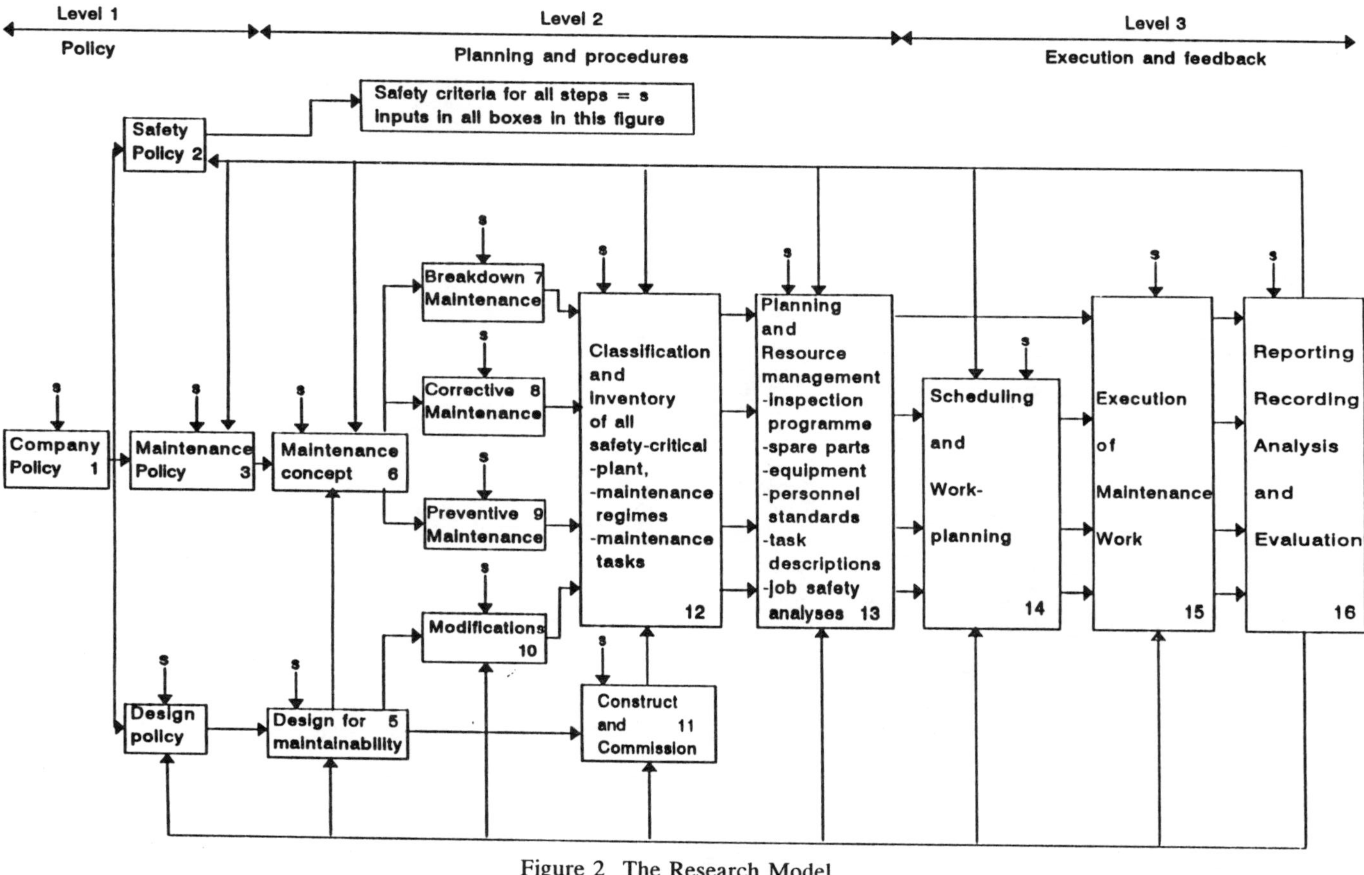


Figure 2 The Research Model

## 2.5 Field testing of the model

The proposed maintenance management model is derived from theoretical considerations. In order to test whether it could form the basis for an assessment of the effectiveness of maintenance management in chemical companies a small scale study was conducted to subject it to peer review. Five organisations were chosen which have a high reputation for the quality of their maintenance. The selection was made in consultation with experts in the field to identify companies and operations where the importance of maintenance for safety was clear and where the record of the company was excellent. The industries represented by the five companies were nuclear power, aircraft operations, railways, oil and gas production and shipping. Each company was visited for one day, for discussions with the senior managers responsible for maintenance and safety. The structure of the model and its implications for policy, activities, responsibilities, supervision, information flow and decision making were discussed with these managers in order to check whether the management systems of these companies could be mapped onto the model, whether they recognised and practised all the steps and whether they could identify other elements which they practised, but were not in the model. As a result of these discussions no fundamental changes in the model were found necessary, though the comments did result in modifications to the detailed level of translating the model into a set of auditing questions.

## 2.6 Development of assessment instruments

The research model was used to develop a set of questions to be used as a checklist in carrying out a detailed audit of a number of companies in the chemical industry, chosen as representative case studies. Each block of the model in figure 2 was turned into a series of questions to discover how and by whom the step was carried out and how safety as a set of criteria was taken account of during each step. Each phase of the life cycle was studied (new construction, modification and direct maintenance), as well as each management level from policy, through plans and procedures to execution, in relation to both maintenance and safety and their interrelationships. The complete checklist is described in Heming et al. [9].

# 3 The Usability of the Maintenance Model

The experience with the use of the maintenance model is positive. The checklist derived from and used during the audit gives a clear framework for the interviews, which was also acknowledged by the companies. The model emphasises the systematic approach towards maintenance management. Human resource management during execution of maintenance is much less emphasised, mainly because this study did not focus on accidents during maintenance. During the audit the importance of the distinction between the presence of a systematic approach and the application in practice became much more clear. The latter was given special attention during the assessment phase [10].

Hence, in practice the model is very useful although the application of some concepts in the model should be reconsidered. It should be pointed out that for a full benefit all of the concepts should be worked out in great detail . At the moment this has been accomplished for only part of the model.

## References


1. Health and Safety Executive. Deadly maintenance. HMSO, London, 1985
2. Koehorst L.J.B. Analyse van een selectie ongevallen in de chemische industry uit de databank FACTS (Analysis of a selection of accidents from the database FACTS). SDU, Den Haag, 1989
3. Hurst N.W., Bellamy L.J., Geyer T.A.W. & Astley J.A. A classification system for pipework failures to include human and socio-technical errors and their contribution to pipework failure frequencies. Journal of Hazardous Materials 1991; 26: 159-186
4. Hale A.R., Heming B.H.J., Rodenburg F.G.Th, & Smit K. Onderhoud en veiligheid: een study naar de relatie tussen onderhoud en veiligheid in de procesindustrie; fase 1 (Maintenance and safety: a study of the relation between maintenance and safety in the process industry; phase 1). Report to the Dutch Ministry of Social Affairs and Employment. Safety Science Group, Delft University of Technology, 1993
5. Heming B.H.J., Rodenburg F.G.Th., Hale A.R., Smit K. & van Leeuwen N.D. Veiligheid in onderhoudsmanagement: verslag van fase 2 (Safety in maintenance management: report of phase 2). Report to the Dutch Ministry of Social Affairs and Employment. Safety Science Group, Delft University of Technology, 1995
6. Rodenburg F., Heming B., Van Leeuwen, D., Smit K. & Hale, A.R. The Quality of Management of Safety in Relation to Maintenance in Dutch Major Hazard Plants: Audit and Survey Results and Analyses (this volume).
7. Smit K. & Slaterus W.H. Information model for maintenance management ISBN 90-71996-56-5). Cap Gemini, Rijswijk, 1992
8. Hale A.R., Heming B.H.J., Carthey J. & Kirwan B. Modelling of safety management systems. Paper submitted to Safety Science, 1995.
9. see 5
10. see 6

# Regulating Safety in Maintenance Activities in High Hazard Plant.

Hale A., Smit K., van Leeuwen D., Rodenburg F. & Heming B.
Delft University of Technology & Rijnconsult, NL

A study of the attention to safety in maintenance management in the Dutch chemical industry indicated areas for improvement. This paper presents a framework for regulatory activities at company, industry and government level to stimulate these improvements and classifies the tools available for doing so

## 1. Introduction

Other papers in this volume describe the structure and results of a study funded by the Dutch Ministry of Social Affairs into the state of safety in relation to maintenance activities in the Dutch chemical industry [1,2]. The study developed an ideal model of the way in which safety should be incorporated into the decision making about maintenance which takes place in all phases of the life cycle of a plant [3]. This model is given in figure 1.

The major weaknesses found in the companies studied were:

- feedback from practice to the design stage to learn from earlier experience
- development of explicit maintenance strategy/programmes for plant, taking into account not only the safety of the plant but also of the maintenance personnel
- concern at the design stage for maintainability
- an explicit safety policy for the Engineering Maintenance Function (EMF)
- management of the planning and resources for maintenance, in particular updating of documentation and communication with contractors personnel
- the process of handing back the plant after completion of work
- organisational learning from breakdowns, incidents and accidents.

The execution phase of the maintenance activity had relatively good management procedures. The study recommended the development of a strong, maintenance engineering function to personify the planning and inspection functions and to ensure feedback and learning (see the dotted line in figure 1).

To meet the objectives of the responsible policy makers a framework was drawn up of the possible factors which could be influenced directly or indirectly by government in order to provide a systematic picture of which actors they should try to activate. Since the study was carried out in a climate of government de-regulation the emphasis of the Ministry was on stimulating industry initiatives, rather than on extending laws and regulations. However, the field work indicated that the aspects of maintenance management which were regulated in detail by the law (e.g. Pressure vessel regulations) were better carried out in the companies studied than those where no specific legal requirements were present. The framework and measures proposed in this paper do not therefore eschew proposals for tightening legal requirements.

## 2. A regulatory framework

The regulatory framework places centrally the company self-regulating safety management system (SMS) which is influenced by external groups exerting pressure on, or providing resources for supporting the SMS. Government can influence the company directly or indirectly by stimulating other actors to perform and coordinate their roles.

To identify how and where these actors could have their influence on the SMS use was made of the Structured Analysis and Design Technique [4] which had been used by our group in Delft for studies of management and regulatory systems in the past [5]. It can be used to analyse decisions at differing levels of detail within the SMS. For this project we used it particularly to look at the major building blocks of policy making, planning and execution in each of the life cycle phases of a chemical plant. The technique conceives of management decisions as transformations of *information*, which is *input* to an activity and translated to form the *output* of the decision box. The transformation is steered by *criteria,* for our purposes the desired level of safety, translated into an appropriate form for the given decision. The transformation is carried out by the use of *resources (people* with knowledge, skills and motivation, *methods or tools* e.g. measuring instruments, analysis techniques). This notation picks up the elements already emphasised in the maintenance management model in figure 1. The "s" feeding into the top of each box there represents the safety criteria and the arrows between boxes the information flows. If these information flows, the criteria and the resources are seen as the levers which can steer the transformation and ensure its quality, we can make an explicit link to regulation. These are the levers which both the company in its own SMS and the outside world, including the government ministries, can use to steer the process and hence, potentially, to improve it. The actors must: ensure that the resources, criteria and information for each part of the SMS are available at the right time and are of the appropriate quality; check the functioning of each step and of the whole system to ensure that it is working and producing the desired output.

The relevant actors in terms of regulation are therefore those organisations which develop, provide or influence the choice of the three types of input and those which check the functioning. The main actors relevant to the regulation of maintenance management and safety in the Dutch chemical industry are the following:

- Ministries (Social Affairs, Environment) and their inspectorates
- European legislators, e.g. concerned with the Major Hazards Directive
- Provincial and Municipal governments responsible for planning approvals.
- The employers' organisation for the chemical industry (VNCI)
- Local associations of companies, e.g. in the Europoort Botlek area of the Rijnmond which houses the major concentration of the Dutch chemical industry (EBB)
- Individual companies, particularly trend-setting multinationals
- Associations of maintenance engineers (NVDO) and safety engineers (NVVK)
- Universities, technical colleges and others training managers, supervisors, maintenance engineers, technical and safety personnel
- Certifying bodies auditing and inspecting management systems and plant
- Designers and design consultants working for chemical companies

- Equipment manufacturers
- Maintenance contractors.
- Research and development organisations developing tools, software etc. and providing advice and consultancy.

We used these frameworks to translate the shortcomings discovered by our studies into proposals for developing, or strengthening action by these actors, aimed at one or more of the essential ingredients of an ideal safety management system.

# 3. Regulatory tools for improvement

## 3.1. Criteria

The model of the ideal safety management system for maintenance, summarised in figure 1 provides one criterion for judging the system existing in companies. Our project provides a preliminary validation of the model, but further research with it is needed to confirm our work and to link it to measures of accident statistics or to detailed analyses of maintenance accidents [6]. The model can be fed into the National Practice Guideline on quality management of maintenance [7], linked to ISO 9001, to provide more explicit attention to safety.

The AVRIM audit [8], which checks compliance with the Plant Safety Report (Post-Seveso) legislation in the Netherlands, already pays considerable attention to maintenance management, but requires amendment to require an explicit maintenance concept based on risk analysis and design for maintainability. The Safety Checklist for Contractors (VCA) [9], which is used by a large number of companies to check on the SMS of contractors, before accepting them onto the bidding list, would also need to be modified to give more emphasis to the feedback from maintenance to the design stage and the link to the maintenance engineering function. It could then be used as a criterion document not only for contractors, but also as an explicit safety policy for the in-house EMF, which was lacking in 75% of the companies surveyed. The VCA is currently a voluntary audit. Its use is expanding under the commercial pressure wielded by large companies. There is strong resistance from them to making it a legal requirement, but government could encourage its use by referring to it as a guideline during inspections. Under the EU fair trading rules (Procurement Directive) its use has already spread outside the Netherlands.

In the design stage there is a need for a much stronger use of safety as a criterion for both design and layout of maintenance-free and -friendly plant and as the basis for an explicit maintenance concept. The Construction Process regulations [10], based on the Temporary and Mobile Workplace Directive of the EU, provide a possible basis for this, as they require the designer to develop a safety and health plan to provide explicit information to the constructor and user on, among other things, the safety of maintenance. A new development would be to develop certification criteria for design consultants and departments (based on ISO 9001) giving more emphasis to design for safe maintenance, and in particular for requirements on designers to establish and maintain systems for securing feedback and learning from practice. There is also a role for manufacturers, in collaboration with industry associations, in developing standard maintenance concepts for certain types of equipment and for developing standards for the type and amount of information which needs to be provided by designers to users in relation to maintenance safety.

### 3.2. Information inputs

Studies in the Norwegian offshore sector [11] show that design contractors have little or no contact with the field situation and its practical safety problems. 50% of the companies in our study did not use inspection and maintenance records, breakdown registration and incident reporting as feedback into the design stage (e.g. via HAZOP). None of the companies surveyed combined the information from their maintenance and safety records. In only two thirds was it standard practice to include someone with maintenance experience in a HAZOP team. Initiatives from the industry (VNCI, EBB, NVDO) could encourage this, as well as promoting the exchange of lessons between companies, perhaps via a confidential data base on incidents and breakdowns for use by all participants.

Explicit attention to safety during maintenance planning and to communication of safety implications to technicians was relatively poorly organised. Existing initiatives by an industry/contractor working group to develop Project Safety Plans in standard and condensed format for use by contractors personnel need to be supported and extended, as do EBB attempts to harmonise Permit-to-Work systems. The process of handing back plant after maintenance was also not formalised in a quarter of the companies surveyed, whilst updating of "as-built" records to incorporate changes made during plant shutdowns was a problem for many. We see a role here for a strong maintenance engineering function, incorporating an internal inspection service (with suitable guarantees of independence provided by formal certification, to prevent undue pressure from production) which can check work done and officially sign it off for resumption of production. Such a user inspectorate could also avoid the delays and expense of calling in external certifying bodies to do this work.

### 3.3. Resources: tools

There is a need for a number of software tools to be developed or improved, particularly for incident and inspection recording and analysis in order to derive design and management lessons, and for software making updating of records easier [12]. Risk analysis tools (e.g. HAZOP) do not currently focus on the safety of maintenance, seeing it more as a measure to reduce the probability of failures than as an opportunity to introduce faults, or as an activity with a high risk for the staff. HAZOP itself requires development by research organisations and industry to fulfil this function. Very few companies had the tools available to make analyses of maintainability at the detailed design stage. The application of job safety analysis to maintenance tasks was also not widespread. The techniques of Reliability Centred Maintenance also require adaptation to incorporate safety more explicitly, in particular the safety of maintenance personnel [13]. In all these cases development of the analysis and assessment tools is needed and government could play a role by stimulating/financing research organisations, industry and professional bodies.

### 3.4. Resources: people

All the above implies the need for skilled and knowledgeable people. Under the auspices of EBB, as part of the VCA audit, all maintenance personnel are required to attend a short safety training course. In the first 3 years of this scheme more than 35,000 contractors' personnel have done so. Initiatives are now being developed to require all maintenance supervisors to attend a longer course. A safety passport, recognised by all participating companies, records the training and refresher courses. However 20% of the surveyed companies still did not give their own staff regular training in safe working procedures, and almost half did not give it to contractors personnel; so there is still major room for improvement.

Training for maintenance engineers and safety engineers from universities and higher technical colleges, for technical and production managers and for inspectors and auditors is much less well organised. The ideal model produced by our study can form a basis for this, as can the existing and to be improved tools mentioned earlier. However, a major initiative from faculties is needed to accept the importance of these topics; both safety and maintenance are undervalued Cinderellas of higher technical education and their combination doubly so. Pressure from government and industry, perhaps via visitation committees, is needed to overcome this current lack of interest.

### 3.5. Functioning checks

The use of the audit tools mentioned in 3.1 by government or certifying bodies, by those letting contracts and by companies for their own activities can provide the necessary checks on the functioning of the management system. Ultimately the motivation to comply with criteria depends strongly on the belief that performance will be checked and suitably rewarded (and non-compliance punished).

## 4. Conclusions

The modelling technique produced a structuring of the regulatory issues which was positively assessed by the policy makers. Plans are being developed to present it to other actors involved.

## 5. References

1. Heming B, Rodenburg F. Smit K. & Hale A. Evaluating the control of safety in maintenance management: models and instruments (This volume)
2. Rodenburg F. Heming B, Smit K. Hale A. & v Leeuwen D. The quality of management of safety in relation to maintenance in Dutch Major Hazards plants: audit and survey results and analyses. (This volume)
3. Hale A.R., Heming B.H.J., Smit K., Rodenburg F.G.Th, & van Leeuwen N.D. Evaluating safety in the management of maintenance activities in the chemical process industry. Safety Science 1996 (in press)
4. Marca D.A. & McGowan C.L. 1988 Structured analysis and design technique. New York: McGraw Hill

5. Hale A.R. Heming B.H.J., Carthey J. & Kirwan B. 1994. Extension of the model of behaviour in the control of danger. Industrial Ergonomics Group, School of Manufacturing & Mechanical Engineering, University of Birmingham.

6. Moll O., Hale A.R. & Smit K. 1994. Prevention of maintenance-related accidents (in dutch) Tijdschrift voor Toegepaste Arbowetenschap 1994; 7(6): 79-86

7. NPR 2720. Draft Guideline for practice: quality management in maintenance. (in Dutch) NVDO. Dutch Standards Institute. Delft.

8. Ministry of Social Affairs & Employment. 1993. Handbook AVRIM: an inspection technique for the plant safety report. (in Dutch). The Hague. Staatsuitgeverij

9. VCA: Safety checklist for contractors. Schiedam. Europoort Botlek Belangen.

10. Construction process regulations (Bouwprocesbesluit). (in Dutch). Ministry of Social Affairs & Employment. The Hague

11. Wulff A-M. NTH Trondheim. Personal communication

12. Smit K. & Slaterus W.H. 1992. Information model for maintenance management. Cap Gemini Publishing. Rijswijk. ISBN 90-71996-56-5

13. Vucinic B. MACAD: Maintenance concept adjustment and design. ISBN 90-370-0112-2.

Figure 1.

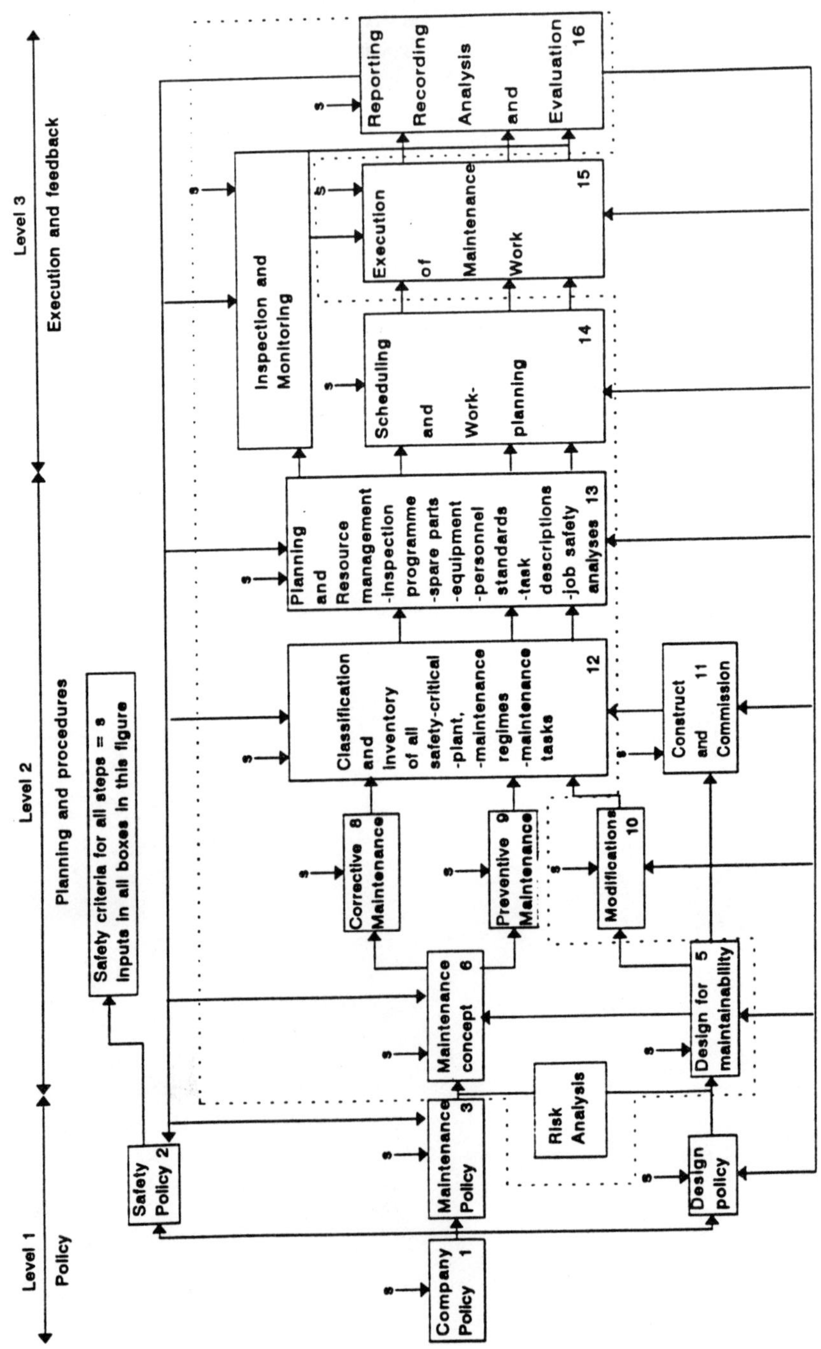

Level 1
Policy
Level 2
Planning and procedures
Level 3
Execution and feedback
Safety criteria for all steps = s
Inputs in all boxes in this figure
Company Policy 1
Safety Policy 2
Maintenance Policy 3
Design policy
Risk Analysis
Maintenance concept 6
Design for maintainability 5
Corrective Maintenance 8
Preventive Maintenance 9
Modifications 10
Construct and Commission 11
Classification and inventory of all safety-critical -plant, -maintenance regimes -maintenance tasks 12
Planning and Resource management -inspection programme -spare parts -equipment -personnel standards -task descriptions -job safety analyses 13
Inspection and Monitoring
Scheduling and Work-planning 14
Execution of Maintenance Work 15
Reporting Recording Analysis and Evaluation 16
s